FOOD PACKAGING AND PRESERVATION

FOOD PACKAGING AND PRESERVATION

Antimicrobial Materials and Technologies

Edited by

AMIT K. JAISWAL

School of Food Science and Environmental Health, Faculty of Sciences and Health, Technological University Dublin-City Campus, Dublin, Ireland

SHIV SHANKAR

Research Laboratories in Sciences Applied to Food, INRS-Armand-Frappier Health Biotechnology Research Centre, Laval, Quebec, Canada

ELSEVIER

ACADEMIC PRESS

An imprint of Elsevier

Academic Press is an imprint of Elsevier
125 London Wall, London EC2Y 5AS, United Kingdom
525 B Street, Suite 1650, San Diego, CA 92101, United States
50 Hampshire Street, 5th Floor, Cambridge, MA 02139, United States
The Boulevard, Langford Lane, Kidlington, Oxford OX5 1GB, United Kingdom

ISBN: 978-0-323-90044-7

For Information on all Academic Press publications
visit our website at https://www.elsevier.com/books-and-journals

Publisher: Nikki P. Levy
Acquisitions Editor: Nina Bandeira
Editorial Project Manager: Suchita Gera
Production Project Manager: Bharatwaj Varatharajan
Cover Designer: Mark Rogers

Typeset by MPS Limited, Chennai, India

Contents

2

Natural antimicrobials for food packaging and preservation

3

Novel non-thermal technology for food preservation

List of contributors

Gulsun Akdemir Evrendilek College of Engineering, Department of Electrical and Computer Engineering, Boston University, MA, United States

Elena Alexandra Alexa Food Safety Department, Teagasc Food Research Centre, Dublin, Ireland

Saeid Alizadeh Department of Food Science and Technology, Ferdowsi University of Mashhad, Mashhad, Iran

Amilcar L. Antonio Centro de Investigação de Montanha (CIMO), Instituto Politécnico de Bragança, Campus de Santa Apolónia, Bragança, Portugal; Laboratório Associado para a Sustentabilidade e Tecnologia em Regiões de Montanha (SusTEC), Instituto Politécnico de Bragança, Campus de Santa Apolónia, Bragança, Portugal

João C.M. Barreira Centro de Investigação de Montanha (CIMO), Instituto Politécnico de Bragança, Campus de Santa Apolónia, Bragança, Portugal; Laboratório Associado para a Sustentabilidade e Tecnologia em Regiões de Montanha (SusTEC), Instituto Politécnico de Bragança, Campus de Santa Apolónia, Bragança, Portugal

Lillian Barros Centro de Investigação de Montanha (CIMO), Instituto Politécnico de Bragança, Campus de Santa Apolónia, Bragança, Portugal; Laboratório Associado para a Sustentabilidade e Tecnologia em Regiões de Montanha (SusTEC), Instituto Politécnico de Bragança, Campus de Santa Apolónia, Bragança, Portugal

María A. Bertuzzi Chemical Industry Research Institute, INIQUI-CONICET, Salta, Argentine; Engineering Department, National University of Salta, Salta, Argentine

Deblina Biswas Department of Electrical and Computer Engineering, Sungkyunkwan University, Suwon, Republic of Korea; Department of Instrumentation and Control Engineering, Dr. B. R. Ambedkar National Institute of Technology Jalandhar, Jalandhar, Punjab, India

Bruna Wendt Böhmer-Maas Department of Agroindustrial Science and Technology, Federal University of Pelotas, Pelotas, Rio Grande do Sul, Brazil

Catherine M. Burgess Food Safety Department, Teagasc Food Research Centre, Dublin, Ireland

Ilaria Cacciotti Department of Engineering, INSTM RU, University of Rome "Niccolò Cusano", Roma, Italy

Prasad Chavan Food Technology and Nutrition, Lovely Professional University, Phagwara, India

Caroline Corrêa de Souza Coelho Food and Nutrition Graduate Program – The Federal University of State of Rio de Janeiro, Rio de Janeiro, Brazil

Ali Demirci Department of Agricultural & Biological Engineering, The Pennsylvania State University, University Park, PA, United States

Ângela Fernandes Centro de Investigação de Montanha (CIMO), Instituto Politécnico de Bragança, Campus de Santa Apolónia, Bragança, Portugal; Laboratório Associado para a Sustentabilidade e Tecnologia em Regiões de Montanha (SusTEC), Instituto Politécnico de Bragança, Campus de Santa Apolónia, Bragança, Portugal

Isabel C.F.R. Ferreira Centro de Investigação de Montanha (CIMO), Instituto Politécnico de Bragança, Campus de Santa Apolónia, Bragança, Portugal; Laboratório Associado para a Sustentabilidade e Tecnologia em Regiões de Montanha (SusTEC), Instituto Politécnico de Bragança, Campus de Santa Apolónia, Bragança, Portugal

Otniel Freitas-Silva Embrapa Agroindústria de Alimentos, The Brazilian Agricultural Research Corporation (EMBRAPA), Rio de Janeiro, Brazil; Food and Nutrition Graduate Program – The Federal University of State of Rio de Janeiro, Rio de Janeiro, Brazil

Jimena E. Gamboni Chemical Industry Research Institute, INIQUI-CONICET, Salta, Argentine; Engineering Department, National University of Salta, Salta, Argentine

Farhad Garavand Department of Food Chemistry and Technology, Teagasc Food Research Centre, Moorepark, Fermoy, Ireland

Tabli Ghosh Department of Food Engineering and Technology, Tezpur University, Tezpur, Assam, India

Hamid Food Technology and Nutrition, Lovely Professional University, Phagwara, India

Carolina Ibarguren Chemical Industry Research Institute, INIQUI-CONICET, Salta, Argentine; Health Sciences Department, National University of Salta, Salta, Argentine

Amit K. Jaiswal School of Food Science and Environmental Health, Faculty of Sciences and Health, Technological University Dublin-City Campus, Dublin, Ireland; Environmental Sustainability and Health Institute (ESHI), Technological University Dublin-City Campus, Dublin, Ireland

Lily Jaiswal Department of Food and Nutrition, BioNanocomposite Research Institute, Kyung Hee University, Dongdaemun-gu, Seoul, Republic of Korea

Swarna Jaiswal School of Food Science and Environmental Health, Faculty of Sciences and Health, Technological University Dublin-City Campus, Dublin, Ireland; Environmental Sustainability and Health Institute (ESHI), Technological University Dublin-City Campus, Dublin, Ireland

Varsha Khare Soft Foundry Institute, Seoul National University, School of Engineering, Kwanak-gu, Seoul, South Korea

Diako Khodaei School of Science and Computing, Galway-Mayo Institute of Technology (GMIT), Galway, Ireland

Aniket Kumar School of Materials Science and Engineering, Chonnam National University, Buk-gu, Gwangju, South Korea

Triona O'Brien Food Safety Department, Teagasc Food Research Centre, Fermoy, Ireland

Angelos Papadochristopoulos Food Safety Department, Teagasc Food Research Centre, Dublin, Ireland

Eliana Pereira Centro de Investigação de Montanha (CIMO), Instituto Politécnico de Bragança, Campus de Santa Apolónia, Bragança, Portugal; Laboratório Associado para a Sustentabilidade e Tecnologia em Regiões de Montanha (SusTEC), Instituto Politécnico de Bragança, Campus de Santa Apolónia, Bragança, Portugal

Kalpani Y. Perera School of Food Science and Environmental Health, Faculty of Sciences and Health, Technological University Dublin-City Campus, Dublin, Ireland; Environmental Sustainability and Health Institute (ESHI), Technological University Dublin-City Campus, Dublin, Ireland

Jack Prendeville School of Food Science and Environmental Health, Faculty of Sciences and Health, Technological University Dublin-City Campus, Dublin, Ireland

Laura Principato DiSTAS Department for Sustainable Food Process, Università Cattolica del Sacro Cuore, Piacenza, Italy

Ruchir Priyadarshi Department of Food and Nutrition, BioNanocomposite Research Center, Kyung Hee University, Seoul, Republic of Korea

Jong-Whan Rhim Department of Food and Nutrition, BioNanocomposite Research Center, Kyung Hee University, Seoul, Republic of Korea

Swarup Roy Department of Food Technology and Nutrition, School of Agriculture, Lovely Professional University, Phagwara, Punjab, India; Department of Food and Nutrition, BioNanocomposite Research Center, Kyung Hee University, Seoul, Republic of Korea

Shiv Shankar Research Laboratories in Sciences Applied to Food, INRS-Armand-Frappier Health Biotechnology Research Centre, Laval, Quebec, Canada

Shubham Sharma School of Food Science and Environmental Health, Faculty of Sciences and Health, Technological University Dublin-City Campus, Dublin, Ireland; Environmental Sustainability and Health Institute (ESHI), Technological University Dublin-City Campus, Dublin, Ireland

Aníbal M. Slavutsky Chemical Industry Research Institute, INIQUI-CONICET, Salta, Argentine; Engineering Department, National University of Salta, Salta, Argentine

Sanjiv Sonkaria Soft Foundry Institute, Seoul National University, School of Engineering, Kwanak-gu, Seoul, South Korea

Andressa Moreira de Souza Embrapa Agroindústria de Alimentos, The Brazilian Agricultural Research Corporation (EMBRAPA), Rio de Janeiro, Brazil

Giorgia Spigno DiSTAS Department for Sustainable Food Process, Università Cattolica del Sacro Cuore, Piacenza, Italy

Amirerfan Taghizadeh Department of Agriculture and Food Science, Islamic Azad University, Science and Research Branch, Tehran, Iran

Felipe Machado Trombete Department of Food Engineering, Federal University of São João del-Rei, Sete Lagoas, MG, Brazil

Gulsad Uslu-Senel Department of Environmental Engineering, Firat University, Elazığ, Turkey

Shima Yousefi Department of Agriculture and Food Science, Islamic Azad University, Science and Research Branch, Tehran, Iran

Preface

Food safety and quality are crucial for ensuring the well-being of consumers, as well as the success of food manufacturers. However, with the increasing demand for fresh and natural foods by consumers, it has become more challenging to ensure the preservation and quality of food products. This challenge has led to the development of innovative food packaging and preservation methods, such as antimicrobial materials and technologies. Antimicrobial materials and technologies have gained attention in recent years due to their potential to improve food safety and extend the shelf life of food products. These materials and technologies, including metallic nanoparticles such as silver, copper, and zinc; natural extracts such as essential oils, plant extracts, nisin, lysozyme; and nonthermal processing methods, such as gamma irradiation, X-ray, cold plasma, ozonation, and pulse electric field, utilize various mechanisms to combat foodborne pathogens.

Nanoparticles have been widely used in food packaging and preservation due to their antimicrobial properties. They have the ability to disrupt the cell walls of microorganisms, thereby preventing their growth and proliferation. Additionally, natural extracts such as essential oils and plant extracts have been used as alternatives to synthetic preservatives, as they are safe and nontoxic. These extracts contain natural antimicrobial compounds that can prevent the growth of microorganisms in food products. Similarly, nonthermal processing methods, such as gamma irradiation, X-ray, cold plasma, ozonation, and pulse electric field, have also gained attention as effective methods for food preservation. These methods utilize various physical and chemical mechanisms to inactivate microorganisms, including the disruption of cell membranes, denaturation of proteins, and DNA damage. One of the key advantages of antimicrobial materials and technologies is their potential for sustainability. They can reduce the use of synthetic preservatives and chemicals, leading to a more sustainable food system. Additionally, some of these technologies can reduce food waste by extending the shelf life of food products.

The book "Food Packaging and Preservation: Antimicrobial Materials and Technologies" provides a comprehensive overview of recent advances in antimicrobial materials and technologies for food packaging and preservation. Divided into three sections, the book covers a wide range of topics, including the principles of food spoilage, microbial contamination, and foodborne illnesses, the development of novel materials such as metallic nanoparticles, natural extracts, and nonthermal processing methods, and the applications of these technologies in food packaging and preservation.

The book has 16 chapters in total and is divided into three sections. The first section of the book focuses on the development and application of antimicrobial materials, including metallic nanoparticles, biopolymer-based nanocomposites, and plastic- and bioplastic-based nanocomposites and their impact on meat, dairy, fruits, and vegetable products. Section two discusses natural antimicrobials, including novel food packaging systems with antimicrobial agents from microbial sources, antimicrobial agents from herbs and spices, and natural antimicrobials from fruits and plant extracts. The third section covers novel nonthermal technologies for food preservation, including irradiation, cold plasma, ozone-based preservation, and pulsed electric field. These technologies offer promising solutions for extending the shelf life of food products and maintaining their safety and quality without the need for thermal processing.

This book is primarily aimed at researchers and industry professionals working in the fields of food technology, materials science, nanotechnology, and polymer science, who are involved in the development of novel antimicrobial technologies for food industry applications. It can also be used by undergraduate and postgraduate students studying food science and technology and bio-nanotechnology.

The content of this book is derived from university-level courses designed for food scientist and nanotechnology scientists that have been successfully taught for decades. We hope that this book will provide a scaffolded introduction to the principles of biological science and nanomaterials, natural antimicrobials, and emerging nonthermal processing methods as antimicrobial technology, enabling researchers and industry professionals to develop sustainable use of these materials and technologies.

We believe that this book will be a valuable resource for researchers, scientists, and industry professionals working in the field of food technology, materials science, nanotechnology, and polymer science. It will provide crucial background information in these arenas and help the readers understand the latest trends and advancements in the field of antimicrobial materials and technologies for food packaging and preservation, thereby contributing to the development of more sustainable and effective food packaging and preservation systems.

Our heartfelt appreciation goes out to the esteemed experts and leaders from academic and research institutions who have provided invaluable and cutting-edge contributions to this book. Their remarkable efforts and contributions have made this book possible. We also extend our gratitude to Elsevier Academic Press for the successful production of this book. We would like to acknowledge the Elsevier editorial team for their exceptional support and guidance throughout the project, particularly Ms. Nina Bandeira, the executive editor, and Ms. Kathrine Esten, the Editorial Project Manager.

We are confident that this book will meet the expectations of its readers and would greatly appreciate any feedback and suggestions for further improvement.

Amit K. Jaiswal and Shiv Shankar

SECTION 1

Antimicrobial materials for food packaging and preservation

CHAPTER

1

Microbial contamination of food

Elena Alexandra Alexa[1], *Angelos Papadochristopoulos*[1], *Triona O'Brien*[2] *and Catherine M. Burgess*[1]

[1]Food Safety Department, Teagasc Food Research Centre, Dublin, Ireland [2]Food Safety Department, Teagasc Food Research Centre, Fermoy, Ireland

1.1 Introduction

Contamination of food makes it unfit for consumption due to the introduction of undesirable elements, and it can have significant impacts on both consumers and food producers. In the case of consumers, food contamination can cause illness and even death, depending on the contaminant. For food businesses, it can cause substantial financial and reputational damage.

Food contaminants generally fall under one of three broad categories. Physical contaminants are physical items not designed for consumption that end up in the product. Examples include hair, jewelry, metal from equipment, broken glass, etc. A number of measures are put in place during food production and processing to prevent the entry of such physical contaminants into products. Chemical contaminants are substances unintentionally present in foods and can arise at different stages of food production and processing. In European legislation (Regulation (EEC) No 315/93, 1993), such a contaminant is defined as any substance unintentionally added to food that is present in such it as a result of the production (including operations carried out in crop husbandry, animal husbandry, and veterinary medicine), manufacturing, processing, preparation, treatment, packaging, transport or holding of such food, or as a result of environmental contamination. Such chemical contaminants can be harmful to humans; examples of chemical contaminants which can occur in foods include natural toxins such as mycotoxins. Environmental contaminants arising from agricultural or industrial activities such as pesticides or heavy metals, or processing contaminants that can naturally form during cooking/processing, such as acrylamide. The third category of food contaminants is represented by biological contaminants and is the focus of this chapter. Biological or microbial contaminants refer to a broad range of microorganisms, including bacteria, viruses, yeasts, molds, and parasites. Biological contaminants can be pathogenic and cause illness, commonly referred to as food poisoning if consumed. Other microbial species can cause food spoilage, impacting product quality, and resulting in food loss and waste. The presence of either is not desirable and therefore food producers must ensure that all steps are taken to prevent food contamination wherever possible.

Different factors influence both the types of microorganisms present in food and their ability to survive, or even multiply, on the product. Intrinsic properties of the food itself, such as pH, nutrient availability, water activity, and antimicrobial components can all influence microbial survival and proliferation of a product. Extrinsic properties imposed by the environment, such as temperature, humidity, gaseous atmosphere, and ultraviolet (UV) irradiation, also play a role in influencing microbial growth in food products.

Microbial contamination of food is diverse and complex and very much influenced by the type of product; therefore, this chapter will not be exhaustive in nature. In order to illustrate how the food itself, its origin, processing, and storage can influence the associated microbial communities and the predominant pathogens and spoilage organisms associated with specific food products, this chapter will focus on three broad product types—meat and meat products, dairy products and fruit and vegetables. For each category the predominant bacterial pathogens and spoilage organisms will be considered, as well as potential routes of contamination.

Food Packaging and Preservation.
DOI: https://doi.org/10.1016/B978-0-323-90044-7.00001-X

1.2 Microbial contamination of meat

Meat has always played an important role in the human diet as it contains fatty acids, vitamins, minerals, and protein (Geiker et al., 2021; Wyness, 2016); proteins from animal sources are more easily digested and provide all the essential amino acids (Elmadfa & Meyer, 2017; FAO, 2011). Global meat consumption is high and continues to increase (Godfray et al., 2018). Nevertheless, a significant amount of meat produced (approximately 23%) is wasted every year (Karwowska, Łaba, & Szczepański, 2021). One of the main reasons for meat waste is spoilage during storage (Lipinski, 2020). Meat is considered a highly perishable food product, as its high nutritional content and high water activity make it ideal for colonization and the growth of microorganisms such as bacteria, yeasts, and molds (Fernandes, 2009; Sofos et al., 2013). Table 1.1 provides an overview of the different bacteria that can be present and grow on meat products.

1.2.1 *Sources of microbial contamination*

Microbial contamination mainly occurs on the surface of meat, which can lead to spoilage but also to the transmission of foodborne pathogens (Singh et al., 2019; Sofos, 2014). This is a significant concern for the meat industry and consumers. The source of meat contamination varies, with the main source being the animals themselves. During slaughter, the carcasses can become contaminated by the hides of cattle, the fleece of sheep, the skin of pigs, or the feathers of poultry that can all carry animal feces, soiling, and animal feed on them, while the spillage of the intestinal material of the slaughtered animals during the evisceration procedure can also lead to contamination of the carcasses (Berends et al., 1997; Brizio et al., 2015; Fernandes, 2009; Sheridan, 1998; Sofos, 2014). Other sources that are responsible for cross-contamination of the carcasses with microorganisms are the processing environment, the equipment used and the workers, while insects, rodents and birds can also carry and transmit microorganisms to meat (Singh et al., 2019; Sofos, 2014). Furthermore, microbial contamination and growth may also take place after processing, and during storage and distribution of meat products (Sung et al., 2013).

Several different microorganisms can be present in meat products and the type depends on the physiological state of the animal at the time of slaughter, the processing methods used, hygiene during processing, surface area, temperature of storage and distribution, and the presence of competitive microorganisms (Rawdkuen, Punbusayakul, & Lee, 2016).

In order to reduce contamination levels, control microbial growth, ensure safety, extend shelf-life, and preserve the quality characteristics of meat products, traditional and novel preservation methods are used. Some of these include drying, thermal processing, irradiation, freezing, refrigeration, fermentation, modified atmosphere packaging, the addition of antimicrobial agents, salting, high hydrostatic pressure, etc. (Quintavalla & Vicini, 2002; Sung et al., 2013).

Generally, improper cooking or inadequate processing of meat products, cross-contamination, and subsequent poor handling of already cooked products are responsible for meat-associated foodborne disease outbreaks (Singh et al., 2019). The cooking of unprocessed meat is very important in order to ensure its safety. In the case of intact pieces of meat, such as steaks or chops, less cooking may be sufficient to ensure its safety because the microorganisms are mainly found on the surface of the meat exposed to the heat source (Singh et al., 2019). On the other hand, for non-intact pieces of meat which have been treated with mechanical tenderization (such as cubing, frenching or pounding devices, blades, needle injectors used for marinating, flavoring, moisture enhancement, or tenderizing) (Sofos, 2014), ground, or comminuted meat products, thorough cooking is essential to ensure the safety of the product. This is because microorganisms have the opportunity to transfer from the surface to the interior part of the meat and are trapped inside the meat tissues (Luchansky et al., 2008; Singh et al., 2019; Sofos, 2014).

1.2.2 *Bacterial pathogens associated with meat and meat products*

The presence of biological contaminants in meat and poultry is of concern because it poses a risk to consumers' health. These include bacteria, viruses, parasites, toxigenic molds, and transmissible spongiform encephalopathies prions (Sofos et al., 2013), but here we focus on bacterial pathogens. The most common foodborne pathogens associated with meat products are *Escherichia coli* O157:H7, non-O157 Shiga toxin-producing *E. coli* (STEC), *Salmonella* spp., *Campylobacter* spp., *Staphylococcus aureus*, *Yersinia enterocolitica*, *Listeria monocytogenes*, *Clostridium* spp., and *Bacillus* spp. Other bacteria that have been reported in meat products include

TABLE 1.1 Bacterial species commonly found in meat and poultry.

Presence on meat and poultry[a]			
Bacterial species	**Fresh**	**Processed**	**Vacuum packaged**
Achromobacter	•		
Acinetobacter	••	•	•
Aeromonas	••	•	•
Alcaligenes	•		
Alteromonas	•	•	
Arthrobacter	•	•	
Bacillus	•	•	••
Brochothrix	•	•	
Buttiauxella	•		
Campylobacter	•		
Carnobacterium	•		••
Chromobacterium	•		
Citrobacter	•		
Clostridium	•		
Corynebacterium	•	•	•
Enterobacter	•	•	•
Enterococcus	••	•	••
Escherichia	•		
Flavobacterium	•		
Hafnia	•	•	
Halobacterium		•	
Janthinobacterium		•	
Klebsiella	•		
Kluyvera	•		
Kocuria	•	•	•
Kurthia	•		•
Lactobacillus	•	••	••
Lactococcus	•		
Leuconostoc	•	•	•
Listeria	•	•	
Microbacterium	•	•	•
Micrococcus	•	•	•
Moraxella	••		
Paenibacillus	•	•	
Pantoea	•		
Proteus	•		
Providencia	•	•	•

(Continued)

TABLE 1.1 (*Continued*)

Presence on meat and poultry[a]			
Bacterial species	**Fresh**	**Processed**	**Vacuum packaged**
Pseudomonas	••	•	
Rahnella	•		
Serratia	•		
Shewanella	•	•	•
Staphylococcus	•	•	•
Streptococcus	•	•	
Vibrio	•		
Weissella	•	•	•
Yersinia	•		•

[a]*• Known to occur; •• Most frequently isolated.*
Source: Adapted from Sofos, J. N. et al. (2013). Meat, poultry, and seafood. In: Food microbiology: Fundamentals and frontiers (pp. 111–168). ASM Press. https://doi.org/10.1128/9781555818463.ch6.

Aeromonas, *Arcobacter*, *Brucella*, *Enterobacter*, *Helicobacter*, *Mycobacterium*, *Plesiomonas*, and *Shigella*, amongst others (Lianou, Panagou, & Nychas, 2017; Singh et al., 2019; Sofos, 2014).

Campylobacter spp. is mostly associated with contaminated poultry but it is also found in other food-producing animals. Campylobacteriosis caused by thermotolerant species of *Campylobacter* is the most common foodborne poisoning in humans in the European Union (EU) since 2007, with 249 outbreaks reported in 2021 (EFSA ECDC, 2022). Moreover, it is estimated that 1.5 million infections every year in the United States are attributable to *Campylobacter* (CDC, 2021). The main source of *Campylobacter* cases is undercooked poultry, but cross-contamination of products that do not require cooking also causes a significant number of cases (Fernandes, 2009). According to the European Food Safety Authority (EFSA), in the period 2017–20, 26% of nonready-to-eat (RTE) meat and meat products, in general, and 33% of non-RTE and meat products from broilers tested positive for *Campylobacter*. The most commonly identified species were *Campylobacter jejuni* and *Campylobacter coli* (EFSA ECDC, 2022).

Salmonella was the EU's second most commonly reported foodborne pathogen in 2021, with 15.7 cases per 100,000 population (EFSA ECDC, 2022). *Salmonella* is found in red meat, poultry products and cooked cured meat products (EFSA ECDC, 2022; Rouger, Tresse, & Zagorec, 2017). According to EFSA, the source of the majority of positive samples from official *Salmonella* monitoring programs were poultry (broiler and turkey meat), while other sources included pig meat and bovine meat. The three most detected serovars were *S.* Enteritidis, which was responsible for most cases in humans, *S.* Typhimurium, and monophasic *S.* Typhimurium (1, 4, [5], 12:i:-) (EFSA ECDC, 2022). A recent outbreak of *Salmonella* linked to the consumption of non-RTE breaded poultry in 9 countries in Europe and the United Kingdom between 2018 and 2020 led to 193 cases and 1 death, with half of the cases in children (ECDC, 2021). In the United States, between 1998 and 2008, *Salmonella* was responsible for 26% of the foodborne disease outbreaks in total, where 145 outbreaks of *Salmonella* linked to poultry were recorded (Gould et al., 2013). A recent outbreak with 176 cases of salmonellosis between 2020 and 2022 was reported in the United Kingdom and was related to the consumption of pork scratching products (FSA, 2022).

STEC is one of the most important pathogens in meat products, especially beef, with cattle considered as its main reservoir (Sofos et al., 2013), with *E. coli* O157:H7 being the most well-known serotype. *E. coli* O157:H7 is most commonly found in red meat and especially in ground beef, non-intact beef products, and fermented meat products (Marsden et al., 2009; Sofos, 2014). The serogroups of STEC, in addition to O157, that are most commonly linked to illness in humans are O26, O103, O111, O145, O121, and O45 (Koutsoumanis, Lianou, & Sofos, 2014), with O157 and O26 serogroups the most detected in humans in the EU in 2021 (EFSA ECDC, 2022). STEC has been implicated in many outbreaks over the years globally. In 2021, STEC was the fourth most reported bacterial foodborne pathogen in the EU, with 31 outbreaks and 6084 cases (EFSA ECDC, 2022). Most were associated with meat products. The first recorded STEC outbreak in meat products dates back to 1982 and it was linked to contaminated hamburgers (Riley et al., 1983), while in 1992–93 in the United States, one of the biggest multistate outbreaks was observed and it was associated with a fast food burger chain and the reservoir of the STEC was ground beef that led to the death of 4 children (CDC, 1993). Another recent outbreak was observed in the United States in 2019, and it was again linked to ground beef, with 29 people hospitalized (CDC, 2019).

Listeriosis is one of the most serious foodborne diseases because of its high fatality rate. According to EFSA, 23 foodborne outbreaks of *L. monocytogenes* were recorded in the EU in 2021. Half of the strong-evidenced outbreaks were caused by the consumption of meat products and poultry contaminated with *L. monocytogenes* (EFSA ECDC, 2022). Some characteristic examples include the multi-country outbreak in Europe from 2017 to 2019 that was linked to sliced RTE meat products (ECDC, 2019) and the biggest outbreak of listeriosis worldwide, in South Africa in 2017. This latter outbreak was associated with the consumption of polony, a RTE processed meat product and it was reported that 27% of the infected people died (Smith et al., 2019; WHO, 2018). *L. monocytogenes* is frequently associated with contaminated RTE meat products such as deli-type meats. It can have a high prevalence in such meat products (Gómez et al., 2015), with studies reporting prevalence levels of up to 40%–45% in fermented sausages (Meloni, 2015). The contamination of RTE meat products can happen at any stage of processing. It can be due to raw materials or ingredients, equipment, employees, poor compartmentalization of the processing lines, cross-contamination by contact with other contaminated materials or raw products, or due to poor cleaning and sanitation procedures (Gómez et al., 2015; Gounadaki et al., 2008; Lunden et al., 2003; Marsden et al., 2009). Of key concern is that *L. monocytogenes* can persist in the processing plant environment for extended periods, create biofilms and grow at low temperatures (Fernandes, 2009; Lundén, Autio, & Korkeala, 2002).

Clostridioides difficile is considered an emerging pathogen and it has been identified in meat products and poultry (Gould & Limbago, 2010). The contamination of meat with this pathogen can occur during slaughtering (Rodriguez et al., 2013). *C. botulimum* and *C. perfringens* are also associated with meat and poultry products, such as bacon and dry-cured ham or canned meat products for *C. botulinum* (Drosinos & Paramithiotis, 2009; Marsden et al., 2009). *C. perfringens* is most commonly found in meat and poultry that have been cooked and kept warm for long periods (Marsden et al., 2009).

S. aureus is present in raw meat and poultry and it is also reported in meat products due to cross-contamination by the employees during processing (Fernandes, 2009; Sofos, 2014). Finally, *Bacillus cereus* has been reported in meat products, such as meatloaf (Marsden et al., 2009) and *Y. enterocolitica* is mostly linked to raw pork meat (Lianou et al. 2017).

1.2.3 Bacterial spoilage organisms associated with meat and meat products

Food spoilage is an important consideration for both consumers and the industry because it results in economic loss for the meat industry, while the consumers' confidence in spoiled products is compromised (Sofos et al., 2013). The presence of bacteria has been connected to meat spoilage and bacterial counts are an indicator of spoilage. When the bacterial counts on meat products reach 10^7 CFU/g, "off odors" are observed, while higher populations (10^9 CFU/g) are linked to a fruity "off odor" and slime (Singh et al., 2019). Generally, spoilage is attributed to a fraction of the microorganisms present on the product and not the entire microbial community (Nychas & Skandamis, 2005), while the type of microorganisms that are present in meat and provoke its spoilage depends on two key factors; first, the conditions of storage of the meat and second, the competition from other microbes (Doulgeraki et al., 2012).

Spoilage varies among different types of meat products, because of the range of factors that influence this phenomenon, including the differing composition of products, the dominant microbial species, pH, water activity, and storage conditions (Sofos, 2014). In general, spoilage of fresh meat is mainly caused by the Gram-negative *Pseudomonas* spp. and Enterobacterales and the Gram-positive *Brochothrix thermosphacta* and lactic acid bacteria (LAB) (Pennacchia, Ercolini, & Villani, 2011). For meat products stored at low temperatures, under aerobic conditions, Gram-negative aerobic or facultative anaerobic bacteria such as *Pseudomonas* spp. (*P. fragi*, *P. fluorescens*, *P. putida* and *P. lundensis*) are the dominant bacteria that cause spoilage (Doulgeraki et al., 2012; Iulietto et al., 2015; Sofos et al., 2013). Other spoilage bacteria commonly found in meat stored aerobically at low temperatures are *Acinetobacter* spp., *B. thermosphacta, Shewanella putrefaciens, Enterococcus* spp., *Moraxella* spp., *Lactobacillus* spp., *Leuconostoc* spp., *Serratia* spp., and *Carnobacterium* spp. (Borch, Kant-Muermans, & Blixt, 1996; Ercolini et al., 2009; Iulietto et al., 2015; Sofos et al., 2013). On the other hand, Gram-positive facultative anaerobic or anaerobic Gram-positive bacteria, such as LAB, are the dominant bacteria that cause spoilage in meat products stored under vacuum or modified atmosphere (MAP) (Doulgeraki et al., 2012; Sofos et al., 2013); the presence of other bacteria such as *B. thermosphacta* has also been reported (Doulgeraki et al., 2012). Moreover, *S. putrefaciens* is one of the most common sources of spoilage in vacuum-packed meat and high pH vacuum-packed meat at chill temperatures (Doulgeraki et al., 2012).

Meat that has a pH higher than 6, dark color, and a glucose deficiency is referred to as dark, firm, dry (DFD) meat. These characteristics of DFD meat favor faster degradation of amino acids by pseudomonads, leading to faster spoilage at lower cell densities (10^6 CFU/cm^2) (Newton & Gill, 1981). Under vacuum or MAP packaging, green discoloration is also observed and has been attributed to the growth of *S. liquefaciens* and *S. putrefaciens* (Newton & Gill, 1981).

Comminuted products such as ground meat are more susceptible to spoilage due to the high initial level of contamination, the higher possibility of cross-contamination with other trimmings, microbial translocation with the grinding process, and the greater availability of nutrients, which have been released from the cells through the grinding process and are available for microbial metabolism (Sofos et al., 2013). The most commonly found bacteria in such meat products stored in aerobic conditions are *Pseudomonas*, *Acinetobacter*, and *Moraxella*. The occurrence of Enterobacterales in comminuted meat products is also more common compared to intact meat (Sofos et al., 2013).

Cooked and processed meat products such as frankfurters, bologna sausages, and luncheon meat are generally considered stable; however, three different types of spoilage caused by bacteria can be observed. First, slime can be formed on the surfaces of meat products by LAB (*Lactobacillus*, *Enterococcus*) and *B. thermosphacta* (Marsden et al., 2009; Sofos et al., 2013). Souring is another characteristic type of spoilage in processed meat products that is attributed to the above bacterial species that can grow under the casings in moist products that are stored at high humidity and they metabolize the sugars producing organic acids, leading to pH reduction (Marsden et al., 2009). Lastly, green discoloration in processed meats is another type of spoilage that is caused mainly by *Lactobacillus viridescens*, *Streptococcus* spp., and *Leuconostoc* spp. which can grow before or after processing of the products (Grant, McCurdy, & Osborne, 1988; Sofos et al., 2013).

For poultry, the main bacterial spoilage agents are *B. thermosphacta*, *P. fluorescens*, and *S. putrefaciens*, while LAB and Enterobacterales are also considered to contribute to spoilage (Rouger et al. 2017).

Psychrophilic and psychrotolerant clostridia can cause spoilage in chilled meat stored under vacuum which is characterized by a putrid smell (H_2S), proteolysis, a metallic sheen on the meat, and massive gas production which is referred to as blown pack spoilage (Bolton, Carroll, & Walsh, 2015; Húngaro et al., 2016). Blown pack spoilage has been observed in several meat products such as beef cuts, lamb, precooked turkey, and sous-vide products (Sofos et al., 2013).

Biogenic amines, such as histamine, putrescine, spermidine, spermine, etc., in spoiled meat products pose a risk to human health. The bacterial groups that are mostly involved in the production of these are Enterobacterales, *Micrococcaceae*, and LAB (Schirone et al., 2022). Biogenic amines can also play an important role as a quality index of fresh meat (Schirone et al., 2022; Triki et al., 2018). Good hygiene practices and proper storage (temperature and time limitation) can control the bacteria involved in the production of biogenic amines (Sofos et al., 2013).

1.3 Microbial contamination of milk and milk products

Milk and milk products play an indispensable role in human nutrition, especially during childhood, providing high-quality proteins and microelements such as calcium, magnesium, zinc, potassium, and phosphorus necessary for the human metabolism and good development and strength of bones (Rozenberg et al., 2016). Evidence has shown that frequent consumption of milk and milk products can contribute to preventing periodontal disease, controlling body weight, blood pressure, and oral microbiota, as well as having antioxidative and antiinflammatory effects, although there have been some debates in recent years regarding longer-term health effects (Zhang et al., 2021). It also serves as an excellent growth medium for a wide range of microorganisms (Quigley et al., 2013).

1.3.1 *Sources of microbial contamination associated with milk and milk products*

Aseptically drawn raw milk from healthy cows is practically sterile and free from microbial contaminants when secreted into the alveoli of the udder. During milking, microorganisms can rapidly contaminate it (Zastempowska, Grajewski, & Twaruzek, 2016).

At the farm and during the milking process, the milk can become contaminated by bacteria from the farm environment entering the milk via the teat skin, air, water, feed, silage, the milking equipment, and the milk tank; therefore, good farm hygiene management is essential for milk quality and safety (Ledo, Hettinga, & Luning, 2020). Contamination can occur due to the use of unhygienic milking machines, improper disinfection of bulk tanks, and insufficient washing of the exterior of the animals' udders which can also be sources of raw milk contamination (Oliver et al., 2009; Zastempowska et al. 2016). The prevalence of beneficial and harmful microorganisms and presumptive interactions that may occur in complex communities present in the raw material can be influenced by other factors such as geographical location, season, farm management practices, farm size, and the number of animals present (Kousta et al., 2010; Skowron et al., 2022).

Dairy processing plants use a clean-in-place system to clean and disinfect the interior of processing equipment, thus ensuring a higher level of safety and increasing productivity since less time and water resources are used for cleaning and disinfection (Memisi et al., 2015). However, dairy plant environments are still at risk as pathogenic microorganisms can reside on various food-contact surfaces such as holding tanks, conveyer belts, cheese molds, brushes used for soft cheese polishing, wooden shelves, and nonfood contact surfaces such as drains, walls, and floors, respectively, therefore persisting for prolonged periods of time (Kousta et al., 2010).

Microbial contamination can occur in a wide range of dairy products, but some are more susceptible to contamination than others. Dairy products' cross-contamination may include post-processing contamination (Kousta et al., 2010). Soft cheeses such as Brie, Camembert, and feta are also at higher risk for contamination due to their high moisture content, which can provide a suitable environment for bacterial growth (Choi et al., 2016). Ice cream and other frozen dairy desserts can also be at risk, as the freezing process can slow the growth of bacteria but may not eliminate them (Marshall, 2001). Processing can effectively control microbial growth in dairy products and reduce the risk of contamination (Boor et al., 2017) to improve product safety and quality, with good manufacturing practices and hygiene control methods to ensure that processing does not introduce new sources of contamination. Some standard processing techniques in the dairy industry are pasteurization, ultra-high temperature (UHT) processing, drying, fermentation, and cheese making. UHT processing and spray drying products can extend the shelf life of dairy products and are commonly used for products such as shelf-stable milk, cream, whey powders, milk-based infant formula, and skim milk powders. Other dairy products at risk for microbial contamination include yogurt, sour cream, and cream cheese, among others. Generally, any dairy product improperly processed, handled, or stored can be at risk of contamination. Following proper food safety practices is essential when handling and preparing dairy products, including refrigerating them promptly and avoiding cross-contamination with other foods.

In recent years there has been an increased interest in artisanal cheese varieties made of raw milk, thus increasing the chances of possible contamination, which may necessitate prolonged cheese ripening to ensure the safety of such products. In certain situations, artisanal cheeses are ripened on wooden shelves that have not been previously cleaned and disinfected, thus enabling the potential transference of resident pathogenic and beneficial microbiota onto the cheese (Settanni et al., 2021). The manufacturing practices may influence the overall safety of milk and milk products, especially those traditionally made. Therefore producers need to ensure appropriate staff attire and hygiene, ongoing monitoring of temperature control, and the strict implementation of cleaning and disinfection protocols.

1.3.2 *Bacterial pathogens associated with milk and milk products*

Raw milk is particularly vulnerable to pathogenic microbial contamination (Verraes et al., 2015) because it is not pasteurized or treated to kill pathogens (Griffiths, 2010). According to EFSA, milk and milk products accounted for 7.9% of total foodborne outbreaks in 2021, compared with products of nonanimal origin (12.7%), eggs and egg products (11.9%), meat and meat products (21.7%), or fish and fishery products (15.5%) (EFSA ECDC, 2022).

Dairy pathogen contamination can come from animal feces, contaminated equipment, poor hygiene, environmental contamination, and cross-contamination during and after processing. Foodborne pathogens which have been reported in dairy products include *L. monocytogenes*, *S. aureus*, *Salmonella* spp., *E. coli* O157:H7, *Cronobacter sakazakii* and *Campylobacter* spp. (EFSA ECDC, 2021; Kousta et al., 2010). It is important to note that these are not the only pathogens that may be present in dairy products, and proper handling, processing, and storage of dairy products are critical to prevent the growth and spread of pathogenic bacteria. Continuous monitoring and testing of dairy products for pathogens are essential to ensure their safety and quality.

Studies examining the cheese processing environments for bacterial pathogens have frequently found *L. monocytogenes* present on food contact and nonfood surfaces, indicating ineffective cleaning and disinfection protocols in place (Fox et al., 2011; Jordan, Hunt, & Fox, 2012). If present in a biofilm, *L. monocytogenes* can persist at low temperatures for months, even years, on non-food contact surfaces such as walls, floors, and drains, becoming a potential source of recontamination for RTE products. Listeriosis cases from consuming contaminated milk and milk products have been reported, particularly affecting immunocompromised people, pregnant women, and the elderly (Kousta et al., 2010; Melo, Andrew, & Faleiro, 2015; Ranasinghe et al., 2021).

Mastitis-related microorganisms can directly contaminate raw milk, further colonizing the dairy processing environment and consequently contaminating the milk products (Kousta et al., 2010; Oliver, Jayarao, & Almeida, 2005), making it unsuitable for human consumption (Zastempowska et al. 2016). The somatic cell count indicates milk quality, where a high number increases the likelihood of mastitis-related pathogenic microorganisms and their toxins, posing health risks if consumed (Oliver et al., 2009). The majority of such cases are associated with *S. aureus*, and its presence in milk and milk products (Kousta et al., 2010) may cause illness, mainly due to intoxication episodes with preformed enterotoxins produced by *S. aureus* that, once present, have a rapid onset of symptoms that may include diarrhea, nausea, and vomiting (Jay, Loessner, & Golden, 2008).

Several cases of salmonellosis have been associated with consuming contaminated cheddar and mozzarella cheeses, while *E. coli* O157:H7 cases have been linked to consuming contaminated unpasteurized cheese, respectively (D'Aoust, Warburton, & Sewell, 1985; Jenkins et al., 2022). *Cronobacter sakazakii* contamination has been highlighted in rehydrated powdered infant formula (Drudy et al., 2006) and milk powders (Jacobs, Braun, & Hammer, 2011), and as contamination can spread in manufacturing facilities, maintaining hygiene is critical in its control (Lindsay et al., 2019). *C. sakazakii* can cause life-threatening invasive diseases, such as necrotizing enterocolitis, meningitis, and sepsis in infants and immunocompromised adults.

Campylobacter spp., a resident of the intestinal tract of animals, has been involved in milk contamination at the processing plant (Davis et al., 2016; Oliver et al. 2005). Consumption of contaminated unpasteurized raw milk with *Brucella* spp. has also been reported (Jansen et al., 2019). Bacterial spore formers *B. cereus* and *C. botulinum* are pathogenic and food spoilage organisms and have been linked with milk contamination (Lopez-Brea, Gómez-Torres, & Arribas, 2017).

1.3.3 Spoilage and beneficial organisms associated with dairy products

Some microorganisms belonging to the same taxonomic group may have roles as starter cultures, probiotics, or both, while others can act as spoilage microorganisms, or be both spoilage and pathogenic (Fernández et al., 2015). Depending on the dairy product, some microorganisms could be considered starter or spoilage microorganisms, like the genus *Propionibacterium* which is considered a starter in Swiss-type cheese but a spoilage microorganism in long-ripened cheeses Parmigiano-Reggiano or Grana Padano.

Spoilage bacteria are responsible for degrading the sensorial quality of dairy products. The microbial spoilage effect can deteriorate texture, color, and odor, causing food waste and shortening product shelf life. Spoilage microorganisms in dairy products include proteolytic and lipolytic bacteria that can be psychrotrophic, mesophilic, and thermoduric. Lipolysis and proteolytic enzymes produced by the bacteria can result in problems such as the "sweet curdling" of pasteurized milk and the formation of flakes in cream when added to hot drinks (bitty cream) (Christiansson, 2011).

Raw milk and raw milk products contain the highest variety of bacteria species, including *Pseudomonas* spp. and other psychrotrophic microorganisms associated with spoilage in cold storage. These are the most common problematic bacteria, which can grow during refrigeration and produce extracellular lipases and proteases, resulting in spoilage. Pasteurized milk and fermented products, such as cheese and yogurt that have fewer bacterial species, can still have *Pseudomonas* spp. and other psychotropic microorganisms that can cause spoilage and thus shorten the shelf life. In fermented products containing higher numbers of LAB, there are fewer spoilage bacteria due to the LAB acting as competing bacteria.

Pasteurization is an effective first step to eliminate and reduce the levels of many spoilage microorganisms; however, spoilage enzymes can impair milk quality and safety even after the thermal inactivation of the vegetative form of microorganisms (Yuan et al., 2018). Therefore reducing spoilage microorganisms is critical before processing, and raw milk quality remains the critical control for all dairy products. While raw milk quality and post-pasteurization contamination remain challenging, processing technologies and preservatives are also needed to prevent spoilage microorganisms' growth and, therefore, extend the shelf life of dairy products. Spoilage of cheese can be caused by psychrotrophic bacteria like *Pseudomonas*, *Alcaligenes*, *Achromobacter*, and *Flavobacterium*, that due to proteolytic and lipolytic reactions taking place a slime is formed on the cheese surface combined with appearance of "off odors" and "off flavors" (de Oliveira et al., 2015).

In cheese, spoilage bacteria can easily gain entry through contaminated water during curd washing or during handling. They may produce gas internally in the cheese thus shortening the shelf life of the product by causing process failures such as inhibiting fermentation. Gas production in fresh cheese, also called "early blowing," is usually caused by Enterobacterales and some *Bacillus* species whereas the "late blowing" defects refer to ripened ones, involving clostridia species (*C. butyricum*, *C. tyrobutyricum*, *C. sporogenes*) responsible for cheese spoilage (Fernández et al., 2015; Yeluri Jonnala et al., 2018). Spore-forming bacteria, such as clostridia, can cause blowout of fermented products (Lopez-Brea et al. 2017). Another spoilage bacteria that produce defects such as pink discoloration in cheese is *Thermus thermophilus*, a species associated with the production of carotenoids (Quigley et al., 2016). Red-brown defects have also occurred in ripened cheeses, mainly associated with *Microbacterium*, *Brevibacterium* and *Corynebacterium*, respectively (Yeluri Jonnala et al., 2018).

Lactic acid bacteria are the most commonly added microbial group in dairy products such as cheese and fermented dairy products, and have a significant impact on increasing the dairy product's shelf life by inhibiting both spoilage and pathogenic organisms, but they can also be a group of bacteria that cause spoilage. Heterofermentative LAB such as lactobacilli and *Leuconostoc* can also develop "off-flavors" and gas in ripened cheeses (Ledenbach & Marshall, 2009).

For high-heat processed products such as UHT milk and milk powders spore-forming *Bacillus* and *Geobacillus* are the main problems. Spoilage microorganisms can be divided into three main groups for these products: spore-forming bacteria, thermoduric, and thermophiles. Spore-forming bacteria can survive high-heat processing and can be problematic in processed and dried dairy products (Li et al., 2019; McHugh et al., 2017). When conditions are favorable, the spores can germinate into vegetative cells. Thermoduric bacteria spores are the most resistant to processing and remain active after milk processing and therefore spoil the milk during prolonged storage, limiting the shelf life of dairy products. Some thermoduric species are spore formers but others are not, such as micrococci, *Corynebacterium*, and enterococci (Fusco et al., 2020; Gleeson, O'Connell, & Jordan, 2013; Gopal et al., 2015). Psychotropic thermoduric species survive pasteurization and grow at refrigeration temperatures, reducing the shelf life of dairy products (Panthi et al., 2017). Thermophilic bacteria grow at high temperatures and can survive in the processing equipment, causing fouling and biofilm formation (Jindal et al., 2016).

Controls in avoiding microbial spoilage include implementing hygiene measurements at each step from cow to processing, packaging, and storage, including the use of inhibitors and preservatives. The continuous monitoring of bacterial levels in the processing environment with good hygiene and sufficient cleaning and disinfection procedures need to be in place. Processing controls include heat treatments, bactofugation, and physical exclusion, such as filtration.

1.4 Microbial contamination of fruits and vegetables

The global consumption of fresh fruits and vegetables has increased in recent years as people adopt a healthier way of living by consuming more fresh produce, including fresh or minimally processed fruits and vegetables. The consumption of fruits and vegetables has been shown to be beneficial for maintaining a healthy diet but also to protect against coronary diseases, diabetes, cancer, and obesity, through the intake of vitamins (such as B, C, and K) and minerals (such as potassium, calcium, and magnesium), as well as dietary fibers (Yahia, García-Solís, & Maldonado Celis, 2019). Additionally, people tend to consume fresh produce for convenience as it is a grab-and-go snack. Herbs and spices also play an important role in diets as people are more inclined to try exotic recipes but also due to their associated health benefits, including anti-inflammatory properties, improving brain function and memory, or lowering blood sugar levels (Yahia et al. 2019).

The largest producers of fruits and vegetables globally are China, India, Brazil, the United States, and Turkey, followed by Mexico, Russia, Nigeria, and Spain, among others (STATISTA, 2021). The tremendous increase in the consumption of fruits and vegetables has put a lot of pressure on importation, with 35% of fresh produce in the United States coming from imports (Balali et al., 2020). With such globalization, ensuring product safety is crucial not only for consumers and importers but also for producers.

Microbial contamination of fruits and vegetables has serious implications from a food safety perspective as such products can become contaminated with foodborne pathogens and spoilage bacteria, which can result in human infection outbreaks, resulting in public health consequences, but also financial repercussions due to possible food recalls (Balali et al., 2020; Lee, Yang, & Yoon, 2021). Special attention needs to be given to herbs and spices, including dried ones, as some microorganisms have the ability to survive in products with a low water activity content for prolonged periods of time (Gurtler & Keller, 2019).

1.4.1 Sources of microbial contamination associated with fruits and vegetables

The quality and safety of fresh produce, minimally processed produce, herbs, spices as well as other categories of products including mixed salads, canned products, desiccated and frozen fruits and vegetables, sprouted seeds and juices, can be influenced by the presence of pathogenic and spoilage microorganisms. Microbial contamination of fresh produce can occur during pre- and post-harvest along the farm-to-fork chain (Fig. 1.1). Potential contamination events at the pre-harvest level are often attributed to crop cultivation, from the use of treated or untreated manure or human biosolids as nutrient sources. However, such fertilizers, if not treated appropriately, can harbor pathogenic microorganisms that can contaminate the crop (Machado-Moreira et al., 2019). In addition, water of poor microbiological quality can be a vehicle for the transmission of foodborne pathogens affecting the safety of crops. Seasonality and the level of precipitation can impact on contamination of fruits and vegetables as extreme weather events such as drought or flooding can lead to the survival of microorganisms that tolerate low water activity and/or cause alterations of soil microbiota, respectively (Machado-Moreira et al., 2019).

Water used for washing is the main source of contamination during post-harvest, as the washing water is generally reused several times before being exhausted and often for different crops successively, limiting the traceability in the case of an outbreak investigation (Balali et al., 2020; Machado-Moreira et al., 2019). In addition, the product-to-water ratio, especially crops carrying debris and small plant fragments, may lead to a higher accumulation of organic matter affecting the efficiency of antimicrobial treatments used for water decontamination purposes (Murray et al., 2017). Therefore the proper design of washing equipment, washing tanks, and processing facilities, as well as proper quality management in place, is critical to limit cross-contamination (Balali et al., 2020; Machado-Moreira et al., 2019). Another important parameter to consider is the maintenance of cold chain conditions at processing, shipping, retail, and in homes or commercial settings to limit the proliferation of microorganisms, especially pathogens, if already present (Ansah, Amodio, & Colelli, 2018).

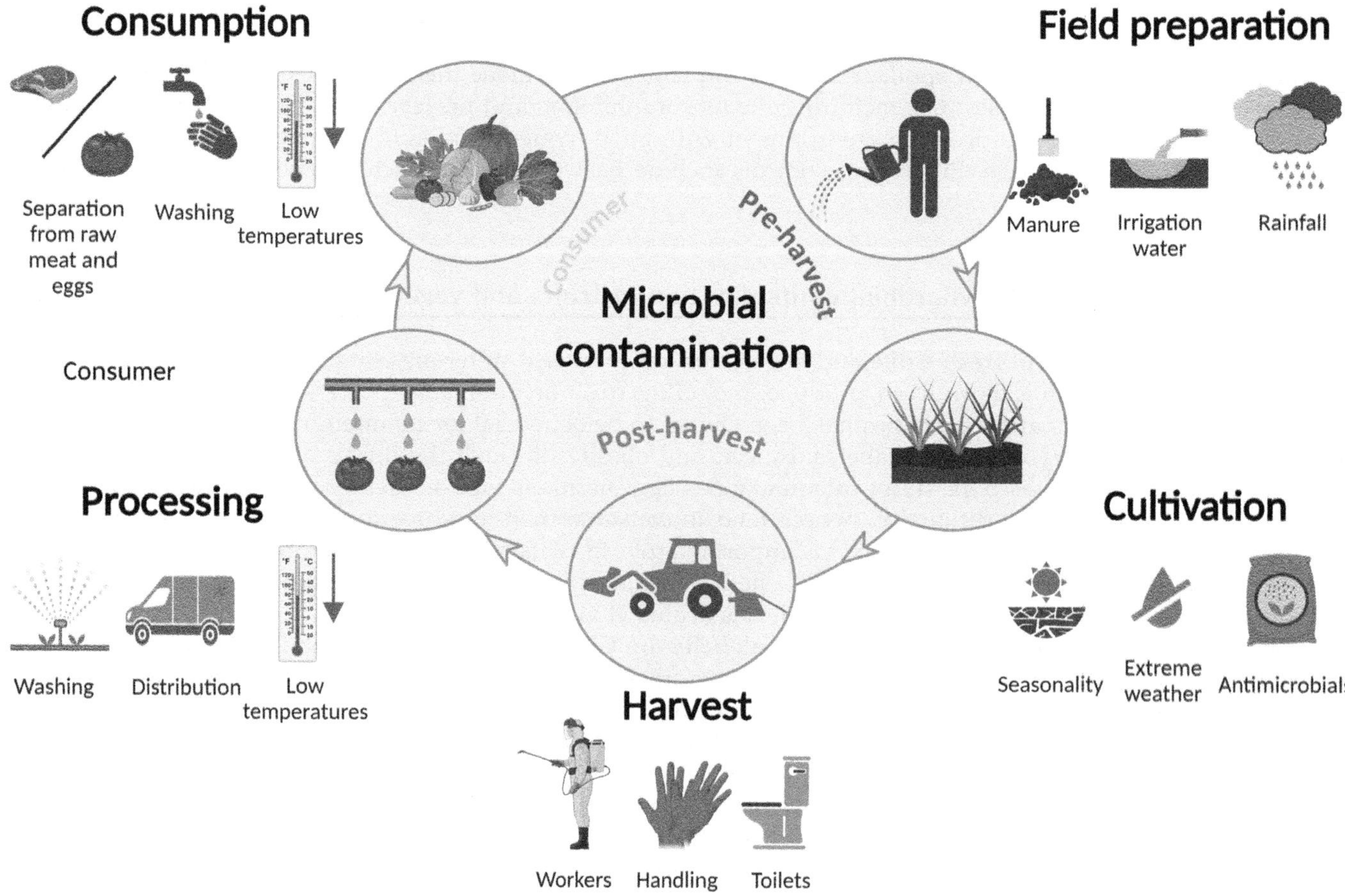

FIGURE 1.1 Microbial contamination of fresh produce from primary production to final consumer. *Source: Created with http://www.biorender.com, 2023.*

Handling contaminated produce can result in the transference of microorganisms to workers' hands or vice versa, with those with underlying conditions becoming a source for crops' contamination (Balali et al., 2020; Julien-Javaux et al., 2019). The risk of contamination increases when gloves, harvesting knives, and other field and processing utensils are not periodically changed and/ or disinfected, and furthermore the risk is amplified in the absence of toilets, hand-washing facilities, disinfectants, respectively (Julien-Javaux et al., 2019). Finally, manipulation of fruits and vegetables, as well as spices and herbs, by the final consumer or commercial settings such as restaurants, canteens, hospitals, or catered events, are also potential risks for cross-contamination if there is not a clear separation from other commodities such as raw meat, eggs, etc. (EFSA ECDC, 2021).

Increasing awareness about the sources of contamination can limit the risks associated with foodborne pathogens and spoilage microorganisms, by minimizing losses and keeping a high level of quality and safety along the farm-to-fork chain. Good hygiene and manufacturing practices are essential to ensure the safety and quality of fresh produce. Furthermore, different interventions can be implemented to decrease microbial loads in produce, thereby limiting the proliferation of microorganisms and thus increasing their shelf life.

1.4.2 *Microbial pathogens associated with fruits and vegetables*

Bacteria, viruses, parasites, and molds have all been implicated in fresh produce-related outbreaks. Often, the soil environment, driven by amendment with manure, human biosolids, and suboptimal water sources, dictates the profile of pathogens, as well as their ability to persist for prolonged periods on the produce surface (Alegbeleye, Singleton, & Sant'Ana, 2018). The increased consumption of fruits, vegetables, herbs, and spices is consistent with the increased number of reported outbreaks in the past years.

A recent review of microbial contamination of fresh produce identified norovirus as being accountable for the highest number of RTE produce-related outbreaks (304 outbreaks, corresponding to 53.2% of the total), followed by *Salmonella* (128 outbreaks) and *E. coli* (41 outbreaks) from 1980 to 2016 (Machado-Moreira et al., 2019). The largest norovirus outbreak occurred in Germany in 2012 when 390 institutions, predominantly schools, and child-care facilities, reported nearly 11,000 cases of gastroenteritis. The likely vehicle was identified as frozen strawberries imported from China being served in the affected institutions (Bernard et al., 2014). Studies have demonstrated the survival of viruses such as norovirus and hepatitis A in different water sources like groundwater, tap water, and surface water droplets. Although their presence is not commonly tested for in fresh produce ([1]), they should not be disregarded as it has been shown that they can survive for up to 10 days in leafy green vegetables and berries, under refrigerated conditions (Lamhoujeb et al., 2008). Other viruses commonly associated with fresh fruits and vegetables include *Rotavirus*, *Sapovirus*, and *Calicivirus* (Balali et al., 2020).

Among bacteria, *E. coli* accounts for a high number of fresh produce-associated illnesses. Although the majority of strains of this species are harmless and normally belong to the gastrointestinal microbiota, several of them are virulent and capable of causing illnesses associated with the gastrointestinal, nervous, and urinary systems (Machado-Moreira et al., 2019). Of particular concern are STEC. Studies have reported survival of *E. coli*, *Salmonella* and *Shigella* resulting in cross-contamination of fresh produce from rain and irrigation water droplets from contaminated fields, respectively (Machado-Moreira et al., 2019). Large STEC outbreaks have been linked to the consumption of fresh produce. Around 10,000 related cases were associated with the consumption of contaminated radish sprout salads in canteens in Japan. Although radish sprouts salads contained other uncooked ingredients such as mayonnaise, the epidemiological findings indicated that contaminated radish sprouts were the main vehicle (Watanabe et al., 1999). Fenugreek seeds contaminated with *E. coli* O104 were the main cause of illness for 3,126 STEC and 773 hemolytic uremic syndrome *E. coli*-associated cases with 47 deaths in Germany and other countries in 2011 (EFSA, 2011).

Different types of pepper and sesame seeds have been associated with multiple serovars of *Salmonella* (Gurtler & Keller, 2019) and it has been shown to be able to survive up to 405 days in manure or fields treated with manure (Machado-Moreira et al., 2019). *Salmonella* spp. is of particular concern as it has the ability to tolerate stressful environmental conditions including products with low water activity such as chocolate, sesame spreads, paprika powders, peanut butter spreads, and other foodstuffs. *Salmonella* spp. includes over 2500 serovars making it difficult to link specific serovars to a possible contamination source by culturing methods alone. For example, *S. Saintpaul* has been involved in two outbreaks related to contaminated peppers and paprika powder (Barton Behravesh et al., 2011; Lehmacher, Bockemühl, & Aleksic, 1995). The outbreak which occurred in 1993 led to approximately 1000 cases involving the consumption of contaminated paprika used to prepare potato snacks and RTE foods (Lehmacher et al. 1995). Similar patterns occurred in another outbreak but this time contaminated jalapeno and serrano peppers were listed as the main source, causing almost 1500 cases in the United States, of which 21% of people were admitted to hospitals and two died (Barton Behravesh et al., 2011). Outbreaks involving *S. javiana* and *S. montevideo* have been linked to contaminated tomatoes where the main source of contamination was the water bath used by the packer (Harris et al., 2003).

Soil fertilization with manure has been linked to *L. monocytogenes* which can then be transferred onto fresh produce, especially crops being in contact with the soil (e.g., melons, celery, cucumber, etc). in one outbreak consumption of contaminated cantaloupe was the source of listeriosis for 147 people, of which 33 (22%) died, and was traced back to improper hygienic conditions identified in the processing facility (McCollum et al., 2013). Contaminated whole apples were the main cause of an *L. monocytogenes* outbreak in 2014. This outbreak caused 35 cases including seven deaths across twelve states in the United States (Angelo et al., 2017).

Several cases of *Campylobacter* spp. contamination in fresh produce has been reported, most likely as a result of cross-contamination from meat sources (Guévremont et al., 2015; Mohammadpour et al., 2018). Guévremont et al. (2015) demonstrated the capacity of *Campylobacter* to survive in spinach leaves for up to 7 days at low temperatures. Campylobacteriosis has been linked with the consumption of fresh produce most likely due to an inadequate separation of raw meat from the fresh produce (Carstens, Salazar, & Darkoh, 2019; Møretrø et al., 2021).

Bacillus spp. (predominantly *B. cereus*, followed by *B. thuringiensis*, *B. licheniformis*, and others), which are widely distributed in soil and plant material, have been particularly associated with the contamination of spices as resistant endospores can survive even after drying and radiation treatments (Gurtler & Keller, 2019). Other reported disease cases included the presence of *C. botulinum* in garlic and cabbage salad (Machado-Moreira et al., 2019). White, red and black pepper, turmeric, and garlic are examples of products linked to outbreaks involving these spore formers (Van Doren et al., 2013).

Protozoan pathogens have been associated with fruit and vegetable-associated illnesses as they can survive without a host for a prolonged period in water and are resistant to many chemical water treatments. An infected individual, either human or animal, excretes the protozoan parasites that are then further ingested through the consumption of contaminated fruits and vegetables. *Cyclospora* and *Cryptosporidium* have been associated with the consumption of raspberries and salads (Machado-Moreira et al., 2019). In the United States, for example, a multistate outbreak of cyclosporiasis occurred in 1996 associated with imported raspberries accounting for 1465 cases, while in 2014, coriander was the main source of an outbreak, being responsible for 304 cases (Herwaldt & Ackers, 1997; Wadamori, Gooneratne, & Hussain, 2017). Nevertheless, other protozoan pathogens such as *Giardia*, *Toxoplasmagondii*, *Ascaris*, and others have been linked to fresh produce, with water as the main source of contamination (Balali et al., 2020).

Fungal contamination can have serious repercussions for consumer health, especially those capable of producing mycotoxins that are heat stable and difficult to eliminate once present (genera belonging to *Penicillium*, *Aspergillus*, *Fusarium*, etc.). Climatic conditions and excessive irrigation of crops can increase fungal colonization by contributing to a gradual increase in mycotoxin production, above acceptable levels (Costa et al., 2019). For example, aflatoxigenic *Aspergillus* and patulin-producing *Penicillium* have been found in smoked paprika samples, despite the low water activity of this product (Casquete et al., 2017).

1.4.3 Spoilage microorganisms associated with fruits and vegetables

Spoilage microorganisms affect the quality of products by making them unappealing for consumption due to a change in taste, odor, color, and texture, often associated with a slime and watery appearance. A rapid increase in visual deterioration occurs in minimally processed fruits and vegetables that are physically injured, older, and/ or wet, becoming more prone to nutrient leaching (Mogren et al., 2018; Mulaosmanovic et al., 2021). The spoilage burden can be reduced by using low temperatures during processing, retail, and at the consumer level. Nonetheless, by their expiration date, products such as RTE salads can be associated with the appearance of defects such as browning and wilting and the development of "off-odors" and "off-flavors" due to an overgrowth of spoilage microorganisms (Xylia et al., 2021).

Despite the fact that microbial composition of crops differs from field to bag (Mulaosmanovic et al., 2021), studies are mostly focused on pathogen detection and methods to increase shelf life by limiting the growth of spoilage microorganisms, as opposed to identifying specific species causing spoilage. Spoilage bacteria such as *Erwinia carotovora*, *Corynebacterium*, *Pseudomonas* spp., and LAB predominantly affect vegetables (Tournas, 2005). Other bacterial species such as *B. mojavensis*, *B. megaterium*, and *P. fluorescens* have been isolated from RTE salads and have been shown to be present in increased numbers at the end of shelf life (Xylia et al., 2021). Certain

Erwinia species can survive and replicate at refrigeration temperatures (Jasper, Elmore, & Wagstaff, 2021). In a study conducted by Mulaosmanovic et al. (2021), an increased abundance of *Pseudomonadaceae* and Enterobacterales, including both spoilage and pathogenic species, were identified after the washing step, indicating the influence of water quality.

Spoilage fungi can cause rotting and wilting of crops, therefore reducing their commercial value and considerably influencing the crops' organoleptic properties. Besides common spoilage fungi isolated such as *Aspergillus* and *Penicillium*, others fungi such as *Mucor, Rhizopus, Harzia, Cladosporium* have been isolated at a lower frequency from crops such as peppers during field cultivation (Costa et al., 2019). Damage to fruit and vegetables makes them much more susceptible to microbial contamination and spoilage. Therefore, careful handling, appropriate packaging, and maintenance of the cold chain are essential to minimize spoilage from microbial contamination.

1.5 Conclusion

Food products are generally not sterile and they have a specific microbial community associated with them. As demonstrated in this chapter, the composition of this community is impacted by the physiochemical properties of the food itself and the environment where it is produced, processed, transported, and stored. Contamination of the food product with pathogenic or spoilage organisms can occur at any of these stages and their ability to survive and grow is similarly impacted by both intrinsic and extrinsic factors. In order to ensure food product safety and quality, it is critical to try to prevent contamination at all stages of production and distribution and minimize microbial proliferation as much as possible. The use of antimicrobial treatments at appropriate points can aid in ensuring product safety and adequate shelf life. The implementation of a robust food safety management system allows the identification of hazards such as microbial contamination and the appropriate controls needed to ensure food safety. Tools such as next-generation sequencing have revolutionized food safety management in recent years and enable the accurate linking of outbreaks to specific foods and contamination sources, as well as providing important information on the virulence of the organism. Such technology also enables the identification of the entire microbial community of foods and their environment and how these change over time. This enables a much more in-depth understanding of the influence of contamination, as well as the impact of any mitigation measures put in place. It also facilitates the identification of suitable biomarkers for monitoring product safety and quality.

Food production faces many challenges, including the need to produce food for a growing global population in a more sustainable manner, as well as the influences of climate change on food production. These challenges will alter the risks posed by microbial contamination of food and such emerging risks need to be considered. These challenges also provide the impetus to develop more sustainable methods to provide food safety assurance and minimize food waste, from production, through processing and packaging.

References

Alegbeleye, O. O., Singleton, I., & Sant'Ana, A. S. (2018). Sources and contamination routes of microbial pathogens to fresh produce during field cultivation: A review. *Food Microbiology, 73*, 177–208. Available from https://doi.org/10.1016/j.fm.2018.01.003.

Angelo, K. M., et al. (2017). Multistate outbreak of Listeria monocytogenes infections linked to whole apples used in commercially produced, prepackaged caramel apples: United States, 2014–2015. *Epidemiology and Infection, 145*(5), 848–856. Available from https://doi.org/10.1017/S0950268816003083.Multistate.

Ansah, F. A., Amodio, M. L., & Colelli, G. (2018). Quality of fresh-cut products as affected by harvest and postharvest operations. *Journal of the Science of Food and Agriculture, 98*(10), 3614–3626. Available from https://doi.org/10.1002/jsfa.8885.

Balali, G. I., et al. (2020). Microbial contamination, an increasing threat to the consumption of fresh fruits and vegetables in today's world. *International Journal of Microbiology, 2020*, 1–13. Available from https://doi.org/10.1155/2020/3029295, ID 3029295.

Barton Behravesh, C., et al. (2011). 2008 outbreak of Salmonella Saintpaul infections associated with raw produce. *New England Journal of Medicine, 364*(10), 918–927. Available from https://doi.org/10.1056/nejmoa1005741.

Berends, B. R., et al. (1997). Identification and quantification of risk factors regarding Salmonella spp. on pork carcasses. *International Journal of Food Microbiology, 36*(2–3), 199–206. Available from https://doi.org/10.1016/S0168-1605(97)01267-1.

Bernard, H., et al. (2014). Large multistate outbreak of norovirus gastroenteritis associated with frozen strawberries, Germany, 2012. *Eurosurveillance, 19*(8), 1–9. Available from https://doi.org/10.2807/1560-7917.ES2014.19.8.20719.

Bolton, D. J., Carroll, J., & Walsh, D. (2015). A four-year survey of blown pack spoilage Clostridium estertheticum and Clostridium gasigenes on beef primal cuts. *Letters in Applied Microbiology, 61*(2), 153–157. Available from https://doi.org/10.1111/lam.12431.

Boor, K. J., et al. (2017). A 100-year review: Microbiology and safety of milk handling. *Journal of Dairy Science, 100*(12), 9933–9951.

Borch, E., Kant-Muermans, M. L., & Blixt, Y. (1996). Bacterial spoilage of meat and cured meat products. *International Journal of Food Microbiology, 33*(1), 103–120. Available from https://doi.org/10.1016/0168-1605(96)01135-X.

Brizio, A. P. D. R., et al. (2015). Visible contamination in broiler carcasses and its relation to the stages of evisceration in poultry slaughter. *International Food Research Journal, 22*(1), 59–63.

Carstens, C. K., Salazar, J. K., & Darkoh, C. (2019). Multistate outbreaks of foodborne illness in the United States associated with fresh produce from 2010 to 2017. *Frontiers in Microbiology, 10*, 1–15. Available from https://doi.org/10.3389/fmicb.2019.02667, November.

Casquete, R., et al. (2017). Occurrence of toxigenic fungi and mycotoxins during smoked paprika production. *Journal of Food Protection, 80*(12), 2068–2077. Available from https://doi.org/10.4315/0362-028X.JFP-17-164.

CDC. (1993). *Multistate outbreak of Escherichia coli O157:H7 infections from hamburgers- Western United States, 1992–1993*. https://www.cdc.gov/.

CDC. (2019). *Outbreak of E. coli infections linked to ground beef*. https://www.cdc.gov/.

CDC. (2021). *Campylobacter (Campylobacteriosis)*. https://www.cdc.gov/.

Choi, K. H., et al. (2016). Cheese microbial risk assessments-a review. *Asian-Australasian Journal of Animal Sciences, 29*(3), 307–314.

Christiansson, A. (2011). *Pathogens in milk: Bacillus cereus. Encyclopedia of dairy sciences* (pp. 24–30). Academic Press.

Costa, J., et al. (2019). Overview of fungi and mycotoxin contamination in capsicum pepper and in its derivatives. *Toxins, 11*(1), 1–16. Available from https://doi.org/10.3390/toxins11010027.

D'Aoust, J. Y., Warburton, D. W., & Sewell, A. M. (1985). Salmonella typhimurium phage-type 10 from Cheddar cheese implicated in a major Canadian foodborne outbreak. *Journal of Food Protection, 48*(12), 1062–1066. Available from https://doi.org/10.4315/0362-028X-48.12.1062.

Davis, K.R. et al. (2016). *Campylobacter jejuni infections associated with raw milk consumption- Utah, 2014, CDC- Morbidity and Mortality Weekly Report*. https://www.cdc.gov.

de Oliveira, G. B., et al. (2015). Psychrotrophic bacteria in milk: How much do we really know? *Brazilian Journal of Microbiology, 46*(2), 313–321. Available from https://doi.org/10.1590/S1517-838246220130963.

Doulgeraki, A. I., et al. (2012). Spoilage microbiota associated to the storage of raw meat in different conditions. *International Journal of Food Microbiology, 157*(2), 130–141. Available from https://doi.org/10.1016/j.ijfoodmicro.2012.05.020.

Drosinos, E. H., & Paramithiotis, S. (2009). *Chapter 4. Cured meats and poultry, including cooked cured meats. Microbiology handbook meat products* (pp. 101–116). Leatherhead Food International Ltd.

Drudy, D., et al. (2006). Enterobacter sakazakii: An emerging pathogen in powdered infant formula. *Clinical Infectious Diseases, 42*(7), 996–1002.

ECDC. (2019). *Rapid outbreak assessment: Multi-country outbreak of Listeria monocytogenes sequence type 6 infections linked to ready-to-eat meat products*. https://www.ecdc.europa.eu/.

ECDC. (2021). *Multi-country outbreak of Salmonella Enteritidis sequence type (ST)11 infections linked to poultry products in the EU/EEA and the United Kingdom*. https://www.ecdc.europa.eu.

EFSA. (2011). Shiga toxin-producing E. coli (STEC) O104:H4 2011 outbreaks in Europe: Taking Stock. *EFSA Journal, 9*(10), 1–22. Available from https://doi.org/10.2903/j.efsa.2011.2390.

EFSA ECDC. (2021). The European Union one health 2020 zoonoses report. *EFSA Journal, 19*(12), 1–324. Available from https://doi.org/10.2903/j.efsa.2021.6971.

EFSA ECDC. (2022). The European Union one health 2021 zoonoses report. *EFSA Journal, 20*(12), 1–273. Available from https://doi.org/10.2903/j.efsa.2022.7666.

Elmadfa, I., & Meyer, A. L. (2017). Animal proteins as important contributors to a healthy human diet. *Annual Review of Animal Biosciences, 5*, 4.1–4.21. Available from https://doi.org/10.1146/annurev-animal-022516-022943.

Ercolini, D., et al. (2009). Mesophilic and psychrotrophic bacteria from meat and their spoilage potential in vitro and in beef. *Applied and Environmental Microbiology, 75*(7), 1990–2001. Available from https://doi.org/10.1128/AEM.02762-08.

FAO. (2011). *Dietary protein quality evaluation in human nutrition. Report of an FAQ Expert Consultation, FAO Food and Nutrition paper*.

Fernandes, R. (2009). *Chapter 1. Chilled and frozen raw meat, poultry and their products. Microbiology handbook meat products* (pp. 1–44). Leatherhead Food International Ltd.

Fernández, M., et al. (2015). Impact on human health of microorganisms present in fermented dairy products: An overview. *BioMed Research International, 2015*, 1–13. Available from https://doi.org/10.1155/2015/412714, ID 412714.

Fox, E., et al. (2011). Listeria monocytogenes in Irish farmhouse cheese processing environments. *International Journal of Food Microbiology, 145*, S39–S45.

FSA. (2022). *FSA issues advice to consumers not to eat several pork scratching products linked to Salmonella poisoning*. https://www.food.gov.uk.

Fusco, V., et al. (2020). Microbial quality and safety of milk and milk products in the 21st century. *Comprehensive Reviews in Food Science and Food Safety, 19*(4), 2013–2049.

Geiker, N. R. W., et al. (2021). Meat and human health- current knowledge and research gaps. *Foods, 10*(7), 1–17. Available from https://doi.org/10.3390/foods10071556.

Gleeson, D., O'Connell, A., & Jordan, K. (2013). Review of potential sources and control of thermoduric bacteria in bulk-tank milk. *Irish Journal of Agricultural and Food Research, 52*, 217–227.

Godfray, H. C. J., et al. (2018). Meat consumption, health, and the environment. *Science (New York, N.Y.), 361*, 1–8. Available from https://doi.org/10.1126/science.aam5324.

Gómez, D., et al. (2015). Occurrence of Listeria monocytogenes in ready-to-eat meat products and meat processing plants in Spain. *Foods, 4*(3), 271–282. Available from https://doi.org/10.3390/foods4030271.

Gopal, N., et al. (2015). The prevalence and control of Bacillus and related spore-forming bacteria in the dairy industry. *Frontiers in Microbiology, 6*, 1–18.

Gould, L. H., & Limbago, B. (2010). Clostridium difficile in food and domestic animals: A new foodborne pathogen? *Clinical Infectious Diseases, 51*(5), 577–582. Available from https://doi.org/10.1086/655692.

Gould, L. H., et al. (2013). Surveillance for foodborne disease outbreaks - United States, 2009-2015. *MMW, 62*(2), 1–34. Available from https://doi.org/10.15585/MMWR.SS6710A1.

Gounadaki, A. S., et al. (2008). Microbial ecology of food contact surfaces and products of small-scale facilities producing traditional sausages. *Food Microbiology, 25*(2), 313–323. Available from https://doi.org/10.1016/j.fm.2007.10.001.

Grant, G. F., McCurdy, A. R., & Osborne, A. (1988). Bacterial greening in cured meats: A review. *Canadian Institute of Food Science and Technology Journal, 21*(1), 50–56.

Griffiths, M. (2010). *Improving the safety and quality of milk: Milk production and processing.*

Guévremont, E., et al. (2015). Survival of Campylobacter jejuni on fresh spinach stored at 4°C or 12°C. *Food Control, 50*, 736–739. Available from https://doi.org/10.1016/j.foodcont.2014.10.023.

Gurtler, J. B., & Keller, S. E. (2019). Microbiological safety of dried spices. *Annual Review of Food Science and Technology, 10*, 409–427. Available from https://doi.org/10.1146/annurev-food-030216-030000.

Harris, L. J., et al. (2003). Outbreaks associated with fresh produce: Incidence, growth, and survival of pathogens in fresh and fresh-cut produce. *Comprehensive Reviews in Food Science and Food Safety*, 2(1 SUPPL.), 78–141. Available from https://doi.org/10.1111/j.1541-4337.2003.tb00031.x.

Herwaldt, B. L., & Ackers, M.-L. (1997). An outbreak in 1996 of cyclosporiasis associated with imported raspberries. *New England Journal of Medicine, 336*(22), 1548–1556. Available from https://doi.org/10.1056/nejm199705293362202.

Húngaro, H. M., et al. (2016). Blown pack spoilage in vacuum-packaged meat: A review on clostridia as causative agents, sources, detection methods, contributing factors and mitigation strategies. *Trends in Food Science and Technology, 52*, 123–138. Available from https://doi.org/10.1016/j.tifs.2016.04.010.

Iulietto, M. F., et al. (2015). Meat spoilage: A critical review of a neglected alteration due to ropy slime producing bacteria. *Italian Journal of Animal Science, 14*(3), 316–326. Available from https://doi.org/10.4081/ijas.2015.4011.

Jacobs, C., Braun, P., & Hammer, P. (2011). Reservoir and routes of transmission of Enterobacter sakazakii (Cronobacter spp.) in a milk powder-producing plant. *Journal of Dairy Science, 94*(8), 3801–3810.

Jansen, W., et al. (2019). Brucella-positive raw milk cheese sold on the inner European market: A public health threat due to illegal import? *Food Control, 100*, 130–137. Available from https://doi.org/10.1016/j.foodcont.2019.01.022, January.

Jasper, J., Elmore, J. S., & Wagstaff, C. (2021). Determining the quality of leafy salads: Past, present and future. *Postharvest Biology and Technology, 180*(111630), 1–13. Available from https://doi.org/10.1016/j.postharvbio.2021.111630.

Jay, J. M., Loessner, M. J., & Golden, D. A. (2008). *Modern food microbiology*. Springer.

Jenkins, C., et al. (2022). Outbreak of STEC O157:H7 linked to a milk pasteurisation failure at a dairy farm in England, 2019. *Epidemiology and Infection, 150*(e114), 1–7. Available from https://doi.org/10.1017/s0950268822000929.

Jindal, S., et al. (2016). Evaluation of modified stainless steel surfaces targeted to reduce biofilm formation by common milk sporeformers. *Journal of Dairy Science, 99*(12), 9502–9513.

Jordan, K. N., Hunt, K., & Fox, E. M. (2012). *Troubleshooting the environmental source of contamination with Listeria monocytogenes in a typical small food manufacturing plant in Ireland. Case studies in food safety and authenticity* (pp. 95–101). Woodhead Publishing.

Julien-Javaux, F., et al. (2019). Strategies for the safety management of fresh produce from farm to fork. *Current Opinion in Food Science, 27*, 145–152. Available from https://doi.org/10.1016/j.cofs.2019.01.004.

Karwowska, M., Łaba, S., & Szczepański, K. (2021). Food loss and waste in meat sector- why the consumption stage generates the most losses? *Sustainability, 13*, 1–13. Available from https://doi.org/10.3390/su13116227.

Kousta, M., et al. (2010). Prevalence and sources of cheese contamination with pathogens at farm and processing levels. *Food Control, 21*(6), 805–815. Available from https://doi.org/10.1016/j.foodcont.2009.11.015.

Koutsoumanis, K. P., Lianou, A., & Sofos, J. N. (2014). Food safety: Emerging pathogens. *Encyclopedia of Agriculture and Food Systems, 3*, 250–272. Available from https://doi.org/10.1016/B978-0-444-52512-3.00049-8, January.

Lamhoujeb, S., et al. (2008). Evaluation of the persistence of infectious human noroviruses on food surfaces by using real-time nucleic acid sequence-based amplification. *Applied and Environmental Microbiology, 74*(11), 3349–3355. Available from https://doi.org/10.1128/AEM.02878-07.

Ledenbach, L. H., & Marshall, R. T. (2009). Microbiological spoilage of dairy products. *Compendium of the microbiological spoilage of foods and beverages*, 41–67. Available from https://doi.org/10.1007/978-1-4419-0826-1_2.

Ledo, J., Hettinga, K. A., & Luning, P. A. (2020). A customised assessment tool to differentiate safety and hygiene control practices in emerging dairy chains. *Food Control, 111*, 107072.

Lee, J. Y., Yang, S. Y., & Yoon, K. S. (2021). Control measures of pathogenic microorganisms and shelf-life extension of fresh-cut vegetables. *Foods, 10*(3). Available from https://doi.org/10.3390/foods10030655.

Lehmacher, A., Bockemühl, J., & Aleksic, S. (1995). Nationwide outbreak of human salmonellosis in Germany due to contaminated paprika and paprika-powdered potato chips. *Epidemiology and Infection, 115*(3), 501–511. Available from https://doi.org/10.1017/S0950268800058660.

Li, F., et al. (2019). Occurrence and identification of spore-forming bacteria in skim-milk powders. *International Dairy Journal, 97*, 176–184.

Lianou, A., Panagou, E.Z. and Nychas, G.J. E. (2017) 'Meat safety-I Foodborne pathogens and other biological issues', in Lawrie's meat science, pp. 521–552. doi: 10.1016/B978-0-323-85408-5.00015-7.

Lindsay, D., et al. (2019). Quantifying the uncertainty of transfer of Cronobacter spp. between fomites and floors and touch points in dairy processing plants. *Food Microbiology, 84*(103256), 1–7.

Lipinski, B. (2020). Why does animal-based food loss and waste matter? *Animal Frontiers, 10*(4), 48–52. Available from https://doi.org/10.1093/af/vfaa039.

Lopez-Brea, S. G., Gómez-Torres, N., & Arribas, M. Á. (2017). *Spore-forming bacteria in dairy products. Microbiology in dairy processing: Challenges and opportunities* (pp. 11–36). Wiley online Library. Available from https://doi.org/10.1002/9781119115007.ch2.

Luchansky, J. B., et al. (2008). Translocation of surface-inoculated Escherichia coli O157:H7 into beef subprimals following blade tenderization. *Journal of Food Protection, 71*(11), 2190–2197. Available from https://doi.org/10.4315/0362-028X-71.11.2190.

Lundén, J. M., Autio, T. J., & Korkeala, H. J. (2002). Transfer of persistent Listeria monocytogenes contamination between food-processing plants associated with a dicing machine. *Journal of Food Protection, 65*(7), 1129–1133. Available from https://doi.org/10.4315/0362-028X-65.7.1129.

Lunden, J., et al. (2003). Persistent and nonpersistent Listeria monocytogenes contamination in meat and poultry processing plants. *Journal of Food Protection, 66*(11), 2062–2069.

Machado-Moreira, B., et al. (2019). Microbial contamination of fresh produce: What, where, and how? *Comprehensive Reviews in Food Science and Food Safety, 18*(6), 1727–1750. Available from https://doi.org/10.1111/1541-4337.12487.

Marsden, J., et al. (2009). *Chapter 2. Cooked meats, poultry and their products. Microbiology handbook meat products* (pp. 53–78). Leatherhead Food International Ltd.

Marshall, R. T. (2001). *Frozen desserts. Applied Dairy Microbiology*. CRC Press.

McCollum, J. T., et al. (2013). Multistate outbreak of listeriosis associated with cantaloupe. *The New England Journal of Medicine, 369*(10), 944–953. Available from https://doi.org/10.1056/NEJMoa1215837.

McHugh, A. J., et al. (2017). Detection and enumeration of spore-forming bacteria in powdered dairy products. *Frontiers in Microbiology, 8*(109), 1–15.

Melo, J., Andrew, P. W., & Faleiro, M. L. (2015). Listeria monocytogenes in cheese and the dairy environment remains a food safety challenge: The role of stress responses. *Food Research International, 67*, 75–90. Available from https://doi.org/10.1016/j.foodres.2014.10.031.

Meloni, D. (2015). Presence of Listeria monocytogenes in mediterranean-style dry fermented sausages. *Foods, 4*(1), 34–50. Available from https://doi.org/10.3390/foods4010034.

Memisi, N., et al. (2015). CIP cleaning processes in the dairy industry. *Procedia Food Science, 5*, 184–186. Available from https://doi.org/10.1016/j.profoo.2015.09.052.

Mogren, L., et al. (2018). The hurdle approach-A holistic concept for controlling food safety risks associated with pathogenic bacterial contamination of leafy green vegetables. A review. *Frontiers in Microbiology, 9*, 1–20. Available from https://doi.org/10.3389/fmicb.2018.01965, AUG.

Mohammadpour, H., et al. (2018). The prevalence of Campylobacter spp. in vegetables, fruits, and fresh produce: A systematic review and meta-analysis. *Gut Pathogens, 10*(1), 1–12. Available from https://doi.org/10.1186/s13099-018-0269-2.

Møretrø, T., et al. (2021). Consumer practices and prevalence of Campylobacter, Salmonella and norovirus in kitchens from six European countries. *International Journal of Food Microbiology, 347*, 1–13. Available from https://doi.org/10.1016/j.ijfoodmicro.2021.109172, March.

Mulaosmanovic, E., et al. (2021). Processing of leafy vegetables matters: Damage and microbial community structure from field to bag. *Food Control, 125*(107894), 1–16. Available from https://doi.org/10.1016/j.foodcont.2021.107894.

Murray, K., et al. (2017). Challenges in the microbiological food safety of fresh produce: Limitations of post-harvest washing and the need for alternative interventions. *Food Quality and Safety, 1*(4), 289–301. Available from https://doi.org/10.1093/fqsafe/fyx027.

Newton, K. G., & Gill, C. O. (1981). The microbiology of DFD fresh meats: A review. *Meat Science, 5*(3), 223–232. Available from https://doi.org/10.1016/0309-1740(81)90005-X.

Nychas, G. J. E., & Skandamis, P. N. (2005). *Fresh meat spoilage and modified atmosphere packaging (MAP). Improving the safety of fresh meat* (pp. 461–502). Woodhead Publishing Ltd. Available from 10.1533/9781845691028.2.461.

Oliver, S. P., Jayarao, B. M., & Almeida, R. A. (2005). Foodborne pathogens in milk and the dairy farm environment: Food safety and public health implications. *Foodborne Pathogens and Disease, 2*(2), 115–129.

Oliver, S. P., et al. (2009). Food safety hazards associated with consumption of raw milk. *Foodborne Pathogens and Disease, 6*(7), 793–806. Available from https://doi.org/10.1089/fpd.2009.0302.

Panthi, R. R., et al. (2017). *Selection and treatment of milk for cheesemaking. Cheese* (pp. 23–50). Academic Press. Available from https://doi.org/10.1016/B978-0-12-417012-4.00002-8.

Pennacchia, C., Ercolini, D., & Villani, F. (2011). Spoilage-related microbiota associated with chilled beef stored in air or vacuum pack. *Food Microbiology, 28*(1), 84–93. Available from https://doi.org/10.1016/j.fm.2010.08.010.

Quigley, L., et al. (2013). The complex microbiota of raw milk. *FEMS Microbiology Reviews, 37*(5), 664–698.

Quigley, L., et al. (2016). Thermus and the pink discoloration defect in cheese. *mSystems, 1*(3), e00023. Available from https://doi.org/10.1128/mSystems.00023-16, Edited by R.J. Dutton.

Quintavalla, S., & Vicini, L. (2002). Antimicrobial food packaging in meat industry. *Meat Science, 62*, 373–380. Available from https://doi.org/10.1016/S0309-1740(02)00121-3.

Ranasinghe, R. A. S. S., et al. (2021). Persistence of Listeria monocytogenes in food commodities: Foodborne pathogenesis, virulence factors, and implications for public health. *Food Research, 5*(1), 1–16. Available from https://doi.org/10.26656/fr.2017.5(1)0.199.

Rawdkuen, S., Punbusayakul, N., & Lee, D. S. (2016). *Antimicrobial packaging for meat products. Antimicrobial food packaging* (pp. 229–238). Elsevier.

Regulation (EEC) No 315/93. (1993). Community procedures for contaminants in food. *Official Journal of the European Communities, 1993*, 1–3, L.

Riley, L. W., et al. (1983). Hemorrhagic colitis associated with a rare Escherichia coli serotype. *The New England Journal of Medicine, 308*(12), 1–4.

Rodriguez, C., et al. (2013). Presence of Clostridium difficile in pigs and cattle intestinal contents and carcass contamination at the slaughterhouse in Belgium. *International Journal of Food Microbiology, 166*(2), 256–262. Available from https://doi.org/10.1016/j.ijfoodmicro.2013.07.017.

Rouger, A., Tresse, O., & Zagorec, M. (2017). Bacterial contaminants of poultry meat: Sources, species, and dynamics. *Microorganisms, 5*(50), 1–16. Available from https://doi.org/10.3390/microorganisms5030050.

Rozenberg, S., et al. (2016). Effects of dairy products consumption on health: Benefits and beliefs-A commentary from the Belgian bone club and the European society for clinical and economic aspects of osteoporosis, osteoarthritis and musculoskeletal diseases. *Calcified Tissue International, 98*(1), 1–17. Available from https://doi.org/10.1007/s00223-015-0062-x.

Schirone, M., et al. (2022). Biogenic amines in meat and meat products: A review of the science and future perspectives. *Foods, 11*(788), 1–14. Available from https://doi.org/10.3390/foods11060788.

Settanni, L., et al. (2021). In-depth investigation of the safety of wooden shelves used for traditional cheese ripening. *Applied and Environmental Microbiology, 87*(23), 1–17. Available from https://doi.org/10.1128/AEM.01524-21.

Sheridan, J. J. (1998). Sources of contamination during slaughter and measures for control. *Journal of Food Safety, 18*, 321–339.

Singh, M., et al. (2019). *Meat and poultry. Food microbiology: Fundamentals and frontiers* (pp. 125–177). ASM Press. Available from 10.1128/9781555819972.ch6.

Skowron, K., et al. (2022). Two faces of fermented foods-The benefits and threats of its consumption. *Frontiers in Microbiology, 13*, 1–17. Available from https://doi.org/10.3389/fmicb.2022.845166, March.

Smith, A. M., et al. (2019). Outbreak of Listeria monocytogenes in South Africa, 2017-2018: Laboratory activities and experiences associated with whole-genome sequencing analysis of isolates. *Foodborne Pathogens and Disease, 16*(7), 524–530. Available from https://doi.org/10.1089/fpd.2018.2586.

Sofos, J. N. (2014). *Meat and meat products. Food safety management: A practical guide for the food industry* (pp. 119–162). Elsevier. Available from 10.32741/fihb.4.meat.

Sofos, J. N., et al. (2013). *Meat, poultry, and seafood. Food microbiology: Fundamentals and frontiers* (pp. 111–168). ASM Press. Available from 10.1128/9781555818463.ch6.

STATISTA. (2021). *Global fresh vegetable production in 2019, by region.* https://www.statista.com/statistics/264066/global-vegegable-production-by-region/.

Sung, S. Y., et al. (2013). Antimicrobial agents for food packaging applications. *Trends in Food Science and Technology, 33*(2), 110–123. Available from https://doi.org/10.1016/j.tifs.2013.08.001.

Tournas, V. H. (2005). Spoilage of vegetable crops by bacteria and fungi and related health hazards. *Critical Reviews in Microbiology, 31*(1), 33–44. Available from https://doi.org/10.1080/10408410590886024.

Triki, M., et al. (2018). Quality assessment of fresh meat from several species based on free amino acid and biogenic amine contents during chilled storage. *Foods, 7*(132), 1–16. Available from https://doi.org/10.3390/foods7090132.

Van Doren, J. M., et al. (2013). Foodborne illness outbreaks from microbial contaminants in spices, 1973-2010. *Food Microbiology, 36*(2), 456–464. Available from https://doi.org/10.1016/j.fm.2013.04.014.

Verraes, C., et al. (2015). A review of the microbiological hazards of dairy products made from raw milk. *International Dairy Journal, 50*, 32–44.

Wadamori, Y., Gooneratne, R., & Hussain, M. A. (2017). Outbreaks and factors influencing microbiological contamination of fresh produce. *Journal of the Science of Food and Agriculture, 97*(5), 1396–1403. Available from https://doi.org/10.1002/jsfa.8125.

Watanabe, Y., et al. (1999). Factory outbreak of Escherichia coli O157:H7 infection in Japan. *Emerging Infectious Diseases, 5*(3), 424–428. Available from https://doi.org/10.3201/eid0503.990313.

WHO. (2018). *Listeriosis- South Africa.* https://www.who.int.

Wyness, L. (2016). The role of red meat in the diet: Nutrition and health benefits. *Proceedings of the Nutrition Society, 75*(3), 227–232. Available from https://doi.org/10.1017/S0029665115004267.

Xylia, P., et al. (2021). Expiration date of ready-to-eat salads: Effects on microbial load and biochemical attributes. *Foods, 10*(941), 1–22. Available from https://doi.org/10.3390/foods10050941.

Yahia, E.M., García-Solís, P. and Maldonado Celis, M.E. (2019) 'Contribution of fruits and vegetables to human nutrition and health', in Postharvest physiology and biochemistry of fruits and vegetables. Elsevier, pp. 19–45. doi: 10.1016/B978-0-12-813278-4.00002-6.

Yeluri Jonnala, B. R., et al. (2018). Sequencing of the cheese microbiome and its relevance to industry. *Frontiers in Microbiology, 9*(1020), 1–12. Available from https://doi.org/10.3389/fmicb.2018.01020.

Yuan, L., et al. (2018). Spoilage potential of psychrotrophic bacteria isolated from raw milk and the thermo-stability of their enzymes. *Journal of Zhejiang University-Science B, 19*(8), 630–642.

Zastempowska, E., Grajewski, J., & Twaruzek, M. (2016). Food-borne pathogens and contaminants in raw milk - A review. *Annals of Animal Science, 16*(3), 623–639. Available from https://doi.org/10.1515/aoas-2015-0089.

Zhang, X., et al. (2021). Milk consumption and multiple health outcomes: Umbrella review of systematic reviews and meta-analyses in humans. *Nutrition and Metabolism, 18*(7), 1–18. Available from https://doi.org/10.1186/s12986-020-00527-y.

C H A P T E R

2

Antimicrobial nanoparticles in active food packaging applications

Swarup Roy[1,2], *Ruchir Priyadarshi*[1], *Deblina Biswas*[3,4] *and Jong-Whan Rhim*[1]

[1]Department of Food and Nutrition, BioNanocomposite Research Center, Kyung Hee University, Seoul, Republic of Korea [2]Department of Food Technology and Nutrition, School of Agriculture, Lovely Professional University, Phagwara, Punjab, India [3]Department of Electrical and Computer Engineering, Sungkyunkwan University, Suwon, Republic of Korea [4]Department of Instrumentation and Control Engineering, Dr. B. R. Ambedkar National Institute of Technology Jalandhar, Jalandhar, Punjab, India

2.1 Introduction

Food is an imperative element for developing and surviving human life; on the other hand, food wastage is a serious issue globally. At present, a large part of the population spends a day deprived of food because of its inadequate supply, increasing population, and severe climate change (Leisner, 2020). According to Food and Agricultural Organization (FAO) and United Nations (UN), >1.3 billion metric tons of consumable food is wasted each year. Wastage of food occurs during harvesting, transportation, and consumption (Nile et al., 2020). Based on data, 20% of food wastage comes from production, 19% from distribution, 1% from processing, and 60% from the consumers (Griffin, Sobal, & Lyson, 2009). A recent report of "Rethink Food Waste Through Economics and Data" suggests that vegetables and fruits are spoiled rapidly and contribute to a larger part of the waste, followed by dairy products, grains, and other supplies (Sridhar, Ponnuchamy, Kumar, & Kapoor, 2020). There are numerous causes of food wastage, but the increasing population and growing demands of consumers make it more provocative for the food processing sector. Every day, enough food is produced to feed the whole population, but due to the wastage and lack of proper food processing, it cannot reach those in need. Food preservation is one of the vital solutions to the food wastage issue. The food preservation technique is not new; our ancestors also used this technology in natural ways to preserve and extend the shelf life of food. The research advancements in this area introduced many modern technologies to extend the food shelf life and restrict its spoilage. Still, food safety is a big concern, and hence day by day, new techniques are evolving for food preservation. The crucial prospect of food preservation is to eliminate the attack of microbes and restrict the chemical alteration in foods due to physical and environmental issues (Sridhar et al., 2020). The commonly used food preservation methods such as freezing, drying, heating, etc., are helping to extend the shelf life of foods, but they have many drawbacks as these methods mainly cause severe nutrient loss, food shrinkage, and as a whole, loss of foods (Nile et al., 2020; Sridhar et al., 2020). Recently, to address this issue, chemical and biological factors such as active packaging, coating, etc., are introduced (Jeevahan & Chandrasekaran, 2019).

In this context, nanotechnology has become an integral part of the food preservation sector. A schematic representation of food nanotechnology is displayed in Fig. 2.1. The superb physio-chemical properties and intense antimicrobial activities of nanoparticles make them useful for food safety and preservation. Nanotechnology can be used in the food sector during food production, processing, storage, and even distribution (Nile et al., 2020). The progress of nanotechnology even makes it possible to detect food contamination through nanosensors (Sridhar et al., 2020). The nanobased food technology thus offers a better alternative to the conventional food preservation methods. In the food packaging sector, the nanobased packaging films offer better barrier, thermal and mechanical performances as well as the functional

Food Packaging and Preservation.
DOI: https://doi.org/10.1016/B978-0-323-90044-7.00002-1

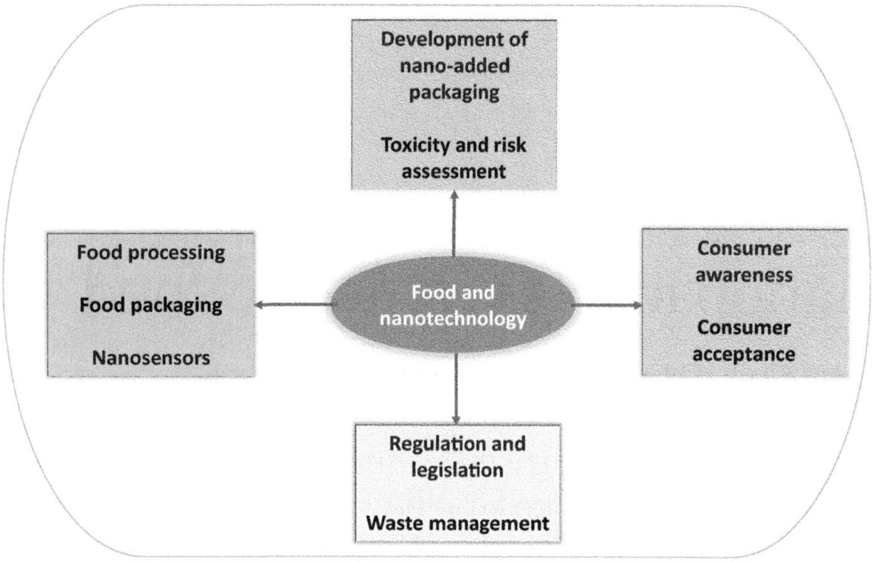

FIGURE 2.1 An overview of food nanotechnology.

ingredients like antimicrobial materials helps in controlling the growth of the microorganism to extend the shelf life of food products (Kuswandi, 2017; Roy & Rhim, 2020b; Roy & Rhim, 2021a; Shankar, Jaiswal, & Rhim, 2020).

Moreover, the nanobased smart packaging offers food safety and security through sensing, detection, and remote controlling of foods (Jeevahan & Chandrasekaran, 2019). The nanobased technology also helps to promote plant growth, increasing the yield of grains and germination of seeds (Bajpai et al., 2018). The antimicrobial property of eco-friendly nanoparticles helps to control plant bugs and reduce pollution. This chapter covers the role and potential of nanotechnology in food preservation with a special mention of food processing and food packaging.

2.2 Antimicrobial nanoparticles

Nanoparticles are widely used in the packaging system because of their potent antimicrobial activity. Ag, ZnO, Cu, CuO, CuS MgO, Se, Pd, Fe, SiO_2, and TiO_2 are the most commonly used inorganic antimicrobial nanoparticles (Basavegowda, Mandal, & Baek, 2019; Hoseinnejad, Jafari, & Katouzian, 2018; Huang, Mei, Chen, & Wang, 2018; Jafarzadeh & Jafari, 2020; Omerović et al., 2021). Organic nanoparticles such as chitosan, nanocelluose, curcumin, melanin, essential oil nanoemulsion, etc., are also actively used in food preservation (Ahankari, Subhedar, Bhadauria, & Dufresne, 2020; Garavand et al., 2020; Jafarzadeh & Jafari, 2020; Kuorwel, Cran, Orbell, Buddhadasa, & Bigger, 2015; Priyadarshi & Rhim, 2020; Roy & Rhim, 2020a,c; Roy, van Hai, Kim, Zhai, & Kim, 2020; Sharma, Barkauskaite, Jaiswal, & Jaiswal, 2020; Silvestre, Duraccio, & Cimmino, 2011).

2.2.1 Inorganic nanoparticles

Metal or their oxides are commonly known for their excellent antimicrobial activity. The antimicrobial properties mainly originate from the reactive oxygen species (ROS), metal ion release, and membrane disruption depending on the metal (Hoseinnejad et al., 2018). Some of the inorganic nanoparticles are discussed here in brief.

So far, silver nanoparticles have been the most popular among all tested antimicrobial nanoparticles (Simbine et al., 2019). Silver nanoparticles possess high thermal stability and broad-spectrum antimicrobial activity. Because of the strong antimicrobial action, they have even been used in commercial consumer products. For instance, Duretham, DuPont, Sumitomo, Metal Ferro, NovaCentrix, US Research Nanomaterials, NANOMAS, Heraeus, and Kodak have used silver nanoparticles in their packaging polymer to improve the shelf life of packaged food products (Hoseinnejad et al., 2018). The function of silver nanoparticles has relied on various factors such as size, shape, concentration, and types of microbes. There are countless other research reports on the use of silver nanoparticles in food packaging (Simbine et al., 2019). The most skeptical question regarding this nanoparticle is whether they are more effective than their counterpart silver ion. Recent reports proved that when the size of particles is <10 nm, the nanoform is effective, while the increase in size dominates the ionic form of silver

(Hoseinnejad et al., 2018). Moreover, the nanoform is more effective in entering the microbial cell than the counterpart. The antibacterial activity of silver nanoparticles rely on various factors such as adsorption and collection in the bacterial cell, the interaction of silver ion and DNA, damage of cell membrane, hindrance of cell respiration, etc. (Barkat et al., 2019; Roy & Das, 2016).

Among the metal oxide nanostructures, ZnO nanoparticles are widely used nanomaterials because of the strong antimicrobial activity (Kim et al., 2020). These nanoparticles are known to be more effective against Gram-negative bacterial strains (Kim, Panicker, Rhim, & Kim, 2021). Reports suggest that ZnO nanoparticles can acts as oxidizing agents for bacterial cells by generating reactive oxygen species (Fu et al., 2015). Zinc ions released from the nanoparticles also have a predominant role in damaging the bacterial cell membrane (Wang et al., 2019).

Copper oxide nanoparticles are another well-known antibacterial agent due to their broad-spectrum antibacterial activity, like the silver nanoparticles (Hasheminya, Rezaei Mokarram, Ghanbarzadeh, Hamishekar, & Kafil, 2018). It is already known that for bacterial cells, copper is an essential element for metabolism and electron transport. Although, when the content exceeds the required limit, it restricts the growth of the cells and the most probable reasons are ion exchange, generation of peroxide-free radicals, attachment in the cell membrane, deactivation of protein and enzymes, etc. (El Sayed, El-Gamal, Morsi, & Mohammed, 2015; Saravanakumar, Sathiyaseelan, Mariadoss, Xiaowen, & Wang, 2020).

Another functional antimicrobial oxide nanoparticle is TiO_2. Although they are very effective antimicrobial agents, cytotoxicity is a concern for practical use (Jovanović, 2015). The antimicrobial activity of TiO_2 nanoparticles is a light-dependent phenomenon since it acts as a photocatalyst (Weir, Westerhoff, Fabricius, Hristovski, & von Goetz, 2012). The strong antimicrobial action of TiO_2 nanoparticles originates from the irradiated nanoparticles under visible light. It is believed that the TiO_2 nanoparticles produce reactive oxygen species (ROS) such as peroxide radicals, superoxide ions, hydroxyl radicals, etc., which are responsible for the oxidation of bacterial cell membrane (Chorianopoulos, Tsoukleris, Panagou, Falaras, & Nychas, 2011; Roy, Zhai, et al., 2021; Siripatrawan & Kaewklin, 2018).

MgO nanoparticles are also known for their good antimicrobial activity. Mg is an essential element for living beings; nevertheless, the nanooxide form of Mg is a highly active agent, and the antimicrobial function is based on ROS production and lipid peroxidation (Wang et al., 2020a; Wang, Cen, Chen, & Fu, 2020b). The MgO nanoparticles possess the capability to adhere to the cell walls and cleave the peptide bonds (Mirtalebi, Almasi, & Alizadeh Khaledabad, 2019).

Silicon dioxide (SiO_2) is also known for its antimicrobial activity in nanoforms. The antimicrobial activity relies on the attachment to the cell surface and restricting bacterial proliferation (Bi et al., 2020). Although SiO_2 nanoparticles are antimicrobials, the activity is not up to the expectation, and as a result, it is often used with commonly used other nanoparticles as a synergistic antimicrobial material (Rukmanikrishnan, Jo, Choi, Ramalingam, & Lee, 2020).

2.2.2 Organic nanoparticles

Organic nanoparticles have also received immense interest nowadays. Some of them are discussed below in brief.

Chitosan nanoparticles are popular polymer-based nanomaterials. The chitosan itself shows antimicrobial activity due to the electrostatic interaction between positively charged chitosan and negatively charged bacterial cell membranes (Garavand et al., 2020; Yanat & Schroën, 2021). Chitosan nanoparticles are more effective compared to bulk chitosan. Moreover, the quaternary chitosan nanoparticles are reported to be far more lethal towards microorganisms due to the presence of quaternary ammonium ions (Garavand et al., 2020).

Melanin nanoparticles are another fascinating nanomaterials that can be effective against the growth of microbes. The exact mechanism of antibacterial activity is not clear. The melanin has a role in damaging the integrity of cell membrane, which results in increased leakage of cell contents, the elevated quantity of NPN uptake, and decreased membrane potential (Xu et al., 2017).

Curcumin is another well-known natural antimicrobial agent used since ancient times. Recently, nanocurcumin has been a topic of interest due to its low water solubility and absorption of curcumin (Basniwal, Buttar, Jain, & Jain, 2011). Moreover, it has been found that the nanocucrumin showed better functional properties compared to its counterpart. Curcumin's antibacterial activity differs depending on the type of bacteria. Its antibacterial action is associated with the inhibition of filamenting temperature-sensitive mutant Z (FtsZ) function, an essential protein necessary for bacterial cell division (Kaur, Modi, Panda, & Roy, 2010). Curcumin can bind to the

FtsZ protein, which thwarts the association of FtsZ protofilaments; as a result, the formation of Z-ring structure is hindered, and this process inhibits the cell division (Kaur et al., 2010).

Essential oils also elucidate strong antimicrobial and antioxidant activity (Aswathanarayan & Ravishankar, 2019; Lu et al., 2018). They are generally recognized as safe and are very useful functional materials. However, their high volatility and stability make them difficult for practical usage (Roy & Rhim, 2021b). The solution to this problem is nanoemulsion or nanoencapsulation. Countless essential oils (such as cinnamon, clove, paprika, ginger, eucalyptus, rosemary, etc.) have been used for this purpose, and many of them showed promising functional properties when they are present as nanoemulsions compared to their original state (Pinzon, Leidy, & Cristian, 2019; Prakash, Revathy, Nithyanand, & Vellingiri, 2018). The presence of diverse types of aldehydes, phenolics, terpenes, and other antimicrobial compounds makes essential oils highly effective against broad spectrum pathogens (Aswathanarayan & Ravishankar, 2019; Pinzon et al., 2019).

2.3 Status on food nanotechnology

Food nanotechnology offers many features of the food sector like packaging, additives, and preservation of food. The role of nanotechnology in various food sectors is presented in schematic illustration (Fig. 2.2), and the steps for using the nanomaterials in the food sector are given in Fig. 2.3. For this purpose, varieties of nanomaterials such as inorganic metal or metallic oxide nanoparticles, organic nanoparticles, and their combinations have already been utilized in the food sector. Metal or metal oxide nanoparticles such as silver, gold, ZnO, TiO_2 nanoparticles are mainly used because of their antimicrobial activity, sensor, or food flavor additives (Bajpai et al., 2018). Organic nanoparticles such as natural product nanoparticles (chitosan, modified nanocelulose, curcumin, essential oil, melanin, etc.) have been used as food supplements and delivery systems. Although the nanobased materials used in the food sector are less toxic, long-term use and exposure to these nanomaterials may cause some adverse effects on health.

Consequently, these risk factors attracted the attention of consumers. The United States Food and Drug Administration (US FDA) and European Commission (EC) are the authorities that control nanotechnology in the food sector. Nanomaterials have been used in the food processing sector as color/flavor additives, nanoemulsion/encapsulations, and carriers of food supplements. In this regard, the US FDA has given zinc oxide, copper oxide, and iron oxide the GRAS grade. In the food packaging part, the nanotechnology-based packaging materials are beneficial over the commonly used packaging films. Various nanomaterials have been used for food packaging. One of the most used ones is nanoclays - thanks to their abundance, cost-effectiveness, and role in improving mechanical, barrier, and thermal properties (Jose et al., 2014). Nanoclays such as montmorillonite and bentonite are considered as GRAS by FDA, although reports suggest the possible risk of migration of nanoclay

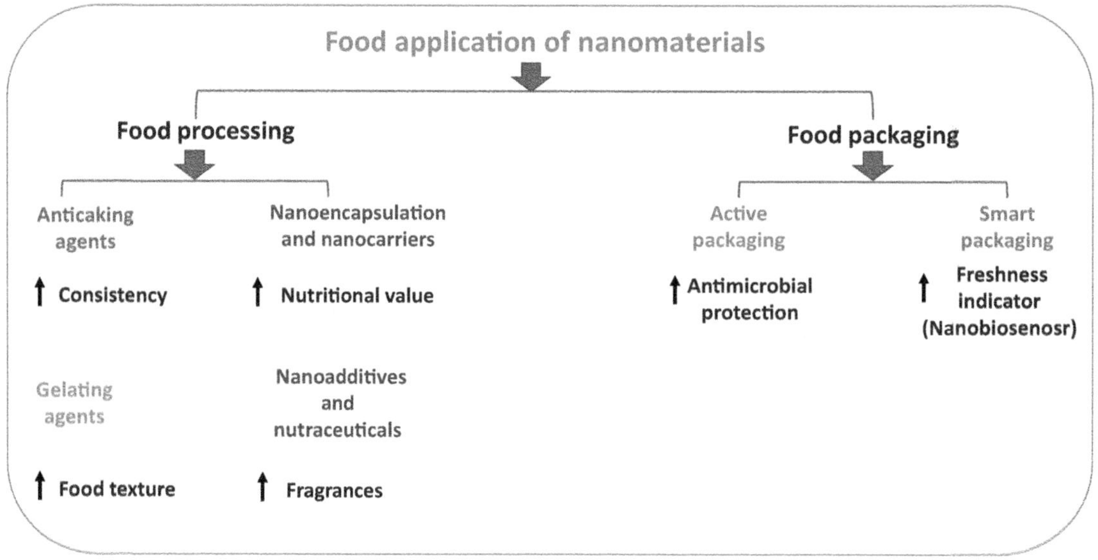

FIGURE 2.2 Schematic illustrations of the function of nanomaterials in food technology.

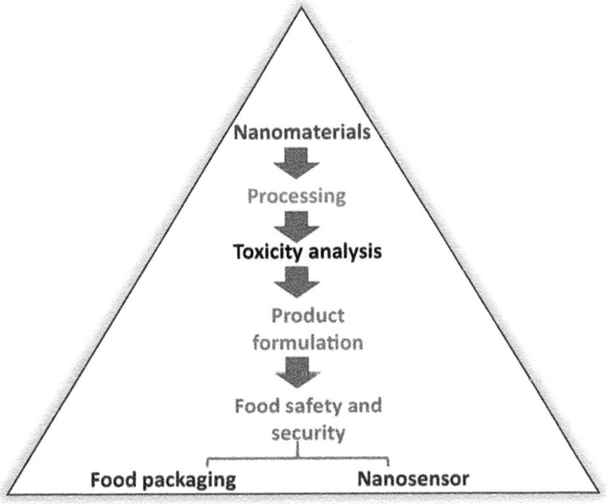

FIGURE 2.3 A schematic for the steps to be followed for using nanomaterials in food sectors.

based packaging (Huang, Li, & Zhou, 2015; Störmer, Bott, Kemmer, & Franz, 2017). The nanocomposites used in the food packaging system can show the gas barrier properties and antimicrobial function (Basumatary, Mukherjee, Katiyar, & Kumar, 2020). So, recently nanofiller incorporated bio-based packaging films have gained much interest, although the main concern is the intake of these nanomaterials along with food and their biodegradability. Hence, it is necessary to conduct a proper study on the release of nanoparticles on human systems and their toxicological effects.

Along with packaging, the edible coating is also a tool for the preservation of food. The coated food products, such as fruits and vegetables, remain fresher for prolonged periods during storage and transportation. To this end, nanomaterials are advantageous, and a layer of nanoedible coating can be used to prevent microbial attacks and as gas and moisture barriers (Bajpai et al., 2018; Störmer et al., 2017). The gas produced inside the packaging is not considered healthy as it promotes microbial growth. In this regard, nanosilica, nanomagnesium oxide, iron nanoparticles, etc., be promising oxygen scavenging materials (Foltynowicz, Bardenshtein, Sängerlaub, Antvorskov, & Kozak, 2017; Swaroop & Shukla, 2018). Active packaging films with included nanomaterials and nanosensor-based intelligent packaging have also become popular nowadays.

2.4 Nanotechnology in food processing

Nanotechnology has been used in plant and animal-based food processing in various ways, such as sensors, flavors, taste, colorants, textures, preservatives, supplements, etc. (He & Hwang, 2016; Sridhar et al., 2020). Nanotechnology enhances the shelf life of various food products and reduces food wastage by providing security towards microbial attacks. Recently nanocarriers have also received attention for carrying food additives without compromising the morphology of the food (Singh et al., 2017). Nanoparticles showed good encapsulation and released proficiency with respect to conventional encapsulators (Singh et al., 2017). The nanoencapsulation provides several functions such as regulated interaction among food and the active materials, controlled release, protection from external factors such as heat, moisture, etc. (Weiss, Takhistov, & McClements, 2006). Polymeric nanomaterials are considered helpful for encapsulating bioactive ingredients like flavonoids, vitamins, etc. (Singh et al., 2017). The encapsulated active compounds help to protect and distribute the ingredients to the specific targets. The diverse application of nanotechnology has shown a significant contribution to food processing.

The nanomagnesium dioxide and silicon dioxide play a role in improving the color and flavor while acting as anticaking agents of food. In contrast, nanotitanium dioxide has been used as an additive, color enhancer, etc. (Sridhar et al., 2020; Weir et al., 2012). Furthermore, zinc oxide, copper oxide, and iron oxide are also considered GRAS grade materials for their use in animal and plant products (He & Hwang, 2016; Roy & Rhim, 2019a,b). Encapsulated lipid nanoparticles with palmitic acid, corn oil, and whey protein isolates showed improved thermal and oxidative stability of beta-carotene (Mehrad, Ravanfar, Licker, Regenstein, & Abbaspourrad, 2018). Bioactive natural compounds are unstable, and so encapsulation can enhance their stability. The nanoencapsulation of anthocyanin improves the stability of the pigments, while the ferritin encapsulation of rutin improves its stability and solubility (Yang et al., 2015; Zhang et al., 2014). SiO_2 nanoparticles are most widely applied in food products as a vehicle of flavors (Singh et al., 2017). Nanoemulsion of fat-soluble active ingredients is another well-known phenomenon due to its role in improving water dispersion and bioavailability (Ozturk, Argin, Ozilgen, & McClements, 2015). Another aspect of food processing is the nutritional value of food products. It is well established that most of the bioactive ingredients like proteins, lipids, carbohydrates, vitamins are easily prone to alteration in an acidic condition of the stomach. So the encapsulation of the active ingredients not only help to remain unaltered in adverse condition allows assistance for integration with food due to water solubility (Yan & Gilbert, 2004). Nanobased edible capsules are also used, useful for transportation of vitamins, micronutrients, etc. (Singh et al., 2017).

2.5 Nanotechnology for food packaging

Packaging is an integral part of food, and it has also been a significant contributor to the global economy. At present, there are two direct pathways in the packaging industry. The first one is improving existing packaging materials' physical (mechanical, barrier, etc.) and functional (antimicrobial, antioxidant, etc.) properties. There are many advancements in this direction to improve food safety and extend the food shelf life. Although these materials are beneficial for the consumers, they generate pollution, which is a drawback of this approach. The second approach is replacing the nonbiodegradable and nonrenewable packaging materials with biodegradable and eco-friendly sources. The petroleum-based plastics pose a severe public challenge since most of the produced plastics are not appropriately disposed of, consequently generating a huge ecological issue, which also applied to biomaterials in case they do not degrade within a suitable period of time. In 2020 only, one trillion food and drink packages are expected to be thrown away only in Europe (Yanat & Schroën, 2021). Throwaway plastic packaging materials accounted for almost 40% of plastic production in 2018 (Iwata, 2015; Pira, 2013). Although eco-friendly materials are advantageous for our ecology, the cost is a big concern for the packaging industry. Reports suggest that the share of biodegradable plastics was less than 1% of the total plastic production in 2019 (Lange, 2019; Plastics Europe, 2019). Hence, more focus should be given to producing cost-effective biodegradable plastics and their use to develop sustainable packaging.

Nowadays, there is a rapid increment in the active and smart packaging of processed food products, which are highly prone to contamination. The primary goal of packaging is to restrict the contamination and spoiling of food and retain its quality without compromising safety (Kerry, O'Grady, & Hogan, 2006). The progress and development of nanotechnology in the food sector play a crucial role in food preservation. In one way, the nanosensor can detect the spoilage and contamination of processed foods, identify the toxin, pesticides, while its use in packaging film or coating purpose gives protection to food over microbial attacks (He & Hwang, 2016). The nanoparticles used in packaging films are generally antimicrobial agents, and as a result, they can act as a vehicle for antimicrobial components and provide protection against food contamination (Cha & Chinnan, 2004).

Moreover, the nanoparticles used in the packaging sector can also act as carriers for biomolecules like antioxidants, enzymes, etc. (Weiss et al., 2006). Metal and their oxide-based nanoparticles such as silver, zinc oxide, carbon, titanium dioxide, iron, etc., are extensively applied as antimicrobial agents to the food packaging field (He & Hwang, 2016). Nanoparticles and polymer-based composites are reported to be the perfect combination to address the food contamination issue. For the controlled release of nanoparticles, the polymer matrix has a good function by regulating the release of active components (Duncan, 2011). Nanoparticles are primarily added to the polymer matrix to create nanocomposites with improved physical and functional properties. For this purpose, to make nanocomposites, thermoplastic polymers (such as polylactic acid, biopolymers like cellulose, agar, gelatin, etc.) and various contents (0–10 wt.%) of nanofillers are commonly used (Duncan, 2011). In polymer-based food packaging films or coatings, the addition of functional nanomaterials in the polymer matrix improves the performance of the packaging materials by imparting antimicrobial and antioxidant activities and therefore extending the shelf life of food products (Rhim, Park, & Ha, 2013). Recently, several inorganic and organic nanoparticles have been induced to different polymer matrices to improve functional properties and enhance the physical properties such as gas barrier, mechanical, thermal, fire-resistant properties, etc. Among the countless reports, some of the results are presented in Table 2.1.

Previous reports suggest that the addition of nanoparticles enhances the oxygen barrier property in PLA-based films, imparted antimicrobial activity, and exhibited better biodegradability of polymers (Brody, Bugusu, Han, Sand, & McHugh, 2008). Besides, the mechanical, thermal, and other physical properties were also profoundly improved by the proper blending ratio of nanofillers. The restriction in oxidation, microbial attacks, and

TABLE 2.1 Use of multifunctional nanofillers in food packaging.

Polymers	Nanofillers	Improved properties	Application	Reference
Sodium alginate	Silver	Antibacterial activity	Carrot and pear packaging	Mohammed Fayaz, Balaji, Girilal, Kalaichelvan, and Venkatesan (2009)
Agar	ZnO	Thermal stability, mechanical properties, and fruit preservation	Green grape	Kumar, Boro, Ray, Mukherjee, and Dutta (2019)
Low-density polyethylene (LDPE)	CuO	Antimicrobial activity	Cheese	Beigmohammadi et al. (2016)
Chitosan	TiO_2	Mechanical and barrier properties	Tomato	Kaewklin, Siripatrawan, Suwanagul, and Lee (2018)
LDPE	SiO_2	Antioxidant activity of fruits	Loquat fruit	Wang et al. (2020a,b)
Gelatin and bacterial cellulose	MgO	Surface hydrophobicity and mechanical properties	Egg	Wang, Luo, Tu, and Zhao (2021)
Gelatin	Fe	Mechanical, antimicrobial, and barrier properties	Grapes	Mehmood, Sadiq, and Khan (2020)
Furcellaran and Gelatin	Se and Ag	Mechanical, antimicrobial, and barrier properties	Fruit (Kiwi) packaging	Jamróz et al. (2019)
Linear low-density polyethylene	Nanoclay	Antibacterial properties	Beef	Tornuk, Hancer, Sagdic, and Yetim (2015)
Chitosan coating	Modified Chitosan	–	Tomato	Meena et al. (2020)
Chitosan	Nanocellulose	Mechanical and barrier properties	Strawberry	Resende, Gonçalves, Reis, Tonoli, and Boas (2018)
Gelatin	Melanin	Oxidative stability	Pork lard	Łopusiewicz, Filip, and Artur (2018)
Zein/potato starch	Curcumin	Mechanical, antioxidant, and barrier properties	Fish fillet	Xin et al. (2020)
Corn starch	Essential oil nanoemulsion	Mechanical and antioxidant properties	Beef patties	Amiri, Aminzare, Azar, and Mehrasbi (2019)

regulation in the gas barrier and respiration properties tailored by nanotechnology significantly impacted the food packaging sector (Chaudhry et al., 2008). Edible coating with nanoparticles is also effective in improving the shelf life of fruits and vegetables. The edible coating is commonly produced from bio-based polymers such as polysaccharides, fats, and proteins which are promising, but the inclusion of nanomaterials can help to improve further the gas barrier property (Díaz-Montes & Castro-Muñoz, 2021). Among nanobased packaging films, active and intelligent packaging is currently being widely used. Compared to traditional packaging, active packaging can act as a passive barrier, and it releases the functional compounds for antimicrobial and antioxidant activity via interaction with foods. Active packaging also assists in removing oxygen or water vapor from the interior of the package. The factor mentioned above of active packaging helped to enhance food stability. The active packaging is based on controlling moisture, gas, and antimicrobial activity, and this type of packaging is used depending on the applications. For instance, wrapping packages are applied for short-span cold storage while the vacuum package, modified atmospheric package are used for long-span cold storage (Kerry et al., 2006). The use of nanomaterials such as nanosilver, nanosilica, etc., in the active food packaging system, is found to be promising to improve the shelf life of meats (Nile et al., 2020).

On the other hand, smart packaging helps detect food contaminants, toxins, pesticides through sensors (Li & Sheng, 2014). The alteration in environment (internal or external) stimulus on food products can be recorded through selective sensors. The advent of nanotechnology in food has exhibited a good role in making intelligent sensors for analysis of food, detection of odors and colors, etc. (Nile et al., 2020). The nanosenosrs have the excellent ability to identify any modification (physical, biological, or chemical) in food during processing, storage, and transportation (Mushtaq & Faizan, 2020). Different types of functional nanomaterials have been utilized in making nanosensors for the food sector, such as freshness indicators, radio frequency identification, time temperature monitoring, etc. (Kerry et al., 2006). Nanosensors are mainly used to detect chemical toxins, gases, pathogens, products generated from microbial degradation, odors, temperature, etc. (King, Osmond-McLeod, & Duffy, 2018). Recently, nanobarcodes have also been developed using nanoparticles, which can be used as identification tags for food products (Branton et al., 2008). In packaging, the use of nanosensors assists in the detection of unwanted biomarkers such as enzymes created from the spoilage of foods, while the packaging helps in improving the shelf-life of food through the internal antimicrobial defense and restricting the entry of air and undesirable materials (Caon, Martelli, & Fakhouri, 2017; Nile et al., 2020). Smart packaging also possesses the ability to reduce the use of artificial food preservatives. As a food nanosensor, the optical sensor illustrates better sensitivity over analytical methods like chromatography, gas analyzer, etc., due to its quick and real-time identification (Kerry et al., 2006; Sridhar et al., 2020).

2.6 Nanotechnology in food safety

Food safety is a concern worldwide. The main aim of the safety process is to certify the safety of food during production and consumption. The food should be protected from any kind of contamination during production, processing, and distribution. Food safety is a serious issue due to the fast progress in food habits. Foodborne contamination can cause serious harm to health, and in this regard, nanotechnology has shown a crucial role in ensuring food safety through nanosensors (Bajpai et al., 2018; He & Hwang, 2016; Stephen Inbaraj & Chen, 2016). Metallic nanoparticles, quantum dots, and other nanoparticles help make biosensors quantify microbes and ensure food safety. Many biosensors have been created to identify the foodborne pathogens, and as a tool, Surface-enhanced Raman scattering (SERS) is used for rapid detection of pathogens (Jarvis & Goodacre, 2004; Kahraman, Yazici, Şahin, & Çulha, 2008). In SERS, for the identification of pathogens, various nanoparticles such as silver, gold, magnetic nanoparticles, carbon nanotubes, etc., have been commonly used (Bajpai et al., 2018; Baranwal, Mahato, Srivastava, Maurya, & Chandra, 2016; Holzinger, Goff, & Cosnier, 2014). Nanomaterials have also been used for various other purposes, such as controlling the food allergens, preventing the reduction of heavy metals, inhibiting biofilm formation, etc. (Bajpai et al., 2018).

2.7 Future prospective and risk associated with nanotechnology

Although nanotechnology has numerous benefits in the food sector, the safety concerns related to nanotechnology cannot be ignored. A lot of advancements have been seen regarding the use of nanotechnology in the food sector in this decade. Even if the nanoparticles have a good influence and role in detecting the pathogens, toxins, and pesticides, they restrict the growth of foodborne pathogens and in other related applications (Fu et al., 2008; Hahn, Keng, & Krauss, 2008; Su & Li, 2004; Yang & Li, 2005), but the potential risk on the toxicity of nanomaterials is still a big concern for the consumers. The effect of nanomaterials on living beings and the environment is still not entirely known. As every nanomaterial has its particular properties, consequently, the toxicity effect of each should be studied. Due to the small size, the nanoparticles can easily reach the cells crossing the blood-brain barrier, leading to a potential concern of bioaggregation of nanoparticles within the body or tissues (Su & Li, 2004). Even inhalation of fine nanoparticles has increased day by day, which is an alarming factor for health safety (Jovanović, 2015). Also, there are some reports regarding the accumulation of commonly used nanomaterials in food and living beings. Therefore these crucial concerns about the risk should be firmly addressed (Bajpai et al., 2018; Yang & Li, 2005). There is no doubt that the advent of nanotechnology in the food sector has a significant role in food processing and preservation. Still, the growing concern of its risk makes it challenging to create a sustainable and healthy food sector. There is the utmost requirement of further research on nanotoxicity, proper testing of nanoadded food before coming to the consumer, and adequate education to the consumers regarding the safety of nanotechnology.

References

Ahankari, S. S., Subhedar, A. R., Bhadauria, S. S., & Dufresne, A. (2020). Nanocellulose in food packaging: A review. *Carbohydrate Polymers, 255*, 117479. Available from https://doi.org/10.1016/j.carbpol.2020.117479.

Amiri, E., Aminzare, M., Azar, H. H., & Mehrasbi, M. R. (2019). Combined antioxidant and sensory effects of corn starch films with nano-emulsion of Zataria multiflora essential oil fortified with cinnamaldehyde on fresh ground beef patties. *Meat Science, 153*, 66–74. Available from https://doi.org/10.1016/j.meatsci.2019.03.004.

Aswathanarayan, J. B., & Ravishankar, R. V. (2019). Nano-emulsions and their potential applications in food industry. *Frontiers in Sustainable Food Systems, 395*. Available from https://doi.org/10.3389/fsufs.2019.00095.

Bajpai, V. K., Kamle, M., Shukla, S., Mahato, D. K., Chandra, P., Hwang, S. K., ... Han, Y. K. (2018). Prospects of using nanotechnology for food preservation, safety, and security. *Journal of Food and Drug Analysis, 26*, 1201–1214. Available from https://doi.org/10.1016/j.jfda.2018.06.011.

Baranwal, A., Mahato, K., Srivastava, A., Maurya, P. K., & Chandra, P. (2016). Phytofabricated metallic nanoparticles and their clinical applications. *RSC Advances, 6*, 105996–106010. Available from https://doi.org/10.1039/c6ra23411a.

Barkat, M. A., Harshita, F. N., Beg, S., Pottoo, F. H., Garg, A., Singh, S. P., & Ahmad, F. J. (2019). Silver nanoparticles and their antimicrobial applications. *Current Nanomedicine, 8*(3), 215–224. Available from https://doi.org/10.2174/2405461503666180806113924.

Basavegowda, N., Mandal, T. K., & Baek, K.-H. (2019). Bimetallic and trimetallic nanoparticles for active food packaging applications: A review. *Food and Bioprocess Technology*. Available from https://doi.org/10.1007/s11947-019-02370-3.

Basniwal, R. K., Buttar, H. S., Jain, V. K., & Jain, N. (2011). Curcumin nanoparticles: Preparation, characterization, and antimicrobial study. *Journal of Agricultural and Food Chemistry, 59*(5), 2056–2061. Available from https://doi.org/10.1021/jf104402t.

Basumatary, I. B., Mukherjee, A., Katiyar, V., & Kumar, S. (2020). Biopolymer-based nanocomposite films and coatings: Recent advances in shelf-life improvement of fruits and vegetables. *Critical Reviews in Food Science and Nutrition. (In Press)*. Available from https://doi.org/10.1080/10408398.2020.1848789.

Beigmohammadi, F., Peighambardoust, S. H., Hesari, J., Azadmard-Damirchi, S., Peighambardoust, S. J., & Khosrowshahi, N. K. (2016). Antibacterial properties of LDPE nanocomposite films in packaging of UF cheese. *LWT-Food Science and Technology, 65*, 106–111. Available from https://doi.org/10.1016/j.lwt.2015.07.059.

Bi, F., Zhang, X., Liu, J., Yong, H., Gao, L., & Liu, J. (2020). Development of antioxidant and antimicrobial packaging films based on chitosan, D-α-tocopheryl polyethylene glycol 1000 succinate and silicon dioxide nanoparticles. *Food Packaging and Shelf Life, 24*, 100503. Available from https://doi.org/10.1016/j.fpsl.2020.100503.

Branton, D., Deamer, D. W., Marziali, A., Bayley, H., Benner, S. A., Butler, T., ... Schloss, J. A. (2008). The potential and challenges of nanopore sequencing. *Nature Biotechnology, 26*, 1146–1153. Available from https://doi.org/10.1038/nbt.1495.

Brody, A. L., Bugusu, B., Han, J. H., Sand, C. K., & McHugh, T. H. (2008). Scientific status summary. *Journal of Food Science, 73*(8), R107–R116. Available from https://doi.org/10.1111/j.1750-3841.2008.00933.x.

Caon, T., Martelli, S. M., & Fakhouri, F. M. (2017). *New trends in the food industry: application of nano-sensors in food packaging. Nanobiosensors* (pp. 773–804). Elsevier. Available from https://doi.org/10.1016/b978-0-12-804301-1.00018-7.

Cha, D. S., & Chinnan, M. S. (2004). Biopolymer-based antimicrobial packaging: A review. *Critical Reviews in Food Science and Nutrition, 44*(4), 223–237. Available from https://doi.org/10.1080/10408690490464276.

Chaudhry, Q., Scotter, M., Blackburn, J., Ross, B., Boxall, A., Castle, L., ... Watkins, R. (2008). Applications and implications of nanotechnologies for the food sector. *In Food Additives and Contaminants - Part A Chemistry, Analysis, Control, Exposure and Risk Assessment, 25*, 241–258. Available from https://doi.org/10.1080/02652030701744538.

Chorianopoulos, N. G., Tsoukleris, D. S., Panagou, E. Z., Falaras, P., & Nychas, G. J. E. (2011). Use of titanium dioxide (TiO2) photocatalysts as alternative means for Listeria monocytogenes biofilm disinfection in food processing. *Food Microbiology, 28*(1), 164–170. Available from https://doi.org/10.1016/j.fm.2010.07.025.

Díaz-Montes, E., & Castro-Muñoz, R. (2021). Edible films and coatings as food-quality preservers: An overview. *Foods, 10*(2), 249. Available from https://doi.org/10.3390/foods10020249.

Duncan, T. v. (2011). Applications of nanotechnology in food packaging and food safety: Barrier materials, antimicrobials and sensors. *Journal of Colloid and Interface Science, 363*(1), 1–24. Available from https://doi.org/10.1016/J.JCIS.2011.07.017.

El Sayed, A. M., El-Gamal, S., Morsi, W. M., & Mohammed, G. (2015). Effect of PVA and copper oxide nanoparticles on the structural, optical, and electrical properties of carboxymethyl cellulose films. *Journal of Materials Science, 50*(13), 4717–4728. Available from https://doi.org/10.1007/s10853-015-9023-z.

Foltynowicz, Z., Bardenshtein, A., Sängerlaub, S., Antvorskov, H., & Kozak, W. (2017). Nanoscale, zero valent iron particles for application as oxygen scavenger in food packaging. *Food Packaging and Shelf Life, 11*, 74–83. Available from https://doi.org/10.1016/j.fpsl.2017.01.003.

Fu, F., Li, L., Liu, L., Cai, J., Zhang, Y., Zhou, J., & Zhang, L. (2015). Construction of cellulose based ZnO nanocomposite films with antibacterial properties through one-step coagulation. *ACS Applied Materials and Interfaces, 7*(4), 2597–2606. Available from https://doi.org/10.1021/am507639b.

Fu, J., Park, B., Siragusa, G., Jones, L., Tripp, R., Zhao, Y., & Cho, Y.-J. (2008). An Au/Si hetero-nanorod-based biosensor for Salmonella detection related content An Au/Si hetero-nanorod-based biosensor for Salmonella detection. *Nanotechnology, 19*(15), 7. Available from https://doi.org/10.1088/0957-4484/19/15/155502.

Garavand, F., Cacciotti, I., Vahedikia, N., Salara, A. R., Tarhan, Ö., Akbari-Alavijeh, S., ... Jafari, S. M. (2020). A comprehensive review on the nanocomposites loaded with chitosan nanoparticles for food packaging. *Critical Reviews in Food Science and Nutrition, 62*. Available from https://doi.org/10.1080/10408398.2020.1843133.

Griffin, M., Sobal, J., & Lyson, T. A. (2009). An analysis of a community food waste stream. *Agriculture and Human Values, 26*(1–2), 67–81. Available from https://doi.org/10.1007/s10460-008-9178-1.

Hahn, M. A., Keng, P. C., & Krauss, T. D. (2008). Flow cytometric analysis to detect pathogens in bacterial cell mixtures using semiconductor quantum dots. *Analytical Chemistry, 80*(3), 864–872. Available from https://doi.org/10.1021/ac7018365.

Hasheminya, S.-M., Rezaei Mokarram, R., Ghanbarzadeh, B., Hamishekar, H., & Kafil, H. S. (2018). Physicochemical, mechanical, optical, microstructural and antimicrobial properties of novel kefiran-carboxymethyl cellulose biocomposite films as influenced by copper oxide nanoparticles (CuONPs). *Food Packaging and Shelf Life, 17*, 196–204. Available from https://doi.org/10.1016/J.FPSL.2018.07.003.

He, X., & Hwang, H.-M. (2016). Nanotechnology in food science: Functionality, applicability, and safety assessment. *Journal of Food and Drug Analysis, 24*(4), 671–681. Available from https://doi.org/10.1016/j.jfda.2016.06.001.

Holzinger, M., Goff, A. le, & Cosnier, S. (2014). Nanomaterials for biosensing applications: A review. *Frontiers in Chemistry, 2*, 63. Available from https://doi.org/10.3389/fchem.2014.00063.

Hoseinnejad, M., Jafari, S. M., & Katouzian, I. (2018). Inorganic and metal nanoparticles and their antimicrobial activity in food packaging applications. *Critical Reviews in Microbiology, 44*(2), 161–181. Available from https://doi.org/10.1080/1040841X.2017.1332001.

Huang, J. Y., Li, X., & Zhou, W. (2015). Safety assessment of nanocomposite for food packaging application. *Trends in Food Science and Technology, 45*, 187–199. Available from https://doi.org/10.1016/j.tifs.2015.07.002.

Huang, Y., Mei, L., Chen, X., & Wang, Q. (2018). Recent developments in food packaging based on nanomaterials. *Nanomaterials (Basel, Switzerland), 8*(10). Available from https://doi.org/10.3390/nano8100830.

Iwata, T. (2015). Biodegradable and bio-based polymers: future prospects of eco-friendly plastics. *Angewandte Chemie International Edition, 54*, 3210–3215. Available from https://doi.org/10.1002/anie.201410770.

Jafarzadeh, S., & Jafari, S. M. (2020). Impact of metal nanoparticles on the mechanical, barrier, optical and thermal properties of biodegradable food packaging materials. *Critical Reviews in Food Science and Nutrition, 61*. Available from https://doi.org/10.1080/10408398.2020.1783200.

Jamróz, E., Kopel, P., Juszczak, L., Kawecka, A., Bytesnikova, Z., Milosavljevic, V., & Makarewicz, M. (2019). Development of furcellaran-gelatin films with Se-AgNPs as an active packaging system for extension of mini kiwi shelf life. *Food Packaging and Shelf Life, 21*, 100339. Available from https://doi.org/10.1016/j.fpsl.2019.100339.

Jarvis, R. M., & Goodacre, R. (2004). Discrimination of bacteria using surface-enhanced Raman spectroscopy. *Analytical Chemistry, 76*(1), 40–47. Available from https://doi.org/10.1021/ac034689c.

Jeevahan, J., & Chandrasekaran, M. (2019). Nanoedible films for food packaging: A review. *In Journal of Materials Science, 54*, 12290–12318. Available from https://doi.org/10.1007/s10853-019-03742-y.

Jose, T., George, S. C., Maya, M. G., Maria, H. J., Wilson, R., & Thomas, S. (2014). Effect of bentonite clay on the mechanical, thermal, and pervaporation performance of the poly(vinyl alcohol) nanocomposite membranes. *Industrial and Engineering Chemistry Research, 53*(43), 16820–16831. Available from https://doi.org/10.1021/ie502632p.

Jovanović, B. (2015). Critical review of public health regulations of titanium dioxide, a human food additive. *Integrated Environmental Assessment and Management, 11*(1), 10–20. Available from https://doi.org/10.1002/ieam.1571.

Kaewklin, P., Siripatrawan, U., Suwanagul, A., & Lee, Y. S. (2018). Active packaging from chitosan-titanium dioxide nanocomposite film for prolonging storage life of tomato fruit. *International Journal of Biological Macromolecules, 112*, 523–529. Available from https://doi.org/10.1016/j.ijbiomac.2018.01.124.

Kahraman, M., Yazici, M. M., Şahin, F., & Çulha, M. (2008). Convective assembly of bacteria for surface-enhanced Raman scattering. *Langmuir: the ACS Journal of Surfaces and Colloids, 24*(3), 894–901. Available from https://doi.org/10.1021/la702240q.

Kaur, S., Modi, N. H., Panda, D., & Roy, N. (2010). Probing the binding site of curcumin in Escherichia coli and Bacillus subtilis FtsZ-a structural insight to unveil antibacterial activity of curcumin. *European Journal of Medicinal Chemistry, 45*, 4209–4214. Available from https://doi.org/10.1016/j.ejmech.2010.06.015.

Kerry, J. P., O'Grady, M. N., & Hogan, S. A. (2006). Past, current and potential utilisation of active and intelligent packaging systems for meat and muscle-based products: A review. *Meat Science, 74*(1), 113–130. Available from https://doi.org/10.1016/j.meatsci.2006.04.024.

Kim, H. C., Panicker, P. S., Rhim, J.-W., & Kim, J. (2021). Cellulose nanofiber-based nanocomposite films reinforced with zinc oxide nanorods and grapefruit seed extract. *Nanomaterials, 11*(4), 877. Available from https://doi.org/10.3390/nano11040877.

Kim, I., Viswanathan, K., Kasi, G., Thanakkasaranee, S., Sadeghi, K., & Seo, J. (2020). ZnO nanostructures in active antibacterial food packaging: Preparation methods, antimicrobial mechanisms, safety issues, future prospects, and challenges. *Food Reviews International, 38*. Available from https://doi.org/10.1080/87559129.2020.1737709.

King, T., Osmond-McLeod, M. J., & Duffy, L. L. (2018). Nanotechnology in the food sector and potential applications for the poultry industry. *Trends in Food Science and Technology, 72*, 62–73. Available from https://doi.org/10.1016/j.tifs.2017.11.015.

Kumar, S., Boro, J. C., Ray, D., Mukherjee, A., & Dutta, J. (2019). Bionanocomposite films of agar incorporated with ZnO nanoparticles as an active packaging material for shelf life extension of green grape. *Heliyon, 5*(6), e01867. Available from https://doi.org/10.1016/j.heliyon.2019.e01867.

Kuorwel, K. K., Cran, M. J., Orbell, J. D., Buddhadasa, S., & Bigger, S. W. (2015). Review of mechanical properties, migration, and potential applications in active food packaging systems containing nanoclays and nanosilver. *Comprehensive Reviews in Food Science and Food Safety, 14*(4), 411–430. Available from https://doi.org/10.1111/1541-4337.12139.

Kuswandi, B. (2017). Environmental friendly food nano-packaging. *Environmental Chemistry Letters, 15*, 205–221. Available from https://doi.org/10.1007/s10311-017-0613-7.

Lange, K.-B. (2019). *Bioplastics market data update*, 1–13. Available from https://www.european-bioplastics.org/market/.

Leisner, C. P. (2020). Review: Climate change impacts on food security- focus on perennial cropping systems and nutritional value. *Plant Science, 293*, 110412. Available from https://doi.org/10.1016/j.plantsci.2020.110412.

Li, Z., & Sheng, C. (2014). Nano-sensors for food safety. *Journal of Nanoscience and Nanotechnology, 14*, 905–912. Available from https://doi.org/10.1166/jnn.2014.8743.

Łopusiewicz, Ł., Filip, J., & Artur, B. (2018). The application of melanin modified gelatin coatings for packaging and the oxidative stability of pork lard. *World Scientific News, 101*, 108–119.

Lu, W.-C., Huang, D.-W., Wang, C.-C., Yeh, C.-H., Tsai, J.-C., Huang, Y.-T., & Li, P.-H. (2018). Preparation, characterization, and antimicrobial activity of nano-emulsions incorporating citral essential oil. *Journal of Food and Drug Analysis, 26*, 82–89. Available from https://doi.org/10.1016/j.jfda.2016.12.018.

Meena, M., Pilania, S., Pal, A., Mandhania, S., Bhushan, B., Kumar, S., … Saharan, V. (2020). Cu-chitosan nano-net improves keeping quality of tomato by modulating physio-biochemical responses. *Scientific Reports, 10*(1), 21914. Available from https://doi.org/10.1038/s41598-020-78924-9.

Mehmood, Z., Sadiq, M. B., & Khan, M. R. (2020). Gelatin nanocomposite films incorporated with magnetic iron oxide nanoparticles for shelf life extension of grapes. *Journal of Food Safety, 40*(4), e12814. Available from https://doi.org/10.1111/jfs.12814.

Mehrad, B., Ravanfar, R., Licker, J., Regenstein, J. M., & Abbaspourrad, A. (2018). Enhancing the physicochemical stability of β-carotene solid lipid nanoparticle (SLNP) using whey protein isolate. *Food Research International, 105*, 962–969. Available from https://doi.org/10.1016/j.foodres.2017.12.036.

Mirtalebi, S. S., Almasi, H., & Alizadeh Khaledabad, M. (2019). Physical, morphological, antimicrobial and release properties of novel MgO-bacterial cellulose nanohybrids prepared by in-situ and ex-situ methods. *International Journal of Biological Macromolecules, 128*, 848–857. Available from https://doi.org/10.1016/J.IJBIOMAC.2019.02.007.

Mohammed Fayaz, A., Balaji, K., Girilal, M., Kalaichelvan, P. T., & Venkatesan, R. (2009). Mycobased synthesis of silver nanoparticles and their incorporation into sodium alginate films for vegetable and fruit preservation. *Journal of Agricultural and Food Chemistry, 57*(14), 6246–6252. Available from https://doi.org/10.1021/jf900337h.

Mushtaq, Z., & Faizan, S. (2020). *Application of nanosensors in agriculture and food processing. Nanobiosensors for agricultural, medical and environmental applications* (pp. 175–186). Singapore: Springer. Available from https://doi.org/10.1007/978-981-15-8346-9_9.

Nile, S. H., Baskar, V., Selvaraj, D., Nile, A., Xiao, J., & Kai, G. (2020). Nanotechnologies in food science: Applications, recent trends, and future perspectives. *Nano-Micro Letters, 12*, 3. Available from https://doi.org/10.1007/s40820-020-0383-9.

Omerović, N., Djisalov, M., Živojević, K., Mladenović, M., Vunduk, J., Milenković, I., ... Vidić, J. (2021). Antimicrobial nanoparticles and biodegradable polymer composites for active food packaging applications. *Comprehensive Reviews in Food Science and Food Safety, 20*. Available from https://doi.org/10.1111/1541-4337.12727.

Ozturk, B., Argin, S., Ozilgen, M., & McClements, D. J. (2015). Formation and stabilization of nanoemulsion-based vitamin E delivery systems using natural biopolymers: Whey protein isolate and gum arabic. *Food Chemistry, 188*, 256–263. Available from https://doi.org/10.1016/j.foodchem.2015.05.005.

Pinzon, M. I., Leidy, T. S., & Cristian, C. V. (2019). *Essential oil nanoemulsions as antimicrobials and antioxidants in composite food packaging. Polymers for agri-food applications* (pp. 545–552). Cham: Springer.

Pira, S. The Future of Bioplastics for Packaging to 2020: Global Market, Forecasts, 2013.

Plastics Europe, (2019). Conversio market & strategy GmbH, plastics - the facts, 2019. https://www.plasticseurope.org/en/resources/market-data.

Prakash, A., Revathy, B., Nithyanand, P., & Vellingiri, V. (2018). Essential oil based nano-emulsions to improve the microbial quality of minimally processed fruits and vegetables: A review. *Food Research International, 111*, 509–523. Available from https://doi.org/10.1016/j.foodres.2018.05.066.

Priyadarshi, R., & Rhim, J. W. (2020). Chitosan-based biodegradable functional films for food packaging applications. *Innovative Food Science & Emerging Technologies, 62*, 102346. Available from https://doi.org/10.1016/j.ifset.2020.102346.

Resende, N. S., Gonçalves, G. A. S., Reis, K. C., Tonoli, G. H. D., & Boas, E. V. B. V. (2018). Chitosan/cellulose nanofibril nanocomposite and its effect on quality of coated strawberries. *Journal of Food Quality, 2018*, 1727426. Available from https://doi.org/10.1155/2018/1727426.

Rhim, J.-W., Park, H.-M., & Ha, C.-S. (2013). Bio-nanocomposites for food packaging applications. *Progress in Polymer Science, 38*(10), 1629–1652. Available from https://doi.org/10.1016/j.progpolymsci.2013.05.008.

Roy, S., & Das, T. K. (2016). Effect of biosynthesized silver nanoparticles on the growth and some biochemical parameters of Aspergillus foetidus. *Journal of Environmental Chemical Engineering, 4*(2), 1574–1583. Available from https://doi.org/10.1016/j.jece.2016.02.010.

Roy, S., & Rhim, J. W. (2019a). Carrageenan-based antimicrobial bionanocomposite films incorporated with ZnO nanoparticles stabilized by melanin. *Food Hydrocolloids, 90*, 500–507. Available from https://doi.org/10.1016/j.foodhyd.2018.12.056.

Roy, S., & Rhim, J.-W. (2019b). Melanin-mediated synthesis of copper oxide nanoparticles and preparation of functional agar/CuO NP nanocomposite films. *Journal of Nanomaterials, 2019*, 1–10. Available from https://doi.org/10.1155/2019/2840517.

Roy, S., & Rhim, J. W. (2020a). Carboxymethyl cellulose-based antioxidant and antimicrobial active packaging film incorporated with curcumin and zinc oxide. *International Journal of Biological Macromolecules, 148*, 666–676. Available from https://doi.org/10.1016/j.ijbiomac.2020.01.204.

Roy, S., & Rhim, J.-W. (2020b). Anthocyanin food colorant and its application in pH-responsive color change indicator films. *Critical Reviews in Food Science and Nutrition, 61*. Available from https://doi.org/10.1080/10408398.2020.1776211.

Roy, S., & Rhim, J.-W. (2020c). Antioxidant and antimicrobial poly(vinyl alcohol)-based films incorporated with grapefruit seed extract and curcumin. *Journal of Environmental Chemical Engineering, 9*, 104694. Available from https://doi.org/10.1016/j.jece.2020.104694.

Roy, S., & Rhim, J.-W. (2021a). New insight into melanin for food packaging and biotechnology applications. *Critical Reviews in Food Science and Nutrition, 62*. Available from https://doi.org/10.1080/10408398.2021.1878097.

Roy, S., & Rhim, J.-W. (2021b). Fabrication of bioactive binary composite film based on gelatin/chitosan incorporated with cinnamon essential oil and rutin. *Colloid Surface B, 204*, 111830.

Roy, S., van Hai, L., Kim, H. C., Zhai, L., & Kim, J. (2020). Preparation and characterization of synthetic melanin-like nanoparticles reinforced chitosan nanocomposite films. *Carbohydrate Polymers, 231*, 115729. Available from https://doi.org/10.1016/j.carbpol.2019.115729.

Roy, S., Zhai, L., Kim, H. C., Pham, D. H., Alrobei, H., & Kim, J. (2021). Tannic-acid-cross-linked and TiO2-nanoparticle-reinforced chitosan-based nanocomposite film. *Polymers, 13*(2), 228. Available from https://doi.org/10.3390/polym13020228.

Rukmanikrishnan, B., Jo, C., Choi, S., Ramalingam, S., & Lee, J. (2020). Flexible ternary combination of gellan gum, sodium carboxymethyl cellulose, and silicon dioxide nanocomposites fabricated by quaternary ammonium silane: Rheological, thermal, and antimicrobial properties. *ACS Omega, 5*(44), 28767–28775. Available from https://doi.org/10.1021/acsomega.0c04087.

Saravanakumar, K., Sathiyaseelan, A., Mariadoss, A. V. A., Xiaowen, H., & Wang, M. H. (2020). Physical and bioactivities of biopolymeric films incorporated with cellulose, sodium alginate and copper oxide nanoparticles for food packaging application. *International Journal of Biological Macromolecules, 153*, 207–214. Available from https://doi.org/10.1016/j.ijbiomac.2020.02.250.

Shankar, S., Jaiswal, L., & Rhim, J.-W. (2020). New insight into sulfur nanoparticles: Synthesis and applications. *Critical Reviews in Environmental Science and Technology, 51*. Available from https://doi.org/10.1080/10643389.2020.1780880.

Sharma, S., Barkauskaite, S., Jaiswal, A. K., & Jaiswal, S. (2020). Essential oils as additives in active food packaging. *Food Chemistry, 343*, 128403. Available from https://doi.org/10.1016/j.foodchem.2020.128403.

Silvestre, C., Duraccio, D., & Cimmino, S. (2011). Food packaging based on polymer nanomaterials. *Progress in Polymer Science, 36*(12), 1766–1782. Available from https://doi.org/10.1016/j.progpolymsci.2011.02.003.

Simbine, E. O., Rodrigues, L. D. A. C., Lapa-guimarães, J., Kamimura, E. S., Corassin, C. H., & Oliveira, C. A. F. (2019). Application of silver nanoparticles in food packages: A review. *Food Science and Technology, 39*. Available from https://doi.org/10.1590/fst.36318.

Singh, T., Shukla, S., Kumar, P., Wahla, V., Bajpai, V. K., & Rather, I. A. (2017). Application of nanotechnology in food science: Perception and overview. *Frontiers in Microbiology, 8*, 1501. Available from https://doi.org/10.3389/fmicb.2017.01501.

Siripatrawan, U., & Kaewklin, P. (2018). Fabrication and characterization of chitosan-titanium dioxide nanocomposite film as ethylene scavenging and antimicrobial active food packaging. *Food Hydrocolloids, 84*, 125–134. Available from https://doi.org/10.1016/j.foodhyd.2018.04.049.

Sridhar, A., Ponnuchamy, M., Kumar, P. S., & Kapoor, A. (2020). *Food preservation techniques and nanotechnology for increased shelf life of fruits, vegetables, beverages and spices: A review. Environmental chemistry letters* (pp. 1–21). Springer Science and Business Media Deutschland GmbH. Available from https://doi.org/10.1007/s10311-020-01126-2.

Stephen Inbaraj, B., & Chen, B. H. (2016). Nanomaterial-based sensors for detection of foodborne bacterial pathogens and toxins as well as pork adulteration in meat products. *Journal of Food and Drug Analysis, 24*(1), 15–28. Available from https://doi.org/10.1016/j.jfda.2015.05.001.

Störmer, A., Bott, J., Kemmer, D., & Franz, R. (2017). Critical review of the migration potential of nanoparticles in food contact plastics. *Trends in Food Science and Technology, 63*, 39–50. Available from https://doi.org/10.1016/j.tifs.2017.01.011.

Su, X. L., & Li, Y. (2004). Quantum dot biolabeling coupled with immunomagnetic separation for detection of Escherichia coli O1517:H7. *Analytical Chemistry, 76*(16), 4806–4810. Available from https://doi.org/10.1021/AC049442.

Swaroop, C., & Shukla, M. (2018). Nano-magnesium oxide reinforced polylactic acid biofilms for food packaging applications. *International Journal of Biological Macromolecules, 113*, 729–736. Available from https://doi.org/10.1016/j.ijbiomac.2018.02.156.

Tornuk, F., Hancer, M., Sagdic, O., & Yetim, H. (2015). LLDPE based food packaging incorporated with nanoclays grafted with bioactive compounds to extend shelf life of some meat products. *LWT-Food Science and Technology, 64*(2), 540–546. Available from https://doi.org/10.1016/j.lwt.2015.06.030.

Wang, H., Gong, X., Miao, Y., Guo, X., Liu, C., Fan, Y. Y., ... Li, W. (2019). Preparation and characterization of multilayer films composed of chitosan, sodium alginate and carboxymethyl chitosan-ZnO nanoparticles. *Food Chemistry, 283*, 397–403. Available from https://doi.org/10.1016/j.foodchem.2019.01.022.

Wang, L., Shao, S., Madebo, M. P., Hou, Y., Zheng, Y., & Jin, P. (2020a). Effect of nano-SiO2 packing on postharvest quality and antioxidant capacity of loquat fruit under ambient temperature storage. *Food Chemistry, 315*, 126295. Available from https://doi.org/10.1016/j.foodchem.2020.126295.

Wang, Y., Cen, C., Chen, J., & Fu, L. (2020b). MgO/carboxymethyl chitosan nanocomposite improves thermal stability, waterproof and antibacterial performance for food packaging. *Carbohydrate Polymers, 236*, 116078. Available from https://doi.org/10.1016/j.carbpol.2020.116078.

Wang, Y., Luo, W., Tu, Y., & Zhao, Y. (2021). Gelatin-based nanocomposite film with bacterial cellulose–MgO nanoparticles and its application in packaging of preserved eggs. *Coatings, 11*(1), 39. Available from https://doi.org/10.3390/coatings11010039.

Weir, A., Westerhoff, P., Fabricius, L., Hristovski, K., & von Goetz, N. (2012). Titanium dioxide nanoparticles in food and personal care products. *Environmental Science and Technology, 46*(4), 2242–2250. Available from https://doi.org/10.1021/es204168d.

Weiss, J., Takhistov, P., & McClements, D. J. (2006). Functional materials in food nanotechnology. *Journal of Food Science, 71*(9), R107–R116. Available from https://doi.org/10.1111/j.1750-3841.2006.00195.x.

Xin, S., Xiao, L., Dong, X., Li, X., Wang, Y., Hu, X., ... Zhu, B. (2020). Preparation of chitosan/curcumin nanoparticles based zein and potato starch composite films for Schizothorax prenati fillet preservation. *International Journal of Biological Macromolecules, 164*, 211–221. Available from https://doi.org/10.1016/j.ijbiomac.2020.07.082.

Xu, C., Li, J., Yang, L., Shi, F., Yang, L., & Ye, M. (2017). Antibacterial activity and a membrane damage mechanism of Lachnum YM30 melanin against Vibrio parahaemolyticus and Staphylococcus aureus. *Food Control, 73*, 1445–1451. Available from https://doi.org/10.1016/j.foodcont.2016.10.048.

Yan, S. S., & Gilbert, J. M. (2004). Antimicrobial drug delivery in food animals and microbial food safety concerns: An overview of in vitro and in vivo factors potentially affecting the animal gut microflora. *Advanced Drug Delivery Reviews, 56*(10), 1497–1521. Available from https://doi.org/10.1016/j.addr.2004.02.010.

Yanat, M., & Schroën, K. (2021). Preparation methods and applications of chitosan nanoparticles; with an outlook toward reinforcement of biodegradable packaging. *Reactive and Functional Polymers, 161*, 104849. Available from https://doi.org/10.1016/j.reactfunctpolym.2021.104849.

Yang, L., & Li, Y. (2005). Quantum dots as fluorescent labels for quantitative detection of Salmonella typhimurium in chicken carcass wash water. *Journal of Food Protection, 68*(6), 1241–1245. Available from https://doi.org/10.4315/0362-028X-68.6.1241.

Yang, R., Zhou, Z., Sun, G., Gao, Y., Xu, J., Strappe, P., ... Ding, X. (2015). Synthesis of homogeneous protein-stabilized rutin nanodispersions by reversible assembly of soybean (Glycine max) seed ferritin. *RSC Advances, 5*(40), 31533–31540. Available from https://doi.org/10.1039/c5ra03542b.

Zhang, T., Lv, C., Chen, L., Bai, G., Zhao, G., & Xu, C. (2014). Encapsulation of anthocyanin molecules within a ferritin nanocage increases their stability and cell uptake efficiency. *Food Research International, 62*, 183–192. Available from https://doi.org/10.1016/j.foodres.2014.02.041.

CHAPTER

3

Biopolymer-based antimicrobial nanocomposite materials for food packaging and preservation

Ruchir Priyadarshi[1], *Tabli Ghosh*[2], *Swarup Roy*[1,3] *and Jong-Whan Rhim*[1]

[1]Department of Food and Nutrition, BioNanocomposite Research Center, Kyung Hee University, Seoul, Republic of Korea [2]Department of Food Engineering and Technology, Tezpur University, Tezpur, Assam, India [3]Department of Food Technology and Nutrition, School of Agriculture, Lovely Professional University, Phagwara, Punjab, India

3.1 Introduction

Packaging is a multidisciplinary approach involving the synergy of art, science, and technology for the safe and economical delivery of goods from the point of initial production to the ultimate consumer. Food packaging is a crucial process that enables the distribution of food produced in one part of the globe to consumers in some other region, even after days, weeks, or months of the initial production. Hence, food packaging is paramount to the globalization of the food industry (Priyadarshi & Rhim, 2020).

Food packaging is not a new concept but dates back to the pre-historic ages. Initially, the nomadic humans gathered food only at the time of consumption. However, with the discovery of agricultural methods, the need for food storage was also generated. The materials that served the purpose were leaves, shells, woven grasses, hollowed logs, animal organs, etc. (Berger, 2003). Gradually, food storage took the form of food packaging, and with time, materials such as metals, glass, paper, and most recently plastics were developed. Plastics and polymers are by far the most utilized and best-considered resources for food packaging applications. They are lightweight, mechanically robust, water-resistant, and easily processable. Due to these advantages, they have intensively been used in rigid as well as flexible packaging.

Currently, the food packaging market is valued at more than 300 billion dollars per annum and is estimated to grow at a compound annual growth rate (CAGR) >5%. Nevertheless, the exponential growth of the food packaging market is also complemented by a proportional upsurge in the generation of municipal solid waste (MSW) due to the substantial use of nonbiodegradable petroleum-based synthetic plastics. According to the United States Environmental Protection Agency (US EPA) report regarding municipal waste generation in 2018, the contribution of plastics is worth pondering. The total MSW generated in 2018 was 292.40 million tons, of which 35.68 million tons were plastic wastes, accounting for 12.2% of the total MSW. This was an increase of 4.3 million tons from 2010 to 2018, and a large part of it came from the containers and packaging categories (US-EPA, 2020). This scenario makes the search for natural biodegradable alternatives indispensable.

3.2 Biopolymers and their classification

Polymers are generally defined as macromolecules having the same or different monomeric units held together by covalent bonds in the form of chains (Shankar & Rhim, 2018). Each polymeric chain may consist of thousands of monomeric units. Polymers can be natural or synthetic based on their origin. If the monomers are synthetically produced and not found in nature, the resulting polymers are synthetic polymers. On the other hand, if the monomers

Food Packaging and Preservation.
DOI: https://doi.org/10.1016/B978-0-323-90044-7.00003-3

are of natural origin, and the resulting polymers are either found in nature or classically synthesized artificially, the resulting macromolecules are known as biopolymers. Biopolymers are naturally occurring macromolecules produced by plants, animals, or microorganisms (Priyadarshi, Kumar, Deeba, Kulshreshtha, & Negi, 2019).

3.2.1 Biopolymers are divided into three broad categories

(i) Biopolymers that are extracted from biomass such as agricultural, animal processing, forest, or ocean wastes (Fabra, López-Rubio, & Lagaron, 2014). These include polysaccharides and proteins of plant or animal origin.

(ii) Polymers that are artificially synthesized using the biooriginated or bioderived monomeric units via classical polymer synthesis routes. These include polylactic acid or polylactide (PLA), bio-based polyethylene terephthalate (Bio-PET), and biopolyolefins.

(iii) Polymers that are produced by wild or genetically modified varieties of microorganisms. These include bacterial cellulose and polyhydroxyalkanoates (PHA) such as polyhydroxybutyrate, polyhydroxyvalerate, and poly(3-hydroxybutyrate-*co*-hydroxyvalerate) which is the copolymer of hydroxybutyrate and hydroxyvalerate. These are generally produced as extracellular secretions by microbes. PHAs are also synthesized artificially using the fermentation of plant sugars and glucose derived from sweet potatoes, soy, or pea starches.

3.2.2 The technical classification of biopolymers has two widely recognized bases

3.2.2.1 Based on methods of production

Based on production methods, the biopolymers can be classified as natural or synthetic biopolymers. Natural biopolymers are obtained or derived from biomass such as agricultural, animal processing, forest, or ocean waste. These include polysaccharides such as alginate, carrageenan, cellulose, chitin and chitosan, curdlan, gellan, pectin, pullulan, starch, and xanthan, and proteins such as collagen, gelatin, zein, soy proteins, and whey proteins. Synthetic biopolymers include polymers artificially synthesized using biooriginated or bioderived monomeric units, such as aliphatic polyesters (PLA, PHA, Bio-PET, etc.) (Priyadarshi et al., 2019).

3.2.2.2 Based on sources

Based on sources, they are classified into plant-based, animal-based, and microbial biopolymers. Plant-based biopolymers are found in species of the plant kingdom and are obtained from different parts of terrestrial or aquatic plants such as leaf, stem, trunk, roots, tubers, or even seeds and fruits. The following biopolymers fall under the plant-based category:

- *Alginate*: It is a copolymer composed of (1 → 4)-linked β-D-mannuronate and α-D-guluronate monomers and is commonly extracted from the cell wall of brown algae (Priyadarshi, Kim, & Rhim, 2021b).
- *Cellulose*: It is the structural component of the plant cell wall and hence is present in all parts of terrestrial and aquatic plants. Besides, it is also obtained from fungi of class oomycetes. Cellulose is the most abundant polysaccharide available on earth (Priyadarshi & Rhim, 2020).
- *Carrageenan*: These are high molecular weight sulfated polysaccharides extracted from certain species of seaweeds, especially Chondrus, Eucheuma, Gigartina, and Hypnea (Saedi, Shokri, & Rhim, 2020).
- *Pectin*: It is a complex heteropolysaccharide composed of (1 → 4)-linked α-D-galacturonate monomeric units obtained from the cell walls of terrestrial plant cells. It is widely used in the food industry as stabilizers, dietary fiber, and gelling agents (Priyadarshi, Kim, & Rhim, 2021a).
- *Starch*: It is a polysaccharide composed of (1 → 4)-linked α-D-glucose monomers, which acts as a form of stored energy in green plants. It is abundantly present in stable foods such as cassava, rice, wheat, maize, and potato (Choi, Lee, Lacroix, & Han, 2017).
- *Soy protein*: It is isolated from soybean and has numerous uses in the food industry, such as non-dairy creamers, infant formulas, soups, cheese, bread, and pasta. It has also been used in the form of natural adhesives (Koshy, Mary, Thomas, & Pothan, 2015).
- *Zein*: It belongs to the class of prolamin proteins, which comprises plant storage proteins rich in proline and glutamine residues. It is generally found in and isolated from maize (Corradini et al., 2014).
- *PLA*: It is a semisynthetic polymer produced by the artificial polymerization of nature-derived lactic-acid monomers. It can be derived from naturally renewable sources such as corn starch, cassava starch, and

sugarcane. It is not actually an acid but a type of aliphatic polyester and is the second-highest consumed bioplastic in the world (Roy & Rhim, 2020).

Animal-based biopolymers are derived from the species of the animal kingdom or the products obtained from them. The following biopolymers fall under this category:

- *Chitin and chitosan*: Chitin is found in the exoskeletons of arthropods and crustaceans, the cell wall of fungi, and scales of fishes. Chemically, it is composed of β-(1→4)-linked 2-(acetylamino)-2-(deoxy-D-glucose). Chitosan is an N-deacetylated derivative of chitin. Chitin and chitosan are the second most abundant biopolymers present on earth after cellulose (Priyadarshi, Sauraj, Deeba, Kulshreshtha, & Negi, 2018).
- *Collagen*: It is a protein of animal origin and is found in the extracellular space of the muscular tissue. Despite its abundance in mammals, fishes are the primary commercial source for its extraction (Alizadeh & Behfar, 2013).
- *Gelatin*: Collagen is irreversibly hydrolyzed to produce gelatin and hence is also a protein. It possesses excellent gelling properties and hence its name. It is widely used in the form of a gelling agent and stabilizer in the food industry (Riahi, Priyadarshi, Rhim, & Bagheri, 2021).
- *Whey protein*: It is a globular protein isolated from whey that is a liquid byproduct of the dairy industry, primarily the cheese production process (Andrade, Ribeiro-Santos, Melo, & Sanches-Silva, 2016).

Microbial biopolymers are produced by microorganisms, generally as extracellular secretions in the growth media. These include:

- *Bacterial Cellulose:* This exopolysaccharide is commonly produced by various bacteria as a cell covering. The most widely studied and used bacteria for this purpose is *Acetobacter xylinum* which produces a high polymer yield utilizing a variety of carbon and nitrogen sources (Bandyopadhyay, Saha, & Saha, 2018).
- *Curdlan*: It is a high molecular weight exopolysaccharide secreted by nonpathogenic bacteria *Alcaligenes faecalis*, and *Agrobacterium biobar*. It is obtained less commonly from other bacteria of *Agrobacterium* genus as well. Chemically, it is composed of β-(1→3)-linked glucose residues (Lee, Lee, Kim, & Park, 1999).
- *Gellan*: This bacterial polysaccharide is obtained from *Sphingomonas elodea*. It is composed of a tetrasaccharide repeating unit which consists of two D-glucose, one L-rhamnose, and one D-glucuronic acid residues (Morris, Nishinari, & Rinaudo, 2012).
- *Pullulan*: It consists of maltotriose units linked by α-(1→3) glycosidic bonds, and after every three consecutive units, α-(1→6) glycosidic linkage is present, which connects their units together. Fungus *Aureobasidium pullulans* converts starch feed into pullulan, which is released as an exopolysaccharide (Priyadarshi et al., 2021a).
- *Xanthan:* This high molecular weight polysaccharide is produced by the bacteria *Xanthomonas campestris*. It is widely used as a thickening agent in the food industry and an emulsion stabilizer. It provides very high viscosity to the solution at relatively lower concentrations (García-Ochoa, Santos, Casas, & Gómez, 2000).
- *Polyhydroxybutyrate*: It belongs to the class of aliphatic polyesters. This polyhydroxyalkanoate is produced when bacteria, such as *Ralstonia eutrophus*, *Bacillus megaterium*, or *Methylobacterium rhodesianum*, are exposed to physiological stress with limited nutrient supply (Plackett & Siró, 2011).

3.3 Biopolymeric nanocomposites

Composites are materials composed of a matrix and a reinforcement agent as constituent materials, both having significantly different physical and chemical properties. The matrix, or the base material, is a continuous phase such as metals, ceramics, or polymers. The most common reinforcement agents are classified into particles, short fibers, and long fibers. The reinforcement in the matrix by these agents may enhance their physical and mechanical attributes and impart several functional properties. Nanocomposites are composite materials where one of the constituents has at least one dimension in the nanometer-size range. They exhibit better mechanical, optical, barrier, and thermal properties with superior dimensional stability than micro- or macro-composites (Sharma, Jafari, & Sharma, 2020).

Bionanocomposites, also known as nanobiocomposites or green nanocomposites, are nanocomposites where biopolymers are used as matrix materials reinforced with nanofillers (Sharma et al., 2020). With the significantly reduced size of the nanofillers, their surface-to-volume ratio increases intensely. When these nanofillers are incorporated into the matrix material, a large interfacial area is shared between the biopolymer and the filler, enabling better molecular mobility and relaxation behavior, resulting in improved mechanical, thermal, and barrier properties of the bionanocomposites (Shankar & Rhim, 2018). Based on the shape of the nanofillers, they are divided into nanoparticles, nanorods, nanotubes, nanowhiskers or nanofibers, and are collectively termed nanostructures. The nanofillers used can be broadly classified into two categories: (1) Organic nanofillers; (2) Inorganic nanofillers (Shankar & Rhim, 2018). Even though a plethora of materials has been studied to date as nanofillers, Fig. 3.1 classifies some cardinal ones paramount to food packaging.

Organic nanofillers are composed of biopolymeric nanostructures of various shapes such as nanosized fibers, crystals, and whiskers. On the other hand, the inorganic nanofillers are more commonly used and are classified into four basic categories: (1) Carbon-based nanostructures, which include CNTs, Fullerenes, graphite, graphene, and carbon fibers; (2) Metal/metal oxide nanostructures such as Ag, Cu, CuO, ZnO, TiO_2, and Au; (3) Nanoclays such as montmorillonite, bentonite, and halloysite; and (4) Other nonmetallic nanostructures such as sulfur and silica nanoparticles. The enhanced physicochemical and structural properties of the base matrix materials are obtained by incorporating almost all of the mentioned fillers. However, not every filler material mentioned here possesses the microbicidal potential to contribute to antimicrobial activity. The basis of the antimicrobial activity is the interaction of these materials with the microbial cell membrane. This interaction is usually chemical in nature. Since most microbes' cell membrane possesses negatively charged groups, the positively charged metal/metal oxide nanoparticles can easily interact with them and cause the antimicrobial effect.

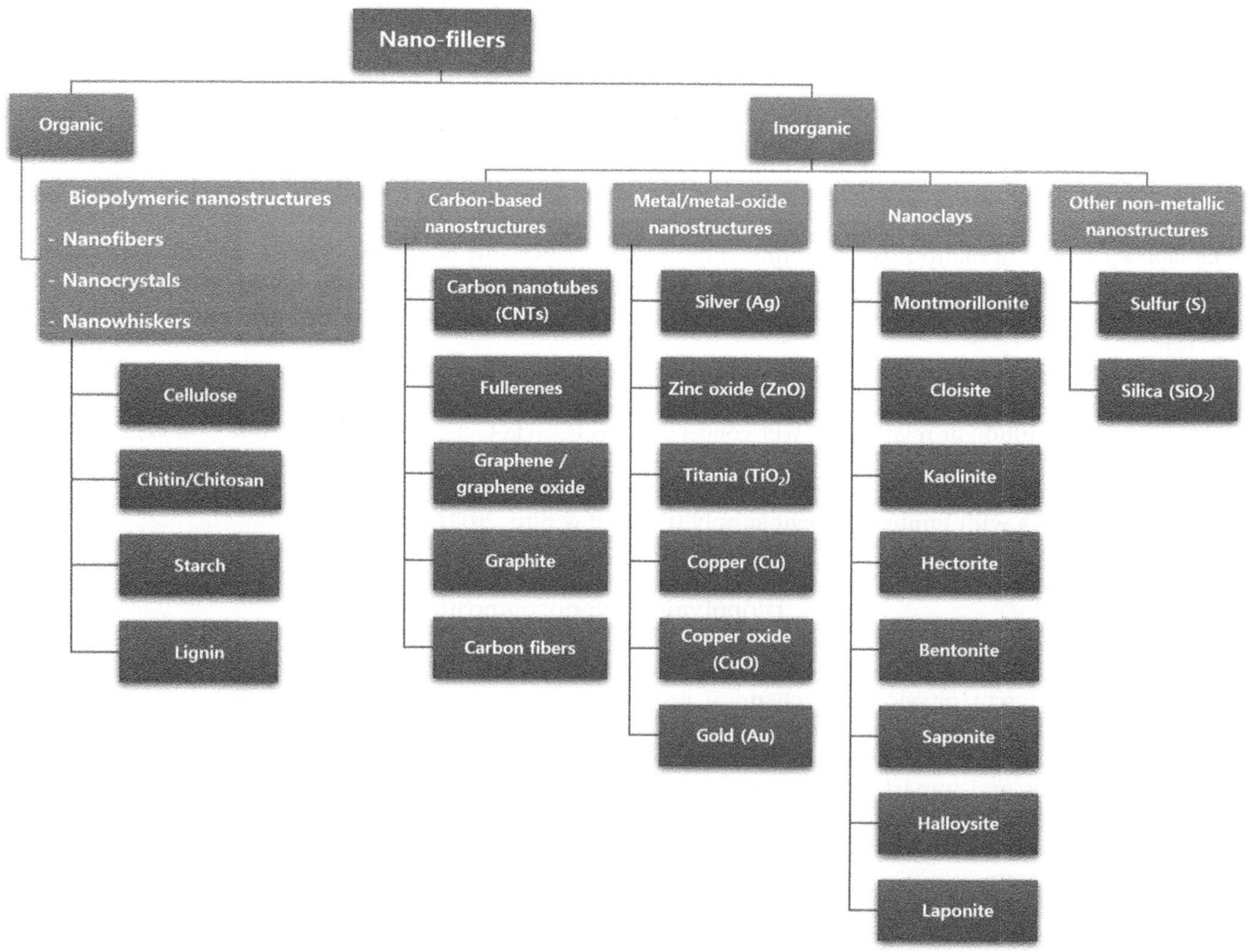

FIGURE 3.1 Various nanofiller types. Nanofillers used in the development of food packaging antimicrobial bionanocomposites materials.

Similarly, chemical interactions are shown by carbon-based materials CNTs, graphene, graphene oxide, etc., biopolymeric chitosan nanostructures, and nonmetallic nanostructures such as nanosulfur. However, clay-based fillers, other biopolymeric nanostructures, and metalloid nanostructures such as silica does not show significant antimicrobial effect. Therefore in the nanocomposites where these nonantimicrobial materials are used only for reinforcement purposes to enhance physicochemical attributes, some other functional materials such as antibiotics, natural extracts, and essential oils polyphenolic compounds can be used for imparting antimicrobial activity.

3.4 Antimicrobial activity of nanofillers

As mentioned above, due to different chemical natures, all the nanofillers do not possess antimicrobial properties and do not interact with the microbial cells. Generally, the metal/metal oxide nanostructures have the highest antimicrobial potential, making them the most widely used antimicrobial nanofillers used in food packaging applications. The antimicrobial potential of metals against bacteria as well as fungi has been known since ancient times. For instance, vessels made up of copper and silver have been used to store and preserve foods and water. Even today, the antimicrobial metal compounds are widely used in industry, agriculture, and healthcare (Lemire, Harrison, & Tuner, 2013).

The production of reactive oxygen species (ROS) by metal/metal oxide nanostructures is presumed to be the most common method of showing antimicrobial activity. ROS, a highly reactive chemical species, is produced by almost all metal/metal oxide nanostructures and can have deleterious effects on the microbial cell structure and its metabolism (Dizaj, Lotfipour, Barzegar-Jalali, Zarrintan, & Adibkia, 2014; Lemire et al., 2013). However, antimicrobial activity does not occur only by this single mechanism but by a combination of several different mechanisms that affect microbial cells in different structural or chemical ways. The other methods by which these nanostructures kill bacteria depend on the type of nanostructure. Each nanostructure possesses unique characteristics and a distinctive way of producing ROS and the overall antimicrobial effect (Dizaj et al., 2014).

Silver (Ag) has been the most widely studied metal for its antimicrobial activity against a wide range of pathogenic microorganisms, including bacteria and fungi. It has been used since times immemorial in different forms to treat bacterial infections, burns, and wounds. When it interacts with bacteria, it acts as an oxidation catalyst and disrupts the electron transfer machinery in bacterial cells. The Ag nanoparticles release Ag^+ ions, which directly interact with the microbial proteins and enzymes and precipitates them, thereby hindering the essential metabolic processes. Also, the Ag^+ ions have a very high affinity towards the sulfhydryl or thiol (–SH) groups present on the surface of the microbial cells. The silver ions replace hydrogen to form stable –S–Ag bonds which prevent the hydrogen transfer and hence inhibit respiration (Martynková & Valášková, 2014).

Zinc oxide (ZnO) and Titania (TiO_2) are semiconducting materials known for their photocatalytic activity. Semiconducting materials absorb light at a particular region of the electromagnetic spectra, which leads to the transfer of electrons from its valence band to the conduction band, initiating a series of chemical reactions. The photocatalytic activity of these metal oxides is the consequence of such chemical reactions. This photocatalytic activity successively leads to the production of ROS, which causes antimicrobial activity. Both ZnO and TiO_2 absorb the light in the UV region of the electromagnetic spectrum. Even though TiO_2 nanostructures depend exclusively on this photocatalytic activity to display antimicrobial effect, ZnO nanostructures can exhibit it in other ways. One such method is the production of Zn^{2+} ions in microbial media, which significantly affects amino acid metabolism and contributes to microbial enzyme system disruption. However, two critical parameters that affect the Zn^{2+} release are: (1) the physicochemical properties of the particles such as porosity, concentration, particle size, and morphology; and (2) The chemistry of the media, including the pH, exposure time, and presence of other elements. Nevertheless, the essential process for antimicrobial activity is ROS production in both ZnO and TiO_2. Specific studies on antimicrobial mechanisms have evidenced that microorganisms exposed to photocatalytic nanostructures exhibited extensive cell wall and membrane alterations, cell inactivation at regulatory network and signaling levels, a significant decrease in the activity of the respiratory chain, and inhibition in the ability to assimilate and transport iron and phosphorous (de Dicastillo, Guerrero Correa, Martínez, Streitt, & José Galotto, 2021).

Copper (Cu) and copper oxides (CuO and Cu_2O) nanostructures, besides ROS production, also have other mechanisms for antimicrobial activity. Under anaerobic conditions, conversion of cupric to cuprous ions takes place, which interacts directly with the cell components and directly damages certain metabolic enzymes. Cu(I) state of copper in Cu_2O might chelate essential DNA binding proteins, hindering metabolism in microbes. Also, just like Ag(I), the Cu(I) bind to thiol groups (–SH), preventing hydrogen transfer and leading to the inhibition of respiration (Meghana, Kabra, Chakraborty, & Padmavathy, 2015). On the other hand, the mechanism of antimicrobial action of CuO was not through the direct release of Cu^{2+} ions released from CuO nanoparticles but via the formation of reactive complexes between the microbial cell components and the nanoparticles (Chatterjee, Chakraborty, & Basu, 2014).

Gold (Au) nanoparticles are considered one of the safest and most potent antimicrobial agents due to their ability to kill a broad spectrum of microbes, including bacteria, fungi, and yeast. Unlike other metal/metal oxide nanostructures, which depend heavily on ROS production for their antimicrobial action, Au nanoparticles' action does not involve any ROS-related process. This ROS-independent mechanism of Au advocates their low toxicity to mammalian cells (Cui et al., 2012). Instead, the antibacterial activity of the Au nanoparticles was accredited to two distinct mechanisms. The first involves their attachment to the bacterial membrane leading to potential membrane alteration and ATP level reduction, while the second is the inhibition of tRNA binding to the ribosome, which halts protein synthesis (Dizaj et al., 2014).

Besides metal/metal oxide nanostructures, some carbon-based nanostructures, such as Carbon Nanotubes (CNT), Fullerenes, and Graphene Oxide (GO), also exhibit potent antimicrobial activity. Among CNTs, the Single-walled CNTs (SWCNT) display higher toxicity towards microorganisms than Multiwalled CNTs (MWCNT), which is attributed to the smaller size of the former. This smaller size of the SWCNT allows them to interact better with the microbial cell surface and easily penetrate them. The interaction of SWCNT with *E. coli* revealed the alteration in the bacteria's cellular membrane integrity, metabolism processes, and morphology. It is also suggested that the formation of cell-CNT aggregates damaged the bacterial cell wall, which released their DNA content, causing cell death. Fullerenes are soccer-ball-shaped carbonaceous molecules that can easily penetrate microbial cells to inhibit bacterial growth by vitiating the respiratory chain due to their smaller size. It is also suggested that the hydrophobic nature of the fullerenes allows them to interact quickly and intercalate themselves into the lipid membranes of the microbial cell walls, subsequently resulting in their disruption. Graphene is defined as a carbon-atom monolayer tightly packed in the form of a 2D structure. The modification of graphene with hydrophilic functional groups, such as hydroxyl and carboxyl, results in GO (Martynková & Valášková, 2014).

Non-metallic nanostructures such as Sulfur have also exhibited decent antimicrobial effects, especially against Gram-positive foodborne pathogenic bacteria such as *Listeria monocytogenes*. Sulfur has been reported to exert antimicrobial action via several mechanisms: (1) disruption of the bacterial metabolic processes resulting from the interaction of SNPs with target molecules on microbial cell wall; (2) production of sulfur ions and H_2S by decomposition of sulfur nanostructures inside the microbial cells, causing degradation of cellular components; (3) microbial cell death resulting from interference in DNA replication by sulfur nanostructures; (4) blocking of the SH-enzymes by sulfur nanostructures in the microbial cell, which is responsible for normal intermediary metabolism of carbohydrates, fats, and proteins, leading to failure of cell machinery and hence death (Priyadarshi et al., 2021b).

Among the organic biopolymeric nanostructures, the only natural polysaccharide that shows antimicrobial activity is chitosan, which exhibits potent activity against bacteria, fungi, and yeast. The main reason for the antimicrobial activity of chitosan is its cationic nature. Nevertheless, the degree of acetylation, concentration, the period of exposure, and the test organism also play a significant role in determining the magnitude of its antimicrobial activity. The commonly proposed mechanisms include: (1) dissolution of the microbial cell wall and leakage of the intracellular fluids resulting from the interaction between the cationic group of chitosan with the anionic peptidoglycans on the microbial cell surface; (2) failure of microbial metabolic machinery resulting from the hindered exchange of essential solutes between the interior and exterior of the cells; (3) binding of low molecular weight chitosan to the DNA molecule and impeding its translation to RNA; (4) chelation of the essential trace metals required for microbial metabolism resulting in inhibition of their metabolic machinery (Priyadarshi & Rhim, 2020).

Clays, in their natural form, are not antimicrobial. However, certain clays, such as montmorillonite when exchanged with cationic surfactants such as cetylpyridinium, develop a surface charge, which makes them capable of interacting with the microbial cell and exhibits strong antimicrobial activity (Martynková & Valášková, 2014). Cloisite 30B, an organically modified MMT, is reported to display a dramatic bactericidal effect against Gram-positive species and a remarkable bacteriostatic effect against Gram-negative organisms (Hong & Rhim, 2008). Even though its effect is not very strong against Gram-negative bacteria, it is one of the prevalent antimicrobial nanoclays and has been widely used to fabricate biopolymeric antimicrobial nanocomposites (Rhim, Hong, & Ha, 2009; Sothornvit, Rhim, & Hong, 2009). Further, to enhance the efficacy of Cloisite 30B, recently Peighambardoust et al. modified the nanoclays with Ag and Cu-ions. The resulting clays displayed a comparatively stronger antimicrobial effect against Gram-negative bacteria and exhibited ameliorated physicochemical properties of the CMC films when used as nanofiller (Peighambardoust, Zahed-Karkaj, Peighambardoust, Ebrahimi, & Peressini, 2020).

3.5 Antimicrobial biopolymeric nanocomposite films

3.5.1 Polysaccharides

3.5.1.1 Alginate

Among available polysaccharides, alginate is derived from brown algae such as *Laminaria digitata, Laminaria hyperborea, Macrocystis pyrifera, Ascophyllum nodosum*, etc., and is a kind of anionic polysaccharide (Lee & Mooney, 2012). Sodium alginate consists of monomeric blocks of (1→4)-linked β-D-mannuronic acid and α-L-guluronic acid (Hammoudi, Ziani Cherif, Borsali, Benmansour, & Meghezzi, 2020; Madhumathi, Jeevana Rekha, & Sampath Kumar, 2018). Alginate exists in the form of sodium, magnesium, and calcium salts of alginic acid. Alginate has the properties of nontoxicity, biodegradability, biocompatibility, easy availability, and others (Lee & Mooney, 2012), which make alginate a potential candidate to be used in developing food packaging materials.

The functional properties of alginate-based films can be modified by incorporating nanofiller materials, and further, the nanofiller materials can control antimicrobial release from nanocomposite materials. Different kinds of essential oils are also added to alginate-based nanocomposites for improved antimicrobial properties. As mentioned earlier, different types of nanoclays include kaolinite, montmorillonite, halloysite, etc. Among available, montmorillonite is a widely used nanoclay (mineral nanofiller) for improved properties of alginate to be used in food packaging (Alboofetileh, Rezaei, Hosseini, & Abdollahi, 2014). Nanoclay as a nanofiller in developing nanocomposite packaging materials can control the release of antimicrobial agents and further retain the aroma compounds in the nanocomposite films. Alginate-montmorillonite/lemon essential oils nanocomposite films exhibited appreciable antimicrobial and antifungal properties (Hammoudi et al., 2020). The films were more effective against Gram-positive bacteria (*Staphylococcus aureus* and *Bacillus cereus*) than Gram-negative bacteria (*Salmonella enteritidis* and *Escherichia coli*).

Similarly, the antimicrobial efficacy of the nanocomposite films based on alginate and MMT, incorporated with essential oils (clove, cinnamon, and marjoram) was tested against Gram-negative bacteria (*E. coli*) and Gram-positive bacteria (*S. aureus* and *L. monocytogenes*) (Alboofetileh et al., 2014). The films provide better effectiveness against Gram-positive bacteria than Gram-negative bacteria. They reported that the marjoram essential oil has the most effective antimicrobial activity among the selected essential oils, and the films can reduce the microbial population by 6.33, 4.52, and 5.80 log CFU/mL for *L. monocytogenes, E. coli*, and *S. aureus*, respectively, compared to control (Alboofetileh et al., 2014).

On the other hand, Halloysite is a kind of aluminosilicate clay available as nanotubes (Shankar, Kasapis, & Rhim, 2018), which can be used as a nanofiller material in developing antimicrobial packaging. The surface activation of halloysite can be obtained by treating it with alkali (sodium hydroxide), where surface functionalization of the alkali-treated halloysite with ZnO NPs (AT-Hal/ZnONP) can be used as nanofiller for developing functional alginate films. In this context, the incorporation of AT-Hal/ZnONP in alginate resulted in films with strong antibacterial activity against *E. coli* and *L. monocytogenes* (Shankar et al., 2018). As shown in Fig. 3.2, the antibacterial property of neat alginate and nanocomposites of alginate reinforced with ZnO nanoparticles functionalized alkali-treated halloysites tested against *L. monocytogenes* and *E. coli*, where the complete sterilization of *L. monocytogenes* and *E. coli* could be obtained after 9 hours and 3 hours, respectively, when nanocomposite films with 7 wt.% of ZnO nanoparticles functionalized alkali-treated halloysite is used. Additionally, the nanocomposite films developed by incorporating Au–TiO_2 in sodium alginate films enhanced the antibacterial activity by ~50% and 60% compared to TiO_2 incorporated sodium alginate films against *E. coli* and *S. aureus*, respectively (Tang et al., 2018). The nanocomposites based on alginate and hydroxyapatite nanoparticles also provide bactericidal effects against foodborne pathogens such as *L. monocytogenes* (Gram-positive bacteria) (Gholizadeh, Buazar, Hosseini, & Mousavi, 2018). In a more recent study, the antimicrobial activity of sulfur nanoparticle-incorporated sodium alginate films was studied. Even though a bacteriostatic effect was observed against Gram-negative *E. coli* bacteria, indicating a weaker antimicrobial effect, the films displayed strong antimicrobial activity against Gram-positive *L. monocytogenes*. The resulting films were hence suggested as suitable for meat packaging under refrigerated conditions, where it is prone to infection by *Listeria* sp (Priyadarshi et al., 2021b).

3.5.1.2 Starch

Starch consists of amylose (20%–25%) and amylopectin (75%–80%), where amylose is responsible for the film-forming property of starch (Mukerjea, Sheets, Gray, & Robyt, 2015). Amylose is a type of linear polysaccharide consisting of D-glucose units linked by α-1,4 glycosidic bonds. In contrast, amylopectin is a branched polysaccharide consisting of D-glucose units linked by α-1,4 glycosidic bonds and α-1,6 glycosidic bonds as the main

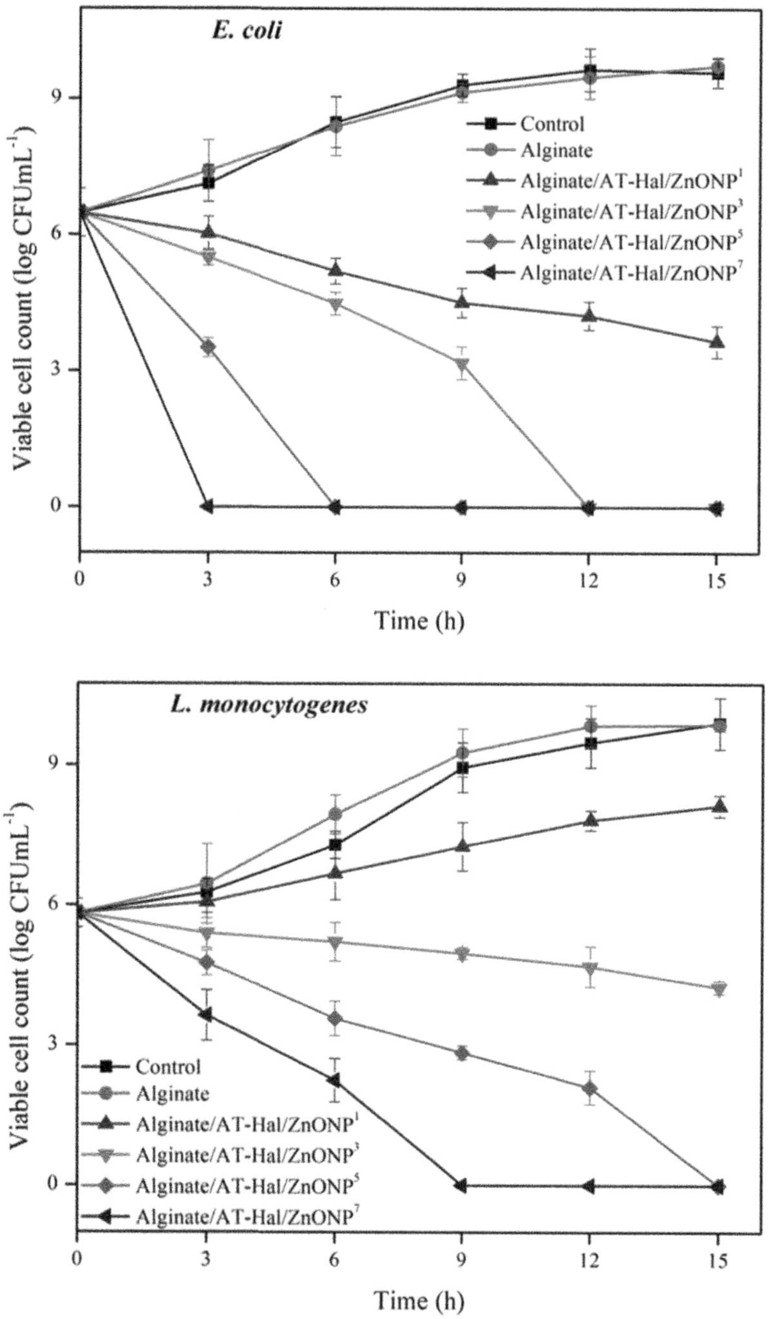

FIGURE 3.2 Alginate-based antimicrobial nanocomposites. Antimicrobial property of alginate nanocomposite reinforced with alkali-treated halloysite with ZnO NPs (AT-Hal/ZnONP) (Shankar et al. 2018). *Source: Data from 1. Shankar, S., Kasapis, S., & Rhim, J. W. (2018). Alginate-based nanocomposite films reinforced with halloysite nanotubes functionalized by alkali treatment and zinc oxide nanoparticles. International Journal of Biological Macromolecules. 118, 1824–1832. https://doi.org/10.1016/j.ijbiomac.2018.07.026.*

chain and branch chain, respectively. Starch is extracted from various sources. The reasons for the extensive utilization of starch in food packaging are cost-effectiveness, biodegradability, biocompatibility, nontoxicity, renewability, edibility, etc. However, the hydrophilic nature, weak mechanical properties, easy degradability, and poor processing stability limit its application in the food packaging sector. The properties of starch-based packaging materials can be modified via different routes such as chemical modifications, nanocomposite formulations, etc. The incorporation of nanofiller materials, including nanocellulose, montmorillonite, inorganic nanomaterials, etc., in the matrix of starch can improve its above-mentioned properties.

The surge in starch-based biomaterials in developing food packaging is increasing due to their biodegradable and water-soluble nature. The antimicrobial starch-based films are obtained by incorporating chitosan, Ag nanoparticles, ZnO, potassium sorbates, lysozyme, hybrid nanomaterials, etc. The antimicrobial nanocomposites based on corn starch (base materials), halloysite clay (nanofiller), and bacteriocins (antimicrobial peptides) have been developed using a solution casting process. The produced films act against foodborne pathogens and exhibit improved packaging properties (Mukerjea, Sheets, Gray, & Robyt, 2015). The existing antimicrobial peptides are bacteriocins isolated from lactic acid bacteria that are considered safe for human ingestion. Some common bacteriocins, such as nisin, pediocin, potassium sorbates, carvacrol, lysozyme, etc., are the potential antimicrobial agents used in developing antimicrobial packaging and further have the status of generally recognized as safe (GRAS).

Additionally, the nanocomposite films based on starch, clay (montmorillonite), and potassium sorbate are more effective in retaining antimicrobial activity than the starch films (Barzegar, Azizi, Barzegar, & Hamidi-Esfahani, 2014). However, the migration rate of potassium sorbates from the packaging materials to the food matrix needs to be controlled for proper functioning as antimicrobial packaging. The migration of antimicrobial agents can be retarded due to silicate layers (clay) in the starch-based nanocomposites providing prolonged antimicrobial activity. The specified nanocomposite films inhibited the growth of *A. niger* when potassium sorbates are incorporated at higher levels (7.5% and above).

On the other hand, graphene oxides have surface functional groups (carbonyl, hydroxyl, and epoxy), which interact with base materials delivering intercalated nanocomposites. The nanocomposite films based on poly (vinyl alcohol) (base polymer), graphene oxide (nanofiller), starch (green, reducing agent), and Ag nanoparticle (nanofiller) were used as a food packaging material with improved mechanical, thermal, and antimicrobial property. Starch was used as a green, reducing agent for graphene oxide and silver salts. The graphene oxide nanofiller displayed uniform dispersion in the poly(vinyl alcohol) matrix due to the presence of polar groups on its surface. The combined effect of graphene oxide and silver nanoparticles provided the improved antimicrobial property in the specified nanocomposite materials (Usman, Hussain, Riaz, & Khan, 2016).

Additionally, the incorporation of hybrid antimicrobial nanoparticles such as ZnO-chitosan nanoparticles (3%) into starch matrix resulted in antimicrobial films that were more effective against Gram-positive bacteria (*S. aureus*) than Gram-negative bacteria (*E. coli*) (Hu, Jia, Zhi, Jin, & Miao, 2019). The nanoparticles of ZnO-chitosan were synthesized using the sol-gel method and exhibited an average size of 25 nm. The antimicrobial property of ZnO-chitosan nanoparticles incorporated in starch bionanocomposite films are positively correlated with the nanofiller loading.

The nanocomposite packaging materials based on tapioca starch, cellulose nanocrystals, and grape pomace extracts are also developed. The sources of grape pomace extracts were cabernet franc (red variety); and viognier (white variety) (Xu, Rehmani, Alsubaie, Kim, & Sismour, 2018). The addition of grape pomace extracts increases the antibacterial activity of starch and cellulose nanocrystal nanocomposites. The antimicrobial activity of grape pomace extracts makes them a good source for developing antimicrobial packaging.

3.5.1.3 *Agar*

Agar is another kind of hydrophilic polysaccharide obtained from red algae (class of *Rhodophyceae*) (Malagurski et al., 2017). Agar is made up of agarose and agaropectin, and possesses good film-forming properties. It is renewable, biobased, biocompatible, and nontoxic. These properties make it a potential source for development and utilization in food packaging. Agarose consists of repeating units of agarobiose and is a linear polysaccharide, whereas agarobiose is a disaccharide composed of D-galactose and 3,6-anhydro-L-galactopyranose monomeric units. Further, agaropectin is a kind of sulfated galactan. Agar is extensively used for developing biodegradable food packaging, including flexible films, edible films, edible coatings, etc. However, agar-based packaging materials have the limitations of low thermal stability, mechanical properties, hydrophilicity, etc. The properties of agar-based films can be modified by incorporating nanoclay, nanocellulose, metallic nanoparticles, and other nanostructures to improve their inherent packaging properties. Several studies report the food packaging potential of agar-based antimicrobial nanocomposite films. The agar-based films with various antimicrobial agents, such as nanoclays, silver NPs, grapefruit seed extract, have been fabricated (Kanmani & Rhim, 2014). The incorporation of metal-based nanofillers in agar films can provide unusual antimicrobial activity. The nanocomposite food packaging materials developed from Zn mineralized (Zn carbonate or Zn phosphate) agar provide antimicrobial properties due to the capability to release Zn(II) ions (Malagurski et al., 2017). The silver nanoparticles incorporated in agar films have also been reported to provide strong antimicrobial activity (Rhim, Wang, & Hong, 2013; Rhim, Wang, Lee, & Hong, 2014). Additionally, the effectiveness of the antimicrobial properties of Ag NPs reinforced agar-based nanocomposites has been represented in Fig. 3.3. Basumatary et al. also reported that the incorporation of silver nanoparticles in agar provided decent antibacterial activity tested against gram-negative bacteria *Aeromonas hydrophilla* and imparted the films with UV light barrier properties (Basumatary, Daimary, Das, Thapa, Singh, Mukherjee, & Kumar, 2018). The nanocomposite films prepared by incorporating bimetallic (Ag-Cu) alloy NPs to Agar provide strong antimicrobial properties tested against *L. monocytogenes* and *Salmonella enterica sv typhimurium*, which preserve the food by inhibiting the growth of

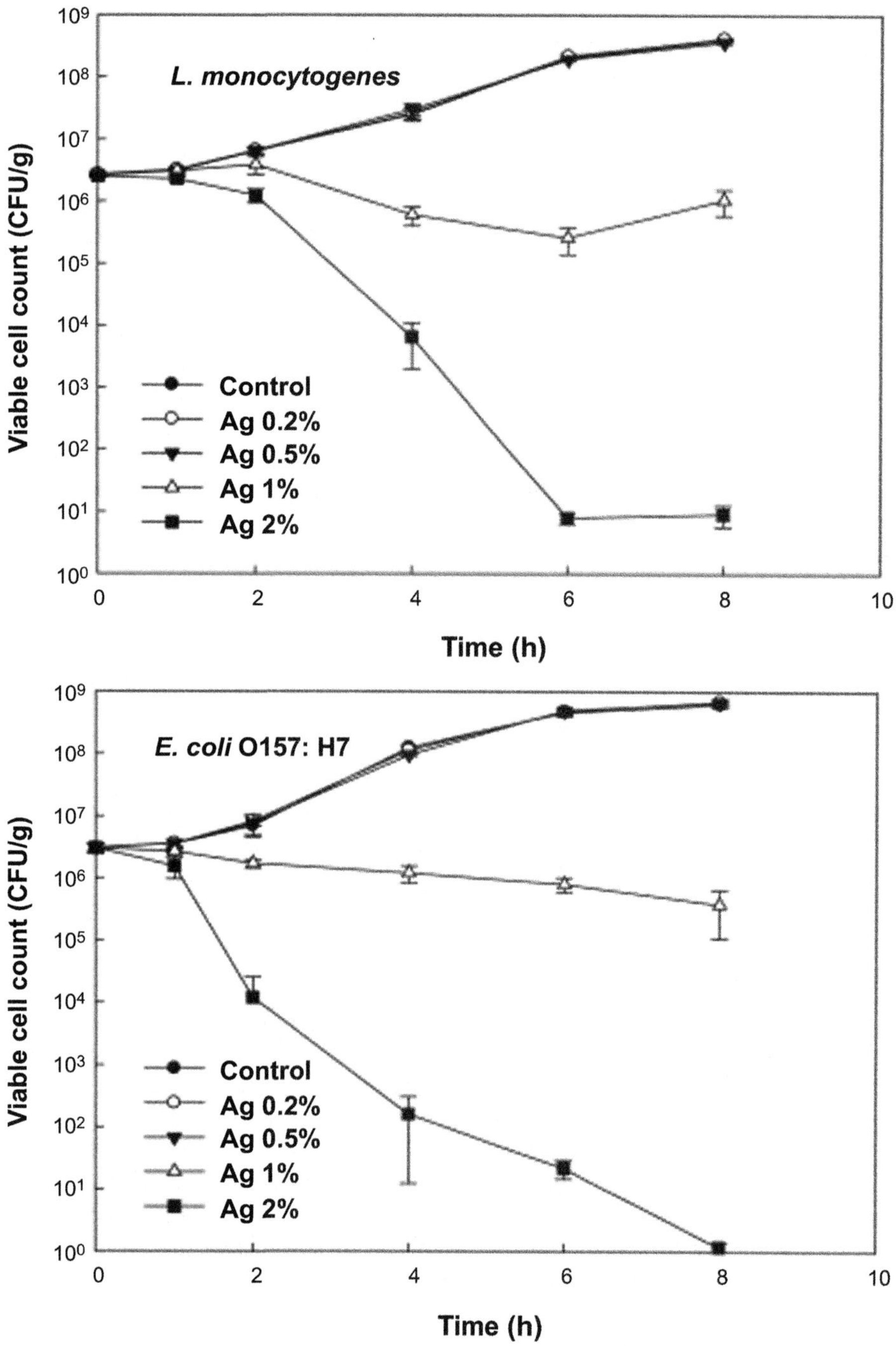

FIGURE 3.3 Agar-based antimicrobial nanocomposites. Antimicrobial activity of agar and agar/AgNPs based nanocomposites against foodborne pathogens (Rhim et al., 2013). *Source: Data from 2. Rhim, J. W., Wang, L. F., & Hong, S. I. (2013). Preparation and characterization of agar/silver nanoparticles composite films with antimicrobial activity. Food Hydrocolloids, 33(2), 327–335. https://doi.org/10.1016/j.foodhyd.2013.04.002.*

food spoilage bacteria (Arfat, Ahmed, & Jacob, 2017). The nanocomposite packaging films based on agar-carboxymethyl cellulose incorporated with silver-modified montmorillonite (MMT, antimicrobial agent) provide strong antibacterial activity against *B. subtilis* (Gram-positive) and *E. coli* (Gram-negative) (Makwana, Castaño, Somani, & Bajaj, 2020).

3.5.1.4 Cellulose

In the current food packaging market, cellulose is one of the most extensively used biomaterials owing to its several remarkable properties. Cellulose is a homopolymer consisting of $\beta(1 \rightarrow 4)$ linked glucose monomeric units. The various sources of cellulose include lignocellulosic biomass such as hemp, wood, ramie, flax, cotton, agricultural byproducts, and others. The cellulose from the lignocellulosic biomass is obtained by removing the hemicellulose and lignin, following the pretreatment methods like pulping and bleaching. On the other hand, the bacterial cellulose obtained from Gram-positive and Gram-negative bacteria has several advantageous properties compared to plant cellulose, such as high purity, low production cost, better yield, and high crystallinity.

The physicochemical properties of cellulose can be altered by synthesizing its derivatives, such as methylcellulose, ethylcellulose, hydroxypropyl cellulose, hydroxypropyl methylcellulose, and carboxymethyl cellulose. Different forms of celluloses have been used to fabricate food packaging materials. The cellulose and its derivatives are used as matrix materials and incorporated with various antimicrobial agents to develop cellulose-based antimicrobial packaging. The nanocomposite films based on methyl cellulose, carvacrol, and montmorillonite are used as active antimicrobial films against *E. coli* and *S. aureus* (Tunç & Duman, 2011). Additionally, the concentration of montmorillonite in the nanocomposite films could control the release of carvacrol. The reduction of *E. coli* counts in carvacrol-added methyl cellulose and methyl cellulose/MMT films are 0.90 and 0.65 log cfu/mL, respectively. Additionally, the reduction of *S. aureus* counts is observed to be 0.90 and 0.70 log cfu/mL, respectively, for both films. It is noteworthy to mention that montmorillonite is widely used as a potential nanofiller to control the release of antimicrobial agents in nanocomposite materials, which makes them effective for a prolonged time (Tunç & Duman, 2011; Tunç, Duman, & Polat, 2016). Besides, the incorporation of metal oxide nanoparticles in bacterial cellulose imparted strong antimicrobial activity to the films. The bacterial cellulose/ZnO nanocomposite films were more effective against *S. aureus* than *E. coli* (Shahmohammadi Jebel & Almasi, 2016). Additionally, the nanocomposite materials based on carboxymethyl cellulose and hybrid chitin nanowhiskers, incorporated with ZnO-Ag nanofillers, possessed multifunctional properties, including antimicrobial activity (Oun & Rhim, 2017). In another study, the in-situ synthesis of silver NPs in the films based on cellulose and banana peel powder was also studied. The nanocomposite films provided antimicrobial activity against *E. coli, Pseudomonas aeroginosa, Bacillus licheniformis,* and *S. aureus* (Thiagamani et al., 2019). Bruna et al. reported the nanocomposite films based on cellulose acetate incorporated with copper-modified montmorillonite for food packaging. The films displayed strong antimicrobial activity, where 98% of the *E. coli* count was reduced by the action of nanocomposite films composed of 5% active nanofiller (Bruna, Galotto, Guarda, & Rodríguez, 2014). Additionally, nanocelluloses obtained from cellulose materials have been used as potential nanofiller in developing nanocomposite materials, which can act as remarkable carriers for antimicrobial agents. The different types of nanocelluloses, that is, cellulose nanocrystals, cellulose nanofibers, and cellulose nanowhiskers, have been widely used to improve the properties of food packaging materials.

3.5.1.5 Chitin and chitosan

Chitosan is the second most available polysaccharide in nature after cellulose having excellent functional properties (Boura-Theodoridou et al., 2020; Priyadarshi & Rhim, 2020). Chitosan is the deacetylated form of chitin, obtained primarily from aquatic fauna. This specific polysaccharide has $-NH_2$ group, which is converted to its protonated form in acidic media, providing chitosan with its intrinsic antimicrobial property. Nevertheless, the antimicrobial property of chitosan is not very strong and depends on several factors, such as the degree of deacetylation (Priyadarshi & Rhim, 2020). Hence, several antimicrobial nanoparticles as fillers in chitosan films have been reported, which helps in a synergistic enhancement of the antimicrobial potential of the resulting nanocomposite films. The metal/metal oxide nanostructures have been the most common filler materials for chitosan-based nanocomposite films. Silver nanoparticle-embedded chitosan films exhibited enhanced antimicrobial activity against Gram-positive bacteria *S. aureus* and *B. subtilis* as well as Gram-negative bacteria *P. aeruginosa* and *E. coli* (Tripathi, Mehrotra, & Dutta, 2011). The in-situ synthesis of silver nanoparticles in the chitosan matrix also improved the antimicrobial activity of the resulting nanocomposite films (Kaur, Choudhary, & Thakur, 2013). Zinc oxide nanoparticles/chitosan nanocomposite films exhibited synergistic antimicrobial activity against *B. subtilis* and *E. coli* bacterial strains (Priyadarshi & Negi, 2017). Youssef et al. prepared chitosan nanocomposite films incorporated with hybrid silver-zinc oxide nanoparticles. The films were effective against a wide range of food pathogens like *S. aureus, E. coli, S. typhimurium, B. cereus,* and *L. monocytogenes* (Youssef, Abou-Yousef, El-Sayed, & Kamel, 2015). Sulfur nanoparticles have also been used as filler materials in chitosan. The resulting films exhibited better physicochemical properties and significantly higher antimicrobial activity than the neat chitosan films. The incorporation of ZnO doped SiO_2 NPs as a nanofiller to the poly vinyl alcohol/chitosan blend films imparted antimicrobial activity against common foodborne pathogens (Al-Tayyar, Youssef, & Al-Hindi, 2020). The nanocomposite films based on chitosan/poly vinyl alcohol/ZnO deliver antimicrobial activity against food-borne pathogens; however, NaOH treatment can reduce the antimicrobial activity of the nanocomposite films due to the reduced polycationic nature of chitosan in alkaline solution, leading to its reduced solubility (Boura-Theodoridou et al., 2020).

Besides inorganic nanoparticles, studies also report the use of organic nanostructures as fillers in chitosan matrix. The use of rod-shaped chitin whiskers at a 3% concentration in chitosan films increased the tensile strength and showed potent antimicrobial activity against foodborne pathogenic bacteria (Ma, Qin, Li, Zhao, & He, 2014). Lignin nanoparticles incorporated chitosan/PVA blend films exhibited remarkable antimicrobial and antioxidant properties making them suitable for multifunctional food packaging applications (Yang, Li, Huang, Yang, & Li, 2016).

3.5.1.6 *Carrageenan*

Carrageenan is an anionic polysaccharide consisting of linear sulfated galactans groups, which is generally extracted from red edible seaweeds (Therkelsen, 2012). The several fractions of carrageenan are kappa, iota, and lambda. The kappa, iota, and lambda carrageenan contain one sulfate group, two sulfate groups, and three sulfate groups per disaccharide. The antimicrobial activity in carrageenan-based food packaging against foodborne pathogens was developed by incorporating organic and inorganic nanomaterials as nanofillers. The nanomaterials used by various researchers include ZnO NPs, Ag NPs, CuO NPs, SiO_2 NPs, and chitin nanofibrils (Meindrawan, Suyatma, Wardana, & Pamela, 2018; Rhim & Wang, 2014; Roy & Rhim, 2019; Shankar, Reddy, Rhim, & Kim, 2015). Besides, using ZnO NPs as a nanofiller in carrageenan-based antimicrobial active food packaging, incorporating ZnO NPs capped with melanin was also reported to provide antibacterial activity against foodborne pathogens (Roy & Rhim, 2019). Additionally, the use of melanin-mediated Ag NPs as a nanofiller in developing carrageenan-based nanocomposite also provided enhanced antimicrobial activity (Roy, Shankar, & Rhim, 2019).

3.5.1.7 *Pullulan*

Pullulan is a linear polysaccharide consisting of maltotriose units, secreted as an exopolysaccharide by the polymorphic fungus *A. pullulans* feeding on starch. The characteristic attributes of pullulan include biodegradable, noncarcinogenic, high flexibility, water solubility, and high stability against sodium chloride. However, the property of pullulan solubility can be altered by etherification, esterification, and crosslinking process. The functional properties of pullulan can be achieved by incorporating active components such as metal oxides, essential oils, and plant extracts. In this context, the incorporation of lysozyme nanofibers in developing pullulan-based nanocomposite provided antimicrobial activity tested against *S. aureus*, making them suitable to be used as edible films (Silva, Vilela, Almeida, Marrucho, & Freire, 2018). Additionally, the various pullulan-based nanocomposites with AgNPs, ZnO NPs, oregano oil, and rosemary oils were also beneficial in maintaining turkey deli meat quality (Khalaf, Sharoba, El-Tanahi, & Morsy, 2013).

3.5.2 *Proteins*

3.5.2.1 *Animal-based proteins*

The important proteins of animal origin that have been used to fabricate antimicrobial nanocomposites for food packaging include gelatin, collagen, and whey protein.

Collagen is a fibrous structural protein found in vertebrate tissues such as skin, ligaments, tendons, and bones. It is a highly versatile, value-added, and abundant proteinaceous biopolymer of animal origin (Ocak, 2018). Due to its compatibility with human tissues, collagen has been widely used in biomedical applications, either alone or in combination with other biopolymers (Parenteau-Bareil, Gauvin, & Berthod, 2010). However, it is not much used in food packaging applications due to its poor film formation, low water barrier, high sensitivity to ambient humidity, low thermal stability, and poor mechanical properties (Ocak, 2018). Due to these drawbacks, it is not commonly used as a standalone material in food packaging applications but has been used to blend with other stronger biopolymers (Wang & Rhim, 2015). Nevertheless, gelatin, which is the derivative of collagen, has been widely used not only in food packaging but have a plethora of important uses in the food industry (Purohit et al., 2020; Riahi et al., 2021).

Gelatin is the partially denatured product of collagen and can form clear, flexible, strong, and biodegradable packaging films. However, the most common commercial source for gelatin extraction is not vertebrate tissues but fish skin. It is the most commonly used biopolymer in the food industry due to its ability to provide elasticity, viscosity, and stability. As the name suggests, gelatin is widely used to make jellies, gums, and soft pharmaceutical capsules. It is a linear polymer that is highly processable and possesses excellent film-forming properties and hence has been used in antimicrobial food packaging in the form of nanocomposite films. Various antimicrobial nanofillers have been used for this purpose. Kanmani and Rhim reported the antimicrobial activity of gelatin film incorporated with AgNPs against *E. coli* and *S. aureus*, exhibiting a 49% and 41% reduction in bacterial cell viability (Kanmani & Rhim, 2014). Arfat et al. reported profound antimicrobial activity of ZnO nanoparticle-incorporated fish gelatin-based film on *L. monocytogenes* and *P. aeruginosa* (Arfat, Benjakul, Prodpran, Sumpavapol, & Songtipya, 2014). The photocatalytic antimicrobial effect of gelatin-TiO_2 nanocomposite film was studied by He et al. against *E. coli* and *S. aureus* after a 120 minutes irradiation under UV-light (365 nm) and obtained a 54% and 45% better activity, respectively (He, Zhang, Cai, & Wang, 2016). Similar results were obtained by Riahi et al., who studied the antimicrobial effect of gelatin-TiO_2 nanocomposite film. The TiO_2 nanoparticle incorporated films displayed a reduction in *E. coli* and *L. monocytogenes* colonies by 7 log cfu/mL and 2 log cfu/mL, respectively, compared to neat gelatin films (Riahi et al., 2021). Besides metal/metal oxide nanoparticles, the antimicrobial effect of the incorporation of bimetallic nanoparticles in gelatin films has also been studied. Arfat et al. embedded bimetallic Ag-Cu nanoparticles in fish gelatin films and tested their antimicrobial activity against *L. monocytogenes* and *S. typhimurium* to observe the reduction in bacterial colonies by 3 log cfu/mL and 7 log cfu/mL, respectively, compared to neat gelatin films (Arfat et al., 2017).

Whey protein, which is produced as waste in large quantities by the dairy industry, is a byproduct of the cheese production process (Said & Sarbon, 2019). It has been used as an edible and biodegradable food packaging material due to its safety for human-contact use and its ability to produce transparent films and coatings that can act as an excellent oxygen barrier (Jain et al., 2021; Karimi, Alizadeh, Almasi, & Hanifian, 2020). There have been several reports of whey protein being used in food packaging. However, very few studies report the fabrication of whey protein-based antimicrobial nanocomposites. Organically modified nanoclays have been the most widely used antimicrobial nanostructures in whey protein films. Sothornvit et al. reported the fabrication of whey protein isolate (WPI)/Cloisite 30B organo-clay composite films and their antimicrobial activity. The nanocomposite films displayed a bacteriostatic effect against Gram-positive *L. monocytogenes* (Sothornvit, Hong, An, & Rhim, 2010; Sothornvit et al., 2009). The bacteriostatic effect of whey protein isolate (WPI)/Cloisite 30B organo-clay nanocomposite films against Gram-positive *L. monocytogenes* is shown in Fig. 3.4. However, very few reports are present on nanostructures as antimicrobials in the whey protein films. Most researchers have used other antimicrobial agents, such as essential oils, plant extracts, antibiotics, either alone or in combination with nanostructures. Sani et al. studied the antimicrobial efficacy of whey protein isolate (WPI)/cellulose nanofibre (CNF) nanocomposite films containing 1.0% (w/w) titanium dioxide (TiO_2) and 2.0% (w/v) rosemary essential oil against inoculated *S. aureus*, *L. monocytogenes*, and *E. coli* O157: H7 on packaged lamb meat. They reported a higher inhibition effect on Gram-positive bacteria than Gram-negative ones (Alizadeh Sani, Ehsani, & Hashemi, 2017). In a more recent study, Karimi et al. fabricated cellulose nanofiber (CNF) and polydextrose incorporated WPI-based nanocomposite film containing *L. plantarum* as the probiotic antimicrobial agent. The incorporation of CNF and polydextrose enhanced the probiotic bacteria's cell viability, and the resulting films exhibited strong antimicrobial action against *P. aeruginosa*, *S. aureus*, and *E. coli* bacteria (Karimi et al., 2020).

3.5.2.2 Plant-based proteins

The important proteins of plant origin that have been used to fabricate antimicrobial nanocomposites for food packaging include soy protein and zein.

Zein is a plant-based protein obtained from *Zea mays* or corn. It is a GRAS protein with the majority of nonpolar amino acid composition, due to which they tend to form highly hydrophobic films (Mei et al., 2017). Zein creates glossy, strong, greaseproof films with relatively good barrier properties. The formation of intermolecular disulfide bonds between sulfur-containing amino acids, such as leucine and alanine, as well as hydrophobic interactions between protein molecules are the bases of film formation (Qu et al., 2019). Various studies report the fabrication of zein-based nanocomposite films using different nanostructures for food packaging applications. However, not much literature is available on the use of these nanostructures for antimicrobial purposes. They are mainly used for the amelioration of the physicochemical properties of the matrix material. Luecha et al. reported the improved physicochemical properties of the zein montmorillonite nanocomposite films with increased clay loading (Luecha, Sozer, & Kokini, 2010). Rouf et al. prepared laponite nanoclay incorporated zein films showing significantly improved mechanical, thermal, and barrier properties, which resulted from the binding of laponite nanoparticles to zein molecules through Si—N bond (Rouf, Schmidt, & Kokini, 2018).

On the other hand, most reports that report zein-based antimicrobial nanocomposites use antimicrobial agents other than nanostructures. Vahedikia et al. prepared zein-based nanocomposite films by incorporating cinnamon essential oil and chitosan nanoparticles and studied the antimicrobial activity against Gram-positive *S. aureus* and Gram-negative *E. coli* bacteria. Although the films displayed strong antimicrobial activity against both the test pathogens, the effect was predominantly due to cinnamon essential oil, while chitosan nanoparticles played a petite synergistic role in slightly enhancing the antimicrobial activity (Vahedikia, Garavand, Tajeddin, Cacciotti, Jafari, Omidi, & Zahedi, 2019). Cui et al. prepared chitosan nanoparticles with antimicrobial pomegranate peel extract in a slightly modified approach and used them as fillers in zein films. The nanocomposite films displayed strong antimicrobial

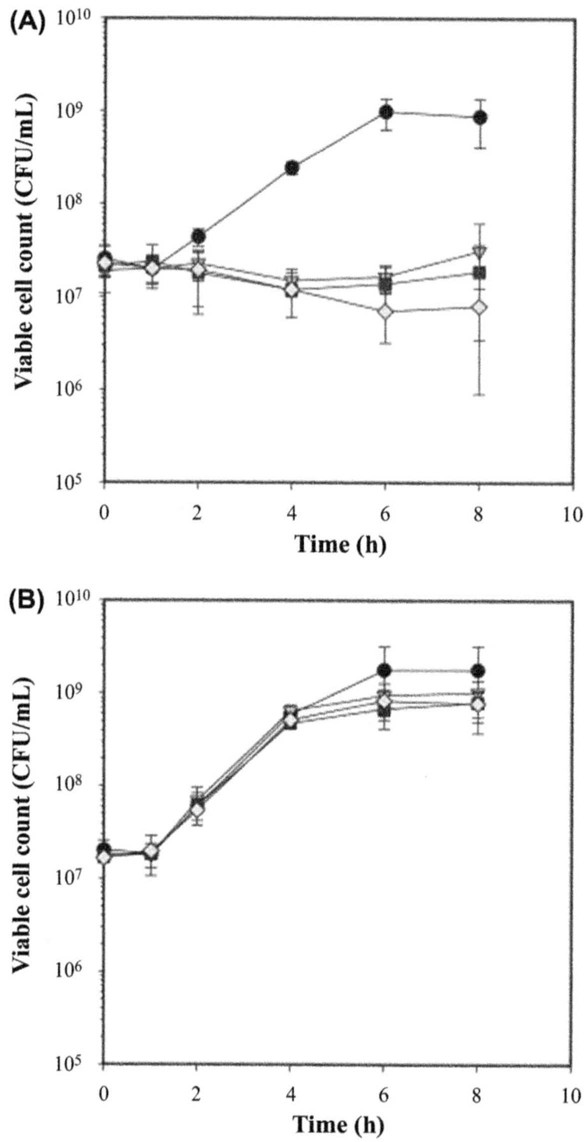

FIGURE 3.4 **Whey protein-based antimicrobial nanocomposites.** The bacteriostatic effect exerted by whey protein isolate/Cloisite 30B organo-clay nanocomposite films against (A) Gram-positive *L. monocytogenes*, and (B) Gram-negative *E. coli* (Sothornvit et al., 2010). *Source: Data from 3. Sothornvit, R., Hong, S. I., An, D. J., & Rhim, J. W. (2010). Effect of clay content on the physical and antimicrobial properties of whey protein isolate/organo-clay composite films.* LWT-Food Science and Technology, *43(2), 279–284. https://doi.org/10.1016/j.lwt.2009.08.010.*

activity against *L. monocytogenes* inoculated on refrigerated pork samples (Cui, Surendhiran, Li, & Lin, 2020). Besides organic chitosan nanoparticles, the use of inorganic antimicrobial nanostructures in zein films has also been reported. In a recent study, zein films with ZnO and ZnO:Mg quantum dots as functional nanofillers were prepared by Schmitz et al., which displayed excellent antimicrobial activity against *S. aureus* (Schmitz, Silva de Albuquerque, Alberton, Riegel-Vidotti, & Zimmermann, 2020).

Soy protein is another plant-based protein used, although less commonly, for fabricating antimicrobial nanocomposite films. Again, the antimicrobial action in these films was not due to the nanostructures but the other antimicrobial agents. Nassar et al. reported the fabrication of soy protein isolate film incorporated with Na^+-MMT nanoclay and eugenol and determined its antibacterial activity against *E. coli*, attributed to antimicrobial characteristics of eugenol (Fernandes Nassar, Dombre, Gastaldi, Touchaleaume, & Chalier, 2018). In another study, Yu et al. studied the antimicrobial action of soy protein-based nanocomposite films incorporated with cellulose nanofibril, pine-needle extract, and lactic acid, against *E. coli* O157:H7, *S. aureus*, *S. typhimurium*, and *L. monocytogenes*. The films with pine-needle extract and lactic acid displayed strong antimicrobial activity against all four pathogenic bacteria (Yu, Dhital, Wang, Sun, Zeng, Mustapha, & Lin, 2019).

3.5.3 *Aliphatic polyesters*

Aliphatic polyesters have recently appeared as a possible solution to the growing environmental concerns due to the accumulation of humungous amounts of non-biodegradable synthetic plastic waste on the planet. Due to their decent biodegradability and properties comparable to petroleum-based plastics, these semisynthetic polymers have become the area of interest for many food packaging researchers. As the name suggests, the monomeric repeating units in the aliphatic polyesters are linked by ester linkages. It is due to the presence of these ester linkages that these semisynthetic polymers possess biodegradability. The enzymes of class "esterase" are ubiquitously present in nature and are responsible for hydrolyzing these ester linkages, hence degrading the polymers. These characteristics have resulted in the commercial production of aliphatic polyesters and their applications in various research areas, the most prevalent being agriculture, biomedicals, and packaging.

The most widely used aliphatic polyester in food packaging research is PLA. Several research groups report the fabrication of PLA-based antimicrobial nanocomposites. Pantani et al. reported antimicrobial activity of PLA–ZnO nanoparticle composite films against *E. coli* and *S. aureus* bacterial strains (Pantani, Gorrasi, Vigliotta, Murariu, & Dubois, 2013). The effect of TiO_2 nanoparticle incorporation in PLA was studied by Shebi et al., who reported strong antimicrobial activity of the nanocomposite films against *S. aureus*. However, the antimicrobial effect was only observed under light conditions due to the photocatalytic nature of the titania nanoparticles (Shebi & Lisa, 2019). Besides the use of metal oxide nanoparticles, PLA-nano clay composites have also been reported. Rhim et al. testified a strong bacteriostatic effect of PLA-Cloisite 30B nanocomposite films against *L. monocytogenes* bacteria (Rhim et al., 2009).

Apart from PLA, several other aliphatic polyester-based antimicrobial nanocomposites have been studied by researchers for food packaging applications. The antimicrobial activity of poly(butylene adipate-co-terephthalate) based nanocomposite films incorporated with ZnO and TiO_2 was reported by Venkatesan et al. in two different studies. Both the nanocomposite films displayed strong antimicrobial activity against *E. coli* and *S. aureus* (Venkatesan & Rajeswari, 2017a, 2017b). Poly(ε-caprolactone) based antimicrobial nanocomposite films incorporated with organically modified montmorillonite were reported by Hadj-Hamou et al. and Yahiaoui et al. in two different studies. They reported the nanocomposite to be highly effective against *E. coli* and *S. aureus* bacterial strains (Hadj-Hamou, Metref, & Yahiaoui, 2017; Yahiaoui et al., 2015).

3.6 Safety aspects and future perspectives

With the advent of nanotechnology and its commercialization, nanocomposites have flourished in various research areas. Food packaging research is also influenced by this and has become one of the rapidly advancing areas of the food industry. Since this research area deals with direct food contact materials, the consequent effect on human health can't be ignored, and therefore stringent identification of their safety aspects and environmental implications becomes paramount. The principal concern in this regard is the leaching of nanomaterials from the package to the foodstuff, which may pose risks to human health (Sharma et al., 2020). The nanometer size of the nanoparticles results in the exhibition of physicochemical properties different from their macro-sized counterparts. This difference in properties estimates their toxicity profiles based on macro-scale particles, a challenging task (Huang, Li, & Zhou, 2015). This entails that the concerns regarding their toxicity differ based on their size (Munro, Haighton, Lynch, & Tafazoli, 2009). This is an intense issue, given the high possibility of allocating the GRAS status to numerous nanoparticles without any size consideration or related safety evaluations (Huang et al., 2015). It has been reported that some nanomaterials are potentially harmful to human health. However, there is still insufficient proof concerning whether this is attributable to any kind of nanomaterial application, especially in the food industry (Sharma et al., 2020).

Risk assessment of nanocomposite packaging materials is the supreme challenge for food researchers and the food industry (Huang et al., 2015). As the use of nanocomposite-based food packaging materials is progressing, not only the common consumers but also the governing organizations are becoming increasingly concerned about their safety effects (Huang et al., 2015). The use of nanomaterials is allowed up to a specific concentration such that they either migrate to food in insignificant quantities or do not migrate at all (EFSA, 2012). The European Union food-contact plastics legislation has registered $10\ mg/dm^2$ as the allowed migration limit of nanoparticles from diverse food packaging nanocomposites to food, and usually, the detected migration values do not exceed this limit (Priyadarshi, Roy, Ghosh, Biswas, & Rhim, 2022). Nevertheless, the European Union Regulation is constantly renewed based on new technical estimations related to food contact nanocomposite plastic materials.

Recently, an extensive assessment of nanocomposites composed of organo-modified MMT nanoclays has been carried out. This evaluation is done through migration tests by using liquid food simulants according to EU Regulation for plastic materials and articles intended to be in contact with food (Priyadarshi, et al., 2022). Bott & Franz reported no migration of laponite from low-density polyethylene (LDPE) in a surfactant solution used as food simulant at 60°C. The analytic method was used with a detection limit of 22 μg laponite per kg of surfactant (Bott & Franz, 2018). In another study, Echegoyen et al. evaluated the release of aluminum (Al^{3+}) from PET bottles in a food simulant media composed of 10 wt.% ethanol and 3 wt.% citric acid at 40°C. They reported 51.65 ng/dm^2 as the maximum migrated amount of aluminum, while the presence of nanoclays in the food simulants was verified by SEM-EDX analyses (Echegoyen, Rodríguez, & Nerín, 2016).

In the continuously competitive food packaging market, the utilization of nanocomposites is regularly increasing due to their various advantages. Hence, the estimation of the risks caused by indirect food contamination through nanomaterial migration from food packaging becomes essential. Therefore the issues that need to be addressed for a complete assessment of the safety of nanocomposites in food packaging are (Huang et al., 2015):

1. Detailed investigation of the physicochemical properties of the developed nanoparticles.
2. Improved protocols for the determination and assessment of nanoparticle migration from packaging to foodstuff.
3. Reliable and apt procedures for identifying, characterizing, and quantifying nanomaterials in complicated food matrices.
4. Estimation of correlations between nanoparticle characteristics and toxicity.
5. Accumulation of data on the toxicokinetic properties of nanoparticles after oral consumption and toxicological dose-response relationships.

Acknowledgments

This research was supported by the Brain Pool Program funded by the Ministry of Science, ICT and Future Planning through the National Research Foundation of Korea (RS-2023-00218228), and the National Research Foundation of Korea (NRF) grant funded by the Korea government (MSIT) (No. 2022R1A2B5B02001422).

References

Alboofetileh, M., Rezaei, M., Hosseini, H., & Abdollahi, M. (2014). Antimicrobial activity of alginate/clay nanocomposite films enriched with essential oils against three common foodborne pathogens. *Food Control*, *36*(1), 1–7. Available from https://doi.org/10.1016/j.foodcont.2013.07.037.

Alizadeh, A., & Behfar, S. (2013). Properties of collagen based edible films in food packaging: A review. *Annals of Biological Research*, 4(2), 253–256.

Alizadeh Sani, M., Ehsani, A., & Hashemi, M. (2017). Whey protein isolate/cellulose nanofibre/TiO_2 nanoparticle/rosemary essential oil nanocomposite film: Its effect on microbial and sensory quality of lamb meat and growth of common foodborne pathogenic bacteria during refrigeration. *International Journal of Food Microbiology*, *251*, 8–14. Available from https://doi.org/10.1016/j.ijfoodmicro.2017.03.018, http://www.elsevier.com/locate/ijfoodmicro.

Al-Tayyar, N. A., Youssef, A. M., & Al-Hindi, R. R. (2020). Antimicrobial packaging efficiency of ZnO-SiO_2 nanocomposites infused into PVA/CS film for enhancing the shelf life of food products. *Food Packaging and Shelf Life*, *25*. Available from https://doi.org/10.1016/j.fpsl.2020.100523, http://www.journals.elsevier.com/food-packaging-and-shelf-life/.

Andrade, M., Ribeiro-Santos, R., Melo, N.R., & Sanches-Silva, A. (2016). The potential of whey protein in food packaging. EC Nutrition Editor's Column-2016 (pp. 25–26).

Arfat, Y. A., Ahmed, J., & Jacob, H. (2017). Preparation and characterization of agar-based nanocomposite films reinforced with bimetallic (Ag-Cu) alloy nanoparticles. *Carbohydrate Polymers*, *155*, 382–390. Available from https://doi.org/10.1016/j.carbpol.2016.08.097, http://www.elsevier.com/wps/find/journaldescription.cws_home/405871/description#description.

Arfat, Y. A., Benjakul, S., Prodpran, T., Sumpavapol, P., & Songtipya, P. (2014). Properties and antimicrobial activity of fish protein isolate/fish skin gelatin film containing basil leaf essential oil and zinc oxide nanoparticles. *Food Hydrocolloids*, *41*, 265–273. Available from https://doi.org/10.1016/j.foodhyd.2014.04.023.

Bandyopadhyay, S., Saha, N., & Saha, P. (2018). Characterization of bacterial cellulose produced using media containing waste apple juice. *Applied Biochemistry and Microbiology*, *54*(6), 649–657. Available from https://doi.org/10.1134/S0003683818060042, http://www.kluweronline.com/issn/0003-6838.

Barzegar, H., Azizi, M. H., Barzegar, M., & Hamidi-Esfahani, Z. (2014). Effect of potassium sorbate on antimicrobial and physical properties of starch-clay nanocomposite films. *Carbohydrate Polymers*, *110*, 26–31. Available from https://doi.org/10.1016/j.carbpol.2014.03.092, http://www.elsevier.com/wps/find/journaldescription.cws_home/405871/description#description.

Basumatary, K., Daimary, P., Das, S. K., Thapa, M., Singh, M., Mukherjee, A., & Kumar, S. (2018). Lagerstroemia speciosa fruit-mediated synthesis of silver nanoparticles and its application as filler in agar based nanocomposite films for antimicrobial food packaging. *Food Packaging and Shelf Life.*, *17*, 99–106. Available from https://doi.org/10.1016/j.fpsl.2018.06.003, http://www.journals.elsevier.com/food-packaging-and-shelf-life/.

Berger, K. R. (2003). A brief history of packaging. *EDIS., 2003*(17). Available from https://doi.org/10.32473/edis-ae206-2002.

Bott, J., & Franz, R. (2018). Investigation into the potential migration of nanoparticles from laponite-polymer nanocomposites. *Nanomaterials, 8* (9), 723. Available from https://doi.org/10.3390/nano8090723.

Boura-Theodoridou, O., Giannakas, A., Katapodis, P., Stamatis, H., Ladavos, A., & Barkoula, N. M. (2020). Performance of ZnO/chitosan nanocomposite films for antimicrobial packaging applications as a function of NaOH treatment and glycerol/PVOH blending. *Food Packaging and Shelf Life, 23*. Available from https://doi.org/10.1016/j.fpsl.2019.100456, http://www.journals.elsevier.com/food-packaging-and-shelf-life/.

Bruna, J. E., Galotto, M. J., Guarda, A., & Rodríguez, F. (2014). A novel polymer based on $MtCu^{2+}$/cellulose acetate with antimicrobial activity. *Carbohydrate Polymers, 102*(1), 317–323. Available from https://doi.org/10.1016/j.carbpol.2013.11.038.

Chatterjee, A. K., Chakraborty, R., & Basu, T. (2014). Mechanism of antibacterial activity of copper nanoparticles. *Nanotechnology, 25*(13). Available from https://doi.org/10.1088/0957-4484/25/13/135101India, http://iopscience.iop.org/0957-4484/25/13/135101/pdf/0957-4484_25_13_135101.pdf.

Choi, I., Lee, J. Y., Lacroix, M., & Han, J. (2017). Intelligent pH indicator film composed of agar/potato starch and anthocyanin extracts from purple sweet potato. *Food Chemistry, 218*, 122–128. Available from https://doi.org/10.1016/j.foodchem.2016.09.050, http://www.elsevier.com/locate/foodchem.

Corradini, E., Curti, P. S., Meniqueti, A. B., Martins, A. F., Rubira, A. F., & Muniz, E. C. (2014). Recent advances in food-packing, pharmaceutical and biomedical applications of zein and zein-based materials. *International Journal of Molecular Sciences, 15*(12), 22438–22470. Available from https://doi.org/10.3390/ijms151222438, http://www.mdpi.com/1422-0067/15/12/22438/pdf.

Cui, H., Surendhiran, D., Li, C., & Lin, L. (2020). Biodegradable zein active film containing chitosan nanoparticle encapsulated with pomegranate peel extract for food packaging. *Food Packaging and Shelf Life, 24*. Available from https://doi.org/10.1016/j.fpsl.2020.100511, http://www.journals.elsevier.com/food-packaging-and-shelf-life/.

Cui, Y., Zhao, Y., Tian, Y., Zhang, W., Lü, X., & Jiang, X. (2012). The molecular mechanism of action of bactericidal gold nanoparticles on Escherichia coli. *Biomaterials, 33*(7), 2327–2333. Available from https://doi.org/10.1016/j.biomaterials.2011.11.057.

de Dicastillo, C. López, Guerrero Correa, M., Martínez, F. B., Streitt, C., & José Galotto, M. (2021). *Antimicrobial effect of titanium dioxide nanoparticles*. IntechOpen. Available from 10.5772/intechopen.90891.

Dizaj, S. M., Lotfipour, F., Barzegar-Jalali, M., Zarrintan, M. H., & Adibkia, K. (2014). Antimicrobial activity of the metals and metal oxide nanoparticles. *Materials Science and Engineering C, 44*, 278–284. Available from https://doi.org/10.1016/j.msec.2014.08.031.

Echegoyen, Y., Rodríguez, S., & Nerín, C. (2016). Nanoclay migration from food packaging materials. Food Additives and Contaminants - Part A Chemistry, Analysis. *Control, Exposure and Risk Assessment, 33*(3), 530–539. Available from https://doi.org/10.1080/19440049.2015.1136844, http://www.tandfonline.com/toc/tfac20/current.

EFSA. (2012). Scientific opinion on the safety evaluation of the substance, titanium nitride, nanoparticles, for use in food contact materials. *EFSA Journal, 10*(3).

Fabra, M. J., López-Rubio, A., & Lagaron, J. M. (2014). *Biopolymers for food packaging applications. Smart polymers and their applications* (pp. 476–509). Spain: Elsevier Ltd.. Available from http://www.sciencedirect.com/science/book/9780857096951, 10.1533/9780857097026.2.476.

Fernandes Nassar, S., Dombre, C., Gastaldi, E., Touchaleaume, F., & Chalier, P. (2018). Soy protein isolate nanocomposite film enriched with eugenol, an antimicrobial agent: Interactions and properties. *Journal of Applied Polymer Science, 135*(10), 45941. Available from https://doi.org/10.1002/app.45941.

García-Ochoa, F., Santos, V. E., Casas, J. A., & Gómez, E. (2000). Xanthan gum: Production, recovery, and properties. *Biotechnology Advances, 18*(7), 549–579. Available from https://doi.org/10.1016/S0734-9750(00)00050-1, http://www.elsevier.com/inca/publications/store/5/2/5/4/5/5/index.htt.

Gholizadeh, B. S., Buazar, F., Hosseini, S. M., & Mousavi, S. M. (2018). Enhanced antibacterial activity, mechanical and physical properties of alginate/hydroxyapatite bionanocomposite film. *International Journal of Biological Macromolecules, 116*, 786–792. Available from https://doi.org/10.1016/j.ijbiomac.2018.05.104, http://www.elsevier.com/locate/ijbiomac.

Hadj-Hamou, A. S., Metref, F., & Yahiaoui, F. (2017). Thermal stability and decomposition kinetic studies of antimicrobial PCL/nanoclay packaging films. *Polymer Bulletin, 74*(9), 3833–3853. Available from https://doi.org/10.1007/s00289-017-1929-y.

Hammoudi, N., Ziani Cherif, H., Borsali, F., Benmansour, K., & Meghezzi, A. (2020). Preparation of active antimicrobial and antifungal alginate-montmorillonite/lemon essential oil nanocomposite films. *Materials Technology, 35*(7), 383–394. Available from https://doi.org/10.1080/10667857.2019.1685292, https://www.tandfonline.com/loi/ymte20.

He, Q., Zhang, Y., Cai, X., & Wang, S. (2016). Fabrication of gelatin-TiO_2 nanocomposite film and its structural, antibacterial and physical properties. *International Journal of Biological Macromolecules, 84*, 153–160. Available from https://doi.org/10.1016/j.ijbiomac.2015.12.012, http://www.elsevier.com/locate/ijbiomac.

Hong, S. I., & Rhim, J. W. (2008). Antimicrobial activity of organically modified nano-clays. *Journal of Nanoscience and Nanotechnology, 8*(11), 5818–5824. Available from https://doi.org/10.1166/jnn.2008.248.

Hu, X., Jia, X., Zhi, C., Jin, Z., & Miao, M. (2019). Improving the properties of starch-based antimicrobial composite films using ZnO-chitosan nanoparticles. *Carbohydrate Polymers, 210*, 204–209. Available from https://doi.org/10.1016/j.carbpol.2019.01.043, http://www.elsevier.com/wps/find/journaldescription.cws_home/405871/description#description.

Huang, J. Y., Li, X., & Zhou, W. (2015). Safety assessment of nanocomposite for food packaging application. *Trends in Food Science and Technology, 45*(2), 187–199. Available from https://doi.org/10.1016/j.tifs.2015.07.002, http://www.elsevier.com/wps/find/journaldescription.cws_home/601278/description#description.

Jain, A., Dasgupta, N., Ranjan, S., Singh, V., Singh, H., Purohit, S. D., ... Samanta, S. K. (2021). Whey protein based electrosprayed nanospheres for encapsulation and controlled release of bioactive compounds from Tinospora cordifolia extract. *Innovative Food Science and Emerging Technologies, 69*. Available from https://doi.org/10.1016/j.ifset.2021.102671, http://www.elsevier.com/locate/ifset.

Kanmani, P., & Rhim, J. W. (2014). Antimicrobial and physical-mechanical properties of agar-based films incorporated with grapefruit seed extract. *Carbohydrate Polymers, 102*(1), 708–716. Available from https://doi.org/10.1016/j.carbpol.2013.10.099.

Karimi, N., Alizadeh, A., Almasi, H., & Hanifian, S. (2020). Preparation and characterization of whey protein isolate/polydextrose-based nanocomposite film incorporated with cellulose nanofiber and L. plantarum: A new probiotic active packaging system. *LWT*, *121*, 108978. Available from https://doi.org/10.1016/j.lwt.2019.108978.

Kaur, P., Choudhary, A., & Thakur, R. (2013). Synthesis of chitosan-silver nanocomposites and their antibacterial activity. *International Journal of Scientific Engineering and Research*, *4*.

Khalaf, H., Sharoba, A., El-Tanahi, H., & Morsy, M. (2013). Stability of antimicrobial activity of pullulan edible films incorporated with nanoparticles and essential oils and their impact on turkey deli meat quality. *Journal of Food and Dairy Sciences*, 557–573. Available from https://doi.org/10.21608/jfds.2013.72104.

Koshy, R. R., Mary, S. K., Thomas, S., & Pothan, L. A. (2015). Environment friendly green composites based on soy protein isolate - A review. *Food Hydrocolloids*, *50*, 174–192. Available from https://doi.org/10.1016/j.foodhyd.2015.04.023.

Lee, J. H., Lee, I. Y., Kim, M. K., & Park, Y. H. (1999). Optimal pH control of batch processes for production of curdlan by Agrobacterium species. *Journal of Industrial Microbiology and Biotechnology*, *23*(2), 143–148. Available from https://doi.org/10.1038/sj.jim.2900714.

Lee, K. Y., & Mooney, D. J. (2012). Alginate: Properties and biomedical applications. *Progress in Polymer Science*, *37*(1), 106–126. Available from https://doi.org/10.1016/j.progpolymsci.2011.06.003.

Lemire, J. A., Harrison, J. J., & Tuner, R. J. (2013). Antimicrobial activity of metals: Mechanisms, molecular targets and applications. *Nature Reviews. Microbiology*, *11*(6), 371–384. Available from https://doi.org/10.1038/nrmicro3028.

Luecha, J., Sozer, N., & Kokini, J. L. (2010). Synthesis and properties of corn zein/montmorillonite nanocomposite films. *Journal of Materials Science*, *45*(13), 3529–3537. Available from https://doi.org/10.1007/s10853-010-4395-6.

Ma, B., Qin, A., Li, X., Zhao, X., & He, C. (2014). Bioinspired design and chitin whisker reinforced chitosan membrane. *Materials Letters*, *120*, 82–85. Available from https://doi.org/10.1016/j.matlet.2014.01.015.

Madhumathi, K., Jeevana Rekha, L., & Sampath Kumar, T. S. (2018). Tailoring antibiotic release for the treatment of periodontal infrabony defects using bioactive gelatin-alginate/apatite nanocomposite films. *Journal of Drug Delivery Science and Technology*, *43*, 57–64. Available from https://doi.org/10.1016/j.jddst.2017.09.015, http://www.editionsdesante.fr/category.php?id_category = 48.

Makwana, D., Castaño, J., Somani, R. S., & Bajaj, H. C. (2020). Characterization of Agar-CMC/Ag-MMT nanocomposite and evaluation of antibacterial and mechanical properties for packaging applications. *Arabian Journal of Chemistry*, *13*(1), 3092–3099. Available from https://doi.org/10.1016/j.arabjc.2018.08.017, http://colleges.ksu.edu.sa/Arabic%20Colleges/CollegeOfScience/ChemicalDept/AJC/default.aspx, ScienceDirect, http://www.sciencedirect.com/science/journal/18785352.

Malagurski, I., Levic, S., Nesic, A., Mitric, M., Pavlovic, V., & Dimitrijevic-Brankovic, S. (2017). Mineralized agar-based nanocomposite films: Potential food packaging materials with antimicrobial properties. *Carbohydrate Polymers*, *175*, 55–62. Available from https://doi.org/10.1016/j.carbpol.2017.07.064, http://www.elsevier.com/wps/find/journaldescription.cws_home/405871/description#description.

Martynková, G. S., & Valášková, M. (2014). Antimicrobial nanocomposites based on natural modified materials: A review of carbons and clays. *Journal of Nanoscience and Nanotechnology*, *14*(1), 673–693, http://docserver.ingentaconnect.com/deliver/connect/asp/15334880/v14n1/s41.pdf?expires = 1391654739&id = 77112110&titleid = 4286&accname = Elsevier + BV&checksum = AB9F6F638C128FC07837504920446E87.10.1166/jnn.2014.8903Czech Republic.

Meghana, S., Kabra, P., Chakraborty, S., & Padmavathy, N. (2015). Understanding the pathway of antibacterial activity of copper oxide nanoparticles. *RSC Advances*, *5*(16), 12293–12299. Available from https://doi.org/10.1039/c4ra12163e, http://pubs.rsc.org/en/journals/journalissues.

Mei, L., Teng, Z., Zhu, G., Liu, Y., Zhang, F., Zhang, J., … Wang, Q. (2017). Silver nanocluster-embedded zein films as antimicrobial coating materials for food packaging. *ACS Applied Materials and Interfaces*, *9*(40), 35297–35304. Available from https://doi.org/10.1021/acsami.7b08152, http://pubs.acs.org/journal/aamick.

Meindrawan, B., Suyatma, N. E., Wardana, A. A., & Pamela, V. Y. (2018). Nanocomposite coating based on carrageenan and ZnO nanoparticles to maintain the storage quality of mango. *Food Packaging and Shelf Life*, *18*, 140–146. Available from https://doi.org/10.1016/j.fpsl.2018.10.006, http://www.journals.elsevier.com/food-packaging-and-shelf-life/.

Meira, S. M. M., Zehetmeyer, G., Werner, J. O., & Brandelli, A. (2017). A novel active packaging material based on starch-halloysite nanocomposites incorporating antimicrobial peptides. *Food Hydrocolloids*, *63*, 561–570. Available from https://doi.org/10.1016/j.foodhyd.2016.10.013.

Morris, E. R., Nishinari, K., & Rinaudo, M. (2012). Gelation of gellan - A review. *Food Hydrocolloids*, *28*(2), 373–411. Available from https://doi.org/10.1016/j.foodhyd.2012.01.004.

Mukerjea, R., Sheets, R. L., Gray, A. N., & Robyt, J. F. (2015). In vitro synthesis of ten starches by potato starch-synthase and starch-branching-enzyme giving different ratios of amylopectin and amylose. *Organic Chemistry Current Research.*, *4*(3).

Munro, I. C., Haighton, L. A., Lynch, B. S., & Tafazoli, S. (2009). Technological challenges of addressing new and more complex migrating products from novel food packaging materials. Food Additives and Contaminants - Part A Chemistry, Analysis. *Control, Exposure and Risk Assessment*, *26*(12), 1534–1546. Available from https://doi.org/10.1080/02652030902995277.

Ocak, B. (2018). Film-forming ability of collagen hydrolysate extracted from leather solid wastes with chitosan. *Environmental Science and Pollution Research*, *25*(5), 4643–4655. Available from https://doi.org/10.1007/s11356-017-0843-z, http://www.springerlink.com/content/0944-1344.

Oun, A. A., & Rhim, J. W. (2017). Characterization of carboxymethyl cellulose-based nanocomposite films reinforced with oxidized nanocellulose isolated using ammonium persulfate method. *Carbohydrate Polymers*, *174*, 484–492. Available from https://doi.org/10.1016/j.carbpol.2017.06.121, http://www.elsevier.com/wps/find/journaldescription.cws_home/405871/description#description.

Pantani, R., Gorrasi, G., Vigliotta, G., Murariu, M., & Dubois, P. (2013). PLA-ZnO nanocomposite films: Water vapor barrier properties and specific end-use characteristics. *European Polymer Journal*, *49*(11), 3471–3482. Available from https://doi.org/10.1016/j.eurpolymj.2013.08.005.

Parenteau-Bareil, R., Gauvin, R., & Berthod, F. (2010). Collagen-based biomaterials for tissue engineering applications. *Materials*, *3*(3), 1863–1887. Available from https://doi.org/10.3390/ma3031863Canada, http://www.mdpi.com/1996-1944/3/3/1863/pdf.

Peighambardoust, S. J., Zahed-Karkaj, S., Peighambardoust, S. H., Ebrahimi, Y., & Peressini, D. (2020). Characterization of carboxymethyl cellulose-based active films incorporating non-modified and Ag or Cu-modified Cloisite 30B and montmorillonite nanoclays. *Iranian Polymer Journal (English Edition)*, *29*(12), 1087–1097. Available from https://doi.org/10.1007/s13726-020-00863-z, http://www.springerlink.com/content/1026-1265.

Plackett, D., & Siró, I. (2011). *Polyhydroxyalkanoates (phas) for food packaging. Multifunctional and nanoreinforced polymers for food packaging* (pp. 498–526). Denmark: Woodhead Publishing Limited. Available from http://www.woodheadpublishing.com/9781845697389, 10.1533/9780857092786.

Priyadarshi, R., Kim, S. M., & Rhim, J. W. (2021a). Pectin/pullulan blend films for food packaging: Effect of blending ratio. *Food Chemistry, 347*. Available from https://doi.org/10.1016/j.foodchem.2021.129022, http://www.elsevier.com/locate/foodchem.

Priyadarshi, R., Kim, H. J., & Rhim, J. W. (2021b). Effect of sulfur nanoparticles on properties of alginate-based films for active food packaging applications. *Food Hydrocolloids, 110*. Available from https://doi.org/10.1016/j.foodhyd.2020.106155, https://www.journals.elsevier.com/food-hydrocolloids.

Priyadarshi, R., Kumar, B., Deeba, F., Kulshreshtha, A., & Negi Y.S. (2019). Food packaging: Natural and synthetic biopolymers. In Encyclopedia of polymer applications (First, pp. 1325–1342). doi:10.1201/9781351019422-140000362.

Priyadarshi, R., & Negi, Y. S. (2017). Effect of varying filler concentration on zinc oxide nanoparticle embedded chitosan films as potential food packaging material. *Journal of Polymers and the Environment, 25*(4), 1087–1098. Available from https://doi.org/10.1007/s10924-016-0890-4, http://www.kluweronline.com/issn/1566-2543/.

Priyadarshi, R., & Rhim, J. W. (2020). Chitosan-based biodegradable functional films for food packaging applications. *Innovative Food Science and Emerging Technologies, 62*. Available from https://doi.org/10.1016/j.ifset.2020.102346, http://www.elsevier.com/locate/ifset.

Priyadarshi, R., Sauraj, K. B., Deeba, F., Kulshreshtha, A., & Negi, Y. S. (2018). Chitosan films incorporated with Apricot (Prunus armeniaca) kernel essential oil as active food packaging material. *Food Hydrocolloids, 85*, 158–166. Available from https://doi.org/10.1016/j.foodhyd.2018.07.003.

Priyadarshi, R., Roy, S., Ghosh, T., Biswas, D., & Rhim, J.-W. (2022). Antimicrobial nanofillers reinforced biopolymer composite films for active food packaging applications - A review. *Sustainable Materials and Technologies, 32*, e00353, In this issue. Available from https://doi.org/10.1016/j.susmat.2021.e00353.

Purohit, S. D., Singh, H., Bhaskar, R., Yadav, I., Chou, C. F., Gupta, M. K., & Mishra, N. C. (2020). Gelatin-alginate-cerium oxide nanocomposite scaffold for bone regeneration. *Materials Science and Engineering C., 116*. Available from https://www.journals.elsevier.com/materials-science-and-engineering-c.10.1016/j.msec.2020.111111.

Qu, L., Chen, G., Dong, S., Huo, Y., Yin, Z., Li, S., & Chen, Y. (2019). Improved mechanical and antimicrobial properties of zein/chitosan films by adding highly dispersed nano-TiO_2. *Industrial Crops and Products, 130*, 450–458. Available from https://doi.org/10.1016/j.indcrop.2018.12.093, http://www.elsevier.com/inca/publications/store/5/2/2/8/2/5.

Rhim, J. W., Hong, S. I., & Ha, C. S. (2009). Tensile, water vapor barrier and antimicrobial properties of PLA/nanoclay composite films. *LWT - Food Science and Technology, 42*(2), 612–617. Available from https://doi.org/10.1016/j.lwt.2008.02.015.

Rhim, J. W., Wang, L. F., & Hong, S. I. (2013). Preparation and characterization of agar/silver nanoparticles composite films with antimicrobial activity. *Food Hydrocolloids, 33*(2), 327–335. Available from https://doi.org/10.1016/j.foodhyd.2013.04.002.

Rhim, J. W., Wang, L. F., Lee, Y., & Hong, S. I. (2014). Preparation and characterization of bio-nanocomposite films of agar and silver nanoparticles: Laser ablation method. *Carbohydrate Polymers, 103*(1), 456–465. Available from https://doi.org/10.1016/j.carbpol.2013.12.075.

Rhim, J. W., & Wang, L. F. (2014). Preparation and characterization of carrageenan-based nanocomposite films reinforced with clay mineral and silver nanoparticles. *Applied Clay Science, 97–98*, 174–181. Available from https://doi.org/10.1016/j.clay.2014.05.025, http://www.elsevier.com/inca/publications/store/5/0/3/3/2/2/index.htt.

Riahi, Z., Priyadarshi, R., Rhim, J. W., & Bagheri, R. (2021). Gelatin-based functional films integrated with grapefruit seed extract and TiO2 for active food packaging applications. *Food Hydrocolloids, 112*. Available from https://doi.org/10.1016/j.foodhyd.2020.106314, https://www.journals.elsevier.com/food-hydrocolloids.

Rouf, T. B., Schmidt, G., & Kokini, J. L. (2018). Zein-Laponite nanocomposites with improved mechanical, thermal and barrier properties. *Journal of Materials Science, 53*(10), 7387–7402. Available from https://doi.org/10.1007/s10853-018-2061-6.

Roy, S., & Rhim, J. W. (2019). Carrageenan-based antimicrobial bionanocomposite films incorporated with ZnO nanoparticles stabilized by melanin. *Food Hydrocolloids, 90*, 500–507. Available from https://doi.org/10.1016/j.foodhyd.2018.12.056.

Roy, S., & Rhim, J. W. (2020). Preparation of bioactive functional poly(lactic acid)/curcumin composite film for food packaging application. *International Journal of Biological Macromolecules, 162*, 1780–1789. Available from https://doi.org/10.1016/j.ijbiomac.2020.08.094, http://www.elsevier.com/locate/ijbiomac.

Roy, S., Shankar, S., & Rhim, J. W. (2019). Melanin-mediated synthesis of silver nanoparticle and its use for the preparation of carrageenan-based antibacterial films. *Food Hydrocolloids, 88*, 237–246. Available from https://doi.org/10.1016/j.foodhyd.2018.10.013.

Saedi, S., Shokri, M., & Rhim, J.-W. (2020). Preparation of carrageenan-based nanocomposite films incorporated with functionalized halloysite using AgNP and sodium dodecyl sulfate. *Food Hydrocolloids, 106*, 105934. Available from https://doi.org/10.1016/j.foodhyd.2020.105934.

Said, N.S., & Sarbon, N.M. (2019). Protein-based active film as antimicrobial food packaging: A. Active antimicrobial food packaging.

Schmitz, F., Silva de Albuquerque, M. B., Alberton, M. D., Riegel-Vidotti, I. C., & Zimmermann, L. M. (2020). Zein films with ZnO and ZnO: Mg quantum dots as functional nanofillers: New nanocomposites for food package with UV-blocker and antimicrobial properties. *Polymer Testing, 91*. Available from https://doi.org/10.1016/j.polymertesting.2020.106709, https://www.journals.elsevier.com/polymer-testing.

Shahmohammadi Jebel, F., & Almasi, H. (2016). Morphological, physical, antimicrobial and release properties of ZnO nanoparticles-loaded bacterial cellulose films. *Carbohydrate Polymers, 149*, 8–19. Available from https://doi.org/10.1016/j.carbpol.2016.04.089, http://www.elsevier.com/wps/find/journaldescription.cws_home/405871/description#description.

Shankar, S., Kasapis, S., & Rhim, J. W. (2018). Alginate-based nanocomposite films reinforced with halloysite nanotubes functionalized by alkali treatment and zinc oxide nanoparticles. *International Journal of Biological Macromolecules, 118*, 1824–1832. Available from https://doi.org/10.1016/j.ijbiomac.2018.07.026, http://www.elsevier.com/locate/ijbiomac.

Shankar, S., Reddy, J. P., Rhim, J. W., & Kim, H. Y. (2015). Preparation, characterization, and antimicrobial activity of chitin nanofibrils reinforced carrageenan nanocomposite films. *Carbohydrate Polymers, 117*, 468–475. Available from https://doi.org/10.1016/j.carbpol.2014.10.010, http://www.elsevier.com/wps/find/journaldescription.cws_home/405871/description#description.

Shankar, S., & Rhim, J.-W. (2018). *Bionanocomposite films for food packaging applications. Reference module in food science* (pp. 1–10). Elsevier. Available from 10.1016/B978-0-08-100596-5.21875-1.

Sharma, R., Jafari, S. M., & Sharma, S. (2020). Antimicrobial bio-nanocomposites and their potential applications in food packaging. *Food Control, 112*. Available from https://doi.org/10.1016/j.foodcont.2020.107086, https://www.journals.elsevier.com/food-control.

Shebi, A., & Lisa, S. (2019). Evaluation of biocompatibility and bactericidal activity of hierarchically porous PLA-TiO2 nanocomposite films fabricated by breath-figure method. *Materials Chemistry and Physics, 230*, 308–318. Available from https://doi.org/10.1016/j.matchemphys.2019.03.045, http://www.journals.elsevier.com/materials-chemistry-and-physics.

Silva, N. H. C. S., Vilela, C., Almeida, A., Marrucho, I. M., & Freire, C. S. R. (2018). Pullulan-based nanocomposite films for functional food packaging: Exploiting lysozyme nanofibers as antibacterial and antioxidant reinforcing additives. *Food Hydrocolloids, 77*, 921–930. Available from https://doi.org/10.1016/j.foodhyd.2017.11.039.

Sothornvit, R., Hong, S. I., An, D. J., & Rhim, J. W. (2010). Effect of clay content on the physical and antimicrobial properties of whey protein isolate/organo-clay composite films. *LWT - Food Science and Technology, 43*(2), 279–284. Available from https://doi.org/10.1016/j.lwt.2009.08.010.

Sothornvit, R., Rhim, J. W., & Hong, S. I. (2009). Effect of nano-clay type on the physical and antimicrobial properties of whey protein isolate/clay composite films. *Journal of Food Engineering, 91*(3), 468–473. Available from https://doi.org/10.1016/j.jfoodeng.2008.09.026.

Tang, S., Wang, Z., Li, P., Li, W., Li, C., Wang, Y., & Chu, P. K. (2018). Degradable and photocatalytic antibacterial Au-TiO2/sodium alginate nanocomposite films for active food packaging. *Nanomaterials, 8*(11). Available from https://doi.org/10.3390/nano8110930, https://www.mdpi.com/2079-4991/8/11/930/pdf.

Therkelsen, G. H. (2012). *CarrageenanIndustrial gums: Polysaccharides and their derivatives* (third ed., pp. 145–180). Denmark: Elsevier Inc.. Available from http://www.sciencedirect.com/science/book/9780080926544, 10.1016/B978-0-08-092654-4.50011-5.

Thiagamani, S. M. K., Rajini, N., Siengchin, S., Varada Rajulu, A., Hariram, N., & Ayrilmis, N. (2019). Influence of silver nanoparticles on the mechanical, thermal and antimicrobial properties of cellulose-based hybrid nanocomposites. *Composites Part B: Engineering, 165*, 516–525. Available from https://doi.org/10.1016/j.compositesb.2019.02.006.

Tripathi, S., Mehrotra, G. K., & Dutta, P. K. (2011). Chitosan-silver oxide nanocomposite film: Preparation and antimicrobial activity. *Bulletin of Materials Science, 34*(1), 29–35. Available from https://doi.org/10.1007/s12034-011-0032-5, http://www.ias.ac.in/matersci/bmsfeb2011/29.pdf.

Tunç, S., Duman, O., & Polat, T. G. (2016). Effects of montmorillonite on properties of methyl cellulose/carvacrol based active antimicrobial nanocomposites. *Carbohydrate Polymers, 150*, 259–268. Available from https://doi.org/10.1016/j.carbpol.2016.05.019, http://www.elsevier.com/wps/find/journaldescription.cws_home/405871/description#description.

Tunç, S., & Duman, O. (2011). Preparation of active antimicrobial methyl cellulose/carvacrol/montmorillonite nanocomposite films and investigation of carvacrol release. *LWT - Food Science and Technology, 44*(2), 465–472. Available from https://doi.org/10.1016/j.lwt.2010.08.018.

US-EPA. (2020). Municipal solid waste generation, recycling, and disposal in the United States: Facts and figures for 2018. Available from http://www.epa.gov/wastes.

Usman, A., Hussain, Z., Riaz, A., & Khan, A. N. (2016). Enhanced mechanical, thermal and antimicrobial properties of poly(vinyl alcohol)/graphene oxide/starch/silver nanocomposites films. *Carbohydrate Polymers, 153*, 592–599. Available from https://doi.org/10.1016/j.carbpol.2016.08.026, http://www.elsevier.com/wps/find/journaldescription.cws_home/405871/description#description.

Vahedikia, N., Garavand, F., Tajeddin, B., Cacciotti, I., Jafari, S. M., Omidi, T., & Zahedi, Z. (2019). Biodegradable zein film composites reinforced with chitosan nanoparticles and cinnamon essential oil: Physical, mechanical, structural and antimicrobial attributes. *Colloids and Surfaces B: Biointerfaces, 177*, 25–32. Available from https://doi.org/10.1016/j.colsurfb.2019.01.045, http://www.elsevier.com/locate/colsurfb.

Venkatesan, R., & Rajeswari, N. (2017a). TiO_2 nanoparticles/poly(butylene adipate-co-terephthalate) bionanocomposite films for packaging applications. *Polymers for Advanced Technologies, 28*(12), 1699–1706. Available from https://doi.org/10.1002/pat.4042, http://onlinelibrary.wiley.com/journal/10.1002/(ISSN)1099-1581.

Venkatesan, R., & Rajeswari, N. (2017b). ZnO/PBAT nanocomposite films: Investigation on the mechanical and biological activity for food packaging. *Polymers for Advanced Technologies, 28*(1), 20–27. Available from https://doi.org/10.1002/pat.3847, http://onlinelibrary.wiley.com/journal/10.1002/(ISSN)1099-1581.

Wang, L. F., & Rhim, J. W. (2015). Preparation and application of agar/alginate/collagen ternary blend functional food packaging films. *International Journal of Biological Macromolecules, 80*, 460–468. Available from https://doi.org/10.1016/j.ijbiomac.2015.07.007, http://www.elsevier.com/locate/ijbiomac.

Xu, Y., Rehmani, N., Alsubaie, L., Kim, C., Sismour, E., & Scales, A. (2018). Tapioca starch active nanocomposite films and their antimicrobial effectiveness on ready-to-eat chicken meat. *Food Packaging and Shelf Life, 16*, 86–91. Available from https://doi.org/10.1016/j.fpsl.2018.02.006, http://www.journals.elsevier.com/food-packaging-and-shelf-life/.

Yahiaoui, F., Benhacine, F., Ferfera-Harrar, H., Habi, A., Hadj-Hamou, A. S., & Grohens, Y. (2015). Development of antimicrobial PCL/nanoclay nanocomposite films with enhanced mechanical and water vapor barrier properties for packaging applications. *Polymer Bulletin, 72*(2), 235–254. Available from https://doi.org/10.1007/s00289-014-1269-0, https://rd.springer.com/journal/289.

Yang, R., Li, H., Huang, M., Yang, H., & Li, A. (2016). A review on chitosan-based flocculants and their applications in water treatment. *Water Research, 95*, 59–89. Available from https://doi.org/10.1016/j.watres.2016.02.068, http://www.elsevier.com/locate/watres.

Youssef, A. M., Abou-Yousef, H., El-Sayed, S. M., & Kamel, S. (2015). Mechanical and antibacterial properties of novel high performance chitosan/nanocomposite films. *International Journal of Biological Macromolecules, 76*, 25–32. Available from https://doi.org/10.1016/j.ijbiomac.2015.02.016, http://www.elsevier.com/locate/ijbiomac.

Yu, Z., Dhital, R., Wang, W., Sun, L., Zeng, W., Mustapha, A., & Lin, M. (2019). Development of multifunctional nanocomposites containing cellulose nanofibrils and soy proteins as food packaging materials. *Food Packaging and Shelf Life, 21*, 100366. Available from https://doi.org/10.1016/j.fpsl.2019.100366.

C H A P T E R

4

Plastic and bioplastic-based nanocomposite materials for food packaging and preservation

Shiv Shankar[1] *and Lily Jaiswal*[2]

[1]Research Laboratories in Sciences Applied to Food, INRS-Armand-Frappier Health Biotechnology Research Centre, Laval, Quebec, Canada [2]Department of Food and Nutrition, BioNanocomposite Research Institute, Kyung Hee University, Dongdaemun-gu, Seoul, Republic of Korea

4.1 Introduction

Packaging materials provide support to preserve, protect, merchandise, market, and distribute foods. They play a significant role in reaching packaged food to consumers in a safe and wholesome form without compromising quality (Shankar & Rhim, 2016a). Since foods are in direct contact with packaging materials, several factors must be considered when choosing the right package for a particular food product. Generally, the packaging material may be rigid or flexible (Shankar & Rhim, 2018). Rigid containers include glass and plastic bottles and jars, cans, pottery, wood boxes, drums, tins, plastic pots, and tubes. They give physical protection to the food inside that is not provided by flexible packaging. Flexible packaging is a major group of materials that includes plastic films, papers, foil, some types of vegetable fibers, and clothes that can be used to make wrappings, sacks, and sealed or unsealed bags. Since early, people first used a variety of locally available natural containers to store and eat foods. Significant developments in food packaging materials have provided protection of food by lowering the growth of microbes and from external microbial contamination. Packaging materials were developed over the years to prevent the deterioration of foods by microbes resulting from exposure to air, moisture, or pH changes associated with the food or its surrounding atmosphere (Raheem, 2012).

For many years, petroleum-based polymer materials have been used as plastic packaging, such as polypropylene (PP), polyester (PET), polyethylene (PE), and polystyrene (PS) (Salwa, Sapuan, Mastura, & Zuhri, 2019). The production of plastic packaging has increased twentyfold since 1964 and represents the most prominent application of polymers with 26% of the total volume. Due to its many good characteristics (e.g., lightweight and good barrier properties), plastics have increasingly replaced other packaging materials. The production is expected to continue growing to double volume within the next 20 years (MacArthur, Waughray, & Stuchtey, 2016). Although plastic packaging has been performing successfully in terms of its functionality, the production of petroleum-based plastics releases greenhouse gases (in particular CO_2). In their disposal, due to the lack of collection or proper handling, plastics tend to end up in landfills, become trash on land and water streams, and ultimately contaminate the oceans (Salwa et al., 2019). The industry is struggling to find a more environmentally friendly way of producing and using plastic. The overall question is how can plastic become (more) sustainable. In this context, food-packaging producers and food industries have been working towards using abundant, low-cost, renewable, and biodegradable alternatives to the traditional, nonrenewable petroleum-based resources (Chi, Wang, & Catchmark, 2020), such as bio-based plastics (Kawashima, Yagi, & Kojima, 2019). They offer the possibility to reduce the use of petroleum resources with the potential subsequent reduction of CO_2 emissions (Shankar & Rhim, 2016a). It is estimated that the eco-friendly food packaging market will increase, with a shift in consumer preference towards recyclable and "eco-friendly" materials. However, there is still significant uncertainty about the potential and possible advantages of bio-based plastics compared to conventional plastics, and

Food Packaging and Preservation.
DOI: https://doi.org/10.1016/B978-0-323-90044-7.00004-5

several misconceptions exist. To reduce plastic waste generation, the EU Commission set an ambitious goal of 55% of plastic packaging recirculation in 2025, and all plastics are recyclable (or reusable) by 2030, following a Circular Economy approach. This puts pressure on increased recycling and reuse of plastic packaging for food, as this represents the largest fraction of all plastic packaging in the EU (Mendes & Pedersen, 2021). Evaluating the sustainability of food packaging requires a broader perspective comprising several aspects (Russell, 2014). This should include the use of materials that create no greenhouse gas emissions, has the potential to be recycled or reused, generate zero landfill waste, reduce water use, built using renewable energy, do not produce air pollution, and do not harm human health, among many other considerations. Although progress has been made towards creating alternative packaging systems, there is not yet a perfect solution that can meet the many criteria for sustainability and ultimately fulfill the functionality of the food packaging: to preserve and deliver the packed foods in good condition (Russell, 2014). Fig. 4.1 shows different categories of packaging materials based on source and biodegradability. Because of the above perspectives, this book chapter describes various types of plastic and bioplastic-based food packaging materials that make it easier to choose specific packaging materials for certain types of food.

4.2 Commonly available food packaging materials

The most common food packaging materials are glass, wood, metal, plastics, paper, and flexible packages such as plastics and bioplastics. Each of these packages offers unique advantages and disadvantages that have to be critically considered in making the right choice by the food processor. Plastic materials are made up of large, organic (carbon-containing) molecules that can be formed into various valuable products. They are fluid, moldable, heat sealable, easy to print, and integrated into production processes where the package is created, filled, and sealed in the same production line (Marsh & Bugusu, 2007). The major disadvantage of plastics is their variable permeability to light, gases, vapors, and low molecular weight molecules. Structural polymers such as polyethylene and polypropylene provide mechanical properties at a low cost. In contrast, barrier polymers such as polyvinylidene chloride and ethylene-vinyl alcohol protect the transfer of gases, flavors, and odors through the package. The use of plastics in packaging has increased worldwide, with an estimated 280 metric tonnes (Shankar & Rhim, 2018). The packaging industry is the largest user of plastics; more than 90% of flexible packaging is made of plastics, compared to only 17% of rigid packaging. Barrier resins are generally employed for plastic containers by modifications to improve product protection and make them more cost-effective.

4.3 Recyclable and recycled plastics

There are more than thirty different plastics in packaging; the most common are polyolefins, polyvinyl, and polyesters. There are possibilities that chemical contaminants in plastic packaging intended for recycling may remain in the recycled material and could migrate into the food. Other aspects of plastics recycling, such as

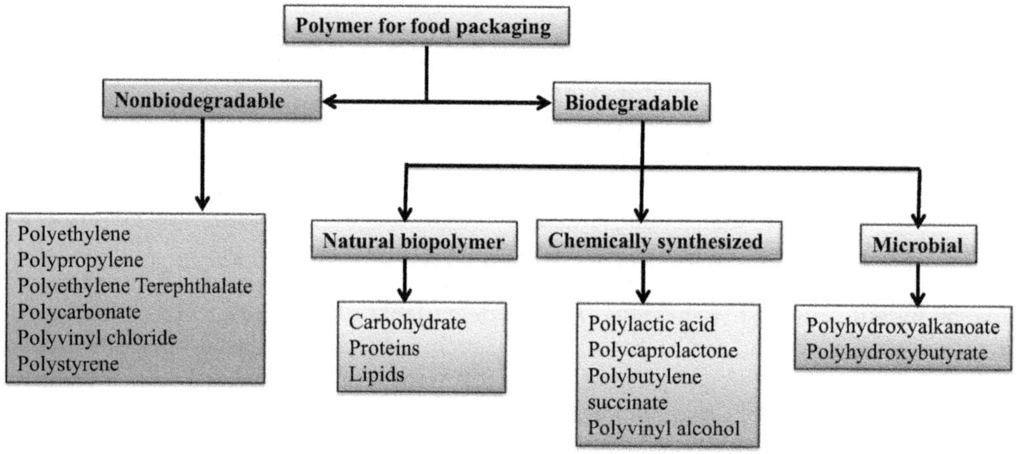

FIGURE 4.1 Types of polymers for food packaging.

microbial contamination and structural integrity of the recycled plastic, are also essential considerations for the safe use of recycled plastics for food contact applications. Plastic recyclers must demonstrate that contaminant levels in the reformed plastic have been reduced to sufficiently low levels to ensure that the resulting packaging is of purity suitable for its intended use. Producing a polymer with the desired qualities will require additional antioxidants, processing aids, or other adjuvants that may need to be added to the recycled polymer (Raheem, 2012). As petroleum reserves become more limited, new varieties of plastics are likely to be made from renewable biomass increasingly. These will contribute to the already extensive array of mechanical and esthetic performance properties that plastics are well known for. The utilization of fossil fuels in the manufacture of plastics accounts for about 7% of worldwide oil and gas (Okada, 2002). These resources will arguably be depleted within the next one hundred years, and the peak in global oil production as estimated by some will occur within the next few decades.

The plastic industry will face fundamental issues associated with using an essentially non-renewable feedstock for most of its products. There is an urgent need to develop new synthetic routes to polymeric materials using renewable resources (Williams & Hillmyer, 2008). Current packaging designs are beginning to incorporate recyclable and recycled plastics, but the search for reuse functions continues. Several factors play into the economic assessment of recycling, including collection, separation, cleaning or reprocessing, and transportation energy. Recycling diverts materials from the waste stream to material recovery. It is different from reuse, which involves using a returned product in its original form. Recycling involves reprocessing the material into new products. The recycling program involves collecting, sorting, processing, manufacturing, and selling recycled materials and products. It was shown that to make recycling economically feasible, recycled products and materials must have a market and the rates of recycling for plastics are on the rise in the United States of America (EPA, 2006). There are several plastic packaging materials for foods, as shown in Fig. 4.2. All thermoplastics are recyclable, that is, they can be melted and reused as raw materials to produce new products. The recycling process requires separation by resin type, as identified by the American Plastics Council. Polyvinyl chloride (PVC) and PS are difficult to recycle. There are concerns that plasticizers such as adipates in PVC may leach to foods, and incineration is a problem because of its chlorine. PS in an expanded form may be used for non-food packaging and cushioning and can then be recycled or incinerated (Marsh & Bugusu, 2007). Table 4.1 shows the application of the most commonly recycled plastic resin.

FIGURE 4.2 The standard type of plastic packaging materials (Raheem, 2012).

TABLE 4.1 Types of plastic resin and their application.

Types of resin	Application
Polyethylene terephthalate	Beverage bottles, mouthwash bottles, boil-in-bag pouches
High-density polyethylene	Milk jugs, trash bags, detergent bottles
Polyvinyl chloride	Cooking oil bottles, packaging around meat
Low-density polyethylene	Grocery bags, food wrap, bread bags
Polypropylene	Yogurt containers, shampoo bottles, straws, margarine tubs, diapers
Polystyrene	Hot beverage cups, take-home boxes, egg cartons, meat trays

4.4 Types of plastic used in packaging

4.4.1 Polyethylene

Polyethylene is structurally the simplest plastic and is made by adding ethylene gas in a high-temperature and pressure reactor (Patel, 2016). Depending on polymerization conditions (temperature, pressure, and catalyst), a range of low, medium, and high-density resins are produced. The processing conditions control the degree of branching in the polymer chain and, therefore, the density and other properties of films and other types of packaging. Polyethylenes are readily heat sealable. They can be made into strong, tough films with a good barrier to moisture and water vapor. They are not a particularly high barrier to oils and fats or gases, such as carbon dioxide and oxygen, compared with other plastics, although barrier properties increase with density. The heat resistance is lower than that of other plastics used in packaging, with a melting point of around 120°C, which increases as the density increases. Polyethylene is not a conductor of electricity and was first used as an insulator in the 1940s. Therefore, PE films are highly susceptible to generating a static charge and need antistatic, slip agents, and anti-blocking compounds added to the resin to assist film manufacturing, conversion, and use. Polymer plants can be found in all countries worldwide, supplying specialist film-making polymers. Low-density polyethylene (LDPE) is easily extruded as a tube and blown to stretch it by a factor of three times the original area (Shankar, Bang, & Rhim, 2019). It is possible to color the films by blending pigment with the polymer before extrusion. Where extruders have more than one die, it is possible to form films with two or more layers of the same material or produce co-extruded films comprised of layers of different plastic materials. With three extruders, it is possible to produce a film where, for example, a moisture-sensitive polymer, EVOH, is sandwiched between protective layers of PE. EVOH provides a gas and odor barrier, and PE offers good heat-sealing properties and a substrate for printing. PE film melts at relatively low temperatures and welds when cut with a hot wire or blade to form effective seals. It is possible to use either premade bags or form/fill/seal machines using flat film in reel form for packaging. A major use of white-pigmented LDPE film is making bags for holding frozen vegetables. By laminating to other substrates with adhesives or extruding the PE polymer onto another material, or web, it is possible to make strong sachets, pouches, and bags with good seal integrity, as the PE flows to fill holes in the sealing area or around contaminants in the seal. PE and other plastics are also combined with paperboard to make the base material for liquid packaging cartons. Major uses of PE film are in shrink and stretch wrapping for collating groups of packs and securing pallet-size loads. Linear low-density PE (LLDPE) film has a density range similar to LDPE. It has short side chain branching and is superior to LDPE in most properties, such as tensile and impact strength and puncture resistance. Major use has been the pillow pack for liquid milk and other liquid foods. LDPE and LLDPE can be used in blends with EVA to improve strength and heat sealing. There is a degree of overlap in application between LDPE and LLDPE. There are differences in both due to the conditions of polymer manufacture and ongoing product development. The thickness used for specific applications can vary, also having commercial implications. Medium-density PE (MDPE) film is mechanically stronger than LDPE and, therefore, used in more demanding situations. LDPE is co-extruded with MDPE to combine the good sealing property of LDPE with the toughness and puncture resistance of MDPE, for example, for the inner extrusion coating of sachets for dry soup mixes. HDPE or high-density PE is the toughest grade and extrudes in the thinnest gauges. This film is used for boil-in-the-bag applications. To improve heat sealing ability, HDPE can be co-extruded with LDPE to achieve peelable seals where the polymer layers can be made to separate easily at the interface of the co-extrusion. A grade of HDPE film is available with either TD mono-axial orientation or biaxial orientation. This film is used for twist-wrapping sugar confectionery and lamination to oriented PP (OPP). The TD-oriented grade easily tears across the web but is more difficult to tear along the web. Being co-extruded, a heat-sealable layer is applied to enable the film to run on conventional form/fill/seal machines. The biaxially oriented film has properties similar to OPP but has a higher moisture vapor barrier. It may be coated in the same way as OPP, including metalizing, to give a high-barrier performance film with the good sealing integrity associated with PE. HDPE is injection molded for closures, crates, pallets, drums, and rotationally molded for intermediate bulk containers (IBCs). A major application of HDPE is for blow-molded milk containers with a 0.5–3 L (Ronca, 2017). PE-based films were also reinforced with antimicrobial compounds to prepare antimicrobial packaging films. Shankar et al. (2019) added grapefruit seed extract (GSE), melanin nanoparticles, and zinc oxide nanoparticles (ZnO NPs) in the LDPE films and coated the films on wrapping paper for sandwich and Korean gimbab packaging.

4.4.2 Polypropylene

PP is a polymer of propylene formed under heat and pressure using Ziegler-Natta type catalysts to produce a linear polymer with protruding methyl (CH_2) groups. The resultant polymer is a more rigid and denser resin than PE and more transparent in its natural form. The use of PP developed from the 1950s onwards. PP has the lowest density and the highest melting point of all the high volume usage thermoplastics and has a relatively low cost. This versatile plastic can be processed in many ways and has many food packaging applications in both flexible film and rigid form (Bang, Shankar, & Rhim, 2020). The high melting point of PP (160°C) makes it suitable for applications where thermal resistance is needed, for example, in hot filling and microwave packaging. In addition, PP may be extrusion laminated to PET or other high-temperature resistant films to produce heat-sealable webs that can withstand up to 115°C−130°C for sterilizing and use in retort pouches. The surfaces of PP films are smooth and have good melting characteristics. PP films are relatively stiff, and the film is glass-clear and heat-sealable when cast. It is used for presentation applications to enhance the appearance of the packaged product. PP is chemically inert and resistant to most commonly found organic and inorganic chemicals. It is a barrier to water vapor and has oil and fat resistance. Aromatic and aliphatic hydrocarbons are, however, able to be dissolved in films and cause swelling and distortion. PP is not subject to environmental stress cracking (ESC). ESC is a surface phenomenon whereby cracks can appear in molded plastic due to contact with materials that affect the surface structure in critical parts of the design. This can lead to cracking without actually degrading the surface. There are specific tests to check for ESC, and shelf life tests with the actual product to be packed should also be carried out. Orientation increases the versatility of PP film. OPP film (BOPP) was the first plastic film to successfully replace regenerated cellulose film in various packaging applications, including biscuit packing. OPP films do not weld or heat seal together easily, as the melting temperature is close to the shrinkage temperature of the film, and the surfaces spring apart when being sealed. However, acrylic-coated OPP has good heat sealing on packing machines designed for RCF, though improved temperature control of the heat-sealing equipment is required. Acrylic coatings also offer good odor-barrier properties. An improved barrier film for both gases and water vapor is produced by coating with PVdC. This film is used as a carton overwrap for assortments of chocolate confectionery and tea bags, providing excellent flavor and moisture barrier protection in both cases. One-side coated EVOH coatings are also available for use in laminations. PP films can be metalized and heat seal coated to produce a high gas and water vapor barrier film. Many of the PP films are used in laminations with other PP and PE films. This allows for the reverse-side printing of one surface buried inside the subsequent laminate. Orienting, or extending, the film, typically, by a factor of 5, in both the machine and transverse direction, increases the ultimate tensile strength and some barrier properties, such as the barrier to water vapor. Orienting in one direction binds the polymer molecules tightly to each other. However, they behave similarly to cotton fibers in a ball of cotton wool in that they demonstrate low mechanical strength. Orienting the polymer fibers biaxially in two directions results in the production of the film with high strength. PP and PE have the lowest surface tension values of the primary packaging plastics and require additional treatment to make them suitable for printing, coating, and laminating. This is achieved with a high-voltage electrical (corona) discharge, ozone treatment, or gas jets (Grieg, Sherman, Pitman, & Barley, 2000). These treatments lightly oxidize the surface by providing aldehydes and ketones that increase the surface energy and improve the adhesion, or keying, of coatings, printing inks, and adhesives. OPP film is produced in widths of up to 10 m or more to achieve cost-effective production. The limiting factors in production are extrusion capacity for the thicker films or winding speed for the very thin films. Most extrusion units today have more than one extruder, enabling production to run at higher rates and using different polymers feeding into one standard die slot. Typically, a film will be made up of three or five layers of resin. The center layer may be a thick core, either opaque or transparent, secondary polymer layers may have unique barrier properties or pigmentation, and the outer layers may be pure PP resin to give gloss to the surface and protect the inner resins should they be moisture sensitive, as is the case with EVOH. In addition, thin layers of special adhesion-promoting resins, known as tie layers, may also be extruded. The range of products packed in PP films includes biscuits, crisps (chips) and snack foods, chocolate and sugar confectionery, ice cream and frozen food, tea, and coffee. Metalized PP film can be used for snacks and crisps (chips) where a higher barrier or longer shelf life is required. White opaque PP films and films with twist-wrapping properties are available. There are several types of heat seal coating, and in addition, converters can apply cold seal coatings on the non-printing side, in register with the print, for wrapping heat-sensitive food products, such as those involving chocolate. Paperboard can be extrusion coated with PP for use as frozen/chilled food trays that the consumer can heat in the microwave and steam-heated ovens. Major food applications of PP are injection-molded pots and tubs for yogurt, ice cream, butter, and margarine. It is also blow-molded for bottles and wide-mouth jars. PP is widely used for the injection molding of closures for bottles and jars. PP can provide a durable living hinge used for flip-top injection molded lids that remain attached to the container when opened, for example, sauce dispensing closure and lid. It is used in thermoforming from PP sheet, as a monolayer, for many food products, such as snacks, biscuits, cheese, and sauces. The functional properties of PP films can also be improved by the addition of certain fillers. Bang et al. (2020) incorporated melanin into PP/polybutylene adipate terephthalate bend films for the prevention of greening of potatoes.

4.4.3 *Polyethylene terephthalate*

Polyesters are condensation polymers formed from ester monomers, resulting from the reaction of a carboxylic acid with an alcohol. There are many different types of polyester, depending on the monomers used. When terephthalic acid reacts with ethylene glycol, it polymerizes into PET. Polyethylene terephthalate (PET) films can be manufactured by extrusion blowing or casting methods. It can be blow-molded, injection-molded, foamed, extrusion coated on paperboard, and extruded as a sheet for thermoforming (Nisticò, 2020). PET can be made into a biaxially oriented range of transparent polyester films produced on essentially the same type of extrusion and Stenter-orienting equipment as OPP. Film thicknesses range from thinner than 12 μm for most polyester films to around 200 μm for laminated composites. No processing additives are used in the manufacture of PET film. Polyesters have much higher heat resistance than many plastics and, when oriented, have very high mechanical strength. The ester has more radicals that may link with other chemicals, making the surface more reactive to inks and not as resistant to chemicals as with polyolefins, such as PE and PP. PET melts at a much higher temperature than PP, typically 260°C, and due to the manufacturing conditions, it does not shrink below 180°C. This means that PET is ideal for high-temperature applications using steam sterilization, boil-in-the-bag, and cooking or reheating in the microwave or conventional radiant heat ovens. The film is also flexible in cold extremes, down to −100°C. Heat sealable versions are available, and it can also be laminated to PE to give good heat-sealing properties. Coated versions, using PVdC, offer an excellent gas barrier and heat-sealing capability. PET is a medium oxygen barrier on its own but becomes a high barrier to oxygen and water vapor when metalized with aluminum. This is used for vacuumed coffee and bag-in-box liquids, where it is laminated with EVA on both sides to produce highly effective seals. It is also used in snack food flexible packaging for products with high-fat content, requiring barriers to oxygen and ultraviolet (UV) light. Metalized PET, either a strip or an elastic laminate, is used as a susceptor in microwaveable packaging. Reverse printed PET film is used as the external ply on f/f/s pouches where it provides a heat-resistant surface for contact with the heat-sealing bars. These packs can be reheated in the microwave and conventional ovens. In the case of PET, the amorphous state is a rigid transparent material, and the crystalline form is white and brittle. PET film is also used as the outer reverse-printed ply in retort pouches, providing strength and puncture resistance, where it is laminated with aluminum foil and either PP or HDPE. PET can be oxide coated with SiO_2 to improve the barrier while remaining transparent, retortable, and microwaveable. Paperboard is extrusion coated with PET for use as ready meal trays that can be reheated in a microwave or conventional radiant heat ovens, that is, dual ovenable. The PET-coated side of the paperboard is on the inside of the tray that is erected by corner heat sealing. PET is the fastest-growing plastic for food packaging applications due to its use in all sizes of carbonated soft drinks and mineral water bottles produced by injection stretch blow molding. PET bottles are also used for edible oils as an alternative to PVC. A foamed-colored PET sheet has been developed under the trade name Escofoam™, which can be laminated and used as the bottom web in thermoformable f/f/s packaging with a printed, peelable seal, top web, for example, with MAP for fresh meat and fish (Kirwan, Plant, & Strawbridge, 2011). PET plastic film is also used as an oxygen barrier in food packaging (Gaikwad, Singh, & Lee, 2018). For packaging the fresh meat as well as fresh produce, a wide range of PET trays are used. Most of the antifogging polymeric films are also made up of PET used in food packaging applications (Gaikwad, Singh, & Negi, 2020).

4.4.4 *Polycarbonate*

Polycarbonate is polyester that contains carbonate groups in its structure. It is formed by the polymerization of the sodium salt of bisphenolic acid with phosgene. It is glass clear, heat resistant, and very tough and durable. Polycarbonate is mainly used as a glass replacement in processing equipment and glazing applications. Its use in packaging is primarily for large, returnable/refillable 3–6 L water bottles (Dhapte et al., 2015). It is used for sterilizable baby feeding bottles and as a replacement in food service. This polycarbonate should not be confused with the thermosetting polycarbonate used in contact lens manufacture. It has been used for returnable milk bottles, ovenable trays for frozen food, and if co-extruded with nylon, could be used for carbonated drinks. Dhapte et al. (2015) prepared transparent ZnO NPs/polycarbonate nanocomposite films with antibacterial and UV-blocking properties for various food packaging applications.

4.4.5 *Polyvinyl chloride*

If one of the hydrogen atoms in ethylene is replaced with a chlorine atom, the resultant molecule is a vinyl chloride monomer (VCM). Addition polymerization of vinyl chloride produces PVC. Unplasticized PVC (uPVC) has beneficial properties but is a rigid, brittle material, and modification is necessary to be used successfully

(Moustafa, Morsy, Ateia, & Abdel-Haleem, 2021). However, flexibility can be achieved by including plasticizers, reduced surface friction with slip agents, various colors by adding pigments, and improved thermal processing by stabilizing agents. Proper care is necessary for selecting additives in food packaging films, which will directly contact food, particularly the migration of packaging components into foodstuffs. Rigid uPVC is generally used for transparent or colored compartmented trays for chocolate assortments and biscuits. It is used with MAP for thermoformed trays to pack salads, sandwiches, and cooked meats. Most PVC films are produced by extrusion, using the bubble process. It can be oriented to make a film with a high degree of shrinking property. Up to 50% shrinkage is possible at relatively low temperatures. The film releases the lowest energy of the commonly used plastic films when heat shrinks around products. It is plasticized, and the high stretch and cling make it suitable for overwrapping fresh produce, for example, apples and meat in rigid trays using semi-automatic and manual methods. Printed PVC film is used for heat-shrinkable sleeve labels for plastic and glass containers. It is also used for tamper-evident shrink bands. Thicker grades are thermoformed to make trays that, after filling, are lidded with a heat seal-compatible top web. PVC has excellent resistance to fat and oil. It can be used to make blow-molded bottles for vegetable oil and fruit drinks. It has good clarity. As a film, it is tough, with high elongation, though with relatively low tensile and tear strength. The moisture vapor transmission rate is relatively high, though adequate for the packaging of mineral water, fruit juice, and fruit drinks in bottles. Depending on its composition, PVC softens at relatively low temperatures (80°C–95°C). PVC easily seals with heat, but heat sealing with a hot wire has the disadvantage of producing HCl gas. The permeability to water vapor and gases depends on the plasticizer's amount used in manufacture. uPVC is a good gas and water vapor barrier, but these properties decrease with increasing plasticizer content. Some grades are used to wrap fresh meat and produce, where an excellent barrier to moisture retards weight loss, but the permeability to oxygen allows the product to breathe. This allows fresh meat to retain its red color and fruits, vegetables, and salads to stay fresh longer by reducing the rate of respiration, especially when packed in a modified atmosphere (MAP). Braga, Rangel, Suarez, and Machado (2018) prepared PVC-containing AgNPs for bread packaging. They found that the bread samples packed using the proposed films with 1 wt.% AgNPs exhibited a total absence of microorganisms in the bread surface compared to the other evaluated films.

4.4.6 *Polystyrene*

PS is a synthetic aromatic hydrocarbon polymer made from styrene. PS has many packaging uses and can be extruded as a monolayer plastic film, co-extruded as a thermoformable plastic sheet, injection-molded, and foamed to give a range of pack types (Chalco-Sandoval, Fabra, López-Rubio, & Lagaron, 2015). It is also co-polymerized to extend its properties. It is less well known as an oriented plastic film, though it has interesting properties. It has high transparency (clarity). It is stiff, with a characteristic crinkle, suggesting freshness, and has a dead fold property. The transparent film is used for carton windows, and the white pigmented film is used for labels. The film is printable. It has a low barrier to moisture and common gases, making it suitable for packaging products, such as fresh produce, which needs to breathe. PS is easily processed by foaming to produce a rigid and lightweight material with good impact protection and thermal insulation properties. It is used in two ways: The blown foam can be extruded as a sheet that can be thermoformed to make trays for meat and fish, egg cartons, a variety of fast food packs, such as the clam shell-shaped containers, as well as cups and tubs. Thin sheets can be used as label stock. The foam can also be produced in pellet or bead form that can be molded with heat and pressure. This is known as expanded PS (EPS). It can be used as a transit case for fresh fish, with thick walls for insulation. PS so far described is general-purpose PS. The main disadvantage of a rigid or semi-rigid container is that it is brittle. This can be overcome by blending with styrene-butadiene co-polymer, SB, or SBC, an elastomeric polymer. The blend is high-impact PS is a more rigid material. It is translucent and is often used in a white-pigmented form. The sheet can be thermoformed for short-shelf life dairy products.

4.5 Nonbiodegradable bioplastics for food packaging

Generally, conventional plastics are polymers produced from monomers as products of the petrochemical industry. The main chemical structures are covalently bound hydrocarbon molecules which are nonbiodegradable. In the last decade, monomers of conventional plastics have been successfully prepared using the fermentation of carbohydrate-based plants, which generates ethanol as a raw material for further conversion to various derivatives such as ethylene and ethylene glycol (EG). Morschbacker (2009) reported the manufacture of ethanol using sugarcane in Brazil as feedstock for the fermentation process, including a catalytic dehydration reaction to prepare an ethylene unit for Bio-PE. This was followed by Braskem launching their "I'm Green Polyethylene" product (i.e., commercial-grade Bio-PE) which retains the same properties and performances of petroleum-based PE with relatively low cost but reduced carbon dioxide (CO_2) emission. Currently, Braskem offers a wide range of bio-based high-density polyethylene (HDPE) and LLDPE. Therefore, Bio-PE facilitates immediate use in the plastics production chain with similar processes to form package and recycling (Byun & Kim, 2014). The manufacturing process of Bio-PP is more complicated than that of Bio-PE and has been commercialized under the name Biopolypropylene 109D, which was utilized for injection-molding packaging. It is made using 20% postindustrial algae biomatter and targets thin-walled applications (Berti et al., 2010). Another nonbiodegradable bioplastic is Bio-PET, having been already produced using resin with further formation into a bottle shape at the industrial scale. PET comprises 30% EG and 70% terephthalic acid (TPA). PlantBottle from Coca-Cola is made of bio-based EG (30% Bio-PET), and similar packaging is used by Dasani, Coca-Cola, Sprite, and Fresca. Berti et al. (2010) developed a viable process to produce bio-based TPA commercially, using isobutanol, n-butanol, isobutylene, limonene, muconic acid, terpenes, and carbohydrates (glucose and fructose) as reactants. Recently, PepsiCo successfully produced 100% Bio-PET from switchgrass, pine bark, and corn husks. The performance of Bio-PET packaging is identical to other petroleum-based PET products (Robertson, 2013). Bio-based-polyethylene furanoate (Bio-PEF), a technology of the Avantium company, has been developed as an alternative to PET bottles. It can be obtained from bio-based EG and furan dicarboxylic acid (FDCA). PEF has the predominantly favorable characteristics of being a better gas and moisture barrier with lower CO_2 emissions than PET. All nonbiodegradable bioplastics as developed have the same structure as those of petroleum-based plastics, allowing the same performance but reducing the carbon footprint of the production process (Nakajima, Dijkstra, & Loos, 2017).

4.6 Biodegradable plastics for food packaging

Biodegradable plastics are classified into two categories according to resources, as presented in Fig. 4.1. Natural biodegradable plastics can be divided into three subgroups by production technologies. The high-potential, natural biodegradable plastics for food packaging are starch, PLA, and Polyhydroxyalkanoates (PHAs), which have been upscaled to industrial manufacturing. Bioplastics sourced from agro-waste have become popular as a nonpolluting and sustainable alternative to their nonbiodegradable counterparts. As defined by European Bioplastics, bioplastics either have biological origin or are biodegradable, or exhibit both properties. In this context, they are divided into two major classes: biodegradable and nonbiodegradable (Plastics Europe & Conversio Market & Strategy GmbH, 2019). Even though sourced from nature, not all bioplastics are biodegradable. Their biodegradability is determined based on the route and rate of degradation (Narancic, Cerrone, Beagan, & O'Connor, 2020). The biodegradable plastics are made from polylactic acid (PLA), starch, cellulose, PHA, and their derivatives (Gómez-Aldapa et al., 2020; Lee et al., 2020; Palanikumar et al., 2020). Quite like fossil-based traditional plastics, bio-based bioplastics can be recycled or incinerated. Alternatively, they can be broken down through microbial activity by composting them at a domestic or industrial scale, creating a paradigm for a zero-waste circular economy (Karan, Funk, Grabert, Oey, & Hankamer, 2019). The carbon footprint of the agricultural biomass, such as jute, banana peels, corn stalks, vegetable waste, wood chips, grain husks, stubble, etc., can be reduced substantially by redirecting the waste towards the fabrication of bioplastics. The raw agro-waste is converted to functional biomaterial by facilitating their breakdown by microbes in a fermentation chamber under specific environmental and nutrition conditions. Natural polyesters like PLA, PHA, and polyhydroxybutyrate (PHB) have also been isolated by mass-producing microorganisms that store these polyesters in the intracellular inclusions (Ali, Zaki, & Obiad, 2017). In both cases, the physiochemical properties of the monomer and the macromolecule produced by this process are affected by the microbial strain and the growth medium being used. To understand its potential for replacing nonbiodegradable plastics, the polymeric resins produced are subjected to life cycle assessment to assess their impact on the environment. These polymers have significant potential as their properties can be altered easily, and the desirable attributes can be imparted with a wide range of disposal options and, as a consequence, reduce their negative environmental impact (Rijpkema et al., 2020).

4.6.1 *Starch-based plastics*

A major source of bioplastics is the storage polysaccharide of cereals, legumes, and tubers, that is, starch, a renewable and widely available raw material. For starch processing, flexibilizer and plasticizer such as sorbitol and glycerin are added. After adding plasticizers and applying thermal and mechanical energy, these constitute thermoplastic starch could be used as a substitute for PS. Starch works as effective packaging material when modified to form films that provide good mechanical properties of high percentage elongation, tensile and flexural strength. Starch is altered by either plasticization, blending with other materials, or chemical modification or combinations of different approaches (Maniglia, Castanha, Rojas, & Augusto, 2021). These starch-based thermoplastic materials such as polyethylene-vinyl alcohol, polyvinyl alcohol, and polycaprolactone (PCL) have found broader industrial applications ranging from extrusion applications, injection molding, blow molding, film blowing, and foaming. In addition to extending the shelf life of food, recent attempts have been made to give starch-based biodegradable food packaging materials new functions. Novel starch-based intelligent food packaging has been prepared, which can be used as a visual indicator of food freshness to monitor the quality changes of seafood, meat, cooking oil, fruit juice, and other food (Luchese, Abdalla, Spada, & Tessaro, 2018; Zhang, Huang, Yan, Hu, & Ren, 2020). Shapi, Othman, Nordin, Basha, and Naim (2020) used starch films incorporated with different concentrations of chitosan nanoparticles for cherry tomato packaging. In vivo study revealed that the starch/chitosan nanoparticle film (15% w/w) was more efficient to inhibit microbial growth in cherry tomatoes (7×10^2 CFU/g) compared to neat starch film (2.15×10^3 CFU/g).

4.6.2 *Chemically synthesized bioplastics*

During the conventional chemical synthesis of polymers, there is also a considerable potential for production of bio-polyesters. A large number of "bio-polyesters" are produced; however, the PLA among them can be commercialized at a larger scale for the production of renewable packaging materials (Endres, 2017). All the classical packaging materials derived from mineral oil in the present time can be obtained from renewable source-based monomers derived from fermentation. However, presently the process is found to be uneconomical due to costs incurred in the production of monomers, and thus a greater insight into knowledge through research is required in this context.

4.6.2.1 Polylactic acid

PLA plastics are derived from the fermentation of agricultural byproducts such as starch-rich substances like maize, wheat, sugar, and corn starch. The process involves the conversion of corn or other carbohydrate sources into dextrose followed by fermentation into lactic acid. PLA derived from lactic acid is thermoplastic, biodegradable aliphatic polyester having ample potential for packaging applications (Shankar, Wang, & Rhim, 2018). The lactic acid monomers are either directly polycondensed or undergo ring-opening polymerization of lactide resulting in the formation of PLA pellets (Shankar, Rhim, & Won, 2018). The properties of PLA as packaging material depend on the ratio between the two optical isomers of the lactic acid monomer. When 100% L-PLA monomers are used, it results in very high crystallinity and melting point, whereas 90/10% D/L co-polymers result in polymerizable melt above its Tg and thus fulfill the requirements of bulk packaging by facilitating its processing. PLA is the first bio-based polymer commercialized on a large scale and can be shaped into injection-molded objects, films, and coatings (Shankar & Rhim, 2016a). PLA has replaced HDPE. While PLA can be synthesized from both renewable (fermentation process) and nonrenewable resources (chemical process), the former is preferred due to its low environmental impact, reduced stress on fossil fuels, and higher optical clarity of the final product. The physical properties like crystallinity, degradation rate, and hydrophobicity of PLA depend on the ratio of Poly(DL-lactide) (PDLA) and poly-l-lactic acid (PLLA) incorporated during polymerization. When compared to PLLA, which has a melting temperature (Tm) of 170°C–180°C, PDLLA (combination of PDLA and PLLA) is more thermostable and has a higher Tm of 220°C–230°C. Techniques to ameliorate the physical properties of PLA include block polymerization method/grafting and physical blending (Nanda, Misra, & Mohanty, 2011). These modifications in the synthesis methods of PLA increase the biodegradation rate of the product while tailoring its toughness according to the final use requirement. Being transparent, PLA has gained wide acceptance for manufacturing packaging films, containers, and stationery products. Biomedical and agricultural sectors find such polymers useful due to their biocompatibility and low carbon footprint after use (Singhvi, Zinjarde, & Gokhale, 2019). The esters in the polymer can be easily hydrolyzed without the use of external agents to release CO_2, water, and partially decomposed organic products that constitute humus and enrich its composting environment. PLA bioplastic in pure form and with additives has been shown to degrade more than 50% of its biomass within 3 months when composted in different environmental soil samples and soil stimulants. The composting of naturals polymers does not affect the diversity and total bacterial and fungal biomass alive per gram of the soil sample, thus, presenting PLA as a viable replacement option for nonbiodegradable plastics such as LDPE, PET, and PS as packaging material (Adhikari et al., 2016). Shankar et al. (2018) prepared PLA-based nanocomposite films incorporated with ZnO NPs with improved mechanical, water vapor barrier, UV-light barrier, and antimicrobial properties.

4.6.2.2 *Polycaprolactone*

PCL is a semi-crystalline, completely biodegradable, easy to process, and cheap fossil-based polymer. It is soluble in many organic and inorganic solvents and has a glass transition temperature (Tg) of −60°C, which increases its application as a compatibilizer in formulations of polyurethane (Mohamed & Yusoh, 2015). Furthermore, the addition of PCL into hydrophilic chitosan polymer increases the overall hydrophobicity of the blend. As a result, low water vapor transmission rate (WVTR) values are observed in the blend compared to pure films. Due to this property, food stored in such films has a longer shelf life (Sarasam, Krishnaswamy, & Madihally, 2006). It is commercially found under the trade names: Tone from Union Carbide, Celgreen from Daicel, and CAPA from Solvay. Bujok et al. (2021) reported environmentally-friendly microwave in-situ synthesis of biodegradable nanocomposite films based on polycaprolactone matrix and ZnO nanoparticles with antimicrobial activity for food packaging applications.

4.6.2.3 *Polybutylene succinate*

Polybutylene succinates (PBSs) are obtained by polycondensation of glycols such as 1,4-butanediol and ethylene glycol with aliphatic dicarboxylic acids, like adipic and succinic acid. It is a white crystalline polymer, with good processibility having a Tg of −45°C to −10°C and a melting point of 90−120°C with 330% elongation at break. PBS has mechanical properties, approximately such as PE and PP (Wang, Zhou, & Li, 2007). They were first invented in 1990 under the trade name Bionolle (Showa Denko) in Japan. Since then, many different copolymers have been prepared, like polybutylene succinate-co-adipate (PBSA) obtained from incorporating adipic acid at a specific concentration. The molecular weight of the polymer can be increased by adding a small amount of coupling agents. Different industries commercialize different PBS by changing the monomeric units, for example, Skygreen by SK Chemicals in Korea. The nature of diols and diacids used for condensation influences these polymers' properties and degradation rate (Wang et al., 2007). Various PBS-based films are reported for their application in food packaging (Łopusiewicz, Zdanowicz, Macieja, Kowalczyk, & Bartkowiak, 2021).

4.6.2.4 *Polybutylene adipate-co-terephthalate*

Polybutylene adipate-co-terephthalate (PBAT) is a linear aromatic co-polyester obtained from the condensation of 1,4-butanediol with a mixture of terephthalic acid and adipic acid. It exhibits excellent properties at a terephthalic acid concentration of more than 35% mol; as the content increases above 55%, the biodegradation rate of PBAT decreases. PBAT is flexible and soft like PCL, so it is used to produce films, filaments, bottles, and molded products. PBAT can be blended with cellulose, starch, and other biodegradable polymers. For improving the hydrophobicity, mechanical, and thermal properties, PBAT is blended with cellulose. The addition of PBAT in polyhydroxybutyrate-valerate (PHBV) decreases the degree of crystallinity (Javadi et al., 2010). It is commercialized under the trade name of Ecoflex developed by BASF Canada Inc, Origo-Bi developed by Novamont, Easter Bio developed by Eastman Chemical. Shankar and Rhim (2016b) prepared antimicrobial PBAT/AgNPs nanocomposite films using tocopherol for food packaging applications.

4.6.2.5 *Polyvinyl-alcohol*

Polyvinyl-alcohol (PVA) is a semi-crystalline polymer comprising mainly amorphous phases with little crystallinity. It consists of 1, 3-diol units or 1, 2-diol units, depending on the hydrolysis degree of poly(vinyl acetate). The properties of PVA generally depend on its molecular weight and degree of hydrolysis, with the molecular weight of PVA usually ranging between 20,000−400,000 and based on the length of vinyl acetate used to produce PVA—the degree of hydrolysis is typically in the range of 80%−99% (Abdullah, Dong, Davies, & Barbhuiya, 2017). Hu et al. (2022) prepared biodegradable polyvinyl alcohol/starch film compatibilized by rod-like ZnO nanofillers for packaging fresh-cut carrot slices to prevent microbial infection and prolong their shelf life.

4.6.2.6 *Polyhydroxyalkanoates*

PHA represents natural polyesters produced by bacterial fermentation of sugar, glucose, or vegetable oil feedstock. It is one of the most recent and widely used biodegradable polymers for food packaging applications. Bacteria accumulate PHAs as a reserve material intracellularly at a 30%−80% dry weight concentration under limited N_2 and abundant carbon (Monroy & Buitrón, 2020). Its Tm ranges from 40–180°C depending on monomers used for synthesis. Depending on the nutrient source for carbon and the organism, PHA may be manufactured from rigid brittle to a rubber-like polymer. A good barrier property film was formed when PHA was blended with zein. An increase of 39%–48% and 27%–35% was seen in the water vapor permeation and oxygen permeation, respectively. A change in blend morphology was also observed due to the incorporation of zein (Fabra, Lopez-Rubio, & Lagaron, 2014). PHAs are completely biodegradable, and biodegradation occurs through the esterase activity that breaks the monomer from the chain ends. The most common PHA is the PHB, formed by the polymerization of 3-hydroxybutyrate. PHB is known for its excellent UV-resistivity, and high optical properties with Tm of 180°C and Tg of 55°C. PHB has a crystallinity of more than 50%. It is well known that PHB is unstable after a melting temperature of 180°C. When kept at a temperature even 10°C below the melting point, it can undergo molecular weight reduction, limiting its processibility (Savenkova et al., 2000). To increase the processibility condition, different strategies such as copolymerization with other alkanoates, adding biodegradable polymer, or blending with a second polymer. Different routes yield different structural polymers. PHB undergoes degradation by various bacteria, fungi, and algae in other environmental conditions. The hydrolytic degradation forms 3-hydroxy butyric acid at a low rate. The co-polymer PHBV is synthesized by adding propionic acid to the feedstock. It is more rigid and less stiff, so it is used as a packaging material. It degrades within 5–6 weeks in a microbiologically active environment, ending with CO_2 and H_2O in aerobic conditions. The degradation is faster, with methane production in anaerobic conditions (Kim, Lee, Yoon, & Chin, 2000). PHB and PHBV are commercialized under different trade names: Biopol from Monsanto, Nodax from Procter & Gamble and Kaneka Corporation, Eamat from Tianan, and Biomer-P™ from Biomer.

4.7 Conclusion

This chapter presented the different types of plastic and bioplastic-based composites for the food packaging applications. The world is dominated by petroleum-based plastic materials for food packaging applications because of easy processing and low cost. However, owing to the nonbiodegradable nature of petroleum-based plastic, they generally end up in landfill which ultimately contaminates land, environment, and marine life. Therefore, researcher and consumers are more focused on developing packaging materials based on bioplastics. Whereas, the higher cost of bioplastic-based packaging materials has become a limiting factor in their commercialization. Moreover, research has to be carried out in developing low-cost bio-plastic using a high percentage of low-cost fillers for the commercialization of sustainable packaging materials.

References

Abdullah, Z. W., Dong, Y., Davies, I. J., & Barbhuiya, S. (2017). PVA, PVA blends, and their nanocomposites for biodegradable packaging application. *Polymer-Plastics Technology and Materials, 56*, 1307–1344.

Adhikari, D., Mukai, M., Kubota, K., Kai, T., Kaneko, N., Araki, K. S., & Kubo, M. (2016). Degradation of bioplastics in soil and their degradation effects on environmental microorganisms. *Journal of Agricultural Chemistry and Environment, 05*, 23–34.

Ali, W. S., Zaki, N. H., & Obiad, S. Y. N. (2017). Prouction of bioplastic by bacteria isolated from local soil and organic wastes. *Current Research in Microbiology and Biotechnology, 5*, 1012–1017.

Bang, Y. J., Shankar, S., & Rhim, J. W. (2020). Preparation of polypropylene/poly (butylene adipate-co-terephthalate) composite films incorporated with melanin for prevention of greening of potatoes. *Packaging Science and Technology, 33*, 433–441.

Berti, C., Binassi, E., Colonna, M., Fiorini, M., Kannan, G., Karanam, S., ... Odeh, I. (2010). Bio-based terephthalate polyesters. US Patent 2,010,168,371.

Braga, L. R., Rangel, E. T., Suarez, P. A. Z., & Machado, F. (2018). Simple synthesis of active films based on PVC incorporated with silver nanoparticles: Evaluation of the thermal, structural and antimicrobial properties. *Food Packaging and Shelf Life, 15*, 122–129.

Bujok, S., Peter, J., Halecký, M., Ecorchard, P., Machálková, A., Medeiros, G. S., ... Beneš, H. (2021). Sustainable microwave synthesis of biodegradable active packaging films based on polycaprolactone and layered ZnO nanoparticles. *Polymer Degradation and Stability, 190*, 109625.

Byun, Y., & Kim, Y. T. (2014). Utilization of bioplastics for food packaging industry. *Innovations in food packaging* (2nd edn, pp. 369–390). London, UK: Academic Press.

Chalco-Sandoval, W., Fabra, M. J., López-Rubio, A., & Lagaron, J. M. (2015). Development of polystyrene-based films with temperature buffering capacity for smart food packaging. *Journal of Food Engineering, 164*, 55–62.

Chi, K., Wang, H., & Catchmark, J. M. (2020). Sustainable starch-based barrier coatings for packaging applications. *Food Hydrocolloids, 103*, 105696.

Dhapte, V., Gaikwad, N., More, P. V., Banerjee, S., Dhapte, V. V., Kadam, S., & Khanna, P. K. (2015). Transparent ZnO/polycarbonate nanocomposite for food packaging application. *Nanocomposites, 1*, 106–112.

Endres, H. J. (2017). In K. Wagemann, & N. Tippkötter (Eds.), *Bioplastics. In Biorefineries. Advances in biochemical engineering/biotechnology* (166, pp. 427–468). Cham, Switzerland: Springer International Publishing.
Environmental Protection Agency (EPA). (2006). Municipal solid waste in the USA: 2005 facts and figures. EPA530-R-06-011. Washington, DC, p. 153.
Fabra, M. J., Lopez-Rubio, A., & Lagaron, J. M. (2014). Nanostructured interlayers of zein to improve the barrier properties of high barrier polyhydroxyalkanoates and other polyesters. *Journal of Food Engineering, 127*, 1–9.
Gómez-Aldapa, C. A., Velazquez, G., Gutierrez, M. C., Rangel-Vargas, E., Castro-Rosas, J., & Aguirre-Loredo, R. Y. (2020). Effect of polyvinyl alcohol on the physicochemical properties of biodegradable starch films. *Materials Chemistry and Physics, 239*, 122027.
Gaikwad, K. K., Singh, S., & Lee, Y. S. (2018). Oxygen scavenging films in food packaging. *Environmental Chemistry Letters, 16*, 523–538.
Gaikwad, K. K., Singh, S., & Negi, Y. S. (2020). Ethylene scavengers for active packaging of fresh food produce. *Environmental Chemistry Letters, 18*, 269–284.
Grieg, S., Sherman, P.B., Pitman, R., & Barley, C. (2000). Adhesion promoters: Corona, flame and ozone: A technology update. In Polymers, laminations and coatings conference. Chicago, IL, USA, 27–31 August 2000, 1.
Hu, W., Zou, Z., Li, H., Zhang, Z., Yu, J., & Tang, Q. (2022). Fabrication of highly transparent and multifunctional polyvinyl alcohol/starch based nanocomposite films using zinc oxide nanoparticles as compatibilizers. *International Journal of Biological Macromolecules, 204*, 284–292.
Javadi, A., Kramschuster, A. J., Pilla, S., Lee, J., Gong, S., & Turng, L. S. (2010). Processing and characterization of microcellular PHBV/PBAT blends. *Polymer Engineering & Science, 50*, 1440.
Karan, H., Funk, C., Grabert, M., Oey, M., & Hankamer, B. (2019). Green bioplastics as part of a circular bioeconomy. *Trends in Plant Science, 24*, 237–249.
Kawashima, N., Yagi, T., & Kojima, K. (2019). How do Bioplastics and fossil-based plastics play in a circular economy? *Macromolecular Materials and Engineering, 304*, 1–14.
Kim, M. N., Lee, A. R., Yoon, J. S., & Chin, I. J. (2000). Biodegradation of poly (3-hydroxybutyrate), Sky-Green® and Mater-Bi® by fungi isolated from soils. *European Polymer Journal, 36*, 1677–1685.
Kirwan, M. J., Plant, S., & Strawbridge, J. W. (2011). Plastics in food packaging. In R. Coles, & M. Kirwan (Eds.), *Food and beverage packaging technology* (pp. 157–211). Hoboken, NJ, USA: Blackwell Publishing Ltd.
Lee, G.-H., Moon, H., Kim, H., Lee, G. H., Kwon, W., Yoo, S., ... Hahn, S. K. (2020). Multifunctional materials for implantable and wearable photonic healthcare devices. *Nature Reviews Materials, 5*, 149–165.
Łopusiewicz, Ł., Zdanowicz, M., Macieja, S., Kowalczyk, K., & Bartkowiak, A. (2021). Development and characterization of bioactive poly (butylene-succinate) films modified with quercetin for food packaging applications. *Polymers, 13*, 1798.
Luchese, C. L., Abdalla, V. F., Spada, J. C., & Tessaro, I. C. (2018). Evaluation of blueberry residue incorporated cassava starch film as pH indicator in different simulants and foodstuffs. *Food Hydrocolloids, 82*, 209–218.
MacArthur, E., Waughray, D., & Stuchtey, M. (2016). *Rethinking plastics, starting with packaging* (pp. 1–206). World Economic Forum.
Maniglia, B. C., Castanha, N., Rojas, M. L., & Augusto, P. E. (2021). Emerging technologies to enhance starch performance. *Current Opinion in Food Science, 37*, 26–36.
Marsh, K., & Bugusu, B. (2007). Food packaging –roles, materials and environmental issues. *Journal of Food Science, 72*, 39–55.
Mendes, A. C., & Pedersen, G. A. (2021). Perspectives on sustainable food packaging:– Is bio-based plastics a solution? *Trends in Food Science & Technology, 112*, 839–846.
Mohamed, R. M., & Yusoh, K. (2015). A Review on the recent research of polycaprolactone (PCL). *Advanced Materials Research, 1134*, 249–255.
Monroy, I., & Buitrón, G. (2020). Production of polyhydroxybutyrate by pure and mixed cultures of purple non-sulfur bacteria: A review. *Journal of Biotechnology, 317*, 39–47.
Morschbacker, A. (2009). Bio-ethanol based ethylene. *Journal of Macromolecular Science: Part C: Polymer Reviews, 49*, 79–84.
Moustafa, H., Morsy, M., Ateia, M. A., & Abdel-Haleem, F. M. (2021). Ultrafast response humidity sensors based on polyvinyl chloride/graphene oxide nanocomposites for intelligent food packaging. *Sensors and Actuators A: Physical, 331*, 112918.
Nakajima, H., Dijkstra, P., & Loos, K. (2017). The recent developments in biobased polymers toward peneral and engineering applications: Polymers that are upgraded from biodegradable polymers, analogous to petroleum-derived polymers, and newly developed. *Polymer, 523*, 1–26.
Nanda, M. R., Misra, M., & Mohanty, A. K. (2011). The effects of process engineering on the performance of PLA and PHBV blends. *Macromolecular Materials and Engineering, 296*, 719–728.
Narancic, T., Cerrone, F., Beagan, N., & O'Connor, K. E. (2020). Recent advances in bioplastics: Application and biodegradation. *Polymers (Basel)* (12, p. 920).
Nisticò, R. (2020). Polyethylene terephthalate (PET) in the packaging industry. *Polymer Testing, 90*, 106707.
Okada, M. (2002). Chemical synthesis of bidegradabl02e polymers. *Progress in Polymer Science, 27*, 87–133.
Palanikumar, L., Al-Hosani, S., Kalmouni, M., Nguyen, V. P., Ali, L., Pasricha, R., ... Magzoub, M. (2020). pH-responsive high stability polymeric nanoparticles for targeted delivery of anticancer therapeutics. *Communications Biology, 3*, 1–17.
Patel, R. M. (2016). Polyethylene. *Multilayer flexible packaging* (pp. 17–34). The Netherlands: Elsevier: Amsterdam.
Plastics Europe, G. M. R., Conversio Market & Strategy GmbH, (2019). Plastics - the Facts.
Raheem, D. (2012). Application of plastics and paper as food packaging materials - An overview. *Emirates Journal of Food and Agriculture, 25*, 177–188.
Rijpkema, S. J., Langens, S. G. H. A., van der Kolk, M. R., Gavriel, K., Toebes, B. J., & Wilson, D. A. (2020). Modular approach to the functionalization of polymersomes. *Biomacromolecules, 21*, 1853–1864.
Robertson, G. L. (2013). Edible, biobased and biodegradable food packaging materials. In G. L. Robertson (Ed.), *Food packaging - Principles and practice* (pp. 49–90). Boca Raton: CRC press.
Ronca, S. (2017). In M. Gilbert (Ed.), *Polyethylene. In Brydson's plastics materials* (8th ed, pp. 247–278). The Netherlands: Elsevier: Amsterdam.
Russell, D. A. M. (2014). Sustainable (food) packaging - an overview. Food Additives & Contaminants Part A Chemistry, Analysis. *Control, Exposure and Risk Assessment, 31*, 396–401.

Salwa, H. N., Sapuan, S. M., Mastura, M. T., & Zuhri, M. Y. M. (2019). Green bio composites for food packaging. *International Journal of Recent Technology and Engineering*, *8*, 450–459.

Sarasam, A. R., Krishnaswamy, R. K., & Madihally, S. V. (2006). Blending chitosan with polycaprolactone: Effects on physicochemical and antibacterial properties. *Biomacromolecules*, *2006*(7), 1131–1138.

Savenkova, L., Gercberga, Z., Nikolaeva, V., Dzene, A., Bibers, I., & Kalnin, M. (2000). Mechanical properties and biodegradation characteristics of PHB-based films. *Process Biochemistry*, *35*, 573–579.

Shankar, S., & Rhim, J.-W. (2016a). Polymer nanocomposites for food packaging applications. *Functional and Physical Properties of Polymer Nanocomposites* (pp. 29–55). John Wiley & Sons, Ltd., Chapter 3.

Shankar, S., & Rhim, J.-W. (2016b). Tocopherol-mediated synthesis of silver nanoparticles and preparation of antimicrobial PBAT/silver nanoparticles composite films. *LWT-Food Science and Technology*, *72*, 149–156.

Shankar, S., & Rhim, J.-W. (2018). Bionanocomposite films for food packaging applications. *Reference Module in Food Science*, 1–10.

Shankar, S., Bang, Y. J., & Rhim, J. W. (2019). Antibacterial LDPE/GSE/Mel/ZnONP composite film-coated wrapping paper for convenience food packaging application. *Food Packaging and Shelf Life*, *22*, 100421.

Shankar, S., Rhim, J. W., & Won, K. (2018). Preparation of poly(lactide)/lignin/silver nanoparticles composite films with UV light barrier and antibacterial properties. *International Journal of Biological Macromolecules*, *107*, 1724–1731.

Shankar, S., Wang, L. F., & Rhim, J. W. (2018). Incorporation of zinc oxide nanoparticles improved the mechanical, water vapor barrier, UV-light barrier, and antibacterial properties of PLA-based nanocomposite films. *Materials Science and Engineering: C*, *93*, 289–298.

Shapi, R. A., Othman, S. H., Nordin, N., Basha, R. K., & Naim, M. N. (2020). Antimicrobial properties of starch films incorporated with chitosan nanoparticles: In vitro and in vivo evaluation. *Carohydrate Polymers*, *230*, 115602.

Singhvi, M. S., Zinjarde, S. S., & Gokhale, D. V. (2019). Polylactic acid: Synthesis and biomedical applications. *Journal of Applied Microbiology*, *127*, 1612–1626.

Wang, X., Zhou, J., & Li, L. (2007). Multiple melting behavior of poly (butylene succinate). *European Polymer Journal*, *43*, 3163–3170.

Williams, C. K., & Hillmyer, M. A. (2008). Polymers from renewable resources. A perspective for a special issue of polymer reviews. *Polymer Review*, *48*, 1–10.

Zhang, K. L., Huang, T. S., Yan, H., Hu, X. Z., & Ren, T. (2020). Novel pH-sensitive films based on starch/polyvinyl alcohol and food anthocyanins as a visual indicator of shrimp deterioration. *International Journal of Biological Macromolecules*, *145*, 768–776.

CHAPTER

5

Impact of nanoparticles on fish and other marine products

Bruna Wendt Böhmer-Maas

Department of Agroindustrial Science and Technology, Federal University of Pelotas, Pelotas, Rio Grande do Sul, Brazil

5.1 Introduction

Several researches have been developed over the years with the aim of maintaining food safety and making food safe for consumers. However, this is a challenge, especially for highly perishable foods, such as fish and other marine products. As is well established, factors such as the rapid installation of rigor mortis, bacterial growth, and autolysis, promoted by the action of cathepsin and calpain enzymes (enzymes naturally present in muscle) and exopeptidases of microbial origin (Bressan & Perez, 2000), are responsible due to the high perishability of these products (Gram & Dalgaard, 2002).

Among the foods of animal origin, fish and other marine products are considered one of the most susceptible to the deterioration process, due to their intrinsic characteristics and also to their natural habitat. The main intrinsic characteristics, which give greater perishability to the fish and other marine products are high metabolic activity of the microbiota, pH close to neutrality (6.2–6.4), high water activity, high content of nutrients that can be easily used by microorganisms, such as proteins, phospholipids, and poly fatty acids—unsaturated omega 3 (eicosapentaenoic acid-EPA and docosahexaenoic-DHA); in addition, crustaceans and mollusks contain carbohydrates in the form of glycogen (Huss, 1997).

The multiplication of microorganisms leads to a decline in the quality of the fish, which is accompanied by changes in sensory characteristics such as flavor, odor, texture, and appearance (Vieira & Saker-Sampaio, 2003). In general, the muscle tissue of freshly caught fish is usually sterile at harvest time. However, the skin, gills, and intestines carry a large number of bacteria (10^2–10^7 cfu/cm^2 on the skin and 10^3–10^9 vcfu/g on the gills and intestine) (Huss, 1997). The deteriorating microflora of fresh fish stored on ice consists mainly of *Pseudomonas* spp. Gram-negative and H_2S-producing bacteria, including *Shewanella putrefaciens*, *Acinobacter*, and *Moraxella* spp., may comprise a portion of the deteriorating microflora (Alcantara, Morais, Cunha, & Souza, 2012). In the post-mortem, bacterial multiplication is favored due to the absence of natural defenses that occurred as a muscle, when it turns into meat, the penetration of bacteria is facilitated (Germano, Oliveira, & Germano, 1993).

In the case of crustaceans, microbial deterioration occurs similarly to fish; however, the greater amount of free amino acids and other soluble nitrogen compounds present leads to rapid deterioration and high levels of volatile basic nitrogen deterioration compounds.

Molluscs meat contains an appreciable amount of carbohydrates in the form of glycogen. Although the microorganisms involved in the deterioration of molluscs are the same as those found in fish and crustaceans (*Pseudomonas*, *Acinetobacter*, and *Moraxella* spp.), the deterioration of the former is mainly glycolytic (contains 1%–5% glycogen) rather than a proteolytic decrease in pH from about 6.5–5.8. Under such acidity conditions, *enterococci*, *lactobacilli*, and yeasts dominate the final stages of deterioration (Adams & Moss, 1999).

Effective conservation methods could delay enzymatic and bacterial multiplication reactions, reducing losses in nutritional value, economic losses, and food waste. In this sense, the irradiation technique has been widely used, as it is effective in reducing the microbial load; however, it can promote adverse reactions in muscle tissue

Food Packaging and Preservation.
DOI: https://doi.org/10.1016/B978-0-323-90044-7.00005-7

due to radiolysis, for example, leading to the formation of superoxide and hydroxyl radicals (Fennema, Damodaran, & Parkin, 2010), as well as representing a high cost to the industry (Koenari et al., 2016; Wolf-Hall, Gibbons, & Bauer, 2009).

In view of this, several studies have been developed in search of alternatives to extend the shelf life of fish and other marine products. Nanostructures such as nanoemulsions (Meral, Ceylan, & Kose, 2019; Ozogul et al., 2020; Yazgan et al., 2017), nanofibers (Ceylan, Sengor, & Yilmaz, 2018; Erbay, Dağtekin, Türe, Yeşilsu, & Torres-Giner, 2017; Shi et al., 2021), nanocomposites (Azizi-Lalabadi, Ehsani, Ghanbarzadeh, & Divband, 2020; Homayonpour, Jalali, Shariatifar, & Amanlou, 2021; Truong, Nguyen, & Nguyen, 2020), and nanocapsules Saloko, Darmadji, Setiaji, & Pranoto, 2014) were produced to reduce the development of bacteria in fish and other marine products. In this chapter, we will cover several applications of antibacterial nanostructures in fish and other marine products.

5.2 Fish production

The increase in the global population and the purchasing power of growing populations will result in a projected increase in food demand. According to the latest statistics published by FAO (FAO, 2020), in 2050, the population will be 9.8 billion, 29% greater than the current one, and the most expressive growth will be in the developing countries, where 70% of the population will be urban and income levels will be higher than today.

To feed this larger urban population with greater purchasing power, food production must increase by 70%, which will lead to greater losses in food, especially perishables. Due to population growth, more and more meat will have to be produced to meet this growing demand.

Aquaculture is one of the sources of animal protein, and fish is meat with excellent nutritional quality. In this way, the production of food derived from aquaculture farming contributes to the nutrition of humanity through the differentiated quality of the protein derived from fish.

In 2018, the average per capita consumption of fish reached 20.5 kg/year in the world, consumption grew by about 1.5% per year. In Latin America, fish consumption will grow by 33% by 2030, according to the FAO of the United Nations.

Sofia is the acronym for "The State of World Fisheries and Aquaculture" and provides society with a summary of statistical and analytical information on aquaculture activities, capture, processing, and consumption of fish around the world. The report is jointly published by the FAO of the United Nations.

According to the latest full report published in Sofia (2020), global fish production is estimated to have reached around 179 million tonnes in 2018, with an estimated total value of US$ 401 billion.

Of the total, 156 million tons were used directly for human consumption, equivalent to an estimated annual supply of 20.5 kg per capita. The remaining 22 million tons were destined for the production of fishmeal and fish oil.

China remained a major producer of fish, accounting for 35% of global fish production in 2018. Excluding China, a significant share of production in 2018 came from Asia (34%), followed by the Americas (14%), Europe (10%), Africa (%), and Oceania (1%).

Global consumption of fish as food increased at an average annual rate of 3.1% from 1961 to 2017, a rate nearly double the annual rate of population growth (1.6%) for the same period, and higher than other foods. proteins (meat, dairy products, milk, etc.), which increased by 2.1% per year. Moreover, per capita food consumption of fish increased from 9.0 kg (live weight equivalent) in 1961 to 20.5 kg in 2018, at about 1.5% per year.

5.3 Fish composition

Foods from the aquatic environment, more specifically the marine, are characterized by the exceptional source of proteins for human consumption, and also of lipids that are fundamental in the prevention of cardiovascular diseases, as well as vascular and hemostatic action and in the intervention of the production of prostaglandins (Ackman, 1989; Badolato et al., 1994; Tan et al., 2018).

Fish is considered a food with ample nutritional support, providing minerals, essential fatty acids, and vitamins of complex B, A, and D (Jiang et al., 2015). The fish has, in its composition, 70% to 80% of water, and 20% to 25% of proteins, which are the sources of essential amino acids such as lysine, methionine, and cysteine. Its lipid contribution is relative and can reach 10%, such as omega 6 and omega 3, as for the mineral content, it has calcium, phosphorus, copper, iron, sulfur, iodine, and selenium, reaching 1.5% of the total body, 55% to 60% of the total fish is considered an edible part, and when 200 g of fish is consumed, at least five essential amino acids recommended for a healthy diet are ingested (Rahimnejad, et al., 2015).

Digestibility varies from 90% to 95%, which is characterized by some factors such as the concentration of proteins linked to connective tissue being lower than proteins linked to muscle structures, facilitating enzymatic activity and, finally, making proteins more unstable in the presence of heat than that of mammals, for example (Rahimnejad, et al., 2015).

5.4 Perishibility of fish

All over the world, improvements in food quality and the prevention of foodborne diseases are sought. According to the WHO (2007), foodborne diseases are those caused by pathogens or toxins present in food. Its transmission occurs through the entry of pathogens and/or toxins into the gastrointestinal tract, where, in general, they cause the first symptoms, such as vomiting, nausea, diarrhea, and abdominal pain.

In some cases, symptoms may manifest late, after a long period of exposure, and can be associated with a multifactorial etiology, not representing a classic foodborne disease (Oliveira, Paula, Capalonga, Cardoso, & Tondo, 2010). The prevention of foodborne diseases is very complex, as numerous pathogens can be inserted along the production chain. Most food poisonings are not reported, as few countries have a foodborne disease reporting system or do not notify inspection bodies or health agencies because their symptoms are usually mild and the patient does not seek medical help.

In view of this, the worldwide incidence of foodborne outbreaks is difficult to estimate, but they lead to the death of at least 600 million people per year (WHO, 2023). The lack of food security threatens not only human health, but entire economies, affecting the most marginalized and vulnerable people.

Fish can be marketed both in nature, including fresh and raw fish, and industrialized, when the fish undergoes a more robust handling and conservation procedure.

According to Penha, Gonçalves, Rosa, Silva, and Bichara (2020), the decomposition initiated after the death of the organism is marked by exogenous and endogenous factors that accelerate its reactions, involving physical, mechanical, biological, and chemical processes. For this reason, the phases that affect fish after slaughter must be known, namely, hyperemia, which is very evident in fish. It consists of the release of hydrophobic mucus produced by the mucous cells, and its molecular composition contributes to the proliferation of microorganisms, another point is rigor mortis which leads to a decrease in the pH of the fish, as a result of the production of lactic acid, giving the meat quick conservation; however, this stage is greatly influenced by the way of slaughter that the fish suffered, due to its energy consumption, which is the main factor that interferes in the duration of rigor mortis.

In sequence, the fish is subjected to chemical and autolydic actions, caused by the action of enzymes such as lipolipidic and proteolytic, generating greater muscle flaccidity as well as increased concentrations of volatile compounds and, finally, degradation by microorganisms (Ackman, 1989).

During rigor mortis, the musculature stiffens, triggering several cellular biochemical changes and for this reason, it is possible to determine the time that the organism took until its death, and this is very important for industries that use fish as raw material, for the guarantee of a fresh product, that is, the fish that receives only the conservative treatment of body temperature control, keeping it close to the ice and less advanced in the degradation phases (Badolato et al., 1994).

In view of this, fish is very susceptible to degradation due to its specific chemical constitution, its pH close to neutral scales, the supply of phospholipids, the high activity of water, and the action of tissue enzymes, which contribute to the elevation of the microbiota and consequently the degradation of the meat (Jiang et al., 2015).

The entry of microorganisms occurs more sharply in the gill portions, integument, and epithelium, as well as in the abdominal cavity, making handling even more cautious in these areas (Cribb & Filho, 2018).

In addition, there may be the presence of toxic substances in fish, precisely due to the action of endogenous enzymes, as mentioned above, and by the action of microorganisms that will act on the amino acids, promoting their decarboxylation and generating organic bases of aliphatic hydrocarbons, aromatic or low molecular weight (MW) heterocyclic structures, called biogenic amines (Neto & Gonçalves, 2011).

In the food field, the term biogenic amines encompasses substances such as histamine, tyramine, spermine, cadaverine, and putrescine, present in low amounts in food and necessary for the physiological functions of man and other animals, but can cause toxic effects if consumed in large amounts (Jiang et al., 2015).

In addition to their toxicological effects, biogenic amines are cause for concern in relation to food hygiene, as high amounts of certain amines can be found in human food as a consequence of using low-quality raw materials, contamination, and inadequate conditions during processing and storage of food.

The amount and type of biogenic amine formed will depend on the composition of the fish and the type of microorganism present (Moura, De Souza, & Filho, 2021). The formation of biogenic amines by bacteria is influenced by temperature, pH, and salt concentration. Low-temperature storage reduces the rate of biogenic amine formation in fish. The amino acid decarboxylase activity is stronger in an acidic environment, with an optimum pH between 4.0 and 5.5 (Soares & Gonçalves, 2012).

The biogenic amine index has been proposed as an indicator of freshness for fish and marine products (Peña, 2006). According to FDA (2001a), histamine is formed as a result of the abuse of the time/temperature binomial in certain fish species, that is, when the fish is exposed to high temperatures for a long time.

Rodrigues (2007) exposed tuna to high temperatures for short periods of time and found potentially toxic concentrations of histamine (67–382 mg/kg) in tuna muscle stored at 0°C–2°C for 12 days and exposed to 30°C for 2 hours daily. When it was stored at 6°C–7°C, and elevated to the same abusive temperature, the levels found were 544.5–4156.6 mg/kg, which is considered highly toxic.

Low levels of histamine are detected in freshly caught fish as its concentration increases with spoilage. In addition, knowledge of histamine levels in fish is useful in assessing its potential to cause histamine intoxication (Leitão, Baldini, Uboldieiroa, & Destro, 1983). The evaluation of the amount of histamine has been routinely used in the quality control procedures adopted by most fishing industries.

Delgado, Gutiérrez, and Hurtado (2003) studied diseases transmitted by the consumption of products of marine origin, fish and shellfish, and identified Staphylococcus aureus, *Vibrio parahaemolyticus*, histamine, paralyzing shellfish toxin, and *Escherichia coli* as the main causal agents.

In the distribution of these cases, intoxications by paralyzing shellfish toxins and histamine stand out. The danger of histamine in fish is intensified by its non-volatility characteristic. Histamine can confer toxicity to the product even before it is considered deteriorated or sensorially unacceptable (Leitão et al., 1983).

To determine the quality of the fish, microbiological, physical-chemical, and/or sensorial methods are used. Bogdanović, Šimat, Frka-Roić, and Marković (2012) reported that sensory evaluation is the main method to measure the quality of fish and is also the most used by both consumers and the industry because it is a quick, objective, and easy to understand.

Traditional conservation techniques such as refrigeration, freezing, salting and drying, continue to be studied and improved to improve the quality and increase the shelf life of fish (Wang, Ran, Ringø, & Zhou, 2018). However, Arancibia, Alemán, López-Caballero, Gómez-Guillén, and Montero (2015) highlight that traditional conservation techniques alone are not able to efficiently maintain fish quality. Thus, new technologies are being developed and improved, such as the use of active packaging.

5.5 Active packaging

Food packaging has the function of exerting protection and barrier to external factors such as light, oxygen, humidity, and microorganisms, providing safety to the consumer (Landim et al., 2016). Packaging has always been of great importance for the development of food commerce and distribution throughout history (Mestriner, 2002).

In the last decade, there has been a growing demand regarding food shelf life extension, product safety, and environmental issues.

Currently, packaging adds new functions, such as awakening the desire to buy, transmitting information, supporting promotional actions, and even more recently, there has been the use of active packaging. In addition to performing the basic functions of a barrier to the external environment, active packaging is designed to incorporate compounds that intentionally interact with food (Fang, Zhao, Warner, & Johnson, 2017), promoting antioxidant, antibacterial, CO_2 emission, ethylene, and oxygen absorption (Fig. 5.1), and thus, extend the shelf life of the products.

In accordance with European regulations 1935/2004/EC and 450/2009/EC, active packaging is designed to deliberately incorporate components that release substances into the food and surrounding space (headspace), as well as components that capture substances from the food and headspace (Restuccia et al., 2010).

The systems may be in contact only with the atmosphere around the food or in contact with the surface of the food (Dainelli, Gontard, Spyropoulos, Zondervan-van den Beuken, & Tobback, 2008). According to the active agent incorporation procedure, active packaging systems can be classified into three groups: Sachets, strips, and labels.

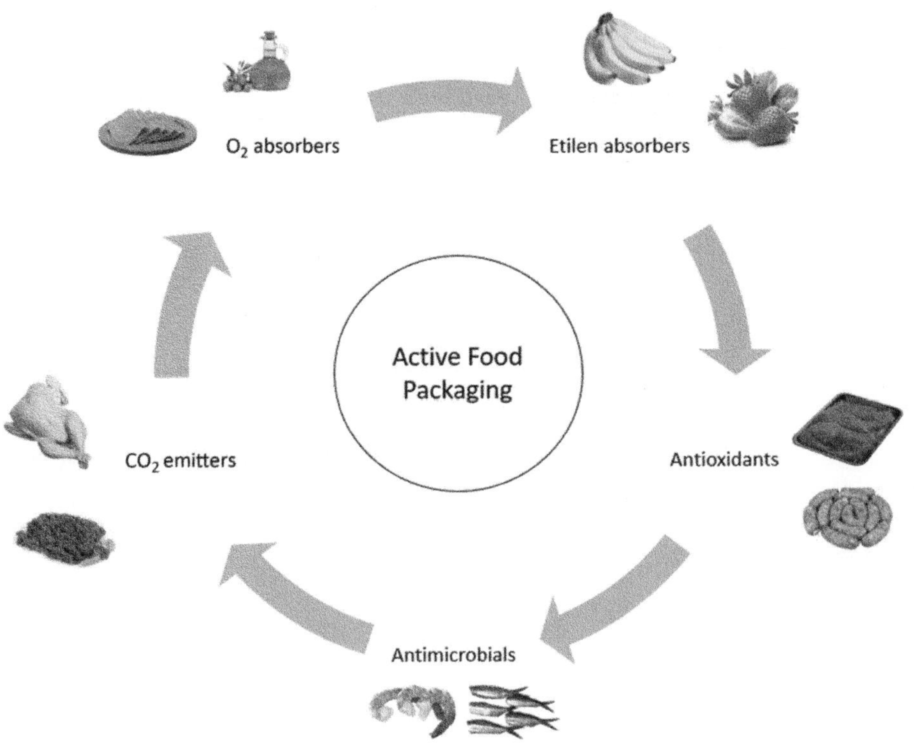

FIGURE 5.1 Different active food packaging. Source: *Adapted from Vilela, C., Kurek, M., Hayouka, Z., Röcker, B., Yildirim, S., Antunes, M. D. C., Nilsen-Nygaard, J., Pettersen, M. K., & Freire, C. S. R. (2018). A concise guide to active agents for active food packaging. Trends in Food Science and Technology, 80, 212–222. https://doi.org/10.1016/j.tifs.2018.08.006.*

5.5.1 *Sachets, strips, and labels*

Sachets, strips, and labels are systems that are not an integral part of the material that makes up the packaging but are found between the packaging and the food independently. They are usually designed with materials with one face permeable to the medium to allow mass transfer, either release or absorption. It is a widespread system, normally used for gas transfers (Ozdemir & Floros, 2004).

5.5.2 *Directly in polymer matrices (monolayer or multilayer)*

The incorporation of substances directly into the polymer matrix takes place at the time of extrusion, in which they fuse with the polymer and are inserted into the polymer network.

5.5.2.1 Added to film surface

In the film system, there is the dispersion of active substances that are in direct contact with the food. In this process, the substances adhere to the polymer when it is already formed, remaining on its surface (and not in the polymeric network). The great advantage of this system, in relation to the incorporation of substances in the matrix, is that the active substances are not exposed to critical conditions, such as high temperatures during packaging formation (Nussinovitch, 2009).

In Table 5.1, some examples of active packaging applied to fish are presented.

5.6 Antimicrobial active packaging

Active packaging with antimicrobial function has great potential, as they prevent or reduces the growth of microorganisms that cause food spoilage and, therefore, the active agent can prolong its shelf life (Realini & Marcos, 2014). This type of system is capable of interfering with the microbial growth phase (lag phase) and inhibiting it, due to the migration of active agents to the food. This type of packaging is very promising, as it avoids the high concentration of preservatives inserted directly into the food since they are incorporated into the polymeric matrix or are placed in sachets or in the form of labels inside the packages and released at a controlled rate (Restuccia et al., 2010). The migration of the antimicrobial active agent (Fig. 5.2) to packaged food can occur by

TABLE 5.1 Examples of active packaging applied to fish.

Fish	Active packaging	Storage temperature	Shelf life	Activity	References
Barracuda fish	Chitosan-based active film embedded with ginger	2°C	20 days	Antioxidant	
Fish fillet	Active film with essential oil of oregano	4°C	10 days	Antioxidant and antimicrobial	
Bluefin tuna fillet	Essential oil-based active films soy-montmorillonite-clove	2°C	12 days	Antioxidant and antimicrobial	Echeverría et al. (2018)
White prawns Pacific	Polyethylene bilayer containing rosemary and cinnamon essential oils	4°C	4 days	Antioxidant and antimicrobial	

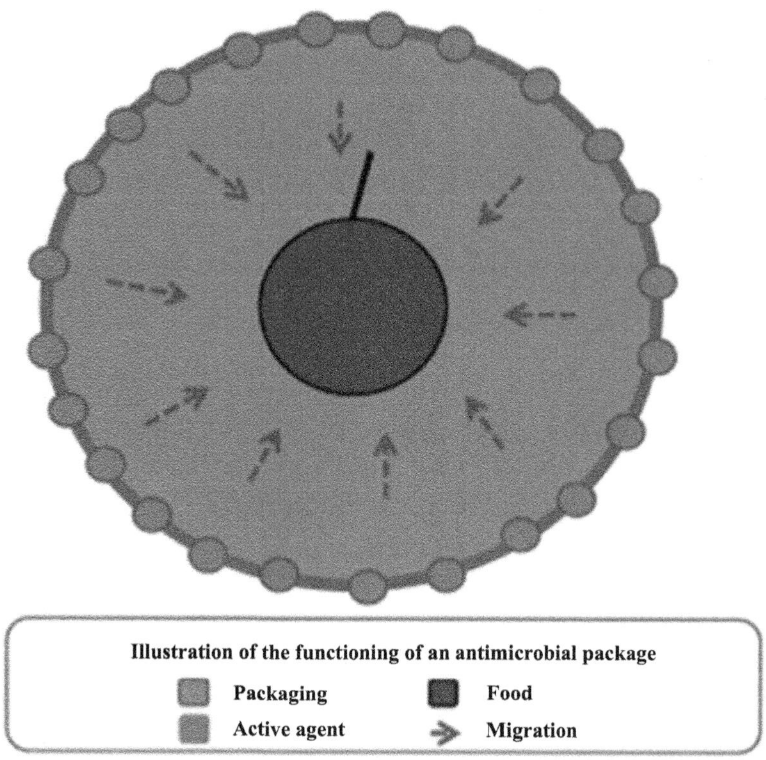

FIGURE 5.2 Illustration of the functioning of an antimicrobial package.

diffusion, dissolution, or equilibrium and can be affected by several factors: film production method, temperature, polarity, volatility of substances, food properties, and the interaction between the active agent and the polymer (Fang et al., 2017). Fig. 5.2 represents the functioning of an antimicrobial active packaging is presented.

5.7 Compounds used in active food packaging

5.7.1 *Zinc oxide*

Inorganic metal oxides have gained greater attention because they offer important advantages, such as lower toxicity and less ability to generate resistance against microorganisms, due to their mechanisms of action and promising bactericidal potential (Lallo da Silva, Caetano, Chiari-Andréo, Pietro, & Chiavacci, 2019; Padmavathy & Vijayaraghavan, 2008). Among the metal oxides, the most studied are ZnO and titanium dioxide (TiO_2) due to their broad spectrum of activity against viruses, bacteria, and fungi (Esmailzadeh, Sangpour, Shahraz, Hejazi, & Khaksar, 2016). Among the metallic oxides, ZnO has gained greater attention for its versatility and multifunctionality. ZnO has interesting physical and chemical properties that provide numerous applications ranging from environmental sciences to biomedical, electronics, or food (Abolghasemi-Fakhri, Ghanbarzadeh, Dehghannya,

Abbasi, & Adun, 2019). The type of nanostructure formed can be controlled through changes in the synthesis parameters. The properties of materials containing ZnO have been the subject of numerous researches and this oxide is widely known to confer excellent properties such as anticancer, anti-inflammatory, drug release modulation, bioimaging, tissue regeneration (Cierech et al., 2019), and UV protection and self-cleaning (Abolghasemi-Fakhri et al., 2019). Among the applications of ZnO–NPs, the antimicrobial activity, which has been widely reported, stands out. A very interesting factor that enables the use of ZnO–NPs is the low cost from the production point of view, in addition, this material has low toxicity, attracting greater attention to biological applications. ZnO–NPs are listed as "generally recognized as safe" by the US FDA. The bactericidal mechanisms of ZnO–NPs are multiple and/or simultaneous against bacterial cells, thus, the resistance of microorganisms against ZnO–NPs would require several simultaneous genetic mutations within the same cell. Therefore, bacterial cells can hardly become resistant to the nonspecific mechanisms of nanoparticles (Yadav, Mehrotra, & Dutta, 2021). Although ZnO–NPs exhibit excellent antibacterial properties, it is not possible to elucidate the main mechanism of action. The works discuss that antibacterial mechanisms occur through chemical or physical interactions. Physical interactions offer biocidal effects through cellular internalization of ZnO–NPs, damage and/or disruption of the cell envelope and chemical interactions of ZnO–NPs lead to the production of reactive oxygen species (ROS) and release of Zn^{2+} ions, causing the death of bacterial cells (Esmailzadeh et al., 2016). The fixation of ZnO–NPs on the surface of Gram-positive and Gram-negative bacteria is responsible for disorganization and membrane changes allowing greater permeability, as well as an increase in the concentration of ZnO–NPs internalized in the bacterial cell (Yadav et al., 2021). Several studies have reported the antibacterial effect of ZnO-incorporated packaging for application in food. Kumar et al. (2020) developed chitosan and gelatin hybrid nanocomposite films containing green synthesized zinc oxide nanoparticles (ZnO–NPs) and evaluated the microstructural properties. Disk diffusion assays showed that the nanocomposite film showed significant antimicrobial activity against *E. coli*. The hybrid nanocomposite films developed showed potential to be developed as a biodegradable alternative for food packaging. Jayakumar et al. (2019) produced starch-PVA-based composite films incorporated with ZnO–NPs and phytochemicals by the solvent casting technique, these have been shown to have improved water barrier, as well as mechanical and antimicrobial properties. The unique characteristics of the nanocomposite with its pH-indicating property demonstrated in the study indicated its potential use in food packaging applications. Al-Naamani, Dobretsov, and Dutta (2016) studied the antimicrobial properties of chitosan and chitosan-ZnO nanocomposite coatings on polyethylene (PE) films. The authors observed that PE coated with chitosan-ZnO nanocomposite films completely inactivated and prevented the growth of food pathogens, while films coated with chitosan without the addition of ZnO showed only a 10-fold decline in viable cell counts of *Salmonella enterica*, *E. coli* and *S. aureus* after 24 hours of incubation compared to the control. Liu et al. (2018) produced ethylcellulose/gelatin nanofibers containing different concentrations of ZnO, and the observed results suggested that the nanofibers containing ZnO showed excellent surface hydrophobicity, water stability, and antimicrobial activity against *E. coli* and *S. aureus*.

Shahvalizadeh et al. (2021) produced antimicrobial nanocomposites based on gelatin and tragacanth with 1%, 3%, and 5% ZnO–NPs. The nanocomposites containing ZnO–NPs had good antimicrobial activity (~10–20 mm) against *S. aureus* and *E. coli*. Baek and Song (2018) produced antibacterial films with extracts of *Gracilaria vermiculophylla* (GVE). Equal amounts of glycerol and sorbitol were added as plasticizers and ZnO–NPs were added to prepare the antibacterial films. Physical and optical properties, ultraviolet (UV) light barrier, scanning electron microscopy (SEM) analysis, and antibacterial activity of the films were evaluated, and the films were applied as packaging material for smoked salmon. The viable cell count indicated that the antimicrobial activity of the films increased with increasing ZnO–NPs content. After 9 h, the population of *Listeria monocytogenes* and *Salmonella typhimurium* decreased by 4.99 and 5.08 log CFU/mL for films containing 5% ZnO–NPs, respectively, compared to films without ZnO–NPs. Smoked salmon packed with GVE films containing 3% ZnO–NPs resulted in increased antibacterial activity against *L. monocytogenes* and *S. Typhimurium* compared to control during storage, as well as a low degree of lipid oxidation. Therefore, these results indicate that GVE films containing ZnO–NPs have potential as active food packaging material to maintain the quality of smoked salmon.

5.7.2 Titanium dioxide

Being considered biologically and chemically inert, TiO_2 is corrosion-resistant and non-toxic (Yoganarasimhan & Rao, 1962). With a wide range of applications, TiO_2 is used in sunscreens, and white paints and is approved by the Food and Drug Administration (FDA) for application in the food industry. TiO_2 has three distinct

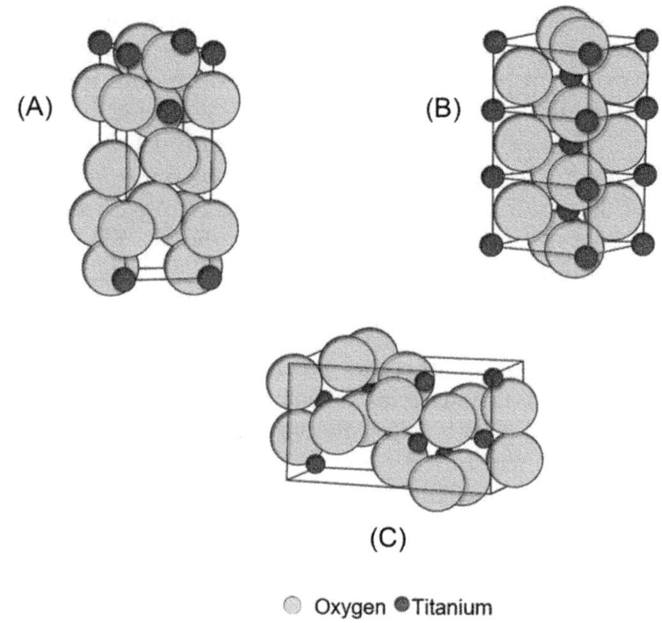

FIGURE 5.3 Structures of the crystalline phases of titanium dioxide: (A) anatase (B) rutile (C) brookite. Source: *Suwa Y., Inagaki, M., & Naka, S. (1984). Polymorphic transformation of titanium dioxide by mechanical grinding. Journal of Materials Science. 19(5), 1397–1405.*

crystalline conformations, namely, anatase, rutile, and brookite, as shown in Fig. 5.3 (Czanderna, Ramachandra Rao, & Honig, 1958).

Both anatase and rutile phases can absorb UV rays. Rutile can also absorb at wavelengths close to the visible, however, the anatase phase has greater photocatalytic activity (Suwa, Inagaki, & Naka, 1984). TiO_2 is considered the most widely used and recognized semiconductor in photocatalysis. This is because it is an excellent photocatalyst in the degradation of organic compounds (Shayegan, Lee, & Haghighat, 2018), viruses (Monmaturapoj, Sri-On, Klinsukhon, Boonnak, & Prahsarn, 2018), bacteria (Feng et al., 2017), fungi (Mukherjee, Acharya, Biswas, & Jana, 2020), and tumor cells (Salahuddin, Abdelwahab, Gaber, & Elneanaey, 2020). The principle of photocatalysis involves activation of a semiconductor (usually TiO_2) by artificial light. A semiconductor is characterized by valence bands (BV) and conduction bands (BC) and the region between them is called a band gap.

When TiO_2 is irradiated at a wavelength capable of activating its photocatalytic action, electrons are transferred from the valence band to the conduction band, producing an unpaired (e-) electron pair. This is followed by a series of reduction and oxidation reactions resulting in the inactivation of bacteria, for example (Almeida, 2011). To form an electron pair, it is necessary that the absorption of photons with energy greater than or equal to the band gap energy occur to move the electron from the valence band to the conduction band, generating an electron-hole (h +) in the valence band (Maneerat & Hayata, 2008).

Currently, some works have evaluated the photocatalytic properties of TiO_2 when applied to food packaging surfaces exposed to UV-C light, mainly with an emphasis on its antibacterial action. As Teymourinia, Salavati-Niasari, Amiri, and Yazdian (2019) developed a nanocomposite of TiO_2/Sb_2S_3/GQDs, it was prepared by the solvothermal method to study its antibacterial properties against *E. coli* and *S. aureus*. The authors observed that this nanocomposite showed antibacterial capacity against *E. coli* and *S. aureus*. Sievers, Pollo, Corção, and Medeiros Cardozo (2020) produced antibacterial polypropylene/titanium dioxide (PP/TiO_2) films through the in situ synthesis of nanostructured TiO_2 directly in the polymer solution. The properties of PP/TiO_2 particles and films were correlated with antibacterial activity against Gram-positive (*L. monocytogenes* and *S. aureus*) and Gram-negative (*S. typhimurium* and *E. coli*) bacteria. The results showed a reduction of about 90% of the bacterial population. The *in situ* method proved to be a simple way to obtain PP/TiO_2 films in a single step and can be used in the manufacture of active packaging with antibacterial action for Gram-positive and Gram-negative bacteria.

Tang et al. (2018) produced an antibacterial sodium alginate film added to Au–TiO_2 nanocomposites. The Au–TiO_2 nanocomposites were hydrothermally synthesized and mixed with the alginate solution to form the film by the casting method. The film showed antibacterial capacity against *E. coli* and *S. aureus*.

5.7.3 Silver

Silver and its compounds have long been used as antimicrobial agents. Its effects have been known since 1000 BC Even before this time, silver and copper vessels were already used to conserve water potability. In addition to silver, many other metals were studied, resulting in the following toxicity scale against microorganisms: Ag > Hg > Cu > Cd > Pb > Co > Au > Zn > Fe > Mn > Mo > Sn (Jafari, Pirouzifard, Khaledabad, & Almasi, 2016).

Duncan (2011) presented evidence that silver ions, when crossing the bacterial membrane and penetrating the interior of cells, interfere with ribosomal activity and interrupt the production of enzymes responsible for energy production in the cell. Possibly, this mechanism occurs through the interaction of silver nanoparticles with the –SH groups of proteins. With action similar to antimicrobial agents commonly classified as sulfonamides. Such proteins project through the bacterial cell membrane, allowing the transport of nutrients through the cell wall.

Monovalent silver ions (Ag^+) replace the hydrogen cation (H^+) of groups (–SH) or thiol groups, inactivating the protein and thus decreasing membrane permeability, which can eventually cause cell death (Abreu et al., 2015).

Fabrega, Fawcett, Renshaw, and Lead (2009) studied the generation of ROSs by silver nanoparticles within bacterial cells, and the generation of these free radicals induces cellular oxidative stress. Fig. 5.4 shows the scheme of the mechanisms of action of silver nanoparticles and silver ions in the fight against microorganisms.

Davies and Etris (1997) demonstrate that only cells containing (–SH) groups on the surface were affected by silver. This would possibly explain why only bacteria and viruses are affected by silver ions, contrary to what happens to the cells of eukaryotic organisms. These cells, because they do not have groups (–SH) exposed on the cell surface, are not affected by AgNP. Therefore, AgNPs are not able to permeate through cell membranes in the absence of –SH groups, making silver relatively non-toxic to humans and animals (Božanić et al., 2011; Cho, Park, Osaka, & Park, 2005).

The possibility of free radical formation being involved in the antibacterial activity of AgNPs was also studied by Soo-Hwan et al. (2011). These authors also evaluated the formation of ROS through the spectrophotometric reading of the fluorescence intensity in order to prove the relationship between the antibacterial activity of AgNP with the formation of free radicals.

In this sense, free radicals generated by AgNP induce cell damage in the bacterial membrane. Under such conditions, high levels of ROS can increase oxidative stress in cells. Oxidative stress can not only damage the cell membrane, but it can also damage proteins, DNA, and intracellular systems such as the respiratory system. After 3 hours of incubation of AgNP solutions with *S. aureus* and *E. coli* bacteria, the authors were able to observe an increase in ROS. Thus, it may indicate that AgNPs can form ROS with water, and then, the cell membrane.

Bumbudsanpharoke, Lee, and Ko (2018) evaluated the effect of silver nanocomposites (AgNPs) on the properties of low-density polyethylene (LDPE) for application in food packaging. The investigation included the study of optical, thermal, mechanical and barrier properties, as well as antimicrobial activity against Gram-positive and Gram-negative bacteria.

The results revealed that spherical AgNPs were distributed both on the surface and inside the LDPE film. While these well-dispersed AgNPs increased the barrier properties of LDPE to oxygen and water vapor, the incorporation of AgNPs affected the thermal properties, reducing the thermal stability and enthalpy of melting of the LDPE film. In addition, the LDPE-AgNP composite film showed effective antibacterial activity against

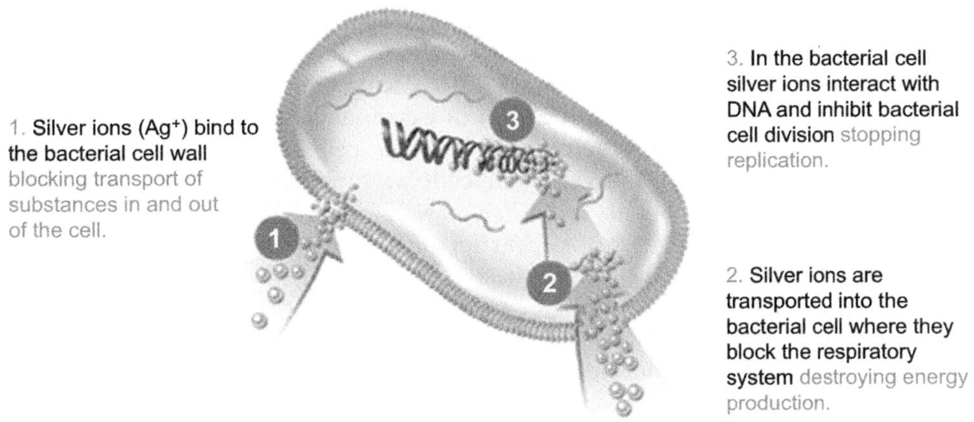

FIGURE 5.4 Silver íons action on the bacterial cell.

S. aureus and *E. coli*. These findings suggest that LDPE-AgNP composite film can be used in a viable way in food packaging to preserve food quality and extend food shelf life.

Trbojevich et al. (2020) evaluated the silver release from commercially available food-contact materials in food simulants (water, acetic acid, ethanol-water, and olive oil) and meats (tuna, ham, and turkey). Additionally, the capacity of migrated silver was examined in meats. Largest silver release was observed in simulants from food touch papers (25 ± 11 mg/kg) as compared to bag, cutting board, and containers. Silver ions and silver nanoparticles were released from food touch paper in food simulants. Food touch paper released the highest amount of silver in tuna (0.5 ± 0.02 mg/kg) than in ham (0.2 ± 0.08 mg/kg) or turkey (0.3 ± 0.08 mg/kg) in the same conditions. Tuna exhibited the lowest pH and highest number of bacterial populations on day 0 compared with other foods. Nonetheless, the significant antibacterial capacity of released silver was noticed predominantly in Turkey for Gram-negative bacteria. These results suggest that silver released in food simulants indicates an overestimation of silver migration; thus, precaution should be maintained when extrapolating such findings to "real" food.

Tripathi, Pudake, Shrivastav, and Shrivastav (2018) promoted the synthesis of a biogenic silver poly (vinyl alcohol)-biogenic antibacterial biodegradable nanocomposite film. Biosynthesized silver nanoparticles were used for the development of the nanocomposite film. *Ficus benghalensis* leaf extract was used to synthesize silver nanoparticles which is an economical, ecological, and fast method. The biogenic silver PVA nanocomposite film was characterized by AFM to determine the surface morphology of the film and confirm the incorporation of silver nanoparticles.

The antibacterial activity of the nanocomposite film was investigated against *S. typhimurium* using the disk diffusion method. The biogenic silver PVA nanocomposite film showed excellent antibacterial activity against *S. typhimurium*.

Ravindran et al. (2020) synthesized silver nanoparticles using rapid biological methods from *Capparis zeylanica* leaf extract, which provides an environmentally friendly, simple, and efficient route for synthesis. The antimicrobial study was investigated against different pathogenic bacteria. This nanocomposite showed antibacterial capacity against *E. coli*, *Pseudomonas aeruginosa*, and *S. aureus*.

5.7.4 Bioactive compounds

Essential oils consist of terpenoids that have gained prominence as antimicrobial agents. These are natural substances of plant origin that can be obtained by physical processes, such as reduced pressure distillation or steam distillation (Zhang, Hortal, Dobon, Bermudez, & Lara-Lledo, 2015). Several studies have related the application of essential oils in active food packaging aiming at antimicrobial action (Alves et al., 2018; Echeverría, López-Caballero, Gómez-Guillén, Mauri, & Montero, 2018; Navarro-Segura, Ros-Chumillas, Martínez-Hernández, & López-Gómez, 2020).

Essential oils are liquid, complex, bioactive, volatile compounds, with characteristic odor and color, formed from secondary plant metabolites, present in all plant organs, such as buds, flowers, leaves, stems, branches, seeds, fruits, and shells (Licciardello, Muratore, Suma, Russo, & Nerín, 2013; Priyadarshi et al., 2018; Ribeiro-Santos, Andrade, Melo, & Sanches-Silva, 2017). They are mainly formed by classes of fatty acid esters, mono and sesquiterpenes, terpenes, phenylpropanones, and aldehyde alcohols (Kaur & Santhiya, 2021; Sharma, Barkauskaite, Duffy, Jaiswal, & Jaiswal, 2020).

More than six thousand years ago, its medicinal properties were already known to the Egyptians, but only from the Middle Ages, with the discovery of its antimicrobial properties, essential oils began to be extracted and marketed by the Arabs, who would have been the first to develop methods such as steam distillation and steam distillation to obtain these compounds (Echeverría et al., 2018; Zhang et al., 2015).

It is important to understand the mechanism of action of essential oils, given that this is exerted by morphological and functional mutations of the cell structure. On bacteria, cell wall degradation occurs, which, with the disruption of this permeable barrier, compromises their cellular functions, including metabolic regulation and maintenance of energy status (Licciardello et al., 2013; Priyadarshi et al., 2018).

For fungi, essential oils can affect the protective barrier of spores or act before their formation, promoting abnormal development and/or rupture of the germ tubes.

Alves et al. (2018) developed chitosan films with grape seed extract and carvacrol microcapsules (CMF), and their physicochemical properties and effect on physicochemical and microbiological parameters in refrigerated salmon were tested. CMF showed higher values of thickness (0.41 ± 0.04 mm), moisture content (13 ± 1 g water/100 g film), a (11 ± 3), b (12 ± 3), opacity ($20 \pm 1\%$), water vapor permeability (WVP) $(4.4 \pm 0.4) \times 10^{-10}$ Pa/s, oxygen permeability (O_2P) $(1.3 \pm 0.3) \times 10^{-12}$ Pa/s and carbon dioxide permeability (CO_2P) $(1.3 \pm 0.3) \times 10^{-12}$ Pa/s compared to chitosan control film (CCF).

CMF showed lower values of L (66 ± 5) and water solubility (17 ± 1%). Salmon packed in CMF showed lower TVB-N value on the 7th day of storage (37 ± 4 mg N/100 g fish) in relation to CCF (42 ± 3 mg N/100 g fish) and control samples (CS; 66 ± 7 mg N/100 g of fish). CMF showed lower pH and luminosity values after 7 days of storage compared to CS and CCF. CMF also showed lower values of mesophilic and psychrophilic bacteria (TPBc) and *Pseudomonas* spp., reaching the maximum limit allowed for the first two only on the 7th day of storage.

CMF increased the shelf life of refrigerated salmon from 4 to 7 days of storage due to the antimicrobial effect of natural agents. The films with microcapsules showed higher permeabilities to water vapor, O_2, and CO_2. The films with microcapsules showed higher values of *a* and *b*, opacity, and moisture content. The microcapsules increased the antimicrobial activity of the chitosan film. The film developed extended the shelf life of chilled salmon.

Echeverría et al. (2018) evaluated the potential of application of active nanocomposite films based on soy protein isolate (SPI)-montmorillonite (MMT)-clove essential oil (CEO) for the preservation of bluefin tuna (*Thunnus thynnus*) muscle fillets during cold storage.

In addition, they analyzed whether the clay diffuses from the packaging into the food. SPI films with CEO (SPI-CEO), MMT (SPI-MMT), or both CEO and MMT (SPI-MMT-CEO), were prepared and used to cover tuna fillets during 17 days of storage at 2°C. Polyethylene films were also used as a control.

Nanoreinforced protein films with 10 g MMT/100 g SPI and activated with CEO were able to decrease microbial growth [assessed by total volatile bases (TVBN) and microorganism count] and lipid autooxidation (assessed according to TBA index, FTIR and color parameters) of fillets of tuna during the storage period studied.

The presence of clay seemed to favor the release of the active principles of clove oil, prolonging its antimicrobial activity (especially effective to inhibit *Pseudomonas* spp.) nanocomposite materials for fish muscle. The nanocomposite films with CEO (SPI-MMT-CEO) improved the quality of tuna in storage at 2°C. The SPI-MMT-CEO films decreased *Pseudomonas* growth and lipid autooxidation. Nanoclay minerals (Si and Al) did not diffuse from the nanocomposite films for fish muscle. MMT favored the release of active ingredients from the CEO, prolonging the shelf life of the tuna.

Navarro-Segura et al. (2020) developed and evaluated the application of a package system of oregano essential oil vapors (OEO) under high vacuum conditions to explore the antimicrobial effect of essential oil vapors applied immediately before packaging fish fillets.

Fresh fillets of farmed sea bream (*Sparus aurata*) have been used as a model to validate this new technology. These fillets, as a refrigerated product under modified atmosphere packaging (MAP), have a relatively short shelf life (12 to 14 days), mainly due to rapid microbial growth.

The effects of conventional OEO dips [pretreatment dip (0.1% OEO) of whole fish (T1) and sea bream fillets (T2)] were compared with the application of vapor phase OEO (67 μL/L) under vacuum (5–10 hPa) immediately before packing the MAP fillet (T3). Samples T3/T2 showed the lowest microbial growth after 28 days at 4°C, with loadings of up to 1/2.6 log units for enterobacteria/lactic acid bacteria compared to untreated samples.

The initial content of trimethylamine nitrogen (TMA-N) (2.6 mg/kg) increased in samples T1 and T2/T3 by 9.6 and 6/7 units, respectively, after 28 days. The quality index method (QIM) better reflected the shelf life of fish fillets than the texture and color measurements. The shelf life for T3/T2 samples was set at least 28 days (4°C), while the QIM threshold (6) was exceeded after 7/21 days in untreated/T1 fillets. Fish shelf life was extended with OEO steam treatment using this new technology, similar to OEO immersion treatment, according to QIM, supported by microbial quality and TMA-N contents.

Maghami, Motalebi, and Anvar (2019) evaluated the effect of chitosan nanoparticles (CNPs) loaded with fennel essential oils together with the MAP system on the chemical, microbial and sensory properties of Huso huso fish fillets during refrigerator storage.

The results showed that coating fish fillets with CNPs and fennel EO significantly reduced the peroxide value, total volatile nitrogen, and thiobarbituric acid value compared to control samples. Microbial analyses showed lower numbers of mesophilic, psychotropic, pseudomonas, and lactic acid bacteria in the coated fillets compared to the control and MAP packaging.

Fish fillets coated with CNPs and EO showed high acceptability in all sensory attributes during storage. It can be concluded that the use of fennel CNPs and EO together with MAP packaging can increase the shelf life of H. huso fillets up to 18 days in the refrigerator. CNPs and fennel EO delayed the chemical deterioration of fish fillets during storage. The use of CNPS and fennel OE reduced the microbial number in the fillets compared to the control. Fennel CNPs and EO improved the sensory properties of Huso huso fillets.

Kunová et al. (2021) added Citrus lemon and *Cinnamomum camphora* essential oils to rainbow trout meat to assess the microbiological quality (bacteria count and identification of the microbiota present) of the fish when vacuum packed and stored for 7 days at 4°C. The results showed that the essential oils of lemon (0.5% and 1%) and *C. camphora* (0.5% and 1%) had a positive effect on the microbiological quality of fish meat, maintaining a high microbial quality of the fish fillets during 7 days of cold storage. Using these essential oils in combination with vacuum packaging has been shown to be beneficial in extending the shelf life of rainbow trout meat.

Del Nobile et al. (2009) analyzed the combined effects of three essential oils [thymol, lemon extract, and grapefruit seed extract (GFSE)] and MAP conditions on quality retention of blue fish burgers were studied and discussed.

In particular, the samples were packed in air and in three different gas mixture compositions: 30:40:30 O_2:CO_2:N_2, 50:50 O_2:CO_2, and 5:95 O_2:CO_2. During a storage period of 28 days at 4°C, the nutritional, microbiological, and sensory quality of the packaged products were evaluated.

The developmental potential of biogenic amines was also evaluated. The results obtained highlight the possibility of improving the microbial quality of bluefish burgers using a very small amount of thymol (110 ppm), GFSE (100 ppm), and lemon extract (120 ppm) in combination with MAP.

Based mainly on the microbiological results, the combined use of the tested natural preservatives and a packaging system characterized by a high concentration of CO_2, was able to guarantee the microbial acceptability of the fish burgers up to the 28th day of storage at 4°C.

On the other hand, the results of the sensory analyzes showed that the sensory quality was the sub-index that limited the shelf life of the hamburgers (about 22 to 23 days), even though the proposed strategy was also effective in minimizing the loss of sensory quality of the hamburgers; product not affecting their nutritional quality.

Ehsani, Jasour, Hashemi, Mehryar, and Khodayari (2014) analyzed the preservative effects of *Zataria multiflora* Boiss essential oil (ZEO) at 0.025%, 0.05%, and 0.1%, sodium acetate (SA) at 2%, and their combination on quality changes in vacuum packed trout burgers during 21 days of refrigerated storage (4°C ± 1°C).

The results showed that both control and ZEO-treated samples reached undesirable levels of rancidity ($P < .05$) at day 15, while samples with the combined effect of SA and ZEOs, especially at higher concentrations, proved to be more stable in terms of sensory, biochemical (PV), TBA, AGL, and pH and microbiological analyzes ($P < .05$).

Therefore, samples treated with ZEO + SA showed good general acceptance until the 21st day. Regarding the results obtained in the samples treated with ZEO, it appears that the combined application of SA and ZEO extended the shelf life of fish burgers during cold storage for 21 days, 1.4 times longer than other treatments, which indicates the potential application of the synergistic activity of these agents in vacuum-packed fish burgers.

Da Rocha et al. (2018) investigated the effect of protein hydrolyzate (PH) and CEO on agar film properties and on improving the shelf life of sole fillets. First, Argentine croaker muscle protein was solubilized in alkali and recovered to obtain a stable substrate. This substrate was hydrolyzed to 20% degree of hydrolysis, by Alcalase (A20) and Protamex (P20).

The MW distribution, amino acids, and in vitro antimicrobial activity of these hydrolysates were tested. Agar films were prepared with and without (control) incorporation of PH (A20 hydrolyzate) or CEO and characterized by mechanical properties, water vapor permeability, solubility, and optical properties. The microbiological, physical, and chemical properties of sole fillets coated with these films were monitored during storage at 5 °C for 15 days.

Compared to P20, the A20 hydrolyzate showed lower MW, higher hydrophobic amino acid content, and the most effective antimicrobial activity against selected pathogens such as *S. aureus*, *Yersinia enterocolitica*, *Aeromonas hydrophila*, *Debaryomyces hansenii*, and *Listeria innocua*.

The presence of PH increased water solubility, water vapor permeability, elongation at break, and yellowing of the films. CEO films showed lower transparency than PH films and control films, due to the presence of oil in the formulations. The fillets covered with the PH and CEO films, especially the latter, presented lower volatile bases and total pH values and considerably retarded the growth of H_2S-producing microorganisms in relation to the control films. CEO can be used as a natural biopreservative to extend the shelf life of the sole. However, more studies are needed to evaluate the antimicrobial effect of PH when used in food packaging.

Phenolic compounds, derived from the secondary metabolism of plants, also have antibacterial activity, they belong to a class of molecules with a great structural diversity, characterized by having one or more aromatic rings that also have one or more hydroxyl groups directly associated with the ring structure, which can vary from simple phenolic molecules to highly polymerized compounds (Fennema et al., 2010).

They are compounds derived from secondary metabolism that do not participate in metabolic pathways responsible for growth and reproduction and, therefore, their nature and concentration vary greatly.

They are divided into two groups: non-flavonoids and flavonoids. Flavonoids are those with the chemical structure described as C6-C3-C6 consisting of two phenolic rings (A and B), linked to a pyran center (containing oxygen) ring (C) (Fonseca et al., 2021; Menzel, González-Martínez, Chiralt, & Vilaplana, 2019).

Variations in C-ring substitution result in different classes of flavonoids, such as flavonols, flavones, flavanones, flavanols, isoflavones, and anthocyanidins. Non-flavonoids are composed of phenolic acids, stilbenes, and hydrolyzable tannins (C6-C2-C6) (Rojas-Lema et al., 2021).

These compounds have a deteriorating effect on the lipid bilayer of the bacterial cell membrane, resulting in the loss of cell structure and, eventually, its function, leading to cell death (Pereira et al., 2007).

The presence of gallic acid esters is responsible for its high affinity to the lipid layer affecting the membrane structure (Menzel et al., 2019; Rojas-Lema et al., 2021). Other phenolic components such as epicatechin, caffeic acid, and benzoic acid can alter cell morphology, influencing the osmotic pressure of the cell, disrupting the cytoplasmic membrane, and causing leakage of cellular constituents (Fu, Lu, & Zhou, 2016).

The use of phenolic extracts from plants, fruits, leaves, and vegetable residues in active packaging may represent an alternative way to maintain food safety and good sensory quality, without resorting to synthetic preservatives.

Basiri, Shekarforoush, Aminlari, and Akbari (2015) evaluated the effect of pomegranate peel extract (PPE) on polyphenol oxidase (PPO) enzyme activity and quality of Pacific white shrimp (*Litopenaeus vannamei*) during 10-day cold storage.

The inhibitory activity of PPE against shrimp PPO appeared to follow a dose-dependent manner. With storage, treatment of shrimp with 1 g/100 mL or 2 g/100 mL of PE solution considerably decreased counts of mesophilic, psychrophilic, lactic acid (LAB), and *enterobacteriaceae* bacteria ($P < .05$).

With storage as well, treatment of shrimp with PE solution at both levels can significantly decrease total volatile basic nitrogen (TVB-N), trimethylamine (TMA), and thiobarbituric acid (TBA) ($P < .05$).

On day 6, PPE-treated shrimp had lower melanosis scores, but on day 10, odor, texture, color, and overall similarity received higher sensory scores. Thus, PPE showed potential as a preservative method and inhibitor of shrimp melanosis and could replace sodium metabisulfite (SMS), commonly used in this type of food.

Zhuang et al. (2019) evaluated the effects of aqueous pomegranate peel extract (APPE) and ethanolic pomegranate peel extract (EPPE) on the microbiota and on changes in the quality of bighead carp (*Aristichthys nobilis*) fillets stored at 4°C.

The results showed that pomegranate peel extract (PPE, which includes both APPE and EPPE) delayed the deterioration of sensory quality and pulp color, inhibited the growth of spoilage bacteria and attenuated the production of biogenic amines, TVB-N, and the degradation of ATP-related compounds.

In addition, EPPE performed better on color attributes and biogenic amines, but APPE was more effective in delaying the rise of TVB-N and K-value. High-throughput sequencing results showed that the microbial composition of all samples become less diverse as storage time increased.

For the control group, *Acinetobacter* was predominant in the middle of storage, while *Pseudomonas*, *Aeromonas*, and *Shewanella* became predominant at the end of storage. Furthermore, PPE decreased the relative abundance of *Acinetobacter* in the middle of the storage period and therefore altered the microbial composition.

Based on our quality assessments and microbial analysis, EPI extended the shelf life of bighead carp fillets by about 2 days and has the potential to become a promising preservative in aquatic products.

Yuan, Lv, Tang, Zhang, and Sun (2016) evaluated the effect of chitosan coating combined with PPE on melanosis and quality of Pacific white shrimp (*L. vannamei*) during 10 days of storage at 0°C.

Melanose and total color difference changes (ΔE values) were significantly delayed, and texture parameters and sensory scores were significantly improved in shrimp treated with chitosan coating, PPE, and chitosan coating combined with PPE, compared with the control. The increase in total aerobic plaque counts, pH, and total volatile basic nitrogen values were significantly inhibited in shrimp treated with chitosan coating, PPE, and chitosan coating combined with PPE.

Melanose and sensory scores, total volatile basic nitrogen values, and total aerobic plaque counts of shrimp treated by chitosan coating combined with PPE were lower than those treated by chitosan coating or PPE alone during the later storage stage at 0°C, suggesting that there is a synergistic effect between the chitosan coating and PPE.

Khaledian and Basiri (2022) analyzed the ability to replace 50% of SMS by coating tragacanth (TG) incorporated with Persian lemon peel extract (PLPE) in extending the shelf life of shrimp during refrigeration.

The incorporation of PLPE and half dose of SMS in the TG coating increased the shelf life of the shrimp, as demonstrated by the delay in microbial growth for about 6 days; reduced production of TVBN, trimethylamine (TMA-N) and index of rancidity (TBA) by about 90%, 75% and 50%, respectively; and delayed melanosis and sensory changes. The authors concluded that the TG coating incorporated with PLPE and half a dose of SMS can serve as an active coating for shrimp conservation.

5.8 Nanotechnology

The prefix nano of the term nanotechnology refers to the 1 to 100 nm scale, but also denotes the techniques used for manipulating elements at the atomic scale (Duncan, 2011). The fields of nanoscience and nanotechnology are relatively recent in the food sector when compared to the area of drug delivery, smart nutrient delivery, rapid sampling of biological contaminants, protein bioseparation, and nanoencapsulation of nutraceuticals (Peidaei, Ahari, Anvar, & Ataee, 2021).

Nanotechnology has already been applied in various fields such as computer electronics, energy production, and communication. Nanomaterials are produced with the aim of mimicking nanodevices found in nature and include DNA, proteins, membranes, and other biomolecules (Mihindukulasuriya & Lim, 2014).

Since most of the nanoparticles that are applied generally belong to the group of colloids (emulsions, micelles, mono, and bi-layers). One of the first colloidal dispersions to be produced was developed by Faraday, in the mid-eighteenth century (Siegrist, Stampfli, Kastenholz, & Keller, 2008); in this context, particles are attracted through Van der Waals forces, which provide colloidal stability. In the case of colloidal particles, steric stabilization is a consequence of the adsorption of polymers and surfactants on the surface.

Nanoparticles can be further stabilized when coating them with molecules that can form chemical bonds. Several studies have pointed out the current and projected short-term applications of nanotechnologies in the food sector, with the main application areas including food packaging and food products that contain nanoencapsulated active ingredients in their formulation (Adeyeye, 2019).

The application of nanotechnology in food has potential that seems to be unlimited, taking into account all facets of the food industry, from ingredients to packaging to analysis methodologies are seeking nanotechnological applications, contributing to the improvement of production, processing, storage, and identification of bacteria and monitoring through smart packaging that has nanosensors that respond to environmental changes such as temperature, oxygen and humidity exposure levels (Alfadul & Elneshwy, 2010).

When the particles have a nano scale, they have different behavior in relation to the properties of macroparticles of the same substance, such as in relation to thermodynamic properties, electrical conductivity, and even color. Nanotechnology can be applied in several stages of the production chain, from agriculture (primary sector), where it can be used in the delivery of pesticides and fertilizers, in the control of water quality, from silver nanoparticles, and in the controlled release of plant growth hormones (Silvestre, Duraccio, & Cimmino, 2011).

In the secondary sector, nanoparticles are generally applied in foods to mask undesirable odors or flavors, mitigate ingredient interactions, control the release, and protect the interest's activness (Emamhadi et al., 2020).

In a survey carried out by Kuang, Burgess, Cuite, Tepper, and Hallman (2020) that evaluated the sensory acceptability and purchase intention of consumers regarding foods using nanotechnology, it can be observed that most consumers were willing to buy products with nanoparticles, claiming that the sensorial and technological improvement are positive factors of the products.

5.9 Nanostructures in the conservation of fish and other marine products

Nanotechnology is an area that has been gaining attention lately, it consists of the production of materials on a nanoscale (1–100 nm), which have different physical-chemical properties compared to materials with conventional size. Due to the size similar to that of bacterial cells, nanomaterials can act as very efficient antibacterial agents with a broad spectrum of activity (Swain et al., 2014).

Throughout this chapter, several works will be cited with an emphasis on the development of antibacterial nanostructures aimed at their application in fish and other marine products. In this context, nanostructures are understood to mean: nanocapsules, nanofibers, nanoemulsions, and nanocomposites.

5.9.1 *Nanocapsules*

Saloko et al. (2014) evaluated the antimicrobial activities of coconut shell liquid smoke nanocapsules using chitosan and maltodextrin and its application on tuna fish preservation. They observed that the application of 5% liquid smoke nanocapsules extends the fish freshness up to 24 hours at room temperature than that of without treatments. The coconut shell liquid smoke nanocapsules are bacteriostatics in nature and show strong inhibition toward bacterial growth. The highest antimicrobial activity was performed against *E. coli* and *P. fluorescens* which represent Gram-negative bacteria.

In another study, da Cunha, de Ávila Scheeren, Fausto, de Melo, Henneman, Frizzo and Baldisserotto (2018) evaluated the antibacterial activity of *Origanum majorana* essential oil and nanocapsules of this essential oil (NOM) in silver catfish (*Rhamdia quelen)* infected with *A. hydrophila*. NOM demonstrated an antibacterial effect against this bacteria.

5.9.2 *Nanofibers*

Polymer nanofibers, an important class of nanomaterials, have attracted increasing attention in recent years. These generally refer to fibers with diameters smaller than 100 nm. This intrinsic characteristic makes nanofibers attractive for many applications where a high specific surface area is highly desirable or necessary.

Ceylan, Sengor, Sağdıç, and Yilmaz (2017) produced electrospun chitosan nanofibers (CN) and thymol liquid smoke-loaded electrospun chitosan nanofibers (LSCN) in terms of limitation of Total Mesophilic Aerobic Bacteria (TMABc), TPBc and Yeast and Mold count (YMc) growth during storage period in sea bass (*Dicentrarchus labrax)* fillets. Microbiological stability tests demonstrated that CN and LSCN were both effective against growth of the tested microorganisms, exhibiting almost 40% and 50% limitation of growth, the results suggested that coating of the fish fillets with the nanofibers would be a promising technique to limit microbial growth.

Erbay et al. (2017), evaluated the quality improvement of rainbow trout fillets by whey protein isolate coatings containing electrospun poly (ε-caprolactone) nanofibers with *Urtica dioica* L. extract during storage. Obtained results demonstrated that the developed bioactive coatings present excellent antimicrobial activity that could extend the shelf life and quality of the fish fillets up to 15 days.

Ceylan, Sengor, Basahel, and Yilmaz (2018) promoted the nanoencapsulation of probiotic bacteria *(L. rhamnosus)* into poly (vinyl alcohol) and sodium alginate-based nanofibers (VSPBe) and also the production of poly (vinyl alcohol) and sodium alginate-based nanofibers (VS), in order to verify the antimicrobial capacity in the fish fillets. The microbiological results revealed that the nanofiber scaffolds VS and VSPBe were significantly effective to delay the growth of TMABc and TPBc in the fish fillets.

Shi et al. (2021) evaluated the antimicrobial mechanism of alkyl gallates against *E. coli* and *S. aureus* and its combined effect with electrospun nanofibers on Chinese Taihu icefish preservation. This work studied the antibacterial properties of enzymatically synthesized alkyl gallates. Octyl gallate (OG) was further incorporated into PLA to develop a novel functionalized nanofiber with outstanding long-term antimicrobial activities, as an active food packaging material to effectively extend the shelf life of Chinese Taihu icefish. The combination of OG and the benefit of electrospun nanofibers can establish a highly efficient multifunctional antibacterial system to address the growing challenges of foodborne infections.

5.9.3 *Nanoemulsions*

The application of nanoemulsions to extend the shelf life of fish and other marine products without changing the sensory characteristics is a relatively new technology.

Nanoemulsions have adverse effects on the structure and function on bacteria by destabilizing the organism's lipid envelope. This is due to the ability of these structures to bind to water, limiting the access of microorganisms to water.

Özogul, Durmus, Ucar, Özogul, and Regenstein (2016) developed nanoemulsions based on commercial oils (sunflower, canola, corn, olive, soybean, and hazelnut oils) and evaluated the effect on microbial, sensory, and chemical qualities of refrigerated farmed sea bass. The study was conducted to investigate the possible use of a number of different commercial oil-based nanoemulsions to determine which would be best for extending the shelf life of fish. The results suggested that nanoemulsion treatment extended the shelf life of the sea bass from 8 days (the control) to 10 days.

Yazgan et al. (2017) evaluated the influence of oil-in-water nanoemulsion based on sunflower oil on the sensory, chemical, and microbiological qualities of sea bream and sea bass fillets stored at 2°C ± 2°C. The results showed that the use of nanoemulsion extended the shelf life of fish 1 or 2 days when compared with the control.

Meral et al. (2019) produced thyme oil nanoemulsions (TONa and TONb having 3.63% and 5.50% thyme oil, respectively) by using the sonication technique; they also investigated the antimicrobial effect of TONs on rainbow trout fillets during the cold storage period. Within 9-day cold storage period, TMAB growth in fish fillets treated with TONb was successfully reduced from 6.42 to 4.62 log CFU/g (change: 28%). In addition, as compared with control group samples, TPB growth of the same group was limited by 1.51 log CFU/g (from 7.93 to

6.42 log CFU/g). The results suggested that treatment of the fish fillets with the TONs would be a useful and practical nano application to delay microbial growth in fish fillets.

In a study by Durmus (2020), nanoemulsions of essential oils (EOs—orange, mandarin, grapefruit, and lemon) were developed and the effect of these was assessed in the shelf life of rainbow trout (*Oncorhynchus mykiss*). The author noted that EOs improved the organoleptic quality of rainbow trout significantly. Mandarin and grapefruit treatment groups were preferred the most by the panelists. Furthermore, compared to the control group, all EOs reduced the chemical parameter values and were effective in slowing bacterial growth.

Ozogul et al. (2017) evaluated antibacterial activities of nanoemulsions based on essential oil of thyme and its purified version on spoilage bacteria of fish (*P. luteola, P. damselae, V. vulnificus, E. faecalis, S. liquefaciens,* and *P. mirabilis*) were evaluated. Damage in bacterial cell membranes after nanoemulsion treatments were also detected.

In a study by Özogul, Özogul, and Kulawik (2021), the antimicrobial properties of grapefruit peel essential oil and its nanoemulsion were investigated. The analyses were carried out on four food-borne pathogens and six fish-spoilage bacteria. Nanoemulsion showed a bacteriostatic effect against some tested strains.

5.9.4 *Nanocomposites*

Truong et al. (2020) developed a nanocomposite of silver nanostructures/diatomite (AgNPs/DA) against pathogenous bacteria for *Pangasianodon hypophthalmus* catfish. The AgNPs/DA nanocomposite inhibited the growth of *A. hydrophila* and *Edwardsiella ictaluristrains,* according to the size of AgNPs deposited in DA. These results demonstrated antibacterial activity of AgNPs/DA nanocomposite against the serious pathogens for *P. hypophthalmus* catfish.

Azizi-Lalabadi et al. (2020) developed a Polyvinyl alcohol/gelatin (PVA/G) nanocomposite containing ZnO, TiO_2, or ZnO/TiO_2 nanostructures doped on zeolite and evaluated the microbial and sensory qualities of packaged white shrimp (*Penaeus vannamei*) during refrigeration. The microbiological and sensory investigation of the efficacy of the PVA/G nanocomposite films as active packaging materials revealed a considerable improvement in shrimp shelf life (12 days) in comparison to the control (6 days). Therefore, these nanocomposite films can be used as novel active packaging in the maintenance of the microbial load and sensory qualities of shrimp.

Luo, Qin, and Ye (2015) evaluated the effect of nano-TiO_2-LDPE packaging on microbiological and physicochemical quality of Pacific white shrimp (*P. vannamei*) during chilled storage. The nano-TiO_2-modified low-density polyethylene (NTLDPE) packaging was prepared by blending LDPE with nano-TiO_2. The overall results indicated that NTLDPE packaging could preserve the quality and extend the shelf-life of Pacific white shrimp. Pacific white shrimp packaged with NTLDPE was able to maintain freshness by inhibiting the growth of spoilage bacteria.

Shankar, Wang, and Rhim (2018) promoted an incorporation of zinc oxide nanostructures improved the mechanical, water vapor barrier, UV-light barrier, and antibacterial properties of PLA-based nanocomposite films. The developed films were applied to the packaging of a minced fish paste and showed strong antibacterial function. They decreased the bacterial count on minced fish paste and increased the shelf life.

Echeverría et al. (2018) evaluated the effect of applying active nanocomposite films based on soy protein isolate (SPI)—montmorillonite (MMT)—CEO on the preservation of muscle fillets of bluefin tuna (*T. thynnus*) during the 17 days of storage at 2°C. Nanorein-enhanced protein films with 10 g MMT/100 g SPI and activated with CEO were able to decrease the microbial growth (evaluated by TVBN and microorganism counts) of tuna fillets during the studied storage period. The presence of clay seemed to favor the release of the active principles of clove oil, by prolonging its antimicrobial activity (especially effective to inhibit *Pseudomonas* spp.)

Alboofetileh, Rezaei, Hosseini, and Abdollahi (2016) evaluated the efficacy of activated alginate-based nanocomposite films to control *L. monocytogenes* and spoilage flora in rainbow trout slices. The results demonstrated that alginate-clay films enriched with 1% MEO significantly delayed the growth of *L. monocytogenes* during the 15-day storage with final counts reaching 6.23 log CFU/g while the counts in control samples were significantly higher reaching 7.38 log CFU/g ($P < .05$). Furthermore, active films efficiently reduced total viable count and psychrotrophic count as well as TVB-N in the fish slice during refrigerated storage.

Xin et al. (2020) produced chitosan/curcumin nanostructures based zein and potato starch composite films for *Schizothorax prenati* fillet preservation. The *S. prenati* fillet preservation experiment showed that curcumin (CCN)/zein/Potato starch (PS) composite films protected the microbial, chemical, and sensorial quality of the treated fish muscle due to the antimicrobial and antioxidant properties of CCN. Therefore, CCN/zein/PS could be an alternative natural bioactive packaging material to improve fish fillet quality during storage and prolong its shelf life.

Abdou, Osheba, and Sorour (2012) studied the effect of different concentrations of chitosan nanostructures as an active coating on microbiological characteristics of fish fingers during frozen storage at −18°C, results indicated that fish fingers coated with chitosan nanostructures had much lower total bacterial count, psychrophilic bacteria, proteolytic bacteria, and coliform bacteria when compared with uncoated fish fingers and that coated with a commercial edible coating.

Homayonpour et al. (2021) evaluated the effects of nano-chitosan coatings incorporating with free/nano-encapsulated cumin (*Cuminum cyminum* L.) essential oil on quality characteristics of sardine fillet. The results of chemical and microbial analyses showed that chitosan coating including nano-encapsulated CCEO on sardine could conduct expansion of shelf-life, enhancement of microbiological safety, and holding of the good quality specification in the storage period at 4°C. All coated samples reduced significantly the counts of microbial than uncoated samples. In general, CCEO encapsulation allows the control of microbial agents on the surface of samples, leading to prolonged antimicrobial activity over 16 days of storage than free CCEO. Coatings of nanochitosan including CCEO-loaded nanoliposomes could be usage in the seafood industry as active packaging.

5.10 Conclusion

Given the above, it can be inferred that the production of antibacterial nanostructures for application in fish and other marine products is a very promising field since it allows to extend the shelf life of these products, avoiding loss of nutritional value and financial losses, in addition to reducing food waste and waste production.

References

Abdou, E. S., Osheba, A. S., & Sorour, M. A. (2012). Effect of chitosan and chitosan-nanostructures as active coating on microbiological characteristics of fish fingers. *International Journal of Applied Science and Technology*, 2(7), 158–169.

Abolghasemi-Fakhri, L., Ghanbarzadeh, B., Dehghannya, J., Abbasi, F., & Adun, P. (2019). Styrene monomer migration from polystyrene based food packaging nanocomposite: Effect of clay and ZnO nanoparticles. *Food and Chemical Toxicology*, *129*, 77–86. Available from https://doi.org/10.1016/j.fct.2019.04.019, http://www.elsevier.com/locate/foodchemtox.

Abreu, A. S., Oliveira, M., De Sá, A., Rodrigues, R. M., Cerqueira, M. A., Vicente, A. A., & Machado, A. V. (2015). Antimicrobial nanostructured starch based films for packaging. *Carbohydrate Polymers*, *129*, 127–134. Available from https://doi.org/10.1016/j.carbpol.2015.04.021, http://www.elsevier.com/wps/find/journaldescription.cws_home/405871/description#description.

Ackman, R. G. (1989). Nutritional composition of fats in seafoods. *Progress in Food and Nutrition Science*, *13*(3–4), 161–289.

Adams, M. R., & Moss, M. O. (1999). Microbiology of primary food commodities. *Food Microbiology* (pp. 103–135). Royal Society of Chemistry.

Adeyeye, S. A. O. (2019). Food packaging and nanotechnology: Safeguarding consumer health and safety. *Nutrition and Food Science*, *49*(6), 1164–1179. Available from https://doi.org/10.1108/NFS-01-2019-0020, http://www.emeraldinsight.com/info/journals/nfs/nfs.jsp.

Alboofetileh, M., Rezaei, M., Hosseini, H., & Abdollahi, M. (2016). Efficacy of activated alginate-based nanocomposite films to control Listeria monocytogenes and spoilage flora in rainbow trout slice. *Journal of Food Science and Technology*, *53*(1), 521–530. Available from https://doi.org/10.1007/s13197-015-2015-9, http://www.springerlink.com/content/121580/.

Alcantara, M., Morais, I. C. L., Cunha, C. M. O., & Souza, C. (2012). Principais microrganismos envolvidos na deterioração das características sensoriais de derivados cárneos. *Revista Brasileira de Higiene e Sanidade Animal*, *6*, 1–18.

Alfadul, S. M., & Elneshwy, A. A. (2010). Use of nanotechnology in food processing, packaging and safety – review. *African Journal of Food, Agriculture, Nutrition and Development*, *10*(6). Available from https://doi.org/10.4314/ajfand.v10i6.58068.

Almeida, M. (2011). Knack, Fotocatálise heterogênea aplicada na degradação do corante rodamina-B utilizando nanopartículas de dióxido de titânio. 2011. 97f. Mestrado (Dissertação em Gestão e Tecnologia Ambiental) - Programa de Pós-Graduação em Tecnologia Ambiental.

Al-Naamani, L., Dobretsov, S., & Dutta, J. (2016). Chitosan-zinc oxide nanoparticle composite coating for active food packaging applications. *Innovative Food Science and Emerging Technologies*, *38*, 231–237. Available from https://doi.org/10.1016/j.ifset.2016.10.010, http://www.elsevier.com/locate/ifset.

Alves, V. L. C. D., Rico, B. P. M., Cruz, R. M. S., Vicente, A. A., Khmelinskii, I., & Vieira, M. C. (2018). Preparation and characterization of a chitosan film with grape seed extract-carvacrol microcapsules and its effect on the shelf-life of refrigerated Salmon (Salmo salar). *LWT - Food Science and Technology*, *89*, 525–534. Available from https://doi.org/10.1016/j.lwt.2017.11.013, http://www.elsevier.com/inca/publications/store/6/2/2/9/1/0/index.htt.

Arancibia, M. Y., Alemán, A., López-Caballero, M. E., Gómez-Guillén, M. C., & Montero, P. (2015). Development of active films of chitosan isolated by mild extraction with added protein concentrate from shrimp waste. *Food Hydrocolloids*, *43*, 91–99. Available from https://doi.org/10.1016/j.foodhyd.2014.05.006.

Azizi-Lalabadi, M., Ehsani, A., Ghanbarzadeh, B., & Divband, B. (2020). Polyvinyl alcohol/gelatin nanocomposite containing ZnO, TiO_2 or ZnO/TiO_2 nanoparticles doped on 4A zeolite: Microbial and sensory qualities of packaged white shrimp during refrigeration. *International Journal of Food Microbiology*, *312*, 108375. Available from https://doi.org/10.1016/j.ijfoodmicro.2019.108375.

Badolato, E. S., Carvalho, J. B., Mello, M. R. P., Tavares, M., Campos, N. C., Auedpimentel, S., & Morais, C. (1994). Centesimal composition of fatty acids and caloric-value of five marine fish species in the different seasons. *Revista Instituto Adolfo Lutz*, *5*(1), 27–35.

Baek, S. K., & Song, K. B. (2018). Development of Gracilaria vermiculophylla extract films containing zinc oxide nanoparticles and their application in smoked salmon packaging. *LWT - Food Science and Technology, 89*, 269–275. Available from https://doi.org/10.1016/j.lwt.2017.10.064, http://www.elsevier.com/inca/publications/store/6/2/2/9/1/0/index.htt.

Basiri, S., Shekarforoush, S. S., Aminlari, M., & Akbari, S. (2015). The effect of pomegranate peel extract (PPE) on the polyphenol oxidase (PPO) and quality of Pacific white shrimp (Litopenaeus vannamei) during refrigerated storage. *LWT - Food Science and Technology, 60*(2), 1025–1033. Available from https://doi.org/10.1016/j.lwt.2014.10.043.

Bogdanović, T., Šimat, V., Frka-Roić, A., & Marković, K. (2012). Development and application of quality index method scheme in a shelf-life study of wild and fish farm affected bogue (Boops boops, L.). *Journal of Food Science, 77*(2), S99–S106. Available from https://doi.org/10.1111/j.1750-3841.2011.02545.x.

Božanić, D. K., Djoković, V., Dimitrijević-Branković, S., Krsmanović, R., McPherson, M., Nair, P. S., ... Radhakrishnan, T. (2011). Inhibition of microbial growth by silver-starch nanocomposite thin films. *Journal of Biomaterials Science, Polymer Edition, 22*(17), 2343–2355. Available from https://doi.org/10.1163/092050610X539532, http://www.tandfonline.com/loi/tbsp20.

Bressan, M. C., & Perez, J. R. O. (2000). Tecnologia de carnes e pescados. *UFLA/FAEPE, 225*.

Bumbudsanpharoke, N., Lee, W., & Ko, S. (2018). A comprehensive feasibility study on the properties of LDPE-Ag nanocomposites for food packaging applications. *Polymer Composites, 39*(9), 3178–3186. Available from https://doi.org/10.1002/pc.24325, http://onlinelibrary.wiley.com/journal/10.1002/(ISSN)1548-0569.

Ceylan, Z., Sengor, G. F. U., Basahel, A., & Yilmaz, M. T. (2018). Determination of quality parameters of gilthead sea bream (Sparus aurata) fillets coated with electrospun nanofibers. *Journal of Food Safety, 38*(6). Available from https://doi.org/10.1111/jfs.12518, http://onlinelibrary.wiley.com/journal/10.1111/(ISSN)1745-4565.

Ceylan, Z., Sengor, G. F. U., & Yilmaz, M. T. (2018). Nanoencapsulation of liquid smoke/thymol combination in chitosan nanofibers to delay microbiological spoilage of sea bass (Dicentrarchus labrax) fillets. *Journal of Food Engineering, 229*, 43–49. Available from https://doi.org/10.1016/j.jfoodeng.2017.11.038, http://www.sciencedirect.com/science/journal/02608774.

Ceylan, Z., Sengor, G. F. U., Sağdıç, O., & Yilmaz, M. T. (2017). A novel approach to extend microbiological stability of sea bass (Dicentrarchus labrax) fillets coated with electrospun chitosan nanofibers. *LWT - Food Science and Technology, 79*, 367–375. Available from https://doi.org/10.1016/j.lwt.2017.01.062, http://www.elsevier.com/inca/publications/store/6/2/2/9/1/0/index.htt.

Cho, K. H., Park, J. E., Osaka, T., & Park, S. G. (2005). The study of antimicrobial activity and preservative effects of nanosilver ingredient. *Electrochimica Acta, 51*(5), 956–960. Available from https://doi.org/10.1016/j.electacta.2005.04.071.

Cierech, M., Wojnarowicz, J., Kolenda, A., Krawczyk-Balska, A., Prochwicz, E., Woźniak, B., ... Mierzwińska-Nastalska, E. (2019). Zinc oxide nanoparticles cytotoxicity and release from newly formed PMMA–ZnO nanocomposites designed for denture bases. *Nanomaterials, 9*(9), 1318. Available from https://doi.org/10.3390/nano9091318.

Cribb A. Y., & Filho, S. C. R. P. (2018). Mello, Manual técnico de manipulação e conservação de pescado. Embrapa.

Czanderna, A. W., Ramachandra Rao, C. N., & Honig, J. M. (1958). The anatase-rutile transition: Part 1. - Kinetics of the transformation of pure anatase. *Transactions of the Faraday Society, 54*, 1069–1073. Available from https://doi.org/10.1039/TF9585401069.

Da Rocha, M., Alemán, A., Romani, V. P., López-Caballero, M. E., Gómez-Guillén, M. C., Montero, P., & Prentice, C. (2018). Effects of agar films incorporated with fish protein hydrolysate or clove essential oil on flounder (*Paralichthys orbignyanus*) fillets shelf-life. *Food Hydrocolloids, 81*, 351–363.

Dainelli, D., Gontard, N., Spyropoulos, D., Zondervan-van den Beuken, E., & Tobback, P. (2008). Active and intelligent food packaging: Legal aspects and safety concerns. *Trends in Food Science and Technology, 19*(1), S103–S112. Available from https://doi.org/10.1016/j.tifs.2008.09.011, http://www.elsevier.com/wps/find/journaldescription.cws_home/601278/description#description.

Davies, R. L., & Etris, S. F. (1997). The development and functions of silver in water purification and disease control. *Catalysis Today, 36*(1), 107–114. Available from https://doi.org/10.1016/S0920-5861(96)00203-9, http://www.sciencedirect.com/science/journal/09205861.

Del Nobile, M. A., Corbo, M. R., Speranza, B., Sinigaglia, M., Conte, A., & Caroprese, M. (2009). Combined effect of MAP and active compounds on fresh blue fish burger. *International Journal of Food Microbiology, 135*(3), 281–287. Available from https://doi.org/10.1016/j.ijfoodmicro.2009.07.024.

Delgado, R., Gutiérrez, C., & Hurtado, Á. (2003). Enfermedades transmitidas por alimentos (ETA) de origen marino en Nueva Esparta II: Características clínicas y etiológicas. *Rev. Inst. Nac. Hig*, 11–16.

Duncan, T. V. (2011). Applications of nanotechnology in food packaging and food safety: Barrier materials, antimicrobials and sensors. *Journal of Colloid and Interface Science, 363*(1), 1–24. Available from https://doi.org/10.1016/j.jcis.2011.07.017.

Durmus, M. (2020). The effects of nanoemulsions based on citrus essential oils (orange, mandarin, grapefruit, and lemon) on the shelf life of rainbow trout (Oncorhynchus mykiss) fillets at $4 \pm 2°C$. *Journal of Food Safety, 40*(1). Available from https://doi.org/10.1111/jfs.12718, http://onlinelibrary.wiley.com/journal/10.1111/(ISSN)1745-4565.

Echeverría, I., López-Caballero, M. E., Gómez-Guillén, M. C., Mauri, A. N., & Montero, M. P. (2018). Active nanocomposite films based on soy proteins-montmorillonite- clove essential oil for the preservation of refrigerated bluefin tuna (Thunnus thynnus) fillets. *International Journal of Food Microbiology, 266*, 142–149. Available from https://doi.org/10.1016/j.ijfoodmicro.2017.10.003, http://www.elsevier.com/locate/ijfoodmicro.

Ehsani, A., Jasour, M. S., Hashemi, M., Mehryar, L., & Khodayari, M. (2014). Zataria multiflora Boiss essential oil and sodium acetate: How they affect shelf life of vacuum-packaged trout burgers. *International Journal of Food Science and Technology, 49*(4), 1055–1062. Available from https://doi.org/10.1111/ijfs.12400.

Emamhadi, M. A., Sarafraz, M., Akbari, M., Thai, V. N., Fakhri, Y., Linh, N. T. T., & Mousavi Khaneghah, A. (2020). Nanomaterials for food packaging applications: A systematic review. *Food and Chemical Toxicology, 146*, 111825. Available from https://doi.org/10.1016/j.fct.2020.111825.

Erbay, E. A., Dağtekin, B. B., Türe, M., Yeşilsu, A. F., & Torres-Giner, S. (2017). Quality improvement of rainbow trout fillets by whey protein isolate coatings containing electrospun poly(ε-caprolactone) nanofibers with Urtica dioica L. extract during storage. *LWT, 78*, 340–351. Available from https://doi.org/10.1016/j.lwt.2017.01.002, http://www.elsevier.com/inca/publications/store/6/2/2/9/1/0/index.htt.

Esmailzadeh, H., Sangpour, P., Shahraz, F., Hejazi, J., & Khaksar, R. (2016). Effect of nanocomposite packaging containing ZnO on growth of Bacillus subtilis and Enterobacter aerogenes. *Materials Science and Engineering C, 58*, 1058–1063. Available from https://doi.org/10.1016/j.msec.2015.09.078.

Fabrega, J., Fawcett, S. R., Renshaw, J. C., & Lead, J. R. (2009). Silver nanoparticle impact on bacterial growth: Effect of pH, concentration, and organic matter. *Environmental Science and Technology, 43*(19), 7285–7290. Available from https://doi.org/10.1021/es803259g, United Kingdom, http://pubs.acs.org/doi/pdfplus/10.1021/es803259g.

FAO. (2020). The State of World Fisheries and Aquaculture. Sustainability in action. Available from https://doi.org/10.4060/ca9229en.

Fang, Z., Zhao, Y., Warner, R. D., & Johnson, S. K. (2017). Active and intelligent packaging in meat industry. *Trends in Food Science and Technology, 61*, 60–71. Available from https://doi.org/10.1016/j.tifs.2017.01.002, http://www.elsevier.com/wps/find/journaldescription.cws_home/601278/description#description.

Feng, Y., Tan, H., Li, C., Wang, Y., Zhang, Y., Wen, P., ... Xu, L. (2017). Preparation and characterization of nano TiO_2 antibacterial corrugating medium. *Journal of Nanoscience and Nanotechnology, 17*(12), 8912–8917. Available from https://doi.org/10.1166/jnn.2017.13896, http://docserver.ingentaconnect.com/deliver/connect/asp/15334880/v17n12/s30.pdf?expires = 1506962195&id = 91516060&titleid = 4286 &accname = Elsevier + BV&checksum = 8015EA849AED2EC9F65BAC485D5A18F4.

Fennema, O. R., Damodaran, S., & Parkin, K. (2010). Química de alimentos de Fennema. Editorial Acribia, 1258, Espanha.

Fonseca, L. M., Silva, F. T. D., Bruni, G. P., Borges, C. D., Zavareze, E. D. R., & Dias, A. R. G. (2021). Aerogels based on corn starch as carriers for pinhão coat extract (Araucaria angustifolia) rich in phenolic compounds for active packaging. *International Journal of Biological Macromolecules, 169*, 362–370. Available from https://doi.org/10.1016/j.ijbiomac.2020.12.110, http://www.elsevier.com/locate/ijbiomac.

Fu, L., Lu, W. Q., & Zhou, X. M. (2016). Phenolic compounds and in vitro antibacterial and antioxidant activities of three tropic fruits: Persimmon, guava, and sweetsop. *BioMed Research International, 2016*, 1–9. Available from https://doi.org/10.1155/2016/4287461.

Germano, P. M. L., Oliveira, J. C. F., & Germano, M. I. S. (1993). O pescado como causa de toxinfecções bacterianas. *Higiene Alimentar, 7*, 40–45.

Gram, L., & Dalgaard, P. (2002). Fish spoilage bacteria - Problems and solutions. *Current Opinion in Biotechnology, 13*(3), 262–266. Available from https://doi.org/10.1016/S0958-1669(02)00309-9, http://www.elsevier.com/locate/copbio.

Homayonpour, P., Jalali, H., Shariatifar, N., & Amanlou, M. (2021). Effects of nano-chitosan coatings incorporating with free/nano-encapsulated cumin (Cuminum cyminum L.) essential oil on quality characteristics of sardine fillet. *International Journal of Food Microbiology, 341*, 109047. Available from https://doi.org/10.1016/j.ijfoodmicro.2021.109047.

Huss, H. H. (1997). Garantia da qualidade dos produtos da pesca. *Food & Agriculture Org, 176*.

Jafari, H., Pirouzifard, M. K., Khaledabad, M. A., & Almasi, H. (2016). Effect of chitin nanofiber on the morphological and physical properties of chitosan/silver nanoparticle bionanocomposite films. *International Journal of Biological Macromolecules, 92*, 461–466. Available from https://doi.org/10.1016/j.ijbiomac.2016.07.051, http://www.elsevier.com/locate/ijbiomac.

Jayakumar, A., Heera, K. V., Sumi, T. S., Joseph, M., Mathew, S., Praveen, G., ... Radhakrishnan, E. K. (2019). Starch-PVA composite films with zinc-oxide nanoparticles and phytochemicals as intelligent pH sensing wraps for food packaging application. *International Journal of Biological Macromolecules, 136*, 395–403. Available from https://doi.org/10.1016/j.ijbiomac.2019.06.018.

Jiang, S., Wu, X., Li, W., Wu, M., Luo, Y., Lu, S., & Lin, H. (2015). Effects of dietary protein and lipid levels on growth, feed utilization, body and plasma biochemical compositions of hybrid grouper (Epinephelus lanceolatus ♂ × Epinephelus fuscoguttatus ♀) juveniles. *Aquaculture (Amsterdam, Netherlands), 446*, 148–155. Available from https://doi.org/10.1016/j.aquaculture.2015.04.034, http://www.journals.elsevier.com/aquaculture/.

Kaur, M., & Santhiya, D. (2021). UV-shielding antimicrobial zein films blended with essential oils for active food packaging. *Journal of Applied Polymer Science, 138*(7). Available from https://doi.org/10.1002/app.49832, http://onlinelibrary.wiley.com/journal/10.1002/(ISSN)1097-4628.

Khaledian, S., & Basiri, S. (2022). The Persian lime peel extract-loaded tragacanth gel coating for sodium metabisulfite reduction in Pacific white shrimp (Litopenaeus vannamei). *Journal of Food Processing and Preservation, 46*(1). Available from https://doi.org/10.1111/jfpp.16191, http://onlinelibrary.wiley.com/journal/10.1111/(ISSN)1745-4549.

Koenari, Z. I., Siagian, C. M., Simanungkalit, B., Nilatany, A., Pratama, I. M., Lasmawati, D., & Nurcahya, C. M. (2016). Potential se of gamma-irradiated ethnic ready-to- eat foods to improve the nutritional status of landslide victims. *Foods, 5*(3), 53.

Kuang, L., Burgess, B., Cuite, C. L., Tepper, B. J., & Hallman, W. K. (2020). Sensory acceptability and willingness to buy foods presented as having benefits achieved through the use of nanotechnology. *Food Quality and Preference, 83*. Available from https://doi.org/10.1016/j.foodqual.2020.103922, https://www.journals.elsevier.com/food-quality-and-preference.

Kumar, S., Mudai, A., Roy, B., Basumatary, I. B., Mukherjee, A., & Dutta, J. (2020). Biodegradable hybrid nanocomposite of chitosan/gelatin and green synthesized zinc oxide nanoparticles for food packaging. *Foods., 9*(9), 1143. Available from https://doi.org/10.3390/foods9091143.

Kunová, S., Sendra, E., Haščík, P., Vukovic, N. L., Vukic, M., & Kačániová, M. (2021). Influence of essential oils on the microbiological quality of fish meat during storage. *Animals, 11*(11), 3145. Available from https://doi.org/10.3390/ani11113145.

Lallo da Silva, B., Caetano, B. L., Chiari-Andréo, B. G., Pietro, R. C. L. R., & Chiavacci, L. A. (2019). Increased antibacterial activity of ZnO nanoparticles: Influence of size and surface modification. *Colloids and Surfaces B: Biointerfaces, 177*, 440–447. Available from https://doi.org/10.1016/j.colsurfb.2019.02.013, http://www.elsevier.com/locate/colsurfb.

Landim, A. P. M., Bernardo, C. O., Martins, I. B. A., Francisco, M. R., Santos, M. B., & De Melo, N. R. (2016). Sustentabilidade quanto às embalagens de alimentos no Brasil. *Polimeros, 26*, 82–92. Available from https://doi.org/10.1590/0104-1428.1897, http://www.scielo.br/pdf/po/v26nspe/0104-1428-po-0104-14281897.pdf.

Leitão, M. F. F., Baldini, V. L. S., Uboldieiroa, M. N., & Destro, M. T. (1983). Bactérias produtoras de histamina em pescado de origem marinha. *Coletânea ITAL, 13*.

Licciardello, F., Muratore, G., Suma, P., Russo, A., & Nerín, C. (2013). Effectiveness of a novel insect-repellent food packaging incorporating essential oils against the red flour beetle (Tribolium castaneum). *Innovative Food Science and Emerging Technologies, 19*, 173–180. Available from https://doi.org/10.1016/j.ifset.2013.05.002.

Liu, Y., Li, Y., Deng, L., Zou, L., Feng, F., & Zhang, H. (2018). Hydrophobic ethylcellulose/gelatin nanofibers containing zinc oxide nanoparticles for antimicrobial packaging. *Journal of Agricultural and Food Chemistry, 66*(36), 9498–9506. Available from https://doi.org/10.1021/acs.jafc.8b03267, http://pubs.acs.org/journal/jafcau.

Luo, Z., Qin, Y., & Ye, Q. (2015). Effect of nano-TiO_2-LDPE packaging on microbiological and physicochemical quality of Pacific white shrimp during chilled storage. *International Journal of Food Science and Technology, 50*(7), 1567–1573. Available from https://doi.org/10.1111/ijfs.12807, http://onlinelibrary.wiley.com/journal/10.1111/(ISSN)1365-2621.

Maghami, M., Motalebi, A. A., & Anvar, S. A. A. (2019). Influence of chitosan nanoparticles and fennel essential oils (Foeniculum vulgare) on the shelf life of Huso huso fish fillets during the storage. *Food Science & Nutrition, 7*(9), 3030–3041. Available from https://doi.org/10.1002/fsn3.1161.

Maneerat, C., & Hayata, Y. (2008). Gas-phase photocatalytic oxidation of ethylene with TiO_2-coated packaging film for horticultural products. *Transactions of the ASABE, 51*(1), 163–168.

Menzel, C., González-Martínez, C., Chiralt, A., & Vilaplana, F. (2019). Antioxidant starch films containing sunflower hull extracts. *Carbohydrate Polymers, 214*, 142–151. Available from https://doi.org/10.1016/j.carbpol.2019.03.022, http://www.elsevier.com/wps/find/journaldescription.cws_home/405871/description#description.

Meral, R., Ceylan, Z., & Kose, S. (2019). Limitation of microbial spoilage of rainbow trout fillets using characterized thyme oil antibacterial nanoemulsions. *Journal of Food Safety, 39*(4). Available from https://doi.org/10.1111/jfs.12644, http://onlinelibrary.wiley.com/journal/10.1111/(ISSN)1745-4565.

Mestriner, F. (2002). Design de embalagem curso básico. Makron Book.

Mihindukulasuriya, S. D. F., & Lim, L. T. (2014). Nanotechnology development in food packaging: A review. *Trends in Food Science and Technology, 40*(2), 149–167. Available from https://doi.org/10.1016/j.tifs.2014.09.009, http://www.elsevier.com/wps/find/journaldescription.cws_home/601278/description#description.

Monmaturapoj, N., Sri-On, A., Klinsukhon, W., Boonnak, K., & Prahsarn, C. (2018). Antiviral activity of multifunctional composite based on TiO_2-modified hydroxyapatite. *Materials Science and Engineering C, 92*, 96–102. Available from https://doi.org/10.1016/j.msec.2018.06.045.

Moura, G. F., De Souza, C. D. O. S. S., & Filho, E. S. D. A. (2021). Salga em pescado: Revisão/Salting in fish: Review. *Brazilian Journal of Development, 7*(12), 121831–121849. Available from https://doi.org/10.34117/bjdv7n12-784.

Mukherjee, K., Acharya, K., Biswas, A., & Jana, N. R. (2020). TiO_2 nanoparticles Co-doped with nitrogen and fluorine as visible-light-activated antifungal agents. *ACS Applied Nano Materials, 3*(2), 2016–2025. Available from https://doi.org/10.1021/acsanm.0c00108, https://pubs.acs.org/journal/aanmf6.

Navarro-Segura, L., Ros-Chumillas, M., Martínez-Hernández, G. B., & López-Gómez, A. (2020). A new advanced packaging system for extending the shelf life of refrigerated farmed fish fillets. *Journal of the Science of Food and Agriculture, 100*(12), 4601–4611. Available from https://doi.org/10.1002/jsfa.10520, http://onlinelibrary.wiley.com/journal/10.1002/(ISSN)1097-0010.

Neto, L., & Gonçalves, A. D. (2011). Tecnologia do pescado: ciência, tecnologia, inovação e legislação. *Atheneu*, 235–245.

Nussinovitch, A. (2009). *Biopolymer films and composite coatings modern biopolymer science* (pp. 295–326). Israel: Elsevier Inc. Available from https://doi.org/10.1016/B978-0-12-374195-0.00010-0, http://www.sciencedirect.com/science/book/9780123741950.

Oliveira, A. B. A. D., Paula, C. M. D. D., Capalonga, R., Cardoso, M. R. D. I., & Tondo, E. C. (2010). Doenças transmitidas por alimentos, principais agentes etiológicos e aspectos gerais: uma revisão. 30.

Ozdemir, M., & Floros, J. D. (2004). Active food packaging technologies. *Critical Reviews in Food Science and Nutrition, 44*(3), 185–193. Available from https://doi.org/10.1080/10408690490441578.

Ozogul, Y., Kuley Boğa, E., Akyol, I., Durmus, M., Ucar, Y., Regenstein, J. M., & Köşker, A. R. (2020). Antimicrobial activity of thyme essential oil nanoemulsions on spoilage bacteria of fish and food-borne pathogens. *Food Bioscience, 36*. Available from https://doi.org/10.1016/j.fbio.2020.100635, http://www.journals.elsevier.com/food-bioscience/.

Özogul, Y., Özogul, F., & Kulawik, P. (2021). The antimicrobial effect of grapefruit peel essential oil and its nanoemulsion on fish spoilage bacteria and food-borne pathogens. *LWT, 136*, 110362. Available from https://doi.org/10.1016/j.lwt.2020.110362.

Ozogul, Y., Yuvka, İ., Ucar, Y., Durmus, M., Kösker, A. R., Öz, M., & Ozogul, F. (2017). Evaluation of effects of nanoemulsion based on herb essential oils (rosemary, laurel, thyme and sage) on sensory, chemical and microbiological quality of rainbow trout (Oncorhynchus mykiss) fillets during ice storage. *LWT - Food Science and Technology, 75*, 677–684. Available from https://doi.org/10.1016/j.lwt.2016.10.009, http://www.elsevier.com/inca/publications/store/6/2/2/9/1/0/index.htt.

Özogul, Y., Durmus, M., Ucar, Y., Özogul, F., & Regenstein, J. M. (2016). Comparative study of nanoemulsions based on commercial oils (sunflower, canola, corn, olive, soybean, and hazelnut oils): Effect on microbial, sensory, and chemical qualities of refrigerated farmed sea bass. *Innovative Food Science and Emerging Technologies, 33*, 422–430. Available from https://doi.org/10.1016/j.ifset.2015.12.018, http://www.elsevier.com/locate/ifset.

Padmavathy, N., & Vijayaraghavan, R. (2008). Enhanced bioactivity of ZnO nanoparticles - An antimicrobial study. *Science and Technology of Advanced Materials, 9*(3). Available from https://doi.org/10.1088/1468-6996/9/3/035004.

Peidaei, F., Ahari, H., Anvar, S. A. A., & Ataee, M. (2021). Nanotechnology in food packaging and storage: A review. *Iranian Journal of Veterinary Medicine, 15*(2), 123–153.

Peña, C. V. M. (2006). Dissertação (Mestre em Ciência e Tecnologia dos Alimentos). *Cento de Ciências Rurais, 84*.

Penha, I. C. D. S., Gonçalves, C. G. E., Rosa, R. M. S. S., Silva, F. E. R. D., & Bichara, C. M. G. (2020). Microbiologia Do Gelo Utilizado Na Conservação Do Pescado Em Um Mercado Municipal De Belém, Pa/The Microbiological Quality Of Ice Used In Fish Conservation In A Municipal Market Of Belém, PA. *Brazilian Journal of Development, 6*(9), 66713–66724. Available from https://doi.org/10.34117/bjdv6n9-200.

Pereira, J. A., Oliveira, I., Sousa, A., Valentão, P., Andrade, P. B., Ferreira, I. C. F. R., ... Estevinho, L. (2007). Walnut (Juglans regia L.) leaves: Phenolic compounds, antibacterial activity and antioxidant potential of different cultivars. *Food and Chemical Toxicology, 45*(11), 2287–2295. Available from https://doi.org/10.1016/j.fct.2007.06.004.

Priyadarshi, R., Sauraj, B., Kumar, F., Deeba, A., Kulshreshtha, Y. S., & Negi. (2018). Chitosan films incorporated with Apricot (Prunus armeniaca) kernel essential oil as active food packaging material. *Food Hydrocolloids, 85*, 158–166. Available from https://doi.org/10.1016/j.foodhyd.2018.07.003.

Rahimnejad, S., Bang, I. C., Park, J. Y., Sade, A., Choi, J., & Lee, S. M. (2015). Effects of dietary protein and lipid levels on growth performance, feed utilization and body composition of juvenile hybrid grouper, Epinephelus fuscoguttatus × E. lanceolatus. *Aquaculture (Amsterdam, Netherlands), 446*, 283–289. Available from https://doi.org/10.1016/j.aquaculture.2015.05.019, http://www.journals.elsevier.com/aquaculture/.

Ravindran, R. E., Subha, V., & Ilangovan, R. (2020). Silver nanoparticles blended PEG/PVA nanocomposites synthesis and characterization for food packaging. *Arabian Journal of Chemistry, 13*(7), 6056–6060.

Realini, C. E., & Marcos, B. (2014). Active and intelligent packaging systems for a modern society. *Meat Science, 98*(3), 404–419. Available from https://doi.org/10.1016/j.meatsci.2014.06.031, http://www.elsevier.com/locate/meatsci.

Restuccia, D., Spizzirri, U. G., Parisi, O. I., Cirillo, G., Curcio, M., Iemma, F., ... Picci, N. (2010). New EU regulation aspects and global market of active and intelligent packaging for food industry applications. *Food Control, 21*(11), 1425–1435. Available from https://doi.org/10.1016/j.foodcont.2010.04.028.

Ribeiro-Santos, R., Andrade, M., Melo, N. R. D., & Sanches-Silva, A. (2017). Use of essential oils in active food packaging: Recent advances and future trends. *Trends in Food Science and Technology, 61*, 132–140. Available from https://doi.org/10.1016/j.tifs.2016.11.021, http://www.elsevier.com/wps/find/journaldescription.cws_home/601278/description#description.

Rodrigues, K. B. (2007). Histamina x pescado. Trabalho de conclusão de curso (Graduação em Medicina Veterinária).

Rojas-Lema, S., Torres-Giner, S., Quiles-Carrillo, L., Gomez-Caturla, J., Garcia-Garcia, D., & Balart, R. (2021). On the use of phenolic compounds present in citrus fruits and grapes as natural antioxidants for thermo-compressed bio-based high-density polyethylene films. *Antioxidants, 10*(1), 14. Available from https://doi.org/10.3390/antiox10010014.

Salahuddin, N., Abdelwahab, M., Gaber, M., & Elneanaey, S. (2020). Synthesis and design of norfloxacin drug delivery system based on PLA/TiO_2 nanocomposites: Antibacterial and antitumor activities. *Materials Science and Engineering: C, 108*, 110337. Available from https://doi.org/10.1016/j.msec.2019.110337.

Saloko, S., Darmadji, P., Setiaji, B., & Pranoto, Y. (2014). Antioxidative and antimicrobial activities of liquid smoke nanocapsules using chitosan and maltodextrin and its application on tuna fish preservation. *Food Bioscience, 7*, 71–79. Available from https://doi.org/10.1016/j.fbio.2014.05.008, http://www.journals.elsevier.com/food-bioscience/.

Shahvalizadeh, R., Ahmadi, R., Davandeh, I., Pezeshki, A., Seyed Moslemi, S. A., Karimi, S., ... Mohammadi, M. (2021). Antimicrobial bio-nanocomposite films based on gelatin, tragacanth, and zinc oxide nanoparticles – Microstructural, mechanical, thermo-physical, and barrier properties. *Food Chemistry, 354*, 129492. Available from https://doi.org/10.1016/j.foodchem.2021.129492.

Shankar, S., Wang, L. F., & Rhim, J. W. (2018). Incorporation of zinc oxide nanoparticles improved the mechanical, water vapor barrier, UV-light barrier, and antibacterial properties of PLA-based nanocomposite films. *Materials Science and Engineering C, 93*, 289–298. Available from https://doi.org/10.1016/j.msec.2018.08.002.

Sharma, S., Barkauskaite, S., Duffy, B., Jaiswal, A. K., & Jaiswal, S. (2020). Characterization and antimicrobial activity of biodegradable active packaging enriched with clove and thyme essential oil for food packaging application. *Foods, 9*(8), 1117. Available from https://doi.org/10.3390/foods9081117.

Shayegan, Z., Lee, C. S., & Haghighat, F. (2018). TiO_2 photocatalyst for removal of volatile organic compounds in gas phase – A review. *Chemical Engineering Journal, 334*, 2408–2439. Available from https://doi.org/10.1016/j.cej.2017.09.153, http://www.elsevier.com/inca/publications/store/6/0/1/2/7/3/index.htt.

Shi, Y. G., Zhang, R. R., Zhu, C. M., Xu, M. F., Gu, Q., Ettelaie, R., ... Leng, X. Y. (2021). Antimicrobial mechanism of alkyl gallates against Escherichia coli and Staphylococcus aureus and its combined effect with electrospun nanofibers on Chinese Taihu icefish preservation. *Food Chemistry, 346*. Available from https://doi.org/10.1016/j.foodchem.2020.128949, http://www.elsevier.com/locate/foodchem.

Siegrist, M., Stampfli, N., Kastenholz, H., & Keller, C. (2008). Perceived risks and perceived benefits of different nanotechnology foods and nanotechnology food packaging. *Appetite, 51*(2), 283–290. Available from https://doi.org/10.1016/j.appet.2008.02.020.

Sievers, N. V., Pollo, L. D., Corção, G., & Medeiros Cardozo, N. S. (2020). In situ synthesis of nanosized TiO_2 in polypropylene solution for the production of films with antibacterial activity. *Materials Chemistry and Physics, 246*. Available from https://doi.org/10.1016/j.matchemphys.2020.122824, http://www.journals.elsevier.com/materials-chemistry-and-physics.

Silvestre, C., Duraccio, D., & Cimmino, S. (2011). Food packaging based on polymer nanomaterials. *Progress in Polymer Science, 36*(12), 1766–1782. Available from https://doi.org/10.1016/j.progpolymsci.2011.02.003.

Soares, A. A., & Gonçalves. (2012). Qualidade e segurança do pescado. *Revista do Instituto Adolfo Lutz, 71*(1), 1–10.

Soo-Hwan, Lee, H. S., Ryu, D. S., Choi, S. J., & Lee, D. S. (2011). Antibacterial activity of silver-nanoparticles against *Staphylococcus aureus and Escherichia coli. Korean Journal of Microbiology and Biotechnology, 39*(1), 77–85.

Suwa, Y., Inagaki, M., & Naka, S. (1984). Polymorphic transformation of titanium dioxide by mechanical grinding. *Journal of Materials Science, 19*(5), 1397–1405. Available from https://doi.org/10.1007/BF00563034.

Swain, P., Nayak, S. K., Sasmal, A., Behera, T., Barik, S. K., Swain, S. K., ... Jayasankar, P. (2014). Antimicrobial activity of metal based nanoparticles against microbes associated with diseases in aquaculture. *World Journal of Microbiology and Biotechnology, 30*(9), 2491–2502. Available from https://doi.org/10.1007/s11274-014-1674-4, http://www.wkap.nl/journalhome.htm/0959-3993.

Tan, X., Sun, Z., Liu, Q., Ye, H., Zou, C., Ye, C., ... Lin, H. (2018). Effects of dietary ginkgo biloba leaf extract on growth performance, plasma biochemical parameters, fish composition, immune responses, liver histology, and immune and apoptosis-related genes expression of hybrid grouper (Epinephelus lanceolatus♂ × Epinephelus fuscoguttatus♀) fed high lipid diets. *Fish & Shellfish Immunology, 72*, 399–409. Available from https://doi.org/10.1016/j.fsi.2017.10.022.

Tang, S., Wang, Z., Li, P., Li, W., Li, C., Wang, Y., & Chu, P. K. (2018). Degradable and photocatalytic antibacterial Au-TiO_2/sodium alginate nanocomposite films for active food packaging. *Nanomaterials, 8*(11). Available from https://doi.org/10.3390/nano8110930, https://www.mdpi.com/2079-4991/8/11/930/pdf.

Teymourinia, H., Salavati-Niasari, M., Amiri, O., & Yazdian, F. (2019). Application of green synthesized TiO_2/Sb_2S_3/GQDs nanocomposite as high efficient antibacterial agent against E. coli and Staphylococcus aureus. *Materials Science and Engineering C, 99*, 296–303. Available from https://doi.org/10.1016/j.msec.2019.01.094.

Trbojevich, R. A., Khare, S., Lim, J. H., Watanabe, F., Gokulan, K., Krohmaly, K., & Williams, K. (2020). Assessment of silver release and biocidal capacity from silver nanocomposite food packaging materials. *Food and Chemical Toxicology, 145*. Available from https://doi.org/10.1016/j.fct.2020.111728, http://www.elsevier.com/locate/foodchemtox.

Tripathi, R. M., Pudake, R. N., Shrivastav, B. R., & Shrivastav, A. (2018). Antibacterial activity of poly (vinyl alcohol)—biogenic silver nanocomposite film for food packaging material. Advances in Natural. *Sciences: Nanoscience and Nanotechnology, 9*(2), 025020. Available from https://doi.org/10.1088/2043-6254/aac4ec.

Truong, H. T., Nguyen, T. D., & Nguyen, H. Q. (2020). Nanocomposite of silver nanostructures/diatomite against pathogenous bacteria for catfish. *Aquaculture Reports, 16*.

Vieira, R. H. S. F., & Saker-Sampaio, S. (2003). O emprego do gelo nos barcos de pesca. Microbiologia, higiene e qualidade do pescado. 37–43.

Vilela, C., Kurek, M., Hayouka, Z., Röcker, B., Yildirim, S., Antunes, M. D. C., … Freire, C. S. R. (2018). A concise guide to active agents for active food packaging. *Trends in Food Science and Technology, 80*, 212–222. Available from https://doi.org/10.1016/j.tifs.2018.08.006, http://www.elsevier.com/wps/find/journaldescription.cws_home/601278/description#description.
Wang, A. R., Ran, C., Ringø, E., & Zhou, Z. G. (2018). Progress in fish gastrointestinal microbiota research. *Reviews in Aquaculture, 10*(3), 626–640. Available from https://doi.org/10.1111/raq.12191, http://onlinelibrary.wiley.com/journal/10.1111/(ISSN)1753-5131.
Wolf-Hall, C. E., Gibbons, W. R., & Bauer, N. A. (2009). Development of a low-cost medium for production of nisin from Lactococcus lactis subsp. lactis. *World Journal of Microbiology and Biotechnology, 25*(11), 2013–2019. Available from https://doi.org/10.1007/s11274-009-0102-7.
World Health Organization-WHO. (2007). Emerging foodborne diseases. Fact Sheet 124. Available at: www.who.int/inffs/en/fact124.html.
World Health Organization-WHO. (2023). Emerging foodborne diseases. Avaiable at: https://www.who.int/activities/estimating-the-burden-of-foodborne-diseases.
Xin, S., Xiao, L., Dong, X., Li, X., Wang, Y., Hu, X., … Zhu, B. (2020). Preparation of chitosan/curcumin nanoparticles based zein and potato starch composite films for Schizothorax prenati fillet preservation. *International Journal of Biological Macromolecules, 164*, 211–221. Available from https://doi.org/10.1016/j.ijbiomac.2020.07.082, http://www.elsevier.com/locate/ijbiomac.
Yadav, S., Mehrotra, G. K., & Dutta, P. K. (2021). Chitosan based ZnO nanoparticles loaded gallic-acid films for active food packaging. *Food Chemistry, 334*, 127605. Available from https://doi.org/10.1016/j.foodchem.2020.127605.
Yazgan, H., Ozogul, Y., Durmuş, M., Balikçi, E., Gökdoğan, S., Uçar, Y., & Aksun, E. T. (2017). Effects of oil-in-water nanoemulsion based on sunflower oil on the quality of farmed sea bass and gilthead sea bream stored at chilled temperature ($2 \pm 2°C$). *Journal of Aquatic Food Product Technology, 26*(8), 979–992. Available from https://doi.org/10.1080/10498850.2017.1366610, http://www.tandfonline.com/toc/wafp20/current.
Yoganarasimhan, S. R., & Rao, C. N. R. (1962). Mechanism of crystal structure transformations. Part 3. - Factors affecting the anatase-rutile transformation. *Transactions of the Faraday Society, 58*, 1579–1589. Available from https://doi.org/10.1039/TF9625801579.
Yuan, G., Lv, H., Tang, W., Zhang, X., & Sun, H. (2016). Effect of chitosan coating combined with pomegranate peel extract on the quality of Pacific white shrimp during iced storage. *Food Control, 59*, 818–823. Available from https://doi.org/10.1016/j.foodcont.2015.07.011.
Zhang, H., Hortal, M., Dobon, A., Bermudez, J. M., & Lara-Lledo, M. (2015). The effect of active packaging on minimizing food losses: Life cycle assessment (LCA) of essential oil component-enabled packaging for fresh beef. *Packaging Technology and Science, 28*(9), 761–774. Available from https://doi.org/10.1002/pts.2135, http://onlinelibrary.wiley.com/journal/10.1002/(ISSN)1099-1522.
Zhuang, S., Li, Y., Jia, S., Hong, H., Liu, Y., & Luo, Y. (2019). Effects of pomegranate peel extract on quality and microbiota composition of bighead carp (Aristichthys nobilis) fillets during chilled storage. *Food Microbiology, 82*, 445–454. Available from https://doi.org/10.1016/j.fm.2019.03.019, http://www.elsevier.com/inca/publications/store/6/2/2/8/3/3/index.htt.

SECTION 2

Natural antimicrobials for food packaging and preservation

CHAPTER

6

Novel food packaging systems with antimicrobial agents from microbial source

María A. Bertuzzi[1,2], *Jimena E. Gamboni*[1,2], *Aníbal M. Slavutsky*[1,2] and *Carolina Ibarguren*[1,3]

[1]Chemical Industry Research Institute, INIQUI-CONICET, Salta, Argentine [2]Engineering Department, National University of Salta, Salta, Argentine [3]Health Sciences Department, National University of Salta, Salta, Argentine

6.1 Background

Food safety is a global priority since it constitutes a permanent and essential need for life. However, food microbial pathogens are still one of the major causes of foodborne illnesses, and food outbreaks remain among the main concerns related to public health (Achi & Halami, 2016; Jagus, Gerschenson, & Ollé, 2016; Nithya & Halami, 2012; Quinto et al., 2019). According to WHO (World Health Organization., 2020), almost 600 million people in the world fall ill after eating contaminated food and 420,000 die every year. In addition, microbiological alterations significantly decrease the food shelf-life stability, causing considerable losses in the food industry (Jayant & Halami, 2020; Kourmentza et al., 2021; Quintavalla & Vicini, 2002). To overcome this complex scenario, it is crucial to adopt measures that ensure food safety throughout the entire food chain, from raw material handling, processing, distribution, and storage, to end-products consumption (Said, Fliss, Offret, & Beaulieu, 2018). Despite several preservation methods that have been developed to assure microbial food safety, globalization in food production, together with the growing consumer demand for minimally processed, more natural, highly nutritional fresh food products, demand major challenges to assure safe, health-promoting, and high-quality food (Jagus, Gerschenson, & Ollé, 2016; Realini & Marcos, 2014).

Packaging is not only a way of protecting and preserving food during handling, transportation, and storage, but also has secondary functions such as sales promotion, customer service, and brand communication that have grown in importance over the years (Contreras, Charles, Toselli, & Strumia, 2017). In packaging development, it must be considered that the chosen design must satisfy the demands related to the product and its conservation, marketing and sale, consumer comfort and needs, and environmental issues related to its disposal. In addition, the design and use of a suitable and convenient packaging system, that is, primary, secondary, and tertiary packages and accessories, are critical issues in the food supply chain. The importance of packaging has increased notably in recent years, associated with the development of electronic commerce operations, which has been strongly boosted by the COVID-19 pandemic.

The European Organization for Packaging and the Environment stated in their New Circular Economy Action Plan that packaging is an integral and essential part of the product supply chain, from the production to the consumption stage. They consider that the packaging supply chain plays a central role in contributing to a resource-efficient and circular economy by optimizing resource use, minimizing waste (food and packaging), and protecting products all along the value chains (EUROPEN, 2021).

Food products require adequate packaging and conservation conditions to move along this supply chain and therefore, the analysis of physical logistics flows and the role of packaging are essential and can influence the definition and design of manufacturing processes (Contreras et al., 2017).

Food Packaging and Preservation.
DOI: https://doi.org/10.1016/B978-0-323-90044-7.00006-9

The characteristics of the packaging systems (e.g., shape, size, design, materials, introduction of active or intelligent components) must respond to the demands of the product (composition, shelf life, and storage conditions) and its supply chain, to improve the performance of companies and minimize their costs. Environmental regulations and consumer demands for more environmentally friendly products are also factors that affect the formulation and design of food packaging systems. Some environmental aspects related to packaging systems should be considered for compliance with sustainability, environmental responsibility, and recycling regulations, such as the use of only the packaging strictly necessary to prevent waste, the minimization of mass and volume of packaging materials, the promotion of reuse (returnable packages) and recycling, and the evaluation of the final disposal (Cha and Chinnan, 2004).

Packaging systems generally consist of primary, secondary, and tertiary levels. The primary packaging provides the structure of the package. It is usually the smallest unit of use and distribution and is in direct contact with the product. The secondary package is related to the visual communication of the product and is used to group the primary packages. Finally, the tertiary package is used for storage and transport shipping. Therefore, packaging systems can fulfill varied general purposes such as physical protection of the content, hygiene and conservation, containment or agglomeration of small parts, information transmission, marketing, and security (traceability, real-time product tracking, etc.) (Robertson, 2012).

6.2 Active packaging

Prof. Gordon L. Robertson defines active packaging as:

Packaging in which subsidiary constituents have been deliberately included in or on either the packaging material or the package headspace to enhance the performance of the package system (Robertson, 2012).

Active packaging is the packaging system that performs, in addition to the basic function of barrier to the external environment, some desired functions that are obtained through the incorporation of active components in the packaging or the use of functional polymers (Han, 2003; Ozdemir & Floros, 2004; Realini & Marcos, 2014; Robertson, 2012; Soltani et al., 2021). These packagings interact with the food or the headspace between the package and the food to achieve goals such as increasing shelf life, improving safety, preserving sensory properties, and maintaining the quality of the product. Active packaging is also used in the pharmaceutical industry and other consumer goods with similar objectives (Han, 2003; Silvestre, Duraccio, & Cimmino, 2011).

The active packaging must fulfill the following conditions (Prasad and Kochhar, 2014):

- The materials must be appropriate and effective for their intended use.
- Good manufacturing practices should be used to manufacture active materials.
- Active components and their released quantity to the food must be informed to the consumer. Its use must be permitted.
- Mandatory "DO NOT EAT" labeling must be provided to enable consumers to distinguish non-edible parts when these may be perceived as edible. This information must be visible, legible, and indelible.

The active compounds commonly incorporated into an active packaging system are antimicrobials, preservatives, oxygen absorbers, water vapor absorbers, and ethylene removers (Brody, Bugusu, Han, Sand, & McHugh, 2008). According to the release mechanism of the active agent, active packaging can be classified into two main types, non-migratory active packaging, acting without intentional migration, and active releasing packaging, allowing controlled migration of non-volatile agents or emission of volatile compounds in the atmosphere surrounding the food (Bastarrachea, Dhawan, & Sablani, 2011; Contreras, Charles, Toselli, & Strumia, 2017).

Active packaging is a widely reviewed topic in the literature from different approaches (Bastarrachea, Wong, Roman, Lin, & Goddard, 2015; Hernandez-Izquierdo & Krochta, 2008; Kim, Kim, Han, & Kimmel, 2008; Ozdemir & Floros, 2004; Prasad & Kochhar, 2014; Realini & Marcos, 2014; Robertson, 2012; Rooney, 1995; Soltani et al., 2021; Brody, 2004) and on which there is growing interest from the food industry as a strategy for food preservation.

6.3 Antimicrobial packaging systems

One of the main causes of the decay in the quality and safety of food is the growth of pathogenic and spoilage microorganisms. Antimicrobial food packaging is a particular and innovative form of active packaging that reduces, inhibits, or stops the growth of food-borne microorganisms on food surfaces and conserves its quality.

In this kind of packaging, the active agent that is deliberately incorporated into the packaging is a compound with antimicrobial activity released into the environment surrounding the packaged food to exert its antimicrobial action (Han, 2003; Malhotra, Keshwani, & Kharkwal, 2015; Sofi et al., 2018). Several substances with antimicrobial activity from different origins, such as essential vegetable oils, antibiotics, silver ions, inorganic gases, organic salts, enzymes, metabolites of bacterial origin (Aider, 2010; Sung et al., 2013; Appendini and Hotchkiss, 2002), and viable cells of probiotic (Espitia, Batista, Azeredo, & Otoni, 2016) have been used for this purpose. This chapter specifically addresses those compounds with antibacterial activity from microorganism sources.

The use of antimicrobial packaging has been driven by consumer interest in more natural and healthy foods with fewer preservatives (Miranda et al., 2016; Appendini and Hotchkiss, 2002). In particular, the advantages of using antimicrobial packaging are:

- It prevents applying antimicrobials directly to food, reducing the total amount of antimicrobial in the product.
- It reduces the amounts of antimicrobials to be applied, as higher antimicrobial concentrations could be maintained on food surfaces compared to direct application to the food itself since it is applied only where it is required.
- It allows antimicrobials to slowly migrate from the packaging material to the food surface maintaining a high concentration of antimicrobials on the food surface, compensating for their consumption during storage.

Different strategies have been studied and implemented to control the amount of antimicrobial released from the container throughout the shelf life of the product, reducing the amount of active agent required (Chawla, Sivakumar, & Kaur, 2021; Sung et al., 2013).

One of the main drawbacks of this technology is that the effectiveness of antimicrobial packaging can be seriously affected by several factors. In general, laboratory conditions cannot be reproduced on an industrial scale and the obtained results change significantly with scale change, (e.g., polymer-embedded antimicrobials). Additionally, research involving antimicrobials at levels well above what is sensory acceptable and feasible and above regulatory limits lacks applicability and is distracting. Besides, the main obstacle to the commercialization and popularization of antimicrobial packaging use is the ability of the package to retain and release antimicrobials with reliable efficacy and compliance with regulatory constraints (López-Carballo et al., 2012). Incorporating antimicrobials into packaging is an important opportunity that requires commitment and joint work from both academia and industry.

6.4 Design of antimicrobial packaging systems

Antimicrobial packaging can be formed as homogeneous or composite materials, micro or nanostructured, that consist of the combination of two or more phases separated by an interface. The active compounds can be integrated or form one of these phases (De Azeredo, 2013). The release kinetics of the active compound is related to the interaction between the active agent and the matrix where it is incorporated as micro or nanostructures (particles, capsules, droplets, or films), and with the material of the other phases (diffusivity, solubility, ionic charge, etc.).

Release kinetics of a compound depends on the packaging materials (polymer composition and glass transition temperature [Tg]), the storage conditions (time, temperature, and relative humidity), and the composition and aggregation state of the food. Antimicrobial release from the polymer can be controlled by the Tg of the material, considering that the diffusivity of the antimicrobial compound in the polymeric material is generally greater in the amorphous than in the vitreous state (Robertson, 2012). This allows antimicrobial release to accelerate as the temperature approaches favorable conditions for microbial growth.

The suitable antimicrobial packaging solution must be selected based on food product characteristics, the target spoilage microorganisms to be controlled, the specific requirements of the product conservation, and its conventional packaging systems. Once the microorganisms of interest for a particular food are identified, specific research must be carried out to obtain antimicrobial packaging that is effective against them. Antimicrobial agents are selected based on food characteristics and a viable antimicrobial packaging solution is evaluated based on feasible modifications of the typical package of the food. Since the antimicrobial compounds can migrate into the food, it is considered an indirect food additive. Therefore, it must be generally recognized as safe (GRAS) or authorized material for contact with food. Also, it must be approved for its use in the estimated concentration that will be incorporated into the food by migration from the container (Robertson, 2012; European Commission, 2009; European Commission, EC 1935, 2004). The choice of the antimicrobial compound is specific to the temperature the food will experience during its shelf life, the microbiota of the product, regulations, and consumer sensory preferences.

As it was stated before, the design of an antimicrobial packaging system depends on the use for which it is intended. Han (2003) proposed the following factors to be considered in the development of antimicrobial packaging (Han, 2003):

1. Chemical nature: Chemical characteristics of the packaging material and the processes associated with its industrial production (e.g., extrusion, lamination, thermo-compression, printing), considering their effect on residual antimicrobial activity of the active compounds after introduction and processing. Compatibility between the active substance and the packaging material (solubility, difference in polarity).
2. Interaction between antimicrobial substance and food: Characteristics of the food (composition, pH, water activity, aggregation state) can have a direct impact on the activity of the antimicrobial agent. The particular microbiota of each food must be considered when selecting and incorporating the antimicrobial agent (Quintavalla and Vicini, 2002).
3. Storage temperature: The temperature can affect the antimicrobial activity of the active compound and its solubility and diffusivity in the food. The temperature has also an important influence on the rate of microbial growth.
4. Functional and physical properties of packaging materials: The incorporation of active substances into the packaging material can modify its barrier and mechanical properties and its final performance.

Antimicrobial packaging is often combined with other active packaging technologies to achieve a synergistic effect (e.g., modified atmosphere, humidity controllers, gas absorbers Brody, 2004) (McManamon, Kaupper, Scollard, & Schmalenberger, 2019; Brody, 2004). Antimicrobial packaging is commonly used in processed meats and seafood, minimally processed fruits and vegetables, cheese, and baked goods. Antimicrobials are mainly incorporated into the material in direct contact with food and that is in turn inside containers that can be made of plastic, cardboard, glass, or multicomponent (Chawla et al., 2021).

Antimicrobial packaging systems can take the following forms (Fang, Zhao, Warner, & Johnson, 2017; Fernández, Cava, Ocio, & Lagarón, 2008; Kim, Kim, Han, & Kimmel, 2008; Robertson, 2012; Appendini and Hotchkiss, 2002; Sofi et al., 2018):

1. Addition of sachets or pads containing volatile antimicrobial agents into packages. The sachets or pockets are enclosed into the inner part of the package and employ volatile substances that inhibit microbial growth when released into the food environment.
2. Incorporation of volatile and non-volatile antimicrobial compounds directly into polymers matrix. The active material is generally processed by extrusion, compression, or injection, which requires thermal and shearing forces, conditions that can affect the stability and activity of the antimicrobial agent.
3. Coating or adsorbing antimicrobials onto polymer surfaces. The polymeric material acts as a carrier of the antimicrobial additives, which are coated onto the film as a thin layer by soaking or spreading. Volatile compounds are released into the package headspace through evaporation and non-volatile substances migrate into the food product through diffusion. Weaker unions such as hydrogen bonds, ionic or hydrophobic interactions, or the Van der Waals forces are responsible for the adsorption onto film surfaces.
4. Immobilization of antimicrobials to polymers by ionic or covalent linkages. This procedure reduces the amount of antimicrobial agent added to the polymer matrix, maintaining its concentration and antimicrobial activity, and moderating migration into the food. The type of bonding controls the release of active agents from the matrix.
5. Inherently antimicrobial polymers: This option requires direct contact between antimicrobial polymer material and the food product for effective inhibition. The solubility and diffusivity of the polymer in the food must be considered, which requires its approval as a food additive.

For each application, the most suitable form must be evaluated.

6.5 Mechanism of antimicrobial action of the package

Films with antimicrobial activity used as a food packaging system can be classified into two groups: (1) films that allow the antimicrobial to migrate into food and (2) films that do not release antimicrobial substances and that inhibit microbial growth on the surface of the food (Contreras, Charles, Toselli, & Strumia, 2017; Suppakul, Miltz, Sonneveld, & Bigger, 2003). The antimicrobial substance must be in contact with the food surface and must be maintained above the minimal inhibitory concentration for the target microorganisms during the shelf life of the product for effective biostatic or biocidal action.

Some factors can affect the performance of the antimicrobial packaging system such as the interaction between the antimicrobial substance and the packaging material or the effect of the film production process on the antimicrobial activity. In addition, the antimicrobial substance can interact with food components (protein, carbohydrate, lipids, etc.) causing the inactivation or elimination of the biocide activity. This was observed even when antimicrobial action was confirmed by in vitro analysis (Ahmed et al., 2020; Ture, Eroglu, Ozen, & Soyer, 2011).

The transference of the antimicrobial from the packaging material to a particular food and its mechanism of action must be studied in detail for each system to obtain a complete description of the phenomena involved in the inactivation of pathogens that can affect that food. Different parameters such as (1) partition coefficient of the active agent in the different phases; (2) diffusion phenomenon of the active substance through the packaging material; (3) volatilization of the antimicrobial into the headspace (if the substance is volatile); (4) solubility and diffusivity of the antimicrobial into the food, especially if it directly interacts with the food, must be considered in this study (López-Carballo et al., 2012).

In the case of packaging materials formed by a film matrix and a coating or multilayer film that releases the antimicrobial agent on the food and the headspace, the mass transport process involves equilibrium and kinetic phenomena. The equilibrium phenomena are related to the partition coefficient of the active compound between the film matrix and the coating, the phases of a multilayer film packaging, and the inner packaging layer or coating and the food. The partition coefficient is a quantitative parameter that represents the relative solubility of a given substance in a system composed of two phases that are immiscible with each other, at a specific temperature. Ideally, materials should be selected to reduce the loss of the antimicrobial agent by retention in the packaging material, to increase the concentration on the food surface (López-Carballo et al., 2012). In addition, the high solubility of the antimicrobial in the food matrix generates a rapid penetration into the food, producing a decrease in the active concentration on the food surface. On the contrary, a low solubility causes the accumulation of the antimicrobial on the food surface, improving protection and reducing the migration of the substance into the food matrix (Cutter et al., 2001).

On the other hand, the kinetic factors associated with the mass transference process involve an appropriate control of the antimicrobial compound diffusion from the packaging system to the food surface. It can help to retard or inhibit the initial growth of the pathogens on the food surface, and can also create a prolonged release or a residual activity that remains during food storage and distribution (Cutter, Willett, & Siragusa, 2001; Han & Floros, 1997; Nerin, Silva, Manso, & Becerril, 2016).

The knowledge of the diffusion process can be used to determine the concentration of antimicrobials in the packaging system to guarantee the necessary amount to maintain the antimicrobial levels above the minimum inhibitory concentration (Contreras et al., 2017). The composition of each phase and other chemical and physical factors affect the diffusion process. The hydrogen bonds, ionic bonds, ionic osmosis, hydrophobic interactions, and electrostatic interactions are some of these factors that influence antimicrobial diffusivity through the packaging systems and the food. The diffusion process also depends on some physical factors such as Tg, free volume, crystallinity, porosity, and the presence of micro or nanoparticles that provide a modification of the tortuosity in the diffusive pathway of the antimicrobial agent through the package. Moreover, due to the complicated composition of foods and the different environmental conditions to which they are exposed during the different stages of their storage (pressure, temperature, and time combinations), a critical study must be performed to evaluate the type and concentration of the appropriate antimicrobial substance for specific food (Chawla et al., 2021). The kinetics of the antimicrobial release can be decisive for the viability and application of the antimicrobial packaging system for a particular food. If the release of the antimicrobial substance is slower than the microbial growth rate, the active packaging system becomes useless. On the contrary, a fast release can exceed allowable levels of the antimicrobial in the food initially and shorten the period of the antimicrobial action of the container.

The migration of antimicrobial agents can be divided into two categories: controlled and uncontrolled release systems. Uncontrolled release packaging systems have widespread use in food applications. However, controlled release systems may be more relevant in these applications due to their ability to prevent sensory or toxicological problems or generate inefficiency of the system, caused by too high or too low concentration of the released substance. The main source of knowledge of controlled release technology comes from the pharmaceutical sector (Mastromatteo, Mastromatteo, Conte, & Del Nobile, 2010). The release mechanics are based on different mechanisms such as diffusion, swelling, dissolution, or degradation of the substance through the matrix to the target point. Depending on whether the mechanism responds to Fick's laws of diffusion, the mechanisms are divided into Fickian and non-Fickian (Chawla et al., 2021).

Generally, the active agent release mechanism can be described through three phenomena (Langer and Peppas, 1983):

Reservoir system: The active agent is contained in reservoirs that work like a rate-controlling barrier of release. The barriers can present different morphologies like micro-porous, macro-porous, or non-porous (dense films). The release rate depends on several variables such as thickness, area, and permeability of the barrier. The rate-limiting step is the diffusion of the active substance through the polymeric barrier.

Swelling-induced release: These systems are characterized by the low diffusivity of the active substance in the polymeric matrix of the antimicrobial packaging. When the matrix is in contact with a compatible liquid medium, the film matrix swells due to the absorption of the fluid. The diffusion coefficient of the active agent in the swollen part of the matrix increases and then diffuses through the packaging system and can act on the food. In these cases, the antimicrobial must have a high solubility in the fluid. Diffusion is possible due to the existence of a swollen zone of the polymer that allows the transfer of the antimicrobial. In swelling-controlled systems, the film must change from a glassy state to a rubbery state when it interacts with the food system. The release rate is determined by the glass-to-rubber transition process.

Degradation-induced release: Two different erosion mechanisms have been proposed: (1) superficial or heterogeneous erosion; (2) bulk or homogeneous erosion. The erosion is superficial when the degradation of the polymer is faster than the absorption of the solvent in the polymer mass. Degradation occurs mainly in the outer layers of the polymer, thus affecting only the surface and not the internal parts of the matrix (heterogeneous process). Instead, bulk erosion occurs when the absorption of the solvent (generally water) by the system is much faster than polymer degradation. In this case, the entire system is rapidly hydrated, and the polymer chains are cleaved. Thus, erosion is not limited to polymer surfaces (a homogeneous process). Generally, polymers that are built from highly reactive functional groups tend to degrade rapidly and to be surface eroded, whereas polymers containing less reactive functional groups tend to be bulk-eroded (Langer and Peppas, 1983).

6.6 Antimicrobial substances from microbial sources

Natural bioactive compounds are produced by living organisms either as primary metabolites, with an essential function in cell life processes (growth, development, and reproduction), or as secondary metabolites, which are not directly involved in primary metabolic processes but usually have other important function, as defense compounds, signaling molecules, among other (Ali, Siddiqui, & Khan, 2018). Some of these naturally occurring metabolites can inhibit microbial growth, being considered natural antimicrobials. They have been recovered from different sources including plants (fruits, vegetables, seeds, herbs, and spices), animals (eggs, milk, and tissues), and microorganisms (bacteria, fungi, and viruses) (Lucera, Costa, Conte, & Del, 2012; Pisoschi et al., 2018; Quinto et al., 2019).

In general, the term "preservatives" refers to food-grade compounds that inhibit or prevent microbial detrimental growth in food products (Jayant and Halami, 2020). Particularly, the use of natural preservatives for the removal of undesirable microorganisms in food with the aim of reducing its level of processing, improving its safety, and extending its useful life, is known as "biopreservation" (Gyawali & Ibrahim, 2014; Pisoschi et al., 2018; Quinto et al., 2019). In turn, about antimicrobial compounds of microbial origin, Montville & Chikindas (2007) define "biopreservation" as the use of microorganisms (including bacteriophages), their metabolic products, or both to preserve foods that are not generally considered fermented (Montville & Chikindas, 2007).

Hence, antimicrobials of natural origin emerge as a new perspective for food preservation, since they can exert specific inhibition against foodborne pathogens, not only ensuring food safety and quality but also extending product shelf-life and reinforcing consumers' confidence. These natural compounds also arise as harmless substitutes to detrimental chemical synthetic preservatives, which in some cases can be carcinogenic and mutagenic, among other negative effects (Achi & Halami, 2016; Pisoschi et al., 2018). Furthermore, natural antimicrobials strengthen as a viable option to microbial resistance caused by antibiotics misuse, which led to the development of multidrug-resistant microorganisms, including foodborne pathogens which become resistant to commonly used antibiotics and to conventional food processing and preservation techniques (Gyawali & Ibrahim, 2014; Pisoschi et al., 2018; Quinto et al., 2019).

The mode of action observed for natural antimicrobials include cell lysis, cell membrane rupture, interference of nucleic acids mechanisms, the decay of the proton motive force, and depletion of adenosine triphosphate (ATP) (Pisoschi et al., 2018; Quinto et al., 2019). When selecting natural antimicrobials as food bio preservatives, their inhibitory spectrum should be considered, since it is desirable, they exert a specific antagonistic effect against foodborne pathogens, with minimal consequences on desirable or beneficial food microorganisms. Additionally, nowadays there is growing concern about bioactive compounds not to influence negatively the consumer's own microbiome (Pisoschi et al., 2018).

The complexity of food matrices can affect or interfere with natural antimicrobial compounds' activity (Quinto et al., 2019). Therefore, biopreservation techniques could render more effective when a mixture of two or more natural antimicrobials is applied in combination with other food preservation techniques, achieving a synergistic effect. "Hurdle technology" has gained attention in the food industry since the final costs of production get diminished and minimal impact on the nutritive value and sensory properties of foods is accomplished, together with enhanced food safety (Deegan, Cotter, Hill, & Ross, 2006).

Although the natural origin of bio preservatives positions them as GRAS, their use in several countries remains unregulated and their incorporation as an ingredient in food will require, in many cases, legal approval (El-Saber Batiha et al., 2021; Lucera, Costa, Conte, & Del, 2012). Hence, proper safety regulations should be established and uniformed worldwide to benefit natural preservatives application with a growing and expanded market in the food industry (Pisoschi et al., 2018).

6.7 Classification, sources, and inhibitory spectrum

Most commonly antimicrobials synthesized by different microorganisms are described below, with special attention to those compounds that have been immobilized in the polymeric matrix and/or applied as food bioactive packaging or coatings.

6.7.1 *Bacteriocins*

Bacteriocins comprise a heterogeneous group of small ribosomal synthesized proteinaceous molecules excreted by a wide variety of Gram-positive or Gram-negative bacteria that inhibit or stop the growth of other bacteria at precise concentrations (Chikindas, Weeks, Drider, Chistyakov, & Dicks, 2018). These antimicrobial peptides have a bacteriostatic or bactericidal effect mainly against bacteria closely related to the producing strain or more rarely against other groups of bacteria (Cotter, Ross, & Hill, 2013; Soltani Firouz, Mohi-Alden, & Omid, 2021).

Bacteriocins are mostly synthesized as non-biologically active precursor peptides that undergo significant post-translational modifications before cleavage of the leader region and extracellular release (Soltani Firouz et al., 2021). These modifications also differentiate their antimicrobial spectrum of activity (Jayant and Halami, 2020). Besides, bacteriocin-producing cells are immune to themselves, since they also set up the production of an "immune protein" (de Freire Bastos, Varella Coelho, & da Silva Santos, 2015).

Most bacteriocins act by forming selective pores or channels in target microorganism membrane cells, resulting in increased permeability and consequently leakage of low molecular mass intracellular components and ions with the disruption of the proton motive force and depletion of intracellular ATP, and eventual death (Kumariya et al., 2019; Pisoschi et al., 2018).

Several bacteriocins have been described and shown to be effective against many foodborne pathogenic bacteria. However, bacteriocins synthesized by lactic acid bacteria (LAB) find particular interest as natural and safe bio preservatives in the food industry, due to the GRAS status of the producing strains (Balciunas et al., 2013). Bacteriocin information has been assembled in different databases, like Bactibase, Bagel, or LAB iocin, which are available online and facilitate comparison between different bacteriocins (peptide sequence, inhibitory spectrum, physicochemical properties, etc.) (Said et al., 2018).

Among the most relevant bacteriocins produced by LAB in food preservation, the following can be mentioned.

6.7.1.1 Nisin

Nisin is synthesized by certain *Lactococcus lactis* subsp. and lactis strains. This positively charged heat-stable peptide belongs to Class I bacteriocins and has 34 amino acids with a 3.5 kDa molecular mass. Nisin has antimicrobial activity against a broad spectrum of Gram-positive bacteria, including LAB, pathogens such as *Listeria*, *Staphylococcus*, and *Mycobacterium*, as well as spore-forming bacteria, such as *Bacillus* and *Clostridium*. However, it is less effective against Gram-negative bacteria, yeasts, and molds (El-Saber Batiha et al., 2021; Jagus, Gerschenson, & Ollé, 2016; Pisoschi et al., 2018). Its antimicrobial action is related to the adsorption of the negatively charged phospholipids on the cell membrane and the subsequent rupture of the membrane by the formation of pores. Thus, only Gram-positive bacteria are affected by nisin. However, Gram-negative bacteria can be inhibited by nisin when they are affected by some processing technologies like heating, freezing, or chelating agents that may cause the permeabilization of their outer membrane (Campos, Schelegueda, Gliemmo, & Barros-Velázquez, 2016).

Nisin A was the first isolated form of this antimicrobial peptide, while other natural variants have been described differing by up to 10 amino acids from the nisin A sequence (nisin Z, Q, F synthesized by *L. lactis*; nisin U1, H, P, produced by *Streptococcus* strains, and nisin O detected in *Blautia obeum*) (Jayant and Halami, 2020). Currently, nisin represents the only bacteriocin approved as a food additive in more than 50 countries (Delves-Broughton and Weber, 2011).

6.7.1.2 *Pediocin*

Pediocins are synthesized by *Pediococcus* strains (*P. acidilactici* and *P. pentosaceus*). They are thermostable small peptides and maintain their functionality over a broad pH range. Pediocins primarily inhibit the growth of Gram-positive pathogens, such as *Listeria monocytogenes*, *Enterococcus faecalis*, *Staphylococcus aureus*, and *Clostridium perfringens* (El-Saber Batiha et al., 2021; Pisoschi et al., 2018; Villalobos-Delgado et al., 2019). Pediocin PA-1/AcH is commercialized as ALTATM2341 and MicroGARDTM, a fermented LAB dried powder with GRAS character, although not approved as a food additive (Balciunas et al., 2013).

6.7.1.3 *Reuterin*

Reuterin (β-hydroxypropionaldehyde), is produced by *Lactobacillus reuteri*. It is water-soluble, thermoresistant, active over a broad pH range, and resistant to proteolytic and lipolytic enzymes (Gyawali & Ibrahim, 2014; Pisoschi et al., 2018; Villalobos-Delgado et al., 2019). Reuterin modifies thiol groups in protein and small molecules, inducing oxidative stress in cells by eventually causing cell death (Gómez-Torres, Ávila, Delgado, & Garde, 2016). This bacteriocin presents a broad spectrum of antimicrobial activity toward several pathogenic and spoilage microorganisms, including Gram-positive and Gram-negative bacteria, yeasts, and molds (Montiel et al., 2016). Inhibitory activity has been reported against *L. monocytogenes*, *Campylobacter jejuni*, *S. aureus*, *Salmonella enterica* serovar Typhimurium, *Escherichia coli* O157:H7, and *Yersinia enterocolitica* (El-Saber Batiha et al., 2021; Pisoschi et al., 2018).

6.7.2 Natamycin

This antimicrobial, also known as pimaricin, is a macrolide antifungal polyene with a molecular weight of 665.7 Da, produced by fermentation of *Streptomyces* species (*S. natalensis*, *S. chmanovgensis*, and *S. Gilvosporeus*) (Duchateau and van Scheppingen, 2018). Natamycin acts by binding irreversibly to ergosterol, thus disrupting specifically the fungal cell membrane leading to cytoplasm efflux (Carocho et al., 2015). Therefore, it is active against almost all foodborne yeasts and molds, but not against bacteria, viruses, or protozoa (Jagus et al., 2016). It has the approval and is used as a food additive in over 40 countries, mainly in dairy-based food products, especially in hard cheese and salami-type sausages for preventive control of fungi spoilage (Carocho, Morales, & Ferreira, 2015; Pisoschi et al., 2018).

6.8 Bacteriocins and lipopeptides from Bacillus

Bacillus genus is known to produce a large scale of bioactive molecules including enzymes, antibiotics, insecticides, and antimicrobial substances, like lipopeptides and bacteriocins (Abriouel, Franz, Omar, & Galvez, 2011; Achi & Halami, 2016; Meena & Kanwar, 2015; Stein, 2005). Despite its technological potential, the genus *Bacillus* has not been widely used in the food industry since not all species are recognized as GRAS. However, some representatives, such as *B. subtilis* and *B. licheniformis* are found safe for use in the food and agricultural industry (Abriouel, Franz, Omar, & Galvez, 2011; Kourmentza et al., 2021; Nithya, Murthy, & Halami, 2013). Moreover, antimicrobials from *Bacillus* are gaining significance and becoming increasingly important due to their broader spectrum of inhibition (in comparison with nisin or other LAB bacteriocins), with activity against Gram-positive and Gram-negative bacteria, yeast, and fungi; also, they have better heat and pH stability (Abriouel, Franz, Omar, & Galvez, 2011; Achi & Halami, 2016; Stein, 2005).

Based on the biosynthetic pathway, *Bacillus* antimicrobials can be classified into two categories. The first group includes ribosomally synthesized peptides, among which are bacteriocins such as subtilin, ericin S, and subtilisin

(Abriouel, Franz, Omar, & Galvez, 2011; Achi & Halami, 2016; Jayant & Halami, 2020; Stein, 2005). While the second group comprises non-ribosomally synthesized peptides, such as lipopeptides, constituted by amphipathic cyclic structures of hydrophilic peptide sequences, usually 7 to 10 amino acids long, with a hydrophobic C13–C18 fatty acid chain (Achi & Halami, 2016; Villalobos-Delgado et al., 2019). The better-characterized lipopeptides are surfactin, iturin, and fengycin (Kourmentza et al., 2021; Meena & Kanwar, 2015). The use of these cyclic ring-structure lipopeptides can reduce protease sensitivity affecting antimicrobial peptides and advantages their future application in the food additives market (Meena and Kanwar, 2015).

6.9 ε-Polylysine

This antimicrobial is a cationic poly-amino acid formed by 25–30 lysine residues produced by different species belonging to *Streptomyces*, *Kitasatospora* genus, among others (Shukla, Singh, Pandey, & Mishra, 2012; Zhang et al., 2019). It has FDA approval as a natural antimicrobial food additive according to GRN N° 336 as an antimicrobial agent in a wide variety of food categories at levels of up to $0.025\%_{w/w}$ since 2010 and with GRN N° 135 for use as an antimicrobial agent in cooked rice and sushi rice at levels from 5 to 50 ppm since 2003 and has been mainly used in Japan, Korea, and USA (Chheda and Vernekar, 2015). ε-Polylysine is common in food applications like boiled rice, cooked vegetables, soups, noodles, and sliced fish (sushi).

The e-polylysine has a strong inhibitory activity against a wide range of Gram-positive and Gram-negative bacteria, including *Bacillus coagulans*, *S. aureus*, *E. coli*, and *S. enterica* serovar Typhimurium, yeasts, and molds, such as *Aspergillus niger*, *Trichophyton mentagrophytes*, *Candida* spp., and *Phaggia rhodozyma*, and against bacteriophages (Lopez-Pena and McClements, 2015). The high cationic charge density of this polymer allows its absorption onto anionic microbial cell surfaces due to electrostatic interactions, damaging the outer membrane of bacteria, leading to the disruption of the cytoplasm, cellular content efflux, and eventually cell death. Nevertheless, of their high water solubility, these highly charged molecules can also interact with various components of complex food matrices, such as binding with acidic polysaccharides, hydrochlorides, phosphates, copper ions, or others, reducing their antibacterial activity and restricting their range of application in the food industry (Lopez-Pena & McClements, 2015; Villalobos-Delgado et al., 2019). On the contrary, its activity is not affected by pH and is stable when heated (120°C for 20 minutes). Therefore, it can be sterilized along with the raw material. ε-Polylysine antibacterial activity can be enhanced with hydrochloric acid, citric acid, malic acid, glycine, and higher fatty glycerides.

6.10 Techniques for measuring antimicrobial packaging action

Different methods have been used to assess the antimicrobial efficiency of active films. Generally, this antimicrobial activity is determined in culture media and/or in food models to establish the effect of the active film against foodborne pathogens and spoilage microorganisms. Even though each author applies different procedures or analytical conditions according to their own antimicrobial-packaging-target microorganism-food system, there are no standards established to test the antimicrobial performance in vitro or in food systems (Abdollahzadeh, Nematollahi, & Hosseini, 2021; Moradi et al., 2021). The most used techniques and some analytical variables are presented in the following sections.

6.10.1 *In vitro assays in culture media*

6.10.1.1 *Agar-based methods*

Disk diffusion and *well diffusion* are among the most common methods, in which the antimicrobial activity is assessed against sensitive or target cells grown on an agar layer (Moradi et al., 2021). These are low-cost, simple, and flexible methods, but they only detect the presence or absence of antimicrobial activity, hence they are considered qualitative procedures (Abdollahzadeh et al., 2021). Briefly, in the *disk diffusion* technique, a sterile plate with a solidified general non-selective agar is inoculated with the target microorganisms, and a piece of sterile disk or square-shaped film, aseptically cut, is placed on the agar (see Fig. 6.1). The plates are incubated and the development of zones of inhibition on the lawn surrounding film diss is observed (Abdollahzadeh, Nematollahi, & Hosseini, 2021; Moradi et al., 2021).

FIGURE 6.1 Agar diffusion technique applied on edible active film Foto by the authors. No Permission Required.

The *well diffusion* technique is preferably used when assessing the antimicrobial activity of antimicrobial polymer casting solutions. The target microorganisms are inoculated on the surface of a sterile agar plate and 5–8 mm diameter wells are cut on the agar with a sterile punch. The antimicrobial studied diffuses through the agar; so, after incubation, inhibition zones are measured around the wells (Abdollahzadeh, Nematollahi, & Hosseini, 2021; Balouiri, Sadiki, & Ibnsouda, 2016).

Also, the *disk volatilization method* can be used when volatile antimicrobials are incorporated in the active films, the *agar spot diffusion* technique is appropriate for high-viscosity antimicrobial solutions, while the *parallel streak method* can be used when handling leaching antimicrobials (Moradi et al., 2021).

The choice of the procedure depends on the antimicrobials and packaging matrix under study. Several parameters should be considered, such as quantity or size of antimicrobial material, incubation time/temperature, target microorganism, culture medium, and inoculum concentration (Dafale, Semwal, Rajput, & Singh, 2016). Regarding the target microorganisms, different culture agar media have been used (tryptic soy (TS), nutrient agar, brain heart infusion (BHI), or Luria-Bertani (LB)), but Muller Hinton (MH) (for bacteria) and yeast extract dextrose chloramphenicol (YGC) (for fungi) agar are preferred for antimicrobial assays (Moradi et al., 2021). The agar plate is generally seeded with an inoculum of ca. 6 log CFU mL^{-1}, however, doses between 5–8 log CFU mL^{-1} have been used (Abdollahzadeh et al., 2021).

Biopolymeric antimicrobial matrices are aseptically cut generally into 6 to 16 mm diameter disks or 1×1 cm squares and sterilized prior to being placed on the agar surface. According to the antimicrobials and polymer characteristics, the active packaging can be sterilized by autoclaving (121°C for 15 minutes), exposing both sides of the films under UV light (254 nm for 2 minutes–10 minutes) or immersing in ethanol (15 minutes), in the case of synthetic polymers. Casting solutions also can be sterilized by filtration (0.45 mm sterile syringe filter) (Abdollahzadeh, Nematollahi, & Hosseini, 2021; Moradi et al., 2021). After incubation, the inhibition zones surrounding the films loaded on the agar surface are observed and measured in millimeters (mm). Some authors subtract the whole inhibition diameter from the diameter of the film or well, while others report the whole inhibition zone diameter (Moradi et al., 2021). Also, the Antimicrobial index calculated as Z–F/F, where Z and F are inhibition zone areas surrounding and under the film disks, respectively, can be determined (Ahmad et al., 2012). The antimicrobial effect of the active films can be classified according to the size of the inhibition diameter, but there is no consensus on this classification and the results will vary according to each experimental condition. For example, the response to edible coatings can be classified as not sensitive (−) for diameters <8 mm; sensitive (+) for diameters 9–14 mm; very sensitive (+ +) for diameters 15–19 mm; and extremely sensitive (+ + +) for diameters >20 mm (Ponce et al., 2003).

6.10.1.2 Viable cell count method

This method allows the study of the antimicrobial release profile as well as the effectiveness of the inhibitory activity in direct contact with target cells in aqueous media. Again, there are several factors to consider (bacterial/fungal type and inoculum, type of culture medium, incubation time, and film characteristics, among others), and although this is an in vitro assay, experimental conditions should be settled as similar as possible to mimic the food matrix in which the active film could find application.

In general, a volume of broth medium (ca. 1–50 mL) and appropriate pieces of antimicrobial film cut aseptically (for example, 1 × 1 cm) are inoculated with ca. 6 log UFC mL^{-1} of target microorganism and incubated at the optimum temperatures and time for the microbial growth (up to 48 hours for bacteria and yeast and 4–7 days for mold). Incubation under agitation is recommended, and samples are taken out at different time intervals to prepare serial dilutions and determine cell viability by agar plate counting (Moradi et al., 2021).

This is considered a semi-quantitative method when it is conducted in a broth culture medium, but more accuracy can be obtained when a neutral solution (e.g., saline solution, peptone water, or PBS buffer) in the absence of a nutrient source is used. Also, the dimension of the antimicrobial film used in the assay, as well as the ratio between the antimicrobial film surface and the culture media volume are important parameters. The assay can be aseptically conducted using tubes, Erlenmeyer flasks, or well plates. The more common ratio used is 1 × 1 cm films per 1–10 mL of broth, but another film/culture media ratio was analyzed; even, the desired amount of active film is sometimes weighted (Moradi et al., 2021). The inclusion of microbial suspension controls, alone and in contact with films without antimicrobial addition, is mandatory. In some cases, free antimicrobial solutions are also included as controls to compare the effect of free against immobilized-released antimicrobial inhibition capacity. This assay renders log CFU mL^{-1} values, and growth/survival curves can be plotted to analyze results. Also, viable cell reductions or antimicrobial activity can be determined as follows (Moradi et al., 2021):

Cell viability reduction (%) = $[(V_0 - V_t)/V_0] \times 100$, where V_0 and V_t are average viable CFU mL^{-1} at t_0 (initial time) and t (time in which the % is calculated), respectively.

Antimicrobial activity (%) = $[(V_c - V_t)/V_c] \times 100$, where V_c and V_t are average viable CFU mL^{-1} in the inoculum control sample and treatment at the time t, respectively.

6.10.1.3 *Optical density-based methods*

In the spectrophotometric method, the growth of the target bacteria in contact with the active film in broth culture media is monitored by optical density measurements (OD; 600 nm). Even though this methodology is considered faster and less expensive than viable cell count, it presents some disadvantages. Both dead and live bacteria are detected by spectrophotometric determinations, and certain colored antimicrobials or polymer film matrices can interfere with the measurements (Abdollahzadeh et al., 2021).

6.11 Food models

Although in vitro tests of the inhibitory effect of bioactive coatings are a necessary first stage of the study, it is desirable to test the application of the coating or packaging directly on the food to analyze all the factors that complex food matrices can introduce on the antimicrobial activity of the metabolite under study.

When designing active food packaging, in advance, it will be oriented to be applied to a specific food product or category and for the inhibition of certain pathogens or spoilage microorganisms. Generally, the antimicrobial film will be applied or introduced to the selected food in several ways, including spraying, immersion, panning, and brushing of the coating solution. The antimicrobial effect can be monitored by artificially contaminating the food systems with pathogens or spoilage microorganisms, applied individually or in pools of strains. Usually, microbial suspensions can be added to food by mixing, dipping, or surface inoculation (Hu and Gurtler, 2017). Although a microbial inoculum of 5–7 log CFU g^{-1} is generally used in this type of essay, to work with a more realistic contamination grade occurring in food, an inoculum of ca. 2 and 3 log CFU g^{-1} is recommended (National Advisory Committee on Microbiological Criteria for Foods, 2010). The incubation conditions and the duration of the experiment will be determined by the active antimicrobial matrix and the food characteristics and shelf life. The procedure basically consists of sampling a portion of the contaminated coated food in time intervals and preparing a homogenate, serially diluting, and determining the survival of each microorganism by plate coating in an adequate agar medium. A reduction of at least 2 log in the viability of the target microorganism is desired and indicative of potential practical significance for the system under study (National Advisory Committee on Microbiological Criteria for Foods, 2010). Controls of food systems without antimicrobial or non-active coatings should be included. Asepsis and biosafety procedures may be maintained in all stages of the experiment (Moradi et al., 2021). Some examples of the application of antimicrobial packaging, including antimicrobials of microbial origin as active components, in food models are listed in Table 6.1.

TABLE 6.1 Applications on foods of antimicrobials food packaging with antimicrobial agents of microbial sources.

Antimicrobial agent	Packaging material/technique	Substrate	Target microorganisms	References
Bacteriocins produced by *Lactobacillus curvatus*	Polyethylene-Oriented polyamide/ soaking, spraying, and coating	Pork steak, ground beef	*Listeria monocytogenes*	Mauriello et al. (2004)
Bacteriocins produced by *L. curvatus*	Polyethylene/coating	Frankfurters	*L. monocytogenes*	Ercolini et al. (2006)
Bacteriocins produced by *L. curvatus*	Polyethylene/soaking	Smoked salmon	*L. monocytogenes*	Ghalfi et al. (2006)
Bacteriocins synthesized by *L. curvatus* CRL705	Plastic multilayer bags (Sealed Air, CRYOVAC) /soaking Wheat gluten/casting	Wieners	*Lactobacillus plantarum* CRL691 and *L. innocua* 7	Blanco Massani et al. (2014)
Enterocins synthesized by *Enterococcus faecium* CTC492	Alginate, zein, Polyvinyl alcohol (PVA)/casting	Sliced cooked ham	*L. monocytogenes*	Marcos et al. (2007)
Enterocins synthesized by *Enterococcus avium* DSMZ17511	Agar film/dipping coating	Cheese	*L. monocytogenes* 01/155	Guitián et al. (2019)
Enterocins produced by *Lactobacillus sakei* CTC494 and *E. faecium* CTC492, and nisin	Interleavers consisting of a non-perforated polypropylene layer, a non-woven polyamide layer in the middle, and a perforated polypropylene layer/ spreading	Sliced cooked ham	*L. monocytogenes* and *Salmonella enterica* subsp. enterica	Jofré, Aymerich, and Garriga (2008)
ε-polylysine				
Natamycin	Chitosan film/spraying and drying	Saloio cheese (semi-hard)	*Aspergillus niger* MUM 03.42, *Penicillium crustosum* MUM 02.30	Fajardo et al. (2010)
Natamycin	Cellulose acetate film/natamycin-alginate nanoparticles	Egyptian Romy cheese	*Aspergillus flavus*	Fayed, Elsayed, and Ali (2021)
Natamycin	Gluten and methylcellulose	Fresh kashar cheese	*A. niger and Penicillium roquefortii*	Ture, Eroglu, Ozen, and Soyer (2011)
Lactic acid bacteria strains and nisin	Alginate-starch-LAB cell films/ casting. Film prayed with nisin	Smoked salmon	*L. monocytogenes*	Concha-Meyer et al. (2011)
Lacticin NK24, Nisin	Polyethylene-polyamide pouches and cellulose-based bioactive inserts	Sliced cheese and ham	*Lactococcus lactis* subsp. lactis HP, *L. innocua* DPC 1770 and *Staphylococcus aureus* MMPR3	Scannell et al. (2000)
Nisin	Cellulose nanofibrils -Nisin nanohybrid films	Ham	*L. monocytogenes*	Yang, Liu, Wu, Ma, and Lu (2020)
Nisin	Polyethylene-Methylcellulose-HPMC	Tofu	*L. monocytogenes*	Cha, Chen, Park, and Chinnan (2003)
Nisin	Calcium alginate/dipping	Smoked salmon	*L. monocytogenes* and *Salmonella anatum*	Datta et al. (2008)
Nisin	LLDPE	Beef burger	*L. monocytogenes*	Ferrocino et al. (2016)
Nisin	Plastic with MC-HPMC	Hot dogs	*L. monocytogenes*	Franklin, Cooksey, and Getty (2004)
Nisin	Polylactic acid (PLA)/casting	Liquid egg, white, orange juice	*L. monocytogenes, Salmonella* *Enteritidis, Escherichia coli*	Jin and Zhang (2008)
Nisin	Glass- PLA	Liquid egg white, skim milk	*L. monocytogenes*	Jin (2010)

(Continued)

TABLE 6.1 (*Continued*)

Antimicrobial agent	Packaging material/technique	Substrate	Target microorganisms	References
Nisin	Pectin and proteins of chicken egg albumin or porcine skin gelatin /extrusion	Turkey meat	*L. monocytogenes*	Jin, Liu, Sommers, Boyd, and Zhang (2009)
Nisin	Pectin-PLA composite film	Orange juice, liquid egg	*L. monocytogenes*	Jin, Liu, Zhang, and Hicks (2009)
Nisin	Paper board with AP or VAE copolymer	Pasteurized milk, orange juice	Total aerobic bacteria, yeast	Kim, An, Park, Park, and Lee (2002)
Nisin	LDPE	Oyster, ground beef	Total aerobic bacteria, Coliform bacteria	Kim, Paik, and Lee (2002)
Nisin	Paper board with VAE copolymer	Milk cream, water/paraffin oil emulsion	*Micrococcus flavus* and Total aerobic bacteria	Lee, An, Lee, Park, and Lee (2004)
Nisin	Paper board with VAE copolymer	Milk, orange juice	Total aerobic bacteria, yeast	Lee, Park, and Lee (2004)
Nisin	PVA films	Fermented sausages	*L. monocytogenes*	Marcos et al. (2013)
Nisin	Galactomannans from *Gleditsia triacanthos*	Ricotta cheese	*L. monocytogenes*	Martins, Cerqueira, Souza, Carmo Avides, and Vicente (2010)
Nisin	LDPE	Milk	*Micrococcus luteus*	Mauriello, De Luca, La Storia, Villani, and Ercolini (2005)
Nisin	LDPE films/coating and drying	Cold-smoked salmon	*L. monocytogenes*	Neetoo et al. (2008)
Nisin	Polyethylene/nisin-impregnated plastic	Beef	*Lactobacillus helveticus* and *Brochothrix thermosphacta*	Siragusa, Cutter, and Willett (1999)
Nisin	Cellophane	Veal meat	Total aerobic bacteria	Guerra, Macias, Agrasara and Castro (2005)
Nisin and chitosan	Surlyns films /coating with chitosan–HPMC solutions	Ham steaks	*L. monocytogenes*	Ye, Neetoo, and Chen (2008b)
Nisin and chitosan	Plastic films/coating and drying	Cold-smoked salmon	*L. monocytogenes*	Ye, Neetoo, and Chen (2008a)
Nisin and EDTA	Polyethylene and polyethylene oxide/ extrusion of all components	Beef	*Brochothrix thermosphacta*	Cutter et al. (2001)
Nisin and EDTA	Co-extruded VDC and EVA as a barrier layer, and LDPE bags (CRYOVAC BB3050, Sealed Air)/ coating internally with nisin solution	Chilled beef	*Brochotrix thermosphacta*, *Carnobacterium* spp., and *Enterobacteriaceae*	Ercolini et al. (2010)
Nisin and natamycin	Tapioca starch films reinforced with rice bran/casting	Argentinian Port Salut cheese	*Saccharomyces cerevisiae* and *L. innocua*	Berti, Flores, and Jagus (2020)
Nisin and pediocin	Cellulose casings and plastic bags/ spraying	Turkey breast, ham, and beef	*L. monocytogenes*	Ming et al. (1997)
Nisin, citric acid, and EDTA	PVC, LDPE, and nylon films/ absorbent tray pads, foam tray pack, and packaging films, all treated with nisin-containing formulations	Broiler drumstick skin	*Salmonella typhimurium*	Natrajan and Sheldon (2000a)

(*Continued*)

TABLE 6.1 (*Continued*)

Antimicrobial agent	Packaging material/technique	Substrate	Target microorganisms	References
Nisin, citric acid, and EDTA	calcium alginate and agar/casting	Broiler drumstick skin	*S. typhimurium*	Natrajan and Sheldon (2000b)
Nisin, lauric acid, and EDTA	Corn zein films	Culture media	*L. monocytogenes* and *Salmonella Enteritidis*	Hoffman, Han, and Dawson (2001)
Nisin, Lauric arginate ester	Chitosan	Deli turkey meat	*L. innocua*	Guo, Jin, Wang, Scullen, and Sommers (2014)
Nisin-Chitosan	Paper board with VAE copolymer/casting	Milk and orange juice	*L. monocytogenes, E. coli, Micrococcus flavus*, Total aerobic bacteria, and yeasts	Lee, An, Park, and Lee (2003)
Pediocin	Cellulose acetate films	Sliced ham	*L. innocua* and *Salmonella spp*	Santiago-Silva et al. (2009)
Peptide produced by *Bacillus licheniformis* Me1	Low-density polyethylene (LDPE) and cellulose films/soaking and spread coating	Paneer	*M. luteus* ATCC 9341, *L. monocytogenes* Scott A, *S. aureus* FRI 722, *S. typhimurium* MTCC 1251, and *Bacillus cereus* F 4433	Nithya et al. (2013)

6.12 Applications

In the scientific literature, there are numerous studies of the performance of antimicrobial substances of microbial origin in materials or containers intended for food packaging, mainly based on the use of authorized and commercially available compounds, such as nisin and natamycin (Table 6.1).

Nisin is a bacteriocin commonly used in food packaging because of its GRAS status. Nisin is frequently used within polymers, however, its antimicrobial effect can be reduced due to the processing temperatures of polymer manufacturing. The decreased mechanical resistance (tensile strength) of polymers containing nisin could be compensated by using stronger outer packaging layers.

Nisin is more effective in processed foods with a low charge of proteolytic enzymes, such as dairy products and processed vegetables. This antimicrobial is a cationic and hydrophobic bacteriocin that is most stable in high-acid foods. Other preservative hurdles such as heat treatment, low water activity, modified atmosphere, refrigeration, low pH, and the presence of other natural or chemical preservatives such as lysozyme, chitosan, and acids, can enhance antimicrobial activity (Delves-Broughton and Weber, 2011).

The use of nisin-activated plastic packaging combined with chill temperatures proved to be effective in enhancing the microbiological quality of beef cuts by reducing the spoilage populations but without affecting the species diversity (Ercolini et al., 2010). Similarly, it was observed that the application of nisin through plastic interleavers combined with 400 MPa high hydrostatic pressure treatment on cooked ham stored at 6°C, appears as an effective combination of hurdles to obtain value-added ready-to-eat products with safe long-term storage (Jofré et al., 2008). On the contrary, Marcos et al. (2013) found that when *L. monocytogenes* was inoculated on the surface of sliced fermented sausages without added sodium salt, antimicrobial packaging (polyvinyl alcohol (PVA) films containing nisin) induced a pronounced reduction in pathogen counts during refrigerated storage. However, the high hydrostatic pressure treatment did not have an antimicrobial effect against *L. monocytogenes* by itself, nor did it improve the performance of the antimicrobial packaging, under the conditions studied. The authors attributed this to the protective effect exerted by the low water activity of the product and its lactate content (Marcos et al., 2013).

Nisin has also shown to be effective in overcoming the problems associated with *L. monocytogenes* contamination after processing vacuum-packaged cold-smoked salmon and reducing initial spoilage microbiota counts (aerobes, anaerobes, and LAB) of samples that were vacuum-packed with nisin-coated Low-density polyethylene (LDPE) plastic films and stored at 4°C (Neetoo et al., 2008). On the other hand, the incorporation of nisin and oyster lysozyme into edible calcium alginate films was able to retain the effectiveness of the antimicrobial agents in the smoked salmon during 35 days of storage, making it more effective than when they have applied alone on the samples (Datta et al., 2008).

Pediocin is a bacteriocin particularly effective against *L. monocytogenes* and has stability across wide pH and temperature ranges. Applying combined bacteriocins onto plastic food-packaging films was an effective way to inhibit *L. monocytogenes* in fresh and processed meat and poultry. Nisin and pediocin were successfully applied to cellulose casings without diffusing out during frankfurter processing. The bags coated with pediocin completely inhibited the growth of inoculated *L. monocytogenes* through 12 weeks of storage at 4°C (Ming et al., 1997). The antimicrobial efficacy of cellulose-based films containing pediocin (25% and 50%) against *Listeria innocua* and *Salmonella* sp. in sliced ham was tested by means of an experiment that consisted of overlapping the slices of contaminated ham with the antimicrobial film. The 50% pediocin-film showed a reduction of 2 log cycles in the growth of *L. innocua* and 0.5 log in the growth of *Salmonella* sp. in relation to the control treatment after 15 days of storage at 12°C in vacuum package (Santiago-Silva et al., 2009). In that regard, a U.S. patent 5,573,797 assigned to Viskase Corp. describes a packaging film formed by a polymeric film or a regenerated cellulosic film, containing a heat-resistant *Pediococcus*-derived bacteriocin in a synergistic combination with a chelating agent to inhibit or kill *L. monocytogenes* on contact with food (Darrel, 1996).

Natamycin is widely used in the food industry. It has reduced the diffusivity from traditional packaging to food. Thus, its use is limited to structures with minimal barrier properties like biopolymers. In this sense, Fajardo et al. (2010) evaluated the effect of the application of chitosan coating containing 0.50 mg/mL of natamycin on the physico-chemical and microbial properties of semi-hard cheese. They found a decrease in molds/yeasts of 1.1 log (CFU g^{-1}) compared to control after 27 days of storage and an increase in O_2 and CO_2 permeabilities (Fajardo et al., 2010). Furthermore, Egyptian Romy cheese wrapping with cellulose sheets fortified with Natamycin-loaded alginate nanoparticles was evaluated as a way of controlling the growth of toxigenic *Aspergillus flavus* and subsequent aflatoxin production. The antimicrobial sheets were sufficient to complete the elimination of 5 log CFU g^{-1} *A. flavus* initially inoculated and the reduction of aflatoxin production by 79%, without affecting the original flavor, color, and overall appearance of traditional Romy cheese (Fayed et al., 2021).

Bacteriocins produced by *Lactobacillus curvatus* were incorporated in plastic packaging for the preservation of different meat products (pork steak, ground beef, frankfurters, wieners, and smoked salmon) to control *L. monocytogenes* contamination (). The results showed that the films activated by soaking, spraying, or coating with these bacteriocins were effective in inhibiting the growth of the pathogen on the food surface (Blanco Massani et al., 2014; Ercolini, La Storia, Villani, & Mauriello, 2006; Ghalfi, Allaoui, Destain, Benkerroum, & Thonart, 2006; Mauriello, Ercolini, La Storia, Casaburi, & Villani, 2004).

Enterocins synthesized by *Enterococcus faecium* and *Enterococcus avium* have also been used to control *L. monocytogenes, S. enterica, S. aureus,* and *Bacillus cereus* in cheese and ham using activated plastic and edible film packaging. The results showed an effective control and delay of the growth of pathogens using enterocin concentrations ranging between 250 and 2000 AU/cm^2 of the film (Guitián et al., 2019; Ibarguren et al., 2015; Jofré, Garriga, & Aymerich, 2007; Marcos, Aymerich, Monfort, & Garriga, 2007).

Another approach to food protection involves incorporating live bacteriocin-producing bacteria into the package. In a study conducted with smoked salmon, films with LAB strains or a combination of both strains and nisin had a bacteriostatic effect on *L. monocytogenes* during 28-day refrigerated storage (Concha-Meyer, Schöbitz, Brito, & Fuentes, 2011).

The overall antimicrobial packaging market size is estimated to grow from USD 9.57 billion in 2018 to USD 17.55 billion in 2026, according to the Fortune Business Insights report on this topic (Fortune Busines Insights., 2019). This is driven by the growth of the packaging market, globally. Antimicrobial packaging is widely used in the food industry because of the rising consumer demand for products that are perishable, preservative-free, and minimally processed, and due to the need to increase the shelf life of the products. The main companies in the market are BASF SE (Germany), The Dow Chemical Company (U.S.), Mondi Plc (South Africa), PolyOne Corporation (U.S.), Biocote Limited (U.K.), Dunmore Corporation (U.S.), Linpac Senior Holdings (U.K.), Microban International (U.S.), Oplon Pure Sciences Ltd. (Israel), and Takex Labo Co. Ltd.

Despite the numerous publications on antimicrobial packaging based on antimicrobials of microbial sources, the commercial applications of this type of active packaging system are limited. Among them, Standa Laboratories offers antifungal coating for the food industry (cheese and sausage rinds) through the Sanico brand. The product is based on natamycin as an active compound (Standa, 2021). Also, JANSSEN PMP (https://www.janssenpmp.com), a division of Janssen Pharmaceutica NV, has announced the signing of a collaboration agreement with Lipofabrik SAS (http://www.lipofabrik.com), a French company specialized in the production, purification, formulation, and commercialization of lipopeptides. Janssen PMP will develop products containing Mycosubtilin, a lipopeptide from Lipofabrik SAS, to preserve the quality of fresh produce in the global market.

6.13 Regulations

In developing a new antimicrobial packaging system, regulatory compliance is almost as important as its effectiveness. The United States (USA) and the European Union (EU) have established strict regulations related to active food packaging. Active packaging is considered active material in the UE and is subjected to regulations EC 1935/2004 (European Commission, EC 1935, 2004) and EC 450/2009 (European Commission, 2009). In addition, the active substances used in the active packaging materials must be identified and the information on their permitted uses must be presented together with the data related to the maximum amount of substance released from the packaging material (European Commission, 2009). In the USA, the FDA must approve the food packaging system prior to marketing when it is intended to be used in contact with food. This normative is also valid when the active substance is considered a food additive. Title 21 of the Code of Federal Regulations (21 CFR) Part 172 and 173, includes approved food additives that are derived from microorganisms (FDA., 2021a).

Nisin and natamycin are among the few antimicrobials of microbial origin accepted for use in food contact applications in the USA and EU. Nisin is considered by the U.S. Food and Drug Administration (FDA) as GRAS since 2000 according to GRAS Notice GRN N° 65 (FDA., 2000) for intended use on casings for frankfurters and on cooked meat and poultry products as an antimicrobial agent. FDA also authorizes the use of nisin preparations as antimicrobial in cheese up to a concentration of 250 ppm (FDA, 2021a). According to the European Union food safety regulation authorities (European Commission), nisin is accepted as a food preservative (E234), and it is recognized as a safe additive in food.

Natamycin has GRAS status from the FDA, with GRN N° 578 for use in ready-to-drink tea beverages, fruit-flavored energy drinks, sport, and isotonic drinks, and fruit-flavored beverages at levels not to exceed 5 ppm; with GRN N° 517 for use as an inhibitor against yeast and mold growth in yogurt at levels not to exceed 5 ppm in the finished product, and authorizes the use of natamycin on cheese as antimycotic in a maximum concentration of 20 ppm in the finished product (FDA., 2021b). Natamycin is also recognized as a natural preservative by the European Union (EEC N° 235).

Regarding antimicrobial content, the Food and Agriculture Organization of the United Nations (FAO) and the WHO allow an acceptable daily intake (ADI) for nisin of 2 mg/kg bw and for natamycin of 0.3 mg/kg bw (FAO, 2013).

For materials in contact with food, special control of the content of the substance that can migrate into the food is required. In addition, an assessment of the potential hazards of ingesting substances that migrate into food should be carried out to ensure consumer safety and protection. In the case of other antimicrobial agents, trials regarding safety and allergy concerns must be applied to assess the safety, delivery mode, and dosage of the antimicrobial agents in contact with food or as food additives. Toxicity is even more important in the case of edible coatings because they should be considered part of the food and not just material in contact with food (Böhme et al., 2016).

Furthermore, active food packaging systems must also comply with environmental legislation related to their disposal. This legislation is related to recycling and reuse and is linked to the composition of materials. In the USA, the packaging waste directive must be complied with, and in the EU, the packaging waste directives of the European Commission.

6.14 Concluding remarks and future trends

In recent years, innovative and disruptive packaging technologies have been implemented that play an important role in food preservation strategies. Antimicrobial packaging systems based on antimicrobial substances of microbial origin have the advantage of their natural origin and highly specific action against other microorganisms, contributing to the development of the concept of hurdle technologies in food preservation. Advances in these antimicrobial packaging are strongly related to biopolymers and biodegradable polymers, along with a more sustainable policy in food packaging.

There are good perspectives for the improvement of the retention or controlled release of the antimicrobial agent in food packaging when used directly in contact with the product or as part of multilayers. Innovative solutions emerged to functionalize polymers, using nanostructured materials such as nanocomposites, nanolaminates, or electrospun fibers that can support the active agent. However, all these technologies must be studied in detail for each application to evaluate their scalability at an industrial level. Furthermore, additional research should be addressed to evaluate the performance of antimicrobial packaging systems in combination with other stress factors, such as high pressure, light pulses, electric fields, among others, in the context of the hurdle technology concept.

A wide variety of studies of possible applications of antimicrobial packaging systems based on substances of microbial origin can be found in the literature. However, the commercial implementation requires exhaustive further assessment. On the one hand, huge and expensive studies are required to demonstrate the safety and compliance with regulatory requirements by the antimicrobial compound. On the other hand, the development of simple, robust, and inexpensive manufacturing processes with a reliable efficacy of the packaging to retain and release the antimicrobial is necessary.

Due to the complexity of the food-antimicrobial packaging system and the variables that can influence its adequate performance, collaborative work between researchers and industry is necessary to achieve useful products and develop cost-effective production methods for them. A multidisciplinary team, including experts in microbiology, food science, sensory evaluation, packaging design and manufacturing, and materials science, is needed for successful implementation.

Acknowledgment

The authors would like to acknowledge the support of the following institutions: INIQUI (Instituto de Investigaciones para la Industria Química), Consejo Nacional de Investigaciones Científicas y Técnicas (CONICET), and Consejo de Investigación de la Universidad Nacional de Salta (CIUNSa).

References

Abdollahzadeh, E., Nematollahi, A., & Hosseini, H. (2021). Composition of antimicrobial edible films and methods for assessing their antimicrobial activity: A review. *Trends in Food Science and Technology*, *110*, 291–303. Available from https://doi.org/10.1016/j.tifs.2021.01.084, http://www.elsevier.com/wps/find/journaldescription.cws_home/601278/description#description.

Abriouel, H., Franz, C. M. A. P., Omar, N. B., & Galvez, A. (2011). Diversity and applications of Bacillus bacteriocins. *FEMS Microbiology Reviews*, *35*(1), 201–232. Available from https://doi.org/10.1111/j.1574-6976.2010.00244.x.

Achi, S., & Halami, P. M. (2016). *Antimicrobial peptides from Bacillus spp.: Use in antimicrobial packaging antimicrobial food packaging* (pp. 527–537). India: Elsevier Inc. Available from http://10.1016/B978-0-12-800723-5.00043-7, http://www.sciencedirect.com/science/book/9780128007235.

Ahmad, M., Benjakul, S., Prodpran, T., & Agustini, T. W. (2012). Physico-mechanical and antimicrobial properties of gelatin film from the skin of unicorn leatherjacket incorporated with essential oils. *Food Hydrocolloids*, *28*(1), 189–199. Available from https://doi.org/10.1016/j.foodhyd.2011.12.003.

Ahmed, S., Sameen, D. E., Lu, R., Li, R., Dai, J., Qin, W., & Liu, Y. (2020). Research progress on antimicrobial materials for food packaging. *Critical Reviews in Food Science and Nutrition*, 1–14. Available from https://doi.org/10.1080/10408398.2020.1863327.

Aider, M. (2010). Chitosan application for active bio-based films production and potential in the food industry: Review. *LWT - Food Science and Technology*, *43*(6), 837–842. Available from https://doi.org/10.1016/j.lwt.2010.01.021.

Ali, S. M., Siddiqui, R., & Khan, N. A. (2018). Antimicrobial discovery from natural and unusual sources. *Journal of Pharmacy and Pharmacology*, *70*(10), 1287–1300. Available from https://doi.org/10.1111/jphp.12976, http://onlinelibrary.wiley.com/journal/10.1111/(ISSN)2042-7158.

Appendini, P., & Hotchkiss J. H. (2002). Review of antimicrobial food packaging. Innovative food science and emerging technologies. <http://www.elsevier.com/locate/ifset>.

Balciunas, E. M., Castillo Martinez, F. A., Todorov, S. D., Franco, B. D. Gd. M., Converti, A., & Oliveira, R. Pd. S. (2013). Novel biotechnological applications of bacteriocins: A review. *Food Control*, *32*(1), 134–142. Available from https://doi.org/10.1016/j.foodcont.2012.11.025.

Balouiri, M., Sadiki, M., & Ibnsouda, S. K. (2016). Methods for in vitro evaluating antimicrobial activity: A review. *Journal of Pharmaceutical Analysis*, *6*(2), 71–79. Available from https://doi.org/10.1016/j.jpha.2015.11.005, http://www.journals.elsevier.com/journal-of-pharmaceutical-analysis.

Bastarrachea, L., Dhawan, S., & Sablani, S. S. (2011). Engineering properties of polymeric-based antimicrobial films for food packaging. *Food Engineering Reviews*, *3*(2), 79–93. Available from https://doi.org/10.1007/s12393-011-9034-8.

Bastarrachea, L. J., Wong, D. E., Roman, M. J., Lin, Z., & Goddard, J. M. (2015). Active packaging coatings. *Coatings.*, *5*(4), 771–791. Available from https://doi.org/10.3390/coatings5040771, https://www.mdpi.com/2079-6412/5/4/771/pdf.

Berti, S., Flores, S. K., & Jagus, R. J. (2020). *Journal of Food Processing and Preservation*, *44*(10), e14827.

Blanco Massani, M., Molina, V., Sanchez, M., Renaud, V., Eisenberg, P., & Vignolo, G. (2014). Active polymers containing Lactobacillus curvatus CRL705 bacteriocins: Effectiveness assessment in Wieners. *International Journal of Food Microbiology*, *178*, 7–12. Available from https://doi.org/10.1016/j.ijfoodmicro.2014.02.013, http://www.elsevier.com/locate/ijfoodmicro.

Brody, A.L. (2004). Commercial uses o f active food packaging and modified atmosphere packaging systems, i..0–12.

Brody, A. L., Bugusu, B., Han, J. H., Sand, C. K., & McHugh, T. H. (2008). Innovative food packaging solutions. *Journal of Food Science*, *73*(8), R107–R116. Available from https://doi.org/10.1111/j.1750-3841.2008.00933.x.

Böhme, K., Caamaño Antelo, S., Fernández-No, I. C., Quintela-Baluja, M., Barros-Velázquez, J., Cañas, B., & Calo-Mata, P. (2016). *Detection of foodborne pathogens using MALDI-TOF mass spectrometryantimicrobial food packaging* (pp. 203–214). Portugal: Elsevier Inc. Available from https://doi.org/10.1016/B978-0-12-800723-5.00015-2, http://www.sciencedirect.com/science/book/9780128007235.

Campos, C. A., Schelegueda, L. I., Gliemmo, M. F., & Barros-Velázquez, J. (2016). *Antimicrobial packaging for seafoodantimicrobial food packaging* (pp. 269–280). Argentina: Elsevier Inc. Available from https://doi.org/10.1016/B978-0-12-800723-5.00020-6, http://www.sciencedirect.com/science/book/9780128007235.

Carocho, M., Morales, P., & Ferreira, I. C. F. R. (2015). Natural food additives: Quo vadis? *Trends in Food Science and Technology, 45*(2), 284–295. Available from https://doi.org/10.1016/j.tifs.2015.06.007, http://www.elsevier.com/wps/find/journaldescription.cws_home/601278/description#description.

Cha, D. S., & Chinnan, M. S. (2004). Biopolymer-based antimicrobial packaging: A review. *Critical Reviews in Food Science and Nutrition, 44*(4), 223–237. Available from https://doi.org/10.1080/10408690490464276.

Cha, D. S., Chen, J., Park, H. J., & Chinnan, M. S. (2003). *International Journal of Food Science and Technology, 38*(4), 499–503.

Chawla, Rekha, Sivakumar, S., & Kaur, Harsimran (2021). Antimicrobial edible films in food packaging: Current scenario and recent nanotechnological advancements- a review. *Carbohydrate Polymer Technologies and Applications, 2*, 100024. Available from https://doi.org/10.1016/j.carpta.2020.100024.

Chheda, A. H., & Vernekar, M. R. (2015). A natural preservative ε-poly-L-lysine: Fermentative production and applications in food industry. *International Food Research Journal, 22*(1), 23–30. <http://www.ifrj.upm.edu.my/22%20(01)%202015/(4).pdf>.

Chikindas, M. L., Weeks, R., Drider, D., Chistyakov, V. A., & Dicks, L. M. (2018). Functions and emerging applications of bacteriocins. *Current Opinion in Biotechnology, 49*, 23–28. Available from https://doi.org/10.1016/j.copbio.2017.07.011, http://www.elsevier.com/locate/copbio.

Concha-Meyer, A., Schöbitz, R., Brito, C., & Fuentes, R. (2011). Lactic acid bacteria in an alginate film inhibit Listeria monocytogenes growth on smoked salmon. *Food Control, 22*(3-4), 485–489. Available from https://doi.org/10.1016/j.foodcont.2010.09.032.

Contreras, C. B., Charles, German, Toselli, Ricardo, & Strumia, Miriam C. (2017). *Antimicrobial active packaging biopackaging* (1, pp. 64–93). CRC Press. <https://www.taylorfrancis.com/chapters/edit/10.1201/9781315152349-3/antimicrobial-active-packaging-cintia-contreras-german-charles-ricardo-toselli-miriam-strumia>.

Cotter, P. D., Ross, R. P., & Hill, C. (2013). Bacteriocins-a viable alternative to antibiotics? *Nature Reviews. Microbiology, 11*(2), 95–105. Available from https://doi.org/10.1038/nrmicro2937.

Cutter, C. N., Willett, J. L., & Siragusa, G. R. (2001). Improved antimicrobial activity of nisin-incorporated polymer films by formulation change and addition of food grade chelator. *Letters in Applied Microbiology, 33*(4), 325–328. Available from https://doi.org/10.1046/j.1472-765X.2001.01005.x.

Dafale, N. A., Semwal, U. P., Rajput, R. K., & Singh, G. N. (2016). Selection of appropriate analytical tools to determine the potency and bioactivity of antibiotics and antibiotic resistance. *Journal of Pharmaceutical Analysis, 6*(4), 207–213. Available from https://doi.org/10.1016/j.jpha.2016.05.006, http://www.journals.elsevier.com/journal-of-pharmaceutical-analysis.

Darrel, L. (1996). United States Patent. Film and method for surface treatment of foodstuffs with antmicrobial compostions.

Datta, S., Janes, M. E., Xue, Q. G., Losso, J., & La Peyre, J. F. (2008). Control of Listeria monocytogenes and Salmonella anatum on the surface of smoked salmon coated with calcium alginate coating containing oyster lysozyme and nisin. *Journal of Food Science, 73*(2), M67–M71. Available from https://doi.org/10.1111/j.1750-3841.2007.00633.x.

De Azeredo, H. M. C. (2013). Antimicrobial nanostructures in food packaging. *Trends in Food Science and Technology, 30*(1), 56–69. Available from https://doi.org/10.1016/j.tifs.2012.11.006.

Deegan, L. H., Cotter, P. D., Hill, C., & Ross, P. (2006). Bacteriocins: Biological tools for bio-preservation and shelf-life extension. *International Dairy Journal, 16*(9), 1058–1071. Available from https://doi.org/10.1016/j.idairyj.2005.10.026.

Delves-Broughton, J., & Weber, G. (2011). *Nisin, natamycin and other commercial fermentates used in food biopreservation* (pp. 63–99). Elsevier BV. Available from https://doi.org/10.1533/9780857090522.1.63.

Duchateau, A. L. L., & van Scheppingen, W. B. (2018). Stability study of a nisin/natamycin blend by LC-MS. *Food Chemistry, 266*, 240–244. Available from https://doi.org/10.1016/j.foodchem.2018.05.121, http://www.elsevier.com/locate/foodchem.

EUROPEN 2020. (2021). 5 29 EUROPEN' s recommendation for a growth - and innovation-oriented Circular Economy, EUROPEN' s recommendation for a growth - and innovation-oriented Circular Economy. <https://europen-packaging.eu/issue-papers.html>.

El-Saber Batiha, G., Hussein, D. E., Algammal, A. M., George, T. T., Jeandet, P., Al-Snafi, A. E., … Cruz-Martins, N. (2021). Application of natural antimicrobials in food preservation: Recent views. *Food Control, 126*. Available from https://doi.org/10.1016/j.foodcont.2021.108066, https://www.journals.elsevier.com/food-control.

Ercolini, D., La Storia, A., Villani, F., & Mauriello, G. (2006). Effect of a bacteriocin-activated polythene film on Listeria monocytogenes as evaluated by viable staining and epifluorescence microscopy. *Journal of Applied Microbiology, 100*(4), 765–772. Available from https://doi.org/10.1111/j.1365-2672.2006.02825.x.

Ercolini, D., Ferrocino, I., La Storia, A., Mauriello, G., Gigli, S., Masi, P., & Villani, F. (2010). Development of spoilage microbiota in beef stored in nisin activated packaging. *Food Microbiology, 27*(1), 137–143. Available from https://doi.org/10.1016/j.fm.2009.09.006.

Espitia, P. J. P., Batista, R. A., Azeredo, H. M. C., & Otoni, C. G. (2016). Probiotics and their potential applications in active edible films and coatings. *Food Research International, 90*, 42–52. Available from https://doi.org/10.1016/j.foodres.2016.10.026, http://www.elsevier.com/inca/publications/store/4/2/2/9/7/0.

European Commission. (2009). Commission Regulation (EC) No. 450/2009 of 29 May 2009 on active and intelligent materials and articles intended to come into contact with food. Official Journal of European Union, 5(450), 3–11.

European Commission, EC 1935. (2004). European Implementation Assessment Study Food Contact Materials Regulation. Official Journal of the European Union. EC 1935. <http://www.europarl.europa.eu/thinktank>.

FAO. (2013). 2021 Evaluations of the Joint FAO/WHO Expert Committee on Food Additives (JECFA). <https://apps.who.int/food-additives-contaminants-jecfa-database/chemical.aspx?chemID = 572>.

FDA. (2000). 2021 5 29 GRAS Notice GRN No 65 for Specified Uses of Nisin. <https://www.cfsanappsexternal.fda.gov/scripts/fdcc/index.cfm?set = GRASNotices&id = 65>.

FDA. (2021a). 5 29 Food additives status list. <https://www.fda.gov/food/food-additives-petitions/food-additive-status-list#additives>.

FDA. (2021b). 2014 5 29 GRAS Notice GRN No 517 for Specified Uses of Natamycin. <https://www.cfsanappsexternal.fda.gov/scripts/fdcc/?set = GRASNotices&id = 517&sort = GRN_No&order = DESC&startrow = 1&type = basic&search = natamycin>.

Fajardo, P., Martins, J. T., Fuciños, C., Pastrana, L., Teixeira, J. A., & Vicente, A. A. (2010). Evaluation of a chitosan-based edible film as carrier of natamycin to improve the storability of Saloio cheese. *Journal of Food Engineering*, *101*(4), 349–356. Available from https://doi.org/10.1016/j.jfoodeng.2010.06.029.

Fang, Z., Zhao, Y., Warner, R. D., & Johnson, S. K. (2017). Active and intelligent packaging in meat industry. *Trends in Food Science and Technology*, *61*, 60–71. Available from https://doi.org/10.1016/j.tifs.2017.01.002, http://www.elsevier.com/wps/find/journaldescription.cws_home/601278/description#description.

Fayed, A., Elsayed, H., & Ali, T. (2021). Packaging fortified with Natamycin nanoparticles for hindering the growth of toxigenic Aspergillus flavus and aflatoxin production in Romy cheese. *Journal of Advanced Veterinary and Animal Research*, *8*(1), 1. Available from https://doi.org/10.5455/javar.2021.h485.

Fernández, A., Cava, D., Ocio, M. J., & Lagarón, J. M. (2008). Perspectives for biocatalysts in food packaging. *Trends in Food Science and Technology*, *19*(4), 198–206. Available from https://doi.org/10.1016/j.tifs.2007.12.004.

Ferrocino, I., Greppi, A., La Storia, A., Rantsiou, K., Ercolini, D., & Cocolin, L. (2016). *Applied and environmental microbiology*, *82*(2), 549–559.

Fortune Busines Insights. (2019). 2021 5 29 Antimicrobial packaging market 2019-2026, <https://www.fortunebusinessinsights.com/antimicrobial-packaging-market-102726.>

Franklin, N. B., Cooksey, K. D., & Getty, K. J. (2004). *Journal of food protection*, *67*(3), 480–485.

de Freire Bastos, M. D. C., Varella Coelho, M. L., & da Silva Santos, O. C. (2015). Resistance to bacteriocins produced by gram-positive bacteria. *Microbiology (United Kingdom)*, *161*(4), 683–700. Available from https://doi.org/10.1099/mic.0.082289-0, http://mic.sgmjournals.org/content/161/Pt_4/683.full.pdf.

Ghalfi, H., Allaoui, A., Destain, J., Benkerroum, N., & Thonart, P. (2006). Bacteriocin activity by Lactobacillus curvatus CWBI-B28 to inactivate Listeria monocytogenes in cold-smoked salmon during 4°C storage. *Journal of Food Protection*, *69*(5), 1066–1071. Available from https://doi.org/10.4315/0362-028X-69.5.1066, http://www.ingentaconnect.com/content/iafp/jfp.

Guerra, N. P., Macias, C. L., Agrasara, T., & Castro, L. P. (2005). *Letters in Applied Microbiology*, *40*, 106–110.

Guitián, M. V., Ibarguren, C., Soria, M. C., Hovanyecz, P., Banchio, C., & Audisio, M. C. (2019). Anti-Listeria monocytogenes effect of bacteriocin-incorporated agar edible coatings applied on cheese. *International Dairy Journal*, *97*, 92–98. Available from https://doi.org/10.1016/j.idairyj.2019.05.016, http://www.elsevier.com/inca/publications/store/4/0/5/8/6/0/index.htt.

Guo, M., Jin, T. Z., Wang, L., Scullen, O. J., & Sommers, C. H. (2014). *Food control*, *40*, 64–70.

Gyawali, R., & Ibrahim, S. A. (2014). Natural products as antimicrobial agents. *Food Control*, *46*, 412–429. Available from https://doi.org/10.1016/j.foodcont.2014.05.047.

Gómez-Torres, N., Ávila, M., Delgado, D., & Garde, S. (2016). Effect of reuterin-producing Lactobacillus reuteri coupled with glycerol on the volatile fraction, odour and aroma of semi-hard ewe milk cheese. *International Journal of Food Microbiology*, *232*, 103–110. Available from https://doi.org/10.1016/j.ijfoodmicro.2016.050.031, http://www.elsevier.com/locate/ijfoodmicro.

Han, J. H. (2003). *Antimicrobial food packaging novel food packaging techniques* (pp. 50–70). Elsevier BV. Available from 10.1533/9781855737020.1.50.

Han, J. H., & Floros, J. D. (1997). Casting antimicrobial packaging films and measuring their physical properties and antimicrobial activity. *Journal of Plastic Film and Sheeting*, *13*(4), 287–298. Available from https://doi.org/10.1177/875608799701300405, http://jpf.sagepub.com/archive/.

Hernandez-Izquierdo, V. M., & Krochta, J. M. (2008). Thermoplastic processing of proteins for film formation - A review. *Journal of Food Science*, *73*(2), R30–R39. Available from https://doi.org/10.1111/j.1750-3841.2007.00636.x.

Hoffman, K. L., Han, I. Y., & Dawson, P. L. (2001). *Journal of Food Protection*, *64*(6), 885–889.

Hu, M., & Gurtler, J. B. (2017). Selection of surrogate bacteria for use in food safety challenge studies: A review. *Journal of Food Protection*, *80*(9), 1506–1536. Available from https://doi.org/10.4315/0362-028X.JFP-16-536, http://jfoodprotection.org/doi/pdf/10.4315/0362-028X.JFP-16-536.

Ibarguren, C., Céliz, G., Díaz, A. S., Bertuzzi, M. A., Daz, M., & Audisio, M. C. (2015). Gelatine based films added with bacteriocins and a flavonoid ester active against food-borne pathogens. *Innovative Food Science and Emerging Technologies*, *28*, 66–72. Available from https://doi.org/10.1016/j.ifset.2015.01.007, http://www.elsevier.com/locate/ifset.

Jagus, R. J., Gerschenson, L. N., & Ollé, C. P. (2016). *Resa, Combinational approaches for antimicrobial packaging: Natamycin and nisinantimicrobial food packaging* (pp. 599–608). Argentina: Elsevier Inc. Available from https://doi.org/10.1016/B978-0-12-800723-5.00049-8, http://www.sciencedirect.com/science/book/9780128007235.

Jayant, D., & Halami, P. M. (2020). *Industrial perspective of food preservatives from microbial origin* (pp. 243–261). Elsevier BV. Available from https://doi.org/10.1016/b978-0-444-64309-4.00011-8.

Jin, T. (2010). *Journal of food science*, *75*(2), M83–M88.

Jin, T., & Zhang, H. (2008). *Journal of Food Science*, *73*, M127–M134.

Jin, T., Liu, L., Sommers, C. H., Boyd, G., & Zhang, H. (2009). *Journal of Food Protection*, *72*(3), 644–649.

Jin, T., Liu, L., Zhang, H., & Hicks, K. (2009). *International Journal of Food Science & Technology*, *44*(2), 322–329.

Jofré, A., Garriga, M., & Aymerich, T. (2007). Inhibition of Listeria monocytogenes in cooked ham through active packaging with natural antimicrobials and high-pressure processing. *Journal of Food Protection*, *70*(11), 2498–2502. Available from https://doi.org/10.4315/0362-028X-70.11.2498, http://www.ingentaconnect.com/content/iafp/jfp.

Jofré, A., Aymerich, T., & Garriga, M. (2008). Assessment of the effectiveness of antimicrobial packaging combined with high pressure to control Salmonella sp. in cooked ham. *Food Control*, *19*(6), 634–638. Available from https://doi.org/10.1016/j.foodcont.2007.06.007.

Kim, Y. T., Kim, K., Han, J. H., & Kimmel, R. M. (2008). *Antimicrobial active packaging for food. smart packaging technologies for fast moving consumer goods* (pp. 99–110). United States: John Wiley & Sons, Ltd. Available from https://doi.org/10.1002/9780470753699.ch6, http://onlinelibrary.wiley.com/book/10.1002/9780470753699.

Kim, Y. M., An, D. S., Park, H. J., Park, J. M., & Lee, D. S. (2002). *Packaging Technology and Science: An International Journal*, *15*(5), 247–254.

Kim, Y. M., Paik, H. D., & Lee, D. S. (2002). *Journal of the Science of Food and Agriculture*, *82*(9), 998–1002.

Kourmentza, K., Gromada, X., Michael, N., Degraeve, C., Vanier, G., Ravallec, R., ... Jauregi, P. (2021). Antimicrobial activity of lipopeptide biosurfactants against foodborne pathogen and food spoilage microorganisms and their cytotoxicity. *Frontiers in Microbiology* (11). Available from https://doi.org/10.3389/fmicb.2020.561060, https://www.frontiersin.org/journals/microbiology#.

Kumariya, R., Garsa, A. K., Rajput, Y. S., Sood, S. K., Akhtar, N., & Patel, S. (2019). Bacteriocins: Classification, synthesis, mechanism of action and resistance development in food spoilage causing bacteria. *Microbial Pathogenesis*, *128*, 171–177. Available from https://doi.org/10.1016/j.micpath.2019.01.002, http://www.elsevier.com/inca/publications/store/6/2/2/9/1/5/index.htt.

Langer, R., & Peppas, N. (1983). Chemical and physical structure of polymers as carriers for controlled release of bioactive agents: A review. *Journal of Macromolecular Science, Part C*, *23*(1), 61–126. Available from https://doi.org/10.1080/07366578308079439.

Lee, C. H., An, D. S., Lee, S. C., Park, H. J., & Lee, D. S. (2004). *Journal of Food Engineering*, *62*(4), 323–329.

Lee, C. H., An, D. S., Park, H. J., & Lee, D. S. (2003). *Packaging technology and science: An international journal*, *16*(3), 99–106.

Lee, C. H., Park, H. J., & Lee, D. S. (2004). *Journal of Food Engineering*, *65*(4), 527–531.

Lopez-Pena, C. L., & McClements, D. J. (2015). Impact of a food-grade cationic biopolymer (ε-polylysine) on the digestion of emulsified lipids: In vitro study. *Food Research International*, *75*, 34–40. Available from https://doi.org/10.1016/j.foodres.2015.05.025, http://www.elsevier.com/inca/publications/store/4/2/2/9/7/0.

Lucera, A., Costa, C., Conte, A., & Del, M. A. (2012). Nobile, Food applications of natural antimicrobial compounds. *Frontiers in Microbiology*, *3*, 1–13. Available from https://doi.org/10.3389/fmicb.2012.00287, http://www.frontiersin.org/Journal/DownloadFile.ashx?pdf = 1&FileId = 11017&articleId = 28087&Version = 1&ContentTypeId = 21&FileName = fmicb-03-00287.pdf.

López-Carballo, G., Gómez-Estaca, J., Catalá, R., Hernández-Muñoz, P., & Gavara, R. (2012). *Active antimicrobial food and beverage packaging. Emerging food packaging technologies: Principles and practice* (pp. 27–54). Spain: Elsevier Ltd.. Available from https://doi.org/10.1016/B978-1-84569-809-6.50003-8, http://www.sciencedirect.com/science/book/9781845698096.

Malhotra, B., Keshwani, A., & Kharkwal, H. (2015). Antimicrobial food packaging: Potential and pitfalls. *Frontiers in Microbiology*, *6*. Available from https://doi.org/10.3389/fmicb.2015.00611, http://journal.frontiersin.org/article/10.3389/fmicb.2015.00611/full.

Marcos, B., Aymerich, T., Monfort, J. M., & Garriga, M. (2007). Use of antimicrobial biodegradable packaging to control Listeria monocytogenes during storage of cooked ham. *International Journal of Food Microbiology*, *120*(1-2), 152–158. Available from https://doi.org/10.1016/j.ijfoodmicro.2007.06.003.

Marcos, B., Aymerich, T., Garriga, M., & Arnau, J. (2013). Active packaging containing nisin and high pressure processing as post-processing listericidal treatments for convenience fermented sausages. *Food Control*, *30*(1), 325–330. Available from https://doi.org/10.1016/j.foodcont.2012.07.019.

Martins, J. T., Cerqueira, M. A., Souza, B. W., Carmo Avides, M. D., & Vicente, A. A. (2010). *Journal of Agricultural and Food Chemistry*, *58*(3), 1884–1891.

Mastromatteo, M., Mastromatteo, M., Conte, A., & Del Nobile, M. A. (2010). Advances in controlled release devices for food packaging applications. *Trends in Food Science & Technology*, *21*(12), 591–598. Available from https://doi.org/10.1016/j.tifs.2010.07.010.

Mauriello, G., Ercolini, D., La Storia, A., Casaburi, A., & Villani, F. (2004). Development of polythene films for food packaging activated with an antilisterial bacteriocin from Lactobacillus curvatus 32Y. *Journal of Applied Microbiology*, *97*(2), 314–322. Available from https://doi.org/10.1111/j.1365-2672.2004.02299.x.

Mauriello, G. D. L. E., De Luca, E., La Storia, A., Villani, F., & Ercolini, D. (2005). *Letters in applied microbiology*, *41*(6), 464–469.

McManamon, O., Kaupper, T., Scollard, J., & Schmalenberger, A. (2019). Nisin application delays growth of Listeria monocytogenes on fresh-cut iceberg lettuce in modified atmosphere packaging, while the bacterial community structure changes within one week of storage. *Postharvest Biology and Technology*, *147*, 185–195. Available from https://doi.org/10.1016/j.postharvbio.2018.10.002, http://www.elsevier.com/inca/publications/store/5/0/3/3/1/3/index.htt.

Meena, K. R., & Kanwar, S. S. (2015). Lipopeptides as the antifungal and antibacterial agents: Applications in food safety and therapeutics. *BioMed Research International*, *2015*. Available from https://doi.org/10.1155/2015/473050, http://www.hindawi.com/journals/biomed/.

Ming, X., Weber, G. H., Ayres, J. W., & Sandine, W. E. (1997). Bacteriocins applied to food packaging materials to inhibit Listeria monocytogenes on meats. *Journal of Food Science*, *62*(2), 413–415. Available from https://doi.org/10.1111/j.1365-2621.1997.tb04015.x, http://www3.interscience.wiley.com/journal/118509799/issueyear?year = 2008.

Miranda, J. M., Mondragón, A. C., Lamas, A., Roca-Saavedra, P., Ibarra, I. S., Rodriguez, J. A., ... Franco, C. M. (2016). Effect of packaging systems on the inactivation of microbiological agents. *Antimicrobial food packaging* (pp. 107–116). Spain: Elsevier Inc. Available from https://doi.org/10.1016/B978-0-12-800723-5.00008-5, http://www.sciencedirect.com/science/book/9780128007235.

Montiel, R., Martín-Cabrejas, I., Peirotén, Á., & Medina, M. (2016). Reuterin, lactoperoxidase, lactoferrin and high hydrostatic pressure treatments on the characteristics of cooked ham. *Innovative Food Science and Emerging Technologies*, *35*, 111–118. Available from https://doi.org/10.1016/j.ifset.2016.04.013, http://www.elsevier.com/locate/ifset.

Moradi, M., Kousheh, S. A., Razavi, R., Rasouli, Y., Ghorbani, M., Divsalar, E., ... Ibrahim, S. A. (2021). Review of microbiological methods for testing protein and carbohydrate-based antimicrobial food packaging. *Trends in Food Science and Technology*, *111*, 595–609. Available from https://doi.org/10.1016/j.tifs.2021.03.007, http://www.elsevier.com/wps/find/journaldescription.cws_home/601278/description#description.

Natrajan, N., & Sheldon, B. W. (2000a). *Journal of Food Protection*, *63*(9), 1189–1196.

Natrajan, N., & Sheldon, B. W. (2000b). *Journal of Food Protection*, *63*(9), 1268–1272.

National Advisory Committee on Microbiological Criteria for Foods. (2010). Parameters for determining inoculated pack/challenge study protocols. *Journal of Food Protection*, *73*(1), 140–202. Available from https://doi.org/10.4315/0362-028X-73.1.140, http://jfoodprotection.org/?code = fopr-site.

Neetoo, H., Ye, M., Chen, H., Joerger, R. D., Hicks, D. T., & Hoover, D. G. (2008). Use of nisin-coated plastic films to control Listeria monocytogenes on vacuum-packaged cold-smoked salmon. *International Journal of Food Microbiology*, *122*(1-2), 8–15. Available from https://doi.org/10.1016/j.ijfoodmicro.2007.11.043.

Nerin, C., Silva, F., Manso, S., & Becerril, R. (2016). *The downside of antimicrobial packaging: Migration of packaging elements into food. Antimicrobial food packaging* (pp. 81–93). Spain: Elsevier Inc. Available from https://doi.org/10.1016/B978-0-12-800723-5.00006-1, http://www.sciencedirect.com/science/book/9780128007235.

Nithya, V., & Halami, P. M. (2012). Antibacterial peptides, probiotic properties and biopreservative efficacy of native bacillus species isolated from different food sources. Probiotics and Antimicrobial. *Proteins*, *4*(4), 279–290. Available from https://doi.org/10.1007/s12602-012-9115-x, http://www.springer.com/new + %26 + forthcoming + titles + %28default%29/journal/12602.

Nithya, V., Murthy, P. S. K., & Halami, P. M. (2013). Development and application of active films for food packaging using antibacterial peptide of Bacillus licheniformis Me1. *Journal of Applied Microbiology*, *115*(2), 475–483. Available from https://doi.org/10.1111/jam.12258.

Ozdemir, M., & Floros, J. D. (2004). Active food packaging technologies. *Critical Reviews in Food Science and Nutrition, 44*(3), 185–193. Available from https://doi.org/10.1080/10408690490441578.

Pisoschi, A. M., Pop, A., Georgescu, C., Turcuş, V., Olah, N. K., & Mathe, E. (2018). An overview of natural antimicrobials role in food. *European Journal of Medicinal Chemistry, 143*, 922–935. Available from https://doi.org/10.1016/j.ejmech.2017.11.095, http://www.journals.elsevier.com/european-journal-of-medicinal-chemistry/.

Ponce, A. G., Fritz, R., Del Valle, C., & Roura, S. I. (2003). Antimicrobial activity of essential oils on the native microflora of organic Swiss chard. *LWT - Food Science and Technology, 36*(7), 679–684. Available from https://doi.org/10.1016/S0023-6438(03)00088-4, http://www.elsevier.com/inca/publications/store/6/2/2/9/1/0/index.htt.

Prasad, P., & Kochhar, A. (2014). Active packaging in food industry: A review. IOSR Journal of Environmental Science. *Toxicology and Food Technology*, 01–07. Available from https://doi.org/10.9790/2402-08530107.

Quintavalla, S., & Vicini, L. (2002). Antimicrobial food packaging in meat industry. *Meat Science, 62*(3), 373–380. Available from https://doi.org/10.1016/S0309-1740(02)00121-3.

Quinto, E. J., Caro, I., Villalobos-Delgado, L. H., Mateo, J., De-Mateo-silleras, B., & Redondo-Del-río, M. P. (2019). Food safety through natural antimicrobials. *Antibiotics., 8*(4). Available from https://doi.org/10.3390/antibiotics8040208, https://www.mdpi.com/2079-6382/8/4/208/pdf.

Realini, C. E., & Marcos, B. (2014). Active and intelligent packaging systems for a modern society. *Meat Science, 98*(3), 404–419. Available from https://doi.org/10.1016/j.meatsci.2014.06.031.

Robertson, G. L. (2012). Food packaging. *Priciples and practice* (Third edition). CRC Press. Available from https://doi.org/10.1201/b21347-27, https://www.routledge.com/Food-Packaging-Principles-and-Practice-Third-Edition/Robertson/p/book/9781439862414.

Rooney, M. L. (1995). *Overview of active food packaging* (pp. 1–37). Springer Science and Business Media LLC. Available from https://doi.org/10.1007/978-1-4615-2175-4_1.

Said, L. B., Fliss, I., Offret, C., & Beaulieu, L. (2018). *Antimicrobial peptides: The new generation of food additives encyclopedia of food chemistry* (pp. 576–582). Canada: Elsevier. Available from http://doi.org/10.1016/B978-0-08-100596-5.21718-6, https://doi.org/10.1016/B978-0-08-100596-5.21718-6.

Santiago-Silva, P., Soares, N. F. F., Nóbrega, J. E., Júnior, M. A. W., Barbosa, K. B. F., Volp, A. C. P., ... Würlitzer, N. J. (2009). Antimicrobial efficiency of film incorporated with pediocin (ALTA® 2351) on preservation of sliced ham. *Food Control, 20*(1), 85–89. Available from https://doi.org/10.1016/j.foodcont.2008.02.006.

Scannell, A. G., Hill, C., Ross, R. P., Marx, S., Hartmeier, W., & Arendt, E. K. (2000). *International Journal of Food Microbiology, 60*(2–3), 241–249.

Shukla, S. C., Singh, A., Pandey, A. K., & Mishra, A. (2012). Review on production and medical applications of ε-polylysine. *Biochemical Engineering Journal, 65*, 70–81. Available from https://doi.org/10.1016/j.bej.2012.04.001.

Silvestre, C., Duraccio, D., & Cimmino, S. (2011). Food packaging based on polymer nanomaterials. *Progress in Polymer Science, 36*(12), 1766–1782. Available from https://doi.org/10.1016/j.progpolymsci.2011.02.003.

Siragusa, G. R., Cutter, C. N., & Willett, J. L. (1999). *Food Microbiology, 16*(3), 229–235.

Sofi, S. A., Singh, J., Rafiq, S., Ashraf, U., Dar, B. N., & Nayik, G. A. (2018). A comprehensive review on antimicrobial packaging and its use in food packaging. *Current Nutrition and Food Science, 14*(4), 305–312. Available from https://doi.org/10.2174/1573401313666170609095732, http://www.eurekaselect.com/612/journal/current-nutrition-amp-food-science.

Soltani, S., Hammami, R., Cotter, P. D., Rebuffat, S., Said, L. B., Gaudreau, H., ... Fliss, I. (2021). Bacteriocins as a new generation of antimicrobials: Toxicity aspects and regulations. *FEMS Microbiology Reviews, 45*(1). Available from https://doi.org/10.1093/femsre/fuaa039, http://femsre.oxfordjournals.org/.

Soltani Firouz, M., Mohi-Alden, K., & Omid, M. (2021). A critical review on intelligent and active packaging in the food industry: Research and development. *Food Research International* 141, 110113. Available from https://doi.org/10.1016/j.foodres.2021.110113, http://www.elsevier.com/inca/publications/store/4/2/2/9/7/0.

Montville, T., & Chikindas, M.L. (2007). Biopreservation of foods food microbiology: Fundamentals and frontiers third edition. American Society for Microbiology, 747–764. Available from: https://doi.org/10.1128/9781555815912.ch34, https://www.asmscience.org/content/book/10.1128/9781555815912.ch34.

Standa. (2021). 5 29 Sanico: Coating and antifungal products for the food industry. http://www.standa-fr.com/eng/laboratoires-standa/sanico/.

Stein, T. (2005). Bacillus subtilis antibiotics: Structures, syntheses and specific functions. *Molecular Microbiology, 56*(4), 845–857. Available from https://doi.org/10.1111/j.1365-2958.2005.04587.x.

Sung, S. Y., Sin, L. T., Tee, T. T., Bee, S. T., Rahmat, A. R., Rahman, W. A. W. A., ... Vikhraman, M. (2013). Antimicrobial agents for food packaging applications. *Trends in Food Science and Technology, 33*(2), 110–123. Available from https://doi.org/10.1016/j.tifs.2013.08.001.

Suppakul, P., Miltz, J., Sonneveld, K., & Bigger, S. W. (2003). Active packaging technologies with an emphasis on antimicrobial packaging and its applications. *Journal of Food Science, 68*(2), 408–420. Available from https://doi.org/10.1111/j.1365-2621.2003.tb05687.x, http://www3.interscience.wiley.com/journal/118509799/issueyear?year = 2008.

Ture, H., Eroglu, E., Ozen, B., & Soyer, F. (2011). *International journal of food science & technology, 46*(1), 154–160.

Villalobos-Delgado, L. H., Nevárez-Moorillon, G. V., Caro, I., Quinto, E. J., Mateo, J. Elsevier BV, 2019 125–157. Available from https://doi.org/10.1016/b978-0-12-817190-5.00004-5.

World Health Organization. (2020). 2020 2021 5 29, Food Safety. Fact Sheet. <https://www.who.int/en/news-room/fact-sheets/detail/food-safety>.

Yang, Y., Liu, H., Wu, M., Ma, J., & Lu, P. (2020). *International Journal of Biological Macromolecules, 161*, 627–635.

Ye, M., Neetoo, H., & Chen, H. (2008a). Effectiveness of chitosan-coated plastic films incorporating antimicrobials in inhibition of Listeria monocytogenes on cold-smoked salmon. *International Journal of Food Microbiology, 127*(3), 235–240. Available from https://doi.org/10.1016/j.ijfoodmicro.2008.07.012.

Ye, M., Neetoo, H., & Chen, H. (2008b). *Food Microbiology, 25*(2), 260–268.

Zhang, Q. Q., Zhang, Y. H., Cai, F. Y., Liu, X. L., Chen, X. H., & Jiang, M. (2019). Comparative antibacterial and antibiofilm activities of garlic extracts, nisin, ε-polylysine, and citric acid on Bacillus subtilis. *Journal of Food Processing and Preservation, 43*(11). Available from https://doi.org/10.1111/jfpp.14179, http://onlinelibrary.wiley.com/journal/10.1111/(ISSN)1745-4549.

CHAPTER

7

Antimicrobial agents from herbs and spices for food packaging applications

Farhad Garavand[1], *Amirerfan Taghizadeh*[2], *Saeid Alizadeh*[3], *Shima Yousefi*[2], *Diako Khodaei*[4] *and Ilaria Cacciotti*[5]

[1]Department of Food Chemistry and Technology, Teagasc Food Research Centre, Moorepark, Fermoy, Ireland [2]Department of Agriculture and Food Science, Islamic Azad University, Science and Research Branch, Tehran, Iran [3]Department of Food Science and Technology, Ferdowsi University of Mashhad, Mashhad, Iran [4]School of Science and Computing, Galway-Mayo Institute of Technology (GMIT), Galway, Ireland [5]Department of Engineering, INSTM RU, University of Rome "Niccolò Cusano", Roma, Italy

7.1 Background

Morbidity and mortality are mainly caused by infectious diseases all over the world. Food pathogens represent a dramatic concern for food consumers and industries, deteriorating and spoiling food. Foodborne diseases represent significant public health, and economic and social burden worldwide. In 2015, the WHO reported the first data related to the global and regional disease burden due to 31 foodborne hazards (Organization, 2015). The collected data highlighted that every year 1 person among 10 people gets sick from food contaminated with microbial or chemical agents, resulting in 600 million illnesses and 420,000 deaths globally (Havelaar et al., 2010). The WHO reported, for example, that worldwide 55 million people died in 2011, one-third due to infectious diseases (Organization, 2010).

The proliferation of microbial pathogens in food may induce spoilage and contribute to foodborne diseases incidence. Furthermore, the actual situation is further aggravated by the emergence of antibiotic therapy, multi-drug and disinfectant-resistant bacteria, for example, *Staphylococcus aureus* (*S. aureus*), *Escherichia coli* (*E. coli*), and *Pseudomonas aeruginosa* (*P. aeruginosa*) can recover and survive after antibiotic drug exposure through their ability to acquire and transmit resistance, leading to increased morbidity and mortality (Marchese et al., 2016; Miladi et al., 2016). Thus, antibiotic resistance has become a public health burden, and there is an urgent need for novel antimicrobial agents to face this problem (Högberg, Heddini, & Cars, 2010). The semi-processed and processed food shelf life can be extended by inhibiting microbial growth and/or decreasing the microbes viability.

A potential strategy to overcome this issue consists in the incorporation of antimicrobials into the food products/ packaging with the final aim to enhance food preservation by controlling food spoilage microorganisms and hindering/suppressing pathogenic microbes (Saeed et al., 2016). For instance, chemical synthetic food additives, such as weak acids (e.g., benzoic and sorbic acids) (Brul & Coote, 1999; Garavand, Razavi, & Cacciotti, 2018), are commonly employed as chemical food preservatives to increase their safety and stability by monitoring pathogenic and spoilage food-related microorganisms (Souza, Stamford, Lima, Trajano, & Barbosa Filho, 2005). However, chemical preservatives cannot completely inhibit the growth of numerous pathogenic bacteria, such as *Listeria monocytogenes* (*L. monocytogenes*), in food products, cannot delay the spoilage microorganisms growth (Tajkarimi, Ibrahim, & Cliver, 2010), can result in the development of microbiological resistance (Silva & Domingues, 2017), are not safe, are commonly carcinogenic, and pose serious risks in terms of food safety (which is the primary demand of the consumers). Furthermore,

Food Packaging and Preservation.
DOI: https://doi.org/10.1016/B978-0-323-90044-7.00007-0

the reduced and limited research about the development of new antibacterial agents, aimed at inhibiting antibiotic-resistant disease-causing microorganisms such as *S. aureus*, aggravates the emerging antibiotic resistance (Högberg et al., 2010). For all these reasons, the current attention is devoted to the use of natural additives able to suppress the growth of different microbes. Indeed, natural products have been increasingly accepted as promising alternatives to synthetic chemical preservatives, because they present high efficacy against pathogens with negligible side effects, being intrinsically better tolerated in the human body (Silva & Domingues, 2017). Different sources of naturally derived antimicrobials can be identified, including plants, animals, and microbes.

The main antimicrobial agents that can be added to the films and coatings are organic acids, bacteriocins, essential oils, and extracts, which maintain food safety during the storage period, by slow release (Abdollahzadeh, Nematollahi, & Hosseini, 2021). Essential oils and plant extracts are natural compounds that are generated from plants and herbs and exhibit a wide application in the food industry as flavoring and antimicrobial agents (Khazaei, Nateghi, Zand, Oromiehie, & Garavand, 2021). Specifically, herbs and spices are considered promising natural antimicrobial additives to be used in foodstuffs, due to their intrinsic self-action against the pathogenic and infectious microorganisms (Kim, Kim, & Ahn, 2003), guaranteeing prime quality food products without safety concerns. Spices and herbs have also played a pivotal role in the breathtaking rise of civilization in the nation's history. Their aroma and tanginess allow for improving the product's sensory features (flavor, color, and taste) and the palatability of different dishes, at the same time masking the off-odors, due to their volatile oils and oleoresins (Saeed et al., 2016).

It has been demonstrated that species and herbs own beneficial and healthy properties and, thus, they could be used for the treatment of different diseases (Garavand, Eghbal, Nooshkam, Miraballes, & Jafari, 2021). For example, some herbs and spices are commonly used for the cure of dyspepsia and gastrointestinal disorders (Silva et al., 2007), and ginger has been used for the treatment of cancer and different nervous system diseases (Choi, Kim, Jeong, & Oh, 2018). Pharmacological and nutraceutical-based industries are involved in the preparation of several medicines based on herbs and spices, and they can be also considered as an alternative to common antibiotics to treat infectious diseases (Liu et al., 2017). Moreover, the use of spices and herbs allows not only to enhance the food product's shelf life but also to improve taste. These natural compounds have been used for several applications, including the medicine and food fields, as drugs for treatments, flavoring agents and agents for food preservation, mainly for fresh fruit juices and processed vegetables. To provide some quantitative data, the range of minimum inhibitory concentration of the herbs and spices extracts lies in the range of $0.2-10\ \mu g/mL$, resulting in a suitable efficiency against different pathogens [e.g., *S. aureus*, *L. monocytogenes*, *Shigella dysenteriae*, *Salmonella typhimurium* (*S. typhimurium*), *E. coli* and *Bacillus subtilis* (*B. subtilis*)] (Brandi, Amagliani, Schiavano, De Santi, & Sisti, 2006; Burt, 2004). Generally, Gram-negative bacteria are less sensitive to antimicrobial agents, due to their lipopolysaccharide outer membrane, which restricts the diffusion of the hydrophobic compounds (Burt, 2004). Furthermore, Gram-negative bacteria are naturally more reluctant to the plant-origin antimicrobial compounds, and no effect is usually revealed concerning Gram-positive bacteria (Stefanello et al., 2008).

Antimicrobial food packaging has recently received a lot of interest due to consumers' growing demand for safe and fresh foods. Therefore, the development of effective and novel antimicrobial packaging should be a solution to overcome the short shelf life of certain food products (Azizi-Lalabadi, Garavand, & Jafari, 2021). The direct application of natural antimicrobial agents from spices and herbs into the food systems is generally limited due to the possible negative impact of these active ingredients on the organoleptic properties of the food, their possible interactions with the food components, and also their low solubility in water (Gahruie, Ziaee, Eskandari, & Hosseini, 2017). Using natural or synthetic polymers as active packaging systems is a novel method to protect these active antimicrobial ingredients during the process and storage of food products and to provide a slow release on the surface of the food system. Different factors, for example, the specific part of the used plant, the method of extraction, the harvesting season, the interaction between the antimicrobial agent and the polymer, and the use of blended polymers, influence the biological activities of these ingredients (Vitoratos, Bilalis, Karkanis, & Efthimiadou, 2013).

Thus, the present chapter is aimed at providing a complete overview of the application of herbs and spices in food packaging materials, including various biopolymeric and polymeric systems, as well as their blends.

7.2 Herbs and spices in food packaging materials

Spices are defined as any aromatic plant substance able to impart flavor and intended for seasoning rather than nutrition. Indeed, they can provide savor or spiciness to food products and drinks. Spices are usually extracted from plants grown in warm or tropical climates and are commonly obtained from the roots, flowers, bark, and seeds. On the other hand, herbs are obtained from the green parts of plants, such as stems and leaves, and can grow in almost all geographical regions around the world and in nearly all conditions (Valdés et al., 2015). Spices have a more intense taste than herbs, so they are primarily used in small amounts (Valdés et al., 2015).

Historically, spices have been used as food and flavoring since ancient times (Lai & Roy, 2004), and as medicine and food preservatives in recent decades (Zheng et al., 2016). Egyptians adopted herbs and spices in their daily food, and their use ascended down to the Asian region including China, India, and Pakistan. For example, in ancient history, around 200 BC, the Chinese population used these spices to refresh their breath during their conversation with the emperor as well as to remove the unpleasant smell and odor. For thousands of years, they have been employed in food preservation, as natural food, and as additives to enhance flavor and aroma. Currently, spices are still very widely and commonly used ingredients in daily diets worldwide and heyday cooking practices. They can provide numerous functional benefits, such as antimicrobial properties, as well as improved flavor and fragrance, and also in very low amounts. They are used in pharmaceutical processing and food industry to provide flavor, aromatic substances, or other functional properties (Valdés et al., 2015). The most frequently used spices and herbs, such as onion, garlic, clove, cinnamon, black paper, ginger, oregano, thyme, mint, and cumin, are shown to provide health benefits. Indeed, they exhibit significant antioxidant and antimicrobial activities against food spoilage bacteria like *B. subtilis* and *Pseudomonas fluorescens (P. fluorescens)*; pathogens like *S. aureus*, *Vibrio parahaemolyticus*, and *S. typhimurium*; harmful fungi like *Aspergillus flavus* (*A. flavus*) and *Aspergillus niger* (*A. niger*); and even antibiotic-resistant microorganisms, such as Methicillin-Resistant *S. aureus* (Lai & Roy, 2004; Liu et al., 2017). Therefore, they can inhibit/reduce the growth of foodborne pathogens, yeast, molds or spoilage bacteria, and could be used to decrease the possibility of food poisoning and spoilage, increase the food safety and shelf life of products and treat some infectious diseases (Liu et al., 2017). Furthermore, different spices and herbs display different antimicrobial and antibacterial activities against different strains within the same species and play therapeutic effects based on the number of the present bioactive components. The spice's antimicrobial activity strictly depends on the spice type and nature (e.g., dried, fresh, or extracted form) and also upon the harvesting seasons and geological sources (Saeed et al., 2016).

Among the most widely applied spices, clove (*Eugenia caryophyllata*) belongs to the *Myrtaceae* family, and has been exploited in medicine as an antiseptic against infectious diseases (e.g., periodontal disease), thanks to its antimicrobial properties against oral bacteria (Chaieb et al., 2007), as well as in food industry as a natural additive or antiseptic to increase shelf life due to the effective antimicrobial activities against some foodborne pathogens (Xu, Liu, Hu, & Cao, 2016).

Bud oil of the clove presents insecticidal, antioxidant, antibacterial, and antifungal properties and has been used in food as a flavoring agent and a preservative antimicrobial substance (Velluti, Sanchis, Ramos, Egido, & Marın, 2003). Clove is a precious source of manganese, dietary fibers, Omega 3 fatty acids, vitamin K, vitamin C, calcium, and magnesium. Additionally, it also contains phosphorus, sodium, potassium, iron, carbohydrates, proteins, and hydrochloric acid (Kim et al., 1998). The main active and most important clove component is phenylpropene eugenol, supplied with substantially high antimicrobial properties against food microorganisms, such as *Lactobacillus* sp., *Aspergillus* sp., *Fusarium* sp., *Alternaria* sp., *Cunninghamella* sp., *Mucor* sp., *Clostridium* sp., *Bacillus* sp., *Salmonella* sp., and *Penicillium* sp. (Soliman & Badeaa, 2002).

Garlic (*Allium sativum*) belongs to the *Liliaceae* family, is characterized by a hot, pungent flavor and has been used for both culinary and medicinal purposes (Nejad, Shabani, Bayat, & Hosseini, 2014). The antimicrobial activities of garlic against gram-positive bacteria, gram-negative bacteria, and foodborne pathogens, such as *Staphylococcus*, *Salmonella*, Coliform, *Streptococcus*, *Lactobacillus*, some anaerobic bacteria (Purwanti & Yuwanta, 2014), and even antibiotic-resistant microorganisms, such asampicillin resistant *E. coli* (*E. coli*^{AmpR}) BL21 strain (Cacciotti et al., 2018) have been demonstrated and ascribed to its main active component, that is, allicin, a diallylthiosulfinate (2-propenyl-2-propenethiol sulfonate) with scavenging antioxidant activity (Chung, 2006). It has been reported that garlic can prevent cardiovascular diseases, high blood pressure, high cholesterol, inhibit tumor cell motility, decrease the cancer metastasis risk, and improve the immune system response, as well as owns anti-inflammatory, anti-atherosclerotic and anti-thrombotic activities, and immunomodulatory effect (Saeed et al., 2016).

Rosemary (*Rosmarinus officinalis*) belongs to the *Lamiaceae* family and consists of a perennial shrub, which grows all over the world and is characterized by a pleasant smell. It has been exploited in pharmaceutical and traditional medicine sectors, and also as a flavoring agent in food products due to its flavor, antioxidant, and antimicrobial features (Azizkhani & Tooryan, 2015). Ginger (*Zingiber officinale*) belongs to the *Zingiberaceae* family and has been commonly used as a food ingredient, as well as in cosmetic and pharmaceutical industries, due to its antimicrobial activity, ascribable to some volatile compounds present in ginger, that is α-pinene, borneol, camphene, and linalool (Sa-Nguanpuag, Kanlayanarat, Srilaong, Tanprasert, & Techavuthiporn, 2011).

Cinnamon belongs to the *genus Cinnamomum* and the *Lauraceae* family. It is a popular tropical spice obtained from the inner bark of trees and used worldwide for many centuries. Almost all parts of the cinnamon tree, such as the root, bark, fruits, flowers, and leaves, have culinary and/or medicinal use (Ranasinghe et al., 2013). Numerous studies have found that cinnamon plays antimicrobial activity against a wide variety of bacteria, fungi, and viruses, such as *B. subtilis, P. aeruginosa, E. coli, Bacillus cereus, A. niger, Aspergillus terreus, Malassezia furfur, Trichophyton mentagrophytes, S. typhimurium, L. monocytogenes,* and *S. aureus* (Ranasinghe et al., 2013). Oregano is a popular herb with high levels of terpenoids, mainly the monoterpenoid phenols, namely thymol and carvacrol. One important property of oregano is its high antimicrobial activity: many studies have reported that oregano has effective antimicrobial properties against G+/G− bacteria, such as *S. aureus, Enterococcus faecalis, B. cereus, E. coli, Salmonella enterica,* and *Streptococcus mutans* (Syafiq et al., 2020).

Similarly, green tea leaves own antimicrobial and antioxidant properties. Their composition is rich in polyphenolic catechins and other compounds, such as minerals and amino acids, which are assumed to be healthy and harmless (De Lacey, López-Caballero, & Montero, 2014; Syafiq et al., 2020). These properties make the green tea extract, supplied with antimicrobial agents, an ideal option for biopolymer food packaging as an active filler. Mint is also a popular herb with a long history of usage. It is used in medicine, the food industry, and cosmetics, so this herb is widely grown worldwide. Mint contains volatile components with quinones, flavonoids, and organic acids. It has anti-inflammatory, antifungal, and antimicrobial properties (Hejna, Kovanda, Rossi, & Liu, 2021).

7.3 Antimicrobial agents from herbs and spices in biopolymers

Appropriate packaging material for use in food packaging should be durable, low cost, flexible, easy to prepare, and able to protect its contents against microbial spoilage, oxidation, water vapor, and oxygen transmission (Garavand, Rouhi, Razavi, Cacciotti, & Mohammadi, 2017; Garavand et al., 2022). Synthetic plastics have all these features and have been widely utilized in the food industry for a long period. However, they slowly break down in nature, and they can require 100 or more years to fully degrade; hence, in addition to the abovementioned ideal properties of food packaging, the effects of the environmental conditions should also be considered in the preparation of packaging material (Mirzaei-Mohkam, Garavand, Dehnad, Keramat, & Nasirpour, 2019). To manage the pollution derived from synthetic nonbiodegradable plastics accumulation, landfill is still an important issue; furthermore, incineration and combustion are also used to manage waste (Alirezalu et al., 2021). Petroleum-based polymers are considered one of the main threats to the integrity of the ecosystem, by generating enormous waste each year, which leads to serious environmental and health issues due to the presence of microplastics in the air, water, seafood, and salts (Khodaei, Oltrogge, & Hamidi-Esfahani, 2020). In recent years, the research and innovation for developing bio-based polymers have gained much attention to reduce the dependency on fossil-based packaging films (Cacciotti, Mori, Cherubini, & Nanni, 2018). These biopolymers can be generated from sustainable natural sources or from agricultural and animal processing by-products which are non-toxic, environmentally friendly, and biodegradable (Kakadellis & Harris, 2020). Natural biopolymers are generally more expensive than synthetic polymers, mainly due to the extraction and purification stages. However, due to the abundance of bio-based polymers in nature, the production cost for these polymers can be relatively low (Sionkowska, 2011). However, recently, a lot of attention has been devoted to the development of biopolymer food packaging that has been proposed to overcome this problem. Biopolymer packaging refers to packaging made of recyclable raw materials obtained from animal or plant origins (Mei et al., 2020). New technologies have enabled biopolymers to be prepared like petroleum-based plastics, such as solvent casting, spinning, extrusion, or injection molding. However, biopolymer films are relatively poor in terms of their barrier, mechanical, and antimicrobial properties, which is not desirable, mainly for packaging some food types (Mirzaei-Mohkam, Garavand, Dehnad, Keramat, & Nasirpour, 2020). In particular, to provide antimicrobial properties, researchers suggest the incorporation of antimicrobial/antioxidant fillers within biopolymer films (Sung et al., 2013; Vahedikia et al., 2019). The fabrication of antimicrobial biopolymer films must contain film-forming agents (biosourced), plasticizers, solvents, and antimicrobial/antioxidant agents with bioactive components (Hermawan et al., 2019). Previous studies have shown that applying plant antimicrobial agents in biopolymer materials could increase packaged foods' shelf life by increasing the lag phase and reducing microorganisms' growth (Fernández-Saiz, Sánchez, Soler, Lagaron, & Ocio, 2013; Kurek, Moundanga, Favier, Galić, & Debeaufort, 2013). An antimicrobial biopolymer packaging system is produced by using antimicrobial fillers so that the agents are slowly dispersed onto the surface of the food. This could help to maintain an adequate concentration of the active agent to efficiently inhibit microbial growth throughout the product's shelf life. Therefore, the antimicrobial agents are uniformly released in the amorphous areas of the biopolymer. In recent years, consumer demand for safer food additives and preservatives has grown, therefore, numerous studies have been conducted to examine the potential of natural antimicrobial/antioxidant agents as food preservatives (Vahedikia et al., 2019; Valdés et al., 2015). Generally, plants that contain phytochemicals are essential, especially those that have medicinal benefits. Plant antimicrobial activity depends on the functional groups, compound level, structure, and composition. Numerous studies have focused on natural compounds due to their natural source, high heat stability, nontoxicity, antioxidant activity, and wide-ranging antimicrobial and antifungal properties towards different microorganisms (Hintz, Matthews, & Di, 2015). A wide range of biopolymers have been used to host natural bioactive compounds from vegetables, oils, tea, fruits and their by-products, cereals, and many other spices and herbs, and their various properties have been investigated (Majidi et al., 2019; Sharma, Jafari, & Sharma, 2020). A growing trend in applying antimicrobial additives obtained from spices and herbs and their beneficial effects to minimize food contamination and enhance food quality and shelf life has been reported and will be discussed in the following parts.

7.3.1 *Herbs in biopolymers*

Owing to the presence of polyphenols in their chemical composition, herbs are known to have beneficial effects on human health, such as digestion stimulants, antimicrobial, anti-inflammatory, antioxidant, and anticarcinogenic activities (Varghese, Siengchin, & Parameswaranpillai, 2020). Over the last few years, biopolymers based on animal or plant sources such as proteins, lipids, starch, gums, cellulose derivatives, and chitosan/chitin have emerged as environmentally friendly sources (Babaee et al., 2022; Hadidi et al., 2022). The use of biopolymer films incorporating natural antimicrobial agents is not only a promising approach to increase food shelf life, quality, and oxidative stability but also an alternative to resolve the problem of synthetic plastic waste (Bahrami et al., 2022; Valdés et al., 2015). It is well-known that herbs, supplied with antimicrobial agents, can be directly incorporated into biopolymer films as active ingredients. In this chapter, we investigate and discuss some studies that have recently demonstrated the antimicrobial activity derived from herbs dispersed within biopolymer films. Some examples of recent studies are summarized in Table 7.1 Fig. 7.1 displays a schematic of the preparation of antimicrobial biopolymer films incorporating herbs or spices.

7.3.2 *Spices in biopolymers*

Numerous studies have shown that spices can be used in biopolymer food packaging as active antimicrobial agents. The antimicrobial effects of spice extracts incorporated into food biopolymer packaging films or coatings in recent years are reviewed in Table 7.1. Turmeric, with its anti-cancer, anti-inflammatory, antibacterial, and antioxidant activity, is widely used as a spice, colorant in the food industry, food preservative, and additive in food biopolymeric packaging (Gul & Bakht, 2015). Kalaycıoğlu, Torlak, Akın-Evingür, Özen, and Erim (2017) added turmeric to chitosan to develop an antimicrobial film, demonstrating an increased antimicrobial activity of chitosan films. Vahedikia et al. (2019) developed antimicrobial zein and cinnamon-based films. Films containing cinnamon showed considerable inhibition against *E. coli* and *S. aureus* growth. Therefore, it was concluded that cinnamon could be used as an active antimicrobial agent for food packaging applications. Han, Yu, and Wang (2018) prepared antimicrobial sodium alginate and carboxymethyl cellulose-based films incorporating cinnamon as an antimicrobial agent. They found that films containing cinnamon exhibited excellent antimicrobial activity against *E. coli* and *S. aureus*, and that the prepared coating could increase the shelf life of bananas. Previous works had shown that cumin inhibited bacterial growth. Alizadeh-Sani et al. prepared films that included sodium caseinate, guar gum, and cumin essential oil as functional agents. Their findings proved that biopolymer films with cumin essential oil exhibited remarkable antibacterial activity against both Gram+ (*L. monocytogenes* and *S. aureus*) and Gram- (*E. coli* and *S. enteritidis*) bacteria (Alizadeh-Sani, Rhim, Azizi-Lalabadi, Hemmati-Dinarvand, & Ehsani, 2020).

7.4 Antimicrobial agents from herbs and spices in biopolymer/polymer blends

Blending is a recent method developed to modify polymers for engineering and industrial applications. Blending natural polymers with synthetic polymers leads to a new class of materials with improved mechanical properties and biodegradability (Yu, Dean, & Li, 2006), taking into account the complementary features, mainly

TABLE 7.1 Antimicrobial agents from herbs and spices in biopolymer films.

Biopolymer matrix	Herb or spice (as an active agent)	Main results	References
Tapioca starch	Cinnamon	*P. fluorescens* growth inhibition	Utami, Khasanah, Manuhara, and Ayuningrum (2019)
Chitosan	Green tea extract	High potential to decrease both foodborne viruses and bacteria	Amankwaah, Li, Lee, and Pascall (2020)
Corn/wheat	Lemon oil	Lemon oil antimicrobial effect in biopolymer film and large inhibition zone in the case of higher lemon oil concentrations	Song, Zuo, and Chen (2018)
Cassava starch	Tea tree essential oil	*S. aureus* and *Candida albicans* growth inhibition	Silveira et al. (2020)
Pea starch	Oregano and neem	Controlled *A. niger* and *Penicillium expansum* growth	Cano, Cháfer, Chiralt, and González-Martínez (2016a)
Egg white protein powder	Cinnamon black cumin	Antimicrobial effect against *E. coli* and *L. monocytogenes*	Kavas (2017)
Chitosan	Cumin, clove, and elecampane	Microbial growth inhibition and improved food preservation in food packaging	Hernández-Ochoa, Macias-Castañeda, Nevárez-Moorillón, Salas-Muñoz, and Sandoval-Salas (2012)
Cassava	Lemon grass	*P. fluorescens* and *A. niger* growth inhibition	Resianingrum et al. (2016)
Sodium alginate	Cinnamon, lavender, tea tree, peppermint	*E. coli* and *C. albicans* growth inhibition	Liakos et al. (2014)

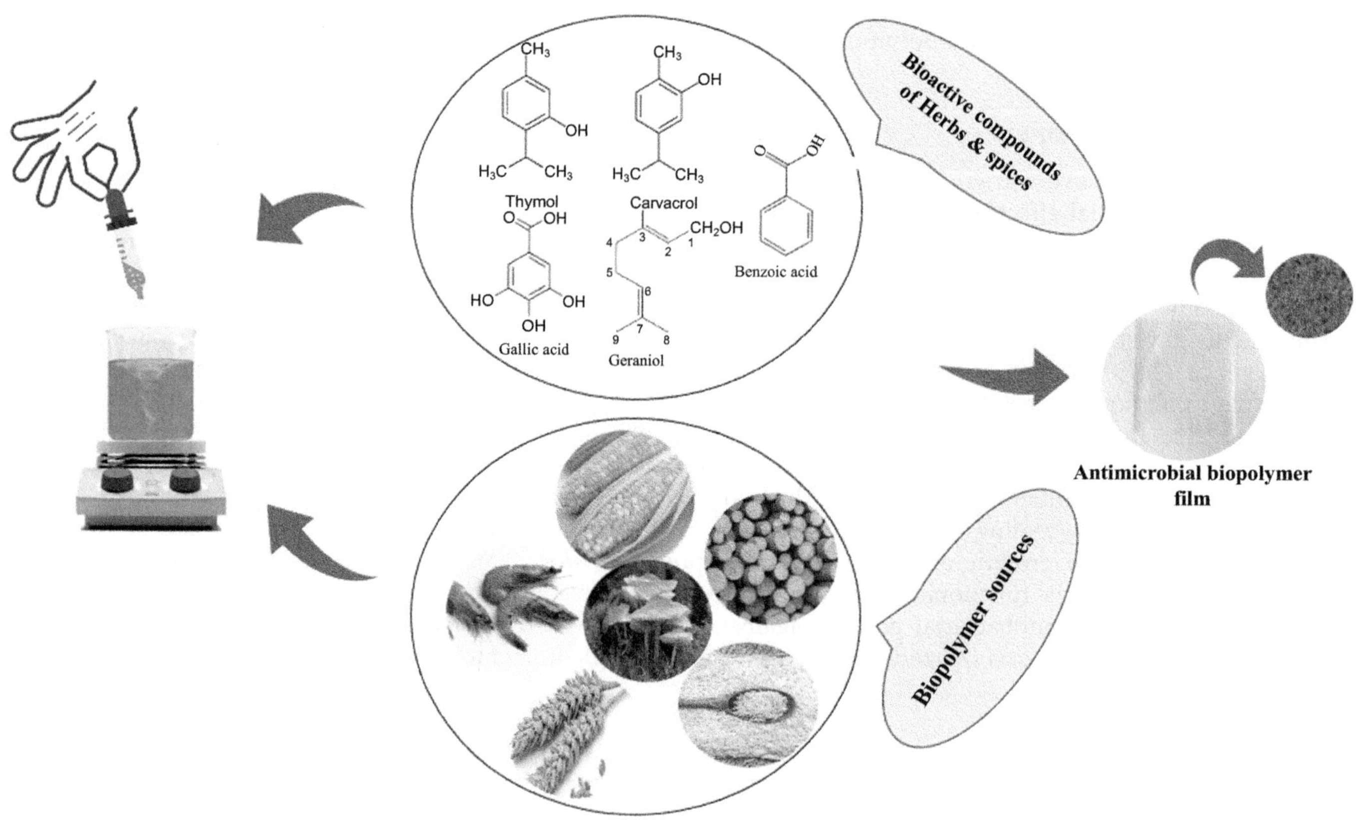

FIGURE 7.1 A schematic of preparation of antimicrobial biopolymer films containing herbs or spices Antimicrobial-loaded biopolymers for food packaging.

the high biocompatibility of natural polymers and the good mechanical and thermal stability of synthetic polymers. The main goals of blending different polymers are to properly adjust the glass transition temperature, improve the material resistance to the fracture, enhance the flexibility, and modify the optical and color characteristics, flammability, and process-ability of polymers (Imre & Pukánszky, 2013). To prepare blended polymers, both natural and synthetic polymers need to be combined in the molten phase (melt mixing) or/and they need to be soluble in the same solvents. Using high pressure along with shear deformation is another method for blending these polymers in solid state (Sionkowska, 2011). Blended polymers have a wide application in packaging and agriculture areas and in the biomedical sector as scaffolds, drug delivery, and controlled release systems. These polymers and their blends can be used as active packaging materials through the incorporation of antimicrobial agents into the film to prohibit the growth of microorganisms on the food surface (Khaneghah, Hashemi, & Limbo, 2018). In fact, by the addition of antimicrobial agents into the films or coatings, the diffusion rate from the packaging material to the product is reduced, helping to maintain high concentrations of the active ingredient where it is required (Rocha, Ferreira, Souza, & Prentice, 2013). Antimicrobial agents from herbs and plants, such as essential oils and plant extracts, have been extensively studied in different biopolymers or synthetic polymers. However, there are few studies about the effect of these antimicrobial agents on the properties of blended films. Some of the recent studies for these antimicrobial blended films are listed in Table 7.2.

Ebrahimzadeh, Bari, Hamishehkar, Kafil, and Lim (2021) developed antimicrobial chitosan thin films laminated with electrospun chitosan/polyvinyl alcohol (PVA) nonwoven membranes which were loaded with *Zataria multiflora* and cinnamon (CEO) essential oils. The results showed no significant difference between antibacterial activity of the films containing EOs against both Gram-negative (*E. coli* and *P. aeruginosa)* and Gram-positive (*B. cereus*, *S. aureus*, and *E. faecalis*) bacteria. However, the films showed antifungal activity against *A. flavus*. The authors pointed out that higher antimicrobial activities of the selected films could be due to the increased hydrophilic property of the film with PVA which led to an enhanced releasing rate of antimicrobial agents. Abbasi, Fahim, and Mahboubi (2021) studied the antimicrobial properties of gelatin-based films blended with cellulose acetate (CA) electrospun fibers containing *Oliveria decumbens Vent* essential oil (OEO). The results from this study showed that neat gelatin film without OEO exhibited no antibacterial properties while the inhibitory properties against *E. coli* increased with the addition of OEO into the film matrix. These blended active films showed a slightly higher antibacterial activity against the Gram-positive bacteria (*S. aureus*). Thus, it is possible to conclude that electrospun fibers can work as controlled release carriers of bioactive compounds, such as OEO, by encapsulating them.

Antimicrobial evaluation of PVA/gum Arabic/chitosan (Amalraj, Haponiuk, Thomas, & Gopi, 2020a) and PVA/gum Arabic/polyethylene glycol (PEG) films (Amalraj, Raj, Haponiuk, Thomas, & Gopi, 2020b), loaded with antimicrobial agents (i.e., black pepper and ginger essential oils), showed a significant inhibitory action against the growth of *B. cereus*, *S. aureus*, *E. coli*, and *S. typhimurium*. These authors reported that the formulated films have great potential to be used in wound dressing or food packaging applications. Antifungal active blended films based on polylactic acid (PLA)/polybutylene-succinate-co-adipate (PBSA) and loaded with thymol EO were manufactured by (Suwanamornlert et al., 2020). In vitro assays showed that the films were highly effective against the growth of *Aspergillus* spp. and *Penicillium* spp. compared to neat PLA film and/or PLA/PBSA blended films. The study of the growth of yeasts and molds on the surface of bread samples packed with these selected films showed that the microbial growth was successfully inhibited by the prepared antifungal films, due to the release of thymol EO in the package headspaces.

Ahmed et al. (2019) studied the antimicrobial properties of PLA/PEG/polycaprolactone (PCL) films incorporated with zinc oxide and clove essential oil, and used them for the scrambled egg packaging. The results showed that these composite films exhibited the highest antimicrobial activity against *S. aureus* and *E. coli* inoculated in scrambled eggs after 21 days of storage at 4 °C. Consequently, the composite film PLA/PEG/PCL/ZnO/CEO completely inhibited the *E. coli* growth, suggesting a synergism between eugenol, the active compound from clove oil, and the ZnO. These authors reported that the hydrophobic compounds from the EO and the cations from zinc nanoparticles would penetrate through the bacterial membrane and rupture the cell structure. The antibacterial activity of grapefruit seed extracts (GSE) incorporated in PLA/polybutylene adipate-co-terephthalate (PBAT) composite films was studied by Shankar and Rhim (2018). The authors reported that the PLA/PBAT/GSE-films containing 7 wt.% of GSE showed high antibacterial activity against *L. monocytogenes*, but only bacteriostatic activity against *E. coli*. Kang, Chandrasekaran, Hwang, Kim, and Lakshmanan (2018) developed PVA/polyvinylpyrrolidone (PVP)/chitosan films loaded with carvacrol and studied the antimicrobial potential of these films against urinary catheters. The results showed that the produced films led to considerable growth inhibition, high resistance to the biofilm formation, and also to the microbial penetration of infectious bacteria isolated from

TABLE 7.2 Blended biopolymer/polymer films incorporated with plant and herbal antimicrobial agents.

Blended polymers	Antimicrobial agent	Studied microorganisms	References
PVA/gum Arabic/ chitosan	Black pepper and ginger essential oils	*B. cereus*, *S. aureus*, *E. coli*, and *S. typhimurium*	Amalraj et al. (2020a)
PVA/gum Arabic/PEG	Black pepper and ginger essential oils	*B. cereus*, *S. aureus*, *E. coli*, and *S. typhimurium*	Amalraj et al. (2020b)
PLA/PBAT	Grapefruit seed extracts	*L. monocytogenes*, *E. coli*	Shankar and Rhim (2018)
PLA/PTMC	Cinnamaldehyde	*E. coli* and *S. aureus*	Qin et al. (2015)
PLA/nano-chitosan	*Polylophium involucratum* essential oil (PIE)	Mesophilic, psychotropic, lactic acid bacteria, and Enterobacteriaceae microorganisms	Javaherzadeh, Tabatabaee Bafroee and Kanjari (2020)
Chitosan/PVA nonwovens	*Zataria multiflora* and cinnamon essential oils.	Gram-positive and Gram-negative bacteria, *A. flavus*	Ebrahimzadeh et al. (2021)
CMC/PVA	Rosemary essential oil.	*Penicillium digitatum*	Fasihi et al. (2017)
PCL/MC/PCL	Antimicrobial formulation A (organic acids, extract of rosmarinic acid and Asian essential oil (EO) mixture), and formulation B (organic acids, extract of rosmarinic acid and Italian EO mixture)	Total aerobic microorganisms and foodborne bacteria (*E. coli* O 157:H7, *S. typhimurium* and *L. monocytogenes*)	Takala et al. (2013)
LDPE/EVA	Clove leaf oil (CL), sweet basil oil, and cinnamon bark oil	*E. coli*, *S. aureus*	Wattananawinrat et al. (2014)
PLA/PEG/PCL	Zinc oxide/clove essential oil	*E. coli*	Ahmed et al. (2019)
PLA/PBSA	Thymol essential oil	*Aspergillus* spp. and *Penicillium* spp.	Suwanamornlert et al. (2020)
Gelatin/CA electrospun fibers	*Oliveria decumbens Vent* essential oil (OEO)	*E. coli*, *S. aureus*	Abbasi et al. (2021)
Starch/PVA	Neem oil, oregano EO, and silver nanoparticles	*L. innocua*,*E. coli*, *A. niger*, and *P. expansum*	Cano et al., 2015; Cano, Cháfer, Chiralt, and González-Martínez (2016b)
PLA/Kenaf fiber	Thymol essential oil	*E. coli*	Tawakkal et al. (2017)
PVA/PVP/Chitosan	Carvacrol	*E. coli*, *B. subtilis*, and *S. typhimurium*	Kang et al. (2018)
PLA/PTC	Oregano essential oil	*E. coli* and *L. monocytogenes*	Liu et al. (2016)
PLA/PTC	Thymol essential oil	*E. coli*, *S. aureus*, *L. monocytogenes*, *B. subtilis*, and *S. typhimurium*	Wu et al. (2014)

hospitals such as *E. coli*, *B. subtilis*, and *S. typhimurium*. Thus, the obtained systems had great potential to be used in biomedical applications.

Fasihi, Fazilati, Hashemi, and Noshirvani (2017) prepared carboxymethyl cellulose (CMC)/PVA blended films loaded with rosemary essential oil (REO). Antimicrobial studies showed a complete inhibitory action against *P. digitatum* in the case of films containing 3% of EO. The presence of PVA, by facilitating the release of EO from the film matrix, improved the antifungal activity of the active film. Moreover, the bread slices covered with the formulated films showed no fungal growth during 60 days of storage, very probably due to the slow and steady release of REO from the blended films. The antimicrobial properties of PLA/kenaf fiber films loaded with thymol EO were evaluated by Tawakkal, Cran, and Bigger (2017). Studying the antimicrobial properties of these blended films in the soy broth system, they demonstrated that the films loaded with 20−30 wt.% of the thymol reduced the *E. coli* growth after two days. These films also showed a great potential to extend the shelf life of sliced chicken samples after 30 days of storage at 10 °C by reducing the number of *E. coli* inoculated on the surface of the product.

Liu et al. (2016) manufactured antimicrobial active packaging films by incorporating oregano EO into PLA/poly(trimethylene carbonate) (PTMC) films. The results from this study showed that PLA/PTMC blend films without the addition of OEO had no bacterial inhibitory effects, while the blended films containing 9 wt.% of EO significantly reduced the growth of *E. coli* and *L. monocytogenes*. Consequently, after 24 hours of incubation, blended PLA/PTMC films reduced the *E. coli* and *L. monocytogenes* from 5 to 1.4 and 1.5 logs, respectively.

Cano, Chafer, Chiralt, and Gonzalez-Martinez (2015) studied the biodegradability of starch/PVA blended films affected by the addition of different antimicrobial agents, such as neem oil, oregano EO, and silver nanoparticles. The results showed that the addition of neem and oregano EO increased the disintegration and biodegradation of the blended films, while the films with higher amounts of silver nanoparticles exhibited a slower biodegradation capacity. Therefore, lower concentrations of silver NPs were recommended to avoid possible changes in the microbial activity of composites. The antimicrobial properties of these films showed that only films containing OEO exhibited antibacterial and antifungal activity. Antibacterial activity against *L. innocua* and *E. coli* appeared at lower doses of OEO (6.7 wt.% in the dried matrix), while the antifungal effect against *A. niger* and *P. expansum* required higher doses of EO in the blended film (Cano et al., 2015). In another study, it was reported that the PLA/PTMC blended films loaded with cinnamaldehyde (9 or 12 wt.%) significantly reduced the number of *E. coli* and *S. aureus* at 16 hours and 24 hours of incubation compared to control blend films without the antimicrobial agent (Qin, Yang, & Xue, 2015).

PLA/PTMC films were manufactured by Wu et al. (2014) and the effect of the thymol EO addition on the antimicrobial properties of the films was evaluated. The antimicrobial study of PLA/PTMC films loaded with different amounts of thymol EO showed that the inhibitory effect of the films against *E. coli, S. aureus, L. monocytogenes, B. subtilis,* and *S. typhimurium* significantly increased with the EO concentration increment. The authors reported that the thymol incorporated into the films can migrate from the film to the agar wells and therefore can exercise antimicrobial activity.

The antimicrobial activity of blended low-density polyethylene (LDPE)/ethyl vinyl acetate (EVA) films containing clove leaf (CL), sweet basil (SB), and cinnamon bark (CB) oils was studied by Wattananawinrat, Threepopnatkul, and Kulsetthanchalee (2014). The antimicrobial activity of the films was increased by EOs lipophilicity and LDPE/EVA film with CL and CB was more efficient against *E. coli* than *S. aureus*. The results showed that CB played the highest antimicrobial activity regardless of its relatively low retention. Indeed blended films containing 2 wt.% of CB showed the highest antimicrobial activity against *S. aureus* and *E. coli*.

Takala et al. (2013) developed bioactive trilayer PCL/methylcellulose (MC)/PCL trilayer films, characterized by two different antimicrobial formulations [A (organic acids, extract of rosmarinic acid and Asian essential oil (EO) mixture) and B (organic acids, extract of rosmarinic acid and Italian EO mixture)] added in MC, by means of compression molding. The authors investigated their controlled release and antimicrobial activity against total aerobic microorganisms and foodborne bacteria (*E. coli* O 157:H7, *S. typhimurium* and *L. monocytogenes*) during the shelf life study of broccoli at 4 °C up to 12 days. The results showed that these films exhibited high antibacterial activity against Gram-negative bacteria such as *E. coli* and *S. typhimurium*, achieving the total inhibition for these microorganisms at 12 and 7 days, respectively. The essential oils formulations also showed a significant influence on the film's antimicrobial properties, and the film containing formulation A exhibited a stronger inhibitory action against *E. coli*, whereas both formulations were efficient against *S. typhimurium*. These films showed a good inhibitory effect against the growth of total aerobic microorganisms in broccoli, up to 10 days of storage, while they showed no antimicrobial activity against *L. monocytogenes*. The authors noted that the diffusion of volatiles might be due to the different mechanisms of action of antimicrobial volatiles as opposed to the direct diffusion of antimicrobials in contact with food.

7.5 Antimicrobial agents from herbs and spices in synthetic polymer

7.5.1 *Synthetic polymers*

Synthetic polymers are derived from petroleum oil through carbon–carbon bonds, while other substances such as oxygen and nitrogen are the major parts of the backbone of hetero-chain polymers (Raja & Mattoso, 2020). Synthetic polymers due to their low costs and easy-to-process properties have been mostly used in the food packaging sector, although they have had problems at the stage of waste disposal. The physical properties of synthetic polymers play a key role in designing packaging structures. These properties can be changed by reactions between polymer chains, such as ionic forces, hydrogen bonding, and crystalline regions (Bolívar-

Monsalve et al., 2021). Elastomers, fibers, and plastics are commonly used in the packaging sector. Elastomers, such as rubber, have high flexibility that gives the capability of returning to the original conformation after stresses. Fibers are typically used to form other materials, and there are two methods to produce synthetic fibers including passing molten thermoplastics through tiny pores in a die and passing thermoplastics dissolved in a volatile solvent through tiny pores in a die (Hacker, Krieghoff, & Mikos, 2019; Ouellette & Rawn, 2015). Plastics have many advantages, including flowability and moldability; they are chemically inert and have appropriate features such as not supporting the growth of microorganisms, the ability to be combined with other plastics, printability, effective costs, lightweight, barrier properties, and heat resistance. Some plastics have special characteristics allowing their use in a wide range of temperatures in food sectors, such as freezing of food and storage, as well as the treatment at high temperatures in processes such as sterilization, and reheating temperature by radiant heat and microwave (Havstad, 2020). Therefore, plastics have been selected as good candidates for use in food products packaging. Common types of plastics that can be used for food packaging are polyethylene (PE), polyesters [polyethylene terephthalate (PET), polyethylene naphthalate (PEN), polycarbonate (PC)], polypropylene (PP), polyamides (PAs), polyvinylidene chloride (PVDC), polyvinyl chloride (PVC), polystyrene (PS), polyvinyl acetate (PVA), fluoropolymers (Polychlorotrifluoroethylene and polytetrafluoroethylene), polymethyl pentene, ionomers, EVA, styrene-butadiene (SB), ethylene vinyl alcohol, acrylonitrile butadiene styrene, high nitrile polymers, and cellulose-based materials. For example, PE, PP, PET, PS, and PVC-based packaging presents higher consumption in the packaging market of Europe (Kirwan & Strawbridge, 2003; Kirwan, Plant, & Strawbridge, 2011).

PVC, the most widely used plastic in the world, has been a common polymer for food packaging design because of several superior advantages, such as softness, flexibility, toughness, chemical inertness, and high permeability (Kanavouras, Hernandez-Munoz, & Coutelieris, 2006). PET can provide a packaging structure with a high crystalline melting temperature, and appropriate mechanical strength, which makes it a potential alternative to PVC (Mangaraj, Goswami, & Mahajan, 2009). PP is resistant to water vapor and has a low density and high melting point, leading to its use for high-temperature packing applications (Maddah, 2016). Low-density polyethylene (LDPE) and high-density polyethylene (HDPE) are two common types of PE that are applicable in food packaging. LDPE is the cheapest plastic polymer with transparency and good resistance to water vapor; nevertheless, its ability is not enough to hold back oxygen and carbon dioxide. HDPE, as well as LDPE, is characterized by good barrier properties against water vapor and low barrier properties against oxygen and carbon dioxide. HDPE is stiffer but less transparent than LDPE (Heldman, 2012). PS is hard and fairly brittle, but, when combined with butadiene, its flexibility is enhanced, and it can be used in food contact applications (Cooper, 2007).

7.5.2 *Approaches for the synthetic polymers functionalization with antimicrobials*

As discussed earlier, synthetic polymers have been widely used in the food packaging sector due to their several advantages. The superior properties of polymers have attracted the attention of researchers to improve their functional properties by the enrichment with functional compounds, to achieve an optimal packaging structure for food products (Garavand et al., 2017). Recently, natural antimicrobial herbs and spices-derived agents have been combined with polymers to produce an antimicrobial packaging structure, resulting in food products' shelf life improvement. Thus, a significant advantage of the packaging structures formulated with polymers embedded with antimicrobials is providing an antimicrobial controlled release from packaging during storage to improve the food shelf life (Vahedikia et al., 2019). Antimicrobial substances derived from herbs and spices can be incorporated into synthetic polymers by using different methods based on the migration of the agents to package (including embedding, non-covalent immobilization, and layer-by-layer deposition) and non-migration of the agents to package (including covalent immobilization, photografting, and layer-by-layer deposition) (Bastarrachea, Dhawan, & Sablani, 2011; Suppakul, Miltz, Sonneveld, & Bigger, 2003). Fig. 7.2 presents the various approaches employed to incorporate active antimicrobials into synthetic polymers.

Antimicrobials can be embedded into polymer matrices by using solution casting (Fig. 7.2A), in which first the polymer and the selected an antimicrobial agent are simultaneously dissolved in a suitable solvent, then the solvent is evaporated from the solubilized polymer and the active agent to promote the formation of a polymer/antimicrobial packaging. Extrusion is another method to incorporate antimicrobials into the polymer, in which a mixture is prepared by the combination of the antimicrobial agent with the melted polymer through heat transfer. The solution casting is commonly used in the laboratory, and in this method high temperatures could be necessary to dissolve the polymer in its solvents, leading to destroy the sensitive active agents. Extrusion can be used to incorporate antimicrobials into the polymer on an industrial scale, but the applied high temperatures can damage the antimicrobials

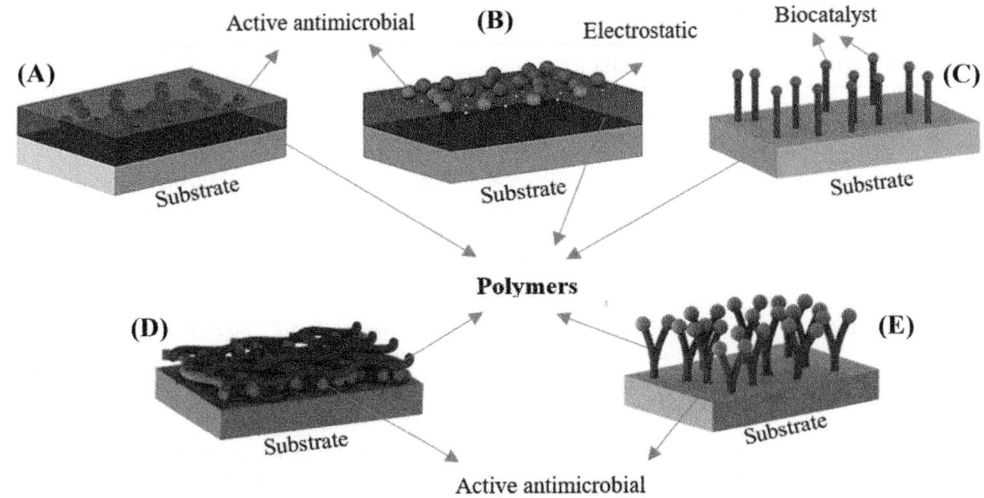

FIGURE 7.2 Different methods to functionalize synthetic polymers with active antimicrobial agents: (A) embedding; (B) surface immobilization (non-covalent); (C) surface immobilization (covalent); (D) layer-by-layer assembly; (E) photografting functionalization.

(Velásquez et al., 2021). Moreover, although active polymers can be obtained by the antimicrobials incorporation within them, this approach can create fundamental modifications in the polymer's structure, resulting in their thermal, mechanical, and gas barrier properties changes. For all these reasons, coating the polymers with the antimicrobial has been introduced as a promising alternative to produce active polymers because this strategy does not create bulk changes in polymer's structure (Ahvenainen, 2003; Bastarrachea et al., 2011). For example, antimicrobial agents can be immobilized on the surface of the polymers in the packaging structure. As some polymers are chemically inert, the agent immobilization on their surface needs some functionalization by wet and/or physical methods. In the wet method, reactive oxygenated species, such as hydroxyl, carbonyl, and carboxylic acid groups, are created on the polymer surface by using corrosive liquids, for example, piranha solution. In order to avoid the use of liquids, handling, and accumulation, it is possible to use physical methods, such as corona discharge, plasma, flame, and UV radiation (Goddard & Hotchkiss, 2007). When the surface of polymers is functionalized, active antimicrobial agents can be immobilized on their surface by using either non-covalent (Fig. 7.2B) or covalent immobilization (Fig. 7.2C). Electrostatic interactions, based on creating opposite charges between the surface of the polymer and active antimicrobial agent, and ligand-receptor interactions are two common non-covalent antimicrobial immobilization approaches (Fig. 7.2B). In case of the covalent immobilization strategy (Fig. 7.2C), active antimicrobials can be directly linked to the polymer surface by using a crosslinker, and, thus the antimicrobials do not migrate to the packaged product matrix (Bastarrachea, Wong, Roman, Lin, & Goddard, 2015; Farkaš & Bystrický, 2010; Goddard & Hotchkiss, 2007). Another method to provide a polymer with antimicrobial features is the layer-by-layer assembly, which is based on the deposition and the attraction of polyelectrolytes via opposite charges onto the polymer surface (Fig. 7.2D). Polyelectrolytes can be attached to the surface of the polymeric substrate by either electrostatic interaction or covalent bonds. Submersion of the substrate into the polyelectrolyte solutions and solutions spraying onto the substrate are two methods of layer-by-layer deposition. Antimicrobial agents can be incorporated between layers and into a polyelectrolyte layer (Bastarrachea et al., 2015). Another approach for immobilizing active antimicrobial agents on the polymer's surface is photografting by using a light-induced mechanism (Fig. 7.2E). In this technique, the surface of the polymer is exposed to UV light free radicals, resulting in the polymerization of the monomers by reactive compounds that produce unpaired electrons. Hydrogen abstraction, electron transfer, and cleavage are three major mechanisms of photografting (Dadashi-Silab, Doran, & Yagci, 2016).

7.5.3 Applications of antimicrobial synthetic polymer-based packaging containing active compounds of herbs and spices

Several antimicrobial active agents derived from herbs and spices have been used to activate food packaging structures based on synthetic polymers (Table 7.3), where a controlled release of antimicrobials can be obtained over time, leading to improved food shelf life (Abedi-Firoozjah et al., 2022). For instance, Gutiérrez-García et al. (2021)

TABLE 7.3 Activated packaging structures based on synthetic polymers loaded with active antimicrobial agents of herbs and spices.

Synthetic polymer	Antimicrobial agent	Results	References
Low-density polyethylene (LDPE)	Epigallocatechin gallate	Morphological changes and *S. aureus* growth inhibition for films with 10 wt.% epigallocatechin gallate (with and without adhesive resin)	Moreno-Vásquez et al. (2017)
LDPE and ethylene vinyl alcohol copolymer (EVOH)	Carvacrol	High antimicrobial activity against *E. coli* and *Alternaria alternata* in *in vitro* micro-atmosphere assays and against *A. alternata* and *Rhizopus* in cherry tomatoes, used as food models in the case of (LDPE/carvacrol)/EVOH and (LDPE/(halloysite nanotubes/carvacrol))/EVOH films	Krepker, Zhang, et al. (2018)
Polypropylene (PP)	Carvacrol	Higher antimicrobial properties against *E. coli and Alternaria alternata*, superior mechanical properties, and prolonged release of carvacrol in the case of PP/(halloysite nanotubes/carvacrol) nanocomposite films	Krepker et al. (2018)
LDPE and thylene vinyl acetate	Clove leaf oil, sweet basil oil, and cinnamon bark oil	The packaging of sliced tomatoes by using antimicrobial films containing essential oil would be able to noticeably withstand the microorganisms' growth.	Wattananawinrat et al. (2014)
Polyurethane	Carvacrol	Sustained release of carvacrol and antibacterial activity against representative pathogens, e.g., *Aeromonas hydrophila*, in the case of carvacrol/halloysite nanotubes/polyurethane films	Hendessi et al. (2016)
PP	Caraway essential oil	Significant antibacterial activity against *S. aureus* and *E. coli* at the essential oil concentration of 4.9 ± 0.2 wt.% and higher.	Strasakova et al. (2021)
Linear low-density polyethylene (LLDPE)	Clove leaf oil	Higher antimicrobial activity against *Salmonella typhimurium* and *L. monocytogenes* in a packed chicken in the case of LLDPE surface chemically modified by chromic acid treatment and coated with clove essential oil	Mulla et al. (2017)
PP/LDPE	*Allium sativum* essence oil	High ability to extend the shelf-life of large yellow croaker by 5 days at 4°C ± 1°C in the case of an active bilayer-films incorporated with Attapulgite loaded with *Allium sativum* essence oil	Dong et al. (2019)
LDPE	Rosemary essential oil (REO)	High efficiency in maintaining the freshness and extending the shelf life of packaged shrimps up to 4 days in the case of active films containing REO	Dong et al. (2018)
PET	Black cumin oil (BCO)	Antimicrobial activity against *S. aureus* and *E. coli*, and provided lower total aerobic mesophilic (TAMC) and psychotropic bacteria counts (PBC) for chicken meat packaged with PET/BCO antimicrobial films	Takma and Korel (2019)

evaluated the capability of LDPE film loaded with *Yucca baccata* butanolic extract to improve the shelf life of ground beef. The produced films presented an inhibitory effect on aerobic mesophilic bacteria in the packed ground beef; the active films improved the shelf life of ground beef from 3 days (observed in the case of using control film for packaging) to 8 days. In another study by Manso, Becerril, Nerín, and Gómez-Lus (2015), cinnamon essential oil as an active substance was applied to PP to produce antimicrobial packaging. They used the method of Garces and de la Puerta (2010) to functionalize PP, which was based on three steps, including the preparation of the main formulation by the combination of the polymer solution obtained in a suitable solvent with the natural extracts containing antimicrobial agents, and finally, the addition of the resulting active liquid to the chosen packaging or support and drying. Then, Manso et al. (2015) evaluated the antifungal activity of the obtained active packaging against *P. roqueforti*, *A. flavus* and *P. expansum* at various temperatures and pH, determining minimal inhibitory concentration (MIC) and minimal fungicidal concentration (MFC). Their results showed that all PP-based packaging delayed the fungus growth. Higher antifungal activity was obtained by enhancing the concentration of cinnamon essential oil and was observed for 30 days. Moreover, the packaging antifungal activity was not decreased at the fridge temperature, showing the very high potential of the produced packaging material to be used for packaging different food products (Manso et al., 2015).

Taking into account that the selection of the optimal methods for the functionalization of the polymers strongly influences their antimicrobial and physical properties, in a study, two methods, i.e. ionizing treatment and extrusion, were applied to incorporate oregano (*Origanum vulgare*) and thyme (*Thymus vulgaris*) EOs into

LDPE (Solano & de Rojas Gante, 2012). In the ionizing treatment, the surface of LDPE films was treated by electronic radiation, and the essential oil was placed on them. In the second method, EOs and resin LDPE were preblended and extruded. The evaluation of the antimicrobial activity of packaging polymers produced by both ionizing and extrusion methods against foodborne pathogens, including *S. typhimurium*, *L. monocytogenes*, and *E. coli* O157:H7, showed that a significantly higher pathogens inhibition was achieved by using extrusion method compared to the ionizing method. This experimental evidence was ascribed to a better antimicrobials distribution into packaging structure by using the extrusion technique. However, the investigation of the polymers physical properties (including tensile strength (TS), thickness, and elongation at break (EAB)) before and after the functionalization with EOs showed that the extrusion method led to an increase in elongation and reduction in tensile strength, as expected, due to EOs plasticizing action (Solano & de Rojas Gante, 2012). Mousavian, Nafchi, Nouri, and Abedinia (2021) produced active systems by immersing orientated polypropylene (OPP) and LDPE films in a thyme EO solution. The produced active films revealed higher antimicrobial activities against *E. coli*, *S. aureus*, and *S. enteritidis*, highlighting that this action was enhanced by increasing the film's immersion time in thyme EOs solution. Moreover, their results indicated that when the dipping time was increased from 1 to 3 days, the tensile strength of the films decreased. Indeed, as already evidenced, the antimicrobial agent could change the internal interactions of the packaging matrix, thus leading to modified mechanical properties. Whereas in general, the TS values were decreased by the EOs loading in polymeric structures, the EAB was enhanced (Mousavian et al., 2021). Indeed, the incorporation of antimicrobials into polymer-based packaging structure can increase its flexibility due to the plasticizer role of antimicrobials (Corrales, Fernández, & Han, 2014). However, it should be considered that the EAB of films can be decreased by increasing the amount of added antimicrobial agents because these agents are placed among polymer chains, leading to the reduction of the polymeric forces (Gamage, Park, & Kim, 2009). The release behavior of EOs incorporated into OPP and LDPE films was also studied in the ethanol as a food model (Mousavian et al., 2021). In this study, the EOs release from both films was increased by increasing immersion temperature from 4°C to 25°C, and the active OPP-based films showed a higher release of EOs compared to LDPE.

Moreover, it is fundamental to consider that further several factors, such as the polymer surface treatment and the addition of other compounds to the film matrix, exert an impact on the properties of the produced polymer-based packaging. For example, Swilem, Stloukal, Abd El-Rehim, Hrabalíková, and Sedlařík (2019) studied the effect of the addition of micro-sized wood flour (WF) and (MS) as carriers for terpene 4-allylanisole (ALY, an antimicrobial agent derived from the basil plant) within the polymeric matrix and the gamma irradiation treatment on the physical and antimicrobial properties of the active packaging based on LDPE, produced by a thermoplastic process. In detail, LDPE films loaded with ALY-carriers were obtained by blending LDPE and antimicrobials at high temperatures, adding the additives, and cooling the mixture to produce active polymers. It was observed that the mechanical properties of the films were modified by the addition of ALY-carriers without the gamma irradiation treatment: the tensile modulus was significantly enhanced, and the tensile strength and EAB were decreased. It indicated that when solid particles such as carriers are added to a polymer matrix, this restricts the mobility of the polymeric chains around each particle and enhances stiffness, increasing the tensile modulus value (Mallik, Barik, & Pal, 2015). In the systems with higher filler amounts, decreased tensile strength and elongation were detected due to the not enough interfacial adhesion between the filler and the polymer matrix under high stress at breakpoint (Huang & Zhang, 2009). Concerning the ALY released from LDPE, a reduced release was ascribed to the network's formation in the polymer structure after the exposure to gamma irradiation (Swilem et al., 2019). The impact of carriers to decrease ALY release was more than that associated with the gamma irradiations, and among the carriers, the highest reduction was revealed in the case of WF compared to MS, which could be related to a more tortuous path created by the addition of WS particle. Moreover, in this study, the antimicrobial activity of the produced active films against *S. aureus* and *E. coli* was investigated, and the results indicated that increasing the concentration of ALY-carriers from 20 wt.% to 30 wt.% did not play a significant impact on the film's antimicrobial activity. In the polymers treated with 30 and 50 kGy gamma irradiations, it was observed that the antimicrobial activity was reduced due to the formation of a more crosslinked structure and slowing release rate of ALY (Swilem et al., 2019). Roy and Rhim (2021) encapsulated grapefruit seed extract (GSE) and curcumin within poly (vinyl alcohol) (PVOH) film using a solvent casting technique. The thickness of the PVOH-films was increased by loading with GSE and curcumin because of the enhancing solid film materials. TS value of the produced films was decreased with adding GSE and curcumin, whereas the GSE addition led to improved flexibility (EB) and decreased stiffness (EM) of PVOH-films, which could be related to the presence of glycerol in GSE. On the contrary, the incorporated curcumin into the films enhanced EM because curcumin may have good compatibility with the polymer at the interface. Moreover, PVOH-films containing GSE and curcumin showed higher antimicrobial activity against *L. monocytogenes* than against *E. coli* with respect to neat polymeric films. In particular, GSE-films showed higher antimicrobial activity against both of the tested bacteria than curcumin-loaded films. Moura, de Souza, de Oliveira, Lemos, and Druzian (2018) used carotenoid (CE) and yerba mate extracts (YME) to functionalize LDPE films and provide them with antimicrobial properties. The investigation of mechanical properties showed that the films TS and EAB values were decreased by adding CE and YME with respect to the control film. For the evaluation of active film's antimicrobial activity, the molds, yeasts, and mesophilic and psychotropic microorganisms of the butter packaged with the active films and stored at refrigerating were counted, and it resulted that the concentration of CE and YME in the structure of the film had an impact on their antimicrobial activities; the lowest antimicrobial activity was observed in the case of active film incorporated with 1.5% CE and 1.5% YME (Moura et al., 2018). Tea tree oil (TTO) with rice bran (RB) was applied to create an antimicrobial LDPE based film (El-Wakil, Moustafa, & Youssef, 2020); first, acrylic acid (AAc) monomer was grafted onto an LDPE using melt blending, and then, the resulted film combined with untreated RB and treated TTO-RB to obtain an active packaging structure. Adding RB and TTO-RB into the film structure did not have a significant effect on the tensile properties; however, an increase in the film's hardness was observed by increasing the content of active agents in the polymer structure. The antimicrobial activity of the produced films was evaluated against *S. aureus*, *E. coli*, *C. albicans*, and *A. nigera*. According to the collected results, LDPE, LDPE/LDPE-g-AAc, and their films containing untreated RB did not show antimicrobial activity against all the tested bacteria. However, bacteria growth inhibition was observed in the case of the films loaded with TTO, due to TTO hydroxyl groups (El-Wakil et al., 2020).

7.6 Concluding remarks and future trends

In recent years, consumers are more aware of the prepared food they eat and are more worried about its safety requirements. Thus, active packaging, based on the use of antimicrobials, is gaining a lot of interest. However, the use of synthetic additives as flavoring agents for a considerable period of time may induce negative health side effects. Additionally, the employment of chemical food additives presents several disadvantages, such as high cost, dangers, health hazards, and environmental pollution. For these reasons, the actual trend is focused on the exploitation of natural additives which are low-cost, easy to use, cost-effective, without side effects, non-hazardous and non-toxic. In this regard, nowadays, the exploitation of plant products, such as essential oils, is a more noteworthy request and an urgent need. Antimicrobials from spices and herbs are widely exploited in active food packaging, but there is not enough knowledge and comprehension about the mechanism of their antimicrobial action, toxicological effects, safety, toxicity, health hazards, proper dose, sensory attributes, and further aspects, such as the interactions of herbs/spices with the considered packaging materials, and how to prevent the transfer of odor/flavor from natural spice extracts loaded packages to the packaged foods. In the future, taking into account that the combinations of several spices demonstrated to play higher inhibitory actions on specific bacteria than the individual spices, it will be pivotal to better and more deeply investigate and comprehend the interactions among the combined spices and their inhibition effect on different microorganisms in different food products. Moreover, actually, most of the biological studies consist of in vitro tests related to the spice and herb's action against human pathogenic bacteria; thus, in vivo studies and clinical trials have to be performed in the future.

References

Abbasi, H., Fahim, H., & Mahboubi, M. (2021). Fabrication and characterization of composite film based on gelatin and electrospun cellulose acetate fibers incorporating essential oil. *Journal of Food Measurement and Characterization, 15*, 2108–2118. Available from https://doi.org/10.1007/s11694-020-00799-1.

Abdollahzadeh, E., Nematollahi, A., & Hosseini, H. (2021). Composition of antimicrobial edible films and methods for assessing their antimicrobial activity: A review. *Trends in Food Science & Technology*.

Abedi-Firoozjah, R., Yousefi, S., Heydari, M., Seyedfatehi, F., Jafarzadeh, S., Mohammadi, R., ... Garavand, F. (2022). Application of red cabbage anthocyanins as pH-sensitive pigments in smart food packaging and sensors. *Polymers, 14*(8), 1629.

Ahmed, J., Mulla, M., Jacob, H., Luciano, G., B, T. B., & Almusallam, A. (2019). Polylactide/poly(ε-caprolactone)/zinc oxide/clove essential oil composite antimicrobial films for scrambled egg packaging. *Food Packaging and Shelf Life*, 21. Available from https://doi.org/10.1016/j.fpsl.2019.100355.

Ahvenainen, R. (2003). *Novel food packaging techniques*. Elsevier.

Alirezalu, K., Pirouzi, S., Yaghoubi, M., Karimi-Dehkordi, M., Jafarzadeh, S., & Khaneghah, A. M. (2021). Packaging of beef fillet with active chitosan film incorporated with ε-polylysine: An assessment of quality indices and shelf life. *Meat Science, 176*, 108475.

Alizadeh-Sani, M., Rhim, J.-W., Azizi-Lalabadi, M., Hemmati-Dinarvand, M., & Ehsani, A. (2020). Preparation and characterization of functional sodium caseinate/guar gum/TiO2/cumin essential oil composite film. *International Journal of Biological Macromolecules, 145*, 835–844.

Amalraj, A., Haponiuk, J. T., Thomas, S., & Gopi, S. (2020a). Preparation, characterization and antimicrobial activity of polyvinyl alcohol/gum arabic/chitosan composite films incorporated with black pepper essential oil and ginger essential oil. *International Journal of Biological Macromolecules, 151*, 366–375.

Amalraj, A., Raj, K. J., Haponiuk, J. T., Thomas, S., & Gopi, S. (2020b). Preparation, characterization, and antimicrobial activity of chitosan/gum arabic/polyethylene glycol composite films incorporated with black pepper essential oil and ginger essential oil as potential packaging and wound dressing materials. *Advanced Composites and Hybrid Materials, 3*, 485–497.

Amankwaah, C., Li, J., Lee, J., & Pascall, M. A. (2020). Antimicrobial activity of chitosan-based films enriched with green tea extracts on murine norovirus, Escherichia coli, and Listeria innocua. *International Journal of Food Science*, 2020.

Azizi-Lalabadi, M., Garavand, F., & Jafari, S. M. (2021). Incorporation of silver nanoparticles into active antimicrobial nanocomposites: Release behavior, analyzing techniques, applications and safety issues. *Advances in Colloid and Interface Science*, 102440.

Azizkhani, M., & Tooryan, F. (2015). Antioxidant and antimicrobial activities of rosemary extract, mint extract and a mixture of tocopherols in beef sausage during storage at 4 C. *Journal of Food Safety, 35*, 128–136.

Babaee, M., Garavand, F., Rehman, A., Jafarazadeh, S., Amini, E., & Cacciotti, I. (2022). Biodegradability, physical, mechanical and antimicrobial attributes of starch nanocomposites containing chitosan nanoparticles. *International Journal of Biological Macromolecules, 195*, 49–58.

Bahrami, R., Zibaei, R., Hashami, Z., Hasanvand, S., Garavand, F., Rouhi, M., ... Mohammadi, R. (2022). Modification and improvement of biodegradable packaging films by cold plasma; a critical review. *Critical Reviews in Food Science and Nutrition, 62*(7), 1936–1950.

Bastarrachea, L., Dhawan, S., & Sablani, S. S. (2011). Engineering properties of polymeric-based antimicrobial films for food packaging: A review. *Food Engineering Reviews, 3*, 79–93.

Bastarrachea, L. J., Wong, D. E., Roman, M. J., Lin, Z., & Goddard, J. M. (2015). Active packaging coatings. *Coatings, 5*, 771–791.

Bolívar-Monsalve, E. J., Álvarez, M. M., Hosseini, S., Espinosa-Hernandez, M. A., Ceballos-González, C. F., Sanchez-Dominguez, M., ... Di Maio, E. (2021). Engineering bioactive synthetic polymers for biomedical applications: A review with emphasis on tissue engineering and controlled release. *Materials Advances*.

Brandi, G., Amagliani, G., Schiavano, G. F., De Santi, M., & Sisti, M. (2006). Activity of *Brassica oleracea* leaf juice on foodborne pathogenic bacteria. *Journal of Food Protection, 69*, 2274–2279.

Brul, S., & Coote, P. (1999). Preservative agents in foods: Mode of action and microbial resistance mechanisms. *International Journal of Food Microbiology, 50*, 1–17.

Burt, S. (2004). Essential oils: Their antibacterial properties and potential applications in foods—A review. *International Journal of Food Microbiology, 94*, 223–253.

Cacciotti, I., Ciocci, M., DI Giovanni, E., Nanni, F., & Melino, S. (2018). Hydrogen sulfide-releasing fibrous membranes: Potential patches for stimulating human stem cells proliferation and viability under oxidative stress. *International Journal of Molecular Sciences, 19*(8), 2368. Available from https://doi.org/10.3390/ijms19082368.

Cacciotti, I., Mori, S., Cherubini, V., & Nanni, F. (2018). Eco-sustainable systems based on poly (lactic acid), diatomite and coffee grounds extract for food packaging. *International Journal of Biological Macromolecules, 112*, 567–575.

Cano, A., Chafer, M., Chiralt, A., & Gonzalez-Martinez, C. (2015). Physical and antimicrobial properties of starch-PVA blend films as affected by the incorporation of natural antimicrobial agents. *Foods*, 5. Available from https://doi.org/10.3390/foods5010003.

Cano, A., Cháfer, M., Chiralt, A., & González-Martínez, C. (2016a). Physical and antimicrobial properties of starch-PVA blend films as affected by the incorporation of natural antimicrobial agents. *Foods, 5*, 3.

Cano, A. I., Cháfer, M., Chiralt, A., & González-Martínez, C. (2016b). Biodegradation behavior of starch-PVA films as affected by the incorporation of different antimicrobials. *Polymer Degradation and Stability, 132*, 11–20. Available from https://doi.org/10.1016/j.polymdegradstab.2016.040.014.

Chaieb, K., Hajlaoui, H., Zmantar, T., Kahla-Nakbi, A. B., Rouabhia, M., Mahdouani, K., & Bakhrouf, A. (2007). The chemical composition and biological activity of clove essential oil, *Eugenia caryophyllata* (*Syzigium aromaticum* L. Myrtaceae): A short review. *Phytotherapy Research: An International Journal Devoted to Pharmacological and Toxicological Evaluation of Natural Product Derivatives, 21*, 501–506.

Choi, J. G., Kim, S. Y., Jeong, M., & Oh, M. S. (2018). Pharmacotherapeutic potential of ginger and its compounds in age-related neurological disorders. *Pharmacology & Therapeutics, 182*, 56–69.

Chung, L. Y. (2006). The antioxidant properties of garlic compounds: Allyl cysteine, alliin, allicin, and allyl disulfide. *Journal of Medicinal Food, 9*, 205–213.

Cooper, I. (2007). Plastics and chemical migration into food. *Chemical Migration and Food Contact Materials*, 228–250.

Corrales, M., Fernández, A., & Han, J. H. (2014). *Antimicrobial packaging systems. Innovations in food packaging* (pp. 133–170). Elsevier.

Dadashi-Silab, S., Doran, S., & Yagci, Y. (2016). Photoinduced electron transfer reactions for macromolecular syntheses. *Chemical Reviews, 116*, 10212–10275.

De Lacey, A. L., López-Caballero, M., & Montero, P. (2014). Agar films containing green tea extract and probiotic bacteria for extending fish shelf-life. *LWT-Food Science and Technology, 55*, 559–564.

Dong, Z., Luo, C., Guo, Y., Ahmed, I., Pavase, T. R., Lv, L., & Lin, H. (2019). Characterization of new active packaging based on PP/LDPE composite films containing attapulgite loaded with Allium sativum essence oil and its application for large yellow croaker (Pseudosciaena crocea) fillets. *Food Packaging and Shelf Life*, 20100320.

Dong, Z., Xu, F., Ahmed, I., Li, Z., & Lin, H. (2018). Characterization and preservation performance of active polyethylene films containing rosemary and cinnamon essential oils for Pacific white shrimp packaging. *Food Control, 92*, 37–46.

Ebrahimzadeh, S., Bari, M. R., Hamishehkar, H., Kafil, H. S., & Lim, L.-T. (2021). Essential oils-loaded electrospun chitosan-poly(vinyl alcohol) nonwovens laminated on chitosan film as bilayer bioactive edible films. *LWT*, 144. Available from https://doi.org/10.1016/j.lwt.2021.111217.

El-Wakil, A. E.-A. A., Moustafa, H., & Youssef, A. M. (2020). Antimicrobial low-density polyethylene/low-density polyethylene-grafted acrylic acid biocomposites based on rice bran with tea tree oil for food packaging applications. *Journal of Thermoplastic Composite Materials*, 0892705720925140.

Farkaš, P., & Bystrický, S. (2010). Chemical conjugation of biomacromolecules: A mini-review. *Chemical Papers, 64*, 683–695.
Fasihi, H., Fazilati, M., Hashemi, M., & Noshirvani, N. (2017). Novel carboxymethyl cellulose-polyvinyl alcohol blend films stabilized by Pickering emulsion incorporation method. *Carbohydrate Polymers, 167*, 79–89. Available from https://doi.org/10.1016/j.carbpol.2017.030.017.
Fernández-Saiz, P., Sánchez, G., Soler, C., Lagaron, J., & Ocio, M. (2013). Chitosan films for the microbiological preservation of refrigerated sole and hake fillets. *Food Control, 34*, 61–68.
Gahruie, H. H., Ziaee, E., Eskandari, M. H., & Hosseini, S. M. H. (2017). Characterization of basil seed gum-based edible films incorporated with Zataria multiflora essential oil nanoemulsion. *Carbohydrate polymers, 166*, 93–103.
Gamage, G. R., Park, H.-J., & Kim, K. M. (2009). Effectiveness of antimicrobial coated oriented polypropylene/polyethylene films in sprout packaging. *Food Research International, 42*, 832–839.
Garavand, F., Cacciotti, I., Vahedikia, N., Rehman, A., Tarhan, Ö., Akbari-Alavijeh, S., ... Jafari, S. M. (2022). A comprehensive review on the nanocomposites loaded with chitosan nanoparticles for food packaging. *Critical Reviews in Food Science and Nutrition, 62*(5), 1383–1416.
Garavand, F., Eghbal, N., Nooshkam, M., Miraballes, I., & Jafari, S. M. (2021). *Salt, spices, and seasonings formulated with nano/microencapsulated ingredients. Application of nano/microencapsulated ingredients in food products* (pp. 435–467). Elsevier.
Garavand, F., Razavi, S. H., & Cacciotti, I. (2018). Synchronized extraction and purification of L-lactic acid from fermentation broth by emulsion liquid membrane technique. *Journal of Dispersion Science and Technology, 39*, 1291–1299.
Garavand, F., Rouhi, M., Razavi, S. H., Cacciotti, I., & Mohammadi, R. (2017). Improving the integrity of natural biopolymer films used in food packaging by crosslinking approach: A review. *International Journal of Biological Macromolecules, 104*, 687–707.
Garces, L., & de la Puerta, C. (2010). *Antimicrobial packaging based on the use of natural extracts and the process to obtain this packaging*. European Patent EP1657181-B1 13.
Goddard, J. M., & Hotchkiss, J. (2007). Polymer surface modification for the attachment of bioactive compounds. *Progress in Polymer Science, 32*, 698–725.
Gul, P., & Bakht, J. (2015). Antimicrobial activity of turmeric extract and its potential use in food industry. *Journal of Food Science and Technology, 52*, 2272–2279.
Gutiérrez-García, G., Quintana-Romero, L., Morales-Figueroa, G., Esparza-Romero, J., Pérez-Morales, R., López-Mata, M., ... Quihui-Cota, L. (2021). Effect of Yucca baccata butanolic extract on the shelf life of chicken and development of an antimicrobial packaging for beef. *Food Control, 127*, 108142.
Hacker, M. C., Krieghoff, J., & Mikos, A. G. (2019). *Synthetic polymers. Principles of regenerative medicine* (pp. 559–590). Elsevier.
Hadidi, M., Jafarzadeh, S., Forough, M., Garavand, F., Alizadeh, S., Salehabadi, A., ... Jafari, S. M. (2022). Plant protein-based food packaging films; recent advances in fabrication, characterization, and applications. *Trends in Food Science & Technology, 120*, 154–173.
Han, Y., Yu, M., & Wang, L. (2018). Physical and antimicrobial properties of sodium alginate/carboxymethyl cellulose films incorporated with cinnamon essential oil. Food Packaging and Shelf. *Life (Chicago, Ill.: 1978), 15*, 35–42.
Havelaar, A., Kirk, M., Torgerson, P., Gibb, H., Hald, T., Lake, R., ... Gargouri, N.World Health Organization Foodborne disease burden epidemiology reference group. (2015). World Health Organization global estimates and regional comparisons of the burden of foodborne disease in 2010. *PLoS Medicine, 12*(12).
Havstad, M. R. (2020). *Biodegradable plastics. Plastic waste and recycling* (pp. 97–129). Elsevier.
Hejna, M., Kovanda, L., Rossi, L., & Liu, Y. (2021). Mint oils: In vitro ability to perform anti-inflammatory, antioxidant, and antimicrobial activities and to enhance intestinal barrier integrity. *Antioxidants, 10*, 1004.
Heldman, D. R. (2012). *Food process engineering*. Springer Science & Business Media.
Hendessi, S., Sevinis, E. B., Unal, S., Cebeci, F. C., Menceloglu, Y. Z., & Unal, H. (2016). Antibacterial sustained-release coatings from halloysite nanotubes/waterborne polyurethanes. *Progress in Organic Coatings, 101*, 253–261.
Hermawan, D., Lai, T. K., Jafarzadeh, S., Gopakumar, D. A., Hasan, M., Owolabi, F. T., ... Khalil, H. A. (2019). Development of seaweed-based bamboo microcrystalline cellulose films intended for sustainable food packaging applications. *BioResources, 14*, 3389–3410.
Hernández-Ochoa, L., Macias-Castañeda, C., Nevárez-Moorillón, G., Salas-Muñoz, E., & Sandoval-Salas, F. (2012). Antimicrobial activity of chitosan-based films including spices' essential oils and functional extracts. *CyTA-Journal of Food, 10*, 85–91.
Hintz, T., Matthews, K. K., & Di, R. (2015). The use of plant antimicrobial compounds for food preservation. *BioMed Research International*, 2015.
Huang, H. X., & Zhang, J. J. (2009). Effects of filler–filler and polymer–filler interactions on rheological and mechanical properties of HDPE–wood composites. *Journal of Applied Polymer Science, 111*, 2806–2812.
Högberg, L. D., Heddini, A., & Cars, O. (2010). The global need for effective antibiotics: Challenges and recent advances. *Trends in Pharmacological Sciences, 31*, 509–515.
Imre, B., & Pukánszky, B. (2013). Compatibilization in bio-based and biodegradable polymer blends. *European Polymer Journal, 49*, 1215–1233.
Javaherzadeh, R., Tabatabaee Bafroee, A. S., & Kanjari, A. (2020). Preservation effect of Polylophium involucratum essential oil incorporated poly lactic acid/ nanochitosan composite film on shelf life and sensory properties of chicken fillets at refrigeration temperature. *LWT*, 118. Available from https://doi.org/10.1016/j.lwt.2019.108783.
Kakadellis, S., & Harris, Z. M. (2020). Don't scrap the waste: The need for broader system boundaries in bioplastic food packaging life-cycle assessment–A critical review. *Journal of Cleaner Production*, 122831.
Kalaycıoğlu, Z., Torlak, E., Akın-Evingür, G., Özen, İ., & Erim, F. B. (2017). Antimicrobial and physical properties of chitosan films incorporated with turmeric extract. *International Journal of Biological Macromolecules, 101*, 882–888.
Kanavouras, A., Hernandez-Munoz, P., & Coutelieris, F. A. (2006). Packaging of olive oil: Quality issues and shelf life predictions. *Food Reviews International, 22*, 381–404.
Kang, T. W., Chandrasekaran, G., Hwang, E. C., Kim, H. S., & Lakshmanan, V.-K. (2018). Characterization and antibacterial activity of PVA–PVP–CS carvacrol-loaded polymer composite films for urinary catheter. *International Journal of Polymeric Materials and Polymeric Biomaterials, 67*, 1016–1027. Available from https://doi.org/10.1080/00914037.2017.1417286.
Kavas, G. (2017). Use of egg white protein powder based films fortified with black cumin and cinnamon essential oils in the storage of Çökelek cheese. *Ege Üniversitesi Ziraat Fakültesi Dergisi, 54*, 439–446.
Khaneghah, A. M., Hashemi, S. M. B., & Limbo, S. (2018). Antimicrobial agents and packaging systems in antimicrobial active food packaging: An overview of approaches and interactions. *Food and Bioproducts Processing, 111*, 1–19.

Khazaei, A., Nateghi, L., Zand, N., Oromiehie, A., & Garavand, F. (2021). Evaluation of physical, mechanical and antibacterial properties of pinto bean starch-polyvinyl alcohol biodegradable films reinforced with cinnamon essential oil. *Polymers, 13,* 2778.
Khodaei, D., Oltrogge, K., & Hamidi-Esfahani, Z. (2020). Preparation and characterization of blended edible films manufactured using gelatin, tragacanth gum and, Persian gum. *LWT, 117,* 108617.
Kim, E.-H., Kim, H.-K., & Ahn, Y.-J. (2003). Acaricidal activity of clove bud oil compounds against *Dermatophagoides farinae* and *Dermatophagoides pteronyssinus* (Acari: Pyroglyphidae). *Journal of Agricultural and Food Chemistry, 51,* 885–889.
Kim, H. M., Lee, E. H., Hong, S. H., Song, H. J., Shin, M. K., Kim, S. H., & Shin, T. Y. (1998). Effect of Syzygium aromaticum extract on immediate hypersensitivity in rats. *Journal of Ethnopharmacology, 60,* 125–131.
Kirwan, M. J., Plant, S., & Strawbridge, J. W. (2011). Plastics in food packaging. *Food and Beverage Packaging Technology,* 157–212.
Kirwan, M. J., & Strawbridge, J. W. (2003). Plastics in food packaging. *Food Packaging Technology, 1,* 174–240.
Krepker, M., Prinz-Setter, O., Shemesh, R., Vaxman, A., Alperstein, D., & Segal, E. (2018). Antimicrobial carvacrol-containing polypropylene films: Composition, structure and function. *Polymers, 10*(1), 79.
Krepker, M., Zhang, C., Nitzan, N., Prinz-Setter, O., Massad-Ivanir, N., Olah, A., & Segal, E. (2018). Antimicrobial LDPE/EVOH layered films containing carvacrol fabricated by multiplication extrusion. *Polymers, 10*(8), 864.
Kurek, M., Moundanga, S., Favier, C., Galić, K., & Debeaufort, F. (2013). Antimicrobial efficiency of carvacrol vapour related to mass partition coefficient when incorporated in chitosan based films aimed for active packaging. *Food Control, 32,* 168–175.
Lai, P., & Roy, J. (2004). Antimicrobial and chemopreventive properties of herbs and spices. *Current Medicinal Chemistry, 11,* 1451–1460.
Liakos, I., Rizzello, L., Scurr, D. J., Pompa, P. P., Bayer, I. S., & Athanassiou, A. (2014). All-natural composite wound dressing films of essential oils encapsulated in sodium alginate with antimicrobial properties. *International Journal of Pharmaceutics, 463,* 137–145.
Liu, D., Li, H., Jiang, L., Chuan, Y., Yuan, M., & Chen, H. (2016). Characterization of active packaging films made from poly(lactic acid)/poly(trimethylene carbonate) incorporated with oregano essential oil. *Molecules (Basel, Switzerland),* 21. Available from https://doi.org/10.3390/molecules21060695.
Liu, Q., Meng, X., Li, Y., Zhao, C.-N., Tang, G.-Y., & Li, H.-B. (2017). Antibacterial and antifungal activities of spices. *International Journal of Molecular Sciences, 18,* 1283.
Maddah, H. A. (2016). Polypropylene as a promising plastic: A review. *American Journal of Polymer Science, 6,* 1–11.
Majidi, H. J., Babaei, A., Bafrani, Z. A., Shahrampour, D., Zabihi, E., & Jafari, S. M. (2019). Investigating the best strategy to diminish the toxicity and enhance the antibacterial activity of graphene oxide by chitosan addition. *Carbohydrate Polymers, 225,* 115220.
Mallik, A., Barik, A. K., & Pal, B. (2015). Comparative studies on physico-mechanical properties of composite materials of low density polyethylene and raw/calcined kaolin. *Journal of Asian Ceramic Societies, 3,* 212–216.
Mangaraj, S., Goswami, T., & Mahajan, P. (2009). Applications of plastic films for modified atmosphere packaging of fruits and vegetables: A review. *Food Engineering Reviews, 1,* 133–158.
Manso, S., Becerril, R., Nerín, C., & Gómez-Lus, R. (2015). Influence of pH and temperature variations on vapor phase action of an antifungal food packaging against five mold strains. *Food Control, 47,* 20–26.
Marchese, A., Barbieri, R., Sanches-Silva, A., Daglia, M., Nabavi, S. F., Jafari, N. J., … Nabavi, S. M. (2016). Antifungal and antibacterial activities of allicin: A review. *Trends in Food Science & Technology, 52,* 49–56.
Mei, L. X., Nafchi, A. M., Ghasemipour, F., Easa, A. M., Jafarzadeh, S., & Al-Hassan, A. (2020). Characterization of pH sensitive sago starch films enriched with anthocyanin-rich torch ginger extract. *International Journal of Biological Macromolecules, 164,* 4603–4612.
Miladi, H., Zmantar, T., Chaabouni, Y., Fedhila, K., Bakhrouf, A., Mahdouani, K., & Chaieb, K. (2016). Antibacterial and efflux pump inhibitors of thymol and carvacrol against food-borne pathogens. *Microbial Pathogenesis, 99,* 95–100.
Mirzaei-Mohkam, A., Garavand, F., Dehnad, D., Keramat, J., & Nasirpour, A. (2019). Optimisation, antioxidant attributes, stability and release behaviour of carboxymethyl cellulose films incorporated with nanoencapsulated vitamin E. *Progress in Organic Coatings, 134,* 333–341.
Mirzaei-Mohkam, A., Garavand, F., Dehnad, D., Keramat, J., & Nasirpour, A. (2020). Physical, mechanical, thermal and structural characteristics of nanoencapsulated vitamin E loaded carboxymethyl cellulose films. *Progress in Organic Coatings, 138,* 105383.
Moreno-Vásquez, M. J., Plascencia-Jatomea, M., Ocaño-Higuera, V. M., Castillo-Yáñez, F. J., Rodríguez-Félix, F., Rosas-Burgos, E. C., & Graciano-Verdugo, A. Z. (2017). Engineering and antibacterial properties of low-density polyethylene films with incorporated epigallocatechin gallate. *Journal of Plastic Film & Sheeting, 33*(4), 413–437.
Moura, L. E., de Souza, C. O., de Oliveira, E. A. S., Lemos, P. V. F., & Druzian, J. I. (2018). Bioactive efficacy of low-density polyethylene films with natural additives. *Journal of Applied Polymer Science, 135,* 46461.
Mousavian, D., Nafchi, A. M., Nouri, L., & Abedinia, A. (2021). Physicomechanical properties, release kinetics, and antimicrobial activity of activated low-density polyethylene and orientated polypropylene films by Thyme essential oil active component. *Journal of Food Measurement and Characterization, 15,* 883–891.
Mulla, M., Ahmed, J., Al-Attar, H., Castro-Aguirre, E., Arfat, Y. A., & Auras, R. (2017). Antimicrobial efficacy of clove essential oil infused into chemically modified LLDPE film for chicken meat packaging. *Food Control, 73,* 663–671.
Nejad, A. S. M., Shabani, S., Bayat, M., & Hosseini, S. E. (2014). Antibacterial effect of garlic aqueous extract on Staphylococcus aureus in hamburger. *Jundishapur Journal of Microbiology,* 7.
Organization, W. H. (2010). *World health statistics 2010.* World Health Organization.
Organization, W. H. (2015). *WHO estimates of the global burden of foodborne diseases: Foodborne disease burden epidemiology reference group 2007–2015.* World Health Organization.
Ouellette, R., & Rawn, J. (2015). *28—Synthetic polymers. Organic chemistry study guide* (pp. 587–601). Boston: Elsevier.
Purwanti, S., & Yuwanta, T. (2014). In vitro study of antibacterial activity of combination of water extract of turmeric and garlic. *International Journal of Pharmacy & Life Sciences,* 5.
Qin, Y., Yang, J., & Xue, J. (2015). Characterization of antimicrobial poly (lactic acid)/poly (trimethylene carbonate) films with cinnamaldehyde. *Journal of Materials Science, 50,* 1150–1158.
Raja, S., & Mattoso, L. H. (2020). *Functionalized polymer-based composite photocatalysts. Green photocatalysts* (pp. 167–188). Springer.
Ranasinghe, P., Pigera, S., Premakumara, G. S., Galappaththy, P., Constantine, G. R., & Katulanda, P. (2013). Medicinal properties of 'true'cinnamon (*Cinnamomum zeylanicum*): A systematic review. *BMC Complementary and Alternative Medicine, 13,* 1–10.

Resianingrum, R., Atmaka, W., Khasanah, L. U., KAWIJI, K., Utami, R., & Praseptiangga, D. (2016). Characterization of cassava starch-based edible film enriched with lemongrass oil (*Cymbopogon citratus*). Nusantara. *Bioscience*, *8*, 278–282.

Rocha, M., Ferreira, F., Souza, M., & Prentice, C. (2013). Antimicrobial films: A review. *Microbial Pathogens and Strategies for Combating Them: Science, Technology and Education*, *1*, 23–31.

Roy, S., & Rhim, J.-W. (2021). Antioxidant and antimicrobial poly (vinyl alcohol)-based films incorporated with grapefruit seed extract and curcumin. *Journal of Environmental Chemical Engineering*, *9*, 104694.

Saeed, M., Yasmin, I., Khan, M. I., Nadeem, M., Shabbir, M. A., & Azam, M. (2016). Herbs and spices as a potential antimicrobial agents for food application. *Pakistan Journal of Food Sciences*, *26*, 153–160.

Sa-Nguanpuag, K., Kanlayanarat, S., Srilaong, V., Tanprasert, K., & Techavuthiporn, C. (2011). Ginger (*Zingiber officinale*) oil as an antimicrobial agent for minimally processed produce: A case study in shredded green papaya. *International Journal of Agriculture & Biology*, 13.

Shankar, S., & Rhim, J.-W. (2018). Preparation of antibacterial poly (lactide)/poly (butylene adipate-co-terephthalate) composite films incorporated with grapefruit seed extract. *International Journal of Biological Macromolecules*, *120*, 846–852.

Sharma, R., Jafari, S. M., & Sharma, S. (2020). Antimicrobial bio-nanocomposites and their potential applications in food packaging. *Food Control*, *112*, 107086.

Silva, F., & Domingues, F. C. (2017). Antimicrobial activity of coriander oil and its effectiveness as food preservative. *Critical Reviews in Food Science and Nutrition*, *57*, 35–47.

Silva, F. G., Oliveira, C., Pinto, J. E. B., Nascimento, V. E., Santos, S. C., Seraphin, J. C., & Ferri, P. H. (2007). Seasonal variability in the essential oils of wild and cultivated *Baccharis trimera*. *Journal of the Brazilian Chemical Society*, *18*, 990–997.

Silveira, M. P., Silva, H. C., Pimentel, I. C., Poitevin, C. G., da Costa Stuart, A. K., Carpiné, D., ... Jorge, R. M. M. (2020). Development of active cassava starch cellulose nanofiber-based films incorporated with natural antimicrobial tea tree essential oil. *Journal of Applied Polymer Science*, *137*, 48726.

Sionkowska, A. (2011). Current research on the blends of natural and synthetic polymers as new biomaterials: Review. *Progress in Polymer Science*, *36*, 1254–1276. Available from https://doi.org/10.1016/j.progpolymsci.2011.050.003.

Solano, A. C. V., & de Rojas Gante, C. (2012). Two different processes to obtain antimicrobial packaging containing natural oils. Food and Bioprocess. *Technology (Elmsford, N.Y.)*, *5*, 2522–2528.

Soliman, K. M., & Badeaa, R. (2002). Effect of oil extracted from some medicinal plants on different mycotoxigenic fungi. *Food and Chemical Toxicology*, *40*, 1669–1675.

Song, X., Zuo, G., & Chen, F. (2018). Effect of essential oil and surfactant on the physical and antimicrobial properties of corn and wheat starch films. *International Journal of Biological Macromolecules*, *107*, 1302–1309.

Souza, E. Ld, Stamford, T. L. M., Lima, Ed. O., Trajano, V. N., & Barbosa Filho, J. M. (2005). Antimicrobial effectiveness of spices: An approach for use in food conservation systems. *Brazilian Archives of Biology and Technology*, *48*, 549–558.

Stefanello, M. É. A., Cervi, A. C., Ito, I. Y., Salvador, M. J., Wisniewski, A., Jr, & Simionatto, E. L. (2008). Chemical composition and antimicrobial activity of essential oils of *Eugenia chlorophylla* (Myrtaceae). *Journal of Essential Oil Research*, *20*, 75–78.

Strasakova, M., Pummerova, M., Filatova, K., & Sedlarik, V. (2021). Immobilization of caraway essential oil in a polypropylene matrix for antimicrobial modification of a polymeric surface. *Polymers*, *13*(6), 906.

Sung, S.-Y., Sin, L. T., Tee, T.-T., Bee, S.-T., Rahmat, A., Rahman, W., ... Vikhraman, M. (2013). Antimicrobial agents for food packaging applications. *Trends in Food Science & Technology*, *33*, 110–123.

Suppakul, P., Miltz, J., Sonneveld, K., & Bigger, S. W. (2003). Active packaging technologies with an emphasis on antimicrobial packaging and its applications. *Journal of Food Science*, *68*, 408–420.

Suwanamornlert, P., Kerddonfag, N., Sane, A., Chinsirikul, W., Zhou, W., & Chonhenchob, V. (2020). Poly (lactic acid)/poly (butylene-succinate-co-adipate)(PLA/PBSA) blend films containing thymol as alternative to synthetic preservatives for active packaging of bread. *Food Packaging and Shelf Life*, *25*, 100515.

Swilem, A. E., Stloukal, P., Abd El-Rehim, H. A., Hrabalíková, M., & Sedlařík, V. (2019). Influence of gamma rays on the physico-chemical, release and antibacterial characteristics of low-density polyethylene composite films incorporating an essential oil for application in food-packaging. *Food Packaging and Shelf Life*, *19*, 131–139.

Syafiq, R., Sapuan, S., Zuhri, M., Ilyas, R., Nazrin, A., Sherwani, S., & Khalina, A. (2020). Antimicrobial activities of starch-based biopolymers and biocomposites incorporated with plant essential oils: A review. *Polymers*, *12*, 2403.

Tajkarimi, M., Ibrahim, S. A., & Cliver, D. (2010). Antimicrobial herb and spice compounds in food. *Food Control*, *21*, 1199–1218.

Takala, P. N., Salmieri, S., Boumail, A., Khan, R. A., Vu, K. D., Chauve, G., ... Lacroix, M. (2013). Antimicrobial effect and physicochemical properties of bioactive trilayer polycaprolactone/methylcellulose-based films on the growth of foodborne pathogens and total microbiota in fresh broccoli. *Journal of Food Engineering*, *116*, 648–655. Available from https://doi.org/10.1016/j.jfoodeng.2013.010.005.

Takma, D. K., & Korel, F. (2019). Active packaging films as a carrier of black cumin essential oil: Development and effect on quality and shelf-life of chicken breast meat. *Food Packaging and Shelf Life*, *19*, 210–217.

Tawakkal, I. S. M. A., Cran, M. J., & Bigger, S. W. (2017). Effect of poly(lactic acid)/kenaf composites incorporated with thymol on the antimicrobial activity of processed meat. *Journal of Food Processing and Preservation*, 41. Available from https://doi.org/10.1111/jfpp.13145.

Utami, R., Khasanah, L. U., Manuhara, G. J., & Ayuningrum, Z. K. (2019). Effects of cinnamon bark essential oil (*Cinnamomum burmannii*) on characteristics of edible film and quality of fresh beef. *Pertanika Journal of Tropical Agricultural Science*, 42.

Vahedikia, N., Garavand, F., Tajeddin, B., Cacciotti, I., Jafari, S. M., Omidi, T., & Zahedi, Z. (2019). Biodegradable zein film composites reinforced with chitosan nanoparticles and cinnamon essential oil: Physical, mechanical, structural and antimicrobial attributes. *Colloids and Surfaces B: Biointerfaces*, *177*, 25–32.

Valdés, A., Mellinas, A., Ramos, M., Burgos, N., Jiménez, A., & Garrigós, Md. C. (2015). Use of herbs, spices and their bioactive compounds in active food packaging. *RSC Advances*, *5*, 40324–40335.

Varghese, S. A., Siengchin, S., & Parameswaranpillai, J. (2020). Essential oils as antimicrobial agents in biopolymer-based food packaging-A comprehensive review. *Food Bioscience*, 100785.

Velluti, A., Sanchis, V., Ramos, A., Egido, J., & Marın, S. (2003). Inhibitory effect of cinnamon, clove, lemongrass, oregano and palmarose essential oils on growth and fumonisin B1 production by Fusarium proliferatum in maize grain. *International Journal of Food Microbiology, 89*, 145–154.

Velásquez, E., Patiño Vidal, C., Rojas, A., Guarda, A., Galotto, M. J., & López de Dicastillo, C. (2021). Natural antimicrobials and antioxidants added to polylactic acid packaging films. Part I: Polymer processing techniques. *Comprehensive Reviews in Food Science and Food Safety*.

Vitoratos, A., Bilalis, D., Karkanis, A., & Efthimiadou, A. (2013). Antifungal activity of plant essential oils against *Botrytis cinerea, Penicillium italicum* and *Penicillium digitatum. Notulae Botanicae Horti Agrobotanici Cluj-Napoca, 41*, 86–92.

Wattananawinrat, K., Threepopnatkul, P., & Kulsetthanchalee, C. (2014). Morphological and thermal properties of LDPE/EVA blended films and development of antimicrobial activity in food packaging film. *Energy Procedia, 56*, 1–9. Available from https://doi.org/10.1016/j.egypro.2014.070.125.

Wu, Y., Yuan, M. W., Yang, J. Y., Qin, Y. Y., Yuan, M. L., & Cao, J. X. (2014). Physical properties and antimicrobial activity of a poly(lactic acid)/poly(trimethylene carbonate) film incorporated with thymol. *Advanced Materials Research, 884–885*, 481–484. Available from https://doi.org/10.4028/http://www.scientific.net/AMR.884-8850.481.

Xu, J.-G., Liu, T., Hu, Q.-P., & Cao, X.-M. (2016). Chemical composition, antibacterial properties and mechanism of action of essential oil from clove buds against Staphylococcus aureus. *Molecules (Basel, Switzerland), 21*, 1194.

Yu, L., Dean, K., & Li, L. (2006). Polymer blends and composites from renewable resources. *Progress in polymer science, 31*, 576–602.

Zheng, J., Zhou, Y., Li, Y., Xu, D.-P., Li, S., & Li, H.-B. (2016). Spices for prevention and treatment of cancers. *Nutrients, 8*, 495.

Further reading

Guerra, I. C. D., de Oliveira, P. D. L., de Souza Pontes, A. L., Lúcio, A. S. S. C., Tavares, J. F., Barbosa-Filho, J. M., ... de Souza, E. L. (2015). Coatings comprising chitosan and Mentha piperita L. or Mentha × villosa Huds essential oils to prevent common postharvest mold infections and maintain the quality of cherry tomato fruit. *International Journal of Food Microbiology, 214*, 168–178.

Perdones, Á., Tur, N., Chiralt, A., & Vargas, M. (2016). Effect on tomato plant and fruit of the application of biopolymer–oregano essential oil coatings. *Journal of the Science of Food and Agriculture, 96*, 4505–4513.

Scartazzini, L., Tosati, J., Cortez, D., Rossi, M., Flôres, S., Hubinger, M., ... Monteiro, A. (2019). Gelatin edible coatings with mint essential oil (*Mentha arvensis*): Film characterization and antifungal properties. *Journal of Food Science and Technology, 56*, 4045–4056.

CHAPTER

8

Natural antimicrobials from fruits and plant extract for food packaging and preservation

Shubham Sharma[1,2], *Kalpani Y. Perera*[1,2], *Amit K. Jaiswal*[1,2] and *Swarna Jaiswal*[1,2]

[1]School of Food Science and Environmental Health, Faculty of Sciences and Health, Technological University Dublin-City Campus, Dublin, Ireland [2]Environmental Sustainability and Health Institute (ESHI), Technological University Dublin-City Campus, Dublin, Ireland

8.1 Introduction

The production of antimicrobial packaging materials containing natural antimicrobial agents is getting popular. Consumer concerns about health-related issues and the use of synthetic antimicrobial agents have driven this curiosity. Because of the high demands of consumers to eat nutritious and safe foods, researchers have turned their attention to developing advanced packaging, which has a significant capacity to preserve the consistency and sensory properties of food products over time. These characteristics are linked to the emergence of new and improved food packaging, such as active, intelligent, and edible systems. Food is perishable by nature since it can be degraded by a variety of environmental factors, including bacterial and fungal contamination. Plant antimicrobial agents, such as polyphenols, essential oils (EOs), etc. are mostly in the form of active components. The natural compounds are extracted from fruits (grapefruit, apricot, orange), rhizomes (ginger), bulbs (garlic, fennel, and onions), flowers or buds (clove), leaves (basil, oregano, thyme, marjoram, rosemary, and sage), bark (cinnamon), seeds (caraway, flax seed, parsley), and other plant parts (Mendonca, Jackson-Davis, Moutiq, & Thomas-Popo, 2018).

Antimicrobials are generally formed in plants to protect them from disease. Many of them are involved in the plants' protection systems against infectious or parasitic agents, either before or after infection (Mendonca et al., 2018; Rauha et al., 2000). Additives, such as lactic bacteria, and antimicrobial agents, such as lactic acid, and bacteriocins, have been shown to inhibit lipid oxidation, prevent color loss, extend storage life, and ensure food safety (Pisoschi et al., 2018). Active packaging has also been demonstrated to be significant in the case of raw and cooked food, requiring the use of various natural extracts derived from herbs, spices, fruits, and vegetables to prevent/preserve color, lipid oxidation, and improve shelf-life. Bioactive compounds such as phenolics (flavonoids and non-flavonoids), phenolic terpenes, and tannins have been assigned such benefits. Several herb and vegetal extracts have also been shown to serve as antioxidants, antimicrobials, anti-diabetics, anti-carcinogens, flavorings, drinks, and repellents, all of which have many applications in food manufacturing.

Encapsulation systems have recently progressed by manufacturing nanocarriers, a process is known as nanoencapsulation, which is one of the most important branches of nanotechnology in the food industry (Bahrami, Delshadi, Assadpour, Jafari, & Williams, 2020; Rostamabadi, Falsafi, & Jafari, 2019; Taheri & Jafari, 2019) The nanoencapsulation process consists of two main steps: loading antibacterial compounds into carriers and minimizing their size to the nanoscale (10^{-9} m). The nanoencapsulation systems have better functional properties such as high absorption, high solubility, increased bioavailability, and controlled release due to their high surface area than the larger particles (Bahrami et al., 2020; Jafari & McClements, 2017). Xavier et al. (2021).

Food Packaging and Preservation.
DOI: https://doi.org/10.1016/B978-0-323-90044-7.00008-2

Encapsulated *Cinnamodendron dinisii Schwanke* oil to produce stable and homogeneous nanoparticles. This active packaging material was able to successfully preserve ground beef by stabilizing oxidation reactions and preserving color (Xavier et al., 2021). López-Gómez, Navarro-Martínez, and Martínez-Hernández (2020) observed that due to enzymatic inhibition, essential oil active sheets also preserved the physicochemical quality of fruits (as measured by titratable acidity, color, soluble solids content, and firmness).

This chapter discusses the various plant-based natural antimicrobials agents such as EOs (Eugenol, Vanillin, Thymol, Carvacrol, Citral, and Cinnamaldehyde) and plant extracts (Tannins, Anthocyanins, Coumarins, Allicin, Alkaloids, and AMPs) and their application for food packaging. In addition, the mechanism of action of these natural antimicrobials is also described. Further, the antimicrobial-loaded nanocarriers (nanoemulsions, nanoliposomes, solid lipid nanoparticles, and biopolymeric nanocarriers) are illustrated for food packaging applications.

8.2 Antimicrobial agents in food packaging

Microorganisms could bring undesirable changes in the food including degrading its organoleptic content and producing toxins. Consumers, regulatory authorities, and the food industry around the world are all concerned about food microbial protection. Application of novel technologies to prevent or delay microbial development, such as controlled atmosphere packaging, activated films, non-thermal treatments, irradiation, and changed atmosphere packaging, can also increase the protection and shelf life of food ingredients. Most of these procedures, on the other hand, can result in the loss of organoleptic properties of foods and a reduction in consumer acceptability. Natural antimicrobial compounds are a strong candidate for green additives in natural and efficient functional foods and packaged food (Fig. 8.1), which would be a key issue for food manufacturers due to the widespread demand for natural foods among customers (Table 8.1).

EOs are antimicrobial agents that are widely dispersed in nature and are commonly used. EOs are volatile oils made up of monoterpenes (such as -pinene, limonene, P-cymene, and sabinene), monoterpenoids (such as citronellal, thymol, carvacrol, carvone, and borneol), and phenylpropanoids (e.g., cinnamaldehyde, eugenol, and vanillin) that are produced by plant limbs or living microorganisms (Bahrami et al., 2020; Pilevar et al., 2020). EO types and targeted foods are two major factors to consider when using EOs as an antimicrobial agent, for example, certain special oils in the meat industry, such as coriander and eugenol, oregano, clove, and thyme oils, have higher antimicrobial activity than others (Burt, 2004). Arana-Sánchez et al. (2010) used a combination of oregano and rosemary EOs as an antimicrobial compound in fresh-cut broccoli, which had shown strong antimicrobial activity against *E. coli*, *Listeria monocytogenes*, *Salmonella choleraesuis*, and *Staphylococcus aureus* (Arana-Sánchez et al., 2010). Sharma, Barkauskaite, Duffy, Jaiswal, and Jaiswal (2020a), Sharma et al., Sharma, Barkauskaite, Jaiswal, Duffy, and Jaiswal (2020b) studied the influence of the incorporation of essential oil (clove oil, cinnamon oil, eucalyptus oil, and thyme oil) in the blend

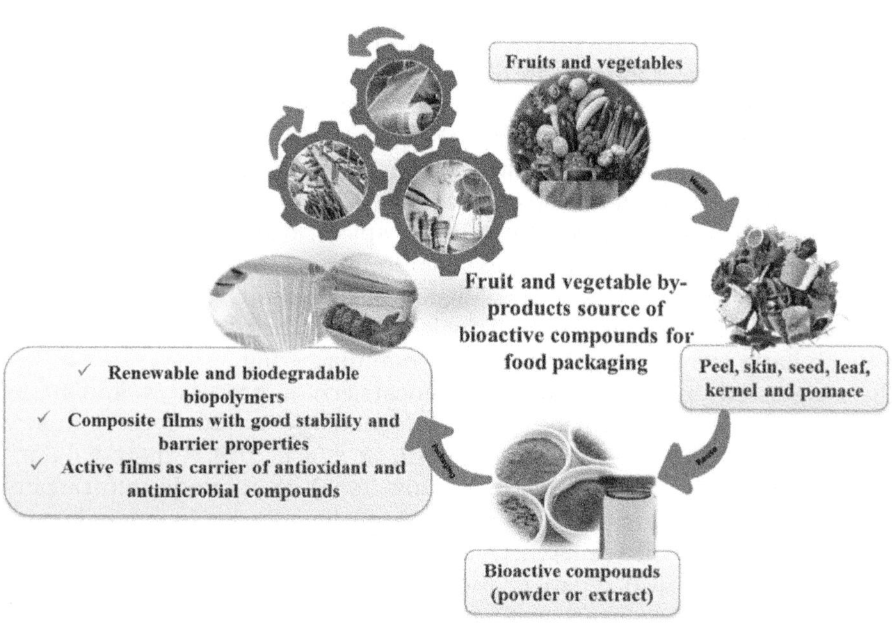

FIGURE 8.1 The potential application of fruit and vegetable-based natural compounds in packaging material. *Source: From Dilucia, F., Lacivita, V., Conte, A., & Del Nobile, M. A. (2020). Sustainable use of fruit and vegetable by-products to enhance food packaging performance.* Foods, *9(7). https://doi.org/10.3390/foods9070857.*

TABLE 8.1 Novel fruits and plant extract incorporated food packaging.

Test food/ media	Packaging materials	Phytochemicals	Results	References
Pork meatballs	Cassava starch film	Cinnamon oil	• Bacterial growth remained below the maximum allowable level until 96 h. • Incorporation of cinnamon oil into packaging based on cassava starch showed antimicrobial efficacy by lowering microbial growth	Iamareerat, Singh, Sadiq, and Anal (2018)
Bread	Chitosan films	Apricot (*Prunus armeniaca*) essential oil	• No fungal growth was observed on bread packed in any of the pouches made up of chitosan-Apricot EO films. • Shown antifungal activity at a very low Apricot EO concentration 0.125 w/v film.	Priyadarshi, Kumar, Deeba, Kulshreshtha, and Negi (2018)
Strawberries	Thermoplastic starch/ montmorillonite	Carvacrol and thymol	• Antimicrobial packaging material balances effectiveness over the control of *Botrytis cinerea*. • No considerable alteration in the quality parameters and organoleptic properties of strawberries.	Campos-Requena et al. (2017)
Strawberries	Chitosan beads	Lavender EO	• Provide an antifungal activity against gray mold rot on strawberries. • Chitosan beads loaded with lavender EO can extend the mold-free storage life of strawberries stored at 7°C from 2 days (control) to 8 days with the acceptable overall sensory scores	Sangsuwan, Pongsapakworawat, Bangmo, and Sutthasupa (2016)
Pork	Polylactic acid nanofilm	Cinnamon essential oil	• Had shown excellent antimicrobial activity against both Gram-positive and Gram-negative bacteria • Shelf life extended from 3 days (Control) to 8 days (PLA/CEO/β-CD nanofilm)	Wen et al. (2016)
Crab stick	Starfish gelatin	Vanillin	• The inhibition zones of the starfish gelatin films containing vanillin increased from 17.24 to 25.24 mm. • After 9 days, the crab stick wrapped with starfish gelatin and vanillin film had a reduction of 1.48 log CFU/g.	Lee et al. (2016)
Chicken fillets	Kafirin	Citral	Citral films showed significant antimicrobial activity ($P < .05$) against the total viable count on chicken fillets stored at 2°C ± 0.5°C for 96 h	Giteru et al. (2017)
Pork	Chitosan	Thymol or thyme essential oil	• Inhibition zones for thyme essential oil and chitosan are *S. aureus* (16.33 ± 1.16 mm) and *E. coli* (14.67 ± 0.58 mm). • Inhibition zones for thymol and chitosan are *S. aureus* (17.33 ± 0.58 mm) and *E. coli* (15.33 ± 1.53 mm). • Total viable counts in pork in 12th day is5.93 and 5.77 log CFU/g respectively for thyme essential oil-chitosan and thymol-chitosan	Liu and Liu (2020)
Apricot fruit	Nanochitosan	Pomegranate peel extract	Addition of pomegranate peel extract significantly reduced total psychrophilic bacterial count, yeast and mold count respectively to 5.3 $\log_{10}$ CFU/g, 5 $\log_{10}$ CFU/g, and 4.7 $\log_{10}$ CFU/g.	Gull et al. (2021)
Agar media	Polylactic acid and poly(butylene adipate-co-terephthalate)	Cinnamon essential oil	• Inhibited *E. coli* biofilm by 89.82%. • Reduced *S. aureus* growth by 4.26 log CFU/mL.	Sharma et al. (2020b)
Nutrient broth	Poly(lactic acid)	Carvacrol	Reduction of 1.02 ± 0.09 log CFU/mL of the *Pseudomonas fluorescens*	Wang et al. (2020)
Agar media	Polyvinyl acetate	Citral	• At 20% of dry matter, Citral showed antimicrobial activity against *E. coli* (2.1 log) and *S. aureus* (4.3 log). • At 3% and 5% of dry matter Citral showed antimicrobial activity against *S. cerevisiae* (4.74 log) and *A. niger* (4.29 log).	Thielmann et al. (2021)
Agar media	Starch nanofibers	Carvacrol	Minimum inhibitory concentration of *S. typhimurium*, *L. monocytogenes*, *S. aureus*, *E. coli* is 0.098 mg/mL	Fonseca et al. (2021)

(Continued)

TABLE 8.1 (*Continued*)

Test food/ media	Packaging materials	Phytochemicals	Results	References
Mueller hinton broth	Polyethylene	Carvacrol or Menthol	Increasing menthol and carvacrol concentration leads to an increase in the antimicrobial activity of films against *E. coli, S. aureus, L. innocua*, and *S. cervicea*.	Jahdkaran et al. (2021)
Agar media	Polylactic acid and poly(butylene adipate-co-terephthalate)	Clove essential oil	• Complete killing of *S. aureus* that is a reduction from 6.5 to 0 log CFU/mL was observed. • Clove oil composite film inhibited *E. coli* biofilm by 93.43%	Sharma et al. (2020a)
Agar media	Sago starch and guar gum	Carvacrol and Citral	• Inhibition zones for the carvacrol and citral incorporated film is *B. cereus* (22.67 ± 2.08–43.33 ± 3.06 mm) and *E. coli* (22.00 ± 2.00–41.67 ± 1.53 mm). • Films incorporated with carvacrol and Citral are extremely sensitive against *B. cereus* and *E. coli*.	Dhumal et al. (2019)
Agar media	Microfibrillated cellulose and Chitosan	Polypyrrole	• Film showed a log reduction bacteria activity against *Escherichia coli* (5.43 log CFU/mL)	Gao et al. (2020)
Agar media	Chitosan nanofiber and Methyl cellulose	Saffron petal anthocyanin	• Inhibition zone for *E. coli* (20.2 ± 3.3 mm) and *S. aureus* (22.8 ± 1.5 mm). • Antimicrobial activity of the films was stronger against *S. aureus* (Gram-positive) than *E. coli* (Gram-negative)	Alizadeh-Sani et al. (2021a)
Agar media	Polylactic acid and poly(butylene adipate-co-terephthalate)	Ferulic acid	• 10% ferulic acid composite film demonstrated 3.5 log CFU/mL growth reduction for *L. monocytogenes*. • It had also demonstrated 2 log CFU/mL reduction in the growth of *E. coli*	Sharma, Jaiswal, Duffy, and Jaiswal (2020)
Agar media	Cellulose nanofiber	Anthocyanin and Oregano essential oil	There was a total bacteria growth inhibition in *E. coli* and *L. monocytogenes*	Chen et al. (2020)

of Poly (Lactide)/Poly (Butylene Adipate-co-Terephthalate) and found clove oil has completely killed *S. aureus*. Cinnamon oil composite films were also found to have higher antimicrobial activity against *S. aureus* than eucalyptus oil-based composite films. Clove oil composite film (10 wt.%) inhibited the *E. coli* biofilm by 93.43%, cinnamon oil (10%) composite film had exhibited by 89.82%, and eucalyptus oil inhibited by 84.37% (Table 8.1).

AMPs are a type of natural antimicrobial agent used in the food industry. They have a wide variety of functions (antibacterial, antifungal, and antiviral properties), are fast-acting, and have anti-inflammatory properties (Bahrami et al., 2020; Narayana & Chen, 2015; Sarabandi et al., 2019). Bacteriocins are produced by bacteria and are divided into four subgroups: class I, which includes nisin (less than 5 kDa, called lantibiotics), class II, which includes pediocin (less than 10 kDa, called non-lantibiotics), class III, which includes lysostaphin (bigger than 10 kDa, responsive to heat), and class IV, which includes *plantaricin S.* (containing different compounds including lipid) (Todorov, 2009). External and internal pathways are used by AMPs like EOs to attack bacteria. Since AMPs damage bacteria's membranes, their vital processes are disrupted. Organic acids, which are formed by plant and animal metabolisms, have a long history as a natural antimicrobial agent. Organic acids are listed as approved antimicrobial agents in European legislation, and their protection is acknowledged. Organic acids' key antimicrobial mechanism is to lower the pH of foods, preventing bacteria from growing. The penetration of non-dissociated (non-ionized) organic acids into the cells of pH-sensitive bacteria (e.g., *L. monocytogenes*, *Salmonella* spp., *Clostridium perfringens*, *Escherichia coli*, and *Campylobacter*) disrupts their essential processes. Since organic acids have no harmful effects on food sensory properties, they can be used in food formulations (Mani-López, García, & López-Malo, 2012).

8.2.1 Essential oils

8.2.1.1 Eugenol

Eugenol is a phenolic component found in a variety of plants, including clove oil, nutmeg oil, cinnamon extract, and a variety of others. It belongs to the Phenylpropanoids class, having the IUPAC name as 4-allyl-2-methoxy phenol. Its molecular mass is 164.2 g/mol. It has effective health-promoting properties, making it a versatile natural ingredient. Eugenol was first isolated from the leaves and buds of clove (*Eugenia caryophyllata*). Eugenol is stated to have various properties such as antimicrobial activity, antioxidant property, anesthetic potential, anti-carcinogenic effects, anti-inflammatory action, anti-diabetic efficacy, and hypolipidemic effectiveness (Khalil et al., 2017). According to the WHO, Eugenol is generally recognized as safe (GRAS) and non-mutagenic. It is isolated in numerous ways such as steam distillation, solvent extraction, hydro-distillation, microwave-assisted extraction, ultra-sound assisted extraction, and supercritical carbon dioxide extraction.

The polyphenol eugenol contained in clove oil has antibacterial properties against a variety of Gram-positive and Gram-negative bacteria. Eugenol damages the cell membrane and cell wall of Gram-negative and Gram-positive bacteria, causing cell lysis and the leakage of intracellular fluid, as well as lipid and protein contents (Oyedemi, Okoh, Mabinya, Pirochenva, & Afolayan, 2009). Gülçin (2011) observed eugenol as a strong antioxidant activity and radical-scavenging activity. Ashrafudoulla, Mizan, Ha, Park, and Ha (2020) observed that different concentrations of eugenol (0.1%–0.6%) significantly reduced biofilm formation, metabolic activities, and extracellular polysaccharide secretion, indicating an efficient antibacterial impact. Jeyakumar and Lawrence (2021) studied the mechanisms of bactericidal action of eugenol against *E. coli* and according to the findings, eugenol changes membrane permeability, causing intracellular contents to leakage.

Natural antioxidants and natural antimicrobials, such as herbs and spices, are of particular interest in active packaging because of their protection and health benefits when compared to synthetic antioxidants. Navikaite-Snipaitiene et al. (2018) have developed an active-packaging material with biopolymers and clove essential oil or eugenol as an active agent. In their results, it has been determined that eugenol has shown promising effects in increasing beef shelf life. Requena, Vargas, and Chiralt (2017) studied the release kinetics of carvacrol and eugenol from poly(hydroxybutyrate-co-hydroxy valerate) films and determined in aqueous simulants, about 50% of the eugenol was delivered, while in fatty systems, 65%–70% was delivered. Encapsulation of eugenol also reduced the loss of eugenol due to an increase in the temperature during processing and transportation (Requena et al., 2017). Woranuch and Yoksan (2013) used ionic gelation of oil in water emulsion for eugenol-loaded chitosan nanoparticles and observed that the use of chitosan particles to encapsulate eugenol could help to minimize eugenol loss during the thermal process. Moreover, their findings indicate that eugenol-loaded chitosan nanoparticles could be used as antioxidants in bioactive plastics for food packaging (Woranuch & Yoksan, 2013).

8.2.1.2 Vanillin

Vanillin (4-hydroxy-3-methoxy benzaldehyde) is a phenolic, volatile, bioactive compound with a distinctive odor which is a major constituent in vanilla beans. Vanillin is naturally extracted from vanilla pods of *Vanilla planifolia*, *Vanilla tahitensis*, and *Vanilla pompon* via a multi-step curing process (Arya, Rookes, Cahill, & Lenka, 2021). Currently, 90% of vanillin used is chemically synthesized from guaiacol, eugenol, or lignin. In addition to the phenolic group, vanillin also contain an aldehyde group making it able to undergo multiple chemical reactions including oxidation. It is a hydrophobic compound with a water solubility of 10 g/L (Arnon-Rips, Cohen, Saidi, Porat, & Poverenov, 2021; Arya et al., 2021; Fitzgerald et al., 2004; Sangsuwan, Rattanapanone, & Pongsirikul, 2015; Yang, Lee, Lee, & Song, 2017). Vanillin is mainly used as a flavoring agent that is generally regarded as safe (FDA-approved); however, it is also used as a food preservative and cross-linking agent (Lee, Lee, Yang, & Song, 2016). Vanillin is an antimicrobial agent with antibacterial activity against bacteria, yeast, and mold. The antimicrobial activity of vanillin depends upon the concentration of bacteria, time of exposure, and the type of microorganism (Sangsuwan et al., 2015; Xavier, Babusha, George, & Ramana, 2015). In addition, vanillin is also having properties such as antioxidant, anticancer, neuroprotective, antibiotic potentiation, and anti-quorum sensing (Arya et al., 2021; Sangsuwan et al., 2015).

The aldehyde group present in vanillin is the most crucial factor supporting antimicrobial activity (Lee et al., 2016). Vanillin can harmfully affect the integrity of the cytoplasmic membrane. This is a result of the loss of pH homeostasis, ion gradients, and respiratory activity inhibition. The energy generation is unaffected by vanillin (Fitzgerald et al., 2004). In *Enterobacter hormaechei* bacteria vanillin disrupt the cell membrane, inhibits biofilm formation, and also kills the cells within biofilms. These changes are caused by the reduction of intracellular Adenosine triphosphate (ATP), pH level, and the change of membrane potential (Qian et al., 2020).

Vanillin is mostly used as a flavor in the food industry as mentioned above; however, its antimicrobial characteristics have prevailed to be used as an antimicrobial agent in food packaging. In recent years, several antimicrobial food packaging materials have been developed with the incorporation of vanillin into biopolymers such as gelatine (Lee et al., 2016), chitosan (Sangsuwan et al., 2015), and Polyhydroxybutyrate (Xavier et al., 2015). These packaging systems demonstrated that vanillin has antimicrobial activity against *E. coli*, *Salmonella typhimurium*, *Shigella flexneri*, *L. monocytogenes*, and *S. aureus*, fungi such as *Aspergillus flavus*, *Aspergillus fumigatus*, *Aspergillus niger*, *Aspergillus parasiticus*, *Aspergillus ochraceus*, *Penicillium viridicatum*, *Penicillium clavigerum*, mold, and yeast.

Xavier et al. (2015) also established that vanillin is more effective against fungi than bacteria through their studies on several bacteria and fungi. However, the combination study of vanillin and trans-cinnamaldehyde by Arnon-Rips et al. (2021) revealed that cinnamaldehyde has better antimicrobial activity against fungi and yeast than vanillin. The study of Yang et al. (2017) focused on the incorporation of vanillin into a *U. pinnatifida* protein and gelatin to increase the shelf-life of smoked chicken breast. Here the vanillin-containing film decreased the *E. coli* by 1.12 log CFU/g when compared to the control film, thus allowing it to act as an antimicrobial packaging material. Further, antimicrobial food packaging materials have been developed with the incorporation of vanillin and cinnamic acid, and had good antimicrobial activity against mesophilic bacteria (1.5 log CFU/g) and *Enterobactericeae* (2.2 log CFU/g). It was also able to reduce the loss of total phenolic content and increase vitamin C retention (Silveira, Moreira, Artes, & Aguayo, 2015).

8.2.1.3 *Thymol*

Thymol is a natural monoterpenoid phenol derivative of p-Cymene, $C_{10}H_{14}O$, found in thyme oil and extracted from *Thymus vulgaris* (common thyme). It is also known as 2-isopropyl-5-methyl phenol. The molecular mass of thymol is 150.221 g/mol. Thymol and carvacrol were found to be highly effective in lowering the minimum inhibitory concentration of many antibiotics against zoonotic pathogens and food spoilage bacteria including *S. typhimurium* SGI 1 and *Streptococcus pyogenes* ermB in an in vitro study (Palaniappan & Holley, 2010).

Thymol is identified as the most active monoterpenoid (Jafri, Ansari, & Ahmad, 2019). Many studies explore the mechanism of action of thymol on microbes. The findings suggest that carvacrol and thymol work by disrupting the cytoplasmic membrane, increasing its permeability, and depolarizing its capacity (Xu, Zhou, Ji, Pei, & Xu, 2008). Furthermore, Gniewosz and Synowiec (2011) suggest thymol's antibacterial effect may be due to the rupture of a lipid layer of the cell's cytoplasmic membrane. Thymol could penetrate the phospholipid layer of the cell membranes, causing it to be disrupted. Thymol causes the cellular membrane to liquefy, resulting in an increase in permeability. An alteration in the potential of the phospholipid bilayer causes the release of ions, ATP molecules, and other critical cytoplasmic contents from the nucleus of a bacterial cell when the structure is disrupted. As a result, apoptosis occurs (Bakkali, Averbeck, Averbeck, & Idaomar, 2008; Gniewosz & Synowiec, 2011; Zhou et al., 2007).

It is used as an active compound in the packaging of food. Aytac, Ipek, Durgun, Tekinay, and Uyar (2017) formed a thymol encapsulated cyclodextrin-inclusion complex (CD-IC). They observed Zein-thymol/γ-CD-IC-nanofibrous webs (2:1) had reduced the bacterial count in beef and extended the stored shelf life for up to 5 days at 4°C (Aytac et al., 2017). Lukic, Vulic, and Ivanovic (2020) used supercritical CO_2 at 40°C and 10 MPa for 5 hours, films of poly (lactic acid) (PLA) with 5% poly(-caprolactone) (PCL), denoted as PLA-5, were filled with thymol, carvacrol, and their mixture, yielding 27.53%, 21.18%, and 21.49% loading, respectively. PLA-5 film filled with thymol and carvacrol mixture had the highest total polyphenol content and antioxidant activity, as well as good storage stability, adequate physical properties, and the lowest release rate (44.51 mg/L released after 6 weeks). In another study, to preserve the consistency, protection, and functional properties of table grapes, changed atmosphere packaging with eugenol or thymol is used (Valero et al., 2006). The berries' consistency losses were greatly reduced in packages containing eugenol or thymol. Furthermore, the active packaging resulted in lower microbial spoilage counts (Valero et al., 2006).

8.2.1.4 *Carvacrol*

Carvacrol (2-methyl-5-(1-methylethyl)-phenol) is a bioactive, volatile, monoterpene phenol that is isolated from aromatic plants and spices such as oregano (*Origanum vulgare*), thyme (*T. vulgaris*), black cumin (*Nigella sativa*), marjoram (*Origanum majorana*), summer savory (*Satureja hortensis* L.) and Ajwain (*Carum copticum*). However, the most common source of carvacrol in Origanum amounting to 75%. Carvacrol is highly soluble in ethanol and ether, while it is insoluble in water (Marchese et al., 2018). It is well known for its antimicrobial and antioxidant activity which is generally recognized as safe by Food and Drug Administration (FDA) as food additives. Carvacrol has high antimicrobial activity against a wide range of microbes including Gram-positive bacteria, Gram-negative bacteria, and fungi (Dhumal, Ahmed, Bandara, & Sarkar, 2019; Jahdkaran, Hosseini, Mohammadi Nafchi, & Nouri, 2021; Łopusiewicz et al., 2021; Lukic et al., 2020; Tunç, Duman, & Polat, 2016; Wang, Dekker, Heising, Fogliano, & Berton-Carabin, 2020). Antimicrobial activity is observed in carvacrol owing to its effects on the structure and function of cytoplasmic membranes of the microorganisms (Jahdkaran et al., 2021). Carvacrol is isomeric with thymol allowing them to have the same path of antimicrobial activity (Marchese et al., 2018). In addition, the combination of carvacrol and thymol has a positive synergetic antimicrobial effect (Lukic et al., 2020).

The antimicrobial activity of carvacrol depends upon the effects on the functions and structure of the cytoplasmic membrane of the microorganism (Jahdkaran et al., 2021) and the hydroxyl group, hydrophobicity of carvacrol plays a crucial role. The cell envelope is the major target of carvacrol, and it decreases in size, length, and diameter after contact. Carvacrol interacts with the lipid bilayer of the microbial cytoplasmic membrane and arranges in a line with the fatty acid chain resulting in the destabilization and the expansion of the structure. This is caused by the inclined permeability and fluidity of ions, cellular material, ATP, nucleic acid, and protons. Further, carvacrol causes the prompt depletion of the intracellular ATP pool by reduction of ATP synthesis or increase of ATP hydrolysis. The membrane potential of the microorganisms also shows an abrupt decrease thus indicating the proton motive force weakening. In addition, the intracellular potassium ions decrease while there is a proportional increase in the extracellular ion amounts. The bacteria cell death is caused by the loss of the ion gradient together with other effects (Burt, 2004; Marchese et al., 2018; Nostro & Papalia, 2012). At lower carvacrol concentration, that is, when not lethal to microbes, carvacrol alters the fatty acid composition of cytoplasm into an unsaturated fatty acid in some bacteria such as *E. coli*, *S. typhimurium*, *Pseudomonas fluorescens*, *Brochothrix thermosphacta* and *S. aureus*. Further, it also affects the toxins and enzymes produced by microorganisms like *S. aureus*. In *E. coli* carvacrol interferes with protein synthesis, inhibits flagellin synthesis, and induces heat shock protein 60 production (Nostro & Papalia, 2012). Carvacrol acts as an antifungal agent by increasing the fungi envelope damage leading to the disruption of membrane integrity and by blocking the biosynthesis of ergosterol (Marchese et al., 2018).

For making active antimicrobial packaging systems carvacrol is incorporated into biopolymers such as methylcellulose (Jahdkaran et al., 2021), PLA (Wang et al., 2020), potato starch nanofibers (Fonseca et al., 2021), chitosan (Mojaddar Langroodi, Nematollahi, & Sayadi, 2021), and zein- poly lactic acid (Altan, Aytac, & Uyar, 2018). These packaging systems showed antimicrobial activity against *E. coli*, *S. aureus*, *Listeria innocua*, *Saccharomyces cervicea*, *Candida albicans*, *S. typhimurium*, *L. monocytogenes*, *Enterobacteriaceae*, *Pseudomonas* spp., lactic acid bacteria yeast, and mold making them suitable active antimicrobial packaging materials. Carvacrol has been blended with the bioactive agent Melanin in a Carboxymethyl Cellulose to form a biodegradable packaging material. These components together showed antimicrobial activity against *E. coli*, *S. aureus*, and *C. albicans*. However, the addition of carvacrol to the packaging material led to the loss of transparency of the film (Łopusiewicz et al., 2021).

8.2.1.5 Citral

Citral (3,7-dimethyl-2, 6-octadienal) is a volatile, monoterpene aldehyde that is found naturally in herbs, plants, and citrus fruits. Citral has two geometrical isomers: geranial/Citral A (trans-isomer) and neral/Citral B (cis-isomer). Citral is currently used in the food, beverage, and cosmetic industry for lemon aroma and flavor. Industrially Citral oil is extracted from fruit, peels, or plants. Citral has properties such as insecticidal, deodorant, expectorant, appetite-stimulating, spasmolytic, anti-inflammatory, and antiseptic in addition to the antimicrobial activity against a wide spectrum of microorganisms including bacteria and fungi (Dhumal et al., 2019; Giteru et al., 2015; Giteru, Oey, Ali, Johnson, & Fang, 2017; Lu et al., 2018; Tamer, Suna, & Özcan-Sinir, 2019; Thielmann et al., 2021). The antimicrobial activity of Citral depends on the reactivity or/and availability of functional groups, biochemical composition, and peripheral components (Dhumal et al., 2019). However, under normal environmental conditions, Citral undergoes oxidative degradation allowing Citral to lose its antimicrobial activity (Lu et al., 2018).

The antimicrobial mechanism of action of Citral is affected by the increased variations of the ATP concentration. This leads to a decrease in cytoplasmic pH levels. The hydrophobic nature of Citral causes increased interaction with the lipid bilayer of the cytoplasmic membrane resulting in disintegrations. This gradually leads to cell permeability and cell death (Dhumal et al., 2019). Citral is further able to suppress the hyphal adhesions and virulent factors in *C. albicans* and genes involved in quorum sensing, peptidoglycan, and fatty acids biosynthesis in *S. aureus*. Thus, involved in cell death in microorganisms with the disruption of cell adhesion factors, virulent genes, and biosynthetic pathways (Gao et al., 2020). In mold, Citral changes the structure of mycelium and cell, destroys the mycelium structural integrity, and distorts the cell wall and membrane structure. This results in the loss of cytoplasm integrity, nucleic acid, protein leakage, destroy pH balance leading to cell death (Zhang et al., 2021).

In recent years, Citral essential oil has been incorporated into kafirin (Giteru et al., 2015a) polyvinyl acetate (Thielmann et al., 2021) that shows antimicrobial activity against *E. coli*, *S. aureus*, *Campylobacter jejuni*, *L. monocytogenes*, and *P. fluorescens*. Further, studies on orange peel powder showed antimicrobial activity against *Botrytis cinerea*, *A. niger*, and *Penicillium* sp. (McKay, Sawant, Fehlberg, & Almenar, 2021). Citral has also been incorporated with other antimicrobial agents Citral-Quercetin-Kafirin (Giteru et al., 2017) and Citral-Carvacrol-Sago Starch—guar gum (Dhumal et al., 2019) to form antimicrobial films for food packaging applications. Giteru et al. (2015a) found that Citral incorporated film has the highest antimicrobial activity, however, these films have decreased the tensile strength and oxygen permeability. Citral-embedded Kafirin film can reduce the total viable bacteria count of fresh chicken fillets during storage at 2°C ± 0.5°C for 96 hours (Giteru et al., 2017). On the other hand, the study by Dhumal et al. (2019) suggested that the combination of Carvacrol and Citral has a strong antimicrobial activity against *B. cereus* and *E. coli*. In addition, Citral and/or oregano EO was incorporated into the Ethylene-vinyl alcohol matrix to improve the shelf-life of packaged salad where the total bacteria count was reduced with storage time in the Citral and Oregano essential oil incorporated film. It was also found by this study that Citral was more effective than Oregano EO in preventing fresh vegetable/fruit spoilage (Muriel-Galet et al., 2013).

8.2.1.6 Cinnamaldehyde

Cinnamaldehyde has the formula $C_6H_5CH{=}CHCHO$ and is an organic compound. Cinnamon's flavor and odor are derived from the trans (E) isomer, which occurs naturally. It is a viscous pale-yellow liquid that can be found in the bark of cinnamon trees and other Cinnamomum plants. Cinnamaldehyde makes up about 90% of the essential oil of cinnamon bark. Also, there are many laboratory synthesis methods available, but the steam distillation of cinnamon bark oil yields the most cost-effective cinnamaldehyde. The FDA has classified cinnamaldehyde as GRAS, as has existing European legislation for products expected to come into contact with food (EU N10/2011 Regulation).

Cinnamaldehyde's antimicrobial activity has been linked to a variety of mechanisms, some of which overlap. The mechanism of cinnamaldehyde's bactericidal action against *L. monocytogenes* and *E. coli* has been studied, and it appears that interaction with the cell membrane causes a rapid inhibition of energy metabolism. The disruption of proton motive forces causes leakage of small ions without leakage of larger components like ATP, as well as inhibition of ATP production and membrane-bound adenosine triphosphatase (ATPase) activity (Friedman, 2017; Gill & Holley, 2006). Nowotarska, Nowotarski, Friedman, and Situ (2014) describe the formation of quantitative parameters using model membranes that suggest antimicrobial compounds like Cinnamaldehyde may change the lipid monolayer structure by forming antimicrobial and lipid aggregates (Nowotarska et al., 2014). After incubation at 37°C for 2 hours, Cinnamon oil and Cinnamaldehyde reduced the intracellular ATP concentration of *Mycobacterium avium* subsp. *paratuberculosis* (Map) cells (Nowotarska et al., 2017). He, Huang, Jiang, and Zhou (2019) observed that Cinnamaldehyde enhanced cell surface hydrophobicity, inhibited acid production, and improved acid tolerance in *S. mutans*. In the presence of Cinnamaldehyde, gene expression in biofilms was down-regulated. As a result, it showed that Cinnamaldehyde suppressed microbial activity in *S. mutans* biofilms by altering aggregation, hydrophobicity, acid tolerance, acid production, and virulence gene expression at sub-MIC levels (He et al., 2019).

It acts as a bacterial, yeast, and filamentous growth inhibitor and is vastly used in the food industry as an active agent. Villegas et al. (2019) super-critically impregnated Thymol and Cinnamaldehyde on PLA/organoclay for active and sustainable food packaging. Because of the plasticizing effect, supercritical impregnation of both active compounds with carbon dioxide changed the thermal and mechanical properties of PLA films. The antimicrobial activity of PLA films infused with Cinnamaldehyde and Thymol, as well as bio-nanocomposites, was demonstrated to be very effective against *E. coli* and *S. aureus*, with viability not detected for both strains. Sanla-Ead, Jangchud, Chonhenchob, and Suppakul (2012) studied the antimicrobial activity of Cinnamaldehyde and Eugenol and the influence of their incorporation into cellulose-based packaging films. According to the findings of this study, Cinnamaldehyde and Eugenol can inhibit a wide range of food pathogenic and contamination microbes. Cellulose-based edible films could serve as reservoirs, releasing Cinnamaldehyde or Eugenol to keep the inhibition activity at the food surface going. Similar results are obtained by Makwana, Choudhary, Dogra, Kohli, and Haddock (2014). Cinnamaldehyde and Eugenol can inhibit a wide variety of food pathogenic and contamination microbes. Edible films made of Cellulose could act as reservoirs, releasing Cinnamaldehyde or Eugenol to maintain an inhibitory effect at the surface of the food (Makwana et al., 2014). Higueras, López-Carballo, Gavara, and Hernández-Muñoz (2015) observed that the procedure utilized and the composition of the suitable solvent had an impact on the amount of Cinnamaldehyde released and the antibacterial activity of the films. The amount of Cinnamaldehyde released increased as the temperature of the treatment increased, improving the efficacy of the films (Higueras et al., 2015).

8.2.2 Plant extracts

8.2.2.1 Tannins

Tannins are water-soluble polyphenols found in a variety of plant foods. Tannic acid (TA) is a type of polyphenol and a specific kind of tannin. Because of the various phenol groups in the structure, it has low acidity (pKa about 6). $C_{76}H_{52}O_{46}$ is a popular chemical formula for commercial tannic acid. Tannins contained in the fruits act as natural protection against microbial infections. Tannic acid's antimicrobial properties can be used in food processing to extend the shelf life of some foods including catfish fillets. The molecular mass of tannic acid is 1,701.19 g/mol. Astringent, antioxidant, hemostatic, and antibacterial properties are among the active properties of TA. Furthermore, the FDA has classified TA as "Generally Recognized as Safe," and it is commonly used in the food and medical industries (Xu, Weng, Gilkerson, Materon, & Lozano, 2015).

Tannic acid has a wide range of chemical and biological properties. Demand for effective antimicrobial agents that reveal more direct bactericidal mechanisms has been affected by the rapid spread of multidrug-resistant bacteria. Tannin-rich plant extracts have been shown to have potent antimicrobial properties. Their antibacterial activity is affected by a variety of factors, including pH, temperature, solvent/matrix form, and action time (Frazier et al., 2010). Tannins are multivalent ligands that bind to proteins primarily through hydrophobic interactions and hydrogen bonds (Jöbstl, Howse, Fairclough, & Williamson, 2006; Theisen et al., 2014). As a result, the metabolic activity of bacteria is inhibited. Tannins isolated from green tea extract have also been shown to have anti-infectious properties. They are effective against *S. pyogenes*, the addition of the extract inhibited the bacteria's attachment to the kidney epithelial cells in a dose-dependent manner, according to the findings (Kaczmarek, 2020; Tucci & Benghuzzi, 2011). TA is found in *Neolamarckia cadamba* fruit extracts, which have antibacterial properties against *E. coli*, *P. aeruginosa*, *Y. enterocolitica*, *S. aureus*, *B. cereus*, and *L. innocua*. One of the key components in the extracts was found to be tannic acid. The findings indicated that the extracts inhibited sugar and amino acid uptake at higher concentrations, which is one of the key mechanisms of bacterial growth inhibition.

Tannin toxicity is thought to be caused by specific inhibition of cell wall synthesis. *Butyrivibrio fibrisolvens A38*, *Streptococcus bovis 45S1*, *Prevotella ruminicola B14*, and *Ruminobacter amylophilus WP225* were found to bind to cells with condensed tannins from *Onobrychis viciifolia* (sainfoin) (Smith, Zoetendal, & Mackie, 2005). In strains with Gram-positive type cell walls, morphological changes were observed in addition to growth and protease activity inhibition. A change in the predominance of Gram-positive to Gram-negative cell-walled bacteria was discovered in the results of an ecological analysis (Smith & Mackie, 2004). According to these findings, studying the effects of tannins on membranes and cell walls may be a promising research subject.

Kamari and Phillip (2018) incorporated TA chitosan, gelatin, and methylcellulose films to synthesize, characterize, and use to preserve cherry tomatoes (*Solanum lycopersicum var. cerasiforme*) and grapes (*Vitis vinifera*). According to antimicrobial research, adding TA to biopolymers strengthened their antibacterial properties against *E. coli* and *S. aureus*. In a 14-day preservation trial, the capacity of films to preserve both fruits was tested. Fruit weight loss and browning index were reduced when biopolymer films treated with TA were used instead of control films. Hazer and Ashby (2021) studied the prevention of autoxidation from light and ambient oxygen, the excellent packaging content polypropylene was functionalized with TA (a natural product). As compared to glass and precursor-chlorinated polypropylene levels, the TA-functionalized polypropylene films demonstrated excellent resistance to autoxidation. Kaczmarek, Owczarek, Nadolna, and Sionkowska (2019) aim to see whether adding a polyphenol like TA to chitosan would result in a new material that could be used in the packaging industry. Chitosan/TA mixtures in various weight ratios were used to make thin films. The research found that combining chitosan and TA results in thin films with surface-free energy suitable for industrial applications such as food packaging. In addition, combining biopolymers enhances mechanical properties (Kaczmarek et al., 2019).

8.2.2.2 Anthocyanins

Anthocyanin is water-soluble, flavonoid polyphenolic pigments that are found in glycosylated forms. Anthocyanins are responsible for the purple, blue, red color in fruits, vegetables, and flowers. Anthocyanins are found in plants such as berries, blackcurrants, black carrots, red cabbage, purple potato, purple mint, purple passion flower, purple sage, common violet, lavender, etc. They are present in the plant as glucose, galactose, arabinose, rhamnose, or xylose. Anthocyanin is pigments that are derived from flavanol, their basic structure is flavylium ion and they are also the glycosylated form of anthocyanidins (e.g., cyanidin, delphinidin, pelargonidin, peonidin, malvidin, and petunidin). Anthocyanins are currently used in the industry as a natural food colorant. However, it has many beneficial properties such as antimicrobial, antioxidative, anticancer, antidiabetic effect, visual health, anti-obesity effect, and neuroprotective effect. The color and stability of the Anthocyanin pigments depend upon pH, light, temperature, and pigment structure. In acidic conditions, Anthocyanin is red while it gradually turns blue as the pH increases (Alizadeh-Sani, Tavassoli, McClements, & Hamishehkar, 2021a; Alizadeh-Sani et al., 2021b; Khoo, Azlan, Tang, & Lim, 2017; Ma et al., 2019; Qin, Liu, Yuan, Yong, & Liu, 2019; Sani, Tavassoli, Hamishehkar, & McClements, 2021). The pH sensitivity effect allows anthocyanin to be used in intelligent packaging in addition to antimicrobial active packaging (Zhang, Liu, Yong, Qin, & Liu, 2019). This effect allows the anthocyanin-incorporated films to determine the antimicrobial function of the packaged food in real-time (Chen et al., 2020).

The antimicrobial activity of Anthocyanins is attributed to the polyphenolic structure and the hydrophobic nature (Sani et al., 2021). Anthocyanins can damage the bacterial cell membrane, resulting in the loss of cellular integrity and thus resulting in the leakage of essential bacterial compounds such as nucleic acids and proteins. Anthocyanin is also able to penetrate through the inner cell membrane and reduce the levels of alkaline phosphatase, adenosine triphosphatase, and superoxide dismutase enzymes thus inhibiting the growth of the microorganism or hindering its self-protection mechanism. The low alkaline phosphatase levels increase the level of calcium ions resulting in the disruption of ion channels. The reduced adenosine triphosphatase levels increase the efflux of ATP and obstruct the respiratory metabolism. Superoxide dismutase enzymes oxide the free radicals and thus the reduced levels result in the lack of eliminating free radicals. Anthocyanins also reduce the rate of the tricarboxylic acid cycle (TCA) associated enzyme production thus blocking the TCA cycle of pathogens. This results in weakened cellular respiration gradually leading to cell death (Ma et al., 2019; Sun et al., 2018).

Anthocyanin is mostly used in the food packaging industry due to its pH-sensitive natural dye that is extremely useful in smart/intelligent packaging. However, in recent years, Anthocyanin has been incorporated into matrixes to form pH-sensitive antimicrobial packaging materials; Chitosan nanofibers-Methyl cellulose (Alizadeh-Sani et al., 2021a, 2021b) and Chitin nanofiber-Methylcellulose (Sani et al., 2021) that showed antimicrobial activity against *E. coli* and *S. aureus*. Further, other antimicrobial agents have been incorporated along with Anthocyanin to form antimicrobial packaging materials; Cellulose nanofiber-Oregano essential oil (Chen et al., 2020), Chitosan-nanosized TiO_2 (Zhang et al., 2019), and Chitosan - silver nanoparticles (Qin et al., 2019) that demonstrated stronger antimicrobial activity than with pure Anthocyanin incorporate. As demonstrated by the above studies anthocyanin dye is pH sensitive and it not only has an antimicrobial effect but also has a strong antioxidant activity as a bioactive agent in food packaging. Thus, the Anthocyanin agent is used in food packaging materials with the potential of monitoring food freshness and spoilage in which the antimicrobial activity is an added advantage.

8.2.2.3 Coumarins

Coumarins (2H-1-benzopyran-2-one) are a class of heterocyclic, phenolic compounds with benzene and an alpha-pyrone ring. There are a few different classes of natural Coumarins present including simple Coumarin derivatives, dihydrofurano Coumarin derivatives, furano Coumarin derivatives, pyrano Coumarin derivatives (linear and angular), phenyl Coumarin derivatives, and bicoumarin derivatives. Coumarins are secondary plant metabolites that are present in various parts of the plant and found in the highest concentration in fruits, roots, stems, and leaves. Coumarins are isolated from hundreds of plants species that are observed in plant families including Angiospermae, Monocotyledoneae, and Dicotyledoneae. Thus, they are present in different concentrations and of different types in plants such as Celery, Coriander, Cumin, Bishop's flower, Dill, Angelica, Fennel, Parsley, etc. Coumarins have a unique sweet odor thus used as a perfume. Coumarins are compounds with bioactivity and photo-reactivity. Thus, they have many beneficial properties such as antimicrobial, antitumor, anticoagulant, antioxidant, anti-inflammatory, anti-carcinogenic, anti-diabetic, analgesic, and anti-neurodegenerative (Cazin, Rossegger, de la Cruz, Griesser, & Schlogl, 2021; Matos et al., 2015).

The free 6-OH and 7-OH in the Coumarin compound play an essential role in antimicrobial activity. Coumarins are inhibitors of DNA topoisomerase type II that are essential for DNA replication, thus Coumarins inhibit nucleic acid formation. In addition, Coumarins also suppress the quorum-sensing network of the microorganism resulting in the lack of biofilm formation and the production of virulence factors. These inhibitions and suppression of Coumarins ultimately lead to the cell death of microorganisms (Khameneh, Iranshahy, Soheili, & Bazzaz, 2019).

Coumarins are currently not in use in pure form for antimicrobial-based food packaging materials. However, Coumarins containing plant extracts have been used as antimicrobial food packaging material such as cumin—*Zataria* (Akrami et al., 2015), Cumin EO—*Shahri Balangu* seed mucilage (Behbahani, Noshad, & Jooyandeh, 2020), coriander oil—Porcine meat and bone meal protein (Lee, Won, & Song, 2015) fennel essential oil—peppermint essential oil—chitosan (Liu, Wang, Chi, Tan, & Liu, 2020). They have antimicrobial activity against a wide range of bacteria such as *Bacillus cereus, L. monocytogenes, Enterococcus faecalis, S. aureus, Salmonella cholerasuis, E. coli, Yersinia enterocolitica*, and *Pseudomonas aeruginosa*.

8.2.2.4 Allicin

Allicin (diallyl thiosulfinate) is the natural antimicrobial agent of garlic that contains sulfur as a natural compound. Allicin is an aromatic compound with properties such as antioxidant, anti-carcinogenic, anti-inflammatory, anti-thrombotic, anti-atherosclerotic, anti-hyperlipidemic, and pro-circulatory other than antimicrobial activity.

Allicin has antimicrobial activity against a wide range of microorganisms including Gram-negative and Gram-positive bacteria, fungi, parasites, and viruses (Esmaeili et al., 2020; Jebali et al., 2013). The antimicrobial activity of allicin is mainly attributed to the sulfur content of the compound (Esmaeili et al., 2020). Allicin is highly unstable and easily decomposed under high pressure and temperature. The decomposition doesn't affect the antimicrobial activity since the breakdown products diallyl sulfide, diallyl disulfide, diallyl trisulphide, and ajoene have antimicrobial activity (Sung et al., 2014).

Allicin is a natural antimicrobial agent that is target-oriented and can penetrate through the cell wall and the cell membrane. Allicin can readily diffuse through the microbial membranes. When in the cell allicin can react with every cysteine residue as long as it has a free −SH-group thus making thymol-containing enzymes a target of inactivation. These activities lead allicin to inactivate several thymol and non-thymol-containing enzymes such as dehydrogenase, hexokinase, triosephosphate dehydrogenase, and alcohol dehydrogenase (Borlinghaus, Albrecht, Gruhlke, Nwachukwu, & Slusarenko, 2014). Allicin also inhibited DNA gyrase enzyme activity which is an important prokaryotic enzyme (Reiter, Hubbers, Albrecht, Leichert, & Slusarenko, 2020). The inactivation of the essential enzymes of the microorganisms leads to the lack of the production of proteins and nucleic acids, in turn leading to reduced cell growth and ultimately cell death.

Antimicrobial packaging material was developed by incorporating allicin into polylactic acid (Radusin et al., 2019), nanocellulose (Jebali et al., 2013), chitosan-whey protein (Esmaeili et al., 2020), and bread aroma (Heras-Mozos et al., 2018) that showed antimicrobial activity against *C. albicans*, *A. niger*, *S. aureus*, *E. coli*, and mold. Additionally, banana flour films Incorporated with garlic essential oil completely inhibited the growth of *A. flavus* showing no fungal activity. The characteristics of this film made it suitable for a potential food packaging system for roasted peanuts or oily food products (Orsuwan & Sothornvit, 2018). However, the study of garlic oil incorporated low-density polyethylene film demonstrated that the antimicrobial effect was observed in packaged meet for *L. monocytogenes* while an insignificant effect was observed for *E. coli* and *B. thermosphacta* (Sung et al., 2014). As per this study, garlic oil significantly reduces the growth rate of *L. monocytogenes* in beef after 3, 6, 9, and 15 days of storage at 4°C.

This discrepancy leaves space for further studies of Allicin incorporated packaging materials for food industry applications. Although Allicin has activity against a wide variety of food spoilage bacteria it is released in higher concentrations at the beginning since it doesn't have a long-lasting antimicrobial activity (Liu, Song, Zhang, & Zhang, 2020). Nanoencapsulation of garlic oil reduces the rate of Allicin release, making the packaging material have a long-lasting antimicrobial activity (Esmaeili et al., 2020). The nano-encapsulated garlic oil was able to increase the shelf life of vacuum-packed sausages making no significant changes to its sensory properties.

8.2.2.5 Alkaloids

Alkaloids are a natural group of heterocyclic structures that contain nitrogen atoms. The alkaloids are classified based on their structure and place of origin. It is broadly classified into two groups (1) non-heterocyclic/atypical alkaloids, such as N-methyltyramine, colchicine, erythromycin, and (2) heterocyclics/ typical alkaloids such as pyrrole and quinine. Alkaloids are found in about 300 plant families. Different types of alkaloids are present in the different parts of the plants (Othman, Sleiman, & Abdel-Massih, 2019). Alkaloids can be found in plants such as coffee, tea, cacao, tomatoes, potatoes, cocaine, etc. The most well-known alkaloids are nicotine, morphine, strychnine, quinine, atropine, caffeine, and ephedrine. Alkaloids have many essential properties such as antibacterial, antimitotic, anti-inflammatory, analgesic, local anesthetic, hypnotic, psychotropic, anti-HIV, insecticidal activities, and antitumor activities. Alkaloids can act as stimulants of the central nervous system and some alkaloids can act as illicit drugs or poisons (Kurek, 2019).

The antimicrobial activity mechanism of the alkaloids depends upon the type of alkaloid because of its structural difference. The different mechanisms of action by which the plant alkaloids are by (1) affecting cell division by inhibiting nucleic acid synthesis through the inhibition of dihydrofolate reductase, (2) inhibition of the respiration activity and enzymes of bacteria, (3) disruption of the bacterial membrane of the bacteria or (4) affecting the virulence genes thus leading to the inhibition of the virulence factors (Othman et al., 2019).

Polypyrrole is a type of Alkaloid that plays an important role as an antimicrobial agent in food packaging. Polypyrrole has a high antimicrobial activity due to its electron donor electron acceptor property in the chain that plays an essential role in bacterial adhesion to a solid surface (Gao et al., 2020). The active food packaging materials have been developed with the incorporation of Polypyrrole; Polypyrrole-Cellulose-Chitosan (Gao et al., 2020), Polypyrrole-Cellulose nanofibers—Polyvinyl alcohol (Bideau, Bras, Saini, Daneault, & Loranger, 2016) and these packaging materials has antimicrobial activity against *E. coli*, and *Bacillus subtilis*.

8.2.2.6 *Antimicrobial peptide*

AMPs are a family of natural antimicrobials that are the first line of defense in plants against many pathogens. AMPs are of positive charge, amphipathic nature, and are rich in cysteine amino acids. The plant-based AMPs are characterized by the presence of a compact spatial structure that is formed by disulfide bonds. They display a peptide promiscuity activity which is its ability to have multiple functions in a single peptide. The plant-based AMPs are categorized into families such as defensins, thionins, α-hairpinins, hevein-like peptides, knottins, snakins, lipid transfer proteins, and cyclotides based on amino acid sequence, cysteine motifs, and the location of disulfide bonds. AMPs are found in all plants and all the parts of the plants. AMPs have many beneficial properties such as antibacterial, antifungal, insecticidal, and antiviral activities (Barashkova & Rogozhin, 2020; Iqbal et al., 2019; Tang, Prodhan, Biswas, Le, & Sekaran, 2018).

The mode of antimicrobial action of AMPs depends on the type of AMPs. The first step of antimicrobial activity is the disruption of the cell membrane integrity. This takes place by the random insertion of the AMPs into the membrane, leading to the hydrophobic sequence alignment that finally leads to the removal of the membrane section and pore formation. In addition to these, the inhibition of DNA synthesis and RNA synthesis takes place in the intracellular matrix of the microbes leading to the loss of nucleic acids. Enzyme inhibition and protein synthesis inhibition also take place which leads to the loss of the functions and the structural properties of the cell. Further, the chaperone proteins that are essential for protein folding are blocked. Finally, the inhibition of cellular respiration and reactive oxygen species formation takes place. This leads to the disruption of mitochondrial membrane integrity and efflux of ATP takes place. All the above changes in the microorganism cell lead to cell death (Amerikova, Pencheva El-Tibi, Maslarska, Bozhanov, & Tachkov, 2019).

AMPs has become an interest in the food packaging industry in recent years due to their antimicrobial properties against microorganism and food preservation ability. AMPs Leg1 and Leg2 from chickpea storage protein showed antimicrobial activity against 16 different bacteria including foodborne pathogens (Heymich et al., 2021). While peptide Snakin-1 from potato tubers has antimicrobial activity against food spoilage yeast *Zygosaccharomyces bailii* (Shwaiki, Arendt, & Lynch, 2020).

8.3 Antimicrobial-loaded nanocarriers for food packaging applications

The process of the incorporation of the bioactive antimicrobial component into the nanocarriers is known as nanoencapsulation. The nanocarriers used are of different shapes such as nanocapsule, nanosphere, nanoshell, branched, nanocube, nanostar, nanorod, and nanocluster. The antimicrobial substances that are incorporated consist of a core containing the antimicrobial agent and a wall consisting of lipids, proteins, or polysaccharides. When produced the nanocarriers can be synthesized in top-down or bottom-up methods. The nanoencapsulation process of the antimicrobial substances provides advantages such as increased solubility and absorption, larger surface area for interaction, protection of antimicrobial agents against environment stress including temperature, pressure, etc., prevention of undesirable interactions, leading towards a target site, and finally controlled release of the antimicrobial agents that have been encapsulated (Bahrami et al., 2020). The different types of nanocarriers that are used in food packaging are illustrated in Fig. 8.2.

8.3.1 *Nanoemulsions*

Nanoemulsions are a lipid-based encapsulation system that contains water and oil as immiscible liquids and are of very small particle size. They consist of a water phase, an oil phase, and an emulsifier. The types of nanoemulsions that are used for encapsulation of antimicrobial agents are; oil in water (O/W), water in oil (W/O), water in oil in water (W/O/W), oil in water in oil (O/W/O), and multilayer emulsions. The no-equilibrium nano-emulsion system is formed through external or internal energy sources and thus fabricated through high-energy or low-energy approaches. Nanoemulsions are of use in food packaging due to their unique properties such as small particle size, increased surface area, less sensitivity to chemical/physical changes, improved digestibility, targeted delivery, increased bioavailability, and efficient encapsulation (Fig. 8.3). Further, nanoemulsions can increase the bioavailability of natural extracts that have been incorporated and they also can incorporate both hydrophilic and hydrophobic compounds. However, there are a few demerits of using nanoemulsions such as regulatory aspects, product safety, and acceptance (Bahrami et al., 2020; McClements, 2021; Salem & Ezzat, 2019; Yousefi, Ehsani, & Jafari, 2019).

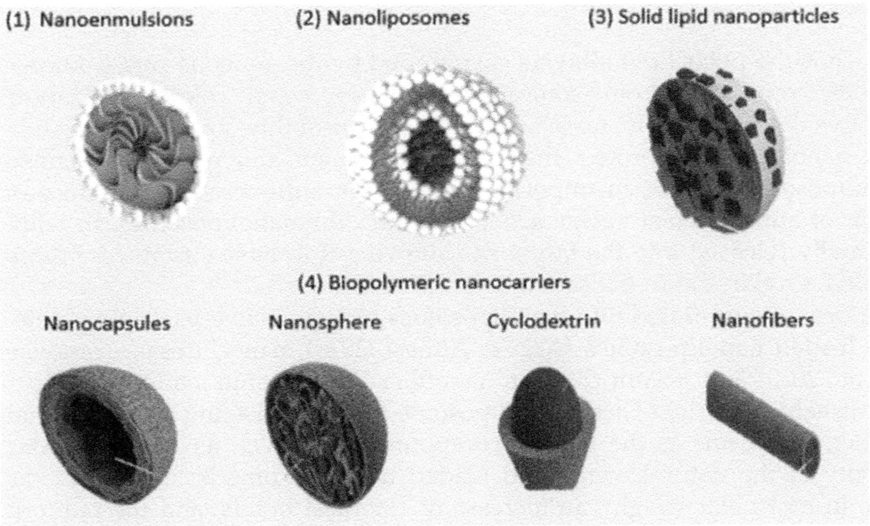

FIGURE 8.2 Antimicrobial-loaded nanocarriers. *Source: From Bahrami, A., Delshadi, R., Assadpour, E., Jafari, S. M., & Williams, L. (2020). Antimicrobial-loaded nanocarriers for food packaging applications.* Advances in Colloid and Interface Science, *278, 102140. https://doi.org/10.1016/j.cis.2020.102140.*

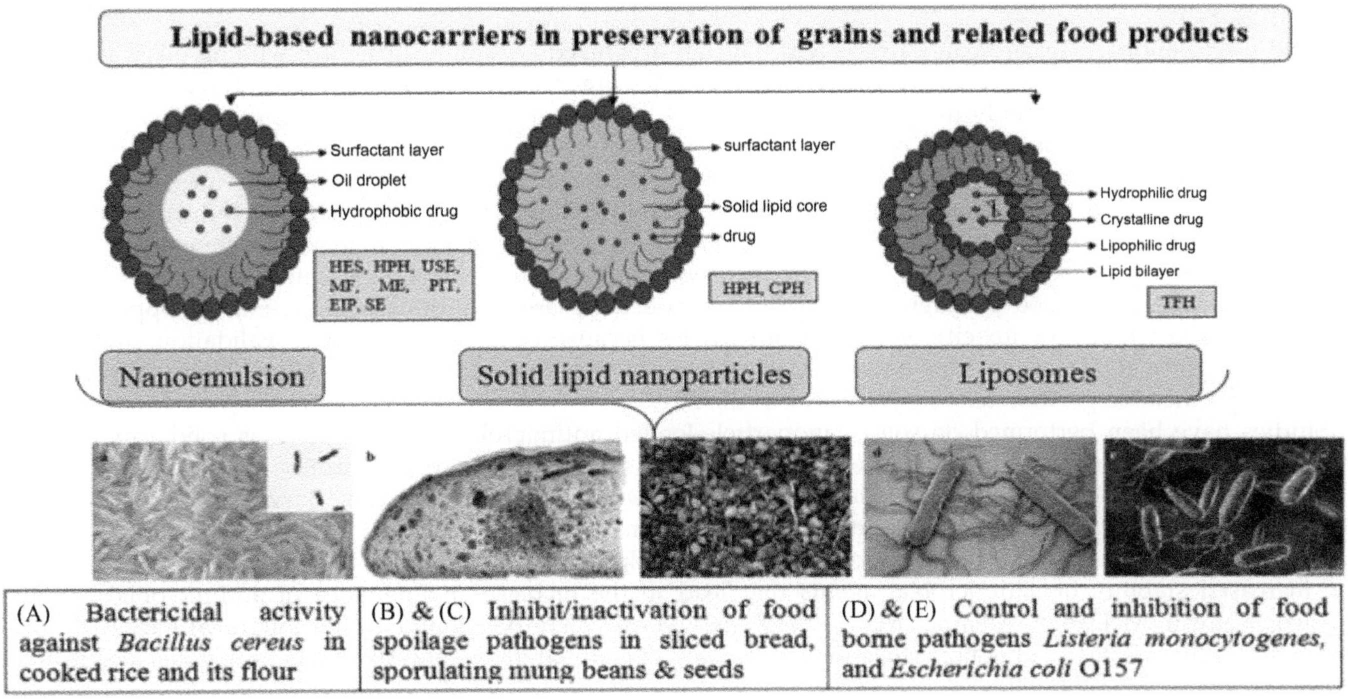

FIGURE 8.3 Lipid-based nanocarriers for food safety. *Source: From Kaliamurthi, S., Selvaraj, G., Hou, L., Li, Z., Wei, Y., Gu, K., & Wei, D. (2019). Synergism of essential oils with lipid based nanocarriers: Emerging trends in preservation of grains and related food products.* Grain & Oil Science and Technology, *2(1), 21–26. https://doi.org/10.1016/j.gaost.2019.04.003.*

Many studies have been conducted on nano-emulsion with the incorporation of natural antimicrobial agents such as pectin—marjoram essential oil nano-emulsion (Almasi, Azizi, & Amjadi, 2020), chitosan-thymol nano-emulsion (Liu & Liu, 2020), pectin–copaiba oil nano-emulsion (Norcino et al., 2020), fish sarcoplasmic protein-chitosan–ginger essential oil nano-emulsion (Cai, Wang, & Cao, 2020) and nano chitosan–pomegranate peel extract nano-emulsion (Gull et al., 2021). These active nano-emulsion antimicrobial packaging materials had antimicrobial activity against the same microorganisms as the natural antimicrobial agent, the antimicrobial activity was not increased, however, antimicrobial activity had a prolonged effect with the slow-release rate making them much more effective.

8.3.2 Nanoliposomes

The nanoscale form of the liposome which is the polar lipid bilayers surrounded by the aqueous core is known as nanoliposomes. Nanoliposomes can be produced from sonication, extrusion, and micro-fluidization. Nanoliposome has the bioavailability of active components, increase solubility, dispensability, and provide stability during stress conditions. Further, the nanoliposomes protect the active component and provide sustained release of the substances (Fig. 8.3) . Nanoliposomes places an important role in the antimicrobial resistance of microorganisms since only a small amount of antimicrobial agents are loaded into the nanoliposomes. In addition, the antimicrobial substances are gradually released into the target site allowing it to have sustained antimicrobial activity for a prolonged time (Demirci, Caglar, Cakir, Gülseren, & Jafari, 2017).

Many natural antimicrobial agents have been incorporated into nanoliposomes to form active packaging materials; whey protein isolate—thyme extract loaded nanoliposomes (Aziz & Almasi, 2018), nanochitosan—caraway seed extract loaded nanoliposome (Wu et al., 2019) and gelatin-chitosan nanofiber-ZnO-betanin loaded nanoliposomes (Amjadi, Nazari, Alizadeh, & Hamishehkar, 2020). These nanoliposome-based packaging materials had antimicrobial activity against the same microorganisms as the natural compounds; however, as stated by Aziz and Almasi (2018), the antimicrobial activity of the natural compound loaded nanoliposome is lower than the natural extract. This is due to the increase in molecular weight, an increase of chemical bonds, and the reduced speed of migration of extract.

8.3.3 Solid lipid nanoparticles

Solid lipid nanoparticles are nanocarriers that have an organized crystalline structure with an oil/water emulsion where the oil phase is solid. The solid nanoparticles are produced from hot homogenization and cold homogenization. The commonly found lipids in the solid lipid nanoparticles are precirol, compritol, cholesterol, cetyl alcohol, glyceryl monostearate, cetylpalmitate, trimyristin, tristearin, and stearic acid. The mobility of the antimicrobial agents is limited in solid lipid when compared to liquid lipid substances; therefore, the release of antimicrobial agents from solid lipid nanoparticles is very low. The antimicrobial agents are incorporated into the solid lipid nanoparticle through different techniques such as homogeneous matrix, bioactive-enriched shell, or bioactive-enriched core models (Fig. 8.3). The incorporation of antimicrobial agents into solid lipid nanoparticles is advantageous in food packaging brcause of controlled release of antimicrobial agent, the opportunity of large-scale production, low toxicity of solid nanoparticles because of the lipids, ease to validation, and getting regulatory approval. However, they have a poor antimicrobial agent encapsulating ability, high water content, and an unpredictable gelation tendency (Yousefi et al., 2019).

Studies have been performed on solid nanoparticle-loaded antimicrobial extracts such as solid nanoparticles loaded with rosmarinic acid and herbal extracts (sage and savory) (Campos, Madureira, Sarmento, Pintado, & Gomes, 2017) and solid lipid nanoparticles containing *E. caryophyllata* essential oil (Bazzaz et al., 2018). These studies confirm that solid nanoparticles increase antimicrobial activity for a longer time. The increase of antimicrobial activity depends upon the ability of solid nanoparticles to interact with the microbial cell membrane and the increased stability of antimicrobial agents that lead to the increase/maintenance of antimicrobial activity. Further, Tonyali, McDaniel, Amamcharla, Trinetta, and Yucel (2020) developed an antimicrobial packaging system with Pullulan-solid lipid nanoparticle-loaded thymol, cinnamaldehyde, and the eugenol that has a controlled release of the active compounds.

8.3.4 Biopolymeric nanocarriers

Biopolymers are used as nanocarriers to encapsulate natural antimicrobial agents. They are effective nanocarriers due to their non-toxicity, antioxidant, antimicrobial properties, gelling, emulsification, and foaming abilities. Polysaccharides such as starch, cellulose, dextran, cyclodextrin, chitosan, pullulan, guar gum, alginate, pectin, hyaluronic acid, and proteins such as casein, whey protein, zein, soy, gelatin are most commonly used as nanocarriers individually or in combination. The most used biopolymeric nanocarriers for encapsulations are nanocapsules, nanosphere, cyclodextrin, and nanofibers (Bahrami et al., 2020). The use of biopolymeric nanocarriers in food packaging has increased and thus natural antimicrobial agents are also been added to these matrixes; chitosan nanoparticles—lime essential oil (Sotelo-Boyas, Correa-Pacheco, Bautista-Banos, & Corona-Rangel, 2017), chitosan nanoparticles—summer savory essential oil (Feyzioglu & Tornuk, 2016), cyclodextrin nanosponges—coriander essential oil (Silva, Caldera, Trotta, Nerín, & Domingues, 2019), zein nanoparticles—Thymbra capitata essential oil (Merino et al., 2019), chitosan nanofiber—methylcellulose (Alizadeh-Sani et al., 2021a, 2021b), and cellulose nanofiber—lemon waste powder (Soofi, Alizadeh, Hamishehkar, Almasi, & Roufegarinejad, 2021). These biopolymeric nanocarrier-based packaging materials had outstanding antimicrobial activity due to the double antimicrobial effects of the biopolymers and the natural antimicrobial agent.

Moreover, the antimicrobial agent is absorbed onto the surface of the nanoparticles therefore the release of the antimicrobial agent is much faster when compared to the lipid-based nanocarriers systems (Sotelo-Boyas et al., 2017). In addition, as per the study of Merino et al. (2019), the antimicrobial activity of the EO encapsulated in zein nanoparticles is higher when compared to pure EO since it can overcome the highly volatile nature and improve the dispersion.

8.4 Conclusion and future trends

In recent years, natural antimicrobial agents extracted from fruits and plants have been incorporated into food packaging materials to form active antimicrobial packaging materials or coating. These packaging materials are eco-friendly, non-toxic, and easy to be made. The mechanism of action of these antimicrobial agents depends on the types and chemical composition of these agents. Loading the antimicrobial agents into nanocarriers is advantageous due to its sustained release of antimicrobial agents leading to prolonged antimicrobial activity. However, the delivery of antimicrobial agents depends upon various factors such as stability, cost, antimicrobial agent efficacy, transportation, and storage.

To improve the utilization of these natural antimicrobials in foods, further research is needed on the toxicological impact and potential synergistic actions of the major and minor natural elements. Moreover, further studies are recommended on the commercial viability of natural antimicrobials in food packaging material.

References

Akrami, F., Rodriguez-Lafuente, A., Bentayeb, K., Pezo, D., Ghalebi, S. R., & Nerin, C. (2015). Antioxidant and antimicrobial active paper based on Zataria (*Zataria multiflora*) and two cumin cultivars (*Cuminum cyminum*). *Lwt-Food Science and Technology, 60*(2), 929–933. Available from https://doi.org/10.1016/j.lwt.2014.09.051.

Alizadeh-Sani, M., Tavassoli, M., McClements, D. J., & Hamishehkar, H. (2021a). Multifunctional halochromic packaging materials: Saffron petal anthocyanin loaded-chitosan nanofiber/methyl cellulose matrices. *Food Hydrocolloids, 111*(13), 106237. Available from https://doi.org/10.1016/j.foodhyd.2020.106237.

Alizadeh-Sani, M., Tavassoli, M., Mohammadian, E., Ehsani, A., Khaniki, G. J., Priyadarshi, R., & Rhim, J. W. (2021b). pH-responsive color indicator films based on methylcellulose/chitosan nanofiber and barberry anthocyanins for real-time monitoring of meat freshness. *International Journal of Biological Macromolecules, 166*, 741–750. Available from https://doi.org/10.1016/j.ijbiomac.2020.10.231.

Almasi, H., Azizi, S., & Amjadi, S. (2020). Development and characterization of pectin films activated by nanoemulsion and Pickering emulsion stabilized marjoram (*Origanum majorana L.*) essential oil. *Food Hydrocolloids, 99*. Available from https://doi.org/10.1016/j.foodhyd.2019.105338.

Altan, A., Aytac, Z., & Uyar, T. (2018). Carvacrol loaded electrospun fibrous films from zein and poly(lactic acid) for active food packaging. *Food Hydrocolloids, 81*, 48–59. Available from https://doi.org/10.1016/j.foodhyd.2018.02.028.

Amerikova, M., Pencheva El-Tibi, I., Maslarska, V., Bozhanov, S., & Tachkov, K. (2019). Antimicrobial activity, mechanism of action, and methods for stabilisation of defensins as new therapeutic agents. *Biotechnology & Biotechnological Equipment, 33*(1), 671–682. Available from https://doi.org/10.1080/13102818.2019.1611385.

Amjadi, S., Nazari, M., Alizadeh, S. A., & Hamishehkar, H. (2020). Multifunctional betanin nanoliposomes-incorporated gelatin/chitosan nanofiber/ZnO nanoparticles nanocomposite film for fresh beef preservation. *Meat Science, 167*. Available from https://doi.org/10.1016/j.meatsci.2020.108161.

Arana-Sánchez, A., Estarrón-Espinosa, M., Obledo-Vázquez, E. N., Padilla-Camberos, E., Silva-Vázquez, R., & Lugo-Cervantes, E. (2010). Antimicrobial and antioxidant activities of Mexican oregano essential oils (*Lippia graveolens* HBK) with different composition when microencapsulated in β-cyclodextrin. *Letters in Applied Microbiology, 50*(6), 585–590.

Arnon-Rips, H., Cohen, Y., Saidi, L., Porat, R., & Poverenov, E. (2021). Covalent linkage of bioactive volatiles to a polysaccharide support as a potential approach for preparing active edible coatings and delivery systems for food products. *Food Chemistry, 338*. Available from https://doi.org/10.1016/j.foodchem.2020.127822.

Arya, S. S., Rookes, J. E., Cahill, D. M., & Lenka, S. K. (2021). Vanillin: A review on the therapeutic prospects of a popular flavouring molecule. *Advances in Traditional Medicine, 21*, 1–17. Available from https://doi.org/10.1007/s13596-020-00531-w.

Ashrafudoulla, M., Mizan, M. F. R., Ha, A. J. W., Park, S. H., & Ha, S. D. (2020). Antibacterial and antibiofilm mechanism of eugenol against antibiotic resistance *Vibrio parahaemolyticus*. *Food Microbiology, 91*, 103500.

Aytac, Z., Ipek, S., Durgun, E., Tekinay, T., & Uyar, T. (2017). Antibacterial electrospun zein nanofibrous web encapsulating thymol/cyclodextrin-inclusion complex for food packaging. *Food Chemistry, 233*, 117–124.

Aziz, S. G. G., & Almasi, H. (2018). Physical characteristics, release properties, and antioxidant and antimicrobial activities of whey protein isolate films incorporated with thyme (*Thymus vulgaris L.*) extract-loaded Nanoliposomes. *Food and Bioprocess Technology, 11*(8), 1552–1565. Available from https://doi.org/10.1007/s11947-018-2121-6.

Bahrami, A., Delshadi, R., Assadpour, E., Jafari, S. M., & Williams, L. (2020). Antimicrobial-loaded nanocarriers for food packaging applications [Review]. *Advances in Colloid and Interface Science, 278*, 12. Available from https://doi.org/10.1016/j.cis.2020.102140, Article 102140.

Bakkali, F., Averbeck, S., Averbeck, D., & Idaomar, M. (2008). Biological effects of essential oils—a review. *Food and Chemical Toxicology, 46*(2), 446–475.

Barashkova, A. S., & Rogozhin, E. A. (2020). Isolation of antimicrobial peptides from different plant sources: Does a general extraction method exist? *Plant Methods, 16*(1). Available from https://doi.org/10.1186/s13007-020-00687-1, Article 143.

Bazzaz, B. S. F., Khameneh, B., Namazi, N., Iranshahi, M., Davoodi, D., & Golmohammadzadeh, S. (2018). Solid lipid nanoparticles carrying Eugenia caryophyllata essential oil: The novel nanoparticulate systems with broad-spectrum antimicrobial activity. *Letters in Applied Microbiology, 66*(6), 506–513. Available from https://doi.org/10.1111/lam.12886.

Behbahani, B. A., Noshad, M., & Jooyandeh, H. (2020). Improving oxidative and microbial stability of beef using Shahri Balangu seed mucilage loaded with Cumin essential oil as a bioactive edible coating. *Biocatalysis and Agricultural Biotechnology, 24*. Available from https://doi.org/10.1016/j.bcab.2020.101563, Article 101563.

Bideau, B., Bras, J., Saini, S., Daneault, C., & Loranger, E. (2016). Mechanical and antibacterial properties of a nanocellulose-polypyrrole multilayer composite [Article]. *Materials Science & Engineering C-Materials for Biological Applications, 69*, 977–984. Available from https://doi.org/10.1016/j.msec.2016.08.005.

Borlinghaus, J., Albrecht, F., Gruhlke, M. C. H., Nwachukwu, I. D., & Slusarenko, A. J. (2014). Allicin: Chemistry and biological properties. *Molecules (Basel, Switzerland), 19*(8), 12591–12618. Available from https://doi.org/10.3390/molecules190812591.

Burt, S. (2004). Essential oils: Their antibacterial properties and potentialapplications in foods—A review. *International Journal of Food Microbiology, 94*, 223–251. Available from https://doi.org/10.1016/j.ijfoodmicro.2004.03.022.

Cai, L., Wang, Y., & Cao, A. (2020). The physiochemical and preservation properties of fish sarcoplasmic protein/chitosan composite films containing ginger essential oil emulsions. *Journal of Food Process Engineering, 43*(10). Available from https://doi.org/10.1111/jfpe.13495.

Campos, D. A., Madureira, A. R., Sarmento, B., Pintado, M. M., & Gomes, A. M. (2017). Technological stability of solid lipid nanoparticles loaded with phenolic compounds: Drying process and stability along storage. *Journal of Food Engineering, 196*, 1–10. Available from https://doi.org/10.1016/j.jfoodeng.2016.10.009.

Campos-Requena, V. H., Rivas, B. L., Pérez, M. A., Figueroa, C. R., Figueroa, N. E., & Sanfuentes, E. A. (2017). Thermoplastic starch/clay nanocomposites loaded with essential oil constituents as packaging for strawberries − In vivo antimicrobial synergy over Botrytis cinerea. *Postharvest Biology and Technology, 129*, 29–36.

Cazin, I., Rossegger, E., de la Cruz, G. G., Griesser, T., & Schlogl, S. (2021). Recent advances in functional polymers containing coumarin chromophores. *Polymers, 13*(1). Available from https://doi.org/10.3390/polym13010056, Article 56.

Chen, S. L., Wu, M., Lu, P., Gao, L., Yan, S., & Wang, S. F. (2020). Development of pH indicator and antimicrobial cellulose nanofibre packaging film based on purple sweet potato anthocyanin and oregano essential oil. *International Journal of Biological Macromolecules, 149*, 271–280. Available from https://doi.org/10.1016/j.ijbiomac.2020.01.231.

Demirci, M., Caglar, M. Y., Cakir, B., Gülseren, İ., & Jafari, S. M. (2017). 3 − Encapsulation by nanoliposomes. *Nanoencapsulation technologies for the food and nutraceutical industries* (pp. 74–113). Academic Press. Available from https://doi.org/10.1016/B978-0-12-809436-5.00003-3.

Dhumal, C. V., Ahmed, J., Bandara, N., & Sarkar, P. (2019). Improvement of antimicrobial activity of sago starch/guar gum bi-phasic edible films by incorporating carvacrol and citral [Article]. *Food Packaging and Shelf Life, 21*, 9. Available from https://doi.org/10.1016/j.fpsl.2019.100380, Article 100380.

Esmaeili, H., Cheraghi, N., Khanjari, A., Rezaeigolestani, M., Basti, A. A., Kamkar, A., & Aghaee, E. M. (2020). Incorporation of nanoencapsulated garlic essential oil into edible films: A novel approach for extending shelf life of vacuum-packed sausages [Article]. *Meat Science, 166*, 9. Available from https://doi.org/10.1016/j.meatsci.2020.108135, Article 108135.

Feyzioglu, G. C., & Tornuk, F. (2016). Development of chitosan nanoparticles loaded with summer savory (*Satureja hortensis L.*) essential oil for antimicrobial and antioxidant delivery applications. *Lwt-Food Science and Technology, 70*, 104–110. Available from https://doi.org/10.1016/j.lwt.2016.02.037.

Fitzgerald, D. J., Stratford, M., Gasson, M. J., Ueckert, J., Bos, A., & Narbad, A. (2004). Mode of antimicrobial action of vanillin against *Escherichia coli, Lactobacillus plantarum* and *Listeria innocua*. *Journal of Applied Microbiology, 97*(1), 104–113. Available from https://doi.org/10.1111/j.1365-2672.2004.02275.x.

Fonseca, L. M., Radünz, M., Crizel, R. L., Camargo, T. M., Gandra, E. A., & Dias, A. R. G. (2021). Effect of carvacrol encapsulation in starch-based nanofibers: Thermal resistance and antioxidant and antimicrobial properties. *Journal of Food Processing and Preservation*. Available from https://doi.org/10.1111/jfpp.15409.

Frazier, R. A., Deaville, E. R., Green, R. J., Stringano, E., Willoughby, I., Plant, J., & Mueller-Harvey, I. (2010). Interactions of tea tannins and condensed tannins with proteins. *Journal of Pharmaceutical and Biomedical Analysis, 51*(2), 490–495.

Friedman, M. (2017). Chemistry, antimicrobial mechanisms, and antibiotic activities of cinnamaldehyde against pathogenic bacteria in animal feeds and human foods. *Journal of Agricultural and Food Chemistry, 65*(48), 10406–10423.

Gao, Q., Lei, M., Zhou, K. M., Liu, X. L., Wang, S. F., & Li, H. M. (2020). Preparation of a microfibrillated cellulose/chitosan/polypyrrole film for Active Food Packaging. *Progress in Organic Coatings, 149*. Available from https://doi.org/10.1016/j.porgcoat.2020.105907, Article 105907.

Gao, S., Liu, G., Li, J., Chen, J., Li, L., Li, Z., ... Zhang, S. (2020). Antimicrobial activity of lemongrass essential oil (Cymbopogon flexuosus) and its active component citral against dual-species biofilms of Staphylococcus aureus and candida species. *Frontiers in Cellular and Infection Microbiology, 10*, 1–14. Available from https://doi.org/10.3389/fcimb.2020.603858.

Gill, A. O., & Holley, R. A. (2006). Disruption of *Escherichia coli, Listeria monocytogenes* and *Lactobacillus sakei* cellular membranes by plant oil aromatics. *International Journal of Food Microbiology, 108*, 1–9.

Giteru, S. G., Coorey, R., Bertolatti, D., Watkin, E., Johnson, S., & Fang, Z. (2015). Physicochemical and antimicrobial properties of citral and quercetin incorporated kafirin-based bioactive films. *Food Chemistry, 168*, 341–347. Available from https://doi.org/10.1016/j.foodchem.2014.07.077.

Giteru, S. G., Oey, I., Ali, M. A., Johnson, S. K., & Fang, Z. (2017). Effect of kafirin-based films incorporating citral and quercetin on storage of fresh chicken fillets. *Food Control, 80*, 37–44. Available from https://doi.org/10.1016/j.foodcont.2017.04.029.

Gniewosz, M., & Synowiec, A. (2011). Antibacterial activity of pullulan films containing thymol. *Flavour and Fragrance Journal, 26*(6), 389–395.

Gull, A., Bhat, N., Wani, S. M., Masoodi, F. A., Amin, T., & Ganai, S. A. (2021). Shelf life extension of apricot fruit by application of nanochitosan emulsion coatings containing pomegranate peel extract. *Food Chemistry, 349*. Available from https://doi.org/10.1016/j.foodchem.2021.129149.

Gülçin, İ. (2011). Antioxidant activity of eugenol: A structure–activity relationship study. *Journal of Medicinal Food, 14*(9), 975–985.

Hazer, B., & Ashby, R. D. (2021). Synthesis of a novel tannic acid-functionalized polypropylene as antioxidant active-packaging materials. *Food Chemistry, 344*, 128644.

He, Z., Huang, Z., Jiang, W., & Zhou, W. (2019). Antimicrobial activity of Cinnamaldehyde on *Streptococcus* mutans biofilms. *Frontiers in Microbiology, 10*, 2241.

Heras-Mozos, R., Muriel-Galet, V., Lopez-Carballo, G., Catala, R., Hernandez-Munoz, P., & Gavara, R. (2018). Active EVOH/PE bag for sliced pan loaf based on garlic as antifungal agent and bread aroma as aroma [Article; Proceedings Paper]. *Food Packaging and Shelf Life, 18*, 125–130. Available from https://doi.org/10.1016/j.fpsl.2018.10.007.

Heymich, M. L., Friedlein, U., Trollmann, M., Schwaiger, K., Bockmann, R. A., & Pischetsrieder, M. (2021). Generation of antimicrobial peptides Leg1 and Leg2 from chickpea storage protein, active against food spoilage bacteria and foodborne pathogens. *Food Chemistry, 347*. Available from https://doi.org/10.1016/j.foodchem.2020.128917, Article 128917.

Higueras, L., López-Carballo, G., Gavara, R., & Hernández-Muñoz, P. (2015). Reversible covalent immobilization of cinnamaldehyde on chitosan films via schiff base formation and their application in active food packaging. *Food and Bioprocess Technology, 8*(3), 526–538.

Iamareerat, B., Singh, M., Sadiq, M. B., & Anal, A. K. (2018). Reinforced cassava starch based edible film incorporated with essential oil and sodium bentonite nanoclay as food packaging material. *Journal of Food Science and Technology, 55*(5), 1953–1959.

Iqbal, A., Khan, R. S., Shehryar, K., Imran, A., Ali, F., Attia, S., … Mii, M. (2019). Antimicrobial peptides as effective tools for enhanced disease resistance in plants. *Plant Cell, Tissue and Organ Culture, 139*(1), 1–15. Available from https://doi.org/10.1007/s11240-019-01668-6.

Jafari, S. M., & McClements, D. J. (2017). Nanotechnology approaches for increasing nutrient bioavailability. *Advances in Food and Nutrition Research, 81*, 1–30.

Jafri, H., Ansari, F. A., & Ahmad, I. (2019). Prospects of essential oils in controlling pathogenic biofilm. *New look to phytomedicine* (pp. 203–236). Academic Press.

Jahdkaran, E., Hosseini, S. E., Mohammadi Nafchi, A., & Nouri, L. (2021). The effects of methylcellulose coating containing carvacrol or menthol on the physicochemical, mechanical, and antimicrobial activity of polyethylene films. *Food Science and Nutrition*. Available from https://doi.org/10.1002/fsn3.2240.

Jebali, A., Hekmatimoghaddam, S., Behzadi, A., Rezapor, I., Mohammadi, B. H., Jasemizad, T., … Sayadi, M. (2013). Antimicrobial activity of nanocellulose conjugated with allicin and lysozyme. *Cellulose, 20*(6), 2897–2907. Available from https://doi.org/10.1007/s10570-013-0084-3.

Jeyakumar, G. E., & Lawrence, R. (2021). Mechanisms of bactericidal action of Eugenol against *Escherichia coli*. *Journal of Herbal Medicine, 26*, 100406.

Jöbstl, E., Howse, J. R., Fairclough, J. P. A., & Williamson, M. P. (2006). Noncovalent cross-linking of casein by epigallocatechin gallate characterized by single molecule force microscopy. *Journal of Agricultural and Food Chemistry, 54*(12), 4077–4081.

Kaczmarek, B. (2020). Tannic acid with antiviral and antibacterial activity as a promising component of biomaterials—A minireview. *Materials, 13*(14), 3224.

Kaczmarek, B., Owczarek, A., Nadolna, K., & Sionkowska, A. (2019). The film-forming properties of chitosan with tannic acid addition. *Materials Letters, 245*, 22–24.

Kamari, A., & Phillip, E. (2018). Chitosan, gelatin and methylcellulose films incorporated with tannic acid for food packaging. *International Journal of Biological Macromolecules, 120*, 1119–1126.

Khalil, A. A., ur Rahman, U., Khan, M. R., Sahar, A., Mehmood, T., & Khan, M. (2017). Essential oil eugenol: Sources, extraction techniques and nutraceutical perspectives. *RSC Advances, 7*(52), 32669–32681.

Khameneh, B., Iranshahy, M., Soheili, V., & Bazzaz, B. S. F. (2019). Review on plant antimicrobials: A mechanistic viewpoint. *Antimicrobial Resistance and Infection Control, 8*. Available from https://doi.org/10.1186/s13756-019-0559-6, Article 118.

Khoo, H. E., Azlan, A., Tang, S. T., & Lim, S. M. (2017). Anthocyanidins and anthocyanins: Colored pigments as food, pharmaceutical ingredients, and the potential health benefits. *Food & Nutrition Research, 61*, 1–21. Available from https://doi.org/10.1080/16546628.2017.1361779, Article 1361779.

Kurek, J. (2019). Alkaloids – Their importance in nature and for human life. *Alkaloids – Their Importance in Nature and Human Life*. Available from https://doi.org/10.5772/intechopen.85400.

Lee, J. H., Won, M., & Song, K. B. (2015). Physical properties and antimicrobial activities of porcine meat and bone meal protein films containing coriander oil [Article]. *Lwt-Food Science and Technology, 63*(1), 700–705. Available from https://doi.org/10.1016/j.lwt.2015.03.043.

Lee, K. Y., Lee, J. H., Yang, H. J., & Song, K. B. (2016). Characterization of a starfish gelatin film containing vanillin and its application in the packaging of crab stick. *Food Science and Biotechnology, 25*(4), 1023–1028. Available from https://doi.org/10.1007/s10068-016-0165-9.

Liu, T., & Liu, L. (2020). Fabrication and characterization of chitosan nanoemulsions loading thymol or thyme essential oil for the preservation of refrigerated pork. *International Journal of Biological Macromolecules, 162*, 1509–1515. Available from https://doi.org/10.1016/j.ijbiomac.2020.07.207.

Liu, T., Wang, J. F., Chi, F. M., Tan, Z. K., & Liu, L. (2020). Development and characterization of novel active chitosan films containing fennel and peppermint essential oils [Article]. *Coatings, 10*(10), 10. Available from https://doi.org/10.3390/coatings10100936, Article 936.

Liu, Y. X., Song, R., Zhang, X. H., & Zhang, D. W. (2020). Enhanced antimicrobial activity and pH-responsive sustained release of chitosan/poly (vinyl alcohol)/graphene oxide nanofibrous membrane loading with allicin. *International Journal of Biological Macromolecules, 161*, 1405–1413. Available from https://doi.org/10.1016/j.ijbiomac.2020.08.051.

Lu, W. C., Huang, D. W., Wang, C. C. R., Yeh, C. H., Tsai, J. C., Huang, Y. T., & Li, P. H. (2018). Preparation, characterization, and antimicrobial activity of nanoemulsions incorporating citral essential oil. *Journal of Food and Drug Analysis, 26*(1), 82–89. Available from https://doi.org/10.1016/j.jfda.2016.12.018.

Lukic, I., Vulic, J., & Ivanovic, J. (2020). Antioxidant activity of PLA/PCL films loaded with thymol and/or carvacrol using $scCO_2$ for active food packaging. *Food Packaging and Shelf Life, 26*, 100578.

López-Gómez, A., Navarro-Martínez, A., & Martínez-Hernández, G. B. (2020). Active paper sheets including nanoencapsulated essential oils: A green packaging technique to control ethylene production and maintain quality in fresh horticultural products—A case study on flat peaches. *Foods, 9*(12), 1904.

Łopusiewicz, Ł., Kwiatkowski, P., Drozłowska, E., Trocer, P., Kostek, M., Śliwiński, M., … Sienkiewicz, M. (2021). Preparation and characterization of carboxymethyl cellulose-based bioactive composite films modified with fungal melanin and carvacrol. *Polymers, 13*(4), 1−17. Available from https://doi.org/10.3390/polym13040499.

Ma, Y., Ding, S. J., Fei, Y. Q., Liu, G., Jang, H. M., & Fang, J. (2019). Antimicrobial activity of anthocyanins and catechins against foodborne pathogens Escherichia coli and Salmonella. *Food Control, 106*. Available from https://doi.org/10.1016/j.foodcont.2019.106712, Article 106712.

Makwana, S., Choudhary, R., Dogra, N., Kohli, P., & Haddock, J. (2014). Nanoencapsulation and immobilization of cinnamaldehyde for developing antimicrobial food packaging material. *LWT-Food Science and Technology, 57*(2), 470−476.

Mani-López, E., García, H. S., & López-Malo, A. (2012). Organic acids as antimicrobials to control *Salmonella* in meat and poultry products. *Food Research International, 45*(2), 713−721.

Marchese, A., Arciola, C. R., Coppo, E., Barbieri, R., Barreca, D., Chebaibi, S., … Daglia, M. (2018). The natural plant compound carvacrol as an antimicrobial and anti-biofilm agent: Mechanisms, synergies and bio-inspired anti-infective materials. *Biofouling, 34*(6), 630−656. Available from https://doi.org/10.1080/08927014.2018.1480756.

Matos, M.J., Santana, L., Uriarte, E., Abreu, O.A., Molina, E., & Yordi, E.G. (2015). Coumarins—An important class of phytochemicals. In V. Rao (Ed.), *Phytochemicals – Isolation, characterisation and role in human health*. https://doi.org/10.5772/59982.

McClements, D. J. (2021). Advances in edible nanoemulsions: Digestion, bioavailability, and potential toxicity. *Progress in Lipid Research, 81*. Available from https://doi.org/10.1016/j.plipres.2020.101081.

McKay, S., Sawant, P., Fehlberg, J., & Almenar, E. (2021). Antimicrobial activity of orange juice processing waste in powder form and its suitability to produce antimicrobial packaging. *Waste Management, 120*, 230−239. Available from https://doi.org/10.1016/j.wasman.2020.11.048.

Mendonca, A., Jackson-Davis, A., Moutiq, R., & Thomas-Popo, E. (2018). Use of natural antimicrobials of plant origin to improve the microbiological safety of foods. *Food and feed safety systems and analysis* (pp. 249−272). Academic Press.

Merino, N., Berdejo, D., Bento, R., Salman, H., Lanz, M., Maggi, F., … Pagan, R. (2019). Antimicrobial efficacy of *Thymbra capitata (L.) Cav.* essential oil loaded in self-assembled zein nanoparticles in combination with heat. *Industrial Crops and Products, 133*, 98−104. Available from https://doi.org/10.1016/j.indcrop.2019.03.003.

Mojaddar Langroodi, A., Nematollahi, A., & Sayadi, M. (2021). Chitosan coating incorporated with grape seed extract and Origanum vulgare essential oil: An active packaging for turkey meat preservation. *Journal of Food Measurement and Characterization*. Available from https://doi.org/10.1007/s11694-021-00867-0.

Muriel-Galet, V., Cerisuelo, J. R., Lopez-Carballo, G., Aucejo, S., Gavara, R., & Hernandez-Munoz, P. (2013). Evaluation of EVOH-coated PP films with oregano essential oil and citral to improve the shelf-life of packaged salad [Article]. *Food Control, 30*(1), 137−143. Available from https://doi.org/10.1016/j.foodcont.2012.06.032.

Narayana, J. L., & Chen, J. Y. (2015). Antimicrobial peptides: Possible anti-infective agents. *Peptides, 72*, 88−94.

Navikaite-Snipaitiene, V., Ivanauskas, L., Jakstas, V., Rüegg, N., Rutkaite, R., Wolfram, E., & Yildirim, S. (2018). Development of antioxidant food packaging materials containing eugenol for extending display life of fresh beef. *Meat Science, 145*, 9−15.

Norcino, L. B., Mendes, J. F., Natarelli, C. V. L., Manrich, A., Oliveira, J. E., & Mattoso, L. H. C. (2020). Pectin films loaded with copaiba oil nanoemulsions for potential use as bio-based active packaging. *Food Hydrocolloids, 106*. Available from https://doi.org/10.1016/j.foodhyd.2020.105862.

Nostro, A., & Papalia, T. (2012). Antimicrobial activity of carvacrol: Current progress and future prospectives. *Recent Patents on Anti-Infective Drug Discovery, 7*(1), 28−35. Available from https://doi.org/10.2174/157489112799829684.

Nowotarska, S. W., Nowotarski, K., Grant, I. R., Elliott, C. T., Friedman, M., & Situ, C. (2017). Mechanisms of antimicrobial action of cinnamon and oregano oils, cinnamaldehyde, carvacrol, 2, 5-dihydroxybenzaldehyde, and 2-hydroxy-5-methoxybenzaldehyde against Mycobacterium avium subsp. paratuberculosis (Map). *Foods, 6*(9), 72.

Nowotarska, S. W., Nowotarski, K. I., Friedman, M., & Situ, C. (2014). Effect of structure on the interactions between five natural antimicrobial compounds and phospholipids of bacterial cell membrane on model monolayers. *Molecules (Basel, Switzerland), 19*, 7497−7515.

Orsuwan, A., & Sothornvit, R. (2018). Active banana flour nanocomposite films incorporated with garlic essential oil as multifunctional packaging material for food application [Article]. *Food and Bioprocess Technology, 11*(6), 1199−1210. Available from https://doi.org/10.1007/s11947-018-2089-2.

Othman, L., Sleiman, A., & Abdel-Massih, R. M. (2019). Antimicrobial activity of polyphenols and alkaloids in middle eastern plants. *Frontiers in Microbiology, 10*. Available from https://doi.org/10.3389/fmicb.2019.00911, Article 911.

Oyedemi, S. O., Okoh, A. I., Mabinya, L. V., Pirochenva, G., & Afolayan, A. J. (2009). The proposed mechanism of bactericidal action of eugenol, ∝-terpineol and g-terpinene against *Listeria monocytogenes, Streptococcus pyogenes, Proteus vulgaris* and *Escherichia coli. African Journal of Biotechnology, 8*(7).

Palaniappan, K., & Holley, R. A. (2010). Use of natural antimicrobials to increase antibiotic susceptibility of drug resistant bacteria. *International Journal of Food Microbiology, 140*(2−3), 164−168.

Pilevar, Z., Hosseini, H., Abdollahzadeh, E., Shojaee-Aliabadi, S., Tajedin, E., Yousefi, M., … Khosroshahi, N. K. (2020). Effect of *Zataria multiflora Boiss.* Essential oil, time, and temperature on the expression of *Listeria monocytogenes* virulence genes in broth and minced rainbow trout. *Food Control, 109*, 106863.

Pisoschi, A. M., Pop, A., Georgescu, C., Turcuş, V., Olah, N. K., & Mathe, E. (2018). An overview of natural antimicrobials role in food. *European Journal of Medicinal Chemistry, 143*, 922−935.

Priyadarshi, R., Kumar, B., Deeba, F., Kulshreshtha, A., & Negi, Y. S. (2018). Chitosan films incorporated with Apricot (*Prunus armeniaca*) kernel essential oil as active food packaging material. *Food Hydrocolloids, 85*, 158−166.

Qian, W. D., Yang, M., Wang, T., Sun, Z. H., Liu, M., Zhang, J. N., … Li, Y. D. (2020). Antibacterial mechanism of vanillic acid on physiological, morphological, and biofilm properties of carbapenem-resistant *Enterobacter hormaechei*. *Journal of Food Protection, 83*(4), 576−583. Available from https://doi.org/10.4315/jfp-19-469.

Qin, Y., Liu, Y. P., Yuan, L. M., Yong, H. M., & Liu, J. (2019). Preparation and characterization of antioxidant, antimicrobial and pH-sensitive films based on chitosan, silver nanoparticles and purple corn extract [Article]. *Food Hydrocolloids, 96*, 102–111. Available from https://doi.org/10.1016/j.foodhyd.2019.05.017.

Radusin, T., Tomsik, A., Saric, L., Ristic, I., Baschetti, M. G., Minelli, M., & Novakovic, A. (2019). Hybrid Pla/wild garlic antimicrobial composite films for food packaging application [Article]. *Polymer Composites, 40*(3), 893–900. Available from https://doi.org/10.1002/pc.24755.

Rauha, J. P., Remes, S., Heinonen, M., Hopia, A., Kähkönen, M., Kujala, T., ... Vuorela, P. (2000). Antimicrobial effects of Finnish plant extracts containing flavonoids and other phenolic compounds. *International Journal of Food Microbiology, 56*(1), 3–12.

Reiter, J., Hubbers, A. M., Albrecht, F., Leichert, L. I. O., & Slusarenko, A. J. (2020). Allicin, a natural antimicrobial defence substance from garlic, inhibits DNA gyrase activity in bacteria. *International Journal of Medical Microbiology, 310*(1). Available from https://doi.org/10.1016/j.ijmm.2019.151359, Article 151359.

Requena, R., Vargas, M., & Chiralt, A. (2017). Release kinetics of carvacrol and eugenol from poly (hydroxybutyrate-co-hydroxyvalerate) (PHBV) films for food packaging applications. *European Polymer Journal, 92*, 185–193.

Rostamabadi, H., Falsafi, S. R., & Jafari, S. M. (2019). Starch-based nanocarriers as cutting-edge natural cargos for nutraceutical delivery. *Trends in Food Science & Technology, 88*, 397–415.

Salem, M. A., & Ezzat, S. M. (2019). Nanoemulsions in food industry. In J. M. Milani (Ed.), *Some new aspects of colloidal systems in foods* (pp. 31–51). . Available from http://doi.org/10.5772/intechopen.79447.

Sangsuwan, J., Pongsapakworawat, T., Bangmo, P., & Sutthasupa, S. (2016). Effect of chitosan beads incorporated with lavender or red thyme essential oils in inhibiting Botrytis cinerea and their application in strawberry packaging system. *LWT, 74*, 14–20.

Sangsuwan, J., Rattanapanone, N., & Pongsirikul, I. (2015). Development of active chitosan films incorporating potassium sorbate or vanillin to extend the shelf life of butter cake. *International Journal of Food Science and Technology, 50*(2), 323–330. Available from https://doi.org/10.1111/ijfs.12631.

Sani, M. A., Tavassoli, M., Hamishehkar, H., & McClements, D. J. (2021). Carbohydrate-based films containing pH-sensitive red barberry anthocyanins: Application as biodegradable smart food packaging materials [Article]. *Carbohydrate Polymers, 255*, 9. Available from https://doi.org/10.1016/j.carbpol.2020.117488, Article 117488.

Sanla-Ead, N., Jangchud, A., Chonhenchob, V., & Suppakul, P. (2012). Antimicrobial activity of cinnamaldehyde and eugenol and their activity after incorporation into cellulose-based packaging films. *Packaging Technology and Science, 25*(1), 7–17.

Sarabandi, K., Jafari, S. M., Mohammadi, M., Akbarbaglu, Z., Pezeshki, A., & Heshmati, M. K. (2019). Production of reconstitutable nanoliposomes loaded with flaxseed protein hydrolysates: Stability and characterization. *Food Hydrocolloids, 96*, 442–450.

Sharma, S., Barkauskaite, S., Duffy, B., Jaiswal, A. K., & Jaiswal, S. (2020a). Characterization and antimicrobial activity of biodegradable active packaging enriched with clove and thyme essential oil for food packaging application. *Foods, 9*(8), 1117.

Sharma, S., Barkauskaite, S., Jaiswal, S., Duffy, B., & Jaiswal, A. K. (2020b). Development of essential oil incorporated active film based on biodegradable blends of poly (lactide)/poly (butylene adipate-co-terephthalate) for food packaging application. *Journal of Packaging Technology and Research, 4*(3), 235–245.

Sharma, S., Jaiswal, A. K., Duffy, B., & Jaiswal, S. (2020). Ferulic acid incorporated active films based on poly (lactide)/poly (butylene adipate-co-terephthalate) blend for food packaging. *Food Packaging and Shelf Life, 24*, 100491.

Shwaiki, L. N., Arendt, E. K., & Lynch, K. M. (2020). Study on the characterisation and application of synthetic peptide Snakin-1 derived from potato tubers – Action against food spoilage yeast. *Food Control, 118*. Available from https://doi.org/10.1016/j.foodcont.2020.107362, Article 107362.

Silva, F., Caldera, F., Trotta, F., Nerín, C., & Domingues, F. C. (2019). Encapsulation of coriander essential oil in cyclodextrin nanosponges: A new strategy to promote its use in controlled-release active packaging. *Innovative Food Science and Emerging Technologies, 56*. Available from https://doi.org/10.1016/j.ifset.2019.102177.

Silveira, A. C., Moreira, G. C., Artes, F., & Aguayo, E. (2015). Vanillin and cinnamic acid in aqueous solutions or in active modified packaging preserve the quality of fresh-cut Cantaloupe melon. *Scientia Horticulturae, 192*, 271–278. Available from https://doi.org/10.1016/j.scienta.2015.06.029.

Smith, A. H., & Mackie, R. I. (2004). Effect of condensed tannins on bacterial diversity and metabolic activity in the rat gastrointestinal tract. *Applied and Environmental Microbiology, 70*(2), 1104–1115.

Smith, A. H., Zoetendal, E., & Mackie, R. I. (2005). Bacterial mechanisms to overcome inhibitory effects of dietary tannins. *Microbial Ecology, 50*(2), 197–205.

Soofi, M., Alizadeh, A., Hamishehkar, H., Almasi, H., & Roufegarinejad, L. (2021). Preparation of nanobiocomposite film based on lemon waste containing cellulose nanofiber and savory essential oil: A new biodegradable active packaging system [Article]. *International Journal of Biological Macromolecules, 169*, 352–361. Available from https://doi.org/10.1016/j.ijbiomac.2020.12.114.

Sotelo-Boyas, M. E., Correa-Pacheco, Z. N., Bautista-Banos, S., & Corona-Rangel, M. L. (2017). Physicochemical characterization of chitosan nanoparticles and nanocapsules incorporated with lime essential oil and their antibacterial activity against food-borne pathogens. *LWT-Food Science and Technology, 77*, 15–20. Available from https://doi.org/10.1016/j.lwt.2016.11.022.

Sun, X. H., Zhou, T. T., Wei, C. H., Lan, W. Q., Zhao, Y., Pan, Y. J., & Wu, V. C. H. (2018). Antibacterial effect and mechanism of anthocyanin rich Chinese wild blueberry extract on various foodborne pathogens. *Food Control, 94*, 155–161. Available from https://doi.org/10.1016/j.foodcont.2018.07.012.

Sung, S. Y., Sin, L. T., Tee, T. T., Bee, S. T., Rahmat, A. R., & Rahman, W. (2014). Control of bacteria growth on ready-to-eat beef loaves by antimicrobial plastic packaging incorporated with garlic oil. *Food Control, 39*, 214–221. Available from https://doi.org/10.1016/j.foodcont.2013.11.020.

Taheri, A., & Jafari, S. M. (2019). Gum-based nanocarriers for the protection and delivery of food bioactive compounds. *Advances in Colloid and Interface Science, 269*, 277–295.

Tamer, C. E., Suna, S., & Özcan-Sinir, G. (2019). Toxicological aspects of ingredients used in nonalcoholic beverages. *Non-alcoholic Beverages: Volume 6. The Science of Beverages*, 441–481. Available from https://doi.org/10.1016/B978-0-12-815270-6.00014-1.

Tang, S. S., Prodhan, Z. H., Biswas, S. K., Le, C. F., & Sekaran, S. D. (2018). Antimicrobial peptides from different plant sources: Isolation, characterisation, and purification. *Phytochemistry, 154*, 94–105. Available from https://doi.org/10.1016/j.phytochem.2018.07.002.

Theisen, L. L., Erdelmeier, C. A., Spoden, G. A., Boukhallouk, F., Sausy, A., Florin, L., & Muller, C. P. (2014). Tannins from *Hamamelis virginiana* bark extract: Characterization and improvement of the antiviral efficacy against influenza A virus and human papillomavirus. *PLoS One, 9*(1), e88062.

Thielmann, J., Theobald, M., Wutz, A., Krolo, T., Buergy, A., Niederhofer, J., ... Muranyi, P. (2021). *Litsea cubeba* fruit essential oil and its major constituent citral as volatile agents in an antimicrobial packaging material [Article]. *Food Microbiology, 96*, 10. Available from https://doi.org/10.1016/j.fm.2020.103725, Article 103725.

Todorov, S. D. (2009). Bacteriocins from *Lactobacillus plantarum* production, genetic organization and mode of action: Produção, organização genética e modo de ação. *Brazilian Journal of Microbiology, 40*(2), 209–221.

Tonyali, B., McDaniel, A., Amamcharla, J., Trinetta, V., & Yucel, U. (2020). Release kinetics of cinnamaldehyde, eugenol, and thymol from sustainable and biodegradable active packaging films. *Food Packaging and Shelf Life, 24*. Available from https://doi.org/10.1016/j.fpsl.2020.100484.

Tucci, M., & Benghuzzi, H. (2011). Evaluation of the antimicrobial efficacy of green tea extract (egcg) against streptococcus pyogenes in vitro-biomed. *Biomedical Sciences Instrumentation, 47*, 177–182.

Tunç, S., Duman, O., & Polat, T. G. (2016). Effects of montmorillonite on properties of methyl cellulose/carvacrol based active antimicrobial nanocomposites. *Carbohydrate Polymers, 150*, 259–268. Available from https://doi.org/10.1016/j.carbpol.2016.05.019.

Valero, D., Valverde, J. M., Martínez-Romero, D., Guillén, F., Castillo, S., & Serrano, M. (2006). The combination of modified atmosphere packaging with eugenol or thymol to maintain quality, safety and functional properties of table grapes. *Postharvest Biology and Technology, 41*(3), 317–327.

Villegas, C., Arrieta, M. P., Rojas, A., Torres, A., Faba, S., Toledo, M. J., ... Valenzuela, X. (2019). PLA/organoclay bionanocomposites impregnated with thymol and cinnamaldehyde by supercritical impregnation for active and sustainable food packaging. *Composites Part B: Engineering, 176*, 107336.

Wang, L., Dekker, M., Heising, J., Fogliano, V., & Berton-Carabin, C. C. (2020). Carvacrol release from PLA to a model food emulsion: Impact of oil droplet size. *Food Control, 114*. Available from https://doi.org/10.1016/j.foodcont.2020.107247.

Wen, P., Zhu, D. H., Feng, K., Liu, F. J., Lou, W. Y., Li, N., & Wu, H. (2016). Fabrication of electrospun polylactic acid nanofilm incorporating cinnamon essential oil/β-cyclodextrin inclusion complex for antimicrobial packaging. *Food Chemistry, 196*, 996–1004.

Woranuch, S., & Yoksan, R. (2013). Eugenol-loaded chitosan nanoparticles: I. Thermal stability improvement of eugenol through encapsulation. *Carbohydrate Polymers, 96*(2), 578–585.

Wu, Z. G., Zhou, W., Pang, C. S., Deng, W. J., Xu, C. L., & Wang, X. Y. (2019). Multifunctional chitosan-based coating with liposomes containing laurel essential oils and nanosilver for pork preservation [Article]. *Food Chemistry, 295*, 16–25. Available from https://doi.org/10.1016/j.foodchem.2019.05.114.

Xavier, J. R., Babusha, S. T., George, J., & Ramana, K. V. (2015). Material properties and antimicrobial activity of polyhydroxybutyrate (PHB) films incorporated with vanillin. *Applied Biochemistry and Biotechnology, 176*(5), 1498–1510. Available from https://doi.org/10.1007/s12010-015-1660-9.

Xavier, L. O., Sganzerla, W. G., Rosa, G. B., da Rosa, C. G., Agostinetto, L., de Lima Veeck, A. P., & Nunes, M. R. (2021). Chitosan packaging functionalized with Cinnamodendron dinisii essential oil loaded zein: A proposal for meat conservation. *International Journal of Biological Macromolecules, 169*, 183–193.

Xu, F., Weng, B., Gilkerson, R., Materon, L. A., & Lozano, K. (2015). Development of tannic acid/chitosan/pullulan composite nanofibers from aqueous solution for potential applications as wound dressing. *Carbohydrate Polymers, 115*, 16–24.

Xu, J., Zhou, F., Ji, B. P., Pei, R. S., & Xu, N. (2008). The antibacterial mechanism of carvacrol and thymol against *Escherichia coli*. *Letters in Applied Microbiology, 47*(3), 174–179.

Yang, H. J., Lee, J. H., Lee, K. Y., & Song, K. B. (2017). Antimicrobial effect of an Undaria pinnatifida composite film containing vanillin against *Escherichia coli* and its application in the packaging of smoked chicken breast. *International Journal of Food Science and Technology, 52*(2), 398–403. Available from https://doi.org/10.1111/ijfs.13294.

Yousefi, M., Ehsani, A., & Jafari, S. M. (2019). Lipid-based nano delivery of antimicrobials to control food-borne bacteria. *Advances in Colloid and Interface Science, 270*, 263–277. Available from https://doi.org/10.1016/j.cis.2019.07.005.

Zhang, J., Du, C., Ailian Hu, Q. L., Peng, R., Sun, F., & Zhang, W. (2021). Inhibition mechanism and antibacterial activity of natural antibacterial agent citral on bamboo mould and its anti-mildew effect on bamboo. *Royal Society Open Science, 8*, 1–12. Available from https://doi.org/10.1098/rsos.202244.

Zhang, X., Liu, Y. P., Yong, H. M., Qin, Y., & Liu, J. (2019). Development of multifunctional food packaging films based on chitosan, TiO_2 nanoparticles and anthocyanin-rich black plum peel extract. *Food Hydrocolloids, 94*, 80–92. Available from https://doi.org/10.1016/j.foodhyd.2019.03.009.

Zhou, F., Ji, B., Zhang, H., Jiang, H. U. I., Yang, Z., Li, J., & Yan, W. (2007). The antibacterial effect of cinnamaldehyde, thymol, carvacrol and their combinations against the foodborne pathogen *Salmonella typhimurium*. *Journal of Food Safety, 27*(2), 124–133.

CHAPTER

9

3D printing of antimicrobial agents for food packaging

Sanjiv Sonkaria and Varsha Khare

Soft Foundry Institute, Seoul National University, School of Engineering, Kwanak-gu, Seoul, South Korea

9.1 Introduction

Microbial ecosystems exist in diverse environments (Canfield, Thamdrup, & Kristensen, 2005) and are unparalleled in their ability to effectively use host systems (Jansma & El Aidy, 2021) to replicate and multiply uncontrollably. However, in the sphere of human-microbial environment, defining the boundary that divides harmful bacterial populations from beneficial microbes is becoming increasingly difficult. In this context, infectious diseases and the mode of transmission of infecting agents via newly emerging species (McArthur, 2019) are an evolving health challenge to the unsuspecting human population. Protection from microbial pathogenicity and infection are long-term prospects that rely on our understanding of molecular processes (Zachary, 2017) in enabling the development of pharmaceutical agents (Cheng et al., 2016) to resist invasion from disease-causing microorganisms.

Humans are exposed to many surfaces and surface contact and transmission of pathogens indiscriminately from surface to surface through human touch (Richard et al., 2017). The lack of microbial presence to human sensing such as touch, smell, audibility, and visibility makes the human population very susceptible to attack. Conventionally, the transmission of infectious microbial contaminants is considered to occur through human-to-human contact or within a transmissible distance mediated by droplets carried through the air which is normally spread by sneezing. The frequent nature of surface-mediated transmission in clinical environments from unsterile environments and wearables, unchecked exposure from hospital personnel, research laboratories, and associated healthcare staff epitomize the diversity of mechanisms microbes utilize to invade hosts may also signify the diversity of material surfaces to which they adhere to. Mechanistic knowledge of surface attachment, detachment, and re-attachment processes although desirable requires long-term commitments to contain the spread of microbial infections within the host environment. The development of pharmaceutical agents to combat host invasion is a lengthy time-consuming, expensive, and complex process, and products may be ineffective in neutralizing the effects of highly adaptive mutational variants of new pathogenic strains. In hindsight, the ability of microbes to survive on diverse surfaces for long periods of time suggests that contaminated surfaces are a prime source for the spread of infection particularly through biofilm formation through repeated touched handling of surfaces (Bhatta et al., 2018) and persistence (Costerton, Stewart, & Greenberg, 1999).

A common route microbes use to replicate themselves is through food packaging (Rodríguez Bernaldo de Quirós, Cardamas, Sendón, & Ibarra, 2019). Common food packaging is highly susceptible to microbial attack (Fig. 9.1). Extending the shelf life of packaged foods is a growing concern both from a food preservation and health point-of-view. Food susceptibility to microbial infection can be air-borne spread through air droplets as small particles driven by air currents or different types of packaging under favorable environmental conditions (Duan et al., 2003). However, how the role surfaces play in mediating or curbing pathogens is a matter of great concern as much remains unknown about the mechanisms involved. It is probable that surfaces converge on a number of physical and chemical factors that is integral to the adhesive and bioactivate properties of

Food Packaging and Preservation.
DOI: https://doi.org/10.1016/B978-0-323-90044-7.00009-4

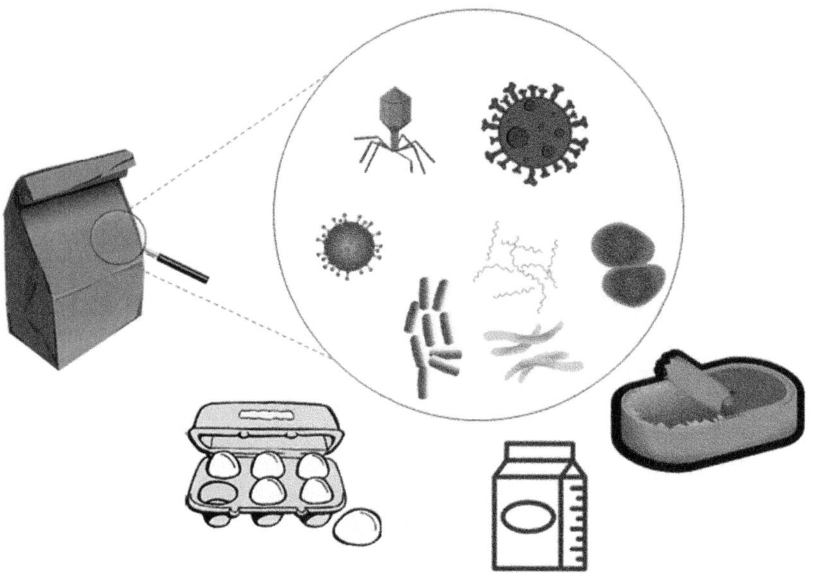

FIGURE 9.1 Schematic representation of contaminated food packaging.

microorganisms as they seek to grow and flourish into disease-spreading colonies. Multi-factors involving environmental conditions such as humidity (Lopez et al., 2013) combined with bacterial biology and surface chemistry play a contributory role in pathogenicity. However, microbes may adopt a more direct approach to contaminating food products possibly through material disintegration.

Surfaces associated with animate materials or living matter are a natural target for microbial growth as they show a tendency to correlate well with biomaterial surface composition both structurally and functionally. This often involves the biodegrading of biocomponents from released toxic products from microbes. The modern-day rationale to combat the fight against existing and emerging bugs, especially for clean and hygienic food packaging applications, is to understand and decipher the molecular basis of adhesive forces, surface charges, elasticity related cell membrane mechanics, and Tuson and adsorption via nanoscale-driven processes (Elbourne et al., 2019), which collectively promote biofilm formation. Clues to predict the design of engineered surfaces from microbial-biological associations to better formulate antimicrobial adhesion properties and prevent or minimize biocontamination are increasingly being sought. While it is proving to be an impossible task to police and routinely disinfect all surfaces susceptible to contamination, a more effective approach would involve the initial prevention of surface contamination after accidental transfer by effectively intervening in adhesion processes to avoid bacterial colonization and microbial spread. Contamination of medical devices poses a considerable risk in the spread pathogen mediated disease. Foreign body-related infections (Brounstein, Yeager, & Labouriau, 2021) are often mediated through surface-associated equipment and easily released via contact with implants. Table 9.1 summarizes commonly found infection-causing agents on contaminated medical devices. Invasion of human organs caused by implants contaminated with foreign bodies is leading to exploratory questions about adhesion processes and the key molecular factors involved that allow microbes to bind inanimate and animate objects with ease. However, different surfaces likely exhibit varying degrees of interaction with microbes with variable degrees of transmission through these associations. This scenario presents new opportunities for existing or emerging technologies to fabricate and supply innovative materials to immobilize, inactivate and destroy disease-causing particles under ambient conditions.

The lack of material resistance to microbes towards the containment of food and their exceptional nature to adapt to diverse environments via their protective capsule poses an engineering challenge to penetrate the sophisticated protective surface layer. However, the problem is considerably more deeply rooted than material composition alone. Since biomaterials are essentially animate, chemistry-driven processes are often secondary to bio-based assessment of their surroundings where the simplest of organisms can intelligently distinguish, select and thrive between the vast spectrum of materials they encounter in the biosphere. This entails the ability of pathogens to mechanistically adjust to environmental cues received by an array of signaling factors prevalent in nature through smarter interaction. Directed growth in this way by decoding and elucidating signals from external stimuli responses (Fig. 9.2), biomimetic materials hold the potential to alter the behavioral properties of microbes to reduce or abolish their pathogenic effects. For example, it has been proposed that electroactive materials may be applied as sensory materials to manipulate microbial behavioral characteristics (Fernandes, Carvalho, & Lanceros-Mendez, 2019).

TABLE 9.1 A summary of fabricated medical implants composed of material types listed against susceptible infections.

Medical implants	Types of materials	Infection causing agents
Orthopedic devices	Stainless steel, cobalt-based alloys, titanium, silicone, polyethylene, polypropylene, and polymethyl methacrylate; aluminum oxide and calcium phosphates.	*Acinetobacter pittii*, *Enterobacter cloacae*, *Micrococcus luteus*, *Staphylococcus epidermidis*, and *Staphylococcus hominis*.
Urologic devices	Polytetrafluoroethylene, rubber, polyurethane, polyamide, silicones, and polyhydroxyalkanoates	*Staphylococcus aureus* and *Staphylococcus epidermis*.
Prosthetic heart valve	Titanium, graphite, pyrolytic carbon, polyethylene, polypropylene, polyamide, and diamond-like carbon	*Mycobacterium tuberculosis* and *hepatitis B virus*.
Bone-anchored hearing systems	Titanium, platinum, aluminum oxide, silicone, teflon, polyethylene, and polyimides.	*Staphylococcus epidermis*
Atheoplasty devices	Cobalt-based alloys, titanium, zirconium, nickel, and ultra-high molecular weight polyethylene.	*Escherichia coli*, *S. aureus*, and *Staphylococcus epidermidis*
Bone allograft	Calcium phosphate ceramic, calcium sulfate, hydroxyapatite, bioactive glasses, and magnesium.	*Hepatitis*, *tuberculosis*, and *human immunodeficiency virus*.
Contact lenses and corneal implants	Silicone hydrogel and polymethyl methacrylate.	*S. epidermidis*, *E. coli*, *P. aeruginosa*, *S. aureus*, *Proteus* spp., *Serratia* spp., and *Candida* spp.
Breast implants	Silicone gel within silicone rubber envelope and inflatable saline.	*S. aureus*, *Enterococcus* spp., *S. epidermidis*, *P. acnes*, and *diphtheroids*
Dental implants	Acrylic resin, titanium and its alloys, zirconia, silver and silver nanoparticles, and ZnO.	*Veillonella* spp., *F. nucleatum*, *A. naeslundii*, *Streptococcus* spp., *C. albicans*, *S. sanguinis*, *P. gingivalis*, *E. timidum*, *E. brachy*, and *P. anerobicus*.

Modified with permission from **Sultana, A., Zare, M., & Luo, H.** (2021). *Surface engineering strategies to enhance the in situ performance of medical devices including atomic scale engineering.*

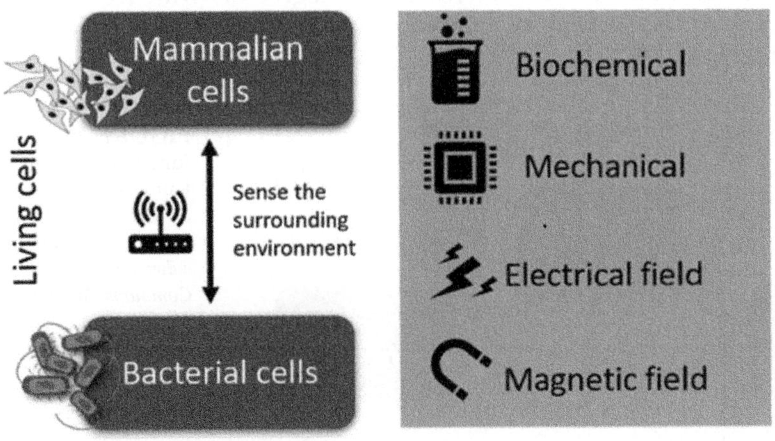

FIGURE 9.2 The use of external sources to control the behavior of cells. *Source: Reproduced with permission from Fernandes, M. M., Carvalho, E. O., & Lanceros-Mendez, S. (2019). Electroactive smart materials: Novel tools for tailoring bacteria behavior and fight antimicrobial resistance.* Frontiers in Bioengineering and Biotechnology, *7*, *277. This work is licensed under a Creative Commons Attribution 4.0 International (CC BY 4.0).*

More recently, researchers have increasingly focused on the 3D printing industry to re-invent antimicrobial materials transpiring at the intersection of important disciplinary fields merging biological and synthetic materials with existing technologies. While conventional approaches have relied on materials with intrinsic antimicrobial properties such as metals and natural biomaterials, micropatterning of surfaces (Mann et al., 2014) is opening up new ways to correlate the anti-adhesive properties of microbes to distinguish and discriminate against the binding signature patterns that may prevail in the microbial world. Indeed, the effects of patterned dimensions relating size, geometrical topographies at the micro level, and surface texture like roughness (Schumacher et al., 2007), and possibly other surface features, can inhibit microbial binding. The effectiveness of antimicrobial materials is now being understood outside the realm of conventional structures which brings into focus surface engineered materials using advanced approaches such as additive manufacturing or 3D printing approaches. In principal, this technology promises to directly impact the ability of microorganisms to colonize surfaces using different material compositions, architectures and shapes. In this chapter, we aim to introduce concepts and discuss 3D printing for applications in food packaging safety.

9.2 Understanding microbial surfaces for rationalizing adhesion processes

Pathogenic microbes by design are surface-adapting entities that live and grow close to the human environments relies on the premise that community-driven colonization facilitated by a surface attachment is a prerequisite to occupying and attainment by growth invasion of its surroundings. Materials and associated chemistries that regulate and determine the degree of adhesion in microbes resulting in both attachment and detachment to surfaces remain to be clearly understood and pose a challenge.

In order to manufacture antimicrobial packaging materials from a knowledge base of effective antimicrobial compositional materials, a better understanding of the molecular interactions that occur between biomolecules (animate) and inert (inanimate) surfaces and the factors that favor their adsorption. In hindsight, exploring surface-active proteins from living sources primarily used by fungi, bacteriophages, and bacteria can be an important source of molecular information from the perspective of understanding intricate chemical interactions that may be complex in nature. Essential questions center on the chemical and biochemical nature of surface binders and the assistance or recruitment of components that may be ubiquitous in nature like water and air.

Decade-old studies have implicated varying topographical signatures as recognition sites for anchoring cells and the size and shape of interacting topographies are capable of aiding cells in modulating metabolic and self-replication (Recum et al., 1996). Such topologies operating at the smallest scales are now considered to give rise to forces (mechanical, electrostatic, Van der Waals, etc.) over a broad range of magnitudes spanning adhesion forces from 10 nN−10 mN (Alam, Kumar, & Varadarajan, 2019) at interfaces with microbes generating precision binding and direction the magnitude of which increases the degree of virulence (Busscher & Mei, 2012). Fig. 9.3 correlates adhesion forces with the severity of virulence which can help predict and screen for colonizing biofilms. The mechanisms that mark the progression from weak to strong adhesion and subsequent acceleration and

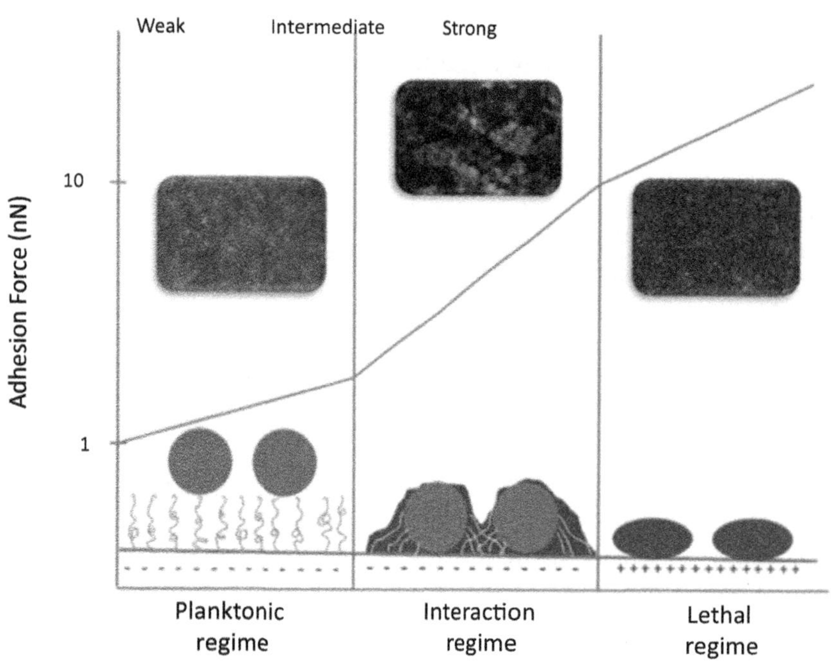

FIGURE 9.3 Adhesion force is defined as a function of weak, intermediate, and strong virulent effects. *Source: Reproduced from Busscher, H. J., & Mei, H. C. V. D. (2012). How do bacteria know they are on a surface and regulate their response to an adhering state?* PLoS Pathogens, 8(1), 1−3. *Creative Commons Attribution License.*

the precipitation of fouling are largely unknown. However, some insight into the microbial surface sensing processes from the initial planktonic state to initiating settlement via signals triggering gene activation to enhance the production of extracellular components for attachment and colony growth.

Central to these queries lie the identity of proteins that co-form into multifunctional microbial coats and mature into biofilm layers at the interfacial boundaries through systematic processes. Well-studied examples of candidate structures are fungal-derived hydrophobins (Berger & Sallada, 2019) and their potential to immobilize cells. Indeed, amphiphilic microbes employ a specific class of hydrophobin proteins to alter the hydrophobicity of the surfaces to enable conditions for cell growth and have been shown to be involved in the aggregation after adsorption (Ley, Christofferson, Penna, & Winkler, 2015). Fig. 9.4 depicts two aspects that begin at the level of microscopic organelles that enable interaction with surfaces adhesion and attachment and are underlined by chemical and topographical considerations resulting in contaminated surfaces through microbial attachment of adhesive proteins. However, microbial adhesive proteins are being reinvented as chimeric structures that combine the adhesive properties of such protein scaffolds with antimicrobials. One example of this strategy which has potential applicability in antimicrobial food packaging involved humanizing a class I hydrophobin protein with a human antimicrobial peptide (Sorrentino et al., 2020). Effective approaches in the future will require the use of materials that can prevent the attachment of proteins that have exclusive surface-binding properties.

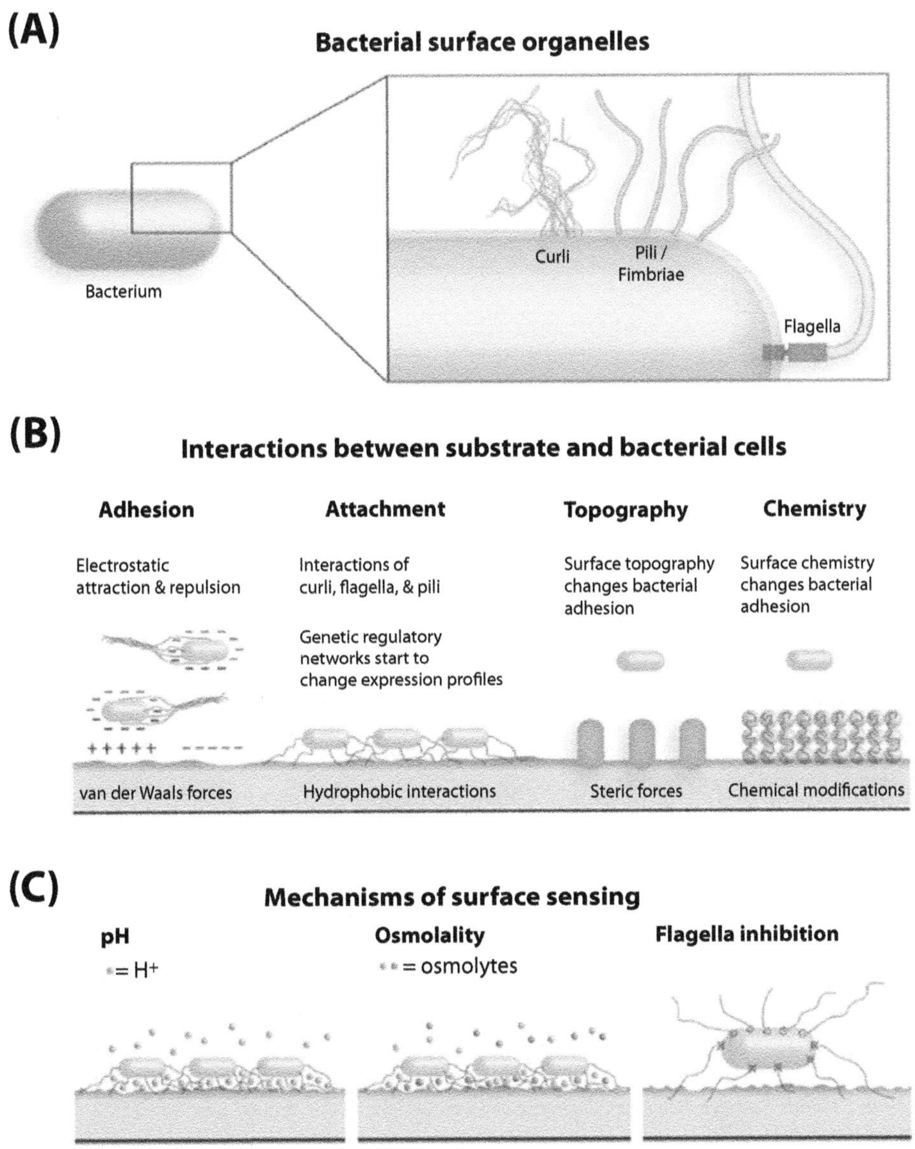

FIGURE 9.4 (A) Bacterial surface organelles enabling (B and C) mechanisms via chemical and topographical processes for attachment and adhesion to surfaces that utilize van der Waals, hydrophobic interactions, steric and chemical interactions. *Source: Reproduced with permission from Tuson, H. H., & Weibel, D. B. (2013). Bacteria-surface interactions.* Soft Matter, *9(17).*

9.3 Conceptualizing "New" materials antimicrobial packaging agents

Microbial persistence is a recognized phenomenon and is well correlated to biofilm formation on surfaces including food packaging. The survival and successful colonization of bacterial populations are determined by the chemical and physicochemical nature of biofilms that form a heterogeneous multiplex of biopolymerized layers from surface and extracellular active secretions. The biofilm supports bacterial growth through the supply of extracellular polymeric substance enabling attachment, growth, and dispersal (Fig. 9.5). The integration of material design with material properties has often been viewed as two separate unrelated aspects. Materials found in nature possess inherent characteristics that from a food hygiene perspective may or may not promote bacterial growth. Biological materials have been a natural choice to combat microbial pathogenicity due to their smart responsive nature to pathogenic attacks. Central to their smart antimicrobial activity can be ascribed to complex shape-responsive or mutable attributes which have the capabilities to change shape and size under different stimulatory environments, exhibit hierarchical nanoscale or sub-nanoscale dimensions, show penetrable or non-penetrable nature, have the ability to assemble and disassemble from other surface contacts and have mechanical robust performance but biodegradable. Animate entities use a diverse range of antifouling mechanisms to discourage colonizers from attaching to surfaces in physical ways to reverse adhesion, secretory antifouling agents and substances and surfaces that act as repellents (Yan, Wu, Yu, Zhao, & Liu, 2020). Shark skin is an example of a self-cleaning naturally occurring biomaterial that optimally utilizes the characteristics of wettability and hydrophobicity (Kong & Kim, 2013).

Fundamental to these collective processes that allow such materials to occupy unique functional roles particularly as defense is to recognize and immobilize unwanted or harmful entities via molecular adhesive processes and engage in their safe and complete removal from vulnerable attack. Processes that enable biologically formulated materials to be cytotoxic toward microbes through different mechanisms while remaining essentially inert to host components and operations is a leading challenging aspect. Incorporating these type of behavioral similarities relate to lessons in understanding biomolecular chemistry such as disulfide hinge-like structures and metal coordinated aligned to bioscaled geometries which impart flexibility to shift between different bond-driven flexible or non-flexible states (Chou et al., 2015).

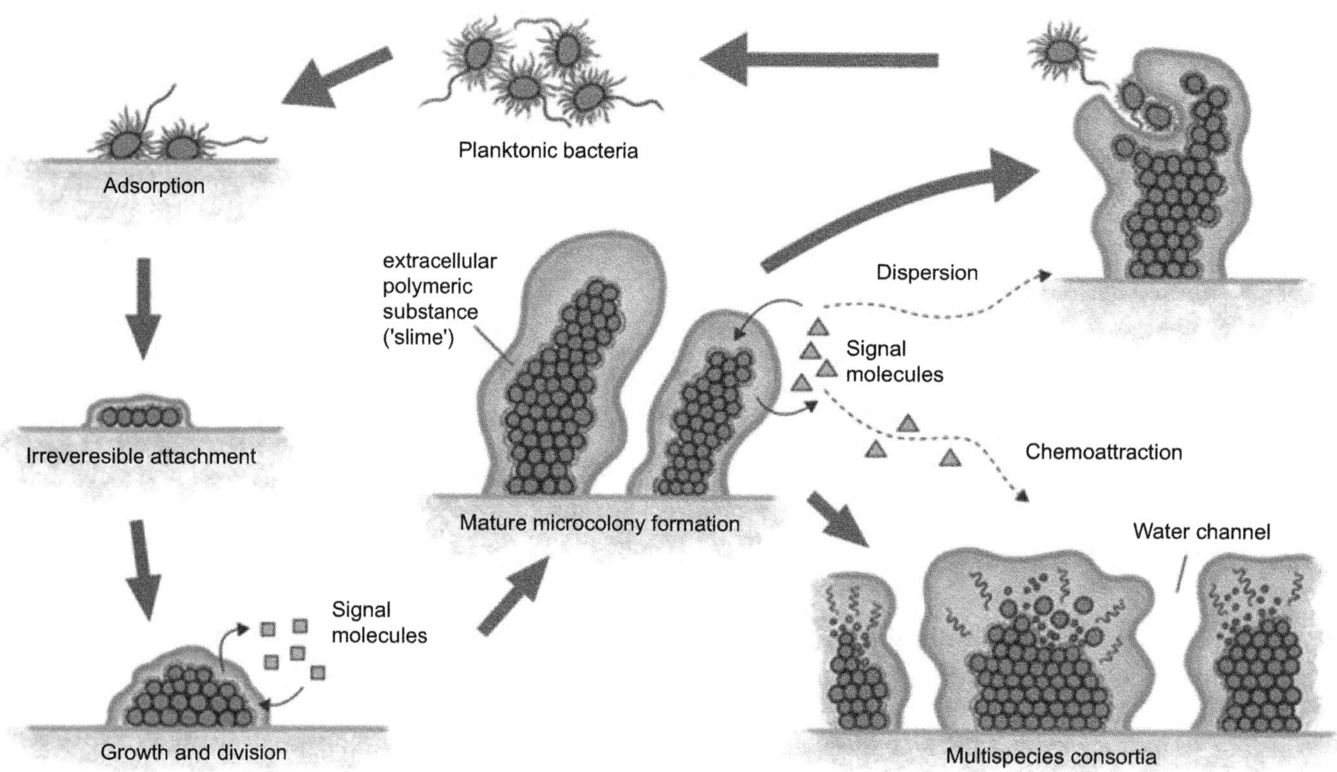

FIGURE 9.5 Evolution of biofilms. *Source: Reproduced with permission from Jacques, M., Aragon, V., & Tremblay, Y. D. N. (2010). Biofilm formation in bacterial pathogens of veterinary importance. Animal Health Research Reviews/Conference of Research Workers in Animal Diseases, 11(2), 97–121 under Creative Commons CC BY. Licensed under Creative Commons CC BY.*

A prerequisite to bacterial invasion is the formation of effective biofilms forming an adhesive polymeric matrix comprising layers of aggregates largely composed of polymerized sugar monomer units having both structural and functional significance. These structures perform key functions and define their capability as surface-adhering molecules demonstrating affinity to polymeric materials on surfaces which vary from species to species (Midelet & Carpentier, 2002). While adhesion processes used by microbes are mediated by natural biopolymers usually in the form of polysaccharides by different mechanisms (Fig. 9.5), synthetically designed polymers that resemble natural polymers in structure and function with added tunable features incorporated in food packages with broad biocidal and food-friendly properties reducing biofilm accumulation. The first line of defense for the implementation of effective antibacterial food packaging is to resist the build-up of densely populated biofilms. Since polymers are core materials that likely regulate adhesive interfacial properties, they are natural targets for controlling the effects of colonization during biofilm assembly by understanding the nature of physicochemical surface-to-surface interactions. Surfaces immobilizing key functional biopolymer units such as polynucleotides, peptides, and proteins. Another strategy might involve employing repulsive forces in the form of electrostatic repulsive charge reducing the affinity of interactions or wetted surfaces that exhibit water-repelling hydrophilic character by avoiding bridging through hydrogen bonding networks.

Some studies have suggested that the host-guest properties between the adhesive nature of microbes relative to host contact surfaces may be independent of surface roughness and more aligned with surface energies. The authors described that adhesion arises from the participation of electron donor (g^-) and electron acceptor (g^+) Lewis acid-base sites expressed in mJ/m^2 of the free energy surface quantified by the Van der Waals interactions, g^{AB}. The study concluded that electron acceptor states residing at the surface of the bacterium in the solid phase were found to positively correlate to adhesion properties of the microbes investigated for a range of polysaccharide natural and synthetic surfaces relative to (Prokopovich & Perni, 2009). Interestingly, borosilicate glass functionalized with a bovine serum protein layer to mimic the natural roughness of bio-surfaces provided a surface inducing a hydrophilic character and supported microbial adhesion of hydrophilic bacteria in which interfacial surface complexation was mediated by bonds that were electron-donating to the recipient strain (Preedy, Perni, Nipiĉ, Bohinc, & Prokopovich, 2014). This result was important to understanding the role of masking primary layers to tune the behavior of surfaces to regulate cell attachment, adhesion, and orientation for further manipulation.

The study also concluded that roughness was not a determining factor in the adhesion process. The contribution of surface roughness however as an adhesive factor is complicated by micro and nanoscale topographic patterning of surfaces and demonstrates nonuniversal effects. In this way, bacterial populations show variable affinities to different topographical impressions at small scales giving rise to preferential binding appropriate to their shape and size. Hence, physical restraints in dimensions, shape, and surface height/depth (Perera-Costa, Bruque, González-Martín, Gómez-García, & Vadillo-Rodríguez, 2014) can alone hinder surface binding and can be implemented as a strategy to potentially discourage microbial settlement's. For example, synthetic surfaces have revealed differences in the interaction of a single homogenous material subject to a grain refinement process compared to the unmodified surface such as TiO_2 at the synthetic-bio interface. In this study, *Staphylococcus aureus* and *Pseudomonas aeruginosa* showed increased sensitivity to the size and skewness of the TiO_2 surface after refinement relative to the less refined surface. Fig. 9.6 shows the different biocidal mechanisms microbes may use to invade host surfaces.

9.4 Bactericidal biomimetic surfaces: clues to modeling antibacterial nanostructured surfaces

The outermost patterned surfaces on microscopic organisms that exist at the micro-nanoscale are formed largely by biopolymers usually in the form of polysaccharides the arrangement of which marks the onset of biofilm formation and antifouling behavior. Bacterial surfaces and their adhesion behavior towards different polysaccharides and the structural arrangements of biopolymer components of repeating units as linear or highly branched chains. However, the adhering or non-adhering ability of bacteria to food packaging relates to the characteristics of the biofilm including its structural properties, binding configuration, and alignment of self-assembling polysaccharides to host polysaccharide surfaces resulting in the formation of a biofilm matrix.

Science is looking toward nature focusing on organisms, insects, plants, and animals in an attempt to model synthetically driven antifouling strategies in the form of new technologies. Some bioinspired designs are shown in Figs. 9.7 and 9.8. At the more intricate level through a microscopic lens, surfaces found in broad environments embody an array of textured exteriors purposely fabricated at the genetic level as barriers against fouling. Lessons from nature reveal a multitude of smart structures with the ability to switch reversibly between by

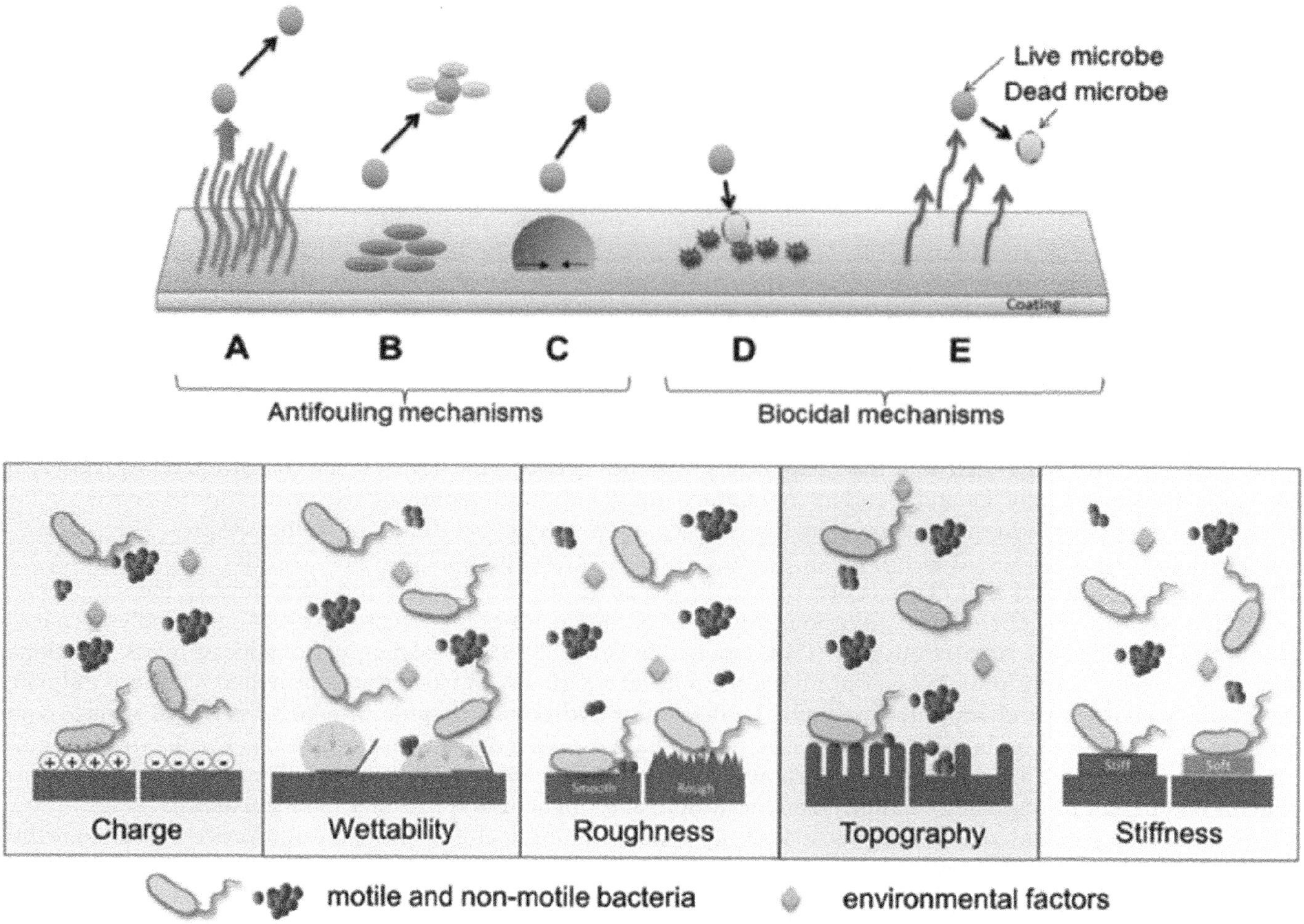

FIGURE 9.6 A schematic representation of possible biocidal mechanisms at the host-microbial surface (Babutan, Lucaci, & Botiz, 2021). *Under Creative Commons CC BY License. Creative Commons CC BY License.*

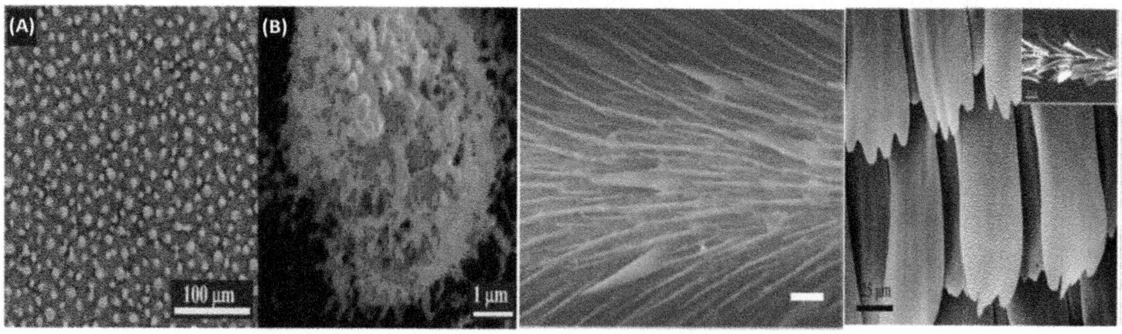

FIGURE 9.7 Nature materials: Targets for biomimetic design. An image of the superhydrophobic leaf pictured in (A) low and (B) high magnification. (C) Water-repellent water strider microsetae with nanogrooves inclined 20°C. *Source: (A, B) Reproduced with permission from Feng, B. X., & Jiang, L. (2006). Design and creation of superwetting / antiwetting surfaces.* Advanced Materials, *18, 3063–3078, (C) Reproduced with permission from Xuefeng, G., & Jiang, L. (2004). Water-repellent legs of water striders.* Nature, *432, 36.*

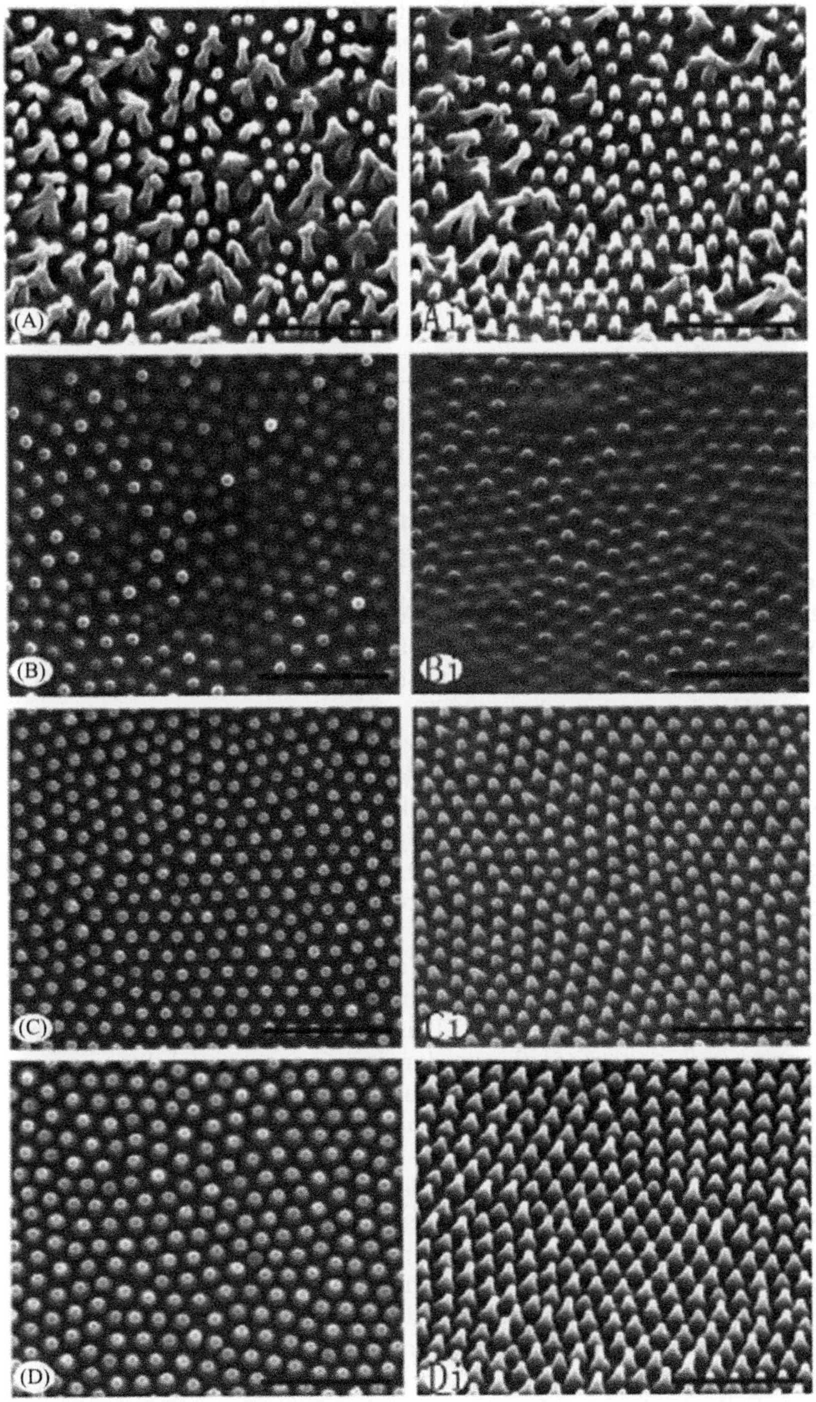

FIGURE 9.8 SEM images showing species-dependent variability of (A–D) four cicada wing membrane types and their comparison to 30°C tilted structures (A_1–D_1). *Source: Reproduced with permission from Sun, G., Fang, Y., Cong, Q., & Ren, L.-Q. (2009). Anisotropism of the non-smooth surface of butterfly wing.* Journal of Bionic Engineering, *6(1), 71–76.*

controlling surface micro- and nano architectures including wetting behavior (Feng & Jiang, 2006). In this context, superhydrophobic structures have been a source of bioinspiration using the material base as a boundary for repelling a microbial attack by interconnecting major parameters like roughness, adhesion, and superhydrophobicity by repelling water (roll-off) by employing epicuticular wax crystalloids applicable to the lotus effect (Barthlott, Neinhuis, Verlot, & Schott, 1997). Further examples are the legs of the water strider comprising tiny microsetae in the shape of nanogrooves with the ability to eject water away from its surface with a volume exceeding many times its own leg volume (Xuefeng & Jiang, 2004). The anisotropic wetting property of superhydrophobic butterfly wings exhibits directional cleaning evidenced by large contact angle at the wing-water interface (Sun, Fang, Cong, & Ren, 2009). Cicada wings are yet another example of anti-wetting materials composed of membranes with nanostructures aligned along hexagonal close-packed structures which display low-surface-energy membranes (Watson, Myhra, Cribb, & Watson, 2008). Species variants have shown distinguishable anti-wetting properties exhibiting both hydrophilic and variable hydrophobic characteristics that differ on the hydrophobicity scale from weak to strong. The irregularity here signifies differences in material order, organization, and chemical nature which was found to correlate with patterning, spacing, and height of varying wing morphologies (Sun et al., 2009).

9.5 Antimicrobial materials: a technological hurdle at the intersection of innovation and materials complexity?

While material organization considers a hierarchical order of assembly as it relates to function, around the same time others ascribed more general properties for the use of polymers as antimicrobial agents. Such polymer characteristics have been identified with reference to the architecture of the bacterial membrane to establish the workability and robustness of polymeric materials based on peptides. Some of the physicochemical properties identified here include the need for strongly adhesive cationic polymers (Grandgirard, Poinsot, Krespi, Nénon, & Cortesero, 2002) combined with hydrophobic elements with the ability to disrupt membranes but deliver a targeted response discriminating between microbial and mammalian cell types (Matsuzaki, 2009). Broader still is the notion that polymers must be fabricated with relative ease, administer potent and sustained antimicrobial activity by acting to reduce prolonged exposure of foods to microbes, and be able to operate via mechanisms that do not generate secondary toxic products (Kenawy, Worley, & Broughton, 2007).

Much of the commercialized antimicrobial products in food packaging have relied on gas phase materials which have included ethanol, chlorine, and sulfur dioxide or aqueous metals like silver ions and nonbiodegradable low-density polyethylene, linear low-density polyethylene, high-density polyethylene, polypropylene, ethylene vinyl acetate, polystyrene, polyethylene terephthalate, polyvinyl chloride, biodegradable materials polycaprolactone, polylactic acid, and thermoplastic starches (Sung et al., 2013). However, fabrication processes that require heating are not compatible with some biodegradable polymers resulting in a substantial loss of efficacy.

The developmental aspect of antimicrobial food packaging is challenging and these challenges at present remain embedded in disengaging multiple events governed by multiscale parameters that utilize different physicochemical concepts which merge into a problem of immense complexity. This has been highlighted by Zeng et al. in a review in which they emphasized aspects such as surface charge density, surface wettability, surface roughness, surface topography, surface stiffness, bacterial motility, biofilm dynamics, and environmental factors (Zheng et al., 2021). The convergence of multiple parameters which in combination blur our understanding of microbial adhesion and biofilm formation must be considered in its entirety in order to design effective antimicrobial surfaces. Conclusions that support the superior performance of multiple antimicrobial films working in synergy with each other are indicative of the cooperation of several mechanisms or more required to combat the growth of biofilms (Joerger, 2007). This finding corroborates the idea of using multi-material to control and inhibit biofouling by employing multiple mechanisms and the need to overcome technological hurdles to their fabrication. Hence, our ability to decipher and control important parametric influences by tuning material properties will likely have significant consequences in improving antibacterial performance with desired outcomes.

With the advent of 3D printing technology, antimicrobial films increase the scope of tuning properties through the implementation of material combinations from both synthetic and natural sources. Considering the structural reinforcement of key components among different materials influencing microbial behavior, technology can play a crucial role in the assembly of chemically and structurally adaptable material configurations to adequately mimic antimicrobial properties, particularly a lower dimension where microbes are able to penetrate and exert considerable biological damage by gaining from surface that aids microbial attachment and growth. In line with this philosophy, a prerequisite to the next generation of antimicrobials is to address multiple surface considerations as functional scaffolds for food packaging. Merging of 3D printed scaffolds layer-by-layer and their homogenous distribution in food packaging via additive manufacturing processes and particularly their availability as filler materials (Cámara-Torres et al., 2021) provides an appreciation of the broad diversity of antimicrobial functionalities that can be generated using this technology.

9.6 3D printing of antimicrobial materials

The current concern of food packaging technology is the incorporation of biocompatible antimicrobial agents which are both cost-effective, environmentally friendly and safe. The treatment of food packaging externally with antimicrobial sprays is the most widely used form of sterilization of containers. Over-utilization may be toxic while under-utilization of antibacterial cleansers may unknowingly favor biofilm accumulation, increasing the rates of infection. A parallel consideration is one that is fundamentally important to the design of smart antimicrobial food packaging. This rests on using a methodology that is sufficiently advanced to technologically drive the fabrication of antifouling materials. Since microbes require surfaces for biofouling during colonization, the shape, texture and geometry of surfaces must be taken into coideration. Melting materials into shapes and patterns have been the hallmark of hot melting extrusion processes for melting pharmaceutical polymers (Maniruzzaman, Boateng, Snowden, & Douroumis, 2012) and creating polymer blends using heat and friction by designing the polymer matrix. The extrusion of polymers at high temperatures is a highly sought route to print nonadhesive surfaces against pathogens. The versatility of using extruded materials under high pressure using software-driven prototyping of multi-materials without creating a physical mold but instead customizing 3D models which can be sliced into layers of "printable" information is termed 3D printing. The fabrication of geometrically defined materials involves the deposition of multilayered materials by the process of additive manufacturing in a layer-by-layer sequence with defined precision. Hence, engineering purpose-built materials remain at the forefront of antimicrobials that may be implemented in food packaging technologies. Here, we describe some of the proofs-of-concept that show the potential as antimicrobial materials.

3D printing offers an attractive route against microbial biofouling as materials can be tailored by adapting structural patterns that discourage biofouling. This was made possible due to the advancement in printing quality substantially improving the resolution of the imprint design at small scales. Since features relating to shape, size, and dimensions of engineered surface scaffolds must be coupled to manufacturing techniques to introduce complex morphologies at the boundary of micro- to nanoscale which holds the potential to design new materials at newly emerging material interfaces. Multifaceted structures through successive layering or structures with dominating patterns can be fabricated using additive manufacturing as a principle tool to generate functional antibacterial materials. The performance of such materials will be determined by the nozzle printing accuracy and the mechanical-driven precision of patterning, texture, and roughness.

3D printing of multi-materials opens up the possibility of designing an array of complex associations of functionally different material types through the customization of digitally created blueprint. This can be particularly useful for combining materials that are inherently antimicrobial in nature demonstrating bio-functionality-like behavior. Here, in the context of new materials development for cleaner food packaging, the widespread importance of the hydrogen bond and its importance in the complexity of interactions that define its energy and bond contribution in dispersed systems, electrostatic, charge-transfer, bond cooperativity, and repulsion/attraction (van der Lubbe & Fonseca Guerra, 2019) can be better aided by 3-dimensional plots and understanding orbital representations can be modeled, therefore, supported by 3D printing technology. In this instance, an alternative to quantum mechanics for mimicking orbital shapes involving hydrogen-bonded associations has been created by 3D printing physical models in a layered pattern from independent modeling tools (Griffith, De Cataldo, & Fogarty, 2016). The method provides a way to visually forecast the different energy terms into a visible construct and the approach may be extended to more complex interfaces. Up until now, large-scale properties of materials originating from 3D printing/additive printing techniques have been explored well. With greater relevance to the micrometer level measurements of micrometer-sized photopolymer droplets, the authors used polyjet printing methodology to develop new composites and examined heterogeneous architectures in relation to layered architectures and interfacial patterns printed along different orientations (Zorzetto et al., 2020). Although the study focused on the mechanical properties of the polymer materials, the importance arises from its correlation with printing techniques which considerably impact the effectiveness of materials against microbial habitats and may lead to the accessibility of material composites of novel design. This can be particularly true for printing new and more complex geometries by overcoming the limitations of physical parameters such as viscosity which is intimately linked with polymer molecular weight of those polymers that are of interest to antifouling food packaging (Scott et al., 2020). Fig. 9.9 demonstrates the reliance on print head orientation along the printing axis which further determines the enhancement or reduction in mechanical parameters such as strength (Lumpe, Mueller, & Shea, 2019).

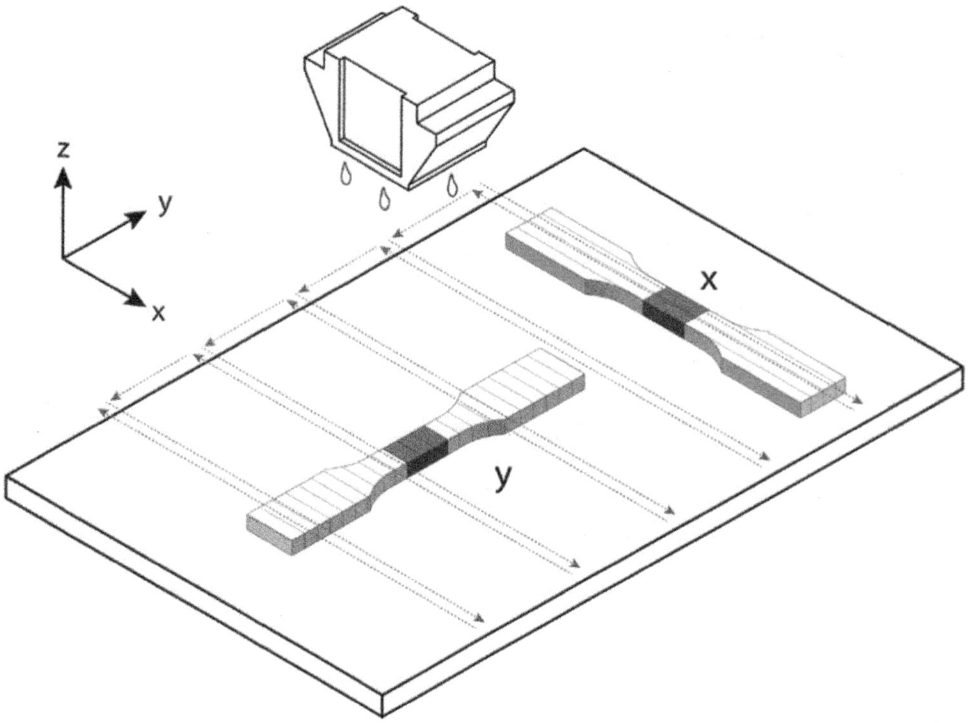

FIGURE 9.9 Printing orientation of materials and manipulation of print head movement to control anisotropy and material properties. *Source: Reproduced with permission from Lumpe, T.S., Mueller, J., & Shea, K. (2019). Tensile properties of multi-material interfaces in 3D printed parts.* Materials and Design, 162, *1–9.*

More recently 3D printing platforms are being explored to fight the build-up of bacterial biofilms by developing materials that have the intrinsic capability of resisting bacterial biolfim build-up without the need for antibiotics and other drugs as part of the formulation. Here, two bioinks differing substantially in moduli properties governing the mechanical properties were combined as a co-printable formulation. The resulting moduli of bioink blend was coded by the relative ratio of the inks in the final mix. This approach gave rise to differing chemical complexities and was modeled by their voxel properties comprising of pixels in the x, y, and z directions, equally distributed along the printing length. The complex interfacial ratios were mapped by a given ratio of the polymers in the bioink tailored to generate the desired voxel positioning in its chemical and mechanical make-up to be as effective as possible as antimicrobial agent (He et al., 2021). The innovative approach is summarized in Fig. 9.10. Another recent study has revealed the potential of 3D printing to control the evolution of different morphologies by modulating material phase separation exemplified by a photosensitive resin. By enabling parametric control over gelation, diffusivity, and kinetics of material phase progression by laser manipulation, key resin components of the system can drive the dominance of specific morphologies. The resulting structures are associated with varying functionalities including antimicrobial agents among others. The process based on polymerization-induced phase separation (PIPS) (Fig. 9.11) has permitted 3D architectures to be manipulated at the nanoscale or indeed accelerating parameters to drive macroscale geometries leaving open the possibility of fabricating surfaces that hinder the adhesion of microbes (Deore et al., 2021). Hence, rationalizing this approach in food packaging may have a significant effect on hygiene food preservation.

On the theme of mechanical properties, hydrogel mimics have been targeted for improving their mechanical properties as antibacterial agents. The incorporation of natural biopolymers such as cellulose nanofibers (CNC) have the ability to enhance the mechanical properties of hydrogels particularly in the presence of photopolymerizable monomers like N-isopropyl acrylamide (NIPAM) and photoinitiator. This element was reasoned to provide a structural constraint on material durability improving the controlled manipulation of the material during printing (Fourmann et al., 2021). The process of fabrication is summarized in Fig. 9.12 from ink formulation to the creation of functional hydrogel. The orientation of components of the hydrogel is by virtue of the CNCs that provide not only structural integrity but also direction to the overall assembly. Besides, implementing a structural role, anisotropic swelling of the hydrogel enables a directional force to the swelling nature of the material.

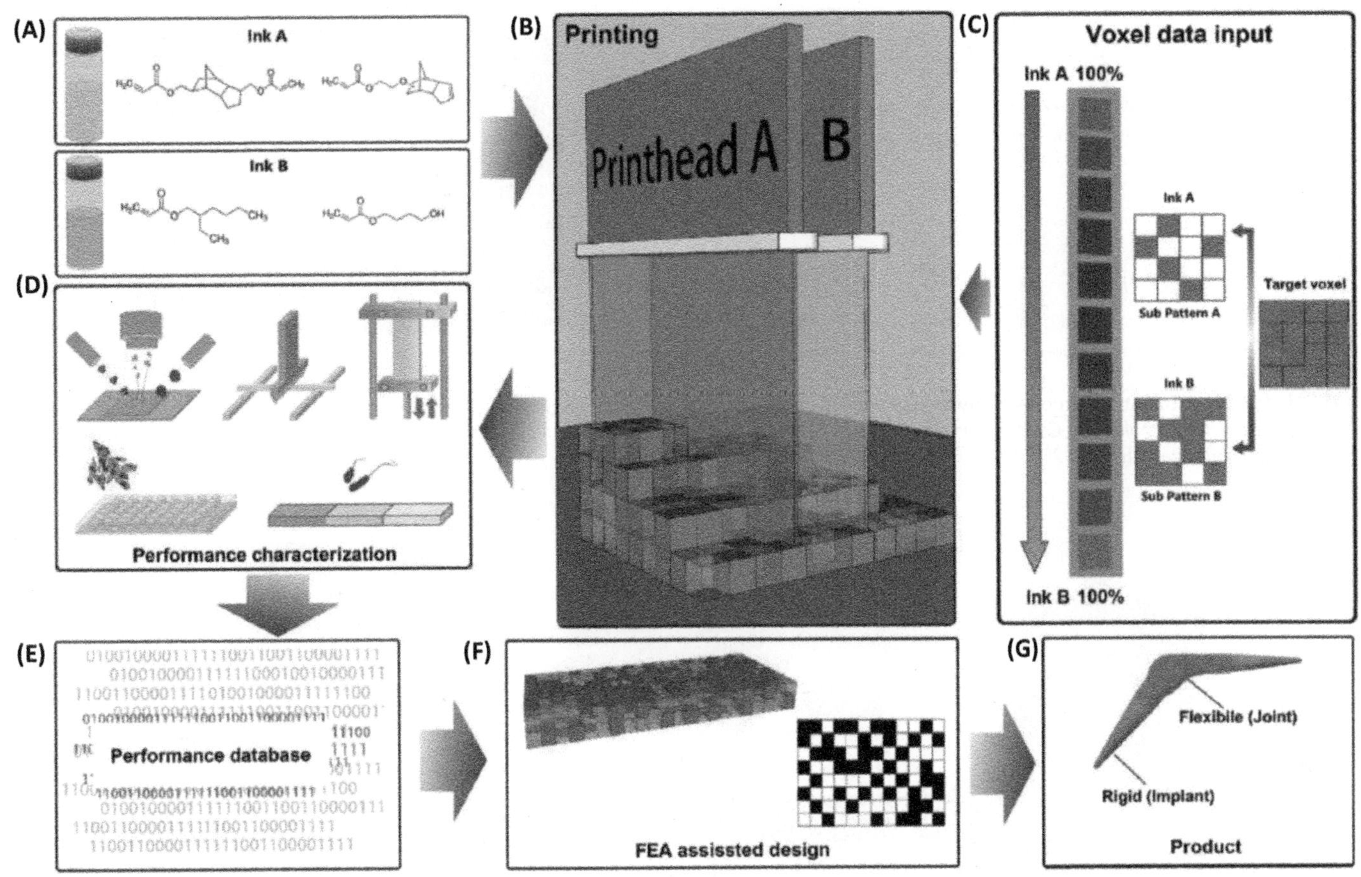

FIGURE 9.10 Strategy to assemble 3D printed antimicrobial formulation from (A–C) inks A and B using a dual print head system targeted to a (D) desirable composition determined by the choice of moduli of separately assembled sub-patterns from the voxel data. (E–G) The printing process was assisted by design from a performance database for producing a dual rigid-flexible materials. *Source: Reproduced with permission from He, Y., Abdi, M., Trindade, G. F., Begines, B., Frédéric Dubern, J., Prina, E., Hook, A. L., et al. (2021). Exploiting generative design for 3D printing of bacterial biofilm resistant composite devices.* Advanced Science, *8(15), 1–11.*

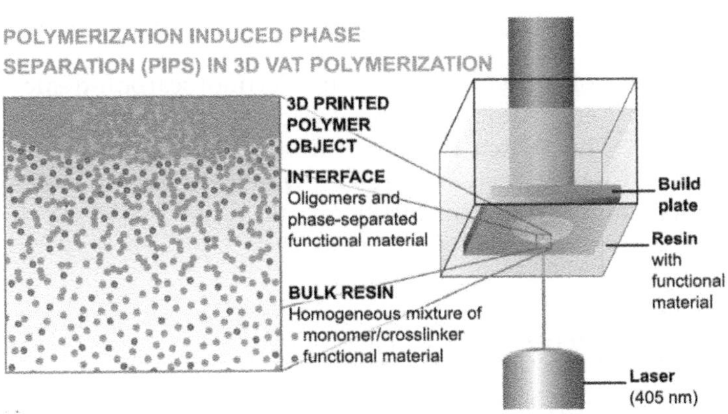

FIGURE 9.11 3D printed manipulation of 2D to 3D geometries *Reproduced from Deore, B., Sampson, K. L., Lacelle, T., Kredentser, N., Lefebvre, J., Young, L. S., Hyland, J., et al. (2021). Direct printing of functional 3D objects using polymerization-induced phase separation. Nature Communications, 12(1), 112.*

3D printed strategies have also been applied to fabricate bioinspired ultra-low-adhesive micro-nanoscale surfaces expelling 10–100 mm water droplets allowing a self-cleaning effect in-between stubs (Watson et al., 2015) mimicking the slippery properties (Yang et al., 2019) of the pitcher plant material. Wu et al. identified that a multifunctional layer is important to the low adhesive properties exhibited by plants and insects as an antibacterial mechanism to inhibit the growth of bacterial colonies. In their effort to construct a good low-adhesive mimic exhibiting a low adhesive energy interface, the slippery surface of the pitcher plant was modeled as a 3D

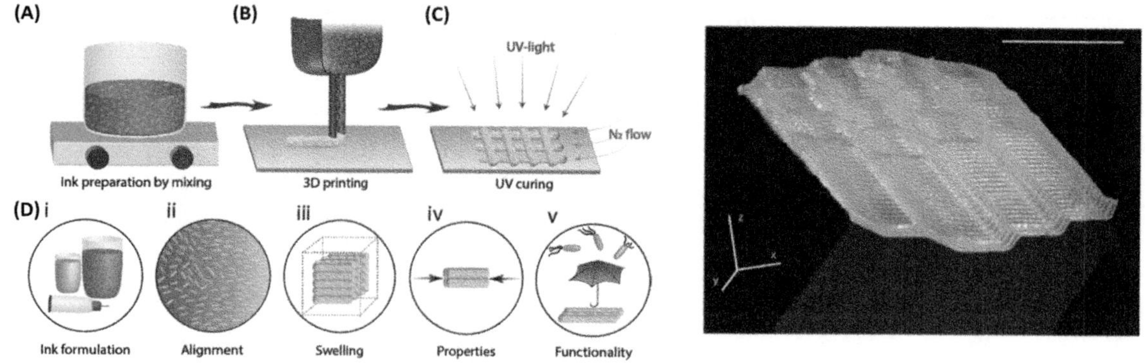

FIGURE 9.12 Fabrication process of hydrogels based on cellulose nano fiber alignment. *Reproduced from Fourmann, O., Hausmann, M. K., Neels, A., Schubert, M., Nyström, G., Zimmermann, T., & Siqueira, G. (2021). 3D printing of shapemorphing and antibacterial anisotropic nanocellulose hydrogels. Carbohydrate Polymers, 259, 1–11.*

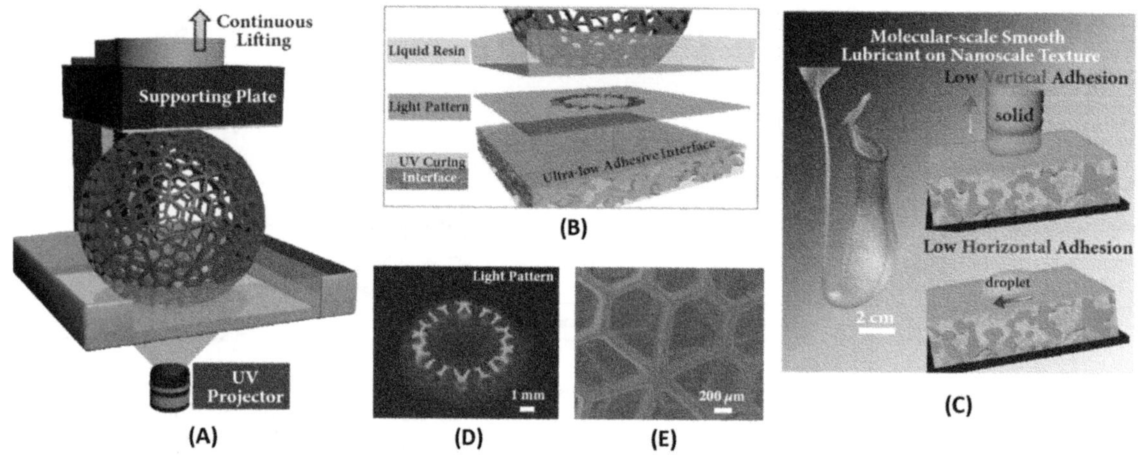

FIGURE 9.13 3D printed mimic of the low adhesive bioinspired anti-microbial materials. *Reproduced from Wu, L., Dong, Z., Du, H., Li, C., Fang, N. X., & Song, Y. (2018). Bioinspired ultra-low adhesive energy interface for continuous 3D printing: Reducing curing induced adhesion. Research; a Journal of Science and its Applications, 2018.*

printable material which comprised of a lubricating layer, a curable polymer, and resin. Perfluorocarbon-treated poly(dimethylsiloxane) provided the ultra-low adhesive energy and curable interface and the underlying polymer was shielded by the perfluorocarbon coating (Fig. 9.13). The switch between the solid-solid adhesion to solid-liquid-solid adhesion in the presence of the resin ensured low adhesive properties which are easily detachable from the cured resin layer providing multilayered functionality (Wu et al., 2018).

9.7 Concluding comments

With increasing uncertainty and unpredictability of rising microbial infections, emerging technologies are expected to play a significant role in the detection and prevention of biofilm-based infections. The changing and adaptive nature of biofilms as actively growing and environmentally responsive biomaterials point towards their heterogeneous character and their ability to persist biofilm degradation emphasizes the importance of material engineering in the context of microbial behavior on food packaging surfaces. While the engineering and advancement of new smart materials have been a core part of recently emerging innovations, future antibacterial-based technologies are driven by the need to apply automation and manufacturing of new materials as an unsolved challenge of the current century. At the heart of this lies the transition from 3D printed materials to multi-4D materials, as new technologies will require materials that can adapt to changing environments.

References

Alam, F., Kumar, S., & Varadarajan, K.M. (2019). Quanti fi cation of adhesion force of bacteria on the surface of biomaterials: Techniques and assays.

Babutan, I., Lucaci, A. D., & Botiz, I. (2021). Antimicrobial polymeric structures assembled on surfaces. *Polymers, 13*(10), 1–35.

Barthlott, W., Neinhuis, C., Verlot, H., & Schott, C.L. (1997). Purity of the sacred lotus, or escape from contamination in biological surfaces. 1–8.

Berger, B.W., & Sallada, N.D. (2019). Hydrophobins: Multifunctional biosurfactants for interface engineering. 1, 1–8.

Bhatta, D. R., Hamal, D., Shrestha, R., Hosuru Subramanya, S., Baral, N., Singh, R. K., ... Gokhale, S. (2018). Bacterial contamination of frequently touched objects in a tertiary care hospital of Pokhara, Nepal: How safe are our hands? *Antimicrobial Resistance and Infection Control, 7*(1), 4–9.

Brounstein, Z., Yeager, C. M., & Labouriau, A. (2021). Development of antimicrobial PLA composites for fused filament fabrication. *Polymers, 13*(4), 1–18.

Busscher, H. J., & Mei, H. C. V. D. (2012). How do bacteria know they are on a surface and regulate their response to an adhering state? *PLoS Pathogens, 8*(1), 1–3.

Cámara-Torres, M., Duarte, S., Sinha, R., Egizabal, A., Álvarez, N., Bastianini, M., Sisani, M., et al. (2021). 3D additive manufactured composite scaffolds with antibiotic-loaded lamellar fillers for bone infection prevention and tissue regeneration. *Bioactive Materials, 6*(4), 1073–1082.

Canfield, D. E., Thamdrup, B., & Kristensen, E. (2005). Microbial ecosystems. *Advances in Marine Biology, 48*(C), 465–506.

Cheng, G., Dai, M., Ahemd, S., Hao, H., Wang, X., & Yuan, Z. (2016). Antimicrobial drugs in fighting against antimicrobial resistance. *Frontiers in Microbiology, 7*, 470.

Chou., et al. (2015). *Mutable mechanics in biomaterials: A study of trichocyte keratin and Nvjp-1 protein (2015) Thesis*. Massachusetts Institute of Technology.

Costerton, J. W., Stewart, P. S., & Greenberg, E. P. (1999). Bacterial biofilms: A common cause of persistent infections. *Science (New York, N.Y.), 284*(5418), 1318–1322.

Deore, B., Sampson, K. L., Lacelle, T., Kredentser, N., Lefebvre, J., Young, L. S., Hyland, J., et al. (2021). Direct printing of functional 3D objects using polymerization-induced phase separation. *Nature Communications, 12*(1), 1–12.

Duan, S. M., Zhao, X. S., Wen, R. F., Huang, J. J., Pi, G. H., Zhang, S. X., ... Dong, X. P. (2003). SARS research team. Stability of SARS coronavirus in human specimens and environment and its sensitivity to heating and UV irradiation. *Biomedical and Environmental Sciences: BES*, 246–255.

Elbourne, A., Chapman, J., Gelmi, A., Cozzolino, D., Crawford, R.J., & Truong, V.K. (2019). Bacterial-nanostructure interactions: The role of cell elasticity and adhesion forces bacterial-nanostructure interactions.

Feng, B. X., & Jiang, L. (2006). Design and creation of superwetting / antiwetting surfaces. *Advanced Materials, 18*, 3063–3078.

Fernandes, M. M., Carvalho, E. O., & Lanceros-Mendez, S. (2019). Electroactive smart materials: Novel tools for tailoring bacteria behavior and fight antimicrobial resistance. *Frontiers in Bioengineering and Biotechnology, 7*, 277.

Fourmann, O., Hausmann, M. K., Neels, A., Schubert, M., Nyström, G., Zimmermann, T., & Siqueira, G. (2021). 3D printing of shape-morphing and antibacterial anisotropic nanocellulose hydrogels. *Carbohydrate Polymers, 259*, 1–11, September 2020.

Grandgirard, J., Poinsot, D., Krespi, L., Nénon, J. P., & Cortesero, A. M. (2002). Costs of secondary parasitism in the facultative hyperparasitoid pachycrepoideus dubius: Does host size matter? *Entomologia Experimentalis et Applicata, 103*(3), 239–248.

Griffith, K. M., De Cataldo, R., & Fogarty, K. H. (2016). Do-it-yourself: 3d models of hydrogenic orbitals through 3D printing. *Journal of Chemical Education, 93*(9), 1586–1590.

He, Y., Abdi, M., Trindade, G. F., Begines, B., Frédéric Dubern, J., Prina, E., Hook, A. L., et al. (2021). Exploiting generative design for 3D printing of bacterial biofilm resistant composite devices. *Advanced Science, 8*(15), 1–11.

Jacques, M., Aragon, V., & Tremblay, Y. D. N. (2010). Biofilm formation in bacterial pathogens of veterinary importance. *Animal Health Research Reviews/Conference of Research Workers in Animal Diseases, 11*(2), 97–121.

Jansma, J., & El Aidy, S. (2021). Understanding the host-microbe interactions using metabolic modeling. *Microbiome, 9*, 16.

Joerger, R. D. (2007). Antimicrobial films for food applications: A quantitative analysis of their effectiveness. *Packaging Technology and Science, 20*(4), 231–273.

Kenawy, E. R., Worley, S. D., & Broughton, R. (2007). The chemistry and applications of antimicrobial polymers: A state-of-the-art review. *Biomacromolecules, 8*(5), 1359–1384.

Kong, Y. S., & Kim, T. W. (2013). Wettability of biomimetic riblet surface like sharkskin. *Journal of the Korean Society of Tribologists and Lubrication Engineers, 29*(5), 304–309.

Ley, K., Christofferson, A., Penna, M., & Winkler, D. (2015). Surface-water interface induces conformational changes critical for protein adsorption: Implications for monolayer formation of EAS hydrophobin. *Frontiers in Molecular Biosciences, 2*, 1–12.

Lopez, G. U., Gerba, C. P., Tamimi, A. H., Kitajima, M., Maxwell, S. L., & Rose, J. B. (2013). Transfer efficiency of bacteria and viruses from porous and nonporous fomites to fingers under different relative humidity conditions. *Applied and Environmental Microbiology, 79*(18), 5728–5734.

Lumpe, T. S., Mueller, J., & Shea, K. (2019). Tensile properties of multi-material interfaces in 3D printed parts. *Materials and Design, 162*, 1–9.

Maniruzzaman, M., Boateng, J. S., Snowden, M. J., & Douroumis, D. (2012). A review of hot-melt extrusion: Process technology to pharmaceutical products. *ISRN Pharmaceutics, 2012*, 1–9, January 2014.

Mann, E. E., Manna, D., Mettetal, M. R., May, R. M., Dannemiller, E. M., Chung, K. K., ... Reddy, S. T. (2014). Surface micropattern limits bacterial contamination. *Antimicrobial Resistance and Infection Control, 3*(1), 1–28.

Matsuzaki, K. (2009). Control of cell selectivity of antimicrobial peptides. *Biochimica et Biophysica Acta – Biomembranes, 1788*(8), 1687–1692.

McArthur, D. B. (2019). Emerging infectious diseases. *Nursing Clinics of North America, 54*, 297–311.

Midelet, G., & Carpentier, B. (2002). Transfer of microorganisms, including Listeria monocytogenes, from various materials to beef. *Applied and Environmental Microbiology, 68*(8), 4015–4024.

Perera-Costa, D., Bruque, J. M., González-Martín, M. L., Gómez-García, A. C., & Vadillo-Rodríguez, V. (2014). Studying the influence of surface topography on bacterial adhesion using spatially organized microtopographic surface patterns. *Langmuir: The ACS Journal of Surfaces and Colloids, 30*(16), 4633–4641.

Preedy, E., Perni, S., Nipič, D., Bohinc, K., & Prokopovich, P. (2014). Surface roughness mediated adhesion forces between borosilicate glass and gram-positive bacteria. *Langmuir: The ACS Journal of Surfaces and Colloids*, *30*(31), 9466–9476.

Prokopovich, P., & Perni, S. (2009). An investigation of microbial adhesion to natural and synthetic polysaccharide-based films and its relationship with the surface energy components. *Journal of Materials Science: Materials in Medicine*, *20*(1), 195–202.

Recum, A. F., Shannon, C. E., Cannon, C. E., Long, K. J., Kooten, T. G., & Meyle, J. (1996). Modifiers of cellular adhesion. *Tissue Engineering*, *2*(4).

Richard, M., Knauf, S., Lawrence, P., Mather, A. E., Munster, V. J., Müller, M. A., … Kuiken, T. (2017). Factors determining human-to-human transmissibility of zoonotic pathogens via contact. *Current Opinion in Virology*, *22*, 7–12.

Rodríguez Bernaldo de Quirós, A., Cardamas, A. L., Sendón, R., & Ibarra, V. G. (2019). *Food contamination by packaging: Migration of chemicals from food contact materials*. Berlin, Boston: De Gruyter.

Schumacher., et al. (2007). Engineered antifouling microtopographies – Effect of feature size, geometry, and roughness on settlement of zoospores of the green alga ulva. *Biofouling*, *23*(1), 55–62.

Scott, P. J., Meenakshisundaram, V., Hegde, M., Kasprzak, C. R., Winkler, C. R., Feller, K. D., … Long, T. E. (2020). 3D printing latex: A route to complex geometries of high molecular weight polymers. *ACS Applied Materials and Interfaces*, *12*(9), 10918–10928.

Sorrentino, I., Gargano, M., Ricciardelli, A., Parrilli, E., Buonocore, C., Pascale, D. D., … Piscitelli, A. (2020). International journal of biological macromolecules development of anti-bacterial surfaces using a hydrophobin chimeric protein. *International Journal of Biological Macromolecules*, *164*, 2293–2300.

Sultana, A., Zare, M., & Luo, H. (2021). *Surface engineering strategies to enhance the in situ performance of medical devices including atomic scale engineering*.

Sun, G., Fang, Y., Cong, Q., & Ren, L.-Q. (2009). Anisotropism of the non-smooth surface of butterfly wing. *Journal of Bionic Engineering*, *6*(1), 71–76.

Sung, S. Y., Sin, L. T., Tee, T. T., Bee, S. T., Rahmat, A. R., Rahman, W. A. W. A., … Vikhraman, M. (2013). Antimicrobial agents for food packaging applications. *Trends in Food Science and Technology*, *33*(2), 110–123.

Tuson, H. H., & Weibel, D. B. (2013). Bacteria-surface interactions. *Soft Matter*, *9*(17).

van der Lubbe, S. C. C. V., & Fonseca Guerra, C. (2019). The nature of hydrogen bonds: A delineation of the role of different energy components on hydrogen bond strengths and lengths. *Chemistry – An Asian Journal*, *14*(16), 2760–2769.

Watson, G. S., Myhra, S., Cribb, B. W., & Watson, J. A. (2008). Putative functions and functional efficiency of ordered cuticular nanoarrays on insect wings. *Biophysical Journal*, *94*(8), 3352–3360.

Watson, G. S., Green, D. W., Schwarzkopf, L., Li, X., Cribb, B. W., Myhra, S., & Watson, J. A. (2015). A gecko skin micro/nano structure – A low adhesion, superhydrophobic, anti-wetting, self-cleaning, biocompatible, antibacterial surface. *Acta Biomaterialia*, *21*, 109–122.

Wu, L., Dong, Z., Du, H., Li, C., Fang, N. X., & Song, Y. (2018). Bioinspired ultra-low adhesive energy interface for continuous 3D printing: Reducing curing induced adhesion. *Research; a Journal of Science and its Applications*, *2018*.

Xuefeng, G., & Jiang, L. (2004). Water-repellent legs of water striders. *Nature*, *432*, 36.

Yan, H., Wu, Q., Yu, C., Zhao, T., & Liu, M. (2020). Recent progress of biomimetic antifouling surfaces in marine. *Advanced Materials Interfaces*, *7*(20), 1–18. Available from https://doi.org/10.1002/admi.202000966.

Yang, X., Huang, Y., Zhao, Y., Zhang, X., Wang, J., Sann, E. E., … Xia, F. (2019). Bioinspired slippery lubricant-infused surfaces with external stimuli responsive wettability: A mini review. *Frontiers in Chemistry*.

Zachary, J. F. (2017). Mechanisms of microbial infections. *Pathologic Basis of Veterinary Disease*, *e1*, 132–241.

Zheng, S., Bawazir, M., Dhall, A., Kim, H. E., He, L., Heo, J., & Hwang, G. (2021). Implication of surface properties, bacterial motility, and hydrodynamic conditions on bacterial surface sensing and their initial adhesion. *Frontiers in Bioengineering and Biotechnology*, *9*, 1–22.

Zorzetto, L., Andena, L., Briatico-Vangosa, F., De Noni, L., Thomassin, J. M., Jérôme, C., Grossman, Q., et al. (2020). Properties and role of interfaces in multimaterial 3D printed composites. *Scientific Reports*, *10*(1), 1–17.

SECTION 3

Novel non-thermal technology for food preservation

CHAPTER

10

Role of irradiation in antimicrobial nanoparticles and nanocomposite material synthesis

Aniket Kumar[1], *Sanjiv Sonkaria*[2] *and Varsha Khare*[2]

[1]School of Materials Science and Engineering, Chonnam National University, Buk-gu, Gwangju, South Korea
[2]Soft Foundry Institute, Seoul National University, School of Engineering, Kwanak-gu, Seoul, South Korea

10.1 Introduction

The presence of nanometer-scale materials designed by thin film deposition methods used in semiconductor process technologies was first pointed out in 1974 by Norio Taniguchi of Tokyo University. This concept was further evolved when understanding the science behind the small-scale structures is clearer and its utility has become essential in many sectors of science and technology such as catalysis, biomedicine, electronics, sensors, energy, and beyond. Scientific and industrial advancement of nanotechnology would be impossible without the workhorse of nanomaterials. Nanoscale materials often have unique properties when compared to their bulk counterparts, which in turn can be attributed to their synthesis and modifications at the outset. There are generally two ways of synthesis and fabrication of nanomaterials which includes a top-down approach where a bulk structure is broken into low dimensional structures and bottom-up approach where physical and chemical forces work together to assemble the molecules/atoms to develop low dimensional structures directly. In nanotechnology, top-down approach to synthesis and fabrication includes conventional methodologies such as lithography, etching, and micro molding. However, with the downsizing in the nanometer range, these approaches are facing challenges due to their limitations in the fabrication of complex and low-dimensional structures. To overcome the limitations the focus was shifted to a "bottom-up" approach. This approach is widely used for nanomaterial preparation for achieving at least 100 nm or less along one dimension. However, contamination issues are unavoidable in both processes owing to the involvement of chemical processes. Chemical processes increase the environmental toxicity, cytotoxicity, and carcinogenicity depending upon the initial materials, byproducts or fumes generated during the reaction process. There are toxicity problems that are brought about by hazardous substances, such as organic solvents, reducing agents, and stabilizers that are used to prevent colloids from agglomerating. Additionally, some materials increase the human toxicity levels at the nanometer range due to surface activity, shape change or composition, and size effects. As a result, clinical and biomedical applications have been prevented due to undesirable by-product formation. There has been a great deal of interest in developing green processes that make nanoparticles that are sustainable, efficient, safe, biocompatible, and environment-friendly. Developing these green processes could help to meet the requirements of green and sustainable chemistry (Jain, Bhargava, Majumdar, Tarafdar, & Panwar, 2011; Kulkarni & Muddapur, 2014).

Nanoparticles and nanomaterials are being created using radiation in recent years, which is an attractive alternative method to traditional synthesis methods (Ekka et al., 2018). The transmission of energy through electromagnetic waves or particle beams is called radiation. A photon or particle can be viewed as a fundamental carrier of radiation, which also acts as a reservoir of energy. As a result, photons and particles interact differently

Food Packaging and Preservation.
DOI: https://doi.org/10.1016/B978-0-323-90044-7.00010-0

with molecules and atoms than Joule heating (wet chemistry), resulting in distinct radiation effects depending on the carrier energy (Radiation, 2019; Radiation Applications, 2018). The use of radiation is controlled by its 12 regulatory principles which are based on an environmentally friendly, chemistry-based approach (Anastas & Eghbali, 2010).

Schiller et al. was the first to utilize the irradiation source to precipitate the solid phase in 1969. Recently, radiation has been a key component in the development of nanomaterials for antimicrobial applications. There are some advantages to radiation-induced processes over common chemical methods, including being largely temperature-independent, the fact that experiments don't need to follow strict protocols and the fact that they produce material with higher purity, with a narrower size distribution. Radiation is one of the simplest methods of preparing nanomaterials according to an increasing number of studies that illustrate the benefits of this method of nanomaterial synthesis (Fig. 10.1).

Nanomaterials can be prepared through the use of ionizing radiation because of their unique properties. There are four categories of nanomaterials:

1. Nanomaterials derived from metals;
2. Nanomaterials derived from carbon;
3. Nanomaterials derived from polymer;
4. Composite nanomaterials.

It is even possible to produce powder materials in large quantities on a micro or nanoscale using this method, making them applicable to larger-scale production. This chapter presents an overview of different alternative routes for synthesizing nanomaterials using ionizing radiation as antimicrobial technologies in a variety of biological applications and points out the benefits of doing so.

1. Nanomaterials derived from metals
2. Metallic Nanoparticles

In order to synthesize metal nanoparticles, chemical reduction routes most commonly use dissolving metal ions in stoichiometric amounts to reduce them to nanoparticles. It is important to note that the size and shape of

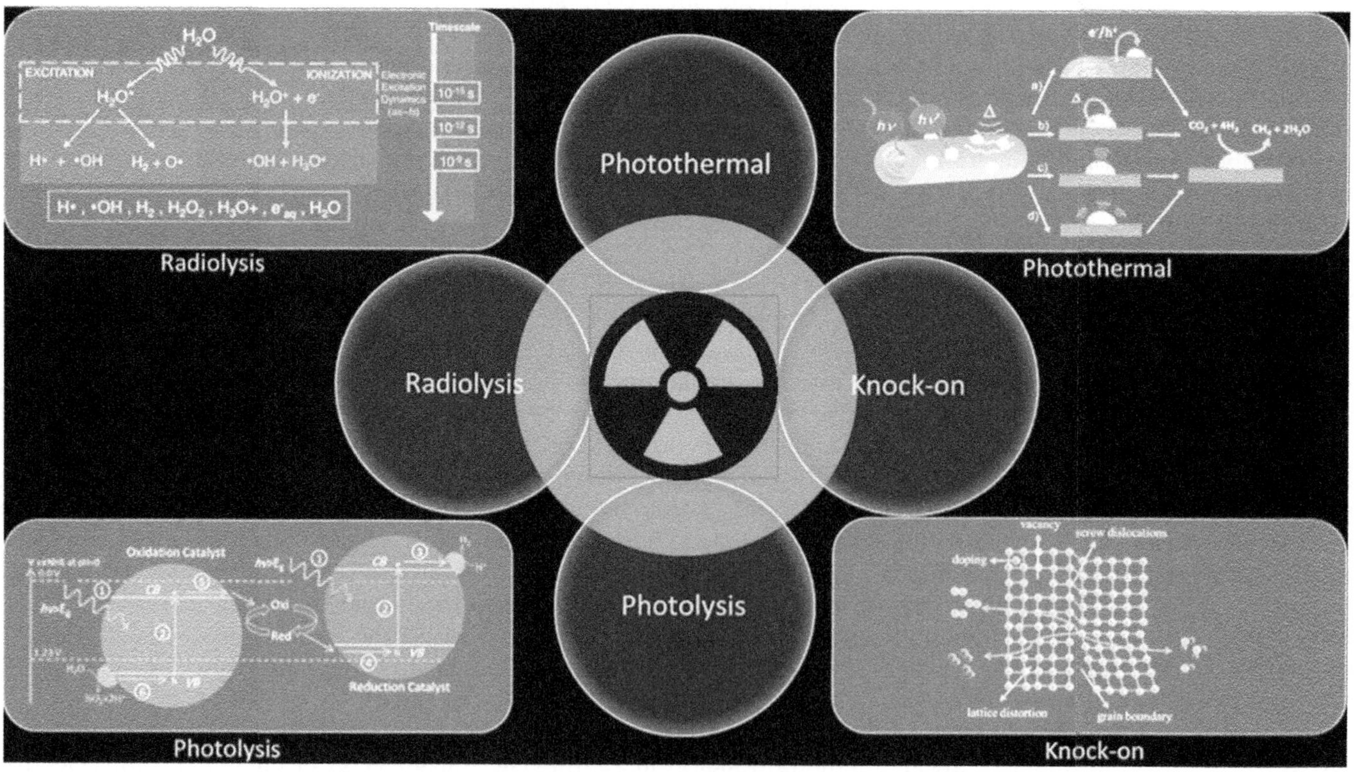

FIGURE 10.1 Mechanisms of radiation-induced synthesis of nanomaterials. Illustrates the four main mechanisms of radiation-induced radiolysis, photolysis, knock-on, and photothermal effects that take place in the chemical reactions for nanomaterial synthesis.

nanoparticles are highly dependent on the characteristics of the reducing agent which largely impacts the nucleation and growth rate of the particles. Recent research has demonstrated that radiolytic reduction is a powerful method for producing mono-sized and highly dispersed metallic clusters (Marignier, Belloni, Delcourt, & Chevalier, 1985). Electron beams, X-rays, gamma rays, and UV light are frequently used electromagnetic waves for metallic nanoparticle synthesis. As part of the radiolytic reduction scheme, metal ions in an aqueous solution with a stabilizing agent are irradiated by the radio frequency, producing hydrated electronic (e^-_{aq}) and $H^\bullet$ atoms are produced (Eq. 10.1). Due to their high reduction potential{($E^o\left(\frac{H_2O}{e^-_{aq}}\right) = -2.87$ VNHE and $E^o\left(\frac{H^+}{H^\cdot}\right) = -2.3$VNHE)}, these radical species produced are highly reducing in nature so much so that they cross the barrier of metal ion reduction to achieve the zero-valent metal particles (Eqs. 10.2 and 10.3) (Rojas & Castano, 2012).

$$H_2O \rightarrow e^-_{aq}, H^+, H^\circ, OH^\circ, H_2O_2, H_3O^+ \tag{10.1}$$

$$M^+ + e^-_{aq} \rightarrow M^o \tag{10.2}$$

$$M^+ + H^\circ \rightarrow M^o + H^+ \tag{10.3}$$

Therefore, mechanistically, additional reducing agents and their side effects are avoided in this approach. Further, the concentration of zero-valent nuclei can be controlled by adjusting the dose of the irradiation.

Furthermore, hydroxyl radicals (OH°) are also strong reducing agents formed in the radiolysis of water according to the equation $E^0 = (OH^\circ/H_2O) = +2.8\ V_{NHE}$ thus, facilitating the oxidation process to achieve a higher oxidation state. At this point prior to irradiation, the introduction of OH° radical scavengers such as primary or secondary alcohols or formate ions into the precursor solutions is important. An example of the above-mentioned mechanism is that the use of isopropanol as a scavenger allows the scavenging of OH° and H° radicals and its conversion into secondary radicals and finally reducing metal ions (M^+) into zero-valent atoms (M^0).

$$OH^\circ + CH_3CH\ OH\ CH_3 \rightarrow H_2O + H_3CC^\circ OH\ CH_3 \tag{10.4}$$

$$H^\circ + CH_3CH\ OH\ CH_3 \rightarrow H_2 + H_3CC^\circ OH\ CH_3 \tag{10.5}$$

$$M^\circ + H_3CC^\circ OH\ CH_3 \rightarrow CH_3COCH_3 + M^0 + H^+ \tag{10.6}$$

It may also include the disproportion of lower valence states in the multistep process of reducing multivalent ions. Fig. 10.2 illustrates these processes schematically.

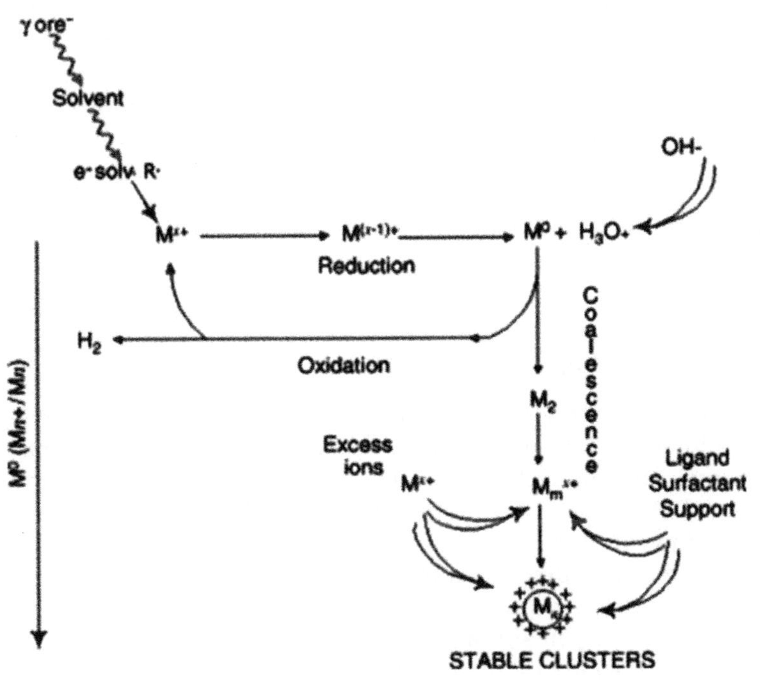

FIGURE 10.2 Radiation reduction of metal ions. A scheme for ionizing radiation reduction of metal ions in solution with a stabilizer. Coalescence occurs between the dispersed atoms M^0. The dispersed atoms are stabilized by ligands, polymers, or supports (Belloni, 2006). *From Belloni, J. (2006). Nucleation, growth and properties of nanoclusters studied by radiation chemistry: Application to catalysis. Catalysis Today, 113(3–4), 141–156. https://doi.org/10.1016/j.cattod.2005.11.082.*

Although in water-saturated electrons environment reduction in metal ions to zero-valent metal (M^0) is a facile process but for multivalent metal ions, a multistep reduction process with intermediate valences makes the process more complex. Radial distribution of the atoms in a radiolytic solution ensures uniform distribution of the atoms throughout the solution. Radiation produces reducing agents that reduce the metal ions in the solution at random. From each newly formed atom, nucleation and coalescence follow. In contrast to bonds between atoms and solvents or atoms and ions, metal atoms and unreduced ions have greater strength (Belloni, 2006). Therefore, metal ions dimerize if they are located near atoms or are bound to the atoms:

$$M^0 + M^0 \rightarrow M_2 \tag{10.7}$$

$$M^0 + M^+ \rightarrow M_2^+ \tag{10.8}$$

Cluster nucleation centers can emerge from the charged dimer clusters. To control the competition between free and absorbed metal ions reduction, the rate at which the reducing agents are formed could be controlled (Rao et al., 2010).

For cluster formation, reducing ions attached to clusters is preferred rather than isolating new atoms. Also, ions may form strong bonds between clusters of unreduced ions, or ions may form strong bonds between clusters of charged ions:

$$M_2^+ + M_2^+ \rightarrow M_4^{2+} \tag{10.9}$$

$$M_m + M^+ \rightarrow M_{m+1}^+ \tag{10.10}$$

$$M_{m+x}^{x+} + M_{p+y}^{y+} \rightarrow M_{n+z}^{z+} \tag{10.11}$$

Here m, n, and p are nuclear ties, and x, y, and z represent a number of associated ions. For example, monodispersed copper nanoparticle was synthesized on the basis of above mention mechanism using gamma irradiation. Pulsed radiolysis, for example, silver, has studied this mechanism in depth (Janata, Henglein, & Ershov, 1994). A study was conducted to determine the optical absorption spectra and reactivity of transient reactions by pulse radiolysis in the nanosecond range to understand the difference between the optical properties of particles synthesized with and without ligand in water. In this paper, transient spectra of Ag^0 and Ag^{2+} with different ligands in water are given in Fig. 10.3 (Mostafavi et al., 1828–1842).

10.1.1 *Multi-metals nanoparticles*

The following types of structures can be identified after irradiating a solution containing two or more metal salts to develop intermetallic or alloyed structures, core/shell structures, and heterostructures. Metallic ions in ionic solutions are reduced by their corresponding redox potentials, which are key factors in determining the resultant particles' structure. The atoms and clusters of the alloy nanoparticles combine with unreacted ions during reduction reactions which are followed by coalescence aggregation resulting in clusters of bimetallic alloys. Using two mixed metal precursors containing solutions as an example, it is possible to identify three kinds of structures: intermetallic or alloyed structures, core/shell structures, and heterostructures (Abedini, Daud, Abdul Hamid, Kamil Othman, & Saion, 2013). Redox potentials, or electrochemical potentials, determine how ionic solutions are reduced, and the particles that result from this are structured as a result.

10.1.2 *Fabrication of Alloy, core/shell, and heterostructured nanoparticles*

Interaction of clusters with unreacted ions with atoms leads to the formation of nanoparticles of alloyed intermetallic materials, for example, clusters of bimetallic alloyed metals are formed by association of reduction with the alternate occurrence of coalescence and aggregation. Abedini and her collaborators have reported that the gamma ray's irradiation on the ionic solution of Al^{3+} and Ni^{2+} suggests that this process leads to the precise formation of alloyed structures however the effect of radiation can vary, for example, reducing Nickel ions is easier than aluminum ions. Therefore, in some cases, the solution of precursor ions requires sequential reduction at the time of irradiation. Further non-reacting ions form charged clusters on their surfaces when they do not react with the newly formed clusters. A hydrated electron reduces these ions in situ at the site to form an alloyed structure (Abedini, Larki, Saion, Zakaria, & Zobir Hussein, 2012a) as indicated by the bright-field TEM images in Fig. 10.4.

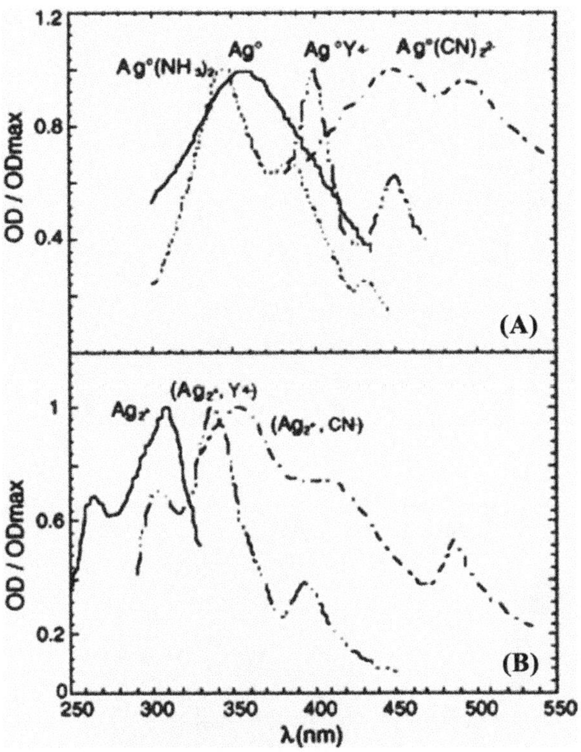

FIGURE 10.3 Optical absorption spectra of (A) Ag ^{0}L and (B) Ag $^{2+}$L. In the following section we present Optical absorption spectra of $Ag^{0}L$ and $Ag^{2+}L$ (L: CN^{-}, NH_3, or EDTA anion = Y^{4-}), as obtained by pulse radiolysis at nanosecond range of a solution containing $Ag^{I}L$, compared with the spectra of the corresponding species without ligand in water (Mostafavi et al., 1828–1842). *From Mostafavi, M., Dey, G. R., François, L., & Belloni, J. (2002). Transient and stable silver clusters induced by radiolysis in methanol. Journal of Physical Chemistry A, 106(43), 10184–10194. https://doi.org/10.1021/jp014257a.*

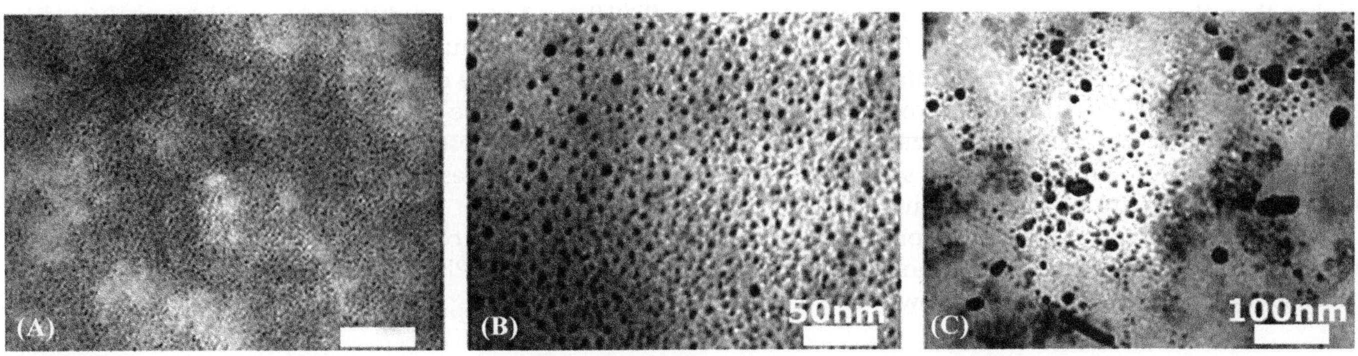

FIGURE 10.4 Binary Al–Ni alloy nanoparticle. Binary Al–Ni alloy nanoparticle TEM images synthesized at 100 kGy with ions concentrations of (A) 5.8×10^{-5} mol/mL and average size of 4.43 nm, (B) 5.7×10^{-5} mol/mL and average size of 6 nm, and (C) 6.4×10^{-5} mol/mL and average size of 9.97 nm. *From Abedini, A., Larki, F., Saion, E., Zakaria, A., & Zobir Hussein, M. (2012). Influence of dose and ion concentration on formation of binary Al-Ni alloy nanoclusters. Radiation Physics and Chemistry, 81(10), 1653–1658. https://doi.org/10.1016/j.radphyschem.2012.05.015.*

It is believed that less noble metals (M_1) are inclined to donate electrons to more noble metals (M_2) when they come in contact with electrons hydrated by water or radicals produced by radioactive decay. In this process, the first step will lead to the formation of single metallic clusters of noble metal (M_n) followed by a decrease in M_2 concentration resulting in the reduction of M_1 ions on the surface of M_n (Belloni, 2006). This concept has been successfully employed in the synthesis of the core-shell clusters, where the more noble metal M_2 is covered by the less noble metal M_1. Realization of this concept can be observed in the formation of Cu–Al_2O_3 core-shell nanoparticles with Cu core and Alumina shell was fabricated using a mixture of $CuCl_2$ and $AlCl_3$ in PVP solutions (Abedini et al., 2012b). The hydration rates of hydrated electrons in solutions containing copper ions were

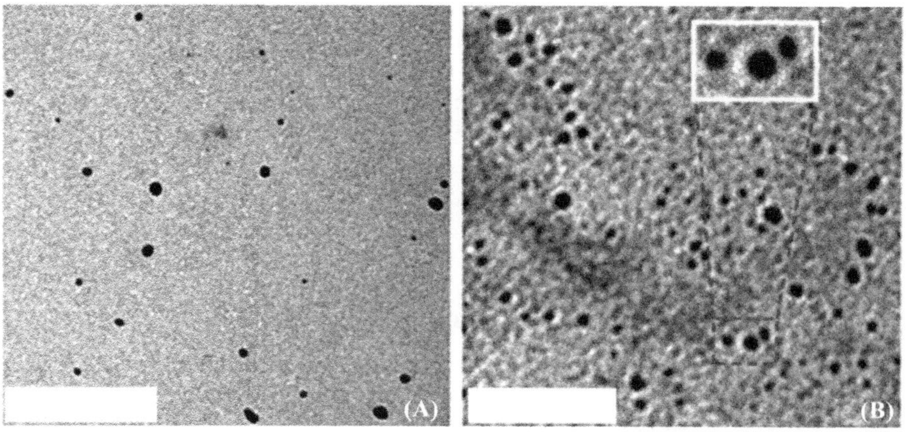

FIGURE 10.5 Core -shell particles. TEM images of (A) Cu and (B) $Cu@CuO_2-Al_2O_3$ nanoparticles. In addition to pure Cu nanoparticles, $Co@CuAlO_2-Al_2O_3$ nanoparticles with core-shell structures (Abedini et al., 2012b). *From Abedini, A., Saion, E., Larki, F., Zakaria, A., Noroozi, M., & Soltani, N. (2012). Room temperature radiolytic synthesized $Cu@CuAlO_2$-Al_2O_3 nanoparticles. International Journal of Molecular Sciences, 13 (9), 11941–11953. https://doi.org/10.3390/ijms130911941.*

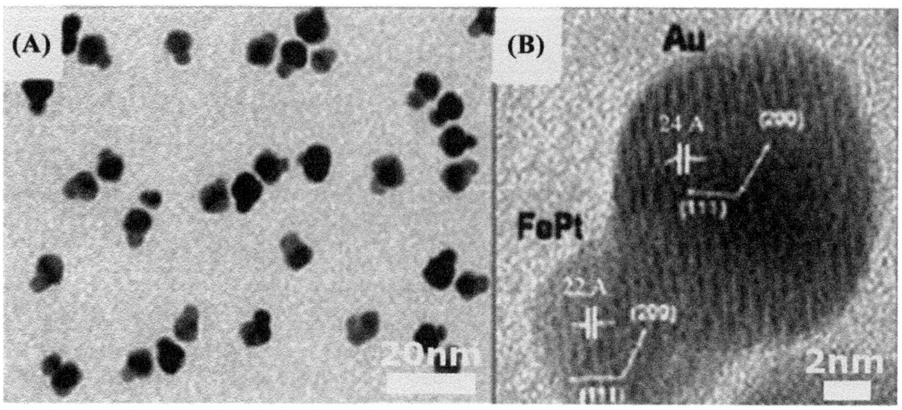

FIGURE 10.6 (A) TEM and (B) HRTEM images of FePt-Auheterodimer nanoparticles indicating the formation via overlaaping of (200) planes of Au and FePt (Choi et al., 2006). *From Choi, J. S., Jun, Y. W., Yeon, S. I., Kim, H. C., Shin, J. S., & Cheon, J. (2006). Biocompatible heterostructured nanoparticles for multimodal biological detection. Journal of the American Chemical Society, 128(50), 15982–15983. https://doi.org/10.1021/ja066547g.*

higher than those in solutions containing aluminum ions ($E^0(V) = -1.66$). The difference could be explained by copper ions' higher redox potentials ($E^0(V) = +0.34$) than those of aluminum ions. Irradiating bivalent Cu ions will therefore reduce their valent content until zero-valent Cu ions become dominant. Next, when the Cu^{2+} ions are depleted, the Al^{3+} reduction increases, resulting in predominantly core-shell structures at the surface of the Cu particles. The core/shell organization of the clusters was clearly observed by TEM, electron diffraction. The boundary between the core and shell is not sharp since $CuAlO_2$ is present in the shell, but Al_2O_3 is not, so the shell contains $CuAlO_2$, as shown in Fig. 10.5.

Several types of metal atoms can form a heterostructure under the right conditions by nucleating and growing in separate locations. When FePt nanoparticles were treated with AuCl-(PPh_3) in 1,2-dichlorobenzene containing 1-hexadecylamine, the successive growth of Au onto FePt was observed (Fig. 10.6) (Choi et al., 2006).

10.1.3 *Metal compounds nanoparticles*

A method of fabrication of organic and inorganic nanoparticles in aqueous solutions based on radiation-induced synthesis has been around for decades (Dispenza, Grimaldi, Sabatino, Soroka, & Jonsson, 2015; Flores-Rojas, López-Saucedo, & Bucio, 2020; Kharisov et al., 2016). By radiolysis of water, the oxidizing radical HO• is converted into the reducing radical e^-_{aq}. By adding proper radical scavengers prior to irradiation, it is possible to shift the redox condition of the solution to either oxidizing or reducing. Radiolytic oxidation or radiolytic reduction changes the valence state of metal ions within irradiated aqueous systems, while precipitation of insoluble products follows.

Radiolytic oxidation involves scavenging a reducing radical e^-_{aq}, from the irradiated system in order to change its redox state to purely oxidative. To convert e^-_{aq} to HO•, nitrous oxide (N_2O) is commonly used. This results in the formation of an equivalent amount of HO•:

$$N_2O + e_a q^- + H_2O \rightarrow HO\bullet + OH^- + N_2 k = 9.1x109\ M^{-1}S^{-1}$$

Therefore, for oxidative radiolysis, the value of G of HO• calculated by the above-mentioned hypothesis is 0.56 mol/J. The reaction pathway for the oxidation of metal ions by HO• is as follows: (Spinks & Woods, 1990)

$$M^{Z+} + OH^{\cdot} \rightarrow M^{(Z+1)+} + OH^{-} \quad (10.12)$$

I would like to emphasize the fact that $M^{(Z+1)+}$ oxidation and hydrolysis happen after its condensation. Hence, metal oxides are created in the process. Using Mn (II) salt as a precursor, MnO_2 has been synthesized using an oxidative radiolytic method (Li et al., 2021; Yadav, Olsson, & Jonsson, 2009). Radiation-induced Nanoparticle Synthesis of "γ-FeOOH" and "CoOOH" Metal Oxyhydroxide Nanoparticles has also been demonstrated (Alrehaily, Joseph, Biesinger, Guzonas, & Wren, 2013; Sutherland et al., 2017). As a result of the oxidation of Fe^{2+} and Co^{2+} by $OH^{\cdot}$, these elements are converted into Fe^{3+} and Co^{3+}. In the following step, the trivalent hydroxide (formed by hydrolysis) was converted to solid-state γ-FeOOH and CoOOH.

Contrastingly, for the radiolytic reduction process, irradiating system's redox condition changes in favor of optimal reducing conditions, thus scavenging of OH is crucial. As mentioned above, isopropanol converts OH to reducing radicals $(CH_3)_2{\bullet}COH$ via the following reaction pathways:

$$OH \times + (CH_3)_2CHOH \rightarrow H_2O + + (CH_3)_2{}^{\circ}CHOH$$

Despite the 2-hydroxyl-2-propyl radical ($(CH_3)_2$ °CHOH) exhibiting powerful reduction properties, it has an $E^0_{((CH3)2CO/(CH3)2\ \bullet COH)} = -1.8\ V_{SHE}$, which makes it a weak redox radical with a strong reduction potential. When isopropanol is applied to scavenge HO•, the system allows the reducing species to bring its G-value to 0.56 mol/J.

There is a tendency for metal ions, in general, to react with the e^-_{aq}. This is the case, however, only when the redox potential is taken into account. Because e^-_{aq} has a lower redox potential than the alkali metal ions (first group) as well as the Be, Mg, Ca, Sr, and Ba ions (second group, alkali earth metal), the alkali metal ions (first group) are unreactive with e^-_{aq}. Reduction by e^-_{aq} is the following for the metal ions:

$$M^{Z+} + e^-_{aq} \rightarrow M^{(Z-1)+} \quad (10.13)$$

Further reduction of $M^{(Z+1)+}$ by e^-_{aq} or by disproportionation can be achieved as follows:

$$2M^{(Z-1)+} \rightarrow M^{(Z-2)+}M^{Z+}$$

$$M^{(Z-1)+} \rightarrow M^{(Z-2)+}M^{Z+}$$

In addition to e^-_{aq} reducing nature, metal ions can also be reduced by $(CH_3)_2$ °CHOH using the following reaction pathway:

$$M^{(Z+)} + (CH_3)_2{}^{\circ}CHOH \rightarrow M^{(z-1)+} + CH_3COCH_3 + H^+$$

Permanganate ions are reductively radiolytically transformed into manganese oxides, MnO_2 and Mn_2O_3, by the oxidative pathway. Manganese oxide composition is affected by the pH of the solution. MnO_2 can be produced at pH values between 6.5 and 8.5, whereas Mn_2O_3 becomes the dominant phase above pH 9.5. Similarly, aqueous $CrO_4^{2-}/Cr_2O_7^{2-}$ solutions under radiolysis produce chromium oxide (Cr_2O_3) nanoparticles. The reduction radiolytic pathway can also be used to produce metal oxide from metal ions and oxometallate. A precursor for the fabrication of Cu_2O particles, for instance, is Cu^{2+}. The combination of metal ions reduction and reducing the radiolytic radicals by irradiation generates many metallic nanoparticles, such as Au, Ag, and Cu.

The formation of metal oxide involves some fundamental steps, in addition to the oxidation state changes caused by the radiolysis products. In addition to hydrolysis and condensation, these steps are discussed. Metal cations M^{z+} act as Lewis acids where the strength of metal cations correlates positively with the electronegativity of the metal. By reacting with M^{z+} as a ligand, water forms metal aquo-complex $[M(OH2)N]^{z+}$ as follows:

$$M^{z+} + NH_2O \rightarrow [M(OH_2)N]^{z+}$$

The hydration number N is equal to the oxidation state z of the metal ion Mz + . In an electron density transfer reaction, the coordinated water molecules transfer their electrons to the metal cation's empty orbitals via the M-OH_2 σ bond. The following reactions might be responsible for water molecule deprotonation:

$$[MOH_2)_N]^{z+} + nH_2O \leftrightarrow [M(OH)_h(OH_2)_{N-h}]^{(z-h)+} + hH_3O^+$$

The hydrolysis ratio is given by h. Depending on the degree of deprotonation and the type of product $[M(OH)_h(OH_2)_{N-h}]^{(z-h)+}$ produced, it may range from the aquo-complex $\left[MOH_2)_N\right]^{z+}$ to the hydroxocomplex $\left[MOH_2)_N\right]^{(z-n)+}$ to the oxo-complex $[MO_N]^{(z-2n)++}$. Further to understand speciation of $[M(OH)_h(OH_2)_{N-h}]^{(z-h)+}$ at any given pH level, one must know the exact metal cation speciation and the chemical evolution during the subsequent condensation process. The partial charge model suggests a very simple principle that while combining atoms of different initial electronegativity (X_i^o), the intermediate electronegativity in the compound will be the same. Therefore in an aqueous-hydroxo complex $[M(OH)_h(OH_2)_{N-h}]^{(z-h)+}$ and in an aqueous solution, the proton exchange will continue until the mean electronegativity of both sides has been achieved:

$$X_h = X_w = 2.732 - 0.035pH \tag{10.14}$$

The X_w is the electronegativity of aqueous solution, with the mean electronegativity of X_h being the electronegativity of $[M(OH)_h(OH_2)_{N-h}]^{(z-h)+}$.

Total charge of the aqueous-hydroxo complex +(z-h) can be obtained by summing up all of the partial charges (δ) of M, O, and H:

$$z - h = \delta_M + N\delta_O + (2N - h)\delta_H$$

thus,

$$h = [z - \delta_M - N\delta_O - 2N\delta_H]/[1 - \delta_H]$$

As a result of the following equation provides the partial charge of M, O, and H:

$$\delta_i = (X_h - X_i^0)/[1.36(X_i^0)^{1/2}]$$

Therefore, at certain pH the hydrolysis ratio h of $[M(OH)_h(OH_2)_{N-h}]^{(z-h)+}$ can be calculated by inserting hydration number N in the equations and electronegativity of atom i χi^0 (obtained from the Allred-Rochow scale) in Eqs. (10.1), (10.3) and (10.4).

Further, z Vs pH plot will be able to provide information about charge vs. pH as shown in Fig. 10.7, which accounts for the hydrolysis product B.

In general, an aquo-, an aquo-hydroxo-, or an aquo-hydroxy-complex are produced upon hydrolysis of metal cations. Hydrolysis of compounds with $z > 4$ results in hydroxo-, oxo-hydroxo-, or oxo-complexes. With $z = 4$, the tetravalent metal ions are on the periphery of the range of possible complex forms and are, therefore, depending on the pH, capable of forming any of the possible complexes.

However, hydrolyzed metal cations are unstable synthesized by this process and are not at equilibrium and there is a high possibility that they may spontaneously condense into their corresponding thermodynamically stable metal oxides as a consequence of olation or oxolation reactions.

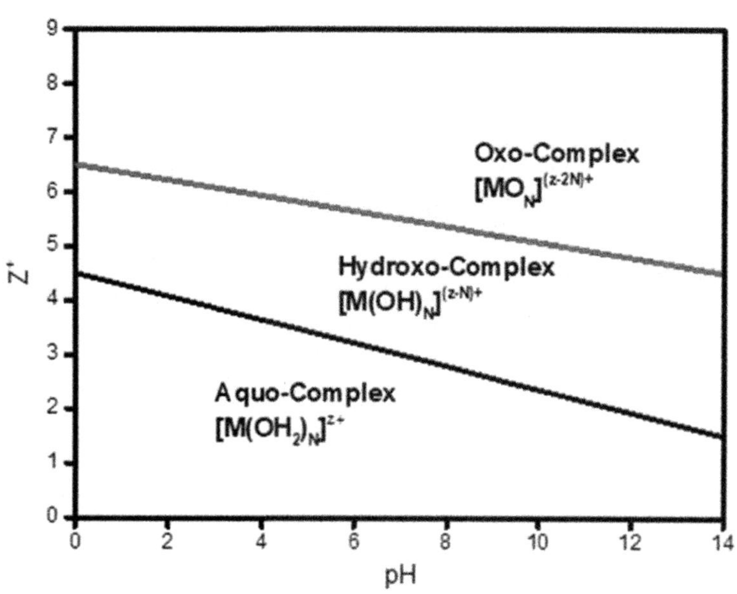

FIGURE 10.7 Metal ion Charge Vs pH. The charge of metal ion (z +) versus pH diagram. The black line corresponds to $h = 1$, and the red line corresponds to $h = 2N - 1$. The two lines separate three domains in which H_2O, OH^- or O^{2-} ligands are formed.

10.1.4 Nanomaterials derived from carbon

Another important area of research is that of carbon materials, which have strong interactions with radiation of high energy. In addition to diamonds, CNTs, graphene, fullerene, and amorphous carbon, carbon exists in a number of other forms. In some cases, radiation has altered the properties of the allotropes as a result of their physical and/or chemical modification. In carbon materials reaction medium is one of the decisive factors for the tenability of reactive pathways to follow oxidation/reduction, defect or surface engineering, surface recreation, etc (Chen, Wu, Liu, & Long, 2006; Gao et al., 2016).

Park et al. have discovered that MWCNTs when exposed to γ-rays in an oxygen atmosphere without solvent then surface functionalization increases drastically. Miao and colleagues have shown that in an airborne experiment, CNT yarns upon y-ray irradiation their surface functionalizes by incorporation of oxygen-containing onto the surface of the CNTs which promotes strong interactions between the CNTs and the polyamide 6,6. Therefore it can be concluded from various studies that in carbon materials after irradiation in air or oxygen the carboxyl content, increases dramatically which enhances the interaction with neighboring groups in contact. This has been evidenced by the improved mechanical properties of CNT yarns due to the hydrogen bond formation by such oxygen groups between neighboring CNTs. Furthermore, some groups have studied γ-ray irradiation onto MWCNTs both in air and epoxy chloropropane (ECP) solvent suggesting the comparative difference in each medium. Graphitized CNTs irradiated in air had a smaller interlayer distance and higher graphitization degree, while CNTs irradiated in ECP produced more defects (Kim et al., 2011; Xu, Chen, Liu, Wu, & Chen, 2011).

Furthermore, apart from CNTs, 2-D materials such as graphene and graphene oxide (GO) have also been studied extensively to understand the effect of irradiation on surface changes and therefore its effect on their properties. As an effect of irradiation reduction of GO and oxidation of graphene produces oxidative as well as reductive species as reported by Li et al. In their study, they observed that in N_2-purged water–alcohol solutions radiation induces reduction of GO to produce reduce graphene oxide (rGO). In order to reduce GO to rGO, the oxygen functional group content was minimized, graphitization was increased, and interlayer spacing was decreased. Zhang and colleagues have also reported GO reduction by irradiating the suspension of GO in N, N-dimethyl formamide where $N(CH_3)_2^+$ groups on the surface of the rGO allow it to be well dispersed in various organic solvents. Reports from Yan et al. suggest that synchronized reduction and hydrogenation of GO where radiolysis of water produced hydrogen radicals simultaneously that hydrogenates rGO (Ansón-Casaos et al., 2014; Cho et al., 2018; Zhang et al., 2012).

Alternatively, a different approach has been used by researchers to modify GO by using radiolytic oxidizing species. Researchers used the radiolysis of N_2O-saturated water to produce HO• radicals to oxidize GO sheets in order to make porous structures. Lacey edges and nonoporous structure formation after removal of edge and basal plane C atom from GO sheets are clearly visible after the irradiation after 23 hours (Fig. 10.8). The process was initiated by the formation of wrinkled sharp-edged rough surface and flat basal plan upon 6 hours irradiation of GO sheet. Further exposure to radiation 17 hours converts the sharp edges a "lacelike" edge and after 23 hours, and Fig. 10.8 the basis plane becomes enriched in pores (Yu et al., 2016).The one-pot reduction method for the deposition of iridium oxide on GO surface by gamma-ray irradiation is proposed by J.V. Rojas as shown in Fig. 10.9. According to their work, oxygenated functionalities of graphene oxide served as locations for the nuclei of iridium atoms to form before their subsequent growth into oxide atoms (Rojas, Molina Higgins, Toro Gonzalez, & Castano, 2015).

10.1.5 Nanomaterials derived from polymer

In top-down nanotechnology, as mentioned earlier, bulk polymers and monomers undergo irradiation by ionized high-energy beams to achieve nanomaterials. Accelerated high-speed electrons from high energy radiation such as γ rays, X-rays, and electrons beam interact with polymers to generate intermediate excited states, ions, or free radicals via a rapid kinetic reaction mechanism involving disproportionation, hydrogen abstraction, or new bond and thus novel structures formation. Further, based on irradiation environment such as air or monomer, the resulting effect will lead to cross-linking, scission of main or side chains, oxidization, or grafting. Various factors such as the characteristics of polymer and conditions before, during, and after the treatment determine the degree of dominance of these transformations. It is possible to tune the polymer under radiation by maneuvering above-mentioned factors.

Exposure to ionizing radiation generally results in degradation or cross-linking of polymers. Radiation processing has been widely used by the polymer industry to crosslink materials and materials, such as thermally shrinkable wires and cable insulations, automobile tires, and PE foams. Polymers of the degrading type are sterically hindered from combining broken chain ends. As a result of disproportionation and the formation of two

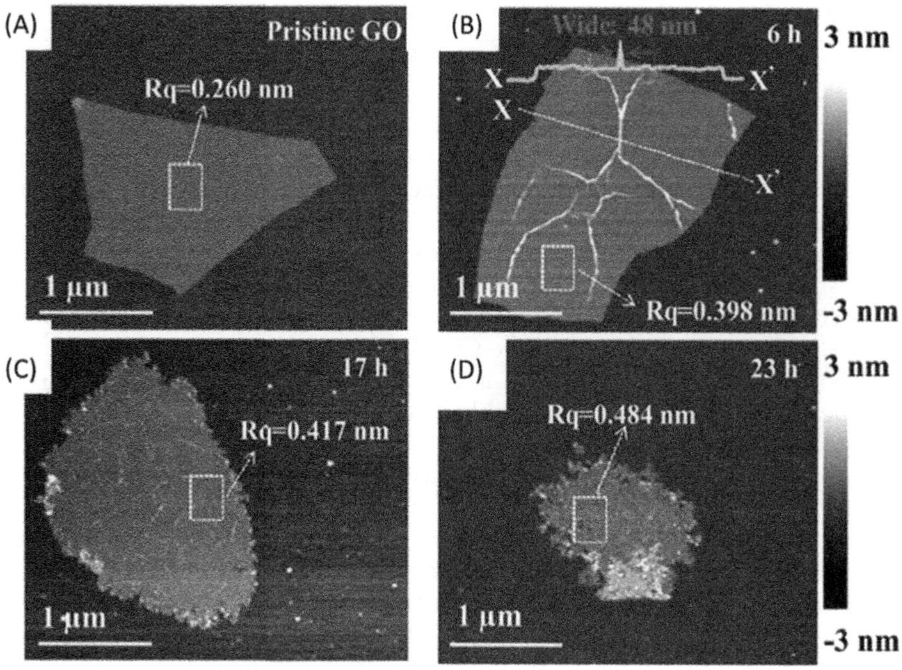

FIGURE 10.8 Effect of irradiation on 2D materials imaged by AFM. Image of pristine GO before and after *radiation using Atomic Force Microscopy (AFM): (A) 12 Pristine GO. (B, C, & D) GO exposed to irradiation for 6 h, 17 h, and 23 h, respectively. Each GO sheet shows the Rq (roughness) of its surface (Rojas et al., 2015). *From Yu, C., Zhang, B., Yan, F., Zhao, J., Li, J., Li, L., & Li, J. (2016). Engineering nano-porous graphene oxide by hydroxyl radicals. Carbon, 105, 291–296. https://doi.org/10.1016/j.carbon.2016.04.050.*

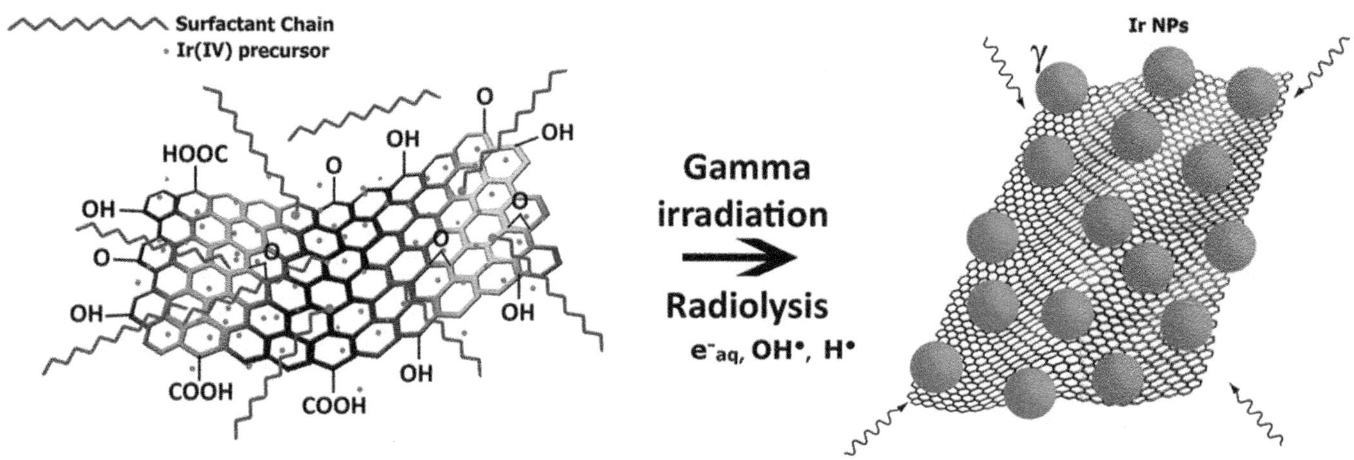

FIGURE 10.9 Irradium nanoparticles on graphene oxide. Schematic representation of the synthesis of iridium nanoparticles on graphene oxide (Karim et al., 2007). *Rojas, J. V, Molina Higgins, M. C., Toro Gonzalez, M., Castano, C. E. (2015). Single step radiolytic synthesis of iridium nanoparticles onto graphene oxide. Applied Surface Science 357, 2087–2093. https://doi.org/10.1016/j.apsusc.2015.09.190.*

stable end groups, polymer radicals are stabilized resulting in resin polymers with reduced chain lengths and lower molecular weights. Chain scission is primarily responsible for degrading polymers becoming more solubilized. In lithography, the degradation of a polymer induced by controlled radiation has long been used to improve its solubility. Generating patterns in resist materials is not possible by a focused beam of energetic particles containing electrons, ions, and X-rays instead photolithography is ideal for this purpose because due to short wavelengths (*0.1 nm) and large depths of focus, these techniques produce extremely high resolutions. Electron beam lithography has demonstrated the use of polymers and a wide variety of materials. A polymer that is subject to extensive chain scissions when exposed to ionizing radiation is poly(methyl methacrylate) PMMA which is highly sensitive to radiation and therefore used frequently as a positive resist in lithography. Using ionizing radiation for photolithography is the first commercial application of controlled degradation of polymers in nanotechnology. A collection of recent works has been published in a thematic issue on photolithography by Huth.

In polymer industry, well-tested ionizing radiations offer properties that are suitable for the fabrication of innovative and functionalized polymers at relatively low dimension. Ionizing radiation produces homogeneously in the media exposed to it, the precursors of chemical reactions namely free radicals as a consequence of very high energies of photons (0.66−1.25 MeV) or electrons (several keV-10 MeV) and their ability to penetrate solid materials. Polymers and monomers in any state can be irradiated in any temperature range above and below ambient. Despite its unique properties, ionizing radiation is a very powerful way to prepare nanostructured materials. Nanoparticles and nanogels are formed when aqueous polymer solutions are irradiated with radiation, as well as polymer-clay nanocomposites and polymer-metal nanocomposites.

10.1.6 *Composite nanomaterials*

Recent advances in hybrid nanomaterials, also known as nanocomposites, have been widely studied due to their interesting properties (Karim et al., 2007; Malta, Louarn, Errien, & Torresi, 2003) in many technological applications, primarily in biomedical fields (Murali Mohan, Lee, Premkumar, & Geckeler, 2007; Satarkar & Zach Hilt, 2008). A single-phase system cannot handle the properties of these NPs, which combine metals and polymers (Fig. 10.10). Physical mixing (Akelah, El-Deen, Hiltner, Baer, & Moet, 1995), sol-gel (Jang, Han, & Im, 2000), emulsions (He, 2005), gamma-irradiation (Yin, Xu, & Zhang, 1998), sonochemical process (Kumar, Diamant, & Gedanken, 2000),

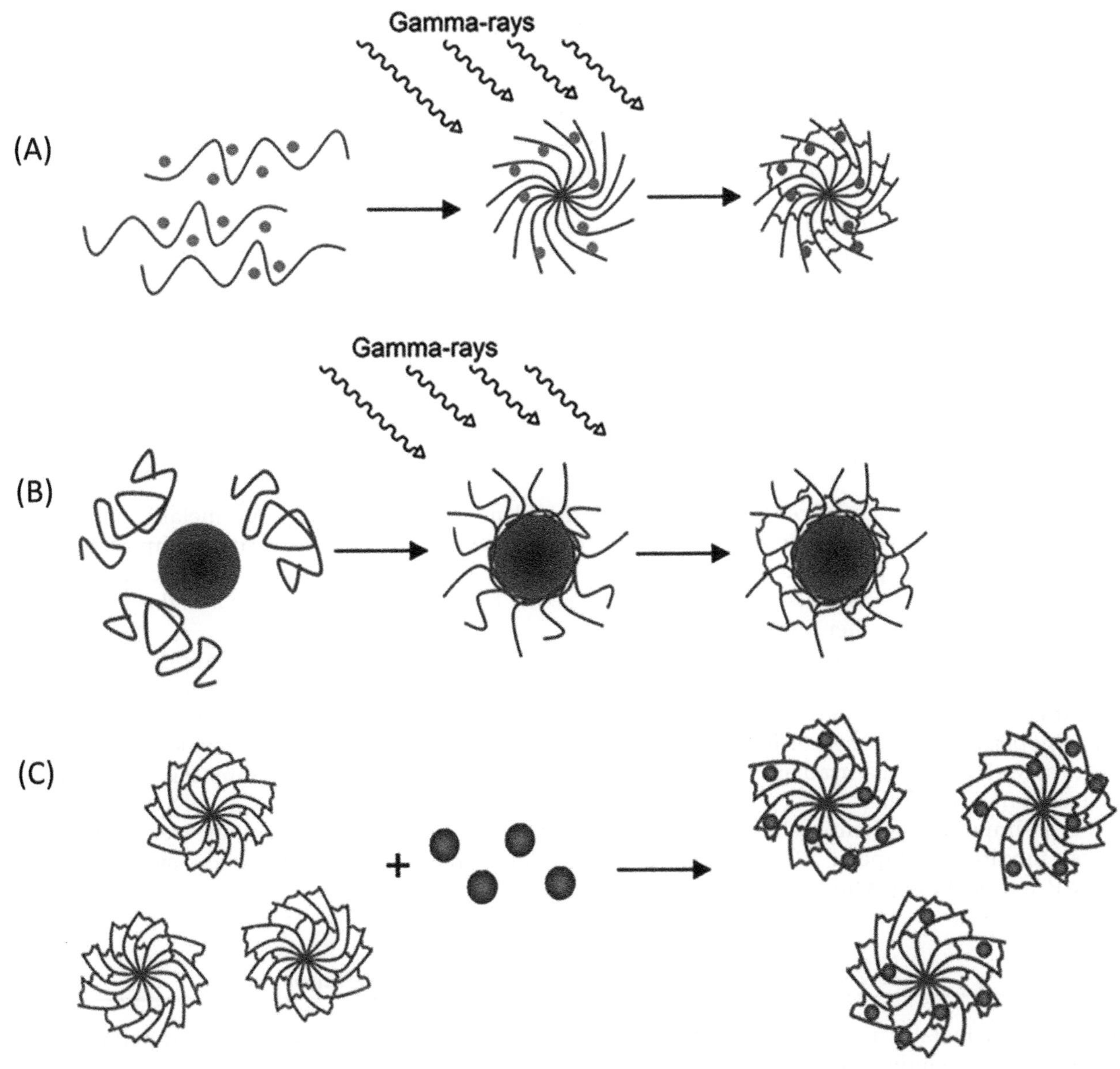

FIGURE 10.10 Nanocomposites synthesis. Nanocomposites are synthesized either in (A) one step, (B) two steps, or (C) by mixing preformed nanoparticles together. *From Flores-Rojas, G. G., López-Saucedo, F., & Bucio, E. (2020). Gamma-irradiation applied in the synthesis of metallic and organic nanoparticles: A short review. Radiation Physics and Chemistry, 169, 107962. https://doi.org/10.1016/j.radphyschem.2018.08.011.*

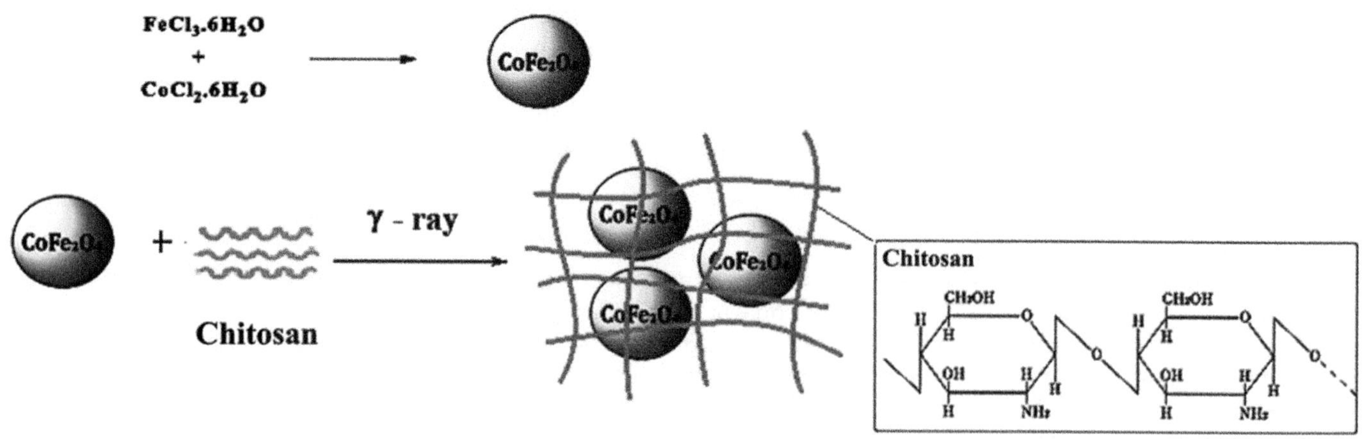

FIGURE 10.11 Nanogel. Metal oxide nanoparticles in nanogel (Kang et al., 2006). *From Taleb, M. F. A. (2014). Adsorption and photocatalytic degradation of 2-CP in wastewater onto CS/$CoFe_2O_4$ nanocomposite synthesized using gamma radiation. Carbohydrate Polymers, 114, 65–72. https://doi.org/10.1016/j.carbpol.2014.07.061.*

as well as cross-linking and polymerization in situ in aqueous solutions have all been employed to synthesize these nanoscale materials (Ray & Biswas, 2000). Nonetheless, irradiation method is extensively utilized for generating metallic and organic nanoparticles at room temperature and normal pressure, adaptability, and little or no impurities (Ershov & Henglein, 1993).

During radiological treatment of an aqueous monomer or polymer and metal ion, two radicals are produced: $H^{\bullet}$ and $OH^{\bullet}$. The e_{aq}^{-}, by contrast, is a powerful reductant that can reduce the metal ions M^{n+} to lower valences and eventually to the metallic state, where these processes can occur simultaneously (Mostafavi, Dey, François, & Belloni, 2002).

In the first case, the nanocomposite is created by combining monomer or polymer with metallic ions, followed by gamma-irradiation in an inert atmosphere. In the second case, the nanocomposites are produced by combining monomer or polymer with the metallic ions. In the second case, the nuclei are first synthesized, either metallic or polymeric, before the shell is synthesized to complete the process of nanocomposite synthesis, during which stages it is possible to use either a chemical or gamma-ray process. In either case, stabilizing agents are often required.

Nanocomposites can be categorized based on the synthesis method, for instance, into metallic nanocomposites with polymeric capping to polymeric nanocomposites containing NPs or polymeric nanocomposites containing metallic nanoparticles. As opposed to this last synthesis method, which produces a nanometric and metallic NPs separately and then combine to make the nanocomposites, the two-step synthesis offers a higher opportunity for selecting the desired nuclei either polymeric or metallic.

PVA, PANI, and poly(N-isopropylacrylamide) as a polymer shell are common combinations of metallic nuclei and polymeric shells. The use of some natural polymers such as chitosan, starch, and cellulose to make nanogels as nanomatrices has also been reported, such as the magnetic nanocomposite synthesized by Abou et al. using magnetic nanoparticles composed of Fe^{3+} and Co^{2+}, A chitosan nanomatrix was then formed using these magnetic nuclei as well as gamma irradiation to complete the synthesis of the nanocomposite (Fig. 10.11), with a size range of 18.2 to 31.7 nm (Taleb, 2014).

Lastly, it is important to emphasize that neither the capping nor the nanomatrix, as well as the synthesis method, can modify the shape of NPs. This behavior was demonstrated by Kang et al., who synthesized Ag NPs covered by PANI obtaining cylindrical and hexagonal shapes (Kang et al., 2006).

10.2 Application of nanoparticles and nanocomposites

In recent years, metallic nanoparticles have been investigated for their antimicrobial properties in various health applications, including water purification, sterile surfaces for biomedical devices, and packaging for food products (Ghosh et al., 2010; González-Sánchez et al., 2015). Various studies have reported strong inhibitory and bactericidal effects of Ag^{+} salts or AgNPs, as well as a wide range of antimicrobial effects on bacteria, fungi, and viruses (González-Sánchez et al., 2015; Yoksan & Chirachanchai, 2009). A higher toxicity than that of other metals, Ag is less toxic to mammals than that of microorganisms (Zhao & Stevens, 1998). At the nanoscale, Ag's antimicrobial properties can be further enhanced. There has been no conclusive proof showing that Ag nanoparticles are an effective weapon against multidrug-resistant (MDR) bacteria as well as multidrug-susceptible strains and that they are effective against MDR bacteria, such as methicillin-resistant *Staphylococcus aureus* and ampicillin-resistant Pseudomonas aeruginosa (Rai, 2012; Loos, 2015). Researchers have explored how Ag NPs can be immobilized on textile fabricsto create antibacterial fabrics.

Although this metal has antimicrobial properties, other metals have also become targets of studies, in an attempt to enhance the properties of these metals for biomedical purposes, by reducing their toxicity and increasing their antimicrobial activity; these include, for instance, gold (Shamaila et al., 2016), copper (Bogdanović et al., 2014), and platinum (Gopal, Hasan, Manikandan, & Wu, 2013), all of which have their own advantages, and titanium and zinc oxide as well (Jesline, John, Narayanan, Vani, & Murugan, 2015; Sirelkhatim et al., 2015). Researchers have also begun studying biomaterials and mixtures in general for their use as antitumor agents, either for imaging diagnostics (Weissleder et al., 1990), drug release (Chen et al., 2013), or by controlling the thermal destruction of cancer cells (Medhat, Mansour, El-sonbaty, Kandil, & Mahmoud, 2017; Wang et al., 2014). A polymeric coating with pH- and thermo-responsive properties (Roy, Patra, Madhuri, & Sharma, 2016) can be applied to metallic nanoparticles, and other types of polymers with thermal (Strong & West, 2011), photo (Gu et al., 2003), or magnetic-responsive (Weinstein et al., 2009) properties are commonly used in biomedical applications, as well as in the creation of sensors for use in various devices, because of interesting combination of physical and chemical properties in nanomaterials.

In terms of the applications of organic nanoparticles (polymeric nanoparticles), among the most common are those related to delivery systems of numerous medicines because of their highly biocompatible characteristics (Singh et al., 2017), but these applications are beyond imagination as researchers are beginning to explore the area of novel nanocomposites which will potentially be used in drug delivery, biomedicine, bioengineering such as self-healing characteristics (Nejadnik et al., 2014), ability to self-assemble (Grzelczak, Vermant, Furst, & Liz-Marzán, 2010; Yang et al., 2017), superconductivity (Tinkham, 2000; Tinkham & Nanowires, 2000), and also in nanoelectronics components in microdevices (Lee & Zhong, 2014). In summary, the future of device development using polymer looks promising with the development and applicability of novel and increasingly complex materials in conjugation with the development of different methods which are compatible with the energy and precision requirements of gamma ray radiation process.

10.3 Conclusions

It takes a high level of engineering efforts to bring nanomaterials to market since most of them are being developed for laboratory applications. In this chapter, we describe how radiation engineering can be used to revolutionize nanomaterial preparation and modification. As ionizing radiation can be used to reduce energy requirements and provide a more environmentally friendly approach than conventional methods, extensive research has been done to prove this. In the future, ionizing radiation will not be limited to conventional methods for preparing nanomaterials but will be used in other applications as well. In many studies, ionizing radiation has been shown to be more energy efficient and environmentally friendly than conventional methods. As nanomaterials are developed, ionizing radiation may replace other conventional techniques, not only for current applications but also for developing new ones in the future.

References

Abedini, A., Daud, A. R., Abdul Hamid, M. A., Kamil Othman, N., & Saion, E. (2013). A review on radiation-induced nucleation and growth of colloidal metallic nanoparticles. *Nanoscale Research Letters*, *8*(1), 474. Available from https://doi.org/10.1186/1556-276X-8-474.

Abedini, A., Larki, F., Saion, E., Zakaria, A., & Zobir Hussein, M. (2012a). Influence of dose and ion concentration on formation of binary Al–Ni alloy nanoclusters. *Radiation Physics and Chemistry*, *81*(10), 1653–1658. Available from https://doi.org/10.1016/j.radphyschem.2012.05.015.

Abedini, A., Saion, E., Larki, F., Zakaria, A., Noroozi, M., & Soltani, N. (2012b). Room temperature radiolytic synthesized Cu@$CuAlO_2$-Al_2O_3 nanoparticles. *International Journal of Molecular Sciences*, *13*(9), 11941–11953. Available from https://doi.org/10.3390/ijms130911941.

Akelah, A., El-Deen, N. S., Hiltner, A., Baer, E., & Moet, A. (1995). Organophilic rubber-montmorillonite nanocomposites. *Materials Letters*, *22*(1), 97–102. Available from https://doi.org/10.1016/0167-577X(94)00167-7.

Alrehaily, L. M., Joseph, J. M., Biesinger, M. C., Guzonas, D. A., & Wren, J. C. (2013). Gamma-radiolysis-assisted cobalt oxide nanoparticle formation. *Physical Chemistry Chemical Physics: PCCP*, *15*(3), 1014–1024. Available from https://doi.org/10.1039/C2CP43094K.

Anastas, P., & Eghbali, N. (2010). Green chemistry: Principles and practice. *Chemical Society Reviews, 39*(1), 301–312. Available from https://doi.org/10.1039/B918763B.

Ansón-Casaos, A., Puértolas, J. A., Pascual, F. J., Hernández-Ferrer, J., Castell, P., Benito, A. M., ... Martínez, M. T. (2014). The effect of gamma-irradiation on few-layered graphene materials. *Applied Surface Science, 301*, 264–272. Available from https://doi.org/10.1016/j.apsusc.2014.02.057.

Belloni, J. (2006). Nucleation, growth and properties of nanoclusters studied by radiation chemistry: Application to catalysis. *Catalysis Today, 113*(3), 141–156. Available from https://doi.org/10.1016/j.cattod.2005.11.082.

Bogdanović, U., Lazić, V., Vodnik, V., Budimir, M., Marković, Z., & Dimitrijević, S. (2014). Copper nanoparticles with high antimicrobial activity. *Materials Letters, 128*, 75–78. Available from https://doi.org/10.1016/j.matlet.2014.04.106.

Chen, S., Wu, G., Liu, Y., & Long, D. (2006). Preparation of poly(acrylic acid) grafted multiwalled carbon nanotubes by a two-step irradiation technique. *Macromolecules, 39*(1), 330–334. Available from https://doi.org/10.1021/ma0520500.

Chen, W.-H., Xu, X.-D., Jia, H.-Z., Lei, Q., Luo, G.-F., Cheng, S.-X., ... Zhang, X.-Z. (2013). Therapeutic nanomedicine based on dual-intelligent functionalized gold nanoparticles for cancer imaging and therapy in vivo. *Biomaterials, 34*(34), 8798–8807. Available from https://doi.org/10.1016/j.biomaterials.2013.07.084.

Cho, J. S., Jang, W., Mun, S. C., Yi, M., Park, J. H., & Wang, D. H. (2018). Tuning surface chemistry and morphology of graphene oxide by γ-ray irradiation for improved performance of perovskite photovoltaics. *Carbon, 139*, 564–571. Available from https://doi.org/10.1016/j.carbon.2018.07.010.

Choi, J., Jun, Y., Yeon, S.-I., Kim, H. C., Shin, J.-S., & Cheon, J. (2006). Biocompatible heterostructured nanoparticles for multimodal biological detection. *Journal of the American Chemical Society, 128*(50), 15982–15983. Available from https://doi.org/10.1021/ja066547g.

Dispenza, C., Grimaldi, N., Sabatino, M. A., Soroka, I. L., & Jonsson, M. (2015). Radiation-engineered functional nanoparticles in aqueous systems. *Journal of Nanoscience and Nanotechnology, 15*(5), 3445–3467. Available from https://doi.org/10.1166/jnn.2015.9865.

Ekka, B., Nayak, S. R., Achary, L. S. K., Sarita., Kumar, A., Mawatwal, S., ... Patel, R. K. (2018). Synthesis of hydroxyapatite-zirconia nanocomposite through sonochemical route: A potential catalyst for degradation of phenolic compounds. *Journal of Environmental Chemical Engineering, 6*(5), 6504–6515. Available from https://doi.org/10.1016/j.jece.2018.09.026.

Ershov, B. G., & Henglein, A. (1993). Optical spectrum and some chemical properties of colloidal thallium in aqueous solution. *The Journal of Physical Chemistry, 97*(13), 3434–3436. Available from https://doi.org/10.1021/j100115a056.

Flores-Rojas, G. G., López-Saucedo, F., & Bucio, E. (2020). Gamma-irradiation applied in the synthesis of metallic and organic nanoparticles: A short review. *Radiation Physics and Chemistry, 169*, 107962. Available from https://doi.org/10.1016/j.radphyschem.2018.08.011.

Gao, S., Zhu, L., Liu, L., Gao, A., Liao, F., & Shao, M. (2016). Improved energy storage performance based on gamma-ray irradiated activated carbon cloth. *Electrochimica Acta, 191*, 908–915. Available from https://doi.org/10.1016/j.electacta.2016.01.151.

Ghosh, S., Kaushik, R., Nagalakshmi, K., Hoti, S. L., Menezes, G. A., Harish, B. N., & Vasan, H. N. (2010). Antimicrobial activity of highly stable silver nanoparticles embedded in agar–agar matrix as a thin film. *Carbohydrate Research, 345*(15), 2220–2227. Available from https://doi.org/10.1016/j.carres.2010.08.001.

González-Sánchez, M. I., Perni, S., Tommasi, G., Morris, N. G., Hawkins, K., López-Cabarcos, E., & Prokopovich, P. (2015). Silver nanoparticle based antibacterial methacrylate hydrogels potential for bone graft applications. *Materials Science & Engineering C-Materials for Biological Applications, 50*, 332–340. Available from https://doi.org/10.1016/j.msec.2015.02.002.

Gopal, J., Hasan, N., Manikandan, M., & Wu, H.-F. (2013). Bacterial toxicity/compatibility of platinum nanospheres, nanocuboids and nanoflowers. *Scientific Reports, 3*(1), 1260. Available from https://doi.org/10.1038/srep01260.

Grzelczak, M., Vermant, J., Furst, E. M., & Liz-Marzán, L. M. (2010). Directed self-assembly of nanoparticles. *ACS Nano, 4*(7), 3591–3605. Available from https://doi.org/10.1021/nn100869j.

Gu, F., Fen Wang, S., Feng Song, C., Kai Lü, M., Xin Qi, Y., Jun Zhou, G., ... Rong Yuan, D. (2003). Synthesis and luminescence properties of SnO_2 nanoparticles. *Chemical Physics Letters, 372*(3), 451–454. Available from https://doi.org/10.1016/S0009-2614(03)00440-8.

He, Y. (2005). Synthesis of polyaniline/nano-CeO_2 composite microspheres via a solid-stabilized emulsion route. *Materials Chemistry and Physics, 92*(1), 134–137. Available from https://doi.org/10.1016/j.matchemphys.2005.01.033.

Jain, N., Bhargava, A., Majumdar, S., Tarafdar, J. C., & Panwar, J. (2011). Extracellular biosynthesis and characterization of silver nanoparticles using Aspergillus flavus NJP08: A mechanism perspective. *Nanoscale, 3*(2), 635–641. Available from https://doi.org/10.1039/C0NR00656D.

Janata, E., Henglein, A., & Ershov, B. G. (1994). First clusters of Ag^+ ion reduction in aqueous solution. *The Journal of Physical Chemistry, 98*(42), 10888–10890. Available from https://doi.org/10.1021/j100093a033.

Jang, S. H., Han, M. G., & Im, S. S. (2000). Preparation and characterization of conductive polyaniline/silica hybrid composites prepared by sol–gel process. *Synthetic Metals, 110*(1), 17–23. Available from https://doi.org/10.1016/S0379-6779(99)00176-9.

Jesline, A., John, N. P., Narayanan, P. M., Vani, C., & Murugan, S. (2015). Antimicrobial activity of zinc and titanium dioxide nanoparticles against biofilm-producing methicillin-resistant Staphylococcus aureus. *Applied Nanoscience, 5*(2), 157–162. Available from https://doi.org/10.1007/s13204-014-0301-x.

Kang, Y.-O., Choi, S.-H., Gopalan, A., Lee, K.-P., Kang, H.-D., & Song, Y. S. (2006). Tuning of morphology of Ag nanoparticles in the Ag/polyaniline nanocomposites prepared by γ-ray irradiation. *Journal of Non-Crystalline Solids, 352*(5), 463–468. Available from https://doi.org/10.1016/j.jnoncrysol.2006.01.043.

Karim, M. R., Lim, K. T., Lee, C. J., Bhuiyan, M. T. I., Kim, H. J., Park, L.-S., & Lee, M. S. (2007). Synthesis of core-shell silver–polyaniline nanocomposites by gamma radiolysis method. *Journal of Polymer Science Part A-Polymer Chemistry, 45*(24), 5741–5747. Available from https://doi.org/10.1002/pola.22323.

Kim, M., Mun, S. C., Lee, C. S., Lee, M. H., Son, Y., & Park, O. O. (2011). Electrical and rheological properties of polyamide 6,6/γ-ray irradiated multi-walled carbon nanotube composites. *Carbon, 49*(12), 4024–4030. Available from https://doi.org/10.1016/j.carbon.2011.05.044.

Kulkarni, N., & Muddapur, U. (2014). Biosynthesis of metal nanoparticles: A review. *Journal of Nanotechnology, 2014*, 510246. Available from https://doi.org/10.1155/2014/510246.

Kumar, R. V., Diamant, Y., & Gedanken, A. (2000). Sonochemical synthesis and characterization of nanometer-size transition metal oxides from metal acetates. *Chemistry of Materials: A Publication of the American Chemical Society, 12*(8), 2301–2305. Available from https://doi.org/10.1021/cm000166z.

Lee, S., & Zhong, Z. (2014). Nanoelectronic circuits based on two-dimensional atomic layer crystals. *Nanoscale, 6*(22), 13283–13300. Available from https://doi.org/10.1039/C4NR03670K.

Li, Z., Yang, Y., Relefors, A., Kong, X., Siso, G. M., Wickman, B., ... Soroka, I. L. (2021). Tuning morphology, composition and oxygen reduction reaction (ORR) catalytic performance of manganese oxide particles fabricated by γ-radiation induced synthesis. *Journal of Colloid and Interface Science, 583*, 71–79. Available from https://doi.org/10.1016/j.jcis.2020.09.011.

Loos, M. (2015). Chapter 1 - Nanoscience and nanotechnology. In M. Loos (Ed.), *Carbon nanotube reinforced composites* (pp. 1–36). Oxford: William Andrew Publishing. Available from https://doi.org/10.1016/B978-1-4557-3195-4.00001-1.

Malta, M., Louarn, G., Errien, N., & Torresi, R. M. (2003). Nanofibers composite vanadium oxide/polyaniline: Synthesis and characterization of an electroactive anisotropic structure. *Electrochemistry Communications, 5*(12), 1011–1015. Available from https://doi.org/10.1016/j.elecom.2003.09.016.

Marignier, J. L., Belloni, J., Delcourt, M. O., & Chevalier, J. P. (1985). Microaggregates of non-noble metals and bimetallic alloys prepared by radiation-induced reduction. *Nature, 317*(6035), 344–345. Available from https://doi.org/10.1038/317344a0.

Medhat, A., Mansour, S., El-sonbaty, S., Kandil, E., & Mahmoud, M. (2017). Evaluation of the antitumor activity of platinum nanoparticles in the treatment of hepatocellular carcinoma induced in rats. *Tumor Biology, 39*(7). Available from https://doi.org/10.1177/1010428317717259, 1010428317717259.

Mostafavi, M., Dey, G. R., François, L., & Belloni, J. (2002). Transient and stable silver clusters induced by radiolysis in methanol. *The Journal of Physical Chemistry. A, 106*(43), 10184–10194. Available from https://doi.org/10.1021/jp014257a.

Mostafavi, M., Remita, S., Delcourt, M. O., & Belloni, J. (1828–1842). Ligand effects on solvated metal cluster properties. *Journal of Chemical Physics, 1996*, 93. Available from https://doi.org/10.1051/jcp/1996931828.

Murali Mohan, Y., Lee, K., Premkumar, T., & Geckeler, K. E. (2007). Hydrogel networks as nanoreactors: A novel approach to silver nanoparticles for antibacterial applications. *Polymer, 48*(1), 158–164. Available from https://doi.org/10.1016/j.polymer.2006.10.045.

Nejadnik, M. R., Yang, X., Bongio, M., Alghamdi, H. S., van den Beucken, J. J. J. P., Huysmans, M. C., ... Leeuwenburgh, S. C. G. (2014). Self-healing hybrid nanocomposites consisting of bisphosphonated hyaluronan and calcium phosphate nanoparticles. *Biomaterials, 35*(25), 6918–6929. Available from https://doi.org/10.1016/j.biomaterials.2014.05.003.

Radiation Applications. (2018). In H. Kudo (Ed.), *An advanced course in nuclear engineering* (07). Singapore: Springer Singapore. Available from https://doi.org/10.1007/978-981-10-7350-2.

Radiation Synthesis of Materials and Compounds. (2016). In B. I. Kharisov, O. V. Kharissova, & U. Ortiz Mendez (Eds.), CRC Press. Available from https://doi.org/10.1201/b14531.

Radiation. (2019). Elsevier. Available from https://doi.org/10.1016/C2016-0-02609-8.

Rai, M. K., Deshmukh, S. D., Ingle, A. P., & Gade, A. K. (2012). Silver nanoparticles: The powerful nanoweapon against multidrug-resistant bacteria. *Journal of Applied Microbiology, 112*(5), 841–852. Available from https://doi.org/10.1111/j.1365-2672.2012.05253.x.

Rao, Y. N., Banerjee, D., Datta, A., Das, S. K., Guin, R., & Saha, A. (2010). Gamma irradiation route to synthesis of highly re-dispersible natural polymer capped silver nanoparticles. *Radiation Physics and Chemistry, 79*(12), 1240–1246. Available from https://doi.org/10.1016/j.radphyschem.2010.07.004.

Ray, S. S., & Biswas, M. (2000). Water-dispersible conducting nanocomposites of polyaniline and poly(N-vinylcarbazole) with nanodimensional zirconium dioxide. *Synthetic Metals, 108*(3), 231–236. Available from https://doi.org/10.1016/S0379-6779(99)00258-1.

Rojas, J. V., & Castano, C. H. (2012). Production of palladium nanoparticles supported on multiwalled carbon nanotubes by gamma irradiation. *Radiation Physics and Chemistry, 81*(1), 16–21. Available from https://doi.org/10.1016/j.radphyschem.2011.08.010.

Rojas, J. V., Molina Higgins, M. C., Toro Gonzalez, M., & Castano, C. E. (2015). Single step radiolytic synthesis of iridium nanoparticles onto graphene oxide. *Applied Surface Science, 357*, 2087–2093. Available from https://doi.org/10.1016/j.apsusc.2015.09.190.

Roy, E., Patra, S., Madhuri, R., & Sharma, P. K. (2016). Stimuli-responsive poly(N-isopropyl acrylamide)-co-tyrosine@gadolinium: Iron oxide nanoparticle-based nanotheranostic for cancer diagnosis and treatment. *Colloids Surfaces B Biointerfaces, 142*, 248–258. Available from https://doi.org/10.1016/j.colsurfb.2016.02.053.

Satarkar, N. S., & Zach Hilt, J. (2008). Hydrogel nanocomposites as remote-controlled biomaterials. *Acta Biomaterialia, 4*(1), 11–16. Available from https://doi.org/10.1016/j.actbio.2007.07.009.

Shamaila, S., Zafar, N., Riaz, S., Sharif, R., Nazir, J., & Naseem, S. (2016). Gold nanoparticles: An efficient antimicrobial agent against enteric bacterial human pathogen. *Nanomaterials, 6*(4). Available from https://doi.org/10.3390/nano6040071.

Singh, Y., Meher, J. G., Raval, K., Khan, F. A., Chaurasia, M., Jain, N. K., & Chourasia, M. K. (2017). Nanoemulsion: Concepts, development and applications in drug delivery. *Journal of Controlled Release: Official Journal of the Controlled Release Society, 252*, 28–49. Available from https://doi.org/10.1016/j.jconrel.2017.03.008.

Sirelkhatim, A., Mahmud, S., Seeni, A., Kaus, N. H. M., Ann, L. C., Bakhori, S. K. M., ... Mohamad, D. (2015). Review on zinc oxide nanoparticles: Antibacterial activity and toxicity mechanism. *Nano-Micro Letters, 7*(3), 219–242. Available from https://doi.org/10.1007/s40820-015-0040-x.

Spinks, J. W. T., & Woods, R. J. (1990). *An introduction to radiation chemistry*. United States: John Wiley and Sons Inc.

Strong, L. E., & West, J. L. (2011). Thermally responsive polymer–nanoparticle composites for biomedical applications. *WIREs Nanomedicine and Nanobiotechnology, 3*(3), 307–317. Available from https://doi.org/10.1002/wnan.138.

Sutherland, T. I., Sparks, C. J., Joseph, J. M., Wang, Z., Whitaker, G., Sham, T. K., & Wren, J. C. (2017). Effect of ferrous ion concentration on the kinetics of radiation-induced iron-oxide nanoparticle formation and growth. *Physical Chemistry Chemical Physics: PCCP, 19*(1), 695–708. Available from https://doi.org/10.1039/C6CP05456K.

Taleb, M. F. A. (2014). Adsorption and photocatalytic degradation of 2-CP in wastewater onto CS/$CoFe_2O_4$ nanocomposite synthesized using gamma radiation. *Carbohydrate Polymers, 114*, 65–72. Available from https://doi.org/10.1016/j.carbpol.2014.07.061.

Tinkham, M. (2000). Limits on superconductivity in nanoparticles and nanowires. *Journal of Superconductivity, 13*(5), 801–804. Available from https://doi.org/10.1023/A:1007847122807.

Tinkham, M. (2000). In I. O. Kulik, & R. Ellialtioğlu (Eds.), *Superconducting nanoparticles and nanowires BT - Quantum mesoscopic phenomena and mesoscopic devices in microelectronics* (pp. 349–360). Dordrecht: Springer Netherlands. Available from https://doi.org/10.1007/978-94-011-4327-1_23.

Wang, Z., Chen, Z., Liu, Z., Shi, P., Dong, K., Ju, E., ... Qu, X. (2014). A multi-stimuli responsive gold nanocage–hyaluronic platform for targeted photothermal and chemotherapy. *Biomaterials, 35*(36), 9678–9688. Available from https://doi.org/10.1016/j.biomaterials.2014.08.013.

Weinstein, J. S., Varallyay, C. G., Dosa, E., Gahramanov, S., Hamilton, B., Rooney, W. D., ... Neuwelt, E. A. (2009). Superparamagnetic iron oxide nanoparticles: Diagnostic magnetic resonance imaging and potential therapeutic applications in neurooncology and central nervous system inflammatory pathologies, a review. *Journal of Cerebral Blood Flow and Metabolism: Official Journal of the International Society of Cerebral Blood Flow and Metabolism, 30*(1), 15–35. Available from https://doi.org/10.1038/jcbfm.2009.192.

Weissleder, R., Elizondo, G., Wittenberg, J., Rabito, C. A., Bengele, H. H., & Josephson, L. (1990). Ultrasmall superparamagnetic iron oxide: Characterization of a new class of contrast agents for MR imaging. *Radiology, 175*(2), 489–493. Available from https://doi.org/10.1148/radiology.175.2.2326474.

Xu, Z., Chen, L., Liu, L., Wu, X., & Chen, L. (2011). Structural changes in multi-walled carbon nanotubes caused by γ-ray irradiation. *Carbon, 49*(1), 350–351. Available from https://doi.org/10.1016/j.carbon.2010.09.023.

Yadav, P., Olsson, R. T., & Jonsson, M. (2009). Synthesis and characterization of MnO_2 colloids. *Radiation Physics and Chemistry, 78*(11), 939–944. Available from https://doi.org/10.1016/j.radphyschem.2009.02.006.

Yang, M., Chan, H., Zhao, G., Bahng, J. H., Zhang, P., Král, P., & Kotov, N. A. (2017). Self-assembly of nanoparticles into biomimetic capsid-like nanoshells. *Nature Chemistry, 9*(3), 287–294. Available from https://doi.org/10.1038/nchem.2641.

Yin, Y., Xu, X., & Zhang, Z. (1998). Synthesis of cadmium sulfide nanoparticles in situ using γ-radiation. *Chemical Communications* (16), 1641–1642. Available from https://doi.org/10.1039/A802910E.

Yoksan, R., & Chirachanchai, S. (2009). Silver nanoparticles dispersing in chitosan solution: Preparation by γ-ray irradiation and their antimicrobial activities. *Materials Chemistry and Physics., 115*(1), 296–302. Available from https://doi.org/10.1016/j.matchemphys.2008.12.001.

Yu, C., Zhang, B., Yan, F., Zhao, J., Li, J., Li, L., & Li, J. (2016). Engineering nano-porous graphene oxide by hydroxyl radicals. *Carbon, 105*, 291–296. Available from https://doi.org/10.1016/j.carbon.2016.04.050.

Zhang, Y., Ma, H.-L., Zhang, Q., Peng, J., Li, J., Zhai, M., & Yu, Z.-Z. (2012). Facile synthesis of well-dispersed graphene by γ-ray induced reduction of graphene oxide. *Journal of Materials Chemistry, 22*(26), 13064–13069. Available from https://doi.org/10.1039/C2JM32231E.

Zhao, G., & Stevens, S. E. (1998). Multiple parameters for the comprehensive evaluation of the susceptibility of Escherichia coli to the silver ion. *Biometals, 11*(1), 27–32. Available from https://doi.org/10.1023/A:1009253223055.

CHAPTER

11

Gamma and electron beam for food processing

Amilcar L. Antonio[1,2], *Ângela Fernandes*[1,2], *Eliana Pereira*[1,2], *João C.M. Barreira*[1,2], *Isabel C.F.R. Ferreira*[1,2] *and Lillian Barros*[1,2]

[1]Centro de Investigação de Montanha (CIMO), Instituto Politécnico de Bragança, Campus de Santa Apolónia, Bragança, Portugal [2]Laboratório Associado para a Sustentabilidade e Tecnologia em Regiões de Montanha (SusTEC), Instituto Politécnico de Bragança, Campus de Santa Apolónia, Bragança, Portugal

11.1 Introduction

Healthy and safe foods contribute to human well-being. Food processing technologies come on this behalf to make more food items available, extending their shelf-life and increasing their quality, eliminating microbiological contaminations, and increasing, in some cases, the bioavailability of some nutrients as well. One of the oldest food processes, cooking, meets these requirements.

However, there are food items that, due to their thermal sensitivity, cannot be processed in this simple way. For example, the elimination of insects (weevils) on cherries or chestnuts consumed fresh, could not be attained by thermal treatment without compromising some of the fruit's properties.

On alternative, there are post-harvest chemical treatments, but these present the inconvenience of not being environment friendly and, in some cases, leaving residues and being dangerous for the operators who apply the process.

As an alternative, the focus of food processing technologies is being directed to physical methods, old or new, adapted to each objective. One of them, hot water treatment, is being used for post-harvest treatment (e.g., disinfestation) of some fruits. Other new physical methods, alone or combined, such as high-pressure processing or modified atmosphere packaging (MAP), are being applied with success on several marketed food products.

Food irradiation, with different technologies (UV, gamma, e-beam, X-rays), comes in this line, to make available a process that contributes to increasing the quality and quantity of food items available in the market. The impact of each process on food attributes should be validated case by case, to document the possibility to use it or not for the target purpose, with less effects on main product characteristics, from the perspective of nutrition, safety, and overall cost.

Following this line, this chapter presents the results of the impact of gamma and e-beam irradiation processing on the main nutritional and bioactive parameters of chestnut fruits, mushrooms, and medicinal and aromatic plants (MAPs).

11.2 Food irradiation technologies: gamma and electron-beam

Non-thermal processes for food preservation, using gamma or electron-beam radiation, are in current use for different types of food products (Eustice, 2018). Gamma radiation comes from the spontaneous emission of photons by an unstable atom, that emits an excess of energy in the nucleus. This energy is generally in the magnitude of mega-electron Volt (MeV, $1\ \text{MeV} = 1.6 \times 10^{-13}$ J). For comparison, UV radiation has an energy of 5 to 10 eV, while X-rays medical apparatus have energies in the range of keV's (kilo-electron Volt) (NASA, 2014).

Food Packaging and Preservation.
DOI: https://doi.org/10.1016/B978-0-323-90044-7.00011-2

Due to its characteristics, gamma radiation could be used to process food boxes of one cubic meter, with high throughput. Also, due to this high energy, it is very efficient in the elimination of bacteria or viruses and lower fungi contamination (Cabo Verde, 2018).

Industrial processing of food by gamma radiation should also guarantee good dose uniformity. For that, the product rotates inside the irradiation chamber. As an analogy, we may think of the microwave apparatus that we have in our kitchens. Similarly, the gamma sources are in a cage and not in contact with the food being processed.

Currently, the Cobalt-60 isotope (^{60}Co) is used as a gamma source in different industrial setups. The sources are stored in a pool, filled with water or dry, on safe conditions for the operators, and always monitored by several instruments and standard procedures, under strict legislation to assure the safety of all the personnel involved in the process (Carreño, 2018). For food processing, the sources are raised and the product is in front of them, rotating to guarantee good dose uniformity (Craven, Schlecht, & Stein, 2018). The dose is expressed in Gray (Gy), which means Joule per kilogram, and the exposure time is adjusted to the target dose, depending on the objective: sprout inhibition (less than 0.5 kGy); insects disinfestation (up to 0.5 kGy); shelf-life extension (1 to 2 kGy); microbial decontamination (up to 5 kGy); or food sterilization (more than 5 kGy) (Antonio, Cabo Verde, & Ferreira, 2018).

Generally, industrial gamma irradiation facilities are multi-purpose, also being used to sterilize medical devices, pharmaceutical products, recipients, or boxes for food. Since this is a nonthermal process with high versatility, the initial high investment costs on the unit are recovered in a short time, making the irradiation food viable (despite its seasonality) at a low competitive price when compared with other processes (Dethier & Mullier, 2018).

Electron-beam (e-beam) radiation processing consists in directing accelerated electrons of high energy to the product, to get the desired effect, with a maximum energy of 10 MeV (mega electron-volt) for food processing. Higher energies allow higher penetration in the product, but even at a maximum energy of 10 MeV, the penetration is limited to a few centimeters (Miller, 2018).

Due to that, products are irradiated on a layer or, in some industrial units, they have two e-beam irradiating the top and bottom of the product, to guarantee a good dose uniformity. The throughput depends on several parameters, namely, target dose and food density. The conveyor velocity is adjusted to these two parameters. For both types of irradiation, gamma and e-beam are used as dosimeters during the process to assure that the product gets the appropriate dose (Kuntz & Strasser, 2018).

The impact of electrons with high energy on food allows us to obtain the same objectives referred to before for gamma radiation preservation processing, with particular characteristics when applied to natural matrices, such as food. Actually, this represents the object of the studies performed by several authors on different food items, particularly on chestnuts, mushrooms, and MAPs.

11.2.1 *Food processing by irradiation*

Research programs for food irradiation were started over the world in the 1950s. Subsequently, several projects were launched to observe and validate the effects of irradiation on the wholesomeness of food and its impact on nutritional content (Ravindran & Jaiswal, 2019). The Joint Food and Agriculture Organization (FAO), International Atomic Energy Agency (IAEA), and World Health Organization (WHO) Expert Committee on Food Irradiation in 1980, based on scientific background, evaluated the results obtained by these projects and concluded that *"the irradiation of any food commodity up to an overall average dose of 10 kGy presents no toxicological hazard"* (Codex Alimentarius Commission, 1999; Koutchma, Keener, & Kotilainen, 2018; Ravindran & Jaiswal, 2019).

In 1983, the Codex Alimentarius Commission (CAC), with support from FAO, IAEA, and the WHO, published the Codex General Standard for Irradiated Foods. This standard remains updated until today and has been represented by other agencies, such as the European Food Safety Authority and Food and Drug Administration (FDA) (Koutchma et al., 2018; Rahman et al., 2018).

Food irradiation is authorized and has been adopted in over 60 countries as a sanitary and phytosanitary processing method with tons of food being processed annually worldwide (Maherani et al., 2016), with China standing out as the country with the highest volume of irradiated foods (particularly spices, garlic, grain, chicken legs, and meat) (Koutchma et al., 2018). Treatment of spices and herbs is the most widely used application of food irradiation, followed by fresh fruit and vegetables. The IAEA and the CAC release the international standards and regulations for the irradiation of food and reveal that the Member States are allowed to convert the standards into national regulations at their preference. Therefore, standards and regulations vary greatly between countries and include the premises about food irradiation, namely, food safety, human health, labeling, packaging, dosimetry, and quality assurance, among others (Maherani et al., 2016; Rahman et al., 2018).

There are only a few countries that have regulations allowing irradiation treatment of any food product to comply with the Codex Standard, namely, Brazil, Mexico, and Singapore have stated that any food can be treated for any purpose. In the United States, the maximum allowed dose authorized for spices is 30 kGy, but in Australia and New Zealand, it varies between 2 and 30 kGy; in India, doses from 6.0 to 14.0 kGy are allowed for the elimination of pathogens and, in Canada, European Union (EU) and China, the maximum allowed dose is 10 kGy. Discordant dose limits and methodologies in dose measurements interrupt the global trade in irradiated foods. Countries that have allowed irradiation of more foodstuffs categories are limited to national markets (Koutchma et al., 2018).

Within the 28 EU Member States, the amount of irradiated foods has been decreasing from approximately 9263 tons in 2010 to 4973 tons in 2016; in 2017, 5237 tons of products were irradiated. Of the treated foods, $\approx$80% were irradiated mainly in two EU Member States, namely Belgium (68%) and the Netherlands (13%). The products treated were frog legs (57%), aromatic herbs, spices and dried vegetable seasoning (21%), poultry (16%), dried vegetables and fruits (5%), and dehydrated blood, plasma, and coagulates (1%) (European Commission, 2019).

Regulation and standardized doses are allocated into three groups:

- *low dose* applications (up to 1 kGy), which can be used to achieve sprouting inhibition in, for example, potatoes, onions, garlic, root ginger (0.05–0.15 kGy); phytosanitation such as insect's disinfestations and parasite disinfection in cereals, pulses, fresh and dried fruits, dried fish and meat (0.15–0.5 kGy); and delay of physiological changes such as ripening in fresh fruits and vegetables (0.25–1.0 kGy).
- *Medium dose* applications (1 to 10 kGy) are recommended to extend shelf-life, for example, of fresh fish, strawberries, and mushrooms (1.0–3.0 kGy); reduce risk of pathogenic and spoilage microbes such as *Salmonella, Campylobacter* and *Escherichia coli* in fresh and frozen seafood, raw or frozen poultry and meat (1.0–7.0 kGy); and improvement the quality of produce, such as juice yield in grapes and cooking time in dehydrated vegetables (2.0–7.0 kGy).
- *High-dose* applications (above 10 kGy) are approved for decontaminating food ingredients like enzymes, natural gum, spices and herbs (10–50 kGy) and sterilization of meat, poultry, seafood, prepared foods, and foods for NASA astronauts and hospital diets at an industrial level (30–50 kGy) (Ihsanullah & Rashid, 2017; Rahman et al., 2018; Ravindran & Jaiswal, 2019).

11.2.2 Legislation aspects and consumer concerns

11.2.2.1 Legislation

The industrial sectors' main concern is to guarantee access to good and safe food products. In this way, a legislative framework around food security is essential in order to cover the entire chain, from cultivation to the arrival of products to the consumer (Balatsas-Lekkas, Arvola, Kotilainen, Meneses, & Pennanen, 2020).

As such, food processing by irradiation also obeys several requirements, being not only efficient in decontamination and food preservation but also safe for the consumer. The industrial sectors' interest in this technique has increased worldwide, being applied in several European, Asian and American countries, Australia, and others. Thus, more than 50 countries have regulations that allow their application in at least one type of food (Roberts, 2014; Roberts, 2016). In order to develop the global regulation of irradiated foods, two important documents were considerably influential: the Codex General Standard for Irradiated Foods and the Recommended International Code of Practice for the Operation of Irradiation Facilities (Nair & Sharma, 2016). The main objective of these regulations is to establish and organize adequate procedures to control food irradiation.

To guarantee that the application of ionizing radiation to food is safe and effective, several legislative documents support this technology; however, its application has some discrepancies according to the country where it is used. In this way, the harmonization of regulations and equivalence of standards, dose, and labeling is one of the biggest challenges faced by the global food sector. So, the legislation needs to be harmonized, revised, and updated constantly to facilitate the efficiency of global food chains and trade, responding to food safety challenges (King et al., 2017).

The control of irradiated foods in the EU has strict legislation and its application is only allowed in dried aromatic herbs, spices, and vegetable seasonings, with a maximum dose of 10 kGy. For that, two directives are followed:

- the *Directive 1999/2/EC*, which deals with all basic aspects of food that have been treated with ionizing radiation; this legislation also makes it necessary for any food item that has undergone irradiation (or has irradiated ingredients) to be labeled as "irradiated" or "treated by ionizing radiation" so that consumers can make an informed choice. In addition, each EU member state is supposed to forward the results of check-ups performed at irradiation facilities.
- the *Directive 1999/3/EC* instructs member states on the foods and food ingredients that can be subjected to ionizing radiation. In the USA, the irradiation of food products is ruled by the FDA. This country follows the legislation 21 CFR 179 *"Irradiation in the production, processing and handling of food,"* which documents the types of ionizing radiation that can be applied (X-rays; gamma, and electron-beam), as well as the food products where it can be applied and the respective dose, and the appropriate labeling (Koutchma et al., 2018).

Otherwise, and with some differences in the applied doses and in the list of foods allowed to be treated, Canada follows the legislation of Food and Drug Regulations Division 26. C.R.C., c.870 and the permitted foods for irradiation include spices, potatoes, onions, and wheat flour (Bashir et al., 2020; Koutchma et al., 2018; Roberts, 2016). In Central and South America, most countries have adopted the Codex Alimentarius standards on food irradiation. Russia also adopted the Codex Alimentarius International Standard for Irradiated Foods in early 2016 and today allows irradiation of several food categories, such as spices and herbs, meat products, and fresh agricultural products (Koutchma et al., 2018). Australia and New Zealand have harmonized their food irradiation legislation under Food Standards Australia New-Zealand and presently, the irradiation treatment is authorized on twelve fruits, spices, herbs, and herbal infusions for phytosanitary purposes (Koutchma et al., 2018).

11.2.2.2 Consumer concerns

As with all technologies, the success of ionizing radiation on the market and its application as a food processing technology is directly related to consumer acceptance (Nair & Sharma, 2016).

Despite the multiple benefits of ionizing irradiation that have been studied and described by several researchers in the decontamination and conservation of different types of food and recognized by different highly credible international organizations (WHO, FAO, IAEA, and CAC), the consumer continues to show some resistance in the consumption of foods processed by this type of technology (Galati, Tulone, Moavero, & Crescimanno, 2019). This translates into one of the reasons why this type of processing is limited, especially among European countries, where the name "irradiation" is still associated with nuclear radiation and radioactivity, and in consequence, with health hazards. Thus, it is in the United States that irradiation has the greatest acceptance (Balatsas-Lekkas et al., 2020).

Considering the importance of consumer opinion and in order to understand not only their knowledge and attitudes toward food irradiation but also to outline ways to positively boost this technology, several studies have been developed, mainly in the United States and Europe (Roberts, 2016). In this sense, these studies focused mainly on two main points: (1) evaluating consumers' beliefs and attitudes toward food irradiation at a general level, and (2) examining factors predicting consumers' acceptance of specific irradiated food products (Balatsas-Lekkas et al., 2020).

According to Galati et al. (2019), it was possible to verify that, although there are scientific studies and competent entities that prove the safety of irradiated products, the consumer still has a negative attitude regarding this type of food. Several factors can explain these negative attitudes, such as the lack of adequate knowledge about its applicability and benefits, as well as the total lack of knowledge about the existence of this type of processing.

In a study performed by Bearth and Siegrist (2019), it was found that consumers' awareness and knowledge about food irradiation is relatively low since only two-thirds of consumers said they had not heard about food irradiation before this study. On the other hand, it is also possible to notice that in previous studies 77% of consumers did not know that irradiation is used for food preservation and 46% believed that this technology attributes radioactivity to food. In addition, it was clear that consumers want to be better informed, and this provision of information can significantly increase the acceptance of this type of food processing.

Overall, it was also notable that families with children under 18 years of age are more opposed to irradiation, and that men are more positive toward irradiation compared to women (Balatsas-Lekkas et al., 2020).

Thus, for irradiation processing to be more widely accepted by the consumer and, consequently, to increase the worldwide applicability rate, it is essential to establish effective strategies to encourage consumers to buy irradiated foods. For this, it is crucial to disseminate information regarding this technology in order to elucidate the consumer of its benefits, as well as to clarify many associated myths that condition the preference for irradiated products (Nishihira, 2020).

11.3 Processing effects on nutrients and bioactive

11.3.1 Chestnut fruits irradiation

11.3.1.1 Chestnut conservation

Chestnuts are edible fruits from chestnut trees that are present in several regions of the planet. In Asia, the main species are *Castanea mollissima*, in China, and *Castanea crenata*, in Korea and Japan. In North America, the predominant specie is *Castanea dentata*, while in Europe region the dominant specie is *Castanea sativa* (Rodrigues, Venâncio, & Lima, 2012). Within *C. sativa* there are different varieties, some of them classified as PDO (*Protected Designation of Origin*), due to their singular characteristics (Neri, Dimitri, & Sacchetti, 2010). As a global food product, present in several countries, it was incorporated into human diet since ancient times (Álvarez-Álvarez, Barrio-Anta, & Diéguez-Aranda, 2006). Nowadays, part of the product goes directly to the food industry for incorporation in other products such as bread, cakes, or sweets (Demirkesen, Mert, Sumnu, & Sahin, 2010; Gomes-Laranjo, Peixoto, Costa, & Ferreira-Cardoso, 2009), while the fruits commercialized in fresh present a defiant task on how to extend their shelf-life and, at the same time, eliminate or minimize the presence of insects and fungi, regarding its conservation for long periods. In the past, several traditional methods were used (cold water, sand storage, etc.) to extend the shelf-life of the product in fresh (FAO, 2003; Jeremini et al., 2006; Kwon, Kwon, Byun, & Kim, 2004).

The rotting of chestnuts is produced by weevils directly in the tree, being in some cases an open door for the entrance of fungi in the fruit. Due to its humidity content, about 50%, and the presence of sugars, the development of different type of fungi is also common and, in some circumstances, lead to the presence of aflatoxins, which are associated with toxicity and carcinogenicity effects in humans (Rodrigues et al., 2012). To overcome these conservation problems, post-harvest fumigation with methyl bromide (CH_3Br) was widely used in European countries, till its banning in 2010, due to environmental concerns (EU, 2008). Other methods were tested, such as sulfuryl fluoride (SO_2F_2) post-harvest fumigation (Vinghes & Ducom, 2000) and MAP using CO_2 (Cecchini, Contini, Massantini, Monarca, & Moscetti, 2011), to overcome conservation problems, with all of these methods presenting advantages and limitations.

In alternative, some physical methods for food preservation are in use to increase the shelf-life and meet the international phytosanitary requirements for the absence of insects, when this product is exported in fresh. One of these physical methods currently in use is hot water treatment, which consists in passing the chestnut fruits in a warm bath for a short period (about 45 minutes) and drying it with forced air (Jeremini et al., 2006). The contact with water and temperature, at about 50°C, are the main problems for this method, mainly due to the possible contribution to increase the presence of fungi in fruits.

Another physical method that could surpass these problems is irradiation since there is no water involved and there is no temperature increase during its use (Ferreira, Antonio, & Cabo Verde, 2018).

11.3.1.2 Effects on chestnut nutrients and bioactives

Food irradiation is being used in several countries to process different type of fruits (mango, papaya, tomatoes, plums, grapes, cherries, etc.) (Eustice, 2018), for insects' disinfestation, microbiological decontamination or shelf-life increase (Cabo Verde, 2018). In chestnut fruits, its impact is minimal and less than storage time effects (Antonio, Carocho et al., 2012; Antonio, Ramalhosa et al., 2012; Carocho et al., 2012, 2013, 2014; Fernandes, Antonio, Barros et al., 2011; Fernandes, Barreira, Antonio et al., 2011; Kwon et al., 2004).

Chestnut fruits of other Korean species (*Castanea bungena*) were treated with gamma radiation at doses up to 0.5 kGy, and the tests to detect DNA fragmentation (DNA Comet assay) and to detect free radicals (ESR—Electron Spin Resonance test) gave no conclusive results to distinguish between irradiated and non-irradiated samples (Chung et al., 2004) (Table 11.1). The same conclusion was obtained for European varieties of *C. sativa*, with no ESR signal difference for chestnuts irradiated up to 1 kGy (Mangiacotti et al., 2009), while the presence and survival of chestnut weevils (*Coleoptera* and *Lepidoptera*) are eliminated for doses above 0.4 kGy (Imamura, Todoriki, Miyanoshita, & Hayashi, 2004).

Color of the fruits is one of the main characteristics valorized by the consumers. Korean chestnuts (*C. crenata*) were gamma irradiated at several doses and stored for 6 months, being the color monitored immediately after processing and along storage (Kwon et al., 2004). In this study, it was observed that for doses up to 1 kGy, there were no significant changes in color, with a more visible change caused by storage time. Only for the 10 kGy dose, the color differences between irradiated and non-irradiated samples were clearly significant.

In 2013, intermediate doses of gamma radiation between 0.5 and 6 kGy were tested in *C. sativa*, concluding that changes in color could be observed for doses above 3 kGy, in lightness (L) parameter, and for doses above

TABLE 11.1 Nutrients and bioactives reported in studies regarding irradiation of chestnut fruits.

Food processing technology	Chestnut specie	Parameter	References
Gamma irradiation	*Castanea molissima*	Sprout inhibition Respiration rate Starch Amylase Catalase Proteins Sugars Ascorbic acid	Iwata and Ogata (1959), Guo-xin et al. (1980), Iwata and Ogata (1959), Guo-xin et al. (1980), Guo-xin et al. (1980), Guo-xin et al. (1980), Guo-xin et al. (1980), Iwata and Ogata (1959), Guo-xin et al. (1980), Iwata and Ogata (1959)
	Castanea bungena	DNA fragmentation Free radicals	Chung et al. (2004), Chung et al. (2004)
	Castanea crenata	Sprout inhibition Respiration rate Sugars Color	Iwata and Ogata (1959), Iwata and Ogata (1961), Iwata and Ogata (1959), Kwon et al. (2004)
	Castanea sativa	Free radicals Color Texture Organic acids Sugars Fatty acids Proteins Fat Ash Carbohydrates Phenols Flavonoids Tocopherols	Mangiacotti et al. (2009), Antonio, Carocho, Bento, Rafalski, and Quintana (2013), Antonio et al. (2013), Carocho et al. (2014), Fernandes, Antonio, Barros et al. (2011), Fernandes, Barreira, Antonio et al. (2011), Antonio et al. (2011), Fernandes, Antonio, Barros et al. (2011), Antonio et al. (2011), Fernandes, Antonio, Barros et al. (2011), Fernandes, Barreira, Antonio et al. (2011), Carocho, Barreira et al. (2012)
E-beam irradiation	*C. sativa*	Color Texture Sugars Sucrose Raffinose Organic acids Fatty acids Proteins Fat Ash Carbohydrates Phenols Flavonoids Tocopherols	Antonio, Carocho et al. (2013), Antonio, Carocho et al. (2012), Antonio, Ramalhosa et al. (2012), Carocho, Barreira et al. (2012), Carocho et al. (2013), Carocho et al. (2014), Carocho et al. (2014), Carocho et al. (2014), Carocho et al. (2014), Carocho et al. (2014), Carocho, Antonio et al. (2012), Carocho et al. (2014)

6 kGy, in blueness (*b*) parameter (Antonio et al., 2013). Four European varieties of *C. sativa*, from Portugal and Italy, were submitted to electron-beam radiation with energy of 10 MeV, with no color changes observed in skins, peeled fruit exterior and interior, after irradiation at 1 kGy (Antonio et al., 2013).

Texture assumes a relevant importance for a fruit that is expected to have a characteristic hardness, as in chestnuts. When irradiated, the softening of the fruit was observed only for doses above 6 kGy for gamma radiation (Antonio et al., 2013). For another type of technology (e-beam radiation), a similar result was observed (Antonio, Carocho et al., 2012; Antonio, Ramalhosa et al., 2012).

Chestnut fresh fruits composition is mainly starch (45%), sugars (8%), proteins (4%), lipids (2%), and with a water content of about 40% (FAO, 2003). Some minerals (potassium, phosphorus, calcium, sodium) are also present in its composition (Corregidor, Antonio, Alves, & Cabo Verde, 2020) and bioactive compounds (organic acids and tocopherols) (Carocho et al., 2013; Fernandes, Antonio, Barros et al., 2011; Fernandes, Barreira, Antonio et al., 2011).

The influence of irradiation processing on chemical parameters of chestnuts has been studied by several authors in different varieties and origins. One of the first studies to assess the impact of gamma radiation on chestnuts was performed in 1973 in Chinese varieties (Guo-xin, Qiong-ling, Yao-chang, & Si-lian, 1980), for doses up to 0.12 kGy (or in old units, 120 kiloRad), proving its efficacy for sprout inhibition. In this study, the effects on amylase and catalase activity were analyzed, and from the available data, there were only slight differences between irradiated and control samples. Regarding starch content, the authors observed a slight reduction for irradiated samples, while for total sugars there was no clear correlation with the dose of irradiation, being lower (for 0.06 kGy) and higher (for 0.12 kGy) than the control. They observed also that the irradiated samples presented lower protein content, 50% less for the dose of 0.10 kGy, while for fat content, the irradiated samples stored for 70 days shows higher values.

Using irradiation as preservation technology for chestnuts has been hypothesized even earlier. As far as in the year 1959, the impact of gamma radiation on sprout inhibition of *C. crenata* and *Castaea molissima* was studied (Iwata & Ogata, 1959). For the latter, authors also analyzed the impact on sugars, concluding that for the tested doses (up to 20 Gy), the chestnuts sugar composition was not affected, observing only a slight difference in ascorbic acid content, compared with the control, with no recognizable difference between the tested doses. A similar conclusion was obtained in another study by the same authors for *C. crenata* at higher doses (up to 600 Gy) (Iwata & Ogata, 1961). In *C. sativa* varieties irradiated at 1 kGy, a slightly higher tendency on total organic acids content in *C. sativa* varieties from Portugal and Italy was observed (Carocho et al., 2014). The same authors have also tested the impact of e-beam radiation for higher doses (up to 6 kGy) stating that "the variance caused by the assayed irradiation doses is slighter, and did not allowed indication of any particular tendency" (Carocho et al., 2013).

The impact of gamma radiation on the respiration rate and CO_2 evolution was tested in *C. molissima*, concluding that irradiation did not affect this parameter after 70 days of storage, for the studied doses (Iwata & Ogata, 1959), while for *C. crenata* they performed a more detailed study, observing that the samples presented a higher respiration rate immediately after processing, becoming closer to control samples value two weeks after irradiation (Iwata & Ogata, 1961).

In *C. sativa* varieties, Fernandes, Antonio, Barros et al. (2011), Fernandes, Barreira, Antonio et al. (2011) observed that gamma irradiation at low doses caused no impact on total sugars, which is in accordance with other studies in Asian varieties for lower doses (Guo-xin et al., 1980; Iwata & Ogata, 1959). In what concerns fatty acids composition in gamma-irradiated *C. sativa* stored for 60 days, authors refer that they were not affected (Fernandes, Antonio, Barros et al., 2011; Fernandes, Barreira, Antonio et al., 2011), while in Asian varieties, it was observed that irradiated samples present a higher fat content when stored till 70 days (Guo-xin et al., 1980). In another study with European chestnut fruits of *C. sativa* irradiated at higher doses (up to 3 kGy), regarding the effects on nutritional composition (proteins, fat, ash, carbohydrates, and energetic value), the authors concluded that irradiation processing at the tested doses "did not affect the nutritional and chemical quality of chestnut fruits" (Fernandes, Antonio, Barros et al., 2011; Fernandes, Barreira, Antonio et al., 2011). In a similar study on *C. sativa* varieties from Turkey, the same conclusions were obtained (Barreira et al., 2012), in line with the results acquired in another study with e-beam irradiated Italian varieties of *C. sativa* (Carocho et al., 2014).

The effects of electron-beam processing on the nutritional parameters of *C. sativa*, irradiated at doses up to 6 kGy, allowed us to conclude that sucrose, total sugars, and raffinose content are higher in irradiated samples (Carocho et al., 2012).

Chestnuts are a source of polyphenolic antioxidants (phenolic acids and flavonoids), and treatment with gamma radiation at low doses (0.5 kGy) has been shown to increase their bioactivity (Antonio et al., 2011). For chestnuts processed with gamma and e-beam radiation (for doses up to 3 kGy), phenolic content was higher for irradiated samples while flavonoids content is lower and, for both types of processing technology, it was observed an increase in antioxidant activity (Carocho et al., 2014).

A study on tocopherols composition of *C. sativa* varieties from Turkey, gamma irradiated up to 3 kGy, concluded that no particular tendency was observed in its content, mainly on its majority isoform (γ-tocopherol) (Barreira et al., 2012). Likewise, when chestnuts were processed with e-beam at higher doses (up to 6 kGy), no significant difference in its content was noted, despite the slight tendency higher contents in irradiated samples (Carocho et al., 2012). In turn, a study with *C. sativa* varieties of Portuguese origin, irradiated at low dose (0.5 kGy), showed an increase in γ-tocopherol after 60 days of storage (Fernandes, Antonio, Barros et al., 2011; Fernandes, Barreira, Antonio et al., 2011).

11.3.2 Edible mushrooms irradiation

11.3.2.1 Mushrooms conservation

Edible mushrooms are attractive from a nutritional point of view, becoming a crucial part of the diet. With the expansion of the global mushroom industry, commercial production and distribution have improved significantly. The demand for fresh mushrooms is increasing across the globe and the global mushrooms market share is dominated by *Agaricus bisporus* (button), *Lentinula edodes* (shiitake), and *Pleurotus ostreatus* (oyster) (Market analysis report, 2020). The production of mushrooms and truffles is controlled by Asia (78.2%), Europe (14.7%), the Americas (6.2%), Oceania (0.6%), and Africa (0.3%) (FAOSTAT, 2020). The market represents medicinal mushrooms (38%), wild (8%), and cultivated edible (54%) mushrooms (Grimm & Wösten, 2018). Wild edible mushrooms are collected and consumed for thousands of years with several million tons harvested annually and valued at up to million euros; the data on wild mushrooms' trade are generally cited from individual people and therefore are not really reliable (Radomir, Mesud, & Žaklina, 2018).

The quality of mushrooms is determined by color, texture, flavor, but the first appealing property perceived by consumers is the color (Lin & Sun, 2019). During postharvest, fresh mushrooms have a short shelf-life of up to 3–4 days at ambient temperature, and quality degradation begins due to their high moisture content, high respiration rate, enzyme activity and bacteria attack, which causes moisture loss, discoloration, texture changes, off flavor and nutrition loss (Nketia, Buckman, Dzomeku, & Akonor, 2020; Zhang, Pu, & Sun, 2018).

Deterioration and loss of quality of mushrooms are affected by many factors, including intrinsic and extrinsic factors, such as browning (e.g., tyrosinase), pathogens (e.g., bacterium *Pseudomonas*, the fungi *Verticillium fungicola, Trichoderma harzianum* and *Lecanicillium fungicola*, among others) and postharvest factors (e.g., water activity, respiration rate, temperature, relative humidity, and mechanical damage) (Lin & Sun, 2019; Zhang et al., 2018).

Many quality preservation methods and technologies have been widely studied and proposed to reduce or inhibit these effects, namely, thermal processes: drying (e.g., hot air, microwave, drying, freeze-drying, and vacuum drying) and cooling (vacuum cooling); physical processes: packaging (MAP), irradiation (gamma, electron-beam, and UV irradiation) and pulsed electric field; and chemical processes: washing (sulfite solutions, citric acid, and hydrogen peroxide), coating (chitosan and alginate), ozone, electrolyzed water and polyphenol oxidase inhibitors (Lin & Sun, 2019; Nketia et al., 2020; Zhang et al., 2018). The quality degradation process of mushrooms is complex and these techniques, despite the advantages, have drawbacks and must be applied depending on the final purpose of the product (Lin & Sun, 2019; Zhang et al., 2018).

11.3.2.2 Effects on mushroom nutrients and bioactives

The effects of irradiation processing on chemical and bioactive properties, namely, nutritional value (moisture, fat, ash, proteins, total carbohydrates, energy), fatty acids, vitamins, soluble sugars, organic acids, phenolic compounds, antioxidant activity, dietary fiber, minerals and enzymes, as well physical parameters and sensory evaluation, were described by different authors in several mushroom species, as summarized in Table 11.2.

Concerning nutritional value (moisture, ash, lipids, proteins, total carbohydrates, and energetic value), different processed *M. procera* samples (frozen, dried and fresh) irradiated with 0.5 and 1 kGy presented very similar results to the control (Fernandes et al., 2014b). In gamma-irradiated *Boletus pinophilus* Pilát & Dermek and *Clitocybe subconnexa* Murrill with 2 kGy, dry matter and carbohydrates revealed significant changes in both mushrooms with less content in irradiated samples. Fat, proteins, and ashes were preserved with 2 kGy and present higher content compared to the non-irradiated samples. The results obtained for *Tuber aestivum* truffles (0.5, 1.0, 1.5, and 2.5 kGy), total carbohydrate content, and protein levels were not affected after irradiation (Tejedor-Calvo et al., 2020). Gamma and electron-beam irradiated *A. bisporus* Portobello (1, 2, and 5 kGy) stored for 8 days revealed a nutritional value similar to non-irradiated samples. Independently of the irradiation source, ash content was higher with 1 and 2 kGy, while the 5 kGy dose was linked with higher protein levels (Cardoso et al., 2019). When *Tuber melanosporum* truffles were submitted to irradiation with electron-beam or gamma rays, no significant changes were noticed in their total carbohydrates, compared to non-irradiated controls. A decrease in protein content was observed, independently of irradiation source and protein degradation was more pronounced with a higher dose (2.5 kGy). This can be explained due to protein fragmentation, denaturalization, and precipitation, as also noticed in other studies with high irradiation doses (Tejedor-Calvo et al., 2019). Gamma-irradiated dried and sliced *Boletus edulis* Bull. (2, 6, and 10 kGy) throughout storage (6 and 12 months) showed identical results to control, but exhibited declines over time during the studied period (Fernandes et al., 2017). The effects of storage time (up to 12 months) and electron-beam irradiated dried *M. procera* (0.5, 1, and 6 kGy), showed a decreasing tendency with time in protein, fat, and ash contents, a result reflected in carbohydrates content, which increased along time. On the hand, the applied irradiation did not cause any marked tendencies, except for lower protein value in samples irradiated with 0.5 kGy (Fernandes et al., 2014c). Dried *B. edulis* and *Russula delica* Fr., treated with electron beam (2, 6, and 10 kGy), the 10 kGy dose exerted the most significant effect in *R. delica*, as it can be concluded from a decrease in fat, proteins, and ash content. For *B. edulis*, the most affected parameter was protein content, which decreased comparatively with the non-irradiated samples (Fernandes et al., 2014a). The same authors reported that electron-beam irradiated *Amanita*

TABLE 11.2 Nutrients and bioactives reported in studies regarding irradiation of mushrooms.

Parameter	Species	Origin	Sample	Radiation technology	Doses (kGy)	References
Nutritional value	*Macrolepiota procera*	Portugal	Frozen, dried and fresh	Gamma	0.5, 1	Fernandes et al. (2014b)
	M. procera	Portugal	Dried	Electron-beam	0.5, 1, 6	Fernandes et al. (2014c)
	Boletus edulis and *Russula delica*	Portugal	Dried	Electron-beam	2, 6, 10	Fernandes et al. (2014a)
	Amanita caesarea and *A. curtipes*	Portugal	Dried	Electron-beam	2, 6, 10	Fernandes, Barreira, Antonio, Rafalski et al. (2015)
	Boletus pinophilus and *Clitocybe subconnexa*	Portugal	Fresh	Gamma	2	Fernandes, Barreira, Antonio, Oliveira et al. (2016)
	B. edulis	Portugal	Dried and sliced	Gamma	2, 6, 10	Fernandes et al. (2017)
	Agaricus bisporus	Portugal	Fresh	Gamma and electron-beam	1, 2, 5	Cardoso et al. (2019)
	Tuber melanosporum	Spain	Fresh	Gamma and electron-beam	1.5, 2.5	Tejedor-Calvo et al. (2019)
	Tuber aestivum	Spain	Fresh	Gamma	0.5, 1.0, 1.5, 2.5	Tejedor-Calvo et al. (2020)
Fatty acids	*M. procera*	Portugal	Frozen, dried and fresh	Gamma	0.5, 1	Fernandes et al. (2014b)
	M. procera	Portugal	Dried	Electron-beam	0.5, 1, 6	Fernandes et al. (2014c)
	B. edulis and *R. delica*	Portugal	Dried	Electron-beam	2, 6, 10	Fernandes et al. (2014a)
	A. caesarea and *A. curtipes*	Portugal	Dried	Electron-beam	2, 6, 10	Fernandes, Barreira, Antonio, Rafalski et al. (2015)
	B. pinophilus and *C. subconnexa*	Portugal	Fresh	Gamma	2	Fernandes, Barreira, Antonio, Oliveira et al. (2016)
	B. edulis	Portugal	Dried and sliced	Gamma	2, 6, 10	Fernandes et al. (2017)
	A. bisporus	Portugal	Fresh	Gamma and electron-beam	1, 2, 5	Cardoso et al. (2019)
Triacylglycerols profiling	*B. edulis*	Portugal	Fresh	Gamma	1, 2	Fernandes, Barreira, Antonio et al. (2014)
	B. edulis	Portugal	Dried	Electron-beam	2, 6, 10	Fernandes, Barreira, Antonio et al. (2014)
	M. procera	Portugal	Frozen, dried and fresh	Gamma	0.5, 1	Fernandes, Barreira, Antonio et al. (2014)
	B. pinophilus	Portugal	Fresh	Gamma	2	Fernandes, Barreira, Antonio et al. (2014)
	R. delica	Portugal	Dried	Electron-beam	2, 6, 10	Fernandes, Barreira, Antonio et al. (2014)

(*Continued*)

TABLE 11.2 (*Continued*)

Parameter	Species	Origin	Sample	Radiation technology	Doses (kGy)	References
Tocopherols	*M. procera*	Portugal	Frozen, dried and fresh	Gamma	0.5, 1	Fernandes et al. (2014b)
	M. procera	Portugal	Dried	Electron-beam	0.5, 1, 6	Fernandes et al. (2014c)
	B. edulis and *R. delica*	Portugal	Dried	Electron-beam	2, 6, 10	Fernandes et al. (2014a)
	A. caesarea and *A. curtipes*	Portugal	Dried	Electron-beam	2, 6, 10	Fernandes, Barreira, Antonio, Rafalski et al. (2015)
	B. edulis	Portugal	Dried and sliced	Gamma	2, 6, 10	Fernandes et al. (2017)
	A. bisporus	Portugal	Fresh	Gamma and electron-beam	1, 2, 5	Cardoso et al. (2019)
Ergosterol	*A. bisporus*	Portugal	Fresh	Gamma and electron-beam	1, 2, 5	Cardoso et al. (2019)
	T. melanosporum	Spain	Fresh	Gamma and electron-beam	1.5, 2.5	Tejedor-Calvo et al. (2019)
	T. aestivum	Spain	Fresh	Gamma	0.5, 1.0, 1.5, 2.5	Tejedor-Calvo et al. (2020)
Soluble Sugars	*M. procera*	Portugal	Frozen, dried and fresh	Gamma	0.5, 1	Fernandes et al. (2014b)
	M. procera	Portugal	Dried	Electron-beam	0.5, 1, 6	Fernandes et al. (2014c)
	B. edulis and *R. delica*	Portugal	Dried	Electron-beam	2, 6, 10	Fernandes et al. (2014a)
	A. caesarea and *A. curtipes*	Portugal	Dried	Electron-beam	2, 6, 10	Fernandes, Barreira, Antonio, Rafalski et al. (2015)
	B. pinophilus and *C. subconnexa*	Portugal	Fresh	Gamma	2	Fernandes, Barreira, Antonio, Oliveira et al. (2016)
	B. edulis	Portugal	Dried and sliced	Gamma	2, 6, 10	Fernandes et al. (2017)
	A. bisporus	Portugal	Fresh	Gamma and electron-beam	1, 2, 5	Cardoso et al. (2019)
Organic acids	*M. procera*	Portugal	Frozen, dried and fresh	Gamma	0.5, 1	Fernandes, Barros et al. (2014)
	B. edulis and *R. delica*	Portugal	Dried	Electron-beam	2, 6, 10	Fernandes et al. (2014a)
	A. caesarea and *A. curtipes*	Portugal	Dried	Electron-beam	2, 6, 10	Fernandes, Barreira, Antonio, Rafalski et al. (2015)
	B. pinophilus and *C. subconnexa*	Portugal	Fresh	Gamma	2	Fernandes, Barreira, Antonio, Oliveira et al. (2016)
	B. edulis	Portugal	Dried and sliced	Gamma	2, 6, 10	Fernandes et al. (2017)
	A. bisporus	Portugal	Fresh	Gamma and electron-beam	1, 2, 5	Cardoso et al. (2019)

Phenolic compounds	*M. procera*	Portugal	Frozen, dried and fresh	Gamma	0.5, 1	Fernandes, Barros et al. (2014)
	B. edulis and *R. delica*	Portugal	Dried	Electron-beam	2, 6, 10	Fernandes et al. (2014a)
	A. caesarea and *A. curtipes*	Portugal	Dried	Electron-beam	2, 6, 10	Fernandes, Barreira, Antonio, Rafalski et al. (2015)
	B. pinophilus and *C. subconnexa*	Portugal	Fresh	Gamma	2	Fernandes, Barreira, Antonio, Oliveira et al. (2016)
	B. edulis	Portugal	Dried and sliced	Gamma	2, 6, 10	Fernandes et al. (2017)
Antioxidant parameters	*M. procera*	Portugal	Frozen, dried and fresh	Gamma	0.5, 1	Fernandes et al. (2014b)
	M. procera	Portugal	Dried	Electron-beam	0.5, 1, 6	Fernandes et al. (2014c)
	B. edulis and *R. delica*	Portugal	Dried	Electron-beam	2, 6, 10	Fernandes et al. (2014a)
	A. caesarea and *A. curtipes*	Portugal	Dried	Electron-beam	2, 6, 10	Fernandes, Barreira, Antonio, Rafalski et al. (2015)
	B. pinophilus and *C. subconnexa*	Portugal	Fresh	Gamma	2	Fernandes, Barreira, Antonio, Oliveira et al. (2016)
	B. edulis	Portugal	Dried and sliced	Gamma	2, 6, 10	Fernandes et al. (2017)
	Volvariella volvacea	China	Fresh	Gamma	0.2, 0.4, 0.6, 0.8, 1.0	Hou et al. (2018)
	A. bisporus	India	Dried and sliced	Gamma	5, 10, 20, 30, 50	
	T. melanosporum	Spain	Fresh	Gamma and electron-beam	1.5, 2.5	Tejedor-Calvo et al. (2019)
	T. aestivum	Spain	Fresh	Gamma	0.5, 1.0, 1.5, 2.5	Tejedor-Calvo et al. (2020)
Dietary fiber	*M. procera*	Portugal	Dried	Electron-beam	0.5, 1, 6	Fernandes, Barreira, Antonio, Morales et al. (2015)
	B. edulis	Portugal	Dried	Electron-beam	2, 6, 10	Fernandes, Barreira, Antonio, Morales et al. (2015)
Macro-and micro-elements	*B. edulis*	Portugal	Fresh	Gamma	1, 2	Fernandes, Barreira, Antonio, Rafalski et al. (2016)
	B. edulis	Portugal	Dried	Electron-beam	2, 6, 10	Fernandes, Barreira, Antonio, Rafalski et al. (2016)
	M. procera	Portugal	Dried	Electron-beam	0.5, 1, 6	Fernandes, Barreira, Antonio, Rafalski et al. (2016)
	Hydnum repandum	Portugal	Fresh	Gamma	1, 2	Fernandes, Barreira, Antonio, Rafalski et al. (2016)

(*Continued*)

TABLE 11.2 (*Continued*)

Parameter	Species	Origin	Sample	Radiation technology	Doses (kGy)	References
Enzymes	*V. volvacea*	China	Fresh	Gamma	0.2, 0.4, 0.6, 0.8, 1.0	Hou et al. (2018)
Physical parameters	*V. volvacea*	China	Fresh	Gamma	0.2, 0.4, 0.6, 0.8, 1.0	Hou et al. (2018)
	Lentinula edodes	China	Fresh	Gamma	1.0, 2.0	Shi et al. (2020)
Sensory evaluation	*V. volvacea*	China	Fresh	Gamma	0.2, 0.4, 0.6, 0.8, 1.0	Hou et al. (2018)
	Pleurotus ostreatus	Ghana	Dried	Gamma	1, 2, 3, 4, 5	Kortei et al. (2020)
	T. aestivum	Spain	Fresh	Gamma	0.5, 1.0, 1.5, 2.5	Tejedor-Calvo et al. (2020)

caesarea (Scop.) Pers. and *A. curtipes* E.-J. Gilbert. (2, 6, and 10 kGy), revealed significant changes in nearly all parameters, except for water content in *A. caesarea*, which was similar to the control. The highest differences among control and irradiated samples for proximate analysis were verified in fat, protein, and ashes content, that have been preserved with all tested doses.

Regarding fatty acids, different processed (frozen, dried, and fresh) gamma-irradiated *M. procera* samples did not show remarkable changes. The percentages obtained for saturated fatty acids (SFA), monounsaturated fatty acids (MUFA), and polyunsaturated fatty acids (PUFA) did not reveal marked changes comparatively to the non-irradiated samples (Fernandes et al., 2014b). The same author reported that fresh *B. pinophilus* is less sensitive to gamma irradiation and, all fatty acids (except palmitoleic acid: C16:1) quantified in gamma-irradiated *C. subconnexa* gave significant changes among control and irradiated samples. These differences were characterized as slight decreases, which may result from oxidative or radiolytic processes. The observed changes were more likely produced by autoxidation, the presence of oxygen increased the formation of free radicals and caused the breakdown of hydrogen peroxide and the destruction of antioxidants that would normally scavenge the free radicals formed (Fernandes, Barreira, Antonio, Oliveira et al., 2016).

Cardoso et al. (2019) reported higher percentages of palmitic acid (C16:0) in fresh *A. bisporus* gamma irradiated with 5 kGy. In the case of electron-beam irradiated fresh Portobello, it was possible to conclude that samples irradiated with 1 kGy presented higher percentages of linoleic acid (C18:2n6) and PUFA and lower percentages of SFA, while non-irradiated ones showed the lowest content of arachidic acid (C20:0). When studying the storage time effect, it was possible to verify that samples stored for 8 days showed the lowest percentage of C18:2n6.

Dried and sliced *B. edulis* gamma irradiated with 2 kGy showed the highest contents in C16:0, C16:1, and SFA. The contents in MUFA, especially due to the vaccenic acid contribution, were favored by the 6 kGy dose and showed also a significant decrease after 12 months of storage. On the other hand, PUFA contents, which were mainly dictated by C18:2n6, were particularly higher in samples irradiated with 10 kGy (Fernandes et al., 2017).

Regarding electron-beam effects in dried samples of different mushrooms, *M. procera* irradiated with 0.5 kGy presented the highest caprylic acid (C8:0) and the lowest tetradecanoic acid (C14:0) levels, while the irradiation with 1 kGy minimizes the pentadecanoic acid (C15:0) content and eicosapentaenoic acid (C20:5) was the only fatty acid with an increase along storage time (up to 12 months). It was also found a decrease in SFA, a similar content of MUFA and an increase in PUFA, in comparison to control samples. Up to one year, SFA increased slightly, MUFA remained constant, and PUFA had a slow decrease (Fernandes et al., 2014c).

After studying *B edulis* and *R. delica* dried samples, the same authors reported C16:0, C18:1, and C18:2n6 as the most abundant fatty acids in both mushrooms. The higher electron-beam irradiation doses (6 and 10 kGy) tended to cause more significant changes in *B. edulis* fatty acids, while *R. delica* was the most affected by the 2 kGy dose. In fact, the general mechanism of lipids radiolysis (primary ionization, followed by migration of the positive charge either toward the carboxyl carbonyl group or double bonds) is more likely to occur in fresh mushrooms (Fernandes et al., 2014a).

In *A. caesarea* and *A. curtipes* samples treated with 2, 6, and 10 kGy, SFA seemed to be more resistant to electron-beam irradiation, while MUFA tended to increase in percentage and PUFA showed the opposite behavior, which is in agreement with its different radiosensitivity. Fatty acid percentages did not suffer any observable change, which might be explained by the fact that irradiation was performed in dried mushrooms (Fernandes, Barreira, Antonio, Rafalski et al., 2015).

Concerning triacylglycerol (TAG) composition, fresh *B. edulis* samples submitted to gamma irradiation presented higher percentages of oleoyl-dilinoleoyl-glycerol (OLL) and 1-palmitoyl-2-oleoyl-3-linoleoyl glycerol (POL) and less dioleoyl-linoleoyl-glicerol (OOL), triolein (OOO) and 1,2-dipalmitoyl-3-oleoyl-rac-glycerol (PPO), especially for 1 kGy dose. The effects of electron-beam in dried samples of the same mushroom were particularly observable for 6 and 10 kGy doses, which produced similar changes: higher percentages in (trilinoleoylglycerol) LLL and OLL and lower percentages in OOL, OOO, and PPO.

Concerning *M. procera*, the effects of gamma irradiation were more evident in dried samples, since the mean value of at least one dose differs from the others for all TAG. Among fresh samples, irradiation with 1 kGy tended to increase LLL, POL, 1-palmitoyl-2-oleoyl-3-oleoyl glycerol (POO), and PPO percentages; the same effect was observable for LLL, OOO, POO, and PPO in dried samples and OLL and POL in frozen samples. Fresh *B. pinophilus* samples submitted to gamma irradiation about OLL, OOL, and POO were shown to be significantly higher with 2 kGy dose. The effects of electron-beam irradiation were significant for most TAG molecules quantified in *R. delica* samples: LLL and POO had maximum percentages in samples irradiated with a 10 kGy dose. The effect of electron-beam irradiation is more pronounced than that obtained for gamma irradiation (except for dried *M. procera* samples). This result might be explained by the higher doses used in electron-beam irradiation (2, 6, and 10 kGy), rather than by the irradiation source (up to 2 kGy). Following the same reasoning, changes in dried samples were more obvious than those verified in fresh or frozen samples (Fernandes, Barreira, Antonio et al., 2014).

Concerning tocopherols content, irradiated dried samples of *M. procera* showed higher levels of β- and γ-tocopherol, while irradiated frozen samples exhibited higher content in δ-tocopherol; generally, the gamma irradiation dose did not cause any noticeable change, either for individual vitamers or for total tocopherols (Fernandes et al., 2014b).

Cardoso et al. (2019) identified two isoforms of vitamin E (α- and β-tocopherol) in irradiated fresh *A. bisporus*, independently of irradiation source; samples stored during 8 days revealed a similar tocopherols content to that in non-irradiated controls.

The tocopherol vitamers tended to be higher in dried and sliced *B. edulis* gamma irradiated samples, namely, α-tocopherol in samples irradiated with 6 kGy, β-tocopherol in those irradiated with 2 kGy and γ-tocopherol in samples irradiated with 10 kGy; up to one year of storage, β- and γ-tocopherols were preserved and α-tocopherol decreased slightly (Fernandes et al., 2017).

γ-tocopherol was also protected with electron beam irradiation (1 kGy) applied to dried samples of *M. procera*; the remaining isoforms (β-, γ-, δ-tocopherol) decline with dose. Up to one year of storage, α- and δ-tocopherol decreased slightly, but the other two isoforms (β- and γ-tocopherol) could not be quantified (Fernandes et al., 2014c).

Regarding dried *B. edulis* and *R. delica*, electron-beam irradiated samples tended to present higher amounts of total tocopherols, particularly for the 2 and 6 kGy doses, respectively (Fernandes et al., 2014a). The same authors applied the same doses and source to *A. caesarea* and *A. curtipes* and verified higher amounts α-, γ- and δ-tocopherol to 10 kGy doses, except for γ-tocopherol in *A. caesarea* (Fernandes, Barreira, Antonio, Rafalski et al., 2015).

A significant increase in ergosterol content was described in fresh *A. bisporus* treated with gamma and electron-beam irradiation (1, 2, and 5 kGy), while those stored for 8 days revealed similar results to non-irradiated samples (Cardoso et al., 2019). Meanwhile, after 21 days of storage, ergosterol levels seemed to grow slowly but continuously up to the end of the storage time (35 days) in fresh *T. melanosporum* with both sources and 1.5 and 2.5 kGy doses (Tejedor-Calvo et al., 2019); in gamma-irradiated fresh *T. aestivum* samples, ergosterol content follows the same tendency up to 42 days (Tejedor-Calvo et al., 2020).

Regarding soluble sugars content, *M. procera* treated with 1 kGy showed an increase in total sugars content; among processed irradiated samples, dried ones showed better results (Fernandes et al., 2014b). Otherwise, total soluble sugars showed a significant decrease in gamma-irradiated (2 kGy) fresh *B. pinophilus* and *C. subconnexa*, except for fructose in *B. pinophilus*. The verified degradation may be explained by a reduction in the optical rotation of sugars, which is a common occurrence when sugars are irradiated (Fernandes, Barreira, Antonio, Oliveira et al., 2016). *A. bisporus* submitted to gamma and electron-beam exhibited smaller rates of decrease of reducing sugars with applied doses and during the storage period (up to 8 days) (Cardoso et al., 2019). Concerning gamma irradiated dried and sliced *B. edulis*, the 2 kGy dose allowed higher mannitol contents; at the end of storage (12 months), none of the applied doses was effective in preventing fructose, mannitol, and trehalose decrease (Fernandes et al., 2017).

Regarding electron-beam dried samples, in *M. procera* samples, the applied irradiation did not cause particular changes, except for higher fructose and trehalose values in samples irradiated with 0.5 and 1 kGy, respectively; on the other hand, trehalose and mannitol presented maximum values in non-stored samples (Fernandes et al., 2014c).

Concerning *B. edulis* treated with 6 and 10 kGy, it was observed an increase in mannitol and trehalose, and, otherwise, a decrease in *R. delica* (Fernandes et al., 2014). Dried *A. caesarea* submitted to electron-beam irradiation, showed an increase in mannitol and trehalose particularly at 10 kGy; *A. curtipes* treated with 10 and 2 kGy showed an increase in mannitol and trehalose, respectively, and a decrease in fructose content, particularly with 6 kGy (Fernandes, Barreira, Antonio, Rafalski et al., 2015).

In what concerns organic acids, Fernandes, Barros et al. (2014) identified oxalic, quinic, malic, citric, and fumaric acids in different processed *M. procera* samples; quinic acid presented the lowest values in samples irradiated with 1 kGy; otherwise, malic and citric present the highest values. In terms of total organic acids, no particular tendency could be observed. Gamma irradiation of *B. pinophilus* at 2 kGy did not affect the organic acid content; in turn, fumaric acid showed a significant increase with the same dose in *C. subconnexa* (Fernandes, Barreira, Antonio, Oliveira et al., 2016). In the case of Portobello, malic acid was higher in gamma-irradiated samples with 5 kGy and was similar to control samples up to 8 days of storage; the same species treated with electron-beam, non-irradiated and samples irradiated with 5 kGy showed the lower and the highest content in malic acid, respectively; oxalic and quinic acids exhibited similar results to controls and also during the storage period (Cardoso et al., 2019).

Gamma irradiated dried and sliced *B. edulis* Bull. at 10 kGy presented the highest oxalic acid content; in turn, fumaric acid showed identical results to control and both oxalic and fumaric acids, exhibited declines over time during the studied year (Fernandes et al., 2017).

Regarding electron-beam effects in dried samples, some slight differences were detected for organic acids; *B. edulis* and *R. delica* irradiated with 2 and 6 kGy doses had similar results, while the 10 kGy dose decreased the organic acids content (Fernandes et al., 2014a). *A. caesarea* and *A. curtipes* irradiated with 6 kGy showed an increase in malic and fumaric acids content (Fernandes, Barreira, Antonio, Rafalski et al., 2015).

Regarding phenolic acids composition, irradiation doses applied to *M. procera* processed samples (particularly frozen and dried) did not cause remarkable changes, except for a slightly lower content of phenolic acids in non-irradiated samples (Fernandes, Barros et al., 2014). Protocatechuic acid (in *C. subconnexa*) and cinnamic acid (in *B. pinophilus*) showed a significant decrease in gamma-irradiated fresh samples with 2 kGy, while *p*-hydroxybenzoic acid (in *C. subconnexa*) increased significantly with irradiation (Fernandes, Barreira, Antonio, Oliveira et al., 2016). Phenolic acids decreased throughout storage time (up to 12 months) in dried and sliced gamma-irradiated *B. edulis*, and this negative effect could not be prevented by any of the assayed gamma irradiation doses. Regardless of storage time, *p*-hydroxybenzoic acid and cinnamic acid did not reveal significant differences for each of the applied doses (2, 6, and 10 kGy) (Fernandes et al., 2017).

With respect to phenolic acids composition, *B. edulis* had interesting levels of *p*-coumaric acid, particularly with the 2 and 10 kGy doses, and its content did not reveal a marked tendency with the increase in electron-beam irradiation. In *R. delica*, the only detected phenolic acid was gallic acid which present higher content with the 10 kGy dose. Both mushrooms presented cinnamic acid (the amounts in *B. edulis* were quite higher with 10 kGy) and decreased with the 6 kGy dose (Fernandes et al., 2014a).

In *A. caesarea* and *A. curtipes*, the only detected phenolic acid was *p*-hydroxybenzoic acid, which was higher with the 2 kGy dose in *A. caesaea*. Some slight differences were detected, despite lacking an identifiable tendency; nevertheless, cinnamic acid suffered a strong decrease in irradiated electron-beam dried samples (Fernandes, Barreira, Antonio, Rafalski et al., 2015).

Concerning antioxidant activity, *M. procera* extracts (frozen, dried, and fresh) maintained the EC_{50} values in different assays, independently of the applied gamma irradiation dose (0.5 and 1 kGy). With respect to processing type, and regardless of the dose applied, fresh samples presented lower DPPH scavenging activity, while frozen samples were less effective inhibitors of β-carotene bleaching and dried samples had less activity against TBARS formation. Concerning total phenolics compounds (TPC) (measured by *Folin-Ciocalteu* assay), the processing type induced again higher changes than the gamma irradiation dose (Fernandes et al., 2014b).

The extracts from irradiated *B. pinophilus* and *C. subconnexa*, showed to be more active in the DPPH radical scavenging and reducing power assays, while *C. subconnexa* extracts were especially active as β-carotene bleaching inhibitors, and *B. pinophilus* in TBARS formation inhibition; TPC increased with irradiation in both mushrooms (Fernandes, Barreira, Antonio, Oliveira et al., 2016). Additionally, gamma-irradiated *Volvariella volvacea* (Bull.) Singer showed a similar trend in the level of lipid peroxidation throughout the 7 days' storage period (Hou et al., 2018). In gamma-irradiated, fresh *T. aestivum* truffles, the levels of TPC after irradiation were almost doubled comparatively to non-irradiated samples (Tejedor-Calvo et al., 2020). The same authors reported that irradiation treatments induced variations in TPC not only immediately after irradiation but also during the storage period. Non-irradiated *T. melanosporum* truffles maintained the same phenolic compound levels during almost 35 storing days. Similar behavior was observed in electron-beam irradiated truffles when 1.5 kGy was applied. Nevertheless, gamma and electron-beam irradiation with 2.5 kGy enhanced the biosynthesis of phenolic compounds up to 21 days, and, in the latter case, an almost threefold higher concentration than its initial levels was observed (Tejedor-Calvo et al., 2020).

Regarding irradiated dried samples, non-stored *B. edulis* samples showed the highest antioxidant activity in all performed assays (DPPH scavenging activity, reducing power, and lipid peroxidation inhibition). On the other hand, irradiation treatment did not induce particularly noticeable effects, despite the slightly lower EC_{50} values found in irradiated samples. This indicates that the effect of a determined gamma irradiation dose is strongly dependent on the period of storage. For the majority of parameters, the applied irradiation doses were not enough to avoid their decrease, probably due to the extended storage time (Fernandes et al., 2017).

Dried *M. procera* irradiated with 6 kGy exhibited lower EC_{50} ferricyanide/Prussian blue assay and TBARS formation inhibition, and maintenance of antioxidant activity (according to all assays) was observed during the storage period (Fernandes et al., 2014c).

In all antioxidant activity assays (DPPH scavenging activity, reducing power, ferricyanide/Prussian blue assay and *Folin–Ciocalteu* assay, lipid peroxidation inhibition, β-Carotene bleaching inhibition, and TBARS formation inhibition), the antioxidant activity was improved in dried irradiated electron-beam samples (except ferricyanide/Prussian blue assay in *B. edulis* and *R. delica*); for *R. delica* it is even possible to point out 6 kGy as the most suitable dose to enhance antioxidant activity (Fernandes et al., 2014a). The same authors reported that, with no exception, electron-beam irradiated samples (especially for the higher doses) were shown to possess more antioxidant capacities, either as DPPH radical scavengers, ferric reducers, and β-carotene bleaching or TBARS formation inhibitors. Except for TBARS formation inhibition, *A. caesarea* showed a higher antioxidant activity than *A. curtipes* and, besides this action, the assayed mushroom proved to be particularly active as reducing agents (Fernandes, Barreira, Antonio, Rafalski et al., 2015).

Regarding dietary fiber assessed in electron-beam irradiated dried samples, *M. procera* irradiated with 6 kGy presented less total fiber, insoluble fiber, and carbohydrates, but the same dose allowed the highest contents in soluble dietary fiber. In *B. edulis* samples, insoluble fiber and total fiber amounts were significantly lower in samples irradiated with 10 kGy, but soluble fiber had no significant changes for any of the assayed doses, while total carbohydrates were significantly lower in non-irradiated samples (Fernandes, Barreira, Antonio, Morales et al., 2015).

The effects of gamma and electron-beam irradiation on the macro- and micro-elements profiles were evaluated in *B. edulis*, *Hydnum repandum* L. and *M. procera*. The effects of irradiation treatment were not significant, as expected due to the chemical nature of micro- and macro-elements, in most cases. Similarly, no particular differences were observed in terms of fresh or dried samples (Fernandes, Barreira, Antonio, Rafalski et al., 2016).

Concerning enzymes, 0.8 and 1.0 kGy of gamma irradiation significantly increased the catalase (CAT) activity of *V. volvacea* (Bul.) Singer samples at 4 and 5 days of storage, respectively; superoxide dismutase (SOD) activities were significantly higher than those of the control (Hou et al., 2018).

γ-Irradiation effects (above 1.0 kGy) on physical parameters of *V. volvacea* and *Lentinus edodes* were evaluated by Hou et al. (2018) and Shi et al. (2020), respectively. Both authors achieved significantly better commercial appearance after 7 days of storage due to maintaining the water state and good appearances, and slower postharvest mushroom softening, browning, weight loss and respiration rate than the control.

Lastly, the sensory quality of mushroom species was evaluated. Fresh *V. volvacea*, stored at 16 °C for 7 days, γ-irradiated with 0.8 kGy did not exhibit watery exudates, showed no evidence of fungal spoilage and discoloration at the end of the storage period, and maintained the best appearance, resulting in the highest sensory evaluation score. However, treatment with 0.4 kGy gave the same sensory evaluation score as the control; the firmness of the *V. volvacea* samples irradiated with 0.8 kGy changed the least over the 7 days of storage period and samples irradiated with 0.4 and 0.6 kGy were softer after 3 days, but showed similar trends to the control during storage (Hou et al., 2018). Kortei et al. (2020) reported that γ-irradiated dried *P. ostreatus* verified no adverse effect on its organoleptic attributes. There were no significant differences recorded for attributes such as appearance, aroma, and taste (irradiated at 2 kGy), and mouthfeel (irradiated at 2 and 5 kGy). Color and overall acceptability recorded some significant differences with 1 and 2 kGy doses, respectively. Also, descriptive textural attributes of hardness (irradiated at 1 and 3 kGy), adhesiveness (at 5 kGy), chewiness (at 1 kGy), and smoothness (at 2 kGy) recorded no significant difference. Cohesiveness, nevertheless, differed significantly (at 3 kGy). Low-dose gamma irradiation (< 10 kGy) of samples caused no adverse effect on its organoleptic attributes (Kortei et al., 2020). The irradiation treatments did not affect negatively *T. aestivum* sensory characteristics, namely, firmness, aspect, aroma, flavor, and general acceptability. The shelf-life limit was established in 21 days for non-irradiated and 0.5 kGy batches, 28 days for 1.0 kGy, 35 days for 1.5 kGy, and 42 days for 2.5 kGy (Tejedor-Calvo et al., 2020).

11.3.3 Dried aromatic and medicinal plants irradiation

11.3.3.1 Dried plants conservation

Throughout human history, approximately 7000 different plant species have been used as food by man, being an essential resource for their well-being (Lambers et al., 2017). MAPs are a very important component of a healthy diet since their consumption in adequate quantities can interfere positively with health, acting in the prevention of some illnesses and diseases (Barata, Rocha, Lopes, & Carvalho, 2016; Bortolin et al., 2019). This capacity is proven scientifically by several studies that highlight their therapeutic properties, namely, inflammatory, cardiovascular, antioxidant, and antimicrobial action (Bahmani, Shirzad, Majlesi, Shahinfard, & Rafieian-Kopaei, 2014; Shayganni, Bahmani, Asgary, & Rafieian-Kopaei, 2016). Thus, the work done by the scientific community on the properties of plants has allowed us to understand that the main benefits of plant species is due to the presence of compounds resulting from their secondary metabolism (mainly phenolic compounds), which can be classified as bioactive molecules (Lima, Turrini, Silva, Melo, & Augusto, 2017). In this sense, taking into account these benefits, its use as an ingredient in the pharmaceutical and food industry has increased significantly over time. Regarding its applicability in the pharmaceutical industry, a quarter of prescription drugs are derived or originated directly from plants and 4 out of 5 people in the world depend on plants for primary health care (Lambers et al., 2017).

However, a major problem is associated with this type of plant matrix. MAPs are, in most cases, grown in an open environment, exposed to chemical and microbiological risks, and unfortunately, no cleaning step can remove all biological and chemical contaminants. Unfortunately, microbial deterioration or insect infestation during storage and transport can reduce the quality and shelf life of plants and also represents a danger to public health, since they are exposed to a high level of natural contamination, namely, bacteria, fungi, molds, and yeasts, which are associated with illness outbreaks (Bearth & Siegrist, 2019; Bortolin et al., 2019; Khawory et al., 2020).

Although one of the biggest health problems in the world is infectious diseases, the need for food continues to increase due to the increase in population, making it essential to create alternatives that allow the consumption of MAPs safely and without changing their organoleptic, nutritional, and bioactive properties (Aleksieva & Yordanov, 2018). In addition, the problems associated with food processing and storage are essential and require the use of effective methods to prevent spoilage (Ortega-Rivas, 2014).

There are several types of processing to decontaminate and conserve food, by eliminating insects, bacteria, and toxins. However, these methodologies must allow to maintain the original appearance and flavor of food products and guarantee the safety of their consumption, avoiding the deposition of chemical residues (Brodowska, & Śmigielski, 2014; Duncan et al., 2017).

Thus, in MAPs, irradiation emerges as an efficient technique in conservation, reducing losses caused by natural physiological processes, by eliminating or reducing microorganisms, making it safer for the consumer, and also by reducing dependence on other technics, especially with chemical fumigants and preservatives used by the industry (Bashir et al., 2020).

In the last two decades, treatment with ionizing radiation has been used in plants, instead of chemical treatments, against microbial deterioration of food during storage and transport, to preserve hygienic quality, guarantee shelf life and, consequently, reduce public health risks (Bortolin et al., 2019).

11.3.3.2 *Effects on dried plants nutrients and bioactives*

When it is intended to apply processing techniques to foods, it is important to verify the maintenance of a wide range of compounds such as fatty acids, tocopherols, organic acids or free sugars, to guarantee their stability in the samples submitted to the decontamination treatments. Likewise, the bioactive properties of the final products should at least maintain the effectiveness of the original products.

Some studies have been done in order to evaluate the effects of the various types of ionizing radiation applied in MAPs. Table 11.3 show several scientific studies in this area, after the year 2013, considering the analyzed plant species, its origin, the type of ionizing radiation applied, and the corresponding dose.

Pereira, Barros, Antonio, Bento, and Ferreira (2014) studied the effects of gamma radiation on nutritional and chemical composition of *Ginkgo biloba* L., as well as, the consequences in their antioxidant action. For that, doses of 1 and 10 kGy were applied and the results proved that irradiation preserved macronutrients, fatty acids, γ- and δ-tocopherols, fructose, trehalose, quinic, and shikimic acids. The 1 kGy, in particular, protected α-tocopherol, oxalic and malic acids contents, while the application of 10 kGy decreased α-tocopherol, glucose, sucrose, oxalic, and malic acids levels. Nevertheless, this dose was the most effective in increasing the antioxidant activity.

Pereira, Barros, Dueñas et al. (2015) also concluded that gamma irradiation (10 kGy) improved the extraction of phenolic compounds in *G. biloba*. In another study, Pereira, Antonio, Barreira et al. (2015) evaluated the effects of gamma irradiation (1 and 10 kGy) on chemical, nutritional and antioxidant properties of several MAPs, namely, *Aloysia citrodora* Paláu, *Melissa officinalis* L., *Melittis melissophyllum* L. and *Mentha piperita* L. The results demonstrated that non-irradiated and irradiated samples were grouped indiscriminately when considering all parameters, indicating that irradiation treatment did not cause sufficient changes to define a specific chemical profile.

The same authors also studied the preserving capacity of gamma radiation (at a dose of 10 kGy) in the same species, stored for 12 and 18 months (Pereira, Antonio, Barreira, Cabo-Verde et al., 2018) and, the results revealed that, despite the maintenance of the main characteristics during the first 12 months of storage, the additional 6 months induced several significant changes in individual compounds. In general, the analyzed species reacted dissimilarly throughout time, but it was possible to verify that the fatty acids, tocopherols and antioxidant

TABLE 11.3 Nutrients and bioactives reported in studies regarding irradiation of dried aromatic and medicinal plants.

Parameters	Species	Origin	Radiation technology	Doses (kGy)	References
Ash	*Ginkgo biloba* L.	China	Gamma	1, 10	Pereira et al. (2014)
Proteins	*Aloysia citrodora* Palau	Portugal	Gamma	1, 10	Pereira, Antonio, Barreira et al. (2015)
				10	Pereira, Antonio, Barreira, Cabo-Verde et al. (2018)
			Electron-beam	1, 10	Pereira, Antonio, Rafalski et al. (2015)
				10	Pereira et al. (2017)
Fat	*Melissa officinalis* L.	Portugal	Gamma	1, 10	Pereira, Antonio, Barreira et al. (2015)
				10	Pereira, Antonio, Barreira, Cabo-Verde et al. (2018)
			Electron-beam	1, 10	Pereira, Antonio, Rafalski et al. (2015)
				10	Pereira et al. (2017)
Carbohydrates	*Melittis melissophyllum* L.	Portugal	Gamma	1, 10	Pereira, Antonio, Barreira et al. (2015)
				10	Pereira, Antonio, Barreira, Cabo-Verde et al. (2018)
			Electron-beam	1, 10	Pereira, Antonio, Rafalski et al. (2015)
				10	Pereira et al. (2017)
Energy	*Mentha piperita* L.	Portugal	Gamma	1, 10	Pereira, Antonio, Barreira et al. (2015)
				10	Pereira, Antonio, Barreira, Cabo-Verde et al. (2018)
			Electron-beam	1, 10	Pereira, Antonio, Rafalski et al. (2015)
				10	Pereira et al. (2017)
	Arenaria montana L.	Portugal	Gamma and Electron-beam	1, 10	Pereira, Barros, Barreira et al. (2016)
Free sugars profile	*G. biloba* L.	China	Gamma	1, 10	Pereira et al. (2014)
Fatty acids profile	*A. citrodora* Palau	Portugal	Gamma	1, 10	Pereira, Antonio, Barreira et al. (2015)
				10	Pereira, Antonio, Barreira, Cabo-Verde et al. (2018)
			Electron-beam	1, 10	Pereira, Antonio, Rafalski et al. (2015)
				10	Pereira et al. (2017)
Tocopherols profile	*M. officinalis* L.	Portugal	Gamma	1, 10	Pereira, Antonio, Barreira et al. (2015)
				10	Pereira, Antonio, Barreira, Cabo-Verde et al. (2018)
			Electron-beam	1, 10	Pereira, Antonio, Rafalski et al. (2015)
				10	Pereira et al. (2017)

Organic acids	*M. melissophyllum* L.	Portugal	Gamma	1, 10	Pereira, Antonio, Barreira et al. (2015)
				10	Pereira, Antonio, Barreira, Cabo-Verde et al. (2018)
			Electron-beam	1, 10	Pereira, Antonio, Rafalski et al. (2015)
				10	Pereira et al. (2017)
	M. piperita L.	Portugal	Gamma	1, 10	Pereira, Antonio, Barreira et al. (2015)
				10	Pereira, Antonio, Barreira, Cabo-Verde et al. (2018)
			Electron-beam	1, 10	Pereira, Antonio, Rafalski et al. (2015)
				10	Pereira et al. (2017)
	A. montana L.	Portugal	Gamma and Electron-beam	1, 10	Pereira, Barros, Barreira et al. (2016)
Phenolic content	*G. biloba* L.	China	Gamma	1, 10	Pereira, Barros, Dueñas et al. (2015)
	A. citrodora Palau	Portugal	Gamma	1, 10	Pereira, Pimenta et al. (2017)
				1, 5, 10	Pereira, Barros et al. (2017)
					Pereira, Pimenta et al. (2018)
	M. piperita L.	Portugal	Gamma	1, 10	Pereira, Pimenta et al. (2017)
				1, 5, 10	Pereira, Pimenta et al. (2018)
	Thymus vulgaris L.	Portugal	Gamma	2, 5, 10	Pereira, Pimenta et al. (2016)
				1, 2	Pereira, Barros, Antonio et al. (2016)
				1, 5, 10	Pereira, Pimenta et al. (2018)
	M. piperita L.	Portugal	Gamma	2, 5, 10	Pereira, Pimenta et al. (2016)
	M. officinalis L.	Portugal	Gamma and Electron-beam	1, 10	Pereira, Antonio, Barreira., Santos-Buelga et al. (2018)
	M. melissophyllum L.	Portugal	Gamma and Electron-beam	1, 10	Pereira, Antonio, Barreira., Santos-Buelga et al. (2018)
	Euodia malayana	Malaysia	Gamma	3–5, 5–10, 6–12, 9–13	Khawory et al. (2020)
	Gnetum gnemon	Malaysia	Gamma	3–5, 5–10, 6–12, 9–13	Khawory et al. (2020)
	Khaya senegalensis	Malaysia	Gamma	3–5, 5–10, 6–12, 9–13	Khawory et al. (2020)
	Nigella sativa	Pakistan	Gamma	1, 3, 5, 7	Naveed et al. (2018)
	Zingiber oficinal	Pakistan	Gamma	1, 3, 5, 7	Naveed et al. (2018)
	Allium sativa	Pakistan	Gamma	1, 3, 5, 7	Naveed et al. (2018)
	Carulluma tubercullata	Pakistan	Gamma	1, 3, 5, 7	Naveed et al. (2018)

(*Continued*)

TABLE 11.3 (*Continued*)

Parameters	Species	Origin	Radiation technology	Doses (kGy)	References
Antioxidant activity	*G. biloba* L.	China	Gamma	1, 10	Pereira et al. (2014)
	A. citrodora Palau	Portugal	Gamma	1, 10	Pereira, Antonio, Barreira et al. (2015)
				10	Pereira, Antonio, Barreira, Cabo-Verde et al. (2018)
			Electron-beam	1, 10	Pereira, Antonio, Rafalski et al. (2015)
				10	Pereira et al. (2017)
	M. officinalis L.	Portugal	Gamma	1, 10	Pereira, Antonio, Barreira et al. (2015)
				10	Pereira, Antonio, Barreira, Cabo-Verde et al. (2018)
			Electron-beam	1, 10	Pereira, Antonio, Rafalski et al. (2015)
				10	Pereira et al. (2017)
	M. melissophyllum L.	Portugal	Gamma	1, 10	Pereira, Antonio, Barreira et al. (2015)
				10	Pereira, Antonio, Barreira, Cabo-Verde et al. (2018)
			Electron-beam	1, 10	Pereira, Antonio, Rafalski et al. (2015)
				10	Pereira et al. (2017)
	M. piperita L.	Portugal	Gamma	1, 10	Pereira, Antonio, Barreira et al. (2015)
				10	Pereira, Antonio, Barreira, Cabo-Verde et al. (2018)
			Electron-beam	1, 10	Pereira, Antonio, Rafalski et al. (2015)
				10	Pereira et al. (2017)
	T. vulgaris L.	Portugal	Gamma	1, 2	Pereira, Barros, Antonio et al. (2016)
	A. montana L.	Portugal	Gamma and Electron-beam	1, 10	Pereira, Barros, Barreira et al. (2016)
	E. malayana	Malaysia	Gamma	3–5, 5–10, 6–12, 9–13	Khawory et al. (2020)
	G. gnemon	Malaysia	Gamma	3–5, 5–10, 6–12, 9–13	Khawory et al. (2020)
	K. senegalensis	Malaysia	Gamma	3–5, 5–10, 6–12, 9–13	Khawory et al. (2020)
	Thapsia garganica L.	Algeria	Gamma	0.1, 0.3, 0.7, 1, 3, 7, 10	Ibrahim et al. (2018)
	N. sativa	Pakistan	Gamma	1, 3, 5, 7	Naveed et al. (2018)
	Z. oficinal	Pakistan	Gamma	1, 3, 5, 7	Naveed et al. (2018)
	A. sativa	Pakistan	Gamma	1, 3, 5, 7	Naveed et al. (2018)
	C. tubercullata	Pakistan	Gamma	1, 3, 5, 7	Naveed et al. (2018)

Cytotoxic activity	*T. vulgaris* L.	Portugal	Gamma	2, 5, 10	Pereira, Pimenta et al. (2016)
				1, 5, 10	Pereira, Pimenta et al. (2018)
	Mentha x piperita L.	Portugal	Gamma	2, 5, 10	Pereira, Pimenta et al. (2016)
				1, 5, 10	Pereira, Pimenta et al. (2018)
	A. citrodora Palau	Portugal	Gamma	1, 10	Pereira, Pimenta et al. (2017)
				1, 5, 10	Pereira, Pimenta et al. (2018)
	M. piperita L.	Portugal	Gamma	1, 10	Pereira, Pimenta et al. (2017)
	E. malayana	Malaysia	Gamma	3–5, 5–10, 6–12, 9–13	Khawory et al. (2020)
	G. gnemon	Malaysia	Gamma	3–5, 5–10, 6–12, 9–13	Khawory et al. (2020)
	K. senegalensis	Malaysia	Gamma	3–5, 5–10, 6–12, 9–13	Khawory et al. (2020)
Microbial activity	*A. citrodora* Palau	Portugal	Gamma	1, 10	Pereira, Pimenta et al. (2017)
				1, 5, 10	Pereira, Pimenta et al. (2018)
	M. piperita L.	Portugal	Gamma	1, 10	Pereira, Pimenta et al. (2017)
				1, 5, 10	Pereira, Pimenta et al. (2018)
	T. vulgaris L.	Portugal	Gamma	1, 5, 10	Pereira, Pimenta et al. (2018)
	E. malayana	Malaysia	Gamma	3–5, 5–10, 6–12, 9–13	Khawory et al. (2020)
	G. gnemon	Malaysia	Gamma	3–5, 5–10, 6–12, 9–13	Khawory et al. (2020)
	K. senegalensis	Malaysia	Gamma	3–5, 5–10, 6–12, 9–13	Khawory et al. (2020)
	N. sativa	Pakistan	Gamma	1, 3, 5, 7	Naveed et al. (2018)
	Z. oficinal	Pakistan	Gamma	1, 3, 5, 7	Naveed et al. (2018)
	A. sativa	Pakistan	Gamma	1, 3, 5, 7	Naveed et al. (2018)
	C. tubercullata	Pakistan	Gamma	1, 3, 5, 7	Naveed et al. (2018)
Virucidal activity	*A. citrodora* Palau	Portugal	Gamma	1, 10	Pereira, Pimenta et al. (2017)
				1, 5, 10	Pereira, Pimenta et al. (2018)
	M. piperita L.	Portugal	Gamma	1, 10	Pereira, Pimenta et al. (2017)
				1, 5, 10	Pereira, Pimenta et al. (2018)
	T. vulgaris L.	Portugal	Gamma	1, 5, 10	Pereira, Pimenta et al. (2018)

capacity presented the most significant changes after 18 months of storage, namely in samples submitted to gamma radiation.

Pereira, Pimenta et al. (2017) continued to study *A. citrodora* and *M. piperita*, evaluating the gamma radiation (1 and 10 kGy) effects on the phenolic composition, cytotoxicity, virucidal and antimicrobial activity. The assays were performed in the infusions of the samples and it became evident that radiation treatment caused a significant increase in total phenolic compounds of both plants. This increase improves the species' antioxidant action, resulting in a benefit for the consumer. The virucidal efficacy was dependent on the plant, the virus, and the absorbed dose.

In addition, in *A. citrodora*, the effect of gamma radiation on the preservation of phenolic compounds and the decontamination capacity in two types of carcinogenic toxins: ochratoxin A (OTA) and aflatoxin B1 (AFB1) were also analyzed. For this purpose, 1, 5, and 10 kGy were applied and, at the highest dose, the mycotoxin reduction observed was 9.6% for OTA and 5.2% for AFB1; a slight degradation of the phenolic compounds in the irradiated samples was also observed (Pereira, Barros et al., 2017).

In addition, these authors (Pereira, Antonio, Rafalski et al., 2015) compared the effects of gamma irradiation and electron beam irradiation in the chemical parameters and bioactive indicators of these species in order to find the most suitable technology in each case. The obtained data showed that several examples of percentage variations specific to a plant species or irradiation condition were found; nevertheless, it was not possible to identify unequivocal trends. But, when evaluated in an integrated way, the parameters with the highest discriminating ability among irradiation conditions or plant species were fatty acids and bioactive activity. Comparing the effects of the two types of irradiation, it might be concluded that the most suitable solution to irradiate aromatic plants would be electron-beam, independently of the applied dose.

As with gamma irradiation, the same authors also studied the preserving power of electron beam irradiation, in the same species, with a dose of 10 kGy and during 18 months of storage (Pereira et al., 2017). The nutritional, chemical, and bioactive properties were evaluated, and the results allowed to prove some heterogeneity among the effects produced in each plant species; however, electron-beam treatment attenuated the reduction of individual compounds (primarily, free sugars, organic acids, tocopherols, and polyunsaturated fatty acids) verified in non-irradiated samples.

Pereira, Pimenta et al. (2018) developed an experimental study on *Mentha* × *piperita* L., *Thymus vulgaris* L. and *A. citrodora* exposed to gamma radiation (1, 5 and 10 kGy) in order to evaluate the effects on phenolic compounds, cytotoxicity, virucity and antimicrobial activity. They observed that the cytotoxic activity varied with plant species and applied dose. The virucidal activity was verified to be dependent on the radiation dose, but this treatment preserved this bioactive action. In turn, in the antimicrobial activity, gamma rays had no effect. Otherwise, the effects of gamma radiation on the phenolic profile were heterogeneous, depending on the species and radiation dose.

Mentha × *piperita* L. and *T. vulgaris* were also studied, namely with regard to the effects of gamma irradiation on cytotoxicity and phenolic compounds, but in methanolic extracts. Different doses of radiation were used: 0, 2, 5, and 10 kGy, and the results showed no statistically significant differences in the phenolic profile of the species when the dose of 10 kGy was applied. However, in the cytotoxic assay, the results are heterogeneous, depending on the plant species, applied dose, and tested cell line (Pereira, Pimenta et al., 2016).

Pereira, Barros, Antonio et al. (2016) developed a study to evaluate the effects of gamma radiation (1 and 10 kGy) on the antioxidant activity and phenolic compounds profile of infusions prepared from irradiated *T. vulgaris*. The results showed that the 10 kGy dose improved the antioxidant capacity and increased the concentrations of some flavonoids content.

Other species, such as *Arenaria montana* L., were also studied by Pereira, Barros, Barreira et al. (2016). Therein, the evaluation of the effects of gamma and electron-beam radiation at different doses (1 and 10 kGy) in nutritional, chemical, and antioxidant parameters were studied. In general, significant variations in response to irradiation treatment were observed; furthermore, it was possible to identify the antioxidant indicators as the most affected parameters in irradiated samples, especially when using the 10 kGy dose and electron-beam irradiation.

M. officinalis and *M. melissophyllum* L. were studied later by Pereira, Antonio, Barreira., Santos-Buelga et al. (2018), specifically to identify potential changes in the phenolic compounds of irradiated samples to further validate electron-beam and gamma irradiation as conservation technologies to be applied to highly consumed plant species. A general increase in individual phenolics was observed as a result of irradiation, especially (despite not for all compounds) when using gamma irradiation.

In a study performed by Khawory et al. (2020), the effects of gamma irradiation on three different methanolic herbal crude extracts and dry plants of *Euodia malayana*, *Gnetum gnemon* and *Khaya senegalensis* on the microbial limit test, quantitative analysis, in vitro cytotoxicity activity and antioxidant activity, were evaluated. The results evidenced that the suitable gamma irradiation dose for leaf extracts was 6–12 kGy and 9–13 kGy for dried leaves, which also induced an increase in the concentration of gallic acid. Regarding the bioactive potential, no significant effect was observed in cytotoxicity, while an increase in free radical scavenging activity (depending on the plant species) and the capacity to reduce the microbial load was proved in samples submitted to gamma radiation treatment.

Naveed, Rizwan, and Sajid (2018) performed a study where several herbs were processed with gamma radiation (1, 3, 5, and 7 kGy) and evaluated considering the microbial loads, and the phytochemical analysis (total phenol contents, total flavonoids contents and antioxidant activity). The data showed that gamma irradiation between 1 to 7 kGy improved the elimination of microbial load, improved the quality of the herbal product, and also enhanced the shelf life of herbal products.

Finally, Ibrahim et al. (2018) evaluated the combined effects of microwave-assisted extraction and gamma irradiation on the extraction yield of the bioactive compounds, as well as on the antioxidant activity of the alcoholic extract, and the results showed that gamma irradiation have no effect on the phytochemical composition of the plant extracts or on their antioxidant properties.

Considering all these scientific studies, it was evident that the effects of ionizing radiation in molecules are heterogeneous, depending on several factors, namely, the plant species, source of radiation, and applied dose. However, it has been proven that the irradiation may be considered a suitable and promising technology to preserve MAPs, considering its effects on the nutritional, chemical, and bioactive potential of several species of MAPs.

11.4 Concluding remarks

Food irradiation emerged as a viable processing technique, applied with the purpose of preserving, decontaminating, and disinfecting several types of food, inhibiting the growth of microorganisms that cause their deterioration and acting in pest control (Balatsas-Lekkas et al., 2020; Mohácsi-Farkas, 2016).

This type of processing has been supported by several national and international organizations, namely the WHO, FAO, IAEA, and CAC. Moreover, it is applied to a wide range of foods, depending on the country and legislation (Ihsanullah & Rashid, 2016).

The scientific community has shown great interest in studying the effects of this technology on nutritional, chemical, physical, and bioactive composition of food (Ravindran & Jaiswal, 2019). Considering the studies carried out by several authors, specifically in chestnuts, mushroom, and MAPs, it is possible to verify that irradiation, with the application of the appropriate doses (duly legislated by the competent entities), is safe and innocuous. In the case of chestnuts, the influence of irradiation process on chemical parameters of chestnuts has been studied over time by several authors in different varieties and origins. The results evidenced some heterogeneity in the effects caused, depending especially on the applied dose and the evaluated parameters. In addition, it was also possible to verify the preserving power of this technology in chestnuts, since in most studies, the nutritional parameters were maintained throughout the storage period.

Likewise, in the case of mushrooms, regardless of the type of ionizing irradiation applied (gamma or electron beam), the studies presented in this chapter proved that the chemical, nutritional, and bioactive profiles were not significantly affected, and several parameters are preserved.

For the aromatic and medicinal plants, several studies were also performed, and the results were heterogeneous. It was possible to verify that the effects of the application of this technology depend on the type of ionizing radiation used, the dose applied, and the plant species. However, it was proved that, although there were changes in the different parameters evaluated, these were not significant, and there was no mischaracterization of the nutritional profile. Considering the particular case of the effects on polyphenols, an increase in the concentration of some molecules was often noticeable. This increase can enhance the intensification of the species' antioxidant action and, consequently, promote benefits in its consumption. In addition, it was also possible to verify the decontaminating capacity of gamma irradiation, by decreasing the concentration of carcinogenic toxins in samples of *A. citrodora*.

Thus, making a general evaluation of the application of irradiation in food products, it can be concluded that it is an apparently safe process. The irradiated foods are nutritionally adequate and equivalent to non-irradiated foods, showing that irradiation does not lead to significant nutrient losses able to reduce their quality. In addition, with regard to its potential as a preservative, several studies have proven that the extension of the shelf life by irradiation is superior to thermal processing, avoiding the need for artificial preservatives, thus maintaining the nutritional value of food (Aylangan, Ic, & Ozyardimci, 2017; Pereira et al., 2017). In this sense, it is expected that this technology will increasingly gain consumer confidence if a clear approach is taken.

References

Aleksieva, K. I., & Yordanov, N. D. (2018). Various approaches in EPR identification of gamma-irradiated plant foodstuffs: A review. *Food Research International*, *105*, 1019–1028.

Álvarez-Álvarez, P., Barrio-Anta, M., & Diéguez-Aranda, U. (2006). Differentiation of sweet chestnut (Castanea sativa Mill.) cultivars by leaf, nut and burr dimensions. *Forestry: International Journal of Forestry Research*, *79*, 149–158.

Antonio, A. L., Carocho, M., Bento, A., Quintana, B., Botelho, M. L., & Ferreira, I. C. F. R. (2012). Effects of gamma radiation on the biological, physico-chemical, nutritional and antioxidant parameters of chestnuts – A review. *Food and Chemical Toxicology*, *50*, 3234–3242.

Antonio, A. L., Carocho, M., Bento, A., Rafalski, A., & Quintana, B. (2013). Influence of e-beam postharvest irradiation in the colour of four European chestnut fruit varieties of *Castanea sativa* Mill. *European Scientific Journal*, *3*, 835–840.

Antonio, A. L., Fernandes, Â., Barreira, J., Bento, A., Botelho, M. L., & Ferreira, I. C. F. R. (2011). Influence of gamma irradiation in the antioxidant potential of chestnuts (*Castanea sativa* Mill.) fruits and skins. *Food and Chemical Toxicology*, *49*, 1918–1923.

Antonio, A.L., Ramalhosa, E., Kaluska, I., Quintana, B., Trigo, M., Ferreira, A., and Bento, A. (2012). Influence of e-beam irradiation processing on chestnuts color and texture. *International Conference on Agricultural Engineering. Valencia* (Spain).

Antonio, A. L., Cabo Verde, S., & Ferreira, I. C. F. R. (2018). Introduction. In I. C. F. R. Ferreira, A. L. Antonio, & S. Cabo Verde (Eds.), *Food irradiation technologies: Concepts, applications and outcomes* (pp. 1–4). UK: Royal Society of Chemistry.

Antonio, A. L., Fernandes, Â., Bento, A., Ferreira, A., Trigo, M., Botelho, M. L., ... Ramalhosa, E. (2013). Gamma irradiation preservation of chestnut fruits: Effects on color and texture. *European Scientific Journal*, *3*, 239–245.

Aylangan, A., Ic, E., & Ozyardimci, B. (2017). Investigation of gamma irradiation and storage period effects on the nutritional and sensory quality of chickpeas, kidney beans and green lentils. *Food Control*, *80*, 428–434.

Bahmani, M., Shirzad, H., Majlesi, M., Shahinfard, N., & Rafieian-Kopaei, M. (2014). A review study on analgesic applications of Iranian medicinal plants. *Asian Pacific Journal of Tropical Medicine*, *7*, S43–S53.

Balatsas-Lekkas, A., Arvola, A., Kotilainen, H. M., Meneses, N., & Pennanen, K. (2020). Effect of labelling fresh cultivated blueberry products with information about irradiation technologies and related benefits on Finnish, German, and Spanish consumers' product acceptance. *Food Control*, *118*, 107387.

Barata, A. M., Rocha, F., Lopes, V., & Carvalho, A. M. (2016). Conservation and sustainable uses of medicinal and aromatic plants genetic resources on the worldwide for human welfare. *Industrial Crops and Products*, *88*, 8–11.

Barreira, J., Antonio, A. L., Günaydi, T., Alkan, H., Bento, A., Botelho, M. L., & Ferreira, I. C. F. R. (2012). Chemometric characterization of gamma irradiated chestnuts from Turkey. *Radiation Physics and Chemistry*, *81*, 1520–1524.

Bashir, K., Jan, K., Kamble, D. B., Maurya, V. K., Jan, S., & Swer, T. L. (2020). *History, status and regulatory aspects of gamma irradiation for food processing* in *reference module in neuroscience and biobehavioral psychology*. Elsevier Inc.

Bearth, A., & Siegrist, M. (2019). "As long as it is not irradiated"–Influencing factors of US consumers' acceptance of food irradiation. *Food Quality and Preference*, *71*, 141–148.

Bortolin, E., Carrat, B., Chiaravalle, A. E., Di Schiavi, M. T., Gargiulo, R., Mangiacotti, M., ... Boniglia, C. (2019). Detection of irradiated plant foods by photo- and thermally-stimulated luminescence. *Food Control*, *105*, 113–122.

Brodowska, A., & Śmigielski, K. (2014). Comparison of methods of herbs and spices decontamination. *CHEMIK*, *68*, 97–102.

Cabo Verde, S. (2018). Food irradiation as sanitary treatment. In I. C. F. R. Ferreira, A. L. Antonio, & S. Cabo Verde (Eds.), *Food irradiation technologies: Concepts, applications and outcomes* (pp. 183–209). UK: Royal Society of Chemistry.

Cardoso, R. V. C., Fernandes, Â., Barreira, J. C. M., Verde, S. C., Antonio, A. L., Gonzaléz-Paramás, A. M., ... Ferreira, I. C. F. R. (2019). Effectiveness of gamma and electron beam irradiation as preserving technologies of fresh *Agaricus bisporus* Portobello: A comparative study. *Food Chemistry*, *278*, 760–766.

Carocho, M., Antonio, A. L., Barros, L., Bento, A., Botelho, M., Kaluska, I., & Ferreira, I. C. F. R. (2012).). Comparative effects of gamma and electron beam irradiation on the antioxidant potential of Portuguese chestnuts (*Castanea sativa* Mill.). *Food and Chemical Toxicology: An International Journal Published for the British Industrial Biological Research Association*, *50*, 3452–3455.

Carocho, M., Barreira, J. C., Antonio, A. L., Bento, A., Kaluska, I., & Ferreira, I. C. F. R. (2012).). Effects of electron-beam radiation on nutritional parameters of Portuguese chestnuts (*Castanea sativa* Mill.). *Journal of Agricultural and Food Chemistry*, *60*(31), 7724–7760.

Carocho, M., Barros, L., Antonio, A. L., Barreira, J., Bento, A., Kaluska, I., & Ferreira, I. C. F. R. (2013). Analysis of organic acids in electron beam irradiated chestnuts (*Castanea sativa* Mill.): Effects of radiation dose and storage time. *Food and Chemical Toxicology: An International Journal Published for the British Industrial Biological Research Association*, *55*, 348–352.

Carocho, M., Barros, L., Antonio, A. L., Barreira, J., Rafalski, A., Bento, A., & Ferreira, I. C. F. R. (2014). Validation of gamma and electron beam irradiation as alternative conservation technology for european chestnuts. *Food and Bioprocess Technology*, *7*, 1917–1927.

Carreño, I. (2018). International standards and regulation on food irradiation. In I. C. F. R. Ferreira, A. L. Antonio, & S. Cabo Verde (Eds.), *Food irradiation technologies: Concepts, applications and outcomes* (pp. 5–27). UK: Royal Society of Chemistry.

Cecchini, M., Contini, M., Massantini, R., Monarca, D., & Moscetti, R. (2011). Effects of controlled atmospheres and low temperature on storability of chestnuts manually and mechanically harvested. *Postharvest Biology and Technology*, *61*(2–3), 131–136.

Chung, H., Delincée, H., Han, S., Hong, J., Kim, H., Kim, M., ... Kwon, J.-H. (2004). Trials to identify irradiated chestnut (*Castaneabungena*) with different analytical techniques. *Radiation Physics and Chemistry*, *71*(1–2), 179–182.

Codex Alimentarius Commission. (1999). *Proposed draft revised general standard for irradiated foods*. Rome. Retrieved from http://www.fao.org/tempref/codex/Meetings/CCFAC/ccfac33/fa01_11e.pdf.

Corregidor, V., Antonio, A. L., Alves, L., & Cabo Verde, S. (2020). *Castanea sativa* shells and fruits: Compositional analysis by proton induced X-ray emission. *Nuclear Instruments & Methods in Physics Research. Section B, Beam Interactions with Materials and Atoms*, 98–103.

Craven, E., Schlecht, J., & Stein, R. (2018). Gamma irradiation plants. In I. C. F. R. Ferreira, A. L. Antonio, & S. Cabo Verde (Eds.), *Food irradiation technologies: Concepts, applications and outcomes* (pp. 28–50). UK: Royal Society of Chemistry.

Demirkesen, I., Mert, B., Sumnu, G., & Sahin, S. (2010). Utilization of chestnut flour in gluten-free bread formulations. *Journal of Food Engineering, 101*(3), 329–336.

Dethier, P., & Mullier, B. (2018). Technical and economical considerations in food irradiation. In I. C. F. R. Ferreira, A. L. Antonio, & S. Cabo Verde (Eds.), *Food irradiation technologies: Concepts, applications and outcomes* (pp. 369–382). UK: Royal Society of Chemistry.

Duncan, S. E., Moberg, K., Amin, K. N., Wright, M., Newkirk, J. J., Ponder, M. A., ... Dickson, J. S. (2017). Processes to preserve spice and herb quality and sensory integrity during pathogen inactivation. *Journal of Food Science, 82*, 1208–1215.

EU. (2008). Commission decision concerning the noninclusion of methyl bromide in Annex I to Council Directive 91/414/EEC and the withdrawal of authorisations for plant protection products containing that substance. *Official Journal of the European Union L, 258*, 68–69.

European Commission. (2019). Report from the commission to the European parliament and the council on food and food ingredients treated with ionising radiation for the years 2016–2017. Brussels. Retrieved from https://eur-lex.europa.eu/legal-content/EN/TXT/?uri=COM:2019:0454:FIN.

Eustice, R. (2018). Global status and commercial applications of food irradiation. In A. L. Antonio, S. Cabo Verde, & I. C. F. R. Ferreira (Eds.), *Food irradiation technologies: Concepts, applications and outcomes* (pp. 397–424). UK: Royal Society of Chemistry.

FAO. (2003). Postharvest handling and storage of chestnuts. Retrieved October 2020, from Food and Agriculture Organization. http://www.fao.org/3/ac645e/ac645e00.htm#Table%20of%20Contents.

FAOSTAT. (2020). Area harvested. Retrieved August 17, 2020, from http://www.fao.org/faostat/en/#data/QC/visualize.

Fernandes, Â., Barreira, J. C. M., Antonio, A. L., Rafalski, A., Morales, P., Férnandez-Ruiz, V., & Ferreira, I. C. F. R. (2016). Gamma and electron-beam irradiation as viable technologies for wild mushrooms conservation: Effects on macro- and micro-elements. *European Food Research and Technology, 242*, 1169–1175.

Fernandes, Â., Barreira, J. C. M., Antonio, A. L., Rafalski, A., Oliveira, M. B. P. P., Martins, A., & Ferreira, I. C. F. R. (2015). How does electron beam irradiation dose affect the chemical and antioxidant profiles of wild dried Amanita mushrooms? *Food Chemistry, 182*, 309–315.

Fernandes, Â., Antonio, A. L., Barros, L., Barreira, J. C., Bento, A., Botelho, M. L., & Ferreira, I. C. F. R. (2011). Low dose γ-irradiation as a suitable solution for chestnut (*Castanea sativa* Miller) conservation: Effects on sugars, fatty acids and tocopherols. *Journal of Agricultural and Food Chemistry, 59*, 10028–10033.

Fernandes, Â., Barreira, J., Antonio, A. L., Bento, A., Botelho, M. L., & Ferreira, I. C. F. R. (2011). Assessing the effects of gamma irradiation and storage time in energetic value and in major individual nutrients of chestnuts. *Food Chemistry and Toxicology, 49*, 2429–2432.

Fernandes, Â., Barreira, J. C. M., Antonio, A. L., Martins, A., Ferreira, I. C. F. R., & Oliveira, M. B. P. P. (2014). Triacylglycerols profiling as a chemical tool to identify mushrooms submitted to gamma or electron beam irradiation. *Food Chemistry, 159*, 399–406.

Fernandes, Â., Barreira, J. C. M., Antonio, A. L., Morales, P., Férnandez-Ruiz, V., Martins, A., ... Ferreira, I. C. F. R. (2015). Exquisite wild mushrooms as a source of dietary fiber: Analysis in electron-beam irradiated samples. *LWT - Food Science and Technology, 60*, 855–859.

Fernandes, A., Barreira, J. C. M., Antonio, A. L., Oliveira, M. B. P. P., Martins, A., & Ferreira, I. C. F. R. (2014a). Feasibility of electron-beam irradiation to preserve wild dried mushrooms: Effects on chemical composition and antioxidant activity. *Innovative Food Science and Emerging Technologies, 22*, 158–166.

Fernandes, Â., Barreira, J. C. M., Antonio, A. L., Oliveira, M. B. P. P., Martins, A., & Ferreira, I. C. F. R. (2014b). Effects of gamma irradiation on chemical composition and antioxidant potential of processed samples of the wild mushroom Macrolepiota procera. *Food Chemistry, 149*, 91–98.

Fernandes, Â., Barreira, J. C. M., Antonio, A. L., Oliveira, M. B. P. P., Martins, A., & Ferreira, I. C. F. R. (2014c). Combined effects of electron-beam irradiation and storage time on the chemical and antioxidant parameters of wild Macrolepiota procera dried samples. *Food and Bioprocess Technology*, 1606–1617.

Fernandes, Â., Barreira, J. C. M., Antonio, A. L., Oliveira, M. B. P. P., Martins, A., & Ferreira, I. C. F. R. (2016). Extended use of gamma irradiation in wild mushrooms conservation: Validation of 2 kGy dose to preserve their chemical characteristics. *LWT - Food Science and Technology, 67*, 99–105.

Fernandes, Â., Barreira, J. C. M., Günaydi, T., Alkan, H., Antonio, A. L., Oliveira, M. B. P. P., & Ferreira, I. C. F. R. (2017). Effect of gamma irradiation and extended storage on selected chemical constituents and antioxidant activities of sliced mushroom. *Food Control, 72*, 328–337.

Fernandes, Â., Barros, L., Antonio, A. L., Barreira, J. C. M., Oliveira, M. B. P. P., Martins, A., & Ferreira, I. C. F. R. (2014). Using gamma irradiation to attenuate the effects caused by drying or freezing in Macrolepiota procera organic acids and phenolic compounds. *Food and Bioprocess Technology, 7*, 3012–3021.

Ferreira, I. C. F. R., Antonio, A. L., & Cabo Verde, S. (Eds.), (2018). *Food irradiation technologies: Concepts, applications and outcomes*. UK: Royal Society of Chemistry.

Galati, A., Tulone, A., Moavero, P., & Crescimanno, M. (2019). Consumer interest in information regarding novel food technologies in Italy: The case of irradiated foods. *Food Research International, 119*, 291–296.

Gomes-Laranjo, J., Peixoto, F., Costa, R., & Ferreira-Cardoso, J. (2009). Portugal. In D. Avanzato, & G. Bounous (Eds.), *Following Chestnut footprints (*Castanea *spp.):* Cultivation and culture, folklore and history, traditions and uses (Vol. 9, pp. 106–111). ISHS Scripta Horticulturae.

Grimm, D., & Wösten, H. A. B. (2018). Mushroom cultivation in the circular economy. *Applied Microbiology and Biotechnology, 102*, 7795–7803.

Guo-xin, Z., Qiong-ling, D., Yao-chang, X., & Si-lian, L. (1980). Effect of ^{60}Co gamma irradiation on the storage of Chinese chestnut fruits. *Acta Botanica Sinica, 22*, 404–406.

Hou, L., Lin, J., Ma, L., Qu, S., Li, H., & Jiang, N. (2018). Effect of 60Co gamma irradiation on postharvest quality and selected enzyme activities of Volvariella volvacea. *Scientia Horticulturae, 235*, 382–390.

Ibrahim, A. M. M., Martinez-Swatson, K. A., Benkaci-Ali, F., Cozzi, F., Zoulikha, F., & Simonsene, H. T. (2018). Effects of gamma irradiation and comparison of different extraction methods on sesquiterpene lactone yields from the medicinal plant *Thapsia garganica* L. (Apiaceae). *Journal of Applied Research on Medicinal and Aromatic Plants, 8*, 26–32.

Ihsanullah, I., & Rashid, A. (2016). Current activities in food irradiation as a sanitary and phytosanitary treatment in the Asia and Pacific Region and a comparison with advanced countries. *Food Control, 72*, 345–359.

Ihsanullah, I., & Rashid, A. (2017). Current activities in food irradiation as a sanitary and phytosanitary treatment in the Asia and the Pacific Region and a comparison with advanced countries. *Food Control, 72*, 345–359.

Imamura, T., Todoriki, S., Miyanoshita, A., & Hayashi, T. (2004). Effects of gamma irradiation on the emergence of larvae of Curculio sikkimensis (Heller) (Coleoptera: *Curculionidae*) and Cydia kurokoi (Amsel) (Lepidoptera: *Tortricidae*). *Food Irradiation, 39*(1–2), 5–7.

Iwata, T., & Ogata, K. (1959). Studies on the storage of chestnuts treated with gamma radiation. *Bulletin of the University of Osaka Prefecture, 59*, 59–65.

Iwata, T., & Ogata, K. (1961). Studies in the Storage of Chestnuts Treated with Gamma Radiation. *Bulletin of the Institute for Chemical Research, Kyoto University, 39*, 112–119.

Jeremini, M., Conedera, M., Sieber, T. N., Sassella, A., Schärer, H., Jelmini, G., & Höhn, E. (2006). Influence of fruit treatments on perishability during cold storage of sweet chestnuts. *Joournal of the Science Food and Agricultural, 86*(6), 877–885.

Khawory, M. H., Sain, A. A., Rosli, M. A. A., Ishak, M. S., Noordin, M. I., & Wahab, H. A. (2020). Effects of gamma radiation treatment on three different medicinal plants: Microbial limit test, total phenolic content, in vitro mytotoxicity effect and antioxidant assay. *Applied Radiation and Isotopes, 157*, 109013.

King, T., Cole, M., Farber, J. M., Eisenbrand, G., Zabaras, D., Fox, E. M., & Hill, J. P. (2017). Food safety for food security: Relationship between global megatrends and developments in food safety. *Trends in Food Science & Technology, 68*, 160–175.

Kortei, N. K., Odamtten, G. T., Obodai, M., Akonor, P. T., Wiafe-Kwagyan, M., Buckman, S., & Mills, S. W. N. O. (2020). Sensory evaluation, descriptive textural analysis, and consumer acceptance profile of steamed gamma-irradiated Pleurotus ostreatus (Ex. Fr.) Kummer kept in two different storage packs. *Scientific African, 8*, e00328.

Koutchma, T., Keener, L., & Kotilainen, H. (2018). Discordant international regulations of food irradiation are a public health impediment and a barrier to global trade. *Global Harmonization Initiative (GHI) Consensus Document on Food Irradiation*, 1–18.

Kuntz, F., & Strasser, A. (2018). Dosimeters for gamma, E-beam, and X-ray irradiation. In I. C. F. R. Ferreira, A. L. Antonio, & S. Cabo Verde (Eds.), *Food irradiation technologies: Concepts, applications and outcomes* (pp. 66–75). UK: Royal Society of Chemistry.

Kwon, J.-H., Kwon, Y.-J., Byun, M.-W., & Kim, K.-S. (2004). Competitiveness of gamma irradiation with fumigation for chestnuts associated with quarantine and quality security. *Radiation Physics and Chemistry, 71*(1), 41–44.

Lambers, H., Yopp, J. H., Dickison, W. C., Schmid, R., Rothwell, G. W., & Woodwell, G. M. (2017). *Plant – Biology*. Encyclopedia Britannica.

Lima, R., Turrini, R., Silva, L. R., Melo, L., & Augusto, S. I. (2017). Popular healing practices and medical plants use for riparian mothers in early childhood care Práticas populares de cura e o uso de plantas medicinais por mães ribeirinhas no cuidado infantil. *Revista de Pesquisa: Cuidado é Fundamental Online, 9*, 1154.

Lin, X., & Sun, D. W. (2019). Research advances in browning of button mushroom (*Agaricus bisporus*): Affecting factors and controlling methods. *Trends in Food Science and Technology, 90*, 63–75.

Maherani, B., Hossain, F., Criado, P., Ben-Fadhel, Y., Salmieri, S., & Lacroix, M. (2016). World market development and consumer acceptance of irradiation technology. *Foods, 5*, 79.

Mangiacotti, M., Chiaravalle, A., Marchesani, G., De Sio, A., Boniglia, C., Bortolin, E., & Onori, S. (2009). Detection of irradiated chestnuts: Preliminary study using three analytical techniques. *Radiation Physics and Chemistry, 78*(7), 695–698.

Market analysis report. (2020). Business insights - Mushroom market size, share & industry analysis, by type, form and regional forecast 2019–2026. Retrieved August 17, 2020, from https://www.fortunebusinessinsights.com/industry-reports/mushroom-ma.%0A1.

Miller, R. B. (2018). Electron beam and X-ray equipment for food irradiation. In I. C. F. R. Ferreira, A. L. Antonio, & S. Cabo Verde (Eds.), *Food irradiation technologies: Concepts, applications and outcomes* (pp. 51–65). UK: Royal Society of Chemistry.

Mohácsi-Farkas, C. (2016). Food irradiation: Special solutions for the immuno-compromised. *Radiation Physics and Chemistry, 129*, 58–60.

Nair, P. M., & Sharma, A. (2016). *Food irradiation*. Elsevier Inc. Available from https://doi.org/10.1016/B978-0-08-100596-5.02950-4.

NASA. (2014). NASA HEASARC Archiver Research Center. Retrieved October 2020, from https://heasarc.gsfc.nasa.gov/docs/heasarc/headates/spectrum.html.

Naveed, S., Rizwan, E., & Sajid, A. (2018). Effect of gamma irradiation on phytochemical content and antimicrobial activities of selected herbs. *ChemXpress, 11*, 132.

Neri, L., Dimitri, G., & Sacchetti, G. (2010). Chemical composition and antioxidant activity of cured chestnuts from three sweet chestnut (*Castanea* sativa Mill.) ecotypes from Italy. *Journal of Food Composition and Analysis, 23*(1), 23–29.

Nishihira, J. (2020). Safety of irradiated food. *Elsevier Inc.* Available from https://doi.org/10.1016/B978-0-12-817240-7.00016-4.

Nketia, S., Buckman, E. S., Dzomeku, M., & Akonor, P. T. (2020). Effect of processing and storage on physical and texture qualities of oyster mushrooms canned in different media. *Scientific African, 9*, e00501.

Ortega-Rivas, E. (2014). Non-thermal processing | Steam Vacuuming. Encyclopedia of Food Microbiology (2nd ed., pp. 982–984).

Pereira, E., Antonio, A. L., Barreira, J. C. M., Barros, L., Bento, A., & Ferreira, I. C. F. R. (2015). Gamma irradiation as a practical alternative to preserve the chemical and bioactive wholesomeness of widely used aromatic plants. *Food Research International, 67*, 338–348.

Pereira, E., Antonio, A. L., Barreira, J. C. M., Cabo-Verde, S., Barros, L., Oliveira, M. B. P. P., & Ferreira, I. C. F. R. (2018). Evaluation of gamma-irradiated aromatic herbs: Hemometric study of samples submitted to extended storage periods. *Food Research International, 111*, 272–280.

Pereira, E., Antonio, A. L., Barreira, J. C. M., Santos-Buelga, C., Barros, L., & Ferreira, I. C. F. R. (2018). How gamma and electron-beam irradiations modulate phenolic profile expression in *Melissa officinalis* L. and *Melittis melissophyllum* L. *Food Chemistry, 240*, 253–258.

Pereira, E., Antonio, A. L., Rafalski, A., Barreira, J. C. M., Barros, L., & Ferreira, I. C. F. R. (2015). Extending the use of irradiation to preserve chemical and bioactiveproperties of medicinal and aromatic plants: A case study with four species submitted to electron beam. *Industrial Crops and Products, 77*, 972–982.

Pereira, E., Antonio, A. L., Rafalski, A., Barreira, J. C. M., Barros, L., Oliveira, M. B. P. P., & Ferreira, I. C. F. R. (2017). Electron beam irradiation as an alternative to preserve nutritional, chemical and antioxidant properties of dried plants during extended storage periods. *LWT - Food Science and Technology, 82*, 386–395.

Pereira, E., Barros, L., Antonio, A., Bento, A., & Ferreira, I. C. F. R. (2014). Analytical methods applied to assess the effects of gamma irradiation on color, chemical composition and antioxidant activity of *Ginkgo biloba* L. *Food Analytical Methods, 8*, 154–163.

Pereira, E., Barros, L., Antonio, A. L., Cabo-Verde, S., Santos-Buelga, C., & Ferreira, I. C. F. R. (2016). Infusions from *Thymus vulgaris* L. treated at different gamma radiation doses: Effects on antioxidant activity and phenolic composition. *LWT - Food Science and Technology, 74*, 34–39.

Pereira, E., Barros, L., Antonio, A. L., Cabo-Verde, S., Santos-Buelga, C., Ferreira, I. C. F. R., & Rodrigues, P. (2017). Is gamma radiation suitable to preserve phenolic compounds and to decontaminate mycotoxins in aromatic plants? A case-study with *Aloysia citrodora* Paláu. *Molecules (Basel, Switzerland)*, *22*, 347.

Pereira, E., Barros, L., Barreira, J. C. M., Carvalho, A. M., Antonio, A. L., & Ferreira, I. C. F. R. (2016). Electron beam and gamma irradiation as feasible conservation technologies for wild *Arenaria montana* L.: Effects on chemical and antioxidant parameters. *Innovative Food Science and Emerging Technologies*, *36*, 269–276.

Pereira, E., Barros, L., Dueñas, M., Antonio, A. L., Santos-Buelga, C., & Ferreira, I. C. F. R. (2015). Gamma irradiation improves the extractability of phenoliccompounds in *Ginkgo biloba* L. *Industrial Crops and Products*, *74*, 144–149.

Pereira, E., Pimenta, A. I., Barros, L., Calhelha, R., Antonio, A. L., Cabo-Verde, S., & Ferreira, I. C. F. R. (2018). Effects of gamma radiation on the bioactivity of medicinal and aromatic plants: *Mentha x piperita* L., *Thymus vulgaris* L. and *Aloysia citrodora* Paláu as case studies. *Food & Function*, *9*, 5150.

Pereira, E., Pimenta, A. I., Calhelha, R., Antonio, A. L., Cabo-Verde, S., Barros, L., ... Ferreira, I. C. F. R. (2016). Effects of gamma irradiation on cytotoxicity and phenolic compounds of *Thymus vulgaris* L. and *Mentha x piperita* L. *LWT - Food Science and Technology*, *71*, 370–377.

Pereira, E., Pimenta, A. I., Calhelha, R. C., Antonio, A. L., Barros, L., Santos-Buelga, C., ... Ferreira, I. C. F. R. (2017). Infusions of gamma irradiated *Aloysia citrodora* L. and *Mentha x piperita* L.: Effects on phenolic composition, cytotoxicity, antibacterialand virucidal activities. *Industrial Crops and Products*, *97*, 582–590.

Radomir, M., Mesud, A., & Žaklina, M. (2018). Conservation and trade of wild edible mushrooms of Serbia - History, state of the art and perspectives. *Nature Conservation*, *25*, 31–53.

Rahman, M. H., Islam, M. S., Begum, S., Ali, M. L., Sutradhar, B. C., Neil, V. O., & Nandwa, C. (2018). Scientific opinion on the standards and regulations of irradiated food. *Journal of Nutrition & Food Sciences*, *08*, 8–13.

Ravindran, R., & Jaiswal, A. K. (2019). Wholesomeness and safety aspects of irradiated foods. *Food Chemistry*, *285*, 363–368.

Roberts, P. B. (2014). Food irradiation is safe: Half a century of studies. *Radiation Physics and Chemistry*, *105*, 78–82.

Roberts, P. B. (2016). *Irradiation of food: Development, status, and prospects. Radiation advisory services*. Lower Hutt, New Zealand: Elsevier Inc, 2016.

Rodrigues, P., Venâncio, A., & Lima, N. (2012). Mycobiota and mycotoxins of almonds and chestnuts with special reference to aflatoxins. *Food Research International*, *48*, 76–90.

Shayganni, E., Bahmani, M., Asgary, S., & Rafieian-Kopaei, M. (2016). Inflammaging and cardiovascular disease: Management by medicinal plants. *Phytomedicine: International Journal of Phytotherapy and Phytopharmacology*, *23*, 1119–1126.

Shi, D., Zhou, R., Feng, X., Dong, X., Gao, H., Fan, X., & Huang, W. (2020). Effects of low-dose γ-irradiation on the water state of fresh *Lentinula edodes*. *Lwt - Food Science and Technology*, *118*, 108764.

Tejedor-Calvo, E., Morales, D., García-Barreda, S., Sánchez, S., Venturini, M. E., Blanco, D., & Marco, P. (2020). Effects of gamma irradiation on the shelf-life and bioactive compounds of *Tuber aestivum* truffles packaged in passive modified atmosphere. *International Journal of Food Microbiology*, *332*, 108774.

Tejedor-Calvo, E., Morales, D., Marco, P., Venturini, M. E., Blanco, D., & Soler-Rivas, C. (2019). Effects of combining electron-beam or gamma irradiation treatments with further storage under modified atmospheres on the bioactive compounds of Tuber melanosporum truffles. *Postharvest Biology and Technology*, *155*, 149–155.

Vinghes, C., & Ducom, P. (2000). Preliminary study on chestnut insect disinfestation with sulfuryl fluoride. In E. Donahaye, S. Navarro, & J. Leesch (Eds.), *Proc. Int. Conf. controlled atmosphere and fumigation in stored products* (pp. 149–153). Fresno, CA (U.S.A.): Executive Printing Services, Clovis, CA, U.S.A.

Zhang, K., Pu, Y. Y., & Sun, D. W. (2018). Recent advances in quality preservation of postharvest mushrooms (*Agaricus bisporus*): A review. *Trends in Food Science and Technology*, *78*, 72–82.

C H A P T E R

12

Cold plasma treatment in food packaging: effects on material properties, sterilization, and safety considerations

Prasad Chavan[1], *Jack Prendeville*[2], *Hamid*[1], *Swarna Jaiswal*[2,3] *and Amit K. Jaiswal*[2,3]

[1]Food Technology and Nutrition, Lovely Professional University, Phagwara, India [2]School of Food Science and Environmental Health, Faculty of Sciences and Health, Technological University Dublin-City Campus, Dublin, Ireland [3]Environmental Sustainability and Health Institute (ESHI), Technological University Dublin-City Campus, Dublin, Ireland

12.1 Introduction

Food packaging is crucial for food marketing and plays a significant role in the industrialization of food. It protects food against physical, chemical, and biological contamination from the outside world, ensuring food quality and extending product shelf life (Sharma, Barkauskaite, Jaiswal, & Jaiswal, 2021; Sharma, Jaiswal, Duffy, & Jaiswal, 2020). Packaging also increases the shelf life of food, prevents contamination, and guarantees food safety in the food supply chain (Carina, Sharma, Jaiswal, & Jaiswal, 2021; Perera, Sharma, Pradhan, Jaiswal, & Jaiswal, 2021). The barrier properties of packaging materials maintain a balance between the environment and the product while protecting food quality and sensory performance (Perera et al., 2021; Suppakul, Miltz, Sonneveld, & Bigger, 2003). Despite industry progress, foodborne infections continue to exist and attract public attention. Modified packaging and antibacterial packaging are crucial, achieved through physicochemical treatments or adding active chemicals to food packaging materials to improve functions or impart new features in food packaging.

In recent years, polymer materials have been increasingly used in food packaging due to their low cost, simple processing, and superior performance. However, the chemical inertness and low surface energy of polymers have limited their use in the food packaging industry (Ozdemir, Yurteri, & Sadikoglu, 1999). To meet market demand and replace expensive multilayer systems composed of different polymers, polymer materials need to continue improving their performance. This requires critical surface modification of polymer packaging materials using chemical and physical procedures (Lee, 2010).

Traditional chemical modification methods involve strong acid and alkali solutions, leading to environmental pollution. New physical alteration techniques include flame treatment, irradiation, laser, ultraviolet light, and cold plasma (CP) treatment, among others. CP treatment is considered the safest and most adaptable option for surface modification. Plasma generated by various devices leads to novel chemical functional groups and surfaces (Chu, Chen, Wang, & Huang, 2002; Karam, Jama, Dhulster, & Chihib, 2013).

Food Packaging and Preservation.
DOI: https://doi.org/10.1016/B978-0-323-90044-7.00012-4

12.2 Plasma

Plasma is commonly known as the fourth state of matter (Grzegorzewski, 2011; Jaeger, Knorr, Meneses, Reineke, & Schlueter, 2014) (Fig. 12.1). When energy is applied to a gas, it ionizes the gas and creates plasma, which is mainly composed of an ionized gaseous mixture (Jaeger et al., 2014; Niemira & Sites, 2008). In addition to electrons and ions, plasma contains various reactive components, such as free radicals, electrons, ions, photons, and gas molecules, that are not in constant contact with each other (Simmons, 2012; Warren, Mather, Neville, & Robson, 2010). Plasma has different temperatures for different components, with electrons having a higher temperature than other reactive components. Plasma can exist in two states: its excited state, where it has a positive charge, and its ground state, where it has a negative charge (Tabares & Junkar, 2021).

Plasma can be classified based on the way it is produced into two types: thermal plasma and low-temperature plasma. Thermal plasma is formed at a temperature of approximately 20,000°C and is in thermodynamic equilibrium with its surrounding environment. Low-temperature plasma, on the other hand, can be divided into two categories: quasi-equilibrium plasma (100°C–150°C) and non-equilibrium plasma (below 60°C). Quasi-equilibrium plasma exhibits local thermodynamic equilibrium between reactive substances like electrons and gas molecules. In non-equilibrium plasma, which is commonly used in the food industry, there is no thermodynamic equilibrium, and the electrons' temperature is higher than that of the other reactive chemicals. As a result, the entire system has a lower temperature, and this type of plasma is referred to as CP in the food sector due to its freezing properties.

12.3 Use of cold plasma in food packaging

CP technology has emerged as a promising method for food packaging decontamination. Packaging materials used in the food industry cannot withstand high temperatures, and the use of chemicals can leave residues and create environmental hazards. CP treatment is applied at low temperatures and does not damage the packaging material structure, making it an attractive option for food packaging decontamination. The principle of CP technology involves the formation of partially ionized positive and negative ions, charged particles of electrons and photons, free radicals, and gas-containing molecules. The CP interacts with bacteria and inactivates gram-positive and gram-negative bacteria, molds, yeasts, and endospores.

The reactive species produced by the breakdown of air, such as hydroxyl radicals, atomic oxygen, peroxides, and superoxide, play a significant role in the inactivation of microorganisms (Deng, Shi, & Kong, 2006). Nitrogen oxide (NO) and nitrogen dioxide (NO_2) inactivate microorganisms by causing damage to their chemical components, such as proteins, lipids, and nucleic acids (Afshari & Hosseini, 2014). CP treatment can be performed directly or indirectly. In direct treatment, the target material is positioned within the plasma discharge region and exposed to the plasma's active species and high-energy photons like UV light. In indirect treatment, the

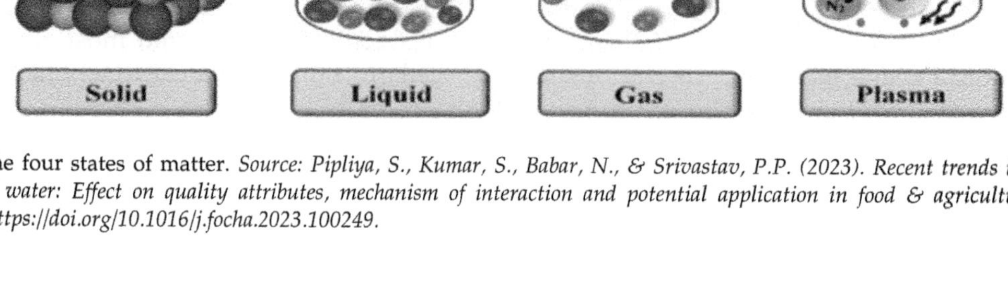

FIGURE 12.1 The four states of matter. *Source: Pipliya, S., Kumar, S., Babar, N., & Srivastav, P.P. (2023). Recent trends in non-thermal plasma and plasma activated water: Effect on quality attributes, mechanism of interaction and potential application in food & agriculture.* Food Chemistry Advances, *100249. https://doi.org/10.1016/j.focha.2023.100249.*

material is positioned outside the plasma discharge region, allowing the active species to leave the plasma region to coat the material's surface.

CP treatment has been effective in decontaminating fresh produce such as fruits and vegetables. The treatment times for in-package CP treatment of fresh produce are comparable to those for other direct plasma treatments, with a 2–7–log reduction achieved in 5 minutes to 15 minutes depending on the contaminant (Schnabel, Niquet, Schlüter, Gniffke, & Ehlbeck, 2015). The need for longer treatment is influenced by the surface texture of the fresh produce, with cherry tomatoes being more effective at being disinfected than strawberries due to their less intricate surface texture (Ziuzina, Patil, Cullen, Keener, & Bourke, 2014). In-package CP treatment has also been effective in reducing bacterial loads in food packages containing cabbage slices, Korean rice cakes, and fish cakes (Kang, Bai, & Min, 2021; Kim & Min, 2021; Lee, Kim, & Min, 2022). The decontamination effects of in-package atmospheric CP treatment depend on various product and process conditions, such as microbial counts, water activity (aw), bacterial growth conditions, and post-treatment storage (Yadav & Roopesh, 2020). CP treatment has also shown its ability to alter the structure of cellulose compounds, leading to better breakdown and potentially improving the development of edible films in food packaging (Zhu, Han, Cheng, & Sun, 2022). Additionally, atmospheric CP has been seen as a preventative measure in inhibiting decay in post-harvest blueberries (Hu et al., 2021).

12.3.1 Impact of cold plasma on food packaging surface sterilization

In recent years, CP technology has emerged as a promising method for the sterilization of food packaging surfaces. The nonthermal and chemical-free process of CP technology provides a safer alternative to traditional sterilization methods. It works by generating plasma discharge and reactive species, such as oxygen and nitrogen, which cause genetic destruction and volatilization of compounds, leading to the elimination of bacteria (Pahwa & Kumar, 2018). This technology has been shown to be effective against several pathogenic bacteria, including *Escherichia coli, Salmonella* spp, *Shigella*, and *Listeria monocytogenes* (Niveditha et al., 2021). The in-package treatment of CP has been found to be the most effective method, where the package is fully sealed with a combination of gases, and a strong electric field is applied, resulting in a breakdown of the gases. In addition, the ionization of these gases act as a sterilizing agent within the packaging, reducing bacteria while sustaining the packaging material structure (Misra, Yepez, Xu, & Keener, 2019).

While CP technology has been found to be effective in reducing microbial loads, there are some concerns about its potential impact on the quality of packaged foods. Laroque, Seó, Valencia, Laurindo, and Carciofi (2022) reported an increase in lipid oxidation after CP treatment, and there is limited research on possible migratory issues into foods. Additionally, the process of surface modification by CP may not be retained over an extended period, and the risk of gas permeation is higher under plasma treatments, which could compromise the shelf-life of packaged foods (Pankaj, Bueno-Ferrer, Misra, Milosavljević, O'donnell, et al., 2014). To mitigate these concerns, incorporating bioactive compounds such as lysozyme, nisin, vanillin, sodium benzoate, and glucose oxidase through plasma treatments has been explored, as it retains a sterile environment for food packaging throughout storage (Pahwa & Kumar, 2018).

It is worth noting that the efficiency of bacterial reduction varies between food products, treatment times, and the type of microorganism. Table 12.1 presents a summary of the inactivation of microorganisms on food packaging, highlighting that this technology may be more applicable to certain foods or reductions in certain microorganisms (Ucar, Ceylan, Durmus, Tomar, & Cetinkaya, 2021). Therefore, more research is needed to determine the effectiveness of CP technology in various food products and the possible migratory issues that could arise.

CP technology has shown promise in the sterilization of food packaging surfaces. Its nonthermal and chemical-free process provides a safer alternative to traditional sterilization methods. In-package treatment using CP has been found to be the most effective method in reducing microbial loads while sustaining the packaging material structure. However, concerns about the impact on the quality of packaged foods and possible migratory issues still need to be addressed. Further research is needed to determine the efficacy of this technology in various food products and its impact on food quality and safety.

12.3.2 Impact of cold plasma on biopolymer-based packaging materials

Biopolymers are known to have inferior mechanical properties, low thermal stability, poor surface functionality, poor printability, and poor adhesiveness compared to plastics (Hoque, McDonagh, Tiwari, Kerry, & Pathania,

TABLE 12.1 Cold plasma inactivation of microorganisms on food packaging surfaces.

Food type	Microorganisms	Plasma source	Treatment times	Log reductions	References
Sushi stored in PET containers.	Aerobic bacteria	DBD with 70 and 80 kV	5 min	1–1.5 log CFU/g	Kulawik et al. (2018)
MAP packed RTE Ham	*Listeria monocytogenes*	DBD with a voltage between 0–3 kV at 3.5 kHz	5 min	2 log CFU/g	Yadav and Roopesh (2020)
Fresh cut apples in polypropylene and PET	COVID-19	Surface DBD	10 min	After 10 min RNA was degraded	Capelli et al. (2021)
Processed chicken breast in a PET container	*E. coli* O157:H7 Salmonella *L. Monocytogenes*	DBD	2 min–5 min	3.9 log CFU/ Cube 3.7 log CFU/ Cube 3.5 log CFU/ Cube	Romani et al. (2019)
Wheat and barley grains in a polypropylene container.	Bacteria and fungi	DBD	20 min	Barley—2.4 log_{10} CFU/g—Bacteria 2.1 log_{10} CFU/g—Fungi Wheat- 1.5 log_{10} CFU/g—Bacteria 2.5 log_{10} CFU/g—Fungi	Los et al. (2018)
Polypropylene PP package containing the strawberry	Aerobic mesophillic bacteria, yeast and mold	Dielectric barrier discharge 60 kV pulsed at 50 Hz	5 min	2.56 log_{10} CFU/g (mesophillic bacteria) 1.56 log_{10} CFU/g (yeast and mold) within 24 h of post-ACP treatment	Misra, Patil et al. (2014)
Orange juice with MA65 (65%O_2 + 30%N_2 + 5%CO2, <5% RH) using a cryovac	*Salmonella enterica* serovar Typhimurium (*S. enteric*)	Dielectric barrier discharge, 90 kV	120 s	3.8- log_{10} CFU/g	Xu, Garner, Tao, and Keener (2017)
Cherry tomatoes in high barrier Cryovac B2630 film	Fungus	Dielectric barrier discharge with 60 kV source, pulsed at 50 Hz	300 s	No visual fungal was detected in any of the treated tomatoes	Misra, Keener, Bourke, Mosnier, and Cullen (2014)
Cherry tomatoes and strawberries in rigid polypropylene container	*Escherichia coli*, *Salmonella enterica Typhimurium* and *L. monocytogenes*	Dielectric barrier discharge system with 70 kV, at 50 Hz frequency	60 s to 300 s	For tomatoes: 3.1- log_{10} CFU/g (Salmonella enteric), 6.3- log_{10} CFU/g (*E. coli*), 6.7- log_{10} CFU/g (*L. monocytogenes*) For strawberries: 3.8- log_{10} CFU/g (S. Enteric), 3.5- log_{10} CFU/g (E. Coli), 4.2- log_{10} CFU/g (L. Monocytogenes)	Ziuzina et al. (2014)
Mandarin in polyethylene terephthalate (PET) containers inserted in composite pouch made of 3% nylon/low-density polyethylene (N-LDPE)	Penicillium digitatum	Dielectric barrier discharge system 27 kV	2 min	Inhibition effect against *P. digitatum* growth in mandarins at 25°C	Bang, Kim, and Min (2021)
Cabbage slices in plastic containers	Salmonella, *E. coli* O157:H7, *L. monocytogenes*, and *Bacillus cereus* spores	Dielectric barrier discharge system, voltage 24.5 kV with Inense Pulse Light treatment 1.5 kV,	2 min	3.1 log CFU/g Salmonella, 2.9 log CFU/g (*E. coli* O157:H7), 3.2 log CFU/g (*L. monocytogenes*), 2.1 log spores/g (*B. cereus*)	Kim and Min (2022)

Mixed vegetables (grape tomato, slices of romaine lettuce, red cabbage, carrot,) in plastic packages (Low density polyethylene polypropylene polyethylene terephthalate)	Salmonella Salmonella enterica subspecies enterica serovar Montevideo (CCARM 8052), S. Enteritidis (CCARM 8040), and S. Typhimurium (DT104)	Dielectric barrier discharge system, 50 kV at 60 Hz	3 min	Grape tomato: 1.5 log CFU/8 g PET and LDPE, 1.2 log CFU/8 g in PP (Salmonella); 0.8 and 0.9 log CFU/8 g PET and LDPE, 1.2 log CFU/8 g in PP (Indigenous mesophilic aerobic bacteria); 1.1 and 1.3 log CFU/8 g PET and LDPE, 1.4 log CFU/8 g in PP (Indigenous yeast and molds); Slices of romaine lettuce, red cabbage, carrot: 1.5 log CFU/8 g PET and LDPE, 1.2 log CFU/8 g in PP (Salmonella); 0.5 log CFU/8 g PET and LDPE, 0.6 log CFU/8 g in PP (Indigenous mesophilic aerobic bacteria); 1.0 log CFU/8 g PET and LDPE, 1.1 log CFU/8 g in PP (Indigenous yeast and molds);	Kim, Bang, and Min (2019)
Chicken breast salad packed in polyethylene terephthalate container	Mesophilic bacteria, Salmonella and Tulane virus	Dielectric barrier discharge system 24 kV	2 min	1.2 log CFU/g (mesophilic bacteria), 1.0−1.5 log CFU/g (Salmonella), 1.0 log PFU/g (Tulane virus)	Lee, Jeon, and Min (2021)
Lettuce packed in clamshell container composed of polyethylene terephthalate and post- consumer recycled plastics	*E. coli* O157:H7	Dielectric barrier discharge system 34.8 kV at 1.1 kHz	5 min	1 log CFU/g	Min et al. (2017)
Grape tomato packed in polyethylene terephthalate clamshell container	Aerobic mesophilic bacteria	Dielectric barrier discharge system 35 kV at 1.1 A	3 min	0.9 ± 0.2 log CFU/tomato to 3.3 ± 0.5 log CFU/tomato	Min et al. (2018)
Korean rice cakes in PET container	Indigenous mesophilic aerobic bacteria, indigenous yeast and molds and Salmonella	Dielectric barrier discharge system 25 kV	3 min	0.8−1.0 log CFU/g 1.7 ± 0.1 log CFU/g 3.9 ± 0.3 log CFU/g	Kang et al. (2021)
Fresh chicken fillets in Cryovac CS977 polymeric tray	Mesophiles, psychrophiles, and pseudomonas spp	Dielectric barrier discharge system 80 kV	3 min	4 logs cfu/g	Wang, Zhuang, Hinton, and Zhang (2016)
Chicken breast meat packed in n polymeric trays sealed with PP films	Campylobacter jejuni and Salmonella Typhimurium	Dielectric barrier discharge system 55 kV	3 min	4.01 log CFU/mL 4.29 log CFU/mL	Zhuang et al. (2019)
Strawberries and spinach in PET bags	*E. coli* *L. innocua*	Dielectric barrier discharge system 100 kV	2.5 min	2.0 and 2.2 log10 CFU/mL reduction s for E. coli and 1.3 and 1.7 log10 CFU/mL reduction s for L. innocua	Ziuzina et al. (2020)
Ready-to-eat chicken breast cubes in plastic containers	Mesophilic bacteria, Salmonella and Tulane virus	Atmospheric dielectric barrier discharge cold plasma 24 kV	3 min	0.7 log CFU/g (mesophilic bacteria), 1.4 log CFU/g (Salmonella), 1.1 log PFU/g (Tulane virus)	Lee et al. (2020)

2022). The application of CP can modify the molecular structure of biopolymer films in various ways, including creating an active site, crosslinking, reordering, and enhancing the structural arrangement of the polymeric network (Hoque et al., 2022). These modifications can have a positive impact on the mechanical, thermal, and barrier properties of biopolymer films.

When plasma is applied to a film, it induces physicochemical phenomena such as etching, degradation, and cross-linking, which can change the mechanical properties of the film (Bahrami et al., 2020). Plasma bombardment of film surfaces cleaves C-C and C-H bonds, generating free radicals that link with surface radicals or participate in chain reactions, thus improving packaging films' mechanical properties (Moosavi et al., 2020).

Dong, Guo, Chen, Jin, and Chen (2020) studied the effect of atmospheric CP on the surface properties of the zein film and reported that a 45-second atmospheric CP treatment significantly increased tensile strength (TS) and elongation at break. This increase may have been caused by strong intermolecular and intramolecular S–S cross-linking in the zein matrix, as the plasma discharge may have caused oxygen to be oxidized into ozone, which changed the free S–H group into S-S. Moosavi et al. (2020) tried to modify protein-based films using CP treatment. They observed that the TS of the control whey protein film was 6.902 MPa, which reached 10.772 and 13.802 MPa after 10 minutes and 15 minutes of plasma treatment, respectively. In contrast, the TS of the control gluten film was 1.854 MPa, which significantly increased to 2.571 MPa after 10 minutes of plasma processing. The differences in mechanical properties were attributed to the different amino acids present in the primary structure of the edible film affected by the plasma treatment.

Dielectric barrier discharge (DBD) plasma can modify the surface properties of biopolymers, such as whey protein concentrate (WPC)/wheat cross-linked starch (WCS) composite films. The surface roughness, TS, barrier characteristics, and thermal stability of edible films can be improved with the right plasma treatment parameters (400 W to 60 seconds), while DBD plasma application reduces the elongation at break (EAB) and the water contact angle (Song et al., 2019). Plasma treatment can cause oxygen-containing groups like –OH, C–O, and –O–C=O to develop on the surface of WPC/WCS films. However, there was no discernible change in the edible film's crystal structure, indicating that DBD plasma treatment is solely a surface modification approach and does not affect the film's overall structure.

Furthermore, the application of CP utilizing ambient air as the gas considerably changes the surface of polycaprolactone (PCL) and polylactic acid (PLA) films, which helps in increasing the films' wettability and hydrophilic nature. The roughness characteristics of the CP-treated films increased over the course of the plasma treatment, with a decrease in water contact angle. Further plasma treatment can help in the improvement of adhesion of starch films onto both PCL and PLA surfaces, with increased delamination resistance over time. The multilayer films with PCL/PLA-starch as an effective packaging material, with higher water vapor barrier than starch films alone and comparative mechanical properties to PCL/PLA films. This presents a promising alternative for reducing synthetic polymers in packaging applications (Heidemann, Dotto, Laurindo, Carciofi, & Costa, 2019).

Atmospheric CP has also been utilized in enhancing the surface hydrophilicity and cytocompatibility of zein films, with a decrease in water contact angles and an improvement in surface free energy. As highlighted in Table 12.2, long-chain aliphatic hydrocarbons (C–H and/or C–C) that were partially transformed into oxygen- and nitrogen-containing bonds (C=O, C–O, and C–N) were the main cause of the change in surface physicochemical features. Additionally, C2C12 cells on zein films could adhere and proliferate more readily after receiving plasma therapy, demonstrating the better cytocompatibility of zein as a biopolymer for packaging applications (Dong et al., 2020).

DBD plasma has shown potential in modifying the surface properties of biopolymers such as WPC/ WCS composite films. The edible film's surface roughness, TS, barrier characteristics, and thermal stability can all be improved with the right plasma treatment parameters (400 W to 60 seconds), while DBD plasma application reduced the EAB and the water contact angle (Song et al., 2019). The plasma treatment can cause oxygen-containing groups like –OH, C–O, and –O–C=O to develop on the surface of WPC/WCS films. However, there was no discernible change in the edible film's crystal structure, demonstrating that DBD plasma treatment is solely a surface modification approach and does not affect the film's overall structure. Thus, plasma technology can provide an eco-friendly and effective technique to improve the surface properties of edible films (Song et al., 2019).

Additionally, glow discharge plasma (GDP) has the potential to be used to modify chitosan films in order to achieve desired qualities in a dose-dependent way. With increased GDP power and improved thermal stability, the GDP treatments on chitosan films have the capacity to change the molecular structure of the chitosan films, as evidenced by the exothermic peaks shifting from 291.5°C (0 W) to 294.1°C (800 W). Furthermore, after GDP treatment, the surface of chitosan films can be altered to become rougher (0–600 W), along with an increase in TS

TABLE 12.2 Application of cold plasma on biopolymer-based packaging materials.

Packaging material	Cold plasma source	Treatment times	Properties of packaging material	References
Pectin—clove essential oil	Dielectric barrier discharge (DBD) system, at a frequency of 10 kHz and a voltage level of 7 kV for	5 min	• High mechanical strength and stretchability • High surface hydrophobicity, mechanical strength, and flexibility • Higher retention of clove essential oil during storage	Jahromi, Niakousari, and Golmakani (2022)
Starch- chitosan	Radio frequency (20 kHz) plasma cleaner system	4, 8 and 12 min	• Tensile strength (TS) increased from 10.59 to 22.09 Mpa after 12 min of treatment • Solubility and hydrophilicity of the films increased • Oxygen permeability (OP) of films improved • Cold plasma treatment did not prolong the shelf life of the fillets	Sheikhi et al. (2021)
Corn starch-based films	Dielectric barrier discharge (DBD) with voltage of 20 kV with frequency of 200 Hz	10, 15, and 20 min	• Higher hydrophobicity, enhanced tensile strength and stiffness after 20 min of treatment • Lowers hydrophilicity and water solubility	Goiana et al. (2021)
Whey protein concentrate/wheat cross-linked starch composite films	Dielectric barrier discharge plasma, 220 V at a frequency of 50 Hz, 400 W	60 s	• Increase in tensile strength, surface roughness (SR), thermal stability, barrier properties of the edible film • Decreases elongation at break (EAB) and the water contact angle • Positive effect on the mechanical and barrier properties and thermal stability	Song et al. (2019)
Casein based films-casein solution treatment	Dielectric barrier discharge cold plasma, 30–70 V for 60 s and 50 V at 15–120 s	15 s–120 s	• Changes in structure of casein solution • Improvement in heat/thermal stability of films • Reported decrease in tensile strength, EAB and barrier to water vapor	Hou et al. (2022)
Banana starch-based films	Dielectric barrier discharge excitation frequency of 400 Hz, voltage of 20 kV	15 min	• Reported better crystalline/amorphous ratio (short-range order structure degree) than untreated films • Reported highest water contact angle, tensile strength and good puncture resistance, resulting highest mechanical strength and hydrophobicity among tested films	Marenco-Orozco, Rosa, and Fernandes (2022)
Zein film	Dielectric barrier discharge plasma, 100 V	30 s	• Enhanced the surface hydrophilic and cytocompatibility of films • Decreased water contact angle (CA) • Change of surface physicochemical properties • Improvement in adhesion rate and proliferation activity of C2C12 cells on zein films	Dong et al. (2020)
Cassava starch multilayer films with PCL and PLA surface	Dielectric barrier discharge plasma50 Hz and 31 kV	10 min	• Film roughness increased throughout plasma treatment time, • Decrease in water contact angle decreased, • Increase in wettability and hydrophilic character of films • Excellent adhesion after 10 min between starch and PCL or PLA layers	Heidemann et al. (2019)
Chitosan film	Glow discharge plasma, 0–800 W	1 min	• Improvement of crystallinity with the increasing power • Increases thermal stability and Tensile strength • TS did not increased at 800 W • Increase in roughness (0–600 W)	Moharram (2010)

(TS) and water vapor permeability (WVP), which also had an impact on the chromatic aberration of chitosan films. As GDP's treatment power increased, the amide II band decreased but crystallinity improved (Li et al., 2019). This reveals that GDP has a broad range of applications for producing contamination-free and desirable chitosan films with convenient methods.

12.3.3 *Impact of cold plasma on modification of food packaging polymers*

CP treatment has been shown to modify the surface properties of various food packaging polymers, such as polyethylene, polystyrene, and polypropylene. This modification can enhance the adhesion between layers, increase surface wettability, and improve barrier properties. The modification can also lead to the introduction of functional groups that can allow for the attachment of active compounds, such as antimicrobial agents, and thus enhance the effectiveness of the packaging material. Additionally, plasma treatment can be used to introduce roughness to the surface of polymers, which can be useful for improving the mechanical properties of packaging materials. Table 12.3 summarizes the impact of CP treatment on the modification of various food packaging polymers, including LDPE, PET, and LDPE.

Polyethylene terephthalate (PET) is a highly versatile plastic used in various industries, including food packaging, due to its strength, flexibility, transparency, and insulation. PET is also easily recyclable, but it has limitations in its ability to withstand high temperatures and certain chemical agents. DBD treatments using CP technology can modify and enhance PET film properties and provide fast and consistent surface treatment. Recent research by Cretu et al. (2021) demonstrated that DBD plasma treatment with atmospheric pressure air can improve adhesion and water repellence of PET films, while DBD treatments on vacuum-packaged food products can reduce bacterial cell counts and oxidation for extended periods (Ucar et al., 2021). However, surface modification is necessary to achieve appropriate adhesive, printing, and dyeing characteristics, as PET has lower surface energy than natural rubber. Crystallinity influences the changes in surface energy induced by CP treatment, which has been investigated for surface characterization using various plasma gases (Almazan-Almazan et al., 2006; Jacobs et al., 2011; Pankaj, Misra, & Cullen, 2013).

Polypropylene (PP) is a widely used flexible polymer in food packaging due to its low density, cost-effectiveness, high melting point, strong heat sealability, and chemical inertness (Pankaj et al., 2013). However, its low surface tension makes it challenging to print, coat, and laminate, necessitating additional surface treatments to increase its surface energy. PP is a saturated hydrocarbon polymer with a carbon backbone containing alternating hydrogen and methyl (CH_3) groups. The reactivity of hydrogen groups for surface reactions in PP depends on the type of C atom they are connected to, with H tert > H sec > H pri having the most reactivity (Dorai & Kushner, 2003).

TABLE 12.3 Impact of cold plasma on modification of food packaging polymers.

Packaging material	Cold plasma source	Treatment	Properties of packaging material	References
LDPE- gallic acid	Cold radiofrequency plasma (13.56 MHz)	Different power (0–90 W) and treatment time (15 s–150 s)	• Reduce growth of *E. coli* and *S. aureus* by about 0.5–0.6 log CFU/mL • Highest tensile strength for the treatment 30 W for 60 s	Wong, Hou, Hsieh, Chang, Wu et al. (2020)
LDPE- Chitosan—savory essential oil	Atmospheric cold plasma (power: 3 kW).	10 s	• Decreased water vapor permeability and oxygen transmission rate • Improved tensile strength and elongation at break of film • The shelf life of fillets extended up to 13 days at refrigerator	Moradi et al. (2020)
LDPE-PET	Atmospheric dielectric barrier discharge cold plasma treatment (ADCPT)	3 min	• Increased water vapor permeability • Tensile properties, transparency, glass transition temperature, and surface morphology are unchanged	Kim et al. (2019)
PET- zinc oxide	High-voltage transformer (220 V, 60 Hz),	3 min	• Higher inactivation of *Salmonella* 2.4 log CFU/g on ZnO-CP treatment than the ZnO-only (0.4 log CFU/g) or CP-only (1.7 logs CFU/g) treatment	Lee et al. (2022)
PET	ADCPT	26 or 27 kV for 1, 2, 3, or 4 min	• Enhanced shelf-life of mandarins in plastic packages by inhibiting the growth of *P. digitatum* on fruit while retaining fruit quality	Bang et al. (2021)

Research has shown that PP can affect the hormones called androgen, which can lead to a stress response in the cells (Zimmermann, Dombrowski, Völker, & Wagner, 2020). CP treatment, such as the Argon PJ process, can be used to improve the properties of PP (Baniya, Guragain, Baniya, Qin, & Subedi, 2020). This process can improve the hydrophilicity of the plastic under atmospheric pressure (Baniya, Guragain, Baniya, Qin, & Subedi, 2020). Additionally, plasma treatment using PJ has been found to inactivate harmful bacteria such as *E. coli K12*, *Staphylococcus aureus*, and *Pseudomonas aeruginosa* (Mandal, Singh, & Pratap Singh, 2018). PP film modification has been extensively researched using high-pressure plasma discharges at high altitudes (Dorai & Kushner, 2003), with the reaction process for PP treatment with air plasma outlined in detail by Akishev et al. (2008). The degradation of PP during plasma treatment occurs mainly through branch-scissions and the creation of low molecular weight organic molecules (LMWOM), with the degradation rate following the order of $N_2 < He < Air << < O_2$ (Poncin-Epaillard, Brosse, & Falher, 1999).

12.4 Effect of cold plasma on packaging properties

12.4.1 *Surface roughness*

Changes to surface roughness are an important consideration in packaging materials as they can impact the adhesion of coatings, printing inks, and other functional layers that are applied to the surface of the packaging. Changes to surface roughness due to CP can be attributed to the "etching effect," which results from the sputtering of high-energy plasma species onto the film surface. These species include electrons, photons, positively or negatively charged ions, free radicals, and atoms in their excited or fundamental states. During the etching process, various changes may occur, such as the cleavage of chemical bonds, scission of polymer chains, rearrangement of low molecular components, and free radical-induced degradation. The extent of roughness obtained after CP treatment depends on the voltage power supplied, exposure time, plasma gas type, discharge gap, and the structural composition of the film.

Honarvar et al. (2017) used atomic force microscopy (AFM) to demonstrate the difference between CP-treated and untreated films. The treated films were composed of polypropylene coated with carboxymethyl cellulose (CMC) infused with *Zataria multiflora* essential oil. The untreated films had a smooth surface layer, while the CP-treated films were significantly rougher, with Ra (average surface roughness) and Rq (root mean square) values increasing from 25.96 and 33.30 nm to 61.54 and 83.17 nm, respectively. The films were exposed to plasma (14 kV at 20 kHz) continuously at a speed of 3.6 m per minute, conveyor belt-style.

Dong et al. (2018) reported that CP treatment of zein films for 30 seconds significantly increased surface roughness. Rq and Ra values increased from 9.4 and 4.2 nm to 82.7 and 67.6 nm, respectively (Fig. 12.2). The increase in surface roughness can impact the adhesion of coatings or printing inks on the film surface. A rougher surface can increase surface energy and wettability, thus improving adhesion. However, excessively rough surfaces can lead to reduced optical clarity and potential defects in the film's barrier properties. Therefore, the appropriate selection of CP treatment parameters is crucial in achieving the desired surface roughness for specific applications.

Pankaj, Bueno-Ferrer, Misra, Bourke, and Cullen (2014) reported that the surface roughness of zein films increased from 16.41 nm (untreated) to 52.04, 126.16, and 132.87 nm when three different voltages of 60, 70, and 80 kV were employed, respectively. Cools et al. (2018) reported that the plasma gas type can influence the extent of roughness during the treatment of "polyethylene-oxide terephthalate (PEOT)/polybutylene terephthalate (PBT) film with DBD plasma involving air, helium (He), argon (Ar), and nitrogen (N_2)." AFM showed that N_2

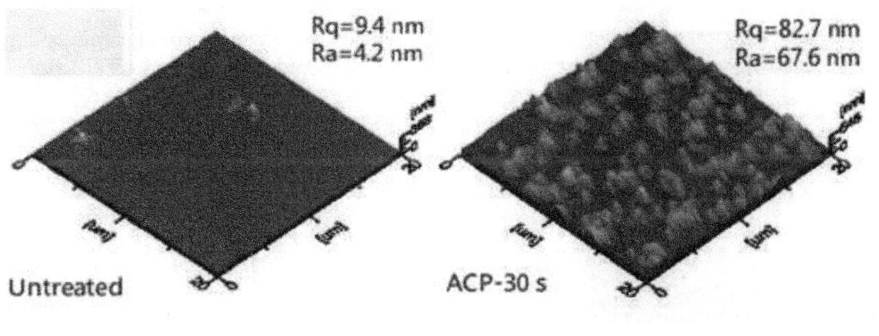

FIGURE 12.2 AFM images of surface roughness of the untreated and ACP treated zein films. Source: *From Dong, S., Guo, P., Chen, Y., Chen, G. y., Ji, H., Ran, Y., Li, S. h., & Chen, Y. (2018). Surface modification via atmospheric cold plasma (ACP): Improved functional properties and characterization of zein film. Industrial Crops and Products, 115, 124–133. https://doi.org/10.1016/j.indcrop.2018.01.080.*

and air led to an increase in surface roughness, whereas He and Ar plasma led to a decrease in roughness. "He and Ar etched away the sharp peaks on the film surface, creating a more uniform surface layer" (Cools et al., 2018). N_2 plasma slightly increased surface roughness, whereas air plasma led to significant changes due to the formation of "hemispherical protrusions on the film surface." Hoque et al. (2022) highlighted that the variation in etching behavior can be due to the unique active species generated by the carrier gas employed to create the plasma discharge.

Increased surface roughness is beneficial for the diffusion of antimicrobial compounds throughout the biopolymer matrix. Karam et al. (2016) reported increased absorption of nisin due to surface roughness modification after exposing low-density polyethylene films to N_2-generated plasma. Significantly lower levels of nisin were absorbed by the untreated films. Pankaj, Bueno-Ferrer, Misra, O'Neill, and Bourke, et al. (2017) prepared antimicrobial films by combining chitosan and thymol using a DBD system. A significant increase in roughness was reported to be responsible for the enhanced diffusion of thymol throughout the film matrix.

Surface roughness can promote adhesion between layers of various polymer materials to create unique multilayered films that exhibit desirable packaging characteristics. Honarvar et al. (2017) found that atmospheric CP treatment enhanced adhesion between CMC and polypropylene films (Fig. 12.2). Likewise, Fazeli, Florez, and Simão (2019) reported air plasma treatment to significantly improve attachment between cellulose fiber and thermoplastic starch.

The increase in surface roughness due to CP treatment enhances the ink printability of films (Hoque et al., 2022). For instance, PLA films are non-absorbent and therefore have poor ink printability properties. Izdebska-Podsiadły (2021) reported that O_2/Ar-generated plasma significantly increased the ink printability of PLA films due to changes in surface roughness. Arolkar, Salgo, Kelkar-Mane, and Deshmukh (2015) and Song, Oh, Roh, Kim, and Min (2015) also reported comparable improvements in ink printability after CP treatment of PLA and starch-poly(ε-caprolactone) films.

12.4.2 Contact angle

Contact angle (CA) is the angle formed between a water droplet and the film surface, and it is a measure of the film's ability to wet or repel water. CP exposure can generally decrease CA, depending on the material. For example, Honarvar et al. (2017) found that the CA of polypropylene films decreased significantly after plasma exposure, resulting in an increase in wettability (Fig. 12.3). This increase was attributed to the deposition of oxygen-containing polar groups on the film surface and the formation of polar functional groups. Etching, which promotes the spreading of water on the film surface, has also been attributed to the resultant increase in wettability by some researchers (Borin, Sbaizero, & Scuor, 2019; Chen, Cheng, & Sun, 2019; Dong et al., 2018).

The CA of a film is influenced by plasma exposure time and voltage applied. In general, higher voltages and extended plasma exposure times result in increased wettability and hydrophilicity due to the formation of polar groups such as –COOH, –OH, and CO on the film surface. The hydrophilicity of food packaging films is crucial as it affects printing, coating, and adhesion properties. However, in some cases, CP treatment can increase CA, resulting in increased hydrophobicity. For example, Honarvar et al. (2017) reported that CP treatment of polypropylene films coated with CMC containing essential oils increased CA, resulting in increased hydrophobicity. Similar findings were reported by Ulbin-Figlewicz, Zimoch-Korzycka, and Jarmoluk (2014) when treating chitosan films with helium gas plasma. Gholamazad, Hosseini, Hosseini, Ramezan, and Rahmanabadi (2022) found that argon-generated plasma increased CA more than oxygen-generated plasma. According to Gholamazad et al. (2022), the wetting properties of films largely depend on the films' wetting nature and the type of gas used for plasma treatment.

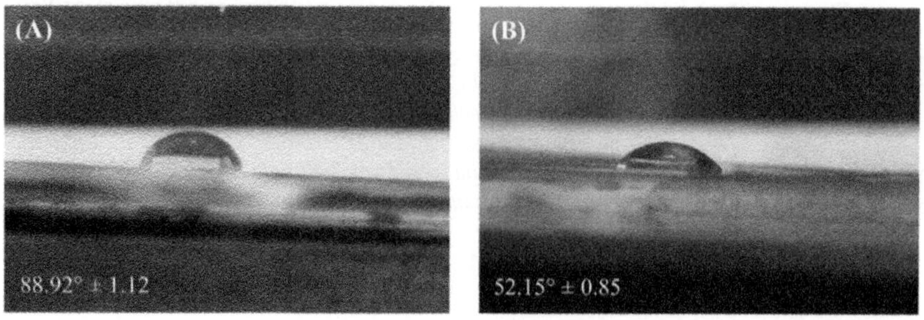

FIGURE 12.3 The water contact angle of untreated (A) and cold plasma treated (B) polypropylene films. Source: *From Honarvar, Z., Farhoodi, M., Khani, M. R., Mohammadi, A., Shokri, B., Ferdowsi, R., & Shojaee-Aliabadi, S. (2017). Application of cold plasma to develop carboxymethyl cellulose-coated polypropylene films containing essential oil. Carbohydrate Polymers, 176, 1–10. https://doi.org/10.1016/j.carbpol.2017.08.054.*

12.4.3 *Mechanical properties*

Mechanical properties such as TS and EAB are crucial factors that determine the structural integrity of packaging materials. TS indicates the maximum tensile stress that the film can withstand, while EAB is the maximum change in the length of the film before rupturing, indicating its stretching capacity (Gholamazad et al., 2022).

Physicochemical changes such as cross-linking, etching, degradation, and molecular rearrangements can impact mechanical properties, and the effect of plasma on TS and EAB depends on the nature of the biopolymer matrix and the type of plasma used (Gholamazad et al., 2022). The bombardment of free radicals on the film surface can lead to stronger intermolecular bonds and increase TS. Free radicals cleave the existing C–C and C–H bonds on the film surface, resulting in cross-linking, which significantly improves the mechanical properties (Bahrami et al., 2020; Sheikhi et al., 2020). The increase in hydrogen and dipole-dipole bonds is another reason for the increase in TS (Gholamazad et al., 2022).

Gholamazad et al. (2022) reported that argon-generated plasma was superior to oxygen gas plasma, as it was more effective at producing the etching effect due to the production of heavier active agents. They also mentioned that argon plasma activates the existing functional groups on the film surface and does not create new reactive functional groups like oxygen-generated plasma. In addition, the presence of glycerol in a sodium alginate film matrix resulted in more free movements of the chains and free interchain space, thereby increasing the film's flexibility.

Studies on different plasma treatments have reported varying effects on TS and EAB values. Dong et al. (2020) reported that TS and EAB of a pristine zein film increased after atmospheric plasma exposure for 45 seconds. High voltages of 100 V and longer exposure times of 60 seconds were found to negatively impact the mechanical properties. Optimal treatment time was reported to be 65 V for 45 seconds. Similar findings have also been reported by Chen et al. (2020) and Dong et al. (2018) with zein-chitosan films and zein-PLA films.

Goiana et al. (2021) used DBD plasma to treat corn starch films and found that as the exposure time increased from 0–20 minutes, TS significantly increased. Sheikhi et al. (2020) reported similar results when using argon gas plasma, as TS increased as exposure time increased from 0–12 minutes. Both studies attributed the TS increase to the etching effect and the introduction of crosslinks such as carbonyl groups, which promote hydrogen bonding and increase structural integrity.

Moosavi et al. (2020) used air and argon plasma to treat whey protein films and found that using air plasma, 15 minutes exposure resulted in a significant increase in TS, whereas argon plasma decreased TS. Hoque et al. (2022) reviewed this and mentioned that this may have been due to the higher ionization yield of argon in comparison to air plasma. Argon ions are heavier and can have a more destructive etching effect on the film surface than other species (Gholamazad et al., 2022). Therefore, overexposure to high-energy plasma can negatively impact mechanical properties. Chen, Cheng, and Sun (2019) reported similar results in which extended treatment times reduced the TS and EAB values of zein-chitosan films.

Ledari, Milani, and Lanbar (2020) used O_2, N_2, air, Ar, and ethanol-argon gas plasma to treat gelatin films. In contrast to Moosavi et al. (2020), no significant changes in TS or EAB were reported compared to the untreated films, regardless of the gas type employed. Various factors such as voltage applied, exposure time, types of reactive species, gas type, and the polymer composition can impact mechanical properties (Hoque et al., 2022). In addition to these factors, the distance of the film from the plasma nozzle can also play a role in the observed mechanical properties (Sheikhi et al., 2020). Dong et al. (2018) reported that TS and EAB of zein/PLA films decreased at nozzle-to-sample distances greater than 10 mm, but increased as the distance decreased. Similar observations have been reported by Sheikhi et al. (2020) and Gholamazad et al. (2022). Therefore, the nozzle-to-sample distance should be optimized for specific plasma treatment applications.

12.4.4 *Thermal properties*

Thermal properties are important as resilience to processing and heat-sealing conditions determines the effectiveness of the packaging. Pankaj, Wan, De León, Mosher, and Colonna et al. (2017) used high voltage (80 kV) atmospheric CP to treat various starch films (corn, potato, tapioca, and rice). They reported CP treatment created a highly ordered structure within the film matrix. The glass transition temperature and enthalpy increased compared to the untreated samples. Interestingly, values were highest for corn starch which contained the most amylose, whereas values were lowest for rice starch, which contained the least amylose. Dong et al. (2018) treated zein films with CP for 60 seconds, they reported an enhanced denaturation temperature. The enhancement was ascribed "to the cross-linking of polar compounds at the film surface." Chen et al. (2020) reported similar results for treatment of zein-PLA films. Treatment for up to 120 seconds led to increased thermal stability. Intermolecular hydrogen bonding between the zein and PLA molecules was identified as the reason for the improvement.

Ledari et al. (2020) investigated the effect of different plasma-forming gases (O_2, N_2, air, Ar, and ethanol-argon [EtOH-Ar]) on the thermal properties of gelatin-corn oil emulsion-based films. O_2, N_2, and EtOH-Ar positively impacted thermal properties, whereas Ar had no impact. Oh, Roh, and Min (2016) found that O_2 plasma treatment decreased glass transition temperature, but Ar and He plasma increased glass transition temperature. Therefore, the plasma gas utilized can impact thermal properties. Oh et al. (2016) discussed that oxygen gas plasma led to a decrease in glass transition temperature due to the "amalgamation of oxygen—containing functional groups at the biopolymer surface," this phenomenon increased molecular mobility which decreased the glass transition temperature.

Some studies have reported CP to have no impact on the thermal properties of PLA, zein, and nisin coated-PLA films (Hu et al., 2018; Pankaj et al., 2014; Pankaj, Bueno-Ferrer, Misra, O'Neill, Jiménez, et al., 2014). Pankaj, Bueno-Ferrer, Misra, O'Neill, Tiwari, et al. (2014) found DBD plasma treatment to have a negative impact on the glass transition temperature of sodium caseinate films compared to the untreated films. Romani et al. (2019) reported similar findings with treatment of fish protein films using GDP. Glass transition temperature, melting temperature, and enthalpy decreased significantly compared to the untreated samples. It is worth noting that for both studies by Pankaj et al. (2014) and Romani et al. (2019), the polymers examined were protein based.

12.4.5 *Water barrier properties*

The effectiveness of packaging materials is largely determined by their ability to prevent moisture from penetrating through the material, which is measured by the WVP. For biopolymers, high WVP can be a significant drawback that impacts packaging effectiveness (Ledari et al. 2020; Wu, Misra, & Mohanty, 2021). While some studies have reported that CP treatment has no impact on WVP (Goiana et al., 2021; Pankaj, Bueno-Ferrer, Misra, O'Neill, Jiménez et al., 2014; Pankaj, Bueno-Ferrer, Misra, O'Neill, Tiwari et al., 2015; Pankaj et al., 2016), others have reported either an increase or a decrease in WVP (Arolkar et al., 2015; Chen et al., 2018, 2020; Dong et al., 2018).

Heidemann et al. (2019) reported a 94% decrease in WVP of starch-PCL and starch-PLA composite films after CP treatment. CP treatment can enhance the adhesion of different polymer materials, leading to improved WVP properties (Hoque et al., 2022). This has been observed with biopolymers such as zein-chitosan (Chen et al., 2019), zein-PLA (Chen et al., 2020), and whey protein concentrate-wheat crosslinked starch (Song et al., 2019). Dong et al. (2018) found that CP treatment reduced the WVP of pristine zein films by 24% compared to the untreated films. Bahrami et al. (2020) and Chen et al. (2019) attribute this reduction in WVP to the increase in surface roughness, which deviates from the flow path of water. In contrast, Arolkar et al. (2015) reported that overexposure to CP increased the water vapor transmission rate (WVTR), suggesting that the etching effect facilitated water vapor transport. Ledari et al. (2020) reported that CP generated with different gases [O_2, N_2, air, Ar, and ethanol-argon (EtOH-Ar)] had no impact on the WVP of gelatin films. Similarly, Sheikhi et al. (2020) reported that both air and oxygen gas plasma had no significant effect on WVP. Factors that may impact WVP include vapor pressure, concentration gradient, diffusion, solubility, thermodynamic, and surface properties (Bahrami et al., 2020; Ledari et al. 2020; Pankaj, Bueno-Ferrer, Misra, O'Neill, Tiwari, et al., 2015; Sheikhi et al., 2020).

12.4.6 *Oxygen permeability*

The rate of oxygen transfer through packaging is a critical factor affecting shelf life and food quality. A low oxygen permeability (OP) value is desirable as it indicates that no oxygen is entering the package. However, the gas barrier properties of biopolymers are generally poor compared to conventional plastic packaging, making them less effective in preserving food quality. One potential solution is the modification of biopolymers through CP treatment.

Moosavi et al. (2020) reported a lower OP after treating whey protein and gelatin films with both argon and air plasma. Sheikhi et al. (2020) used O_2 plasma to treat starch films and found that OP significantly decreased. Hoque et al. (2022) discussed that improvements in OP could be attributed to enhanced cross-linking between polymer chains, which compresses the surface layer and prevents oxygen from entering. Other researchers have similarly described that plasma treatment increases polymer crosslinking, leading to a denser surface layer and an enhanced OP (Ledari et al. 2020; Moosavi et al., 2020; Sheikhi et al., 2020). As the network of polymer chains becomes tightly packed after plasma treatment, the rate at which the permeant (gas) diffuses through the film is significantly reduced (Hoque et al., 2022).

Ledari et al. (2020) reported a decrease in OP for gelatin-based emulsion films using various plasma gases [O_2, N_2, air, Ar, and ethanol-argon (EtOH-Ar)]. A significant decrease in OP was reported with each plasma gas, with O_2 and argon found to be the most effective at reducing OP. However, some studies have reported CP to negatively impact or have no effect on OP. Arolkar et al. (2015) reported that OP increased (worsened) after treatment of corn starch-poly(ε-caprolactone) films using a radio frequency generated plasma reactor. Pankaj et al. (2014) reported no change in OP after DBD plasma treatment of gelatin films. In a follow-up study by Pankaj et al. (2016), using chitosan films, they also reported no change in OP after DBD treatment. Overall, the effect of CP on OP is dependent on various factors, including the type of biopolymer and the plasma gas used.

12.4.7 *Antimicrobial properties of cold plasma treated films*

CP has been found to enhance the effectiveness of active packaging films in several studies. Hoque et al. (2022) discussed that films loaded with bioactive compounds like essential oils, peptides, or functional components show higher antimicrobial activity when exposed to plasma. Honarvar et al. (2017) combined polypropylene films with CMC loaded with *Z. multiflora* essential oil using plasma. Antimicrobial activity was recorded against Gram-positive (*S. aureus*) and Gram-negative (*E. coli, S. typhimurium,* and *P. aeruginosa*) bacteria. They reported a significant increase in antimicrobial activity compared to the control films. Plasma treatment facilitated adhesion between the polypropylene and CMC layers. The improvement in antimicrobial activity was attributed to the polypropylene layer acting as a barrier to the loss of the Zataria essential oil.

Lei et al. (2014) prepared plasma-treated films from PET/polypropylene film loaded with chitosan, sodium benzoate (1.0%–5.0%), potassium sorbate (1.0%–5.0%), calcium propionate (1.0%–5.0%), and ε-poly-lysine (0.2%–1.0%). The plasma-treated films showed high antimicrobial activity. They discussed that CP has no direct impact on antimicrobial activity. However, CP increases adhesion between the polymer layers, which prevents the loss of antimicrobial components, thereby increasing antimicrobial activity.

Wong et al. (2020) coated polyethylene films with gallic acid using plasma (30 W for 60 seconds). A 0.5–1.1 log reduction was reported for *E. coli* and *S. aureus*. Hu et al. (2018) coated PLA films with nisin. The films demonstrated a 3.23 log reduction against *L. monocytogenes*. Hu et al. (2018) found pristine PLA films without the nisin coating to be ineffective at inhibiting the growth of *L. monocytogenes*. As plasma exposure time was increased, more nisin was absorbed by the PLA, thereby increasing the antimicrobial activity. In a similar study by De Vietro, Conte, Incoronato, Del Nobile, and Fracassi (2017), polycarbonate films were coated with copper using low-pressure aerosol-assisted plasma to promote adhesion. A three-log reduction from 10^8 to 10^5 CFU/mL was reported against *Pseudomonas fluorescens* and *P. putida*. The antimicrobial effect was attributed to the release of Cu^{2+} ions, which penetrated the bacterial cells.

Pankaj et al. (2014) found DBD plasma treatment increased the diffusion of thymol throughout a zein film matrix. This enhanced diffusion was credited to the etching effect. Karam et al. (2016) showed that plasma treatment enhanced the release of nisin from an LDPE-nisin film. The film was effective at inhibiting the growth of *L. innocua*, and a 0.9 log reduction was reported.

Chen et al. (2018) reported that CP had a negative impact on the antimicrobial activity of chitosan-ciprofloxacin hydrochloride antimicrobial films compared to the untreated control. The CP treatment significantly reduced the release of ciprofloxacin hydrochloride, thereby reducing the antimicrobial effect. The reduction was ascribed to crosslinking, which may have over-enhanced the adhesion of the ciprofloxacin hydrochloride to the polymer layers, thus reducing its release. Aranaz, Harris, Navarro-García, Heras, and Acosta (2016) reported similar results after CP treatment of films composed of sodium tripolyphosphate, chitosan, and ciprofloxacin. The authors suggested that cross-linking prevented the diffusion of ciprofloxacin. However, these findings do not negate the potential for CP treatment to improve the antimicrobial properties of packaging materials. The optimization of CP treatment parameters, such as exposure time and voltage applied, should be further investigated to optimize antimicrobial effectiveness.

12.4.8 *Biodegradability*

Bacteria, yeast, and fungi are known to produce enzymes responsible for the degradation of biodegradable materials, limiting the use of synthetic non-biodegradable materials in the food packaging industry due to environmental hazards (Hoque et al., 2022). Plasma surface modification has emerged as a promising method for increasing biodegradability. Arolkar et al. (2015) conducted a study on the biodegradation of CP-treated corn starch/poly(ε-caprolactone) films under soil burial conditions in the presence of *Bacillus subtilis*. The results showed a significant loss in mechanical properties of the air-plasma treated films compared to untreated films after 56 days of soil burial. They observed a 35% loss in TS and a 42% loss in elongation at break. Moreover, films exposed to plasma for longer periods degraded faster, and a higher level of bacterial colonization was observed in the CP-treated films. Therefore, Arolkar et al. (2015) concluded that air plasma is an attractive option for the development of biodegradable packaging.

Chen et al. (2020) investigated the biodegradability of zein films with porous PLA coatings after CP treatment. Compared to untreated films, a high degree of degradation was reported, and microbial growth was observed after 7 days of compost burial (Fig. 12.4). They suggested that the development of pores on the surface layer due to CP treatment facilitated swift degradation as microorganisms gained entry to the inner packaging layers. Oh et al. (2016) reported a similar result after treating defatted soybean meal films. The etching effect increased the roughness of the film, thereby enhancing the ability of microbes to adhere to the surface layer. Song et al. (2015) found that plasma-treated PLA films exhibited higher levels of degradation than non-treated films after 28 days of soil burial. In summary, CP treatment has shown potential in increasing the biodegradability of packaging materials, and further investigation is necessary to optimize this process.

12.5 Safety of cold plasma for food packaging applications

The use of CP in food processing and packaging applications is a unique approach that still requires permission. However, the use of this latest plasma technology has not received FDA approval. This is because reactive species produced by plasma come into direct contact with food surfaces, and it is unclear exactly how these reactive organisms interact with the food matrix. Limited studies have been conducted on the interactions of food components at the molecular level with CP and their impact on human health. Han, Suh, Hong, Kim, and Min (2016) examined the acute and sub-acute oral toxicity of edible films treated with CP in a rat model as part of their investigation into the safety of such materials. Their findings showed that the edible film treated with CP had a very low level of toxicity, indicating that CP did not produce any negative byproducts in the edible film.

Lee, Cheigh, Kang, Lee, and Min (2020) evaluated the impact of in-package atmospheric dielectric barrier discharge CP (ADCP) treatment on the inactivation of microorganisms, nitrates and nitrite contents, oral toxicity, and keeping quality of protein-coated boiling chicken breast cubes. Additionally, the toxicity tests' hematological and blood biochemical characteristics showed that ADCP-treated chicken breast cubes were toxicologically safe. Chicken breast cubes' pH, color, total volatile basic nitrogen, lipid oxidation, and softness were unaffected by ADCP treatment when stored at 4°C and 24°C. They also retained their original sensory characteristics after three days at 4°C.

Although there are various drawbacks to plasma processing, such as a rise in lipid oxidation, a drop in color, a loss in fruit firmness, an increase in acidity, etc. (Thirumdas, Sarangapani, & Annapure, 2015), the use of CP technology has also shown promise in disinfecting the surface of walnuts and peanuts (Thirumdas et al., 2015). However, the most significant issue is the 20% rise in peroxide value in walnuts with greater power and treatment times, likely due to radicals oxidizing lipid molecules, which increases the peroxide value. Because the

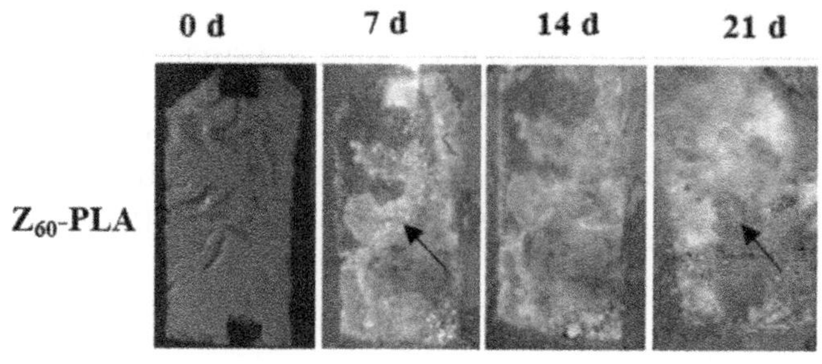

FIGURE 12.4 The biodegradability of zein-PLA films (CP treated for 60 s) after 0, 7, 14, and 21 days. Films were exposed to homemade compost at 37°C and 90% RH. *Source: From Chen, G., Chen, Y., Jin, N., Li, J., Dong, S., Li, S., Zhang, Z. & Chen, Y. (2020). Zein films with porous polylactic acid coatings via cold plasma pre-treatment. Industrial Crops and Products, 150, 112382. https://doi.org/10.1016/j.indcrop.2020.112382.*

plasma effect is a surface phenomenon, it is not viable to employ this technique to inactivate endogenous enzymes present in entire fruits in their natural state. Another drawback is that using direct plasma has been shown to result in a considerable loss in firmness in strawberry fruit after 24 hours of treatment (Misra, Patil, et al., 2014).

In another study, Kim et al. (2015) assessed the microbiological and physicochemical properties of milk treated with encapsulated DBD plasma. Milk samples were treated with DBD plasma for durations of 5 minutes and 10 minutes after being created in a plastic container using 250 W, 15 kHz, and ambient air. After the 10-minutes plasma treatment, the pH of the sample milk had decreased, and the hunter color L^* and b^* values of milk increased, whereas the a^* value dropped. After plasma treatment, the generation of 2-thiobarbituric acid reactive compounds was marginally reported but not considerably increased. This concludes that CP may generate reactive species that will be directly related to the dose and plasma exposure time.

Vleugels et al. (2005) reported that DBD treatments of bell peppers with helium and oxygen as process gas led to slight color losses. Therefore, there is little discoloration due to plasma treatment, and the change in the hue angle for yellow peppers was smaller than that seen for green peppers. Both Grzegorzewski et al. (2009) and Grzegorzewski et al. (2010) conducted extensive research on the effects of atmospheric pressure plasma on polyphenols. After 120 seconds of plasma application, the polyphenols quercetin and kaempferol exhibited significant degradation; however, myricetin and taxifolin, which lack a C=C double bond of the C-ring, demonstrated less deterioration. The presence or absence of structural elements affects the stability differences. In both cases, the plasma treatment causes the compounds to degrade over time. This process is likely mediated by ROS and radicals in the plasma effluent, which cause oxidation and breakage of the flavonoid skeleton. In a relatively short exposure period, reactive species in the plasma were able to completely disintegrate large layers of quercetin and kaempferol on the substrate surface. Studies that have investigated the potentially harmful effects of plasma-generated reactive species on human health are limited. Therefore, further risk assessments are necessary to determine the potential toxicity of plasma-treated food or food packaging for both in vitro and in vivo situations.

12.6 Conclusion and future perspectives

In conclusion, CP treatment is a promising technology for the food packaging industry. It has shown potential in improving surface sterilization, modifying biopolymer-based packaging materials, and enhancing packaging properties such as surface roughness, contact angle, water barrier properties, and biodegradability. However, its impact on the nutritional quality of food and the potential toxicity of plasma-treated food or packaging materials needs further investigation. Overall, the optimization of CP treatment parameters, such as exposure time and voltage applied, should be further investigated to enhance its efficacy and safety. In the future, further research is needed to optimize the CP treatment parameters and develop efficient and cost-effective plasma-based techniques for food packaging applications. The impact of CP on the nutritional quality of food and its potential toxicity to human health should also be investigated. In addition, the potential for combining CP treatment with other antimicrobial agents or preservation techniques, such as high-pressure processing or modified atmosphere packaging, should be explored. Furthermore, the development of plasma-based sensors for real-time monitoring of food quality and safety is an exciting avenue for future research in the field of food packaging. With further research and development, CP technology has the potential to become a key technology in the development of innovative, sustainable, and safe food packaging solutions.

References

Afshari, R., & Hosseini, H. (2014). Non-thermal plasma as a new food preservation method, Its present and future prospect. *Journal of Paramedical Sciences (JPS) Winter*, *5*(1), 2008–4978.

Akishev, Y., Grushin, M., Dyatko, N., Kochetov, I., Napartovich, A., Trushkin, N., ... Descours, S. (2008). Studies on cold plasma–polymer surface interaction by example of PP-and PET-films. *Journal of Physics D: Applied Physics*, *41*(23), 235203.

Almazan-Almazan, M. C., Paredes, J. I., Perez-Mendoza, M., Domingo-García, M., Fernandez-Morales, I., Martínez-Alonso, A., & Lopez-Garzon, F. J. (2006). Surface characteristics of activated carbons obtained by pyrolysis of plasma pretreated PET. *Journal of Physical Chemistry B*, *110*(23), 11327–11333.

Aranaz, I., Harris, R., Navarro-García, F., Heras, A., & Acosta, N. (2016). Chitosan based films as supports for dual antimicrobial release. *Carbohydrate Polymers*, *146*, 402–410. Available from https://doi.org/10.1016/j.carbpol.2016.03.064.

Arolkar, G. A., Salgo, M. J., Kelkar-Mane, V., & Deshmukh, R. R. (2015). The study of air-plasma treatment on corn starch/poly(ε-caprolactone) films. *Polymer Degradation and Stability*, *120*, 262–272. Available from https://doi.org/10.1016/j.polymdegradstab.2015.07.016.

Bahrami, R., Zibaei, R., Hashami, Z., Hasanvand, S., Garavand, F., Rouhi, M., ... Mohammadi, R. (2020). Modification and improvement of biodegradable packaging films by cold plasma; a critical review. *Critical Reviews in Food Science and Nutrition*, 1–15. Available from https://doi.org/10.1080/10408398.2020.1848790.

Bang, I. H., Kim, Y. E., & Min, S. C. (2021). Preservation of mandarins using a microbial decontamination system integrating calcium oxide solution washing, modified atmosphere packaging, and dielectric barrier discharge cold plasma treatment. *Food Packaging and Shelf Life*, *29*(April), 100682. Available from https://doi.org/10.1016/j.fpsl.2021.100682.

Baniya, H. B., Guragain, R. P., Baniya, B., Qin, G., & Subedi, D. P. (2020). Improvement of hydrophilicity of polyamide using atmospheric pressure plasma jet. *Bibechana*, *17*, 133–138.

Borin, D., Sbaizero, O., & Scuor, N. (2019). Application of a dielectric barrier discharge plasma for heating plastic materials. *Plasma Research Express*, *1*(2), 025009. Available from https://doi.org/10.1088/2516-1067/ab1038.

Capelli, F., Tappi, S., Gritti, T., de Aguiar Saldanha Pinheiro, A. C., Laurita, R., Tylewicz, U., ... & Rocculi, P. (2021). Decontamination of food packages from SARS-CoV-2 RNA with a cold plasma-assisted system. *Applied Sciences*, *11*(9), 4177. Available from: https://doi.org/10.3390/app11094177.

Carina, D., Sharma, S., Jaiswal, A. K., & Jaiswal, S. (2021). Seaweeds polysaccharides in active food packaging: A review of recent progress. *Trends in Food Science & Technology*, *110*, 559–572. Available from https://doi.org/10.1016/j.tifs.2021.02.022.

Chen, G., Ali, F., Dong, S., Yin, Z., Li, S., & Chen, Y. (2018). Preparation, characterization and functional evaluation of chitosan-based films with zein coatings produced by cold plasma. *Carbohydrate Polymers*, *202*, 39–46. Available from https://doi.org/10.1016/j.carbpol.2018.08.122.

Chen, G., Chen, Y., Jin, N., Li, J., Dong, S., Li, S., ... Chen, Y. (2020). Zein films with porous polylactic acid coatings via cold plasma pre-treatment. *Industrial Crops and Products*, *150*, 112382. Available from https://doi.org/10.1016/j.indcrop.2020.112382.

Chen, Y.-Q., Cheng, J.-H., & Sun, D.-W. (2019). Chemical, physical and physiological quality attributes of fruit and vegetables induced by cold plasma treatment: Mechanisms and application advances. *Critical Reviews in Food Science and Nutrition*, *60*(16), 2676–2690. Available from https://doi.org/10.1080/10408398.2019.1654429.

Chu, P. K., Chen, J., Wang, L., & Huang, N. (2002). Plasma-surface modification of biomaterials. *Materials Science and Engineering: R: Reports*, *36*(5–6), 143–206.

Cools, P., Asadian, M., Nicolaus, W., Declercq, H., Morent, R., & De Geyter, N. (2018). Surface treatment of PEOT/PBT (55/45) with a dielectric barrier discharge in air, helium, argon and nitrogen at medium pressure. *Materials*, *11*(3), 391.

Cretu, D. E., Rusu, C. C., Burlica, R., Beniuga, O., Astanei, D., & Tesoi, D. (2021). DBD non-thermal plasma used on surface treatment of polymeric film for food packaging. In *2021 International Conference on Electromechanical and Energy Systems (SIELMEN)*, (pp. 454–457). IEEE.

De Vietro, N., Conte, A., Incoronato, A. L., Del Nobile, M. A., & Fracassi, F. (2017). Aerosol-assisted low pressure plasma deposition of antimicrobial hybrid organic-inorganic Cu-composite thin films for food packaging applications. *Innovative Food Science & Emerging Technologies*, *41*, 130–134. Available from https://doi.org/10.1016/j.ifset.2017.02.010.

Deng, X., Shi, J., & Kong, M. G. (2006). Physical mechanisms of inactivation of Bacillus subtilis spores using cold atmospheric plasmas. *IEEE Transactions on Plasma Science*, *34*(4 II), 1310–1316. Available from https://doi.org/10.1109/TPS.2006.877739.

Dong, S., Guo, P., Chen, G. Y., Jin, N., & Chen, Y. (2020). Study on the atmospheric cold plasma (ACP) treatment of zein film: Surface properties and cytocompatibility. *International Journal of Biological Macromolecules*, *153*, 1319–1327. Available from https://doi.org/10.1016/j.ijbiomac.2019.10.268.

Dong, S., Guo, P., Chen, Y., Chen, G., Ji, H., Ran, Y., ... Chen, Y. (2018). Surface modification via atmospheric cold plasma (ACP): Improved functional properties and characterization of zein film. *Industrial Crops and Products*, *115*, 124–133. Available from https://doi.org/10.1016/j.indcrop.2018.01.080.

Dorai, R., & Kushner, M. J. (2003). A model for plasma modification of polypropylene using atmospheric pressure discharges. *Journal of Physics D: Applied Physics*, *36*(6), 666.

Fazeli, M., Florez, J. P., & Simão, R. A. (2019). Improvement in adhesion of cellulose fibers to the thermoplastic starch matrix by plasma treatment modification. *Composites Part B: Engineering, [online]*, *163*, 207–216. Available from https://doi.org/10.1016/j.compositesb.2018.11.048.

Gholamazad, A., Hosseini, S., Hosseini, S., Ramezan, Y., & Rahmanabadi, A. (2022). Effect of low-pressure cold plasma on the properties of edible film based on alginate enriched with *Dunaliella salina* powder. *Plasma Processes and Polymers*, *19*(5), 2100118. Available from https://doi.org/10.1002/ppap.202100118.

Goiana, M. L., de Brito, E. S., Alves Filho, E. G., Miguel, E., de, C., Fernandes, F. A. N., ... Rosa, M. D. F. (2021). Corn starch based films treated by dielectric barrier discharge plasma. *International Journal of Biological Macromolecules*, *183*(April), 2009–2016. Available from https://doi.org/10.1016/j.ijbiomac.2021.05.210.

Grzegorzewski, F. (2011). Influence of non-thermal plasma species on the structure and functionality of isolated and plant-based 1, 4-benzopyrone derivatives and phenolic acids. *Journal of Agricultural and Food Chemistry*, *59*(6), 2396–2401.

Grzegorzewski, F., Rohn, S., Quade, A., Schröder, K., Ehlbeck, J., Schlüter, O., & Kroh, L. W. (2010). Reaction chemistry of 1,4-benzopyrone deriwates in non-equilibrium low-temperature plasmas. *Plasma Processes and Polymers*, *7*(6), 466–473. Available from https://doi.org/10.1002/ppap.200900140.

Grzegorzewski, F., Schlüter, O., Ehlbeck, J., Weltmann, K. D., Geyer, M., Kroh, L. W., & Rohn, S. (2009). Plasma-oxidative degradation of polyphenolics - Influence of non-thermal gas discharges with respect to fresh produce processing. *Czech Journal of Food Sciences*, *27*(SPEC. ISS.), 35–39. Available from https://doi.org/10.17221/1107-cjfs.

Han, S. H., Suh, H. J., Hong, K. B., Kim, S. Y., & Min, S. C. (2016). Oral toxicity of cold plasma-treated edible films for food coating. *Journal of Food Science*, *81*(12), T3052–T3057. Available from https://doi.org/10.1111/1750-3841.13551.

Heidemann, H. M., Dotto, M. E. R., Laurindo, J. B., Carciofi, B. A. M., & Costa, C. (2019). Cold plasma treatment to improve the adhesion of cassava starch films onto PCL and PLA surface. *Colloids and Surfaces A: Physicochemical and Engineering Aspects*, *580*(July). Available from https://doi.org/10.1016/j.colsurfa.2019.123739.

Honarvar, Z., Farhoodi, M., Khani, M. R., Mohammadi, A., Shokri, B., Ferdowsi, R., & Shojaee-Aliabadi, S. (2017). Application of cold plasma to develop carboxymethyl cellulose-coated polypropylene films containing essential oil. *Carbohydrate Polymers, 176*, 1–10. Available from https://doi.org/10.1016/j.carbpol.2017.08.054.

Hoque, M., McDonagh, C., Tiwari, B. K., Kerry, J. P., & Pathania, S. (2022). Effect of cold plasma treatment on the packaging properties of biopolymer-based films: A review. *Applied Sciences, [online], 12*(3), 1346. Available from https://doi.org/10.3390/app12031346.

Hou, Y., Liu, Q., Mian, S. M., Luo, Y., Mu, G., Jiang, S., ... Wu, X. (2022). Effects of treatment of dielectric barrier discharge cold plasma (DBD-CP) on mechanical, barrier and functional characteristics of casein-based films. *International Journal of Food Science and Technology, 57*, 705–718. Available from https://doi.org/10.1111/ijfs.15451.

Hu, S., Li, P., Wei, Z., Wang, J., Wang, H., & Wang, Z. (2018). Antimicrobial activity of nisin-coated polylactic acid film facilitated by cold plasma treatment. *Journal of Applied Polymer Science, 135*(47), 46844. Available from https://doi.org/10.1002/app.46844.

Hu, X., Sun, H., Yang, X., Cui, D., Wang, Y., Zhuang, J., ... Jiao, Z. (2021). Potential use of atmospheric cold plasma for postharvest preservation of blueberries. *Postharvest Biology and Technology, 179*, 111564.

Izdebska-Podsiadły, J. (2021). Effect of plasma surface modification on print quality of biodegradable PLA films. *Applied Sciences, [online], 11* (17), 8245. Available from https://doi.org/10.3390/app11178245.

Jacobs, T., De Geyter, N., Morent, R., Van Vlierberghe, S., Dubruel, P., & Leys, C. (2011). Plasma modification of PET foils with different crystallinity. *Surface and Coatings Technology, 205*, S511–S515.

Jaeger, H., Knorr, D., Meneses, N., Reineke, K., & Schlueter, O. (2014). Food safety: Shelf life extension technologies. In N. K. Van Alfen (Ed.), *Encyclopedia of agriculture and food systems* (pp. 289–303). Oxford: Academic Press. Available from https://doi.org/10.1016/B978-0-444-52512-3.00050-4.

Jahromi, M., Niakousari, M., & Golmakani, M. T. (2022). Fabrication and characterization of pectin films incorporated with clove essential oil emulsions stabilized by modified sodium caseinate. *Food Packaging and Shelf Life, 32*(August 2021), 100835. Available from https://doi.org/10.1016/j.fpsl.2022.100835.

Kang, J. H., Bai, J., & Min, S. C. (2021). Inactivation of indigenous microorganisms and Salmonella in Korean rice cakes by in-package cold plasma treatment. *International Journal of Environmental Research and Public Health, 18*, 3360. Available from https://doi.org/10.3390/ijerph18073360.

Karam, L., Casetta, M., Chihib, N. E., Bentiss, F., Maschke, U., & Jama, C. (2016). Optimization of cold nitrogen plasma surface modification process for setting up antimicrobial low density polyethylene films. *Journal of the Taiwan Institute of Chemical Engineers, 64*, 299–305. Available from https://doi.org/10.1016/j.jtice.2016.04.018.

Karam, L., Jama, C., Dhulster, P., & Chihib, N.-E. (2013). Study of surface interactions between peptides, materials and bacteria for setting up antimicrobial surfaces and active food packaging. *Journal of Materials and Environmental Science, 4*(5), 798–821.

Kim, H. J., Yong, H. I., Park, S., Kim, K., Choe, W., & Jo, C. (2015). Microbial safety and quality attributes of milk following treatment with atmospheric pressure encapsulated dielectric barrier discharge plasma. *Food Control, 47*, 451–456. Available from https://doi.org/10.1016/j.foodcont.2014.07.053.

Kim, S. Y., Bang, I. H., & Min, S. C. (2019). Effects of packaging parameters on the inactivation of Salmonella contaminating mixed vegetables in plastic packages using atmospheric dielectric barrier discharge cold plasma treatment. *Journal of Food Engineering, 242*, 55–67. Available from https://doi.org/10.1016/j.jfoodeng.2018.08.020.

Kim, Y. E., & Min, S. C. (2021). Inactivation of Salmonella in ready-to-eat cabbage slices packaged in a plastic container using an integrated in-package treatment of hydrogen peroxide and cold plasma. *Food Control, 130*, 108392. Available from https://doi.org/10.1016/j.foodcont.2021.108392.

Kim, Y. E., & Min, S. C. (2022). Consecutive treatments of cold plasma and intense pulsed light for microbial decontamination of fresh cabbage slices in plastic containers. *International Journal of Food Microbiology, 369*, 109626. Available from https://doi.org/10.1016/j.ijfoodmicro.2022.109626.

Kulawik, P., Alvarez, C., Cullen, P. J., Aznar-Roca, R., Mullen, A. M., & Tiwari, B. (2018). The effect of non-thermal plasma on the lipid oxidation and microbiological quality of sushi. *Innovative Food Science & Emerging Technologies, 45*, 412–417. Available from: https://doi.org/10.1016/j.ifset.2017.12.011.

Laroque, D. A., Seó, S. T., Valencia, G. A., Laurindo, J. B., & Carciofi, B. A. M. (2022). Cold plasma in food processing: Design, mechanisms, and application. *Journal of Food Engineering, 312*, 110748. Available from https://doi.org/10.1016/j.jfoodeng.2021.110748.

Ledari, S. A., Milani, J. M., & Lanbar, F. S. (2020). Improving gelatin-based emulsion films with cold plasma using different gases. *Food Science & Nutrition, [online], 8*(12), 6487–6496. Available from https://doi.org/10.1002/fsn3.1939.

Lee, E. S., Jeon, Y. J., & Min, S. C. (2021). Microbial inactivation and quality preservation of chicken breast salad using atmospheric dielectric barrier discharge cold plasma treatment. *Foods, 10*, 1214. Available from https://doi.org/10.3390/foods10061214.

Lee, E. S., Cheigh, C.-I., Kang, J. H., Lee, S. Y., & Min, S. C. (2020). Evaluation of in-package atmospheric dielectric barrier discharge cold plasma treatment as an intervention technology for decontaminating bulk ready-to-eat chicken breast cubes in plastic containers. *Applied Sciences, 10*(18), 6301. Available from https://doi.org/10.3390/app10186301.

Lee, H. S., Kim, N., & Min, S. C. (2022). Inactivation of Salmonella in steamed fish cake using an in-package combined treatment of cold plasma and ultraviolet-activated zinc oxide. *Food Control, 135*, 108772. Available from https://doi.org/10.1016/j.foodcont.2022.108772.

Lee, K. T. (2010). Quality and safety aspects of meat products as affected by various physical manipulations of packaging materials. *Meat Science, 86*(1), 138–150. Available from https://doi.org/10.1016/j.meatsci.2010.04.018.

Lei, J., Yang, L., Zhan, Y., Wang, Y., Ye, T., Li, Y., ... Li, B. (2014). Plasma treated polyethylene terephthalate/polypropylene films assembled with chitosan and various preservatives for antimicrobial food packaging. *Colloids and Surfaces B: Biointerfaces, [online], 114*, 60–66. Available from https://doi.org/10.1016/j.colsurfb.2013.09.052.

Li, Y., Wu, C., Bai, Y., Liu, S., Yuan, C., Ding, T., & Hu, Y. (2019). Effect of glow discharge plasma on surface modification of chitosan film. *International Journal of Biological Macromolecules, 138*, 340–348. Available from https://doi.org/10.1016/j.ijbiomac.2019.07.039.

Los, A., Ziuzina, D., Akkermans, S., Boehm, D., Cullen, P. J., Van Impe, J., & Bourke, P. (2018). Improving microbiological safety and quality characteristics of wheat and barley by high voltage atmospheric cold plasma closed processing. *Food Research International, 106*, 509–521. Available from: https://doi.org/10.1016/j.foodres.2018.01.009.

Mandal, R., Singh, A., & Pratap Singh, A. (2018). Recent developments in cold plasma decontamination technology in the food industry. *Trends in Food Science and Technology, 80*, 93–103. Available from https://doi.org/10.1016/j.tifs.2018.07.014.

Marenco-Orozco, G. A., Rosa, M. F., & Fernandes, F. A. N. (2022). Effects of multiple-step cold plasma processing on banana (Musa sapientum) starch-based films. *Packaging Technology and Science*. Available from https://doi.org/10.1002/pts.2636.

Min, S. C., Roh, S. H., Boyd, G., Sites, J. E., Uknalis, J., Fan, X., et al. (2017). Inactivation of Escherichia coli O157: H7 and aerobic microorganisms in romaine lettuce packaged in a commercial polyethylene terephthalate container using atmospheric cold plasma. *Journal of Food Protection, 80*(1), 35–43. Available from https://doi.org/10.4315/0362-028X.JFP16-148.

Min, S. C., Roh, S. H., Niemira, B. A., Boyd, G., Sites, J. E., Fan, X., ... Jin, T. Z. (2018). In-package atmospheric cold plasma treatment of bulk grape tomatoes for microbiological safety and preservation. *Food Research International, 108*, 378–386. Available from https://doi.org/10.1016/j.foodres.2018.03.033.

Misra, N. N., Keener, K. M., Bourke, P., Mosnier, J. P., & Cullen, P. J. (2014). In-package atmospheric pressure cold plasma treatment of cherry tomatoes. *Journal of Bioscience and Bioengineering, 118*(2), 177–182. Available from https://doi.org/10.1016/j.jbiosc.2014.02.005.

Misra, N. N., Patil, S., Moiseev, T., Bourke, P., Mosnier, J. P., Keener, K. M., & Cullen, P. J. (2014). In-package atmospheric pressure cold plasma treatment of strawberries. *Journal of Food Engineering, 125*(1), 131–138. Available from https://doi.org/10.1016/j.jfoodeng.2013.10.023.

Misra, N. N., Yepez, X., Xu, L., & Keener, K. (2019). In-package cold plasma technologies. *Journal of Food Engineering, 244*, 21–31. Available from https://doi.org/10.1016/j.jfoodeng.2018.09.019.

Moharram, H. A. (2010). Application of laser in food technology: an overview. *Alexandria Journal of Food Science and Technology, 7*(1), 35. Available from https://doi.org/10.21608/ajfs.2010.20177.

Moosavi, M. H., Khani, M. R., Shokri, B., Hosseini, S. M., Shojaee-Aliabadi, S., & Mirmoghtadaie, L. (2020). Modifications of protein-based films using cold plasma. *International Journal of Biological Macromolecules, 142*, 769–777. Available from https://doi.org/10.1016/j.ijbiomac.2019.10.017.

Moradi, E., Moosavi, M. H., Hosseini, S. M., Mirmoghtadaie, L., Moslehishad, M., Khani, M. R., ... Shojaee-Aliabadi, S. (2020). Prolonging shelf life of chicken breast fillets by using plasma-improved chitosan/low density polyethylene bilayer film containing summer savory essential oil. *International journal of biological macromolecules, 156*, 321–328.

Niemira, B. A., & Sites, J. (2008). Cold plasma inactivates Salmonella Stanley and Escherichia coli O157: H7 inoculated on golden delicious apples. *Journal of Food Protection, 71*(7), 1357–1365.

Niveditha, A., Pandiselvam, R., Prasath, V. A., Singh, S. K., Gul, K., & Kothakota, A. (2021). Application of cold plasma and ozone technology for decontamination of Escherichia coli in foods-A review. *Food Control, 130*, 108338.

Oh, Y. A., Roh, S. H., & Min, S. C. (2016). Cold plasma treatments for improvement of the applicability of defatted soybean meal-based edible film in food packaging. *Food Hydrocolloids, 58*, 150–159. Available from https://doi.org/10.1016/j.foodhyd.2016.02.022.

Ozdemir, M., Yurteri, C. U., & Sadikoglu, H. (1999). Physical polymer surface modification methods and applications in food packaging polymers. *Critical Reviews in Food Science and Nutrition, 39*(5), 457–477.

Pahwa, A., & Kumar, H. (2018). Influence of cold plasma technology on food packaging materials: A review. *International Journal of Chemical Studies, 6*(2), 594–603. Available from https://doi.org/10.1039/c3cs60148e.

Pankaj, S. K., Bueno-Ferrer, C., Misra, N., Milosavljević, V., O'donnell, C., Bourke, P., ... Cullen, P. (2014). Applications of cold plasma technology in food packaging. *Trends in Food Science & Technology, 35*(1), 5–17. Available from https://doi.org/10.1016/j.tifs.2013.09.007.

Pankaj, S. K., Bueno-Ferrer, C., Misra, N. N., Bourke, P., & Cullen, P. J. (2014). Zein film: Effects of dielectric barrier discharge atmospheric cold plasma. *Journal of Applied Polymer Science, 131*(18), n/a–n/a. Available from https://doi.org/10.1002/app.40803.

Pankaj, S. K., Bueno-Ferrer, C., Misra, N. N., O'Neill, L., Bourke, P., & Cullen, P. J. (2017). Effects of cold plasma on surface, thermal and antimicrobial release properties of chitosan film. *Journal of Renewable Materials, 5*(1), 14–20. Available from https://doi.org/10.7569/jrm.2016.634105.

Pankaj, S. K., Bueno-Ferrer, C., Misra, N. N., O'Neill, L., Jiménez, A., Bourke, P., & Cullen, P. J. (2014). Characterization of polylactic acid films for food packaging as affected by dielectric barrier discharge atmospheric plasma. *Innovative Food Science & Emerging Technologies, 21*, 107–113. Available from https://doi.org/10.1016/j.ifset.2013.10.007.

Pankaj, S. K., Bueno-Ferrer, C., Misra, N. N., O'Neill, L., Jiménez, A., Bourke, P., & Cullen, P. J. (2014). Surface, thermal and antimicrobial release properties of plasma-treated zein films. *Journal of Renewable Materials, 2*(1), 77–84. Available from https://doi.org/10.7569/jrm.2013.634129.

Pankaj, S. K., Bueno-Ferrer, C., Misra, N. N., O'Neill, L., Tiwari, B. K., Bourke, P., & Cullen, P. J. (2014). Physicochemical characterization of plasma-treated sodium caseinate film. *Food Research International, 66*, 438–444. Available from https://doi.org/10.1016/j.foodres.2014.10.016.

Pankaj, S. K., Bueno-Ferrer, C., Misra, N. N., O'Neill, L., Tiwari, B. K., Bourke, P., & Cullen, P. J. (2015). Characterization of dielectric barrier discharge atmospheric air cold plasma treated gelatin films. *Food Packaging and Shelf Life, 6*, 61–67. Available from https://doi.org/10.1016/j.fpsl.2015.09.002.

Pankaj, S. K., Bueno-Ferrer, C., Misra, N. N., O'Neill, L., Tiwari, B. K., Bourke, P., & Cullen, P. J. (2015). Dielectric barrier discharge atmospheric air plasma treatment of high amylose corn starch films. *LWT - Food Science and Technology, [online], 63*(2), 1076–1082. Available from https://doi.org/10.1016/j.lwt.2015.04.027.

Pankaj, S. K., Bueno-Ferrer, C., O'Neill, L., Tiwari, B. K., Bourke, P., & Cullen, P. J. (2016). Characterization of dielectric barrier discharge atmospheric air plasma treated chitosan films. *Journal of Food Processing and Preservation, 41*(1), e12889. Available from https://doi.org/10.1111/jfpp.12889.

Pankaj, S. K., Misra, N. N., & Cullen, P. J. (2013). Kinetics of tomato peroxidase inactivation by atmospheric pressure cold plasma based on dielectric barrier discharge. *Innovative Food Science & Emerging Technologies, 19*, 153–157. Available from https://doi.org/10.1016/j.ifset.2013.03.001.

Pankaj, S. K., Wan, Z., De León, J. E., Mosher, C., Colonna, W., & Keener, K. M. (2017). High-voltage atmospheric cold plasma treatment of different types of starch films. *Starch - Stärke, 69*(11–12), 1700009. Available from https://doi.org/10.1002/star.201700009.

Perera, K. Y., Sharma, S., Pradhan, D., Jaiswal, A. K., & Jaiswal, S. (2021). Seaweed polysaccharide in food contact materials (active packaging, intelligent packaging, edible films, and coatings. *Foods, 10*(9), 2088. Available from https://doi.org/10.3390/foods10092088.

Poncin-Epaillard, F., Brosse, J. C., & Falher, T. (1999). Reactivity of surface groups formed onto a plasma treated poly (propylene) film. *Macromolecular Chemistry and Physics, 200*(5), 989–996.

Romani, V. P., Olsen, B., Pinto Collares, M., Meireles Oliveira, J. R., Prentice-Hernández, C., & Guimarães Martins, V. (2019). Improvement of fish protein films properties for food packaging through glow discharge plasma application. *Food Hydrocolloids, 87*, 970–976. Available from https://doi.org/10.1016/j.foodhyd.2018.09.022.

Schnabel, U., Niquet, R., Schlüter, O., Gniffke, H., & Ehlbeck, J. (2015). Decontamination and sensory properties of microbiologically contaminated fresh fruits and vegetables by microwave plasma processed air (PPA). *Journal of Food Processing and Preservation, 39*(6), 653–662. Available from https://doi.org/10.1111/jfpp.12273.

Sharma, S., Barkauskaite, S., Jaiswal, A. K., & Jaiswal, S. (2021). Essential oils as additives in active food packaging. *Food Chemistry, 343*, 128403. Available from https://doi.org/10.1016/j.foodchem.2020.128403.

Sharma, S., Jaiswal, A. K., Duffy, B., & Jaiswal, S. (2020). Ferulic acid incorporated active films based on poly(lactide)/poly(butylene adipate-co-terephthalate) blend for food packaging. *Food Packaging and Shelf Life, 24*, 100491. Available from https://doi.org/10.1016/j.fpsl.2020.100491.

Sheikhi, Z., Hosseini, S. M., Khani, M. R., Farhoodi, M., Abdolmaleki, K., Shokri, B., ... Mirmoghtadaie, L. (2020). Treatment of starch films with a glow discharge plasma in air and O_2 at low pressure. *Food Science and Technology International, 27*(3), 276–285. Available from https://doi.org/10.1177/1082013220948641.

Sheikhi, Z., Mirmoghtadaie, L., Abdolmaleki, K., Khani, M. R., Farhoodi, M., Moradi, E., ... Shojaee-Aliabadi, S. (2021). Characterization of physicochemical and antimicrobial properties of plasma-treated starch/chitosan composite film. *Packaging Technology and Science, 34*(7), 385–392. Available from https://doi.org/10.1002/pts.2559.

Simmons, A. (2012). 11 - Future trends for the sterilisation of biomaterials and medical devices. In S. Lerouge, & A. Simmons (Eds.), *Sterilisation of biomaterials and medical devices* (pp. 310–320). Woodhead Publishing. Available from https://doi.org/10.1533/9780857096265.310.

Song, A. Y., Oh, Y. A., Roh, S. H., Kim, J. H., & Min, S. C. (2015). Cold oxygen plasma treatments for the improvement of the physicochemical and biodegradable properties of polylactic acid films for food packaging. *Journal of Food Science, 81*(1), E86–E96. Available from https://doi.org/10.1111/1750-3841.13172.

Song, J., Jiang, B., Wu, Y., Chen, S., Li, S., Sun, H., & Li, X. (2019). Effects on surface and physicochemical properties of dielectric barrier discharge plasma-treated whey protein concentrate/wheat cross-linked starch composite film. *Journal of Food Science, 84*(2), 268–275. Available from https://doi.org/10.1111/1750-3841.14387.

Suppakul, P., Miltz, J., Sonneveld, K., & Bigger, S. W. (2003). Active packaging technologies with an emphasis on antimicrobial packaging and its applications. *Journal of Food Science, 68*(2), 408–420. Available from https://doi.org/10.1111/j.1365-2621.2003.tb14146.x.

Tabares, F. L., & Junkar, I. (2021). Cold plasma systems and their application in surface treatments for medicine. *Molecules (Basel, Switzerland), 26*(7), 1903. Available from https://doi.org/10.3390/molecules26071903.

Thirumdas, R., Sarangapani, C., & Annapure, U. S. (2015). Cold plasma: A novel non-thermal technology for food processing. *Food Biophysics, 10*(1), 1–11. Available from https://doi.org/10.1007/s11483-014-9382-z.

Ucar, Y., Ceylan, Z., Durmus, M., Tomar, O., & Cetinkaya, T. (2021). Application of cold plasma technology in the food industry and its combination with other emerging technologies. *Trends in Food Science & Technology, 114*, 355–371. Available from https://doi.org/10.1016/j.tifs.2021.01.047.

Ulbin-Figlewicz, N., Zimoch-Korzycka, A., & Jarmoluk, A. (2014). Antibacterial activity and physical properties of edible chitosan films exposed to low-pressure plasma. *Food and Bioprocess Technology, 7*(12), 3646–3654. Available from https://doi.org/10.1007/s11947-014-1379-6.

Vleugels, M., Shama, G., Deng, X. T., Greenacre, E., Brocklehurst, T., & Kong, M. G. (2005). Atmospheric plasma inactivation of biofilm-forming bacteria for food safety control. *IEEE Transactions on Plasma Science, 33*(2 II), 824–828. Available from https://doi.org/10.1109/TPS.2005.844524.

Wang, J., Zhuang, H., Hinton, A., & Zhang, J. (2016). Influence of in-package cold plasma treatment on microbiological shelf life and appearance of fresh chicken breast fillets. *Food Microbiology, 60*, 142–146. Available from https://doi.org/10.1016/j.fm.2016.07.007.

Warren, J. M., Mather, R. R., Neville, A., & Robson, D. (2010). Gas plasma treatment of polypropylene dental tape. In S. C. Anand, J. F. Kennedy, M. Miraftab, & S. Rajendran (Eds.), *Medical and healthcare textiles* (pp. 423–429). Woodhead Publishing. Available from https://doi.org/10.1533/9780857090348.423.

Wong, L.-W., Hou, C.-Y., Hsieh, C.-C., Chang, C.-K., Wu, Y.-S., & Hsieh, C.-W. (2020). Preparation of antimicrobial active packaging film by capacitively coupled plasma treatment. *LWT, 117*, 108612. Available from https://doi.org/10.1016/j.lwt.2019.108612.

Wu, F., Misra, M., & Mohanty, A. K. (2021). Challenges and new opportunities on barrier performance of biodegradable polymers for sustainable packaging. *Progress in Polymer Science, [online], 117*, 101395. Available from https://doi.org/10.1016/j.progpolymsci.2021.101395.

Xu, L., Garner, A. L., Tao, B., & Keener, K. M. (2017). Microbial inactivation and quality changes in orange juice treated by high voltage atmospheric cold plasma. *Food and Bioprocess Technology, 10*(10), 1778–1791. Available from https://doi.org/10.1007/s11947-017-1947-7.

Yadav, B., & Roopesh, M. S. (2020). In-package atmospheric cold plasma inactivation of Salmonella in freeze-dried pet foods: Effect of inoculum population, water activity, and storage. *Innovative Food Science and Emerging Technologies, 66*(June), 102543. Available from https://doi.org/10.1016/j.ifset.2020.102543.

Zhu, H., Han, Z., Cheng, J.-H., & Sun, D.-W. (2022). Modification of cellulose from sugarcane (Saccharum officinarum) bagasse pulp by cold plasma: Dissolution, structure and surface chemistry analysis. *Food Chemistry, 374*, 131675. Available from https://doi.org/10.1016/j.foodchem.2021.131675.

Zhuang, H., Rothrock, M. J., Jr., Hiett, K. L., Lawrence, K. C., Gamble, G. R., Bowker, B. C., & Keener, K. M. (2019). In-package air cold plasma treatment of chicken breast meat: treatment time effect. *Journal of Food Quality, 2019*, 1–7. Available from https://doi.org/10.1155/2019/1837351.

Zimmermann, L., Dombrowski, A., Völker, C., & Wagner, M. (2020). Are bioplastics and plant-based materials safer than conventional plastics? In vitro toxicity and chemical composition. *Environment International*, *145*, 106066.

Ziuzina, D., Misra, N. N., Han, L., Cullen, P. J., Moiseev, T., Mosnier, J. P., ... Bourke, P. (2020). Investigation of a large gap cold plasma reactor for continuous in-package decontamination of fresh strawberries and spinach. *Innovative Food Science and Emerging Technologies*, *59*, 102229. Available from https://doi.org/10.1016/j.ifset.2019.102229.

Ziuzina, D., Patil, S., Cullen, P. J., Keener, K. M., & Bourke, P. (2014). Atmospheric cold plasma inactivation of Escherichia coli, Salmonella enterica serovar Typhimurium and Listeria monocytogenes inoculated on fresh produce. *Food Microbiology*, *42*, 109–116. Available from https://doi.org/10.1016/j.fm.2014.02.007.

CHAPTER

13

Current development in ozone-based food preservation

Otniel Freitas-Silva[1,2], *Caroline Corrêa de Souza Coelho*[2], *Felipe Machado Trombete*[3] and *Andressa Moreira de Souza*[1]

[1]**Embrapa Agroindústria de Alimentos, The Brazilian Agricultural Research Corporation (EMBRAPA), Rio de Janeiro, Brazil** [2]**Food and Nutrition Graduate Program – The Federal University of State of Rio de Janeiro, Rio de Janeiro, Brazil** [3]**Department of Food Engineering, Federal University of São João del-Rei, Sete Lagoas, MG, Brazil**

13.1 Introduction

The search for safe and quality products portrays consumers as increasingly aware of the limitations of common sanitization techniques, while it acts directly, guaranteeing safer food products that have gone through processes that maintain their final quality (Nath, Kale, & Bhushan, 2019; Yu, Neal, & Sirsat, 2018). To satisfy these requirements, the food industry is currently studying non-thermal techniques such as ozone-based treatments, ultraviolet radiation, pulsed light, cold plasma, ultrasound, and novel packaging practices, such as smart and active packaging (Guo, Zhao, Ma, & Ji, 2021; Kalpana, Priyadarshini, Maria Leena, Moses, & Anandharamakrishnan, 2019). All these technologies have great potential in the field of minimally processed foods. However, there is still a lack of available information about the advantages and limitations of these technologies when applied to food processing. The efficiency of the processes depends directly on the combination of food/contaminant/process and the fresh characteristics that are important to maintain.

Several studies report the use of ozone as a bactericide and fungicide in the decontamination of surfaces, indoor environment, food preservation, and water treatment. Such applications due to the high oxidizing capacity of ozone leave no residue after the decontamination process since oxygen is the final product of its decomposition (Joseph et al., 2021; Li et al., 2019; Mohammadi, Mazloomi, Eskandari, Aminlari, & Niakousari, 2017). The direct effects of ozonation in reducing the levels of different mycotoxins have also been reported in the scientific literature and found significant results that qualify ozone as a powerful agent of degradation of fungal toxins (Freitas-Silva & Venâncio, 2010; Trombete, Freitas-Silva, Saldanha, Venâncio, & Fraga, 2016).

On the other hand, with an improvement in food quality, ozone can also cause undesirable effects. In some cases, it may promote oxidative degradation of chemical constituents, causing loss of color, change in aromaticity, and mischaracterization of the initial quality of the food (Pandiselvam et al., 2019). Thus, the field of ozone as an alternative technology for foods has much to be explored. In this way, there is a need for further studies to improve existing techniques or the development of new technologies that help to improve the quality of the food, preventing microbiological contamination and the formation of toxic products.

13.2 Ozone general aspects and properties

Ozone, which is the triatomic form of oxygen (O_3), is an extremely unstable gas that is partially soluble in water. Ozone gas has a pungent odor and is easily detectable at very low concentrations (0.01–0.05 ppm)

Food Packaging and Preservation.
DOI: https://doi.org/10.1016/B978-0-323-90044-7.00013-6

(Lapolli Rubens, 2003). When compared with other oxidizing agents, ozone stands out for its high oxidation potential (2.07 mV). Ozone is the second most powerful oxidizing agent, only behind fluorine (3.06 mV) (Freitas-Silva & Moreira, 2016).

The overall reaction for the generation of ozone from oxygen can be described as:

$$3O_2 \leftrightarrow 2O_3 \Delta H = +284.5 \text{ kJ/mol} \tag{13.1}$$

Reaction (1) is highly endothermic and is not spontaneous ($\Delta G = +161.3$ kJ/mol). This is because ozone cannot be generated by the thermal activation of oxygen since ozone decomposes rapidly when heated (Javier & Vidal, 2003). For the generation of ozone, one must first divide a molecule of diatomic oxygen. The free oxygen, resulting from the breakdown of the oxygen molecule, can react with other oxygen molecules to form ozone molecules. However, to break the oxygen molecule, a large amount of energy is required (Freitas-Silva & Moreira, 2016).

Ozone can be produced by photochemical methods, and electric discharge can also be produced by thermal, electrochemical, and radiochemical methods, such as UV radiation (Conte et al., 2020; Kim, Yousef, & Dave, 1999).

The commercial production of ozone is performed by the electrical discharge process, also called the corona process (Russel, Hugo, & Avliffe, 1999; USEPA, 1999). An ozone generator that uses the corona process consists of two electrodes subjected to a potential difference higher than approximately 1000 V. Ozone is generated by passing air or pure oxygen between the two electrodes. When the electrons have enough energy to dissociate the oxygen molecules, these molecules begin to collide, causing oxygen dissociation and the subsequent formation of ozone (USEPA, 1999). Because of the instability of ozone, it is necessary to generate it where it will be used since the gas decays rapidly when exposed to air (Freitas-Silva & Moreira, 2016). In the corona process, ozone production varies depending on the voltage, frequency of the electrical current, dielectric constant, and separation between the electrodes (Kim et al., 1999).

The corona process can convert molecular oxygen into ozone at concentrations up to 4% by weight when generated from the air and up to 14% from pure oxygen (Kunz et al., 1999). Pretreatment sets are usually necessary, to generate ozone from air, including filtration, compression, cooling, and dehumidification. In turn, the generation of oxygen is carried out by feeding the ozone generator from a liquid oxygen tank preceded by an evaporator. The advantages of using ozonation fed with oxygen are lower maintenance costs due to the simplicity of the equipment and higher yield. However, the disadvantage is the high cost of oxygen (Lapolli Rubens, 2003).

In another technique to produce ozone, the system produces ozone at concentrations that are three to four times higher (10%–18% in the production of ozone gas mixture) than the levels attainable by corona discharge. In this technique, ozone is produced by an electrochemical method in which water is converted into hydrogen and oxygen and then the oxygen atoms are converted into ozone through electrolysis. The hydrogen molecules are separated from the gas and water and the oxygen atoms combine to form ozone and diatomic oxygen (Kim et al., 1999).

Ozone was discovered in 1839 by Schonbein and used for the first time as a food preservative agent in 1902. However, at that time, its use in the food industry did not reach greater proportions. Initially, ozone was declared a recognized safe substance (GRAS—"Generally Recognized as Safe"), by the Food and Drug Administration (FDA) with permitted use for the sanitization of bottled water, and only a few years later its use was extended to food. Since then, interest in the application of ozone in food processing has emerged as an alternative technology to the traditional chlorine-based treatment due to its effectiveness at low concentrations, short contact time, and its decomposition into nontoxic products (Conte et al., 2020; Freitas-Silva & Venâncio, 2010; Guzel-Seydim, Greene, & Seydim, 2004; Joseph et al., 2021). Fig. 13.1 presents the ozone timeline referring to the main events till it reaches application in the food industry.

Ozone is widely used in the handling and processing of fruits, meats, and poultry as well as for washing and sterilization of plant foods and dairy products. Ozone is also used in silos, storage chambers, and tanks to protect and preserve grains, fruits, vegetables, poultry, meats, and cheeses. The direct application of ozone during storage keeps the environment clean and sterile and can ensure the hygiene and improvement in color, smell, and visual appearance of these products and this will generally allow an increase in storage of the food (Brodowska, Nowak, & Śmigielski, 2018; Conte et al., 2020; Luz & Rickes, 2022; Tiwari, O'Donnell, Patras, Brunton, & Cullen, 2009a; Wen et al., 2020).

At low concentrations, ozone is colorless at room temperature, whereas at high concentrations it acquires a bluish color. With increasing temperatures, in water the ozone has low solubility, making it less stable (Langlais, Reckhow, & Brink,1991; Lapolli Rubens, 2003; USEPA, 1999). Galdeano et al. (2018), Melicia Cintia Galdeano et al. (2018) demonstrated that when ozone gas in a concentration of 22.3 mg/L is bubbled in water at 8°C (pH 6.0) it is possible to obtain saturation corresponding to 7.81 mg/L. In water at 25°C, under the same ozonation conditions,

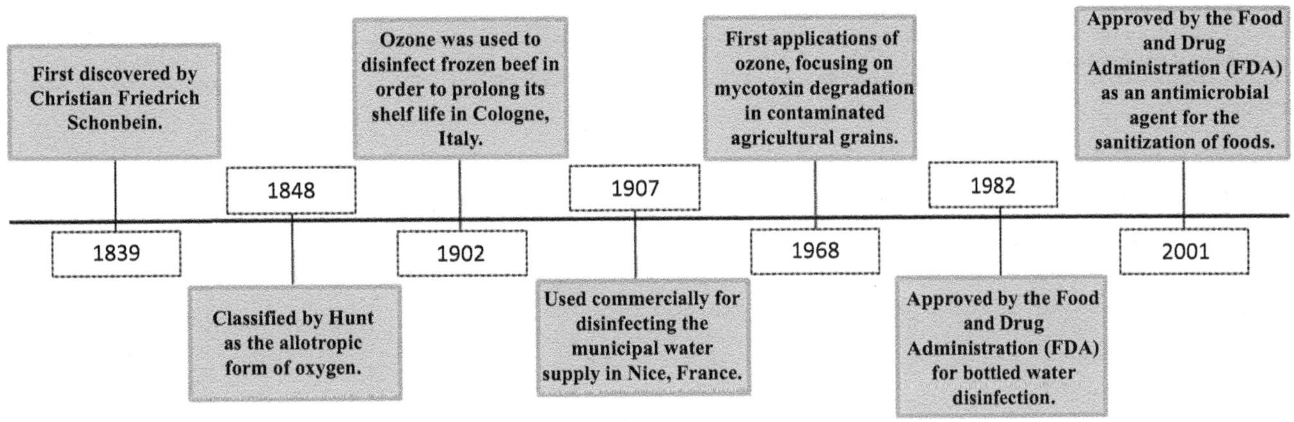

FIGURE 13.1 Timeline related to the O_3 discovery and its use in the food industry.

the saturation corresponded to 6.03 mg/L. However, the increase in temperature does not significantly change the rate of ozone disinfection, which is relatively independent of temperature (Langlais et al., 1991; Lapolli Rubens, 2003; USEPA, 1999).

According to Prado and Esplugas (1999), there are two forms of ozone degradation. In the first, ozone reacts directly with the solute as an electrophile and the reaction is quite selective. The second reaction occurs via OH radicals (indirect reaction) and is a nonselective and faster reaction. The OH radicals generated from ozone in an aqueous solution can lead to the opening of aromatic rings (Chu and Ma, 2000).

The ripening process of fruits and vegetables is retarded by ozone through a chemical reaction with ethylene, removing this compound naturally produced by the respiratory process. According to Sripong, Uthairatanakij, and Jitareerat (2022), the application of gaseous ozone at 900 mg/L for 3 minutes in fresh-cut durian (*Durio zibethinus* Murr.) and stored at 4°C for 14 days and later at 10°C and 25°C for 1 day was able to maintain pulp firmness, reduce the respiration rate and ethylene production of the fruit, increase the antioxidant capacity and the total phenolic content without negatively affecting the pulp color and sensory properties such as flavor and appearance of the product.

High pH favors the decomposition of ozone and the formation of different kinds of oxidizing compounds with different reactivity, such as hydroxyl radicals (high oxidation power) (Lapolli Rubens, 2003). Aside from temperature and pH, ultraviolet radiation or the presence of palladium-based catalysts, manganese, and nickel oxide, as well as metals, metal oxides, hydroxides, and peroxides, accelerate the decomposition of ozone (Langlais et al., 1991; Lapolli Rubens, 2003). Ozone is relatively unstable in an aqueous solution and has a half-life ranging from 20 to 30 minutes in distilled water at 20°C (Khadre, Yousef, & Kim, 2001). In the gas phase, ozone is more stable, with a half-life in atmospheric air, as measured by the Environmental Protection Agency of the United States, at approximately 12 hours (Russel et al., 1999). The main physicochemical properties of ozone are shown in Table 13.1.

13.3 Ozone uses for food safety and quality

13.3.1 *Use of ozone for microbial inactivation*

Ozone is a powerful oxidizing agent that can inactivate various bacteria, including gram-negative and gram-positive vegetative and sporulated forms, as well as components of the cell envelope, spores, fungal or viral capsids at relatively low concentrations, and short contact times (Khadre et al., 2001; Kim et al., 1999). The reduction or inactivation of microbes due to ozonation has been studied by some authors (Gorito et al., 2021; Ketteringham, Gausseres, James, & James, 2006; Oyom et al., 2022; Sripong et al., 2022; Wen et al., 2020). It also depends on ozone concentration, time of application, and the microorganism involved. This allows the extension of the potential use of ozone against microorganisms, as well as surface sterilization of tools and contaminated containers used in agriculture, greenhouses, and packing houses.

The pH has no direct effect on the disinfection of bacteria, viruses, and protozoa. Changes in the efficiency of the disinfection process are related to changes in the rate of decomposition of ozone as a function of pH. It is

TABLE 13.1 Main physicochemical properties of ozone.

Physicochemical properties	
Molar mass	48 Da
Density (0°C and 101.3 kPa)	2.514 kg/m^3
Boiling point	−111.9°C ± 0.3°C
Melting point	−192.5°C ± 0.3°C
Water solubility at 0°C	20 g/m^3
Water solubility at 30°C	1.5 g/m^3
Critical temperature	−12.1°C
Critical pressure	5.53 MPa

Adapted from Vidal (2003), Guzel-Seydim et al. (2004), Silva, Luvielmo, Geyer, and Prá (2011).

known that pH values higher than neutral allow the formation of hydroxyl radicals with high oxidation power, while at pH values less than neutral, the disinfection efficiency has been credited to molecular ozone (Langlais et al., 1991). Some studies involving the use of ozone in postharvest treatments of fruits and vegetables show that both aqueous and gas ozone is effective to control plant pathogens (Table 13.2). According to Martínez (2002), aqueous ozone (2.2 mg/L) reduced the germination of fungal spores commonly associated with postharvest decay of mangoes (*Colletotrichum gloeosporioides, Lasiodiplodia theobromae, Fusarium oxysporum*, and *Pestalotiopsis mangiferae*) and also extended the fruits' shelf life.

Arnal, Salvador, and Martínez-jávega (2005) studied the benefits of postharvest treatments of persimmons. They reported that the fruits treated with ozone had a commercial life of about 16−35 days, due to the reduction of postharvest pathogens. Freitas-Silva, Morales-Valle, and Venâncio (2013) demonstrated the effect of aqueous ozone in reducing the population of fungi (*Aspergillus* section Flavi) associated with the production of aflatoxins in Brazil nuts. An ozone concentration of 20 mg/L was effective to reduce populations of native and inoculated *A. flavus*. Similarly, Scussel et al. (2011) demonstrated the effect of ozone gas alone or in combination with a modified atmosphere for the reduction of *A. flavus*. Beirao-da-Costa, Moura-Guedes, Ferreira-Pinto, Empis, and Moldao-Martins (2014) reported that the microorganism count was reduced in post-harvested kiwi fruits with 0.2 ppm aqueous ozone.

Ong, Kazi, Forney, and Ali (2013) studied the effect of aqueous ozone on the control of anthracnose on papaya fruits. Different ozone concentrations (0 (control) 0.04, 1.6, and 4 ppm) were applied for three exposure periods (48, 96, and 144 hours). The authors reported that the radial growth of the mycelium and conidia germination of *C. gloeosporioides* was significantly inhibited in all treatments with ozone when compared to the control for up to 8 days of incubation at 25°C ± 3°C. In turn, Kechinski et al. (2012) studied the use of 4 ppm of ozone in aqueous thermotherapy applied to "Golden" papaya fruits. The combination of treatments did not alter the cuticular surface of the fruit, did not form crystals, and reduced rottenness-causing fungi. Finally, the reduction of *C. musae* on banana cv "Nanicão" was demonstrated by de Alencar, Faroni, Pinto, da Costa, and da Silva (2013), who showed that aqueous ozone was more effective than the gaseous form, not only to reduce *C. musae* but also to preserve the sensory appeal and extend the shelf life of the fruits. Coelho, Freitas-Silva, Alcantara, da. Silva, and Cabral (2015) demonstrated that treatments with aqueous ozone (0.5 and 1.0 mg/L for 5 minutes) were more effective than treatments with chlorine (150 mg/L) in removing filamentous fungi and yeasts in fresh strawberries.

Talbot, Martinelli, Talvy, Chauveheid, and Haut (2012) studied the ozone inactivation of resistant microorganisms and a method to assess the efficiency of a drinking water plant to inactivate resistant microorganisms. To evaluate the duration of the lag phase of the ozone inactivation of resistant pathogenic microorganisms, an experimental procedure was proposed and applied to *Bacillus subtilis* spores. The procedure characterized the ozone inactivation kinetics of *B. subtilis* spores for different temperature and ozone concentration conditions.

Trombete et al. (2017) reported reductions in total fungal count in wheat grains of approximately 3.0 cycles of log CFU/g, using 60 mg/L of O_3. Similar results were obtained by Porto et al. (2019) with ozonation of corn grits (at 60 mg/L of O^3). The total fungal count also was reduced around 3.0 cycles log CFU/g, and total mesophilic counts were reduced to non-detectable levels.

Freitas-Silva et al. (2013) investigated the reduction of mycotoxigenic *Aspergillus flavus* on Brazil nuts and other fungi associated with the shell using aqueous ozone. They applied O_3 on nuts with high and lower conidia

TABLE 13.2 Microbial degradation by ozone in different foods.

Food product	O_3 application form	Target microbial Population	O_3 effect	References
Carrots	An intermittent flush of ozone (0, 7.5, 15, 30, or 60 mL/L) into chambers for 8 h daily for 28 days at storage temperatures of 2, 8, and 16°C.	*Botrytis cinerea* *S. sclerotiorum*	Results suggested that an ozone supply of 15 mL L for 8 h a day at 2°C could provide some disease protection with a minimum of physical and physiological damage.	Liew Chiam and Prange (1994)
Blackberries	0.2 to 0.6 g/m^3 of gaseous O_3, for 12 days, at 2°C.	Fungal	Reduction of fungal decay	Barth, Zhou, Mercier, and Payne (1995)
Grapes (table)	Ozone application (16 ppm) in a cylinder and storage at 20°C and 85%–90% RH for 6 days.	Natural microflora of fungi, yeasts, and bacteria grape decay caused by *Rhizopus stolonifer*	All the microflora naturally present on the berry surface was greatly reduced by 20 min ozone exposure A significant decrease in decay was observed in berries that were treated with ozone either before or after being inoculated with *Rhizopus stolonifer* spores.	Sarig et al. (1996)
Strawberries	Gaseous O_3.	Reduction of fungal decay		Pérez, Sanz, Ríos, Olías, and Olías (1999)
Lettuce	Ozone flush into a chamber with and without prior application of vacuum.	Microbial counts	When the vacuum was applied before the ozone flush, a higher decrease in the total microbial counts was achieved.	Kim et al. (1999)
Lemons	Exposure to an intermittent day-night ozone cycle (0.3 ppm) at 4.5°C for 9 weeks and then continuous ozone exposure (1.0 ppm) at 10°C for 2 weeks.	*P. digitatum* *P. italicum*. Development of postharvest green and blue molds.	The incidence of blue mold was delayed, but not reduced. In contrast, the sporulation of blue mold was greatly reduced by a day-night ozone cycle. Green mold incidence was significantly lower among fruit exposed to 1.0 ppm ozone during storage.	Palou, Crisosto, Smilanick, Adaskaveg, and Zoffoloi (2001)
Oranges	Continuous exposure to 0.3 ppm ozone at 5°C for 4 weeks, and then ozone continuous exposure (1.0 ppm) at 10°C for 2 weeks.	*P. digitatum and P. italicum.* Development of postharvest green and blue molds.	Growth of *P. italicum*, but not of *P. digitatum*, was significantly reduced by 0.3 ppm ozone exposure.	Palou et al. (2001)
Peaches	Continuous ozone gaseous exposure at 0.3 ppm and storage for 4 weeks at 5°C and 90% RH.	*Monilinia fructicola; B. cinerea; M. piriformis; Penicillium expansum*	Reduction of mycelia growth and fungi sporulation	Palou, Crisosto, Smilanick, Adaskaveg, and Zoffoli (2002)
Table grapes	Continuous ozone gaseous exposure at 0.3 ppm and storage for 7 weeks at 5°C and 90% RH.	*M. fructicola; B. cinérea;* M. piriformis; *P. expansum*	Gray mold nesting was nearly prevented under 0.3 ppm ozone.	Palou et al. (2002)
Peppers (green)	Samples were treated with 2–8 ppm for 10–40 min, under 60%–90% RH and at 22°C	*E. coli* O157:H7	Ozone gas concentration, RH, and time were all significant factors for the inactivation of *E. coli*, ozone concentration being the most important.	Han, Floros, Linton, Nielsen, and Nelson (2002)
Lettuce	Exposure to 2.1, 5.2, and 7.6 ppm ozone for 5, 10, and 15 min, under 80% RH and 22°C, respectively.	*E. coli* O157:H7	Ozone treatments inactivated *E. coli* by 0.79–1.79 log CFU/g. The bactericidal effect increased with concentration and length of exposure to gaseous ozone.	Singh, Singh, Bhunia, and Stroshine (2002)
Carrots (baby)	Exposure to 2.1, 5.2, and 7.6 ppm ozone for 5, 10, and 15 min, under 80% RH and 22°C, respectively.	*E. coli* O157:H7	Ozone treatments inactivated *E. coli* by 1.11–2.64 log CFU/g The bactericidal effect increased with concentration and length of exposure to gaseous ozone.	Singh et al. (2002)

(*Continued*)

TABLE 13.2 (*Continued*)

Food product	O_3 application form	Target microbial Population	O_3 effect	References
Strawberries	Storage for 3 days at 2°C in an atmosphere with 1.5 μL/L ozone and transferred to room temperature	*B. cinereal*	Visible mycelial growth developed more slowly on strawberries previously stored in an ozone-enriched cold atmosphere.	Nadas, Olmo, and García (2003)
Orange juice	Gaseous O_3 (pumped into juice).	Yeast counts (*S. cerevisiae*)	Reduction of yeast counts (*S. cerevisiae*)	Angelino (2002)
Apple cider	Gaseous O_3 (pumped into juice).	Molds and yeast Counts	Reduction of molds and yeast Counts	Choi and Nielsen (2005)
Spinach	Exposure to 40 ppm ozone for 1 h.	Total bacteria count	0.28 log reductions were observed.	Qiang, Demirkol, Ercal, and Adams (2005)
Peppers (red)	Exposure to 40 ppm ozone for 1 h.	Total bacteria count	1.0 log reductions were observed.	Qiang et al. (2005)
Cucumbers	Exposure to 40 ppm ozone for 1 h.	Total bacteria count.	1.04 log reductions on bacterial counts were observed	Qiang et al. (2005)
Green beans	Exposure to 40 ppm ozone for 1 h.	Total bacteria count.	0.38 log reductions were observed.	Qiang et al. (2005)
Asparagus	Exposure to 40 ppm ozone for 1 h.	Total bacteria count.	0.75 log reductions were observed.	Qiang et al. (2005)
Apple juice	Ozone exposure (9 g/h) for (1) 90 min or (2) 60 min, followed by 24 h storage at 4°C	*E. coli* O157:H7; *Salmonella*.	Ozone treatment was more effective in reducing. Salmonella microbial load than *E. coli*	Williams, Sumner, and Golden (2005)
Longan fruits	Ozone exposure (200 μL/L) alone or combined with citric, ascorbic, or oxalic acid for 0, 15, 30, 60, and 120 min and then storage at 25°C.	Natural microflora of fungi, yeasts, and bacteria.	Exposing fruits to ozone for 60 and 120 min significantly reduced the microbial population on the fruit surface.	Whangchai, Saengnil, and Uthaibutra (2006)
Figs (dried)	Storage for 3 h and 5 h in a chamber with gaseous ozone at 5 and 10 ppm.	Total bacterial counts; Coliforms; Yeasts/molds	A minimum 3 h treatment at 5 ppm was required to reduce microbial counts. The decrease in total aerobic mesophilic microorganisms and yeasts/molds counts was 38% and 72% at 5 ppm. All coliforms were inactivated.	Öztekin, Zorlugenç, and Zorlugenç (2006)
Sliced tomatoes	Gaseous O_3		Reduction of microbial counts	Aguayo, Escalona, and Artés (2006)
Cherry tomatoes	Exposure to 5, 10, 20, and 30 ppm ozone for 5, 10, 15, and 20 min and then packaging in a modified atmosphere and storage at 7°C and 22°C	*S. enteritidis* PT4 E10	Complete reduction of *S. enteritidis* at 20 ppm after 15 min treatment. *S. enteritidis* could survive storage, depending on the location of the pathogen on the fruit, the load of contamination, and the storage temperature.	Daş, Gürakan, and Bayindirli (2006)
Carrots	Carrots treated with 1-methylcyclopropene were exposed to an atmosphere with 0, 300, and 1000 nL/L ozone at 10°C for 0, 1, 2, or 4 days, and stored at 0°C for up to 24 weeks.	*B. cinereal*; *S. sclerotiorum*	Ozone treatments (1000 nL/L) induced decay resistance to *B. cinerea*, but not to *S. sclerotiorum*.	Forney, Song, Hildebrand, Fan, and McRae (2007)
Plums	Exposure to 0.1 μmol/mol ozone and maintenance at 13°C under 95% RH for 13 days.	*B. cinerea*	Ozone treatment suppressed spore production by as much as 20% in comparison with "control" fruit. Ozone treatment resulted in no effects on lesion development.	Tzortzakis, Borland, Singleton, and Barnes (2007)

Raspberries	Four different gaseous treatments were applied: (1) continuous ozone exposure (50 g/kg O_3) and (2) pressurized ozone exposure (83 kPa; 50 g/kg O_3), for 2, 4, 8, 16, 32, and 64 min; (3) continuous ozone exposure followed by pressurized ozone (64 + 64 min); (4) vacuum followed by pressurized ozone for 64 min.	*E. coli* O157:H7; *Salmonella*	Maximum reductions were achieved with treatment (3). 3.55 and 3.75 log reductions were achieved for *Salmonella* and *E. coli*, respectively.	Bialka and Demirci (2007)
Strawberries	Four different gaseous treatments were applied: (1) continuous ozone exposure (50 g/kg O_3) and (2) pressurized ozone exposure (83 kPa; 50 g/kg O_3), for 2, 4, 8, 16, 32, and 64 min; (3) continuous ozone exposure followed by pressurized ozone (64 + 64 min); (4) vacuum followed by pressurized ozone for 64 min.	*E. coli* O157:H7; *S. enterica*	Maximum reductions were achieved with treatment (3). 2.60 and 2.96 log reductions were achieved for *Salmonella* and *E. coli*, respectively.	Bialka and Demirci (2007)
Blueberries	Four different gaseous treatments were applied: (1) continuous ozone exposure (50 g/kg O_3) and (2) pressurized ozone exposure (83 kPa; 50 g/kg O_3), for 2, 4, 8, 16, 32, and 64 min; (3) continuous ozone exposure followed by pressurized ozone (64 + 64 min); (4) vacuum followed by pressurized ozone for 64 min.	*E. coli* O157:H7; *Salmonella*	Maximum reduction of Salmonella (3.0 log) was achieved after 64 min of pressurized ozone. For *E. coli*, the maximum reduction (2.2 logs) was attained after 64 min of continuous ozone exposure. The combined ozone treatments (continuous followed by pressurized and vacuum followed by pressurized ozone) did not increase log reductions for either *Salmonella* or *E. coli*.	Bialka and Demirci (2007)
Clementines	Exposure to 0.1 μmol/mol ozone and maintenance at 13°C under 95% RH for 13 days.	*B. cinerea*	Ozone treatment suppressed spore production by as much as 95% in comparison with "control" fruit. Lesion development was also significantly suppressed by ozone.	Tzortzakis et al. (2007)
Tomatoes	Exposure to 0.05–5.0 μmol/mol ozone, and maintenance at 13°C under 95% RH for 13 days.	*B. cinerea*	Ozone treatment suppressed spore production by as much as 50% in comparison with "control" fruit. Lesion development was significantly suppressed by ozone.	Tzortzakis et al. (2007)
Grapes (table)	Exposure to 0.1 L mol/mol ozone and maintenance at 13°C under 95% RH for 13 days	*B. cinerea*	Ozone treatment suppressed spore production by as much as 75% in comparison with "control" fruit.	Tzortzakis et al. (2007)
Figs (dried)	Inoculated figs were dried for 1 h at 25°C and then exposed to gaseous ozone (0.1, 0.5, and 1.0 ppm up to 360 min for *E. coli* and *B. cereus*; 1.0, 5.0, 7.0, and 9.0 ppm for 360 min for *B. cereus* spores) in a chamber at 20°C and 70% RH.	*E. coli* ATCC 8739; *B. cereus* ATCC 6633; *B. cereus* spores	*E. coli* and *B. cereus* counts were decreased by 3.5 log numbers at 1.0 ppm ozone concentration for 360 min. Up to 2 log reductions in the number of *B. cereus* spores were observed above 1.0 ppm ozone concentration at the end of 360 min of ozonation.	Akbas and Ozdemir (2008)
Figs (dried)	Dried figs were exposed to 13.8 ppm of gaseous ozone during 7.5, 15, and 30 min for microbial studies and during 30, 60, and 180 min for aflatoxin level determination.	aerobic mesophilic bacteria, *E. coli*; Coliforms; Yeasts/molds	Microbial counts were decreased by 0.81–1.42, and 0.46–1.84. 0.16–2.09 and 0.59 log for aerobic mesophilic bacteria, coliforms, total yeasts, and total molds, respectively. *E. coli* was not detected after a 7.5 min ozone treatment.	Zorlugenç, Kiroğlu Zorlugenç, Öztekin, and Evliya (2008)
Cantaloupes	Exposure to 5000, 10,000, and 20,000 ppm ozone for 30 min under 90%–95% RH and 11°C, and then storage at 5°C for up to 7 days.	*S. poona*; *E. coli*; *Pseudomonas fluorescens*; Coliforms	Ozone treatment greatly reduced the initial counts of all microbial groups tested. It was effective against *Salmonella* not only initially but also during the	Selma, Ibáñez, Cantwell, and Suslow (2008)

(*Continued*)

TABLE 13.2 (*Continued*)

Food product	O_3 application form	Target microbial Population	O_3 effect	References
Tomatoes	Exposure to 0.05, 0.2, 1.0, or 5.0 ppm ozone for up to 312 h, under dark at 13°C, 95% RH, and then storage for up to 13 days.	*Alternaria alternata; Colletotrichum coccodes*	A low level of ozone resulted in a modest, but statistically significant, reduction in fungal lesion development. Higher concentrations of the gas resulted in greater effects.	Tzortzakis, Singleton, and Barnes (2008)
Carrots	Exposure to 450 ppb ozone for 48 h, and storage for 12 days at 20°C with 95%–97% RH.	*B. cinereal; Sclerotinia sclerotiorum*	Lesion size and height of the aerial mycelium were significantly reduced by the ozone treatment	Sharpe et al. (2009)
Spinach	Prepackaged samples were exposed to ozone (1.6 and 4.3 ppm) for 5 min and then stored at 5°C and 22°C for 0.5, 2, and 24 h in the air or oxygen atmosphere.	*E. coli* O157:H7	All treated samples showed reductions in *E. coli* populations with the largest reductions (3–5 log CFU/leaf) after 24 h of storage. Treated spinach showed discoloration. Only leaves stored or treated in an oxygen atmosphere showed relative humidity changes, indicating that ozone generation in oxygen (and not in the air) affects relative humidity levels inside the package.	Klockow and Keener (2009)
Spinach	Treatment in a pilot-scale system with combinations of vacuum cooling (4°C) and high gaseous ozone levels (935 ppm) for 30 min. A separate treatment used low ozone levels (5–10 ppm) for 3 days. The combination of the two treatments was also studied.	*E. coli* O157:H7	The combination of vacuum cooling with high gaseous ozone levels decreased *E. coli* load by 1.8 log CFU/g with no apparent damage to the quality of spinach. Refrigerated samples treated with low ozone levels for 3 days, in a system that simulated sanitization during transportation, decreased *E. coli* load by 1.4 log CFU/g. When the treatments were applied sequentially, *E. coli* was undetectable (i.e., >5 log reduction) after the 3 days of the process.	Vurma, Pandit, Sastry, and Yousef (2009)
Orange juice	Exposure to 75–78 ppm ozone for different periods (0–18 min).	*E. coli* ATCC 25922 and NCTC 12900	Generally, the ozone treatment of orange juice resulted in a population reduction of 5 log cycles.	Patil, Bourke, Frias, Tiwari, and Cullen (2009)
Apples	Exposure to 450 ppb ozone for 48 h and storage for 12 days at 20°C with 95%–97% relative humidity (RH).	*B. cinerea*	Lesion size was reduced on treated apples.	Sharpe et al. (2009)
Date fruits	Exposure to ozone (1, 3, and 5 ppm) for four different periods (15, 30, 45, and 60 min)	Total bacterial counts; Coliforms; Staphylococcus aureus; Yeasts/molds	A 1 h ozone treatment at 5 ppm could be successfully used for reducing coliforms and S. aureus of date fruits, but longer exposure times are required for the elimination of the total mesophilic bacteria as well as yeasts/molds	Habibi Najafi, Haddad Khodaparast (2009)
Grapes (green table)	Exposure to 450 ppb ozone for 48 h and storage for 12 days at 20°C with 95%–97% RH	*B. cinerea*	The decay incidence of treated grapes was reduced. Ozone treatment had no significant effect on chlorophyll fluorescence.	Sharpe et al. (2009)
Table grapes	2.5, 5, and 10 g/m^3 of gaseous O_3 for 1 h.	Reduction of *B. cinerea* Contamination	Reduction of residues of fenhexamid, cyprodinil, pyrimethanil, and pyraclostrobin. No changes in iprodione and boscalid residues.	Gabler, Smilanick, Mansour, and Karaca (2010)

kiwifruits	0.0003 g/m^3 of aqueous O_3 for 2, 8, 24, 72, and 144 h.	Reduction of stem-end rot disease, caused by *Botrytis cinerea.*	Negative correlation between disease incidence and antioxidant substances and antioxidant activity.	Minas, Karaoglanidis, Manganaris, and Vasilakakis (2010)
Apple juice	Exposure to 0.4 ppm ozone at 20°C during 0–10 min.	*E. coli* ATCC 25922 and NCTC 12900	A desired 5-log reduction of *E. coli* load was achieved within 5 min of ozone treatment.	Patil, Valdramidis, Cullen, Frias, and Bourke (2010)
Apple juice	Exposure to 33,40 ppm ozone for 8 min at 15°C, 18°C, and storage at 4, 8, 12, and 16°C for 30 days. A dynamic storage temperature study was also carried out.	*Saccharomyces cerevisiae* ATCC 9763	A lag phase was observed for all ozone-processed samples, probably due to the oxidizing action of the applied ozone treatment, which may exert additional stress before allowing growth. The authors reported the effectiveness of ozone in the extension of the shelf life of apple juice.	Patil, Valdramidis, Tiwari, Cullen, and Bourke (2011)
"Magnum" peaches	3×10^{-5} g/m^3 of gaseous O_3 for 1 h.	*M. fructicola*	Increase of browning on the fruit skin. O_3 reduced the growth of lesions caused by *M. fructicola.*	Sautter et al. (2011)
Cantaloupe	Ozone exposure (4.3 ppm) for 5 min and storage at room temperature for 24 h.	*Salmonella poona*	Ozone gas treatment resulted in 4.29 log CFU/g reduction.	Trinetta, Vaidya, Linton, and Morgan (2011)
Tables Grapes	Exposure to 200, 250, or 350 ppm ozone, at 95%, 75%, or 35% RH, respectively.	Conidia of *Penicillium digitatum, Penicillium italicum,* and *B. cinerea*	Conidia died more rapidly during ozone exposure at a higher humidity than at lower humidity *P. digitatum* and *P. italicum* were more resistant to ozone than *B. cinerea.*	Ozkan, Smilanick, and Karabulut (2011)
Red bell peppers, strawberries, and watercress	6.34×10^{-4} and 0.004 g/m^3 of aqueous O_3 for 1, 2, and 3 min.	*Listeria innocua*, total mesophiles, and total coliforms.	Combining blanching and ozone did not generate synergistic effects. O_3 reduction counts on *L. innocua*, total mesophiles, and total coliforms.	Alexandre, Santos-Pedro, Brandão, and Silva (2011)
Tomatoes	Ozone exposure (4.3 ppm) for 5 min and storage at room temperature for 24 h.	*S. poona*	Ozone gas treatments resulted in 4.06 log CFU/g reduction.	Trinetta et al. (2011)
Lettuce	Ozone exposure (4.3 ppm) for 5 min and storage at room temperature for 24 h.	*E. coli* O157:H7	Ozone gas treatments resulted in a 3.53 log CFU/g reduction.	Trinetta et al. (2011)
Tangerines	Ozone exposure (200 ppm) for 0, 2, 4, or 6 h and storage for 3 days at 25°C with 75%–80% RH.	*P. digitatum*	Ozone treatment for 4 and 6 h reduced fungi growth. The activities of antioxidant enzymes increased after ozone fumigation and remained significantly higher through storage.	Boonkorn et al. (2012)
Grape tomato	Exposure to 1.71, 3.43, and 6.85 mg/L gaseous ozone for 2 or 4 h. and stored at 10°C for 21 days.	*Salmonella enterica* serovar Typhimurium	Ozone (6.85 mg/L) treatment reduced Salmonella populations by 2 log CFU per fruit.	Wang, Fan, Sokorai, and Sites (2019)
Cherry tomato	Combination of ultrasound and ozonated water. Exposure to 0.85 ± 0.2 mg/L of aqueous O_3 by 5, 10, and 15 min; process. Ultrasound condition: 25°C ± 2°C, 20, 40, 20/40 kHz, 300 W, 10 s (pulsation on/off time) and 5–15 min (treatment time).	Mesophilic bacteria and molds & yeasts	Washing with aqueous ozone showed an increasing microbial reduction with time. However, the combination of ultrasound and ozonated water, resulted in 2.09–3.42 and 2.30–3.72 log CFU/g reduction in both Mesophilic bacteria and molds & yeasts, respectively.	Mustapha (2020)

(*Continued*)

TABLE 13.2 (*Continued*)

Food product	O_3 application form	Target microbial Population	O_3 effect	References
Strawberries, raspberries, and blueberries	The combined effect of modified atmosphere packaging (MAP), and a gaseous ozone pretreatment. Exposure to 13 mg/m^3 of gaseous ozone (about 6 ppm) for 16 h at 1°C (±0.5°C) and then stored for 15 d under MAP (10 kPa O_2 and 40 kPa CO_2) or in the air at 4°C.	Mesophilic bacteria and molds & yeasts	The gaseous ozone reduced the yeast and mold loads in the strawberries and raspberries by one order of magnitude, while the total molds on blueberries were reduced by 2.5 log CFU/g and the total yeasts count by 1.2 log CFU/g. Indeed, the exposure of the raspberries and blueberries to gaseous ozone reduced the total mesophilic bacteria by 1.7 and 3.8 log CFU/g, respectively, compared to those stored in the air. The reduction in the total bacterial mesophilic load on the strawberries was less than one order of magnitude	Pinto et al. (2020)
Garlic bulbs	Ozone gas was applied by continuous flow (2.14 μg/m^3) for 4 days, 20 h a day in garlic bulbs. After treatment, the samples were stored at 22°C.	*Fusarium proliferatum*	The exposure of garlic bulbs to 2.14 μg/m^3 of O_3 for 4 days, 20 h a day, significantly decreased the decay incidence after 2 weeks of storage at 22°C.	Santis, Garzoli, and Vettraino (2021)
Clementine fruit	Application of ozonated water (3 mg/L) in combination with 3% sodium bicarbonate (SBC), on Clementine fruits stored at 5°C ± 1°C with 90% RH for 30 days.	Green mold, blue mold, and minor rots	The integrated treatments reduced the total fungal pathogen count (more than 1 logarithmic unit). In addition, it significantly reduced the incidence of decomposition during all storage compared to the control.	Strano et al. (2021)
Onion	Application of aqueous ozone sanitization with a concentration of 1 to 5 mg/L, exposure time from 2 to 8 min, and pH 3 to 5, for whole peeled onions.	Natural microflora of fungi, yeasts, and bacteria.	The application of aqueous ozone at 4.51 mg/L for 8 min at a pH of 3 provided sanitization with reduced microbial load (3.74 logs CFU/g) and extended its safe storage compared to the control by a margin of 2 days under refrigerated conditions.	Aslam, Alam, Singh, and Kumar (2021)
Potato	Application of gaseous ozone (2 mg/L ozone for 0 s, 30 s, 1 min, 2 min) on potato tubers.	*Fusarium sulphureum*	The application of ozone gas at 2 mg/L was able to suppress the development of dry rot and significantly inhibited the mycelial growth of *F. sulphureum*.	Li et al. (2022)
Apple	Combination of 1.0 mg/L of 1-MCP for 24 h, refrigeration (0.6°C), controlled atmosphere (2% O_2, 1% CO_2, 0.6°C) and application of 51–87 μg/L of ozone gas during storage (36 weeks).	*L. innocua*, yeast/mold, and bacteria	Continuous application of ozone gas (51 μg/L) together with a controlled atmosphere was able to reduce 5.0 $\log^{10}$ CFU/apple of *L. innocua* after 36 weeks of storage. Just as it was efficient in mitigating apple resident bacteria and Y/M counts.	Sheng et al. (2022)

Adapted from Freitas-Silva, O., & Venâncio, A. (2010). Ozone applications to prevent and degrade mycotoxins: A review. Drug Metabolism Reviews, *42(4), 612–620.*

concentrations of *A. flavus* at different exposure times (2.5, 5, 10, 20, 30, or 60 minutes) and four ozone concentrations (0.1, 10, or 20 mg/L) to assess whether the time factor was affected by the ozone concentration (interaction between effects). Their results showed no differences in O_3 exposure time on *A. flavus* and the best O_3 concentration for mycobiota control was achieved with 20 mg/L. In another work on Brazil nuts, de Oliveira et al. (2020) evaluated the potential of gas ozone in mitigating contamination caused by *A. flavus* and the possible effects on the color and quality of crude oil extracted from nuts. The authors reported that the concentrations of 8.88 and 13.24 mg/L for 240 minutes were efficient in inactivating potentially aflatoxigenic microorganisms in the Brazil nuts. They also observed a greenish color in non-ozonated nuts, characteristic of the colonies of *A. flavus*.

These examples demonstrate the great potential of ozone for the reduction and control of postharvest pathogens, mycotoxigenic, and storage microorganisms. The use of ozone can have a greater extent, depending on the expected reduction of microorganisms, pesticides, and other contaminants in agriculture. Since it is GRAS as a food-processing aide, ozone has been approved for postharvest use on organic produce (Freitas-Silva & Venâncio, 2010).

13.3.2 *Use of ozone for mycotoxin degradation*

Early studies of ozone application on the degradation of mycotoxins in contaminated agriculture appeared in the late 1960s. Dwarakanath, Rayner, Mann, and Dollear (1968) reported that ozone gas (25 mg/min) reduced total Aflatoxin (AFT) in flour, cottonseed meal, and peanut meal by 91% (from 214 to 20 mg/kg) after 2 hours and 78% (82–18 mg/kg) after 1 hour. Aflatoxin B1 (AFB1) and Aflatoxin G1 (AFG1) were found to be more sensitive to ozone than Aflatoxin B2 (AFB2) and aflatoxin G2 (AFG2). At this same time, biological experiments with mice showed that ozone (0.025 g/min) was able to reduce levels of Aflatoxin (AF) and toxicity (Dollear et al., 1968), indicating that the AF residues were less toxic after degradation.

Although these studies showed promising preliminary results, the use of ozone for mycotoxin degradation was considered only after 20 years (Table 13.3). Later studies confirmed ozone application, against other groups of mycotoxins, such as Zearalenone (ZEN), Deoxynivalenol (DON), Deoxynivalenol 3-glucoside (D3G), Ochratoxin A (OTA), HT-2 toxin (HT-2), T-2 toxin (T-2) and citrinin, as reviewed by Conte et al. (2020).

Recent research has shown that some degradation by-products generated during the ozonation process may persist after the total elimination of the parent mycotoxin. This fact highlights the need to characterize these degradation products to identify persistent and possibly toxic compounds formed therein (Xu, Ji, et al., 2019; Xu, Wang, et al., 2019). Li, Guan, and Bian (2019) evaluated the degradation products of ozonolysis of DON by mass spectrometry (UHPLC Q-Orbitrap) and the toxicology was investigated using structure-activity relationships. Ozone (8 mg/L for 15 seconds) resulted in 95.7% degradation, and the toxicities of the ozonized products were significantly decreased due to de-epoxidation and the attack of ozone at the C9–10 double bond in DON. Xu, Ji, et al. (2019) and Xu, Wang, et al. (2019) evaluated the degradation products of ZEN (20 μg/mL) by ozonized water at 2.25 mg/L and identified 4 ozonolytic products using a mass spectrometry system (UPLC-ESI-QTOF-Ms). The toxicity of the ozonolytic products was evaluated using BALB/c mice and demonstrated less toxic effects than the original mycotoxin. Xu, Ji, et al. (2019) and Xu, Wang, et al. (2019) applied saturated aqueous ozone to DON solution for different amounts of time, and by-products were collected for cytotoxicity evaluation using Caco-2 cells and untargeted GC-TOFMS-based metabolomics. Authors demonstrated that ozone treatment could degrade DON structure; however, its degradation products and cellular toxicity existed under different treatment times of ozonation.

In Brazil nuts, de Oliveira et al. (2020) evaluated the application of O_3 gas to preserve the nuts and its efficiency in mitigating contamination caused by *Aspergillus flavus* of crude oil extracted from ozonated nuts. Regarding the effect of O_3 on *A. flavus*, the authors reported that the concentrations of 8.88 and 13.24 mg/L, for 240 minutes were efficient in inactivating potentially aflatoxigenic microorganisms in Brazil nuts.

13.3.3 *Use of ozone for storage insects' control*

Ozone concentrations and exposure times have been established for killing stored grain insects such as *Tribolium castaneum* (Herbst), *Ryzopertha dominica*, *Oryzaephilus surinamensis* (Linnaeus), *Sitophilus oryzae* (L), *S. zeamais* (Motsch.), *Ephestia elutella*, *E. kuehniella* (Zeller), *Tribolium confusum* (du Val), *Plodia interpunctella* (Hübner), *Lasioderma serricorne* (F.), *Cryptolestes ferrugineus* and *Psocids* (Amoah & Mahroof; 2018; Paul, Radhakrishnan, Anandakumar,

TABLE 13.3 Mycotoxins degraded by ozone in different foods.

Food	O_3 concentration/application	O_3 effect	References
Cottonseed meal and peanut meal	0.025 g/min of gaseous O_3	Reduction in AFT in cottonseed meal and peanut meal, by 91% (from 214 to 20 ppb) in 2 h and by 78% (from 82 to 18 ppb) in 1 h, respectively.	Dwarakanath et al. (1968)
Feed	1.1 g/m^3 of gaseous O_3, 5 min	Inactivation of AF.	Maeba, Takamoto, Kamimura, and Miura (1988)
Corn and rice powder	25 g/m^3 gaseous O_3, 15 s	Rapid degradation of AFB1 and AFG1, while AFB2 and AFG2 were more resistant to oxidation.	McKenzie et al. (1997)
Corn	18 g/m^3 gaseous O_3, for 92 h at a flow rate of 200 mg/min	Reduction of AF by 95%.	McKenzie et al. (1998)
Corn kernels	12–16 g/m^3 of gaseous O_3	Reduction of AF by 92%.	Prudente and King (2002)
Peanut kernels and flour	5 g/m^3 of gaseous O_3, at various temperatures (25°C, 50°C, 75°C) and exposure times (5, 10, 15 min)	Reduction of AFB2 and AFG2 by 51% in peanut kernels Reduction of AFB2 and AFG2 by 20% and 30% for AFB2 and AFG2 in peanut flour, respectively.	Proctor, Ahmedna, Kumar, and Goktepe (2004)
Aqueous solutions	0.25 to 25 g/m^3 of aqueous O_3	Reduction of toxicity after treatment for most trichothecenes except for fumonisins, where degradation products were still toxic.	Young, Zhu, and Zhou (2006)
Kernels and ground pistachios	5.0, 7.0 and 9.0 g/m^3 of gaseous O_3, for 140 and 420 min, at 20°C and 70% RH	Reduction of AFT and AFB1 in pistachio kernels by 24% and 23%, respectively. Reduction of AFT and AFB1 in ground pistachio kernels by 5%	Akbas and Ozdemir (2006)
Flaked red pepper	16, 33, and 66 g/m^3 of gaseous O_3 at 7.5, 15, 30, and 60 min.	Reduction of AFB1 by 93%, in 60 min.	Inan, Pala, and Doymaz (2007)
Dried figs	13.8 g/m^3 of gaseous O_3 and 1.7 g/m^3 aqueous O_3, for 7.5, 15, and 30 min.	Reduction of AFB1 in 15 min Gaseous O_3 was more effective than aqueous O_3 in AFB1 reduction.	Zorlugenç et al. (2008)
The aqueous solution and diluted apple juice	15 g/m^3 of gaseous O_3	Reduction of patulin by 92% and 100% in water and apple juice, respectively.	Cataldo (2008)
Brazil nuts	31 g/m^3 of gaseous O_3 for 5 h	Total degradation of AF. Inhibition of fungal viability. No changes in sensory characteristics.	Giordano et al. (2010)
Brazil nuts	10.0 g/m^3 of gaseous O_3 for 90 min	Total degradation of AF (from 15 μg/kg to not detected).	Scussel et al. (2011)
Peanuts	13 and 21 g/m^3 of gaseous O_3, at 18°C, for 0, 24, 48, 72, and 96 h.	Reduction of AFB1 and AFT.	de Alencar, Faroni, Martins, da. Costa, and Cecon (2011)
		Reduction of AF levels in kernels.	
Wheat	Various concentrations and exposure times of O_3 (g)	Reduction of AFB1 by 96.6% with an exposure time of 20 min.	El-Desouky, Sharoba, El-Desouky, El-Mansy, and Naguib (2012)
Wheat	60 g/m^3 of gaseous $O_3$for 300 min	Reductions of DON and total aflatoxins corresponded to 64.3% and 48.0%, respectively,	Trombete et al. (2017)
Corn grits	60 g/m^3 of gaseous $O_3$for 480 min	Reducion of total aflatoxins by 57%	Porto et al. (2019)
Brazil nuts/oil	2.42, 4.38, 8.88, and 13.24 mg/L of gaseous O_3 for 0, 60, 120, 180, and 240 min at 25°C.	At a concentration of 8.88 mg/L for 240 min, the gas was able to reduce the count of *Aspergillus flavus* to greater than 3.10 log cycles.	de Oliveira et al. (2020)
Hazelnuts	5, 10, and 20 ppm of gaseous O_3 for 10 and 20 min.	The highest degradation for AFB1(38.96%) and AF total (31.35%) were obtained with the use of 20 ppm. 20 min^{-1} ozonation. In the 5 μg/kg contaminated groups, the AFB1 reduction rate was 19.88% after 5 min and 32.23% after 20 min.	Atakan and Caner (2021)
Potato	Application of gaseous ozone (2 mg/L ozone for 0 s, 30 s, 1 min, 2 min) on potato tubers.	The treatment effectively suppressed the development of postharvest diseases, and the accumulation of diacetoxyscirpenol *in vivo* and *in vitro* and significantly inhibited the mycelial growth of *F. sulphureum*.	Li et al. (2022)

Adapted from Freitas-Silva, O., & Venâncio, A. (2010). Ozone applications to prevent and degrade mycotoxins: A review. Drug Metabolism Reviews, *42(4), 612–620.*

Shanmugasundaram, & Anandharamakrishnan, 2020; Tiwari et al., 2010). Reductions as high as 3-log in microorganisms associated with stored grain have been achieved with ozone treatments as well as significant reductions in mycotoxin levels (Tiwari et al., 2010). Furthermore, investigations on ozone-treated grain indicate that ozone does not impact negatively the intrinsic quality of the grain. When applied to grains under suitable conditions, the ozonation can strengthen dough, increase bread volume, enhance the shelf-life of grain products such as noodles, and modify starch and protein properties (Trombete et al., 2016; Zhu, 2018).

McDonough et al. (2011) investigated whether humidified ozone was more effective than dry ozone. The screw conveyor system had a 76L tank designed to add moisture to the ozone stream by simply bubbling the ozone through the tank. The amount of water vapor in the gas flowing into the manifold was not determined. For both humidified and dry ozone treatments, 100% mortality of *S. zeamais* and *T. castaneum* was achieved after three passes through the screw conveyor with an exposure time of about 5.4 minutes. Statistical analysis indicated no significant difference between humidified and dry ozone treatments. Significant differences were observed between the ozone treatments and their respective controls (no ozone). Total mortality for *T. castaneum* and *S. zeamais* with the non-ozone (air) treatment was less than 20% and 33%, respectively. These authors reported that the effect of humidified ozone on the viability of fungi on corn kernel surfaces was similar to the effect shown on insects. Three passes through the screw conveyor without ozone resulted in a 78% reduction of CFU for humidified air and 83% for dry air. When ozone was applied to the screw conveyor system, a 99% reduction in CFU was observed for humidified ozone and a 97% reduction for dry ozone.

13.3.4 Use of ozone for pesticide degradation

Ozone also can reduce the levels of pesticides in foods. Studies show the potential to reduce the level of fungicide residues, dithiocarbamate, and mancozeb in blackcurrants (Antos, Kurdziel, Sadło, & Balawejder, 2013). These authors applied two different ozone treatment procedures. The first one entailed washing the plant material with an aqueous solution of ozone. This ozone-enriched water solution caused a 59% reduction of mancozeb residues compared with the initial concentration. The latter method was based on the utilization of ozone in a gaseous phase combined with a drying process. In that procedure, samples of blackcurrant fruit were exposed to a 19 ppm ozone concentration, and then the blackcurrants were dried. The utilization of ozone in a gaseous phase promoted a 38% reduction of mancozeb residues in comparison with the initial concentration. The combination of both processes; ozonation and drying, produced a 58% reduction of mancozeb residues.

Chen, Lin, and Kuo (2013) reported that the primary function of the proposed ozone vegetable cleaner is to remove or oxidize the pesticide residues on vegetables. Pesticide residue is removed from the leaf surfaces into the water and oxidized or decomposed via pump recirculation and ozone oxidation by the generator when ozone is injected into the reaction tank. From an environmental protection perspective, pesticides are toxic compounds that can be decomposed by ozone, partially transformed into wastewater or dissolved into water (depending on pesticide solubility), or oxidized by ozone into another derivative.

Chen et al. (2013) conducted pesticide analyses in which the R^2 values of the calibration curves were higher than 0.99 (the standard curves were recalibrated every other week). The 209 Polychlorinated Biphenyl (PCB) solution with 1 mg/L was spiked as the internal standard with the detection limit of 0.1 μg/L. The recovery rates of 100% ± 10% were typically observed, including the errors of extraction and gas chromatography (GC) as analysis. At an ozone production rate of 250 mg/h, pesticide removal efficiencies improved over time. Taking 15 minutes as the time, a household would likely use the proposed ozone vegetable cleaner, average removal efficiencies for the three pesticides were poor at roughly 10%. In addition, the average time-weighted removal efficiencies for permethrin, chlorfluazuron, and chlorothalonil were 24.5%, 12.7%, and 14.2%, respectively. When the ozone production rate was 500 mg/h, removal efficiencies by ozone treatment for the three pesticide solutions increased as treatment time increased. The study leads the investigators to conclude that removal efficiency for these pesticide solutions improved as ozone treatment time increased. Further, average time-weighted removal efficiencies for permethrin, chlorfluazuron, and chlorothalonil were 33.0%, 15.0%, and 24.2%, respectively.

In another study, by Zhang, Liu, and Hong (2007), Chinese white cabbages sprayed with excessive amounts of permethrin were harvested in the first stage of the field experiment. Samples were picked on day 5 (safe harvest day is day 4) after spraying for a preliminary two-group comparison. Samples were soaked for 15 minutes, followed by operation of the ozone generator and water pump recirculation for 15 minutes under an ozone output rate of 160 mg/h. This study indicated that the ozone output rate was 160 mg/h and the pump water flow rate was 12 L/min to recirculate water in the tank, facilitating the removal of impurities on vegetables and distributing the ozone evenly in the water, thereby improving removal efficiency for pesticide residues.

Rodrigues et al. (2021) evaluated different washing procedures (ozonized water, water bubbling continuously with ozone, washing with distilled water, detergent, acetic acid, sodium bicarbonate, and sodium hypochlorite solutions) to remove fungicides (azoxystrobin, difenoconazole, and chlorothalonil) in peppers. According to the authors, water bubbled continuously with ozone at a concentration of 3 mg/L was the most efficient treatment with the removal of fungicide residues ranging from 67% to 87%, however, it negatively affected some physicochemical parameters such as mass losses, color, and vitamin C content. Despite lower rates of removal of fungicide residues (between 53% and 75%), the application of 1 mg/L also preserved the sensory quality of the pepper over 13 days of storage.

Li et al. (2022) studied the application of high concentrations of gaseous ozone (0.83 mg/L at 0.0 to 2 hours exposure) to remove pesticide residues from apple surfaces. Trichlorfon is an organophosphate insecticide of high efficiency and low toxicity, suitable for the control of aphids during fruit growing. However, according to the WHO, organophosphate pesticides account for 70% of poisoning incidents (World Health Organization, 2018). According to the authors, ozone was used to remove residual pesticides on the surface of the apple and the degradation rate of trichlorfon after 120 minutes of treatment reached 73.3%. In addition, ozone treatment did not affect the surface color of the apple.

13.3.5 *Effects of ozone application on food quality*

Some authors have reported results from the experimental application of ozone employing fumigation, spraying, or immersion in ozonated water on fruits and vegetables, evaluating the results of ozonation on some types of infection, with or without inoculation, and its effects on some of the physicochemical properties of treated products (Freitas-Silva & Venâncio, 2010; Miller, Silva, & Brandão, 2013). The application of ozone can cause effects other than the microbial degradation of the unwanted compound and removing pesticide residues.

Nutrient losses or impaired sensory qualities of food can occur (Table 13.4), but these can be insignificant when compared to the benefits. In addition, ozone gas reacts with atmospheric water, reducing the relative humidity. When the reduction of the relative humidity is not adjusted, the product may dry during storage, as observed by Raila et al. (2006).

Antos et al. (2018) demonstrated that the initial quality of the product and the treatments applied, as a fungicide, have a direct impact on the extension of the shelf life of products treated by ozonation. Fruits that were damaged during harvesting or transport and produced with a chemical treatment did not respond well to the modification of the ozone-enriched atmosphere; on the other hand, ozone had a positive impact on the undamaged fruits, extending the shelf life.

13.4 Aspects of toxicity and safety

13.4.1 *Effect of ozone on human health*

The presence of ozone can be detected by the human sense of smell due to its characteristic odor. The concentration of ozone, which kills bacteria has been variously reported to be 0.04 to 0.1 ppm (volume), whereas the toxicity in small animals is 3–12 ppm. Humans experience headaches, and most people can detect the odor at 0.02 to 0.04 ppm (Prabakaran, Selvi, Merinal, & Panneerselvam, 2012). According to Bocci (2005), for the safety of patients and personnel, not a trace of ozone should be present, and a suitable exchange of air can be insured by an aspirator supplied with an ozone destructor. Moreover, a monitor analyzer with warning lights and a loud alarm must be turned on all the time to immediately sound an alert in case of contamination. The odor perception threshold for ozone is about 0.01 ppmv (0.02 mcg/L), ten times lower than the maximum worksite

TABLE 13.4 Effect of ozone on the quality and nutritional parameters of foods.

Food product	O_3 application form	Quality and nutritional Attributes	References
Blackberries	0.2 to 0.6 g/m^3 of gaseous O_3, for 12 days, at 2°C.	No effect on anthocyanin content Reduction in peroxidase activity. Reduction of fungal decay	Barth et al. (1995)
Strawberries	Gaseous O_3.	Reduction of sucrose Increase of glucose, fructose, vitamin C, and aroma quality. Reduction of fungal decay	Pérez et al. (1999)
Peaches (cv Elegant Lady)	0.6 g/m^3 of gaseous O_3 for 4 weeks; storage conditions at 5°C and 90% relative humidity.	Increase of water loss after 5 weeks of storage. No changes in respiration and ethylene production rates. No phytotoxic injuries. Reduction of mycelia growth and fungi sporulation	Palou et al. (2002)
Orange juice	Gaseous O_3 (pumped into juice).	Reduction of ascorbic acid. Color changes.	Angelino (2002)
Apple cider	Gaseous O_3 (pumped into juice).	Increase of sediments. Color change.	Choi and Nielsen (2005)
Longan fruits	Ozone exposure (200 μL/L) alone or combined with citric, ascorbic, or oxalic acid for 0, 15, 30, 60, and 120 min and then storage at 25°C.	Fruit treated with ozone in combination with oxalic or citric acid had less browning and a reduction of PPO activity.	Whangchai et al. (2006)
Tomatoes	Gaseous O_3	Reduction of aroma. Increase of texture, ascorbic acid, glucose, and fructose. No changes in appearance, taste, or overall quality. Reduction of microbial counts	Artes et al. (2006)
Kernels and ground pistachios	5.0, 7.0, and 9.0 g/m^3 of gaseous O_3 for 2.3 and 7 h.	No changes in pH, color, moisture content, free fatty acid values, fatty acid compositions, sweetness, rancidity, flavor, appearance, and palatability	Akbas and Ozdemir (2006)
Carrots	Carrots treated with 1-methylcyclopropene were exposed to an atmosphere with 0, 300, and 1000 nL/L ozone at 10°C for 0, 1, 2, or 4 days, and stored at 0°C for up to 24 weeks.	Reduced carrot quality through the increase of respiration rates and the reduction of firmness and sucrose content. Ozone treatments also stimulated the production of volatile compounds.	Forney et al. (2007)
Flaked red Pepper	16, 33, and 66 g/m^3 of gaseous ozone for 7.5, 15, 30, and 60 min.	No changes in color.	Inan et al. (2007)
Figs (dried)	Inoculated figs were dried for 1 h at 25°C and then exposed to gaseous ozone (0.1, 0.5, and 1.0 ppm up to 360 min for *E. coli* and *B. cereus*; 1.0, 5.0, 7.0, and 9.0 ppm for 360 min for *B. cereus* spores) in a chamber at 20°C and 70% RH.	No significant changes in color, pH, moisture content, sweetness, rancidity, flavor, appearance, and overall palatability of ozonated dried figs were observed.	Akbas and Ozdemir (2008)
Cantaloupes	Exposure to 5000, 10,000, and 20,000 ppm ozone for 30 min under 90%–95% RH and 11°C, and then storage at 5°C for up to 7 days.	No impact of ozone treatment on visual quality, aroma, firmness, and soluble solids content after treatment and storage was observed. Translucency and lightness were slightly increased and decreased, respectively, after ozone treatments.	Selma et al. (2008)
Apple juice	15.38 g/m^3 of gaseous O_3.	No changes in total sugars.	Cataldo (2008)
Orange juice	Ozone exposure (0.6%–10.0%) at 20°C up to 10 min.	No significant changes in pH, Brix, titratable acidity, cloud value, and non-enzymatic browning were found. However, ozonation was found to have a significant effect on juice color and ascorbic acid content.	Tiwari, Muthukumarappan, O' Donnell, and Cullen (2008)
Tomato juice	Ozone exposure (1.6%–7.8%) at 20°C up to 10 min.	No significant changes in pH, Brix, titratable acidity, cloud value, and non-enzymatic browning were found with ozonation. Color values and ascorbic acid content changed significantly. For 10 min at the highest ozone concentration, an ascorbic acid reduction of 96% was observed.	Tiwari et al., 2009

(*Continued*)

TABLE 13.4 (*Continued*)

Food product	O_3 application form	Quality and nutritional Attributes	References
Blackberry juice	Ozone exposure (0%–7.8%) at 20°C up to 10 min.	Ozone concentration and treatment time were found to be critical factors influencing both anthocyanins and color degradation.	Tiwari et al. (2009)
Grape juice	Ozone exposure (0%–7.8% w/w) at 20°C up to 10 min.	No significant changes in pH, Brix, and titratable acidity of ozonated samples were observed. However, significant changes in the juice color and anthocyanin content were observed during ozone treatment.	Tiwari et al. (2009)
Strawberry juice	Ozone exposure (1.6%–7.8%) at 20°C up to 10 min.	Significant reductions in anthocyanins (98.2%) and ascorbic acid (85.8%) were observed at an ozone concentration of 7.8% w/w and a treatment time of 10 min.	Tiwari et al. (2009)
Apples	Exposure to 450 ppb ozone for 48 h and storage for 12 days at 20°C with 95%–97% relative humidity (RH).	Lesion size was reduced on treated apples. Ozone treatment had no significant effect on color or chlorophyll fluorescence.	Sharpe et al. (2009)
Bananas	Ozone exposure at a flow rate of 8 mL/s for 0, 10, 20, and 30 min.	Results showed that ozone treatment for up to 20 min would be useful in enhancing the nutritional properties. Total phenol and total flavonoid content Vitamin C. Antioxidant activity.	Alothman, Kaur, Fazilah, Bhat, and Karim (2010)
Tomatoes	0.01 g/m^3 of gaseous; O_3 for 10 min	No changes in color, sugar content, acidity, antioxidant capacity, or hemicellulose solubilization. Reduction in fruit damage, pectin, and weight loss. Increase of phenolic accumulation. Delay in softening.	Rodoni, Casadei, Concellón, Chaves Alicia, and Vicente (2010)
kiwifruits	0.0003 g/m^3 of aqueous O_3 for 2, 8, 24, 72, and 144 h.	Negative correlation between disease incidence and antioxidant substances and antioxidant activity.	Minas et al. (2010)
Apple juice	Exposure to 0.4 ppm ozone at 20°C during 0–10 min.	Apple juice color and total phenols were significantly affected by ozone concentration and treatment time.	Patil et al. (2010)
Date palm	Exposure to 2 mg/L of aqueous O_3, at 25°C ± 0.1°C for 15, 30, 45, and 60 min.	The application of aqueous ozone removed the acaricides from the date fruit surface and did not affect sugars, antioxidant capacity, and phenolic compounds	Osman (2015)
Apple	Ozone-enriched atmosphere. Exposure to 1 ppm every 12 h for 1 min between 4°C and 6°C for 84 days.	Results showed that ozonated apples had better firmness but showed a loss of mass similar to non-ozonized apples.	Antos et al. (2018)
Cherry tomato	Combination of ultrasound and ozonated water. Exposure to 0.85 ± 0.2 mg/L of aqueous O_3 by 5, 10, and 15 min; process. Ultrasound condition: 25°C ± 2°C, 20, 40, 20/40 kHz, 300 W, 10 s (pulsation on/off time) and 5–15 min (treatment time).	The application of ultrasound and aqueous ozone had no detrimental effect on the physicochemical properties (except the firmness), bioactive compounds, and antioxidants.	Mustapha (2020)
Strawberries, raspberries, and blueberries	The combined effect of modified atmosphere packaging (MAP), and a gaseous ozone pretreatment. Exposure to 13 mg/m^3 of gaseous ozone (about 6 ppm) for 16 h at 1°C (± 0.5°C) and then stored for 15 d under MAP (10 kPa O_2 and 40 kPa CO_2) or in the air at 4°C.	The main quality parameters of the three types of fruit were not affected by combined ozone and MAP storage. The combined effects led to better preservation of the total and individual anthocyanins in the blueberries during cold storage.	Pinto et al. (2020)
Garlic bulbs	Ozone gas was applied by continuous flow (2.14 $\mu g/m^3$) for 4 days, 20 h a day in garlic bulbs. After treatment, the samples were stored at 22°C.	The gaseous ozone treatment, slightly modifies the aromatic profile of the garlic, without, however, producing an evident deviation, and therefore, without causing the loss of its flavoring capacity.	Santis et al. (2021)

(*Continued*)

TABLE 13.4 (*Continued*)

Food product	O_3 application form	Quality and nutritional Attributes	References
Clementine fruit	Application of ozonated water (3 mg/L) in combination with 3% sodium bicarbonate (SBC), on Clementine fruits stored at 5°C ± 1°C with 90% RH for 30 days.	No perceptible changes were observed between treatments in terms of fruit weight loss, chemical composition, and physiological disorders. The sensory characteristics of the fruits were not compromised by the combination of treatments, since the off-flavor levels remained low during storage, and no perceptible differences were detected by the tasters.	Strano et al. (2021)
Onion	Application of aqueous ozone sanitization with a concentration of 1 to 5 mg/L, exposure time from 2 to 8 min, and pH 3 to 5, for whole peeled onions.	The application of aqueous ozone at 4.51 mg/L for 8 min at a pH of 3 retained better quality attributes in terms of pyruvate content (10.9% higher), color change (16.5% lower), and firmness (3.1% higher) compared to chlorinated samples.	Aslam et al. (2021)
Blueberries	Application of aqueous ozone (5.1 mg/L ozone for 10 min) on blueberries stored at 4°C for 15 d.	No significant changes in color were observed in ozone-treated samples, showing that neither changes in epicuticular waxes, nor their partial removal nor changes in the ultrastructure of the epidermis affected the quality of blueberries.	Li et al. (2022)

Adapted from Aslam et al. (2021), Freitas-Silva andVenâncio (2010), Miller et al. (2013), and Li et al. (2022).

concentration (WSC) of 0.1 ppmv (0.2 mcg/L) over a breathing period of 1 hour. WHO permits working for 8 hours when the ozone concentration is 0.06 ppmv (0.12 mcg/L), which is easily perceived as a fairly strong ozone smell. Needless to say, people should never trust their noses because olfactory receptors become quickly tolerant, and, in any case, the air in the clinic must be ozone-free. It is, unfortunately, confusing that ozone concentrations are reported as either ppmv or as mcg/mL in the USA or Europe, respectively. The conversion is as follows: 1 ppmv = 0.002 mcg/mL.

13.4.2 Toxicity and safety

Toxicity is a criterion of utmost importance to adopting ozone in food processing. The primary effects of ozone in humans are associated with the respiratory tract. Symptoms of toxicity include headache, dizziness, burning sensation in the eyes and throat, a feeling of pungent odor, and cough. At low concentrations, ozone is not an extremely toxic gas, but at high concentrations, it can be lethal (Russel et al., 1999). The accidental inhalation of ozone causes rapid breathing, and the degree will depend on the content and duration of inhalation. The affected person should rest immediately after the accident to reduce the choking effect and must be taken to the hospital as quickly as possible (Langlais et al., 1991).

An ozone concentration of approximately 0.1 ppm can irritate the nose, throat, and eyes. Researchers claim that ozone concentrations from 0.02 to 0.04 ppm can be detected by humans and that prolonged exposure to concentrations equal to or greater than 1.0 ppm can cause death. A 1-hour exposure to concentrations of 2, 4, 15, and 95 ppm can cause symptomatic, irritating, toxic, and lethal effects, respectively. The toxic effects of ozone by inhalation are manifested in the lungs. In addition, a variety of extrapulmonary damages can also result from the action of ozone and its products (Kim et al., 1999).

Ozone is a toxic gas with a distinctive odor and is a normal constituent of the earth's atmosphere. Because of its strong tendency to decompose and release oxygen, ozone is extremely reactive and is a powerful oxidizing agent which reacts explosively with oil and grease. Nevertheless, it can be used safely in food industries. Since it readily oxidizes organic matter, it is used with great efficiency and efficacy to control microorganisms and reduce levels of pesticides and mycotoxins (Freitas-Silva & Venâncio, 2010).

Ozone cannot be stored or transported in vessels because it decomposes spontaneously in the presence of oxidizable impurities, humidity, and solid surfaces. The rate of decomposition increases with temperature. Ozone is an irritant to mucous membranes of the eyes and respiratory tract, and high concentrations can cause pulmonary edema. There can be secondary sites of reaction to ozone characterized by a defect in the oxygen dissociation from oxyhemoglobin in the tissues. Even at an exposure level of 0.1 ppm ozone, premature aging can result in people if exposure is sufficiently prolonged (Bocci, 2005).

13.4.3 *Effect of ozone on safe working systems*

According to Health and Safety Executive (HSE) (1996), areas with an ozonation system need to be equipped with adequate ventilation and extraction facilities. In ozone plant rooms, it is recommended that ten changes of room air per hour be achieved to enable dangerous accumulations of gas to be dispersed within a few minutes. Ozone should be prevented from entering the workroom air by the use of exhaust appliances placed close to the source of emission. The ozone may then be passed through appropriate filters before discharge to a safe place in the open air. In the case of ozonators operating under negative pressure, the process acts as its exhaust ventilator. In the case of atmospheric pressure ozonators, effective local exhaust ventilation is necessary because the ozonators are not enclosed. To ensure that plants and processes are properly operated and controlled to minimize health risks, satisfactory safe systems of work need to be established and maintained by employing appropriate training and supervision. All people operating ozone plants should be given full training in all aspects of the operation of the ozonator and associated equipment and should be trained in emergency and first aid procedures. Emergency action plans should be prepared for all sites where ozone is generated deliberately in potentially hazardous quantities.

13.4.4 *Regulation*

Gaseous ozone treatment needs some adjustments in its application to comply with regulations since O_3 concentrations cannot exceed 0.075 μL/L during an 8-hour workday (USEPA, 2008). Since ozone becomes a toxic gas above certain concentrations, maximum exposure limits are set and people working in ozonation plants must undergo regular medical reviews. The reference limits for human exposure to ozone have been established by some U.S. regulators (Langlais et al., 1991).

The recommended exposure limit for ozone is 0.1 ppm (0.2 mg/m^3), calculated as an 8-hour time-weighted average concentration. There is also a short-term exposure limit for ozone of 0.3 ppm (0.6 mg/m^3) calculated as a 15-minute time-weighted average concentration (HSE, 1996). According to NIOSH (2004) and ACGIH (2005), the Threshold Limit Values (TLV) and Immediately Dangerous to Life and Health (IDLH) values for ozone are 0.05 and 5 ppm, respectively.

Due to the strongly oxidizing properties of ozone, it is a primary irritant, affecting especially the eyes and respiratory systems, and can be hazardous at even low concentrations (OSHA, 2004). The potential symptoms are eye and mucous membrane irritation, headache, fatigue, dizziness, drowsiness, anorexia, nausea, vomiting, cough, sore throat, shortness of breath, pulmonary edema, chronic respiratory disease, conjunctivitis, eye and skin burns, frostbite (on contact with liquid), rapid heart rate (tachycardia), and low blood pressure (hypotension). In the United States, the current permissible level for ozone exposure in the workplace is 0.2 g/m^3, as adopted by the Occupational Safety and Health Administration for a breathing period of 1 hour (OSHA, 2004). According to the World Health Organization, the concentration at which a susceptible individual may be continuously exposed to ozone under normal working conditions for 8 hours a day or 40 hours a week without adverse effects is 0.12 g/m^3 (Bocci, 2005). The short-term exposure limit is 0.6 g/m^3: Short-term means exposure for less than 15 minutes, not more than four times a day, with intervals of at least 1 hour between each short-term exposure (Pascual, Llorca, & Canut, 2007).

Diao, Hou, and Dong (2013) describe that in addition, ozone has high reactivity, penetrability, and spontaneous decomposition into a nontoxic product (oxygen) without forming any hazardous residues on treated products (Karaca & Velioglu, 2007). These features make ozone an important alternative for the food industry. WHO, Food and Agriculture Organization (FAO), and FDA regard ozone as a safe and effective chemical applied in the food industry (FAO, 2003; FDA, 2002; WHO, 2007) (Fig. 13.2).

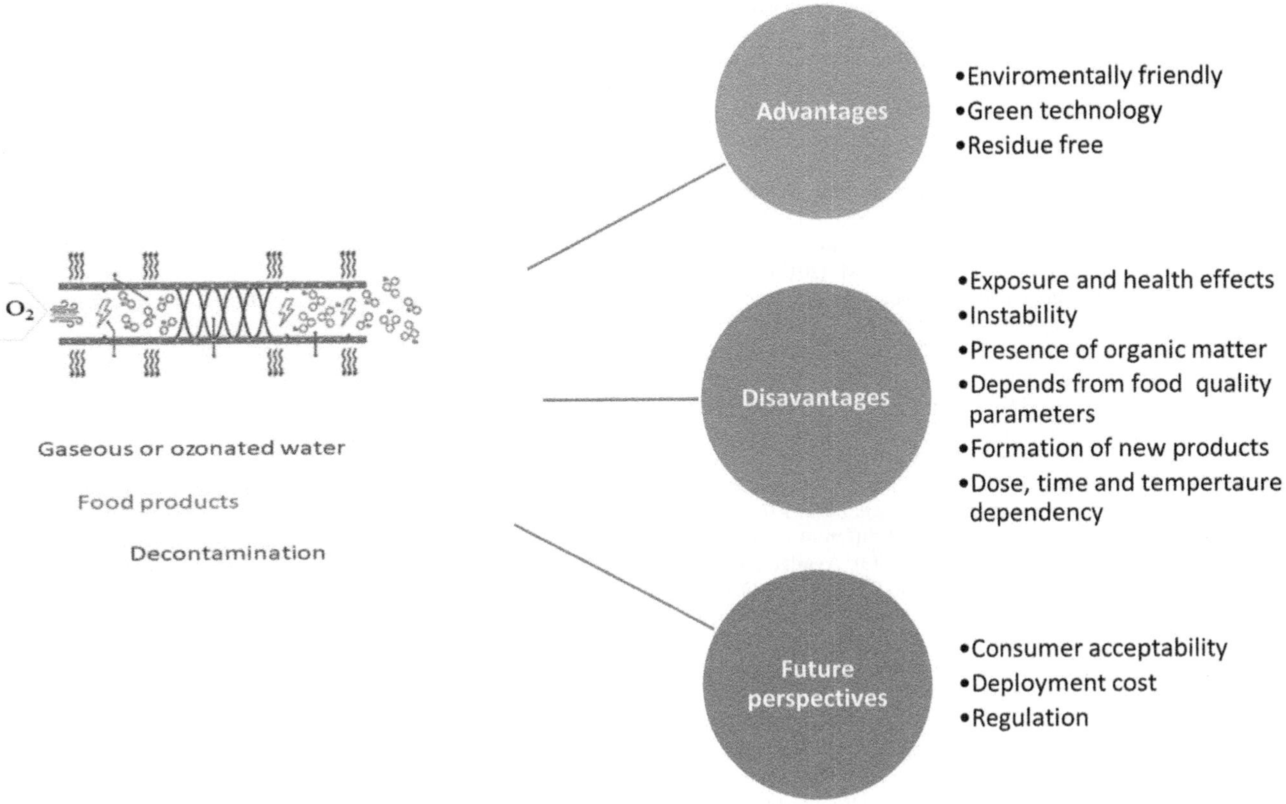

FIGURE 13.2 Ozone technology in food: advantages, disadvantages, and future perspectives.

13.5 Concluding remarks and trends

From the environmental point of view, one of the facts that arouse interest in the use of ozone as a sanitizing agent is that its self-decomposition does not generate toxic waste. Rather, its decomposition products (radicals OH^-, O_2^- e HO_2^-) act to increase its oxidizing effect. This latter feature, combined with its high oxidation potential, makes it effective against many species of pathogenic microorganisms and spoilage, including spores and protozoa, at low concentrations and contact time when compared to sanitizers used commercially. The reduction in the time of sanitization is attractive to processing and packing houses seeking to optimize the time on the production line.

Ozone also has the particularity of decomposing pesticides and other chemical residues in foods, as well as chlorine by-products and toxic organic compounds, by hastening their biodegradation and flocculation. It also can be used to precipitate minerals, remove color and odor in water, and reduce mycotoxin and pesticides. It is important for more research aiming to characterize the degradation products generated after the ozonation of pesticide residues and mycotoxins, as well as to verify their toxicities in vivo. Such studies have been published recently, but due to the diversity of pesticides and mycotoxins molecules, the results are still limited.

Therefore, ozone can be used not only as a sanitizing agent but as a tool to increase the shelf life of many agricultural products, since it minimizes the potential of postharvest infection, mainly caused by fungi.

References

Aguayo, E., Escalona, V. H., & Artés, F. (2006). Effect of cyclic exposure to ozone gas on physicochemical, sensorial and microbial quality of whole and sliced tomatoes. *Postharvest Biology and Technology, 39*(2), 169–177. Available from https://doi.org/10.1016/j.postharvbio.2005.11.005.

Akbas, M. Y., & Ozdemir, M. (2006). Effect of different ozone treatments on aflatoxin degradation and physicochemical properties of pistachios. *Journal of the Science of Food and Agriculture, 86*(13), 2099–2104. Available from https://doi.org/10.1002/jsfa.2579.

Akbas, M. Y., & Ozdemir, M. (2008). Application of gaseous ozone to control populations of Escherichia coli, Bacillus cereus and Bacillus cereus spores in dried figs. *Food Microbiology, 25*(2), 386–391. Available from https://doi.org/10.1016/j.fm.2007.09.007.

Alexandre, E. M. C., Santos-Pedro, D. M., Brandão, T. R. S., & Silva, C. L. M. (2011). Influence of aqueous ozone, blanching and combined treatments on microbial load of red bell peppers, strawberries and watercress. *Journal of Food Engineering, 105*(2), 277–282. Available from https://doi.org/10.1016/j.jfoodeng.2011.02.032.

Alothman, M., Kaur, B., Fazilah, A., Bhat, R., & Karim, A. A. (2010). Ozone-induced changes of antioxidant capacity of fresh-cut tropical fruits. *Innovative Food Science & Emerging Technologies, 11*(4), 666–671. Available from https://doi.org/10.1016/j.ifset.2010.08.008.

Amoah, B. A., & Mahroof, R. M. (2018). Susceptibility of the life stages of cigarette beetle, Lasioderma serricorne (F.) (Coleoptera: Anobiidae) to ozone. *Journal of Stored Products Research, 78*, 11–17. Available from https://doi.org/10.1016/j.jspr.2018.05.009.

Angelino, P.D. (2002). IFT annual meeting book of abstract "effect of ozone treatment on quality oforange juice."

Antos, P., Kurdziel, A., Sadło, S., & Balawejder, M. (2013). Preliminary study on the use of ozonation for the degradation of dithiocarbamate residues in the fruit drying process: Mancozeb residue in blackcurrant is the example used. *Journal of Plant Protection Research, 53*(1), 48–52. Available from https://doi.org/10.2478/jppr-2013-0007.

Antos, P., Piechowicz, B., Gorzelany, J., Matłok, N., Migut, D., Józefczyk, R., ... Balawejder, M. (2018). Effect of ozone on fruit quality and fungicide residue degradation in apples during cold storage. *Ozone: Science and Engineering, 40*(6), 482–486. Available from https://doi.org/10.1080/01919512.2018.1471389, http://www.tandf.co.uk/journals/titles/01919512.asp.

Arnal, L., Salvador, A., & Martínez-jávega, J. M. (2005). Efecto Del Ozono En La Mantenimiento de La Calidad de Caqui 'Rojo Brillante. *Revista Iberoamericana de Tecnología Postcosecha, 6*(2), 99–106.

Aslam, R., Alam, M. S., Singh, S., & Kumar, S. (2021). Aqueous ozone sanitization of whole peeled onion: Process optimization and evaluation of keeping quality during refrigerated storage. *LWT, 151*, 112183. Available from https://doi.org/10.1016/j.lwt.2021.112183.

Atakan, O., & Caner, C. (2021). Evaluation of different ozonation on aflatoxin degradation and physicochemical characteristics of hazelnuts. *Journal of Food Processing and Preservation, 45*(3). Available from https://doi.org/10.1111/jfpp.15276.

Barth, M. M., Zhou, C., Mercier, J., & Payne, F. A. (1995). Ozone storage effects on anthocyanin content and fungal growth in blackberries. *Journal of Food Science, 60*(6), 1286–1288. Available from https://doi.org/10.1111/j.1365-2621.1995.tb04575.x.

Beirao-da-Costa, S., Moura-Guedes, M. C., Ferreira-Pinto, M. M., Empis, J., & Moldao-Martins, M. (2014). Alternative sanitizing methods to ensure safety and quality of fresh-cut kiwifruit. *Journal of Food Processing and Preservation, 38*(1), 1–10. Available from https://doi.org/10.1111/j.1745-4549.2012.00730.x.

Bialka, K. L., & Demirci, A. (2007). Utilization of gaseous ozone for the decontamination of Escherichia coli O157:H7 and salmonella on raspberries and strawberries. *Journal of Food Protection, 70*(5), 1093–1098. Available from https://doi.org/10.4315/0362-028x-70.5.1093.

Boonkorn, P., Gemma, H., Sugaya, S., Setha, S., Uthaibutra, J., & Whangchai, K. (2012). Impact of high-dose, short periods of ozone exposure on green mold and antioxidant enzyme activity of tangerine fruit. *Postharvest Biology and Technology, 67*, 25–28. Available from https://doi.org/10.1016/j.postharvbio.2011.12.012.

Brodowska, A. J., Nowak, A., & Śmigielski, K. (2018). Ozone in the food industry: Principles of ozone treatment, mechanisms of action, and applications: An overview. *Critical Reviews in Food Science and Nutrition, 58*(13), 2176–2201. Available from https://doi.org/10.1080/10408398.2017.1308313.

Cataldo, F. (2008). Ozone decomposition of patulin – A micotoxin and food contaminant. *Ozone: Science and Engineering, 30*(3), 197–201. Available from https://doi.org/10.1080/01919510801925930.

Chen, J. Y., Lin, Y. J., & Kuo, W. C. (2013). Pesticide residue removal from vegetables by ozonation. *Journal of Food Engineering, 114*(3), 404–411. Available from https://doi.org/10.1016/j.jfoodeng.2012.08.033, http://www.sciencedirect.com/science/journal/02608774.

Choi, L. H., & Nielsen, S. S. (2005). The effects of thermal and nonthermal processing methods on apple cider quality and consumer acceptability. *Journal of Food Quality, 28*(1), 13–29. Available from https://doi.org/10.1111/j.1745-4557.2005.00002.x.

Chu, W., & Ma, C. W. (2000). Quantitative prediction of direct and indirect dye ozonation kinetics. *Water Research, 34*(12), 3153–3160. Available from https://doi.org/10.1016/S0043-1354(00)00043-9, http://www.elsevier.com/locate/watres.

Coelho, C. C. de S., Freitas-Silva, O., Alcantara, I., da. Silva, J. P. L., & Cabral, L. M. C. (2015). Ozônio em morangos minimamente processados, uma alternativa ao uso do cloro na segurança de alimentos. *Vigilância Sanitária em Debate, 0*(0). Available from https://doi.org/10.3395/2317-269x.00437.

Conte, G., Fontanelli, M., Galli, F., Cotrozzi, L., Pagni, L., & Pellegrini, E. (2020). Mycotoxins in feed and food and the role of ozone in their detoxification and degradation: An update. *Toxins., 12*(8), 486. Available from https://doi.org/10.3390/toxins12080486.

Daş, E., Gürakan, G. C., & Bayindirli, A. (2006). Effect of controlled atmosphere storage, modified atmosphere packaging and gaseous ozone treatment on the survival of Salmonella Enteritidis on cherry tomatoes. *Food Microbiology, 23*(5), 430–438. Available from https://doi.org/10.1016/j.fm.2005.08.002.

de Alencar, E. R., Faroni, L. R. D. A., Pinto, Md. S., da Costa, A. R., & da Silva, T. A. (2013). Postharvest quality of ozonized \nanicão\ cv. Bananas. *Revista Ciencia Agronomica, 44*(1), 107–114. Available from https://doi.org/10.1590/s1806-66902013000100014, http://www.ccarevista.ufc.br/seer/index.php/ccarevista/article/download/1076/779.

de Alencar, E. R., Faroni, L. R. D., Martins, M. A., da. Costa, A. R., & Cecon, P. R. (2011). Decomposition kinetics of gaseous ozone in peanuts. *Engenharia Agrícola., 31*(5), 930–939. Available from https://doi.org/10.1590/s0100-69162011000500011.

de Oliveira, J. M., de Alencar, E. R., Blum, L. E. B., de Souza Ferreira, W. F., Botelho, S. D. C. C., Racanicci, A. M. C., . . . da Silva, C. R. (2020). Ozonation of Brazil nuts: Decomposition kinetics, control of Aspergillus flavus and the effect on color and on raw oil quality. *LWT, 123*109106. Available from https://doi.org/10.1016/j.lwt.2020.109106, https://www.journals.elsevier.com/lwt.

Diao, E., Hou, H., & Dong, H. (2013). Ozonolysis mechanism and influencing factors of aflatoxin B1: A review. *Trends in Food Science and Technology, 33*(1), 21–26. Available from https://doi.org/10.1016/j.tifs.2013.06.002.

Dollear, F. G., Mann, G. E., Codifer, L. P., Gardner, H. K., Koltun, S. P., & Vix, H. L. E. (1968). Elimination of aflatoxins from peanut meal. *Journal of the American Oil Chemists' Society, 45*(12), 862–865. Available from https://doi.org/10.1007/BF02540171.

Dwarakanath, C. T., Rayner, E. T., Mann, G. E., & Dollear, F. G. (1968). Reduction of aflatoxin levels in cottonseed and peanut meals by ozonization. *Journal of the American Oil Chemists' Society, 45*(2), 93–95. Available from https://doi.org/10.1007/BF02890715.

El-Desouky, T. A., Sharoba, A. M. A., El-Desouky, A. I., El-Mansy, H. A., & Naguib, K. (2012). Effect of ozone gas on degradation of aflatoxin B1 and Aspergillus flavus fungal. *Journal of Environmental and Analytical Toxicology, 2*(2).

Forney, C. F., Song, J., Hildebrand, P. D., Fan, L., & McRae, K. B. (2007). Interactive effects of ozone and 1-methylcyclopropene on decay resistance and quality of stored carrots. *Postharvest Biology and Technology, 45*(3), 341–348. Available from https://doi.org/10.1016/j.postharvbio.2007.03.006.

Freitas-Silva, O., Morales-Valle, H., & Venâncio, A. (2013). Potential of aqueous ozone to control aflatoxigenic fungi in Brazil nuts. *ISRN Biotechnology., 2013*, 1–6. Available from https://doi.org/10.5402/2013/859830.

Freitas-Silva, O., & Moreira, Ad. S. (2016). *Ozone in food processing: Impact on food products attributes* (pp. 519–552). Brazil: CRC Press. Available from https://www.crcpress.com/Food-Processing-Technologies-Impact-on-Product-Attributes/Jaiswal/p/book/9781482257540, https://doi.org/10.1201/9781315372365.

Freitas-Silva, O., & Venâncio, A. (2010). Ozone applications to prevent and degrade mycotoxins: A review. *Drug Metabolism Reviews, 42*(4), 612–620. Available from https://doi.org/10.3109/03602532.2010.484461.

Gabler, F. M., Smilanick, J. L., Mansour, M. F., & Karaca, H. (2010). Influence of fumigation with high concentrations of ozone gas on postharvest gray mold and fungicide residues on table grapes. *Postharvest Biology and Technology, 55*(2), 85–90. Available from https://doi.org/10.1016/j.postharvbio.2009.09.004.

Galdeano, M. C., Wilhelm, A. E., Goulart, I. B., Tonon, R. V., Freitas-Silva, O., Germani, R., ... Chávez, D. W. H. (2018). Effect of water temperature and pH on the concentration and time of ozone saturation. *Brazilian Journal of Food Technology, 21*(0). Available from https://doi.org/10.1590/1981-6723.15617.

Giordano, B. N. E., Simao, V., Manfio, D., Galvao, S., Scussel, J. N., & Scussel, V. M. (2010). Reduction of in-shell Brazil nut (*Bertholletia excelsa* H.B.K.) aflatoxin contamination by ozone gas application during storage. *Julius-Kühn-Archiv, 425*, 566–572. Available from https://core.ac.uk/download/pdf/235686264.pdf.

Gorito, A. M., Pesqueira, J. F. J. R., Moreira, N. F. F., Ribeiro, A. R., Pereira, M. F. R., Nunes, O. C., ... Silva, A. M. T. (2021). Ozone-based water treatment (O3, O3/UV, O3/H2O2) for removal of organic micropollutants, bacteria inactivation and regrowth prevention. *Journal of Environmental Chemical Engineering, 9*(4), 105315. Available from https://doi.org/10.1016/j.jece.2021.105315.

Guo, Y., Zhao, L., Ma, Q., & Ji, C. (2021). Novel strategies for degradation of aflatoxins in food and feed: A review. *Food Research International, 140*, 109878. Available from https://doi.org/10.1016/j.foodres.2020.109878.

Guzel-Seydim, Z. B., Greene, A. K., & Seydim, A. C. (2004). Use of ozone in the food industry. *LWT – Food Science and Technology, 37*(4), 453–460. Available from https://doi.org/10.1016/j.lwt.2003.10.014.

Habibi Najafi, M. B., & Haddad Khodaparast, M. H. (2009). Efficacy of ozone to reduce microbial populations in date fruits. *Food Control, 20*(1), 27–30. Available from https://doi.org/10.1016/j.foodcont.2008.01.010.

Han, Y., Floros, J. D., Linton, R. H., Nielsen, S. S., & Nelson, P. E. (2002). Response surface modeling for the inactivation of Escherichia coli O157:H7 on green peppers (Capsicum annuum) by ozone gas treatment. *Journal of Food Science, 67*(3), 1188–1193. Available from https://doi.org/10.1111/j.1365-2621.2002.tb09475.x, http://www3.interscience.wiley.com/journal/118509799/issueyear?year = 2008.

Inan, F., Pala, M., & Doymaz, I. (2007). Use of ozone in detoxification of aflatoxin B1 in red pepper. *Journal of Stored Products Research, 43*(4), 425–429. Available from https://doi.org/10.1016/j.jspr.2006.11.004.

Javier, F., & Vidal, R. (2003). Procesos de Potabilización Del Agua E Influencia Del Tratamiento de Ozonización. ISBN 847978587X. 1ª edition.

Joseph, C. G., Farm, Y. Y., Taufiq-Yap, Y. H., Pang, C. K., Nga, J. L. H., & Li Puma, G. (2021). Ozonation treatment processes for the remediation of detergent wastewater: A comprehensive review. *Journal of Environmental Chemical Engineering, 9*(5), 106099. Available from https://doi.org/10.1016/j.jece.2021.106099.

Kalpana, S., Priyadarshini, S. R., Maria Leena, M., Moses, J. A., & Anandharamakrishnan, C. (2019). Intelligent packaging: Trends and applications in food systems. *Trends in Food Science and Technology, 93*, 145–157. Available from https://doi.org/10.1016/j.tifs.2019.09.008, http://www.elsevier.com/wps/find/journaldescription.cws_home/601278/description#description.

Kechinski, C. P., Montero, C. R. S., Guimarães, P. V. R., Noreña, C. P. Z., Marczak, L. D. F., Tessaro, I. C., & Bender, R. J. (2012). Effects of ozonized water and heat treatment on the papaya fruit epidermis. *Food and Bioproducts Processing, 90*(2), 118–122. Available from https://doi.org/10.1016/j.fbp.2011.01.005.

Ketteringham, L., Gausseres, R., James, S. J., & James, C. (2006). Application of aqueous ozone for treating pre-cut green peppers (Capsicum annuum L.). *Journal of Food Engineering, 76*(1), 104–111. Available from https://doi.org/10.1016/j.jfoodeng.2005.05.019.

Khadre, M. A., Yousef, A. E., & Kim, J. G. (2001). Microbiological aspects of ozone applications in food: A review. *Journal of Food Science, 66*(9), 1242–1252. Available from https://doi.org/10.1111/j.1365-2621.2001.tb15196.x, http://www3.interscience.wiley.com/journal/118509799/issueyear?year = 2008.

Kim, J.-G., Yousef, A. E., & Dave, S. (1999). Application of ozone for enhancing the microbiological safety and quality of foods: A review. *Journal of Food Protection, 62*(9), 1071–1087. Available from https://doi.org/10.4315/0362-028x-62.9.1071.

Klockow, P. A., & Keener, K. M. (2009). Safety and quality assessment of packaged spinach treated with a novel ozone-generation system. *LWT – Food Science and Technology, 42*(6), 1047–1053. Available from https://doi.org/10.1016/j.lwt.2009.02.011.

Kunz, A., Freire, R. S., Rohwedder, J. J. R., Duran, N., Mansilla, H., & Rodriguez, J. (1999). Construção e otimização de um sistema para produção e aplicação de ozônio em escala de laboratório. *Quimica Nova, 22*(3), 425–428.

Langlais, B., Reckhow, D.A., & Brink, D.R. (1991). Process technologies for water treatment 145-56 ozone in water treatment processes: Application and engineering.

Lapolli Rubens, F. (2003). Desinfecção de Efluentes Sanitários Por Meio Da OzonizaçãoDesinfecção de Enfluentes Sanitários. ABES.

Liew Chiam, L., & Prange, R. K. (1994). Effect of ozone and storage temperature on postharvest diseases and physiology of carrots (Daucus carota L.). *Journal of the American Society for Horticultural Science, 119*(3), 563–567. Available from https://doi.org/10.21273/jashs.119.3.563, http://journal.ashspublications.org/.

Li, C., Chi, K., Yu, H., Guo, Y., Ya, W., & Qian, H. (2022). Degradation, migration, and removal of trichlorfon on harvested apples during storage at room temperature. *Food Chemistry, 381*, 132243. Available from https://doi.org/10.1016/j.foodchem.2022.132243.

Li, M., Guan, E., & Bian, K. (2019). Structure elucidation and toxicity analysis of the degradation products of deoxynivalenol by gaseous ozone. *Toxins, 11*(8), 474. Available from https://doi.org/10.3390/toxins11080474.

Li, H., Xiong, Z., Gui, D., Pan, Y., Xu, M., Guo, Y., ... Li, X. (2019). Effect of ozonation and UV irradiation on aflatoxin degradation of peanuts. *Journal of Food Processing and Preservation, 43*(4), e13914. Available from https://doi.org/10.1111/jfpp.13914.

Li, L., Xue, H., Bi, Y., Zhang, R., Kouasseu, C. J., Liu, Q., ... Prusky, D. (2022). Ozone treatment inhibits dry rot development and diacetoxyscirpenol accumulation in inoculated potato tuber by influencing growth of Fusarium sulphureum and ergosterol biosynthesis. *Postharvest Biology and Technology, 185*, 111796. Available from https://doi.org/10.1016/j.postharvbio.2021.111796.

Luz, S.R. da, & Rickes (2022). Reduced of mycotoxin levels in parboiled rice by using ozone and its effects on technological and chemical properties.

Maeba, H. I., Takamoto, Y. U., Kamimura, M. I., & Miura, T. O. (1988). Destruction and detoxification of aflatoxins with ozone. *Journal of Food Science, 53*(2), 667–668. Available from https://doi.org/10.1111/j.1365-2621.1988.tb07786.x.

Martínez, B. (2002). Effects of ozone, iodine and chlorine on spore germination of fungi isolated from mango fruits. *Revista Mexicana de Fitopatología, 20*(1), 60–65.

McDonough, M. X., Campabadal, C. A., Mason, L. J., Maier, D. E., Denvir, A., & Woloshuk, C. (2011). Ozone application in a modified screw conveyor to treat grain for insect pests, fungal contaminants, and mycotoxins. *Journal of Stored Products Research, 47*(3), 249–254. Available from https://doi.org/10.1016/j.jspr.2011.04.001.

McKenzie, K. S., Kubena, L. F., Denvir, A. J., Rogers, T. D., Hitchens, G. D., Bailey, R. H., ... Phillips, T. D. (1998). Aftatoxicosis in Turkey poults is prevented by treatment of naturally contaminated corn with ozone generated by electrolysis. *Poultry Science, 77*(8), 1094–1102. Available from https://doi.org/10.1093/ps/77.8.1094, https://www.journals.elsevier.com/poultry-science/.

McKenzie, K. S., Sarr, A. B., Mayura, K., Bailey, R. H., Miller, D. R., Rogers, T. D., ... Phillips, T. D. (1997). Oxidative degradation and detoxification of mycotoxins using a novel source of ozone. *Food and Chemical Toxicology, 35*(8), 807–820. Available from https://doi.org/10.1016/S0278-6915(97)00052-5.

Miller, F. A., Silva, C. L. M., & Brandão, T. R. S. (2013). A review on ozone-based treatments for fruit and vegetables preservation. *Food Engineering Reviews, 5*(2), 77–106. Available from https://doi.org/10.1007/s12393-013-9064-5.

Minas, I. S., Karaoglanidis, G. S., Manganaris, G. A., & Vasilakakis, M. (2010). Effect of ozone application during cold storage of kiwifruit on the development of stem-end rot caused by Botrytis cinerea. *Postharvest Biology and Technology, 58*(3), 203–210. Available from https://doi.org/10.1016/j.postharvbio.2010.07.002.

Mohammadi, H., Mazloomi, S. M., Eskandari, M. H., Aminlari, M., & Niakousari, M. (2017). The effect of ozone on aflatoxin M 1, oxidative stability, carotenoid content and the microbial count of milk. *Ozone: Science & Engineering, 39*(6), 447–453. Available from https://doi.org/10.1080/01919512.2017.1329647.

Mustapha, A. T. (2020). Sonozonation: Enhancing the antimicrobial efficiency of aqueous ozone washing techniques on cherry tomato. *Ultrasonics Sonochemistry, 64*.

Nadas, A., Olmo, M., & García, J. M. (2003). Growth of Botrytis cinerea and strawberry quality in ozone-enriched atmospheres. *Journal of Food Science, 68*(5), 1798–1802. Available from https://doi.org/10.1111/j.1365-2621.2003.tb12332.x, http://www3.interscience.wiley.com/journal/118509799/issueyear?year = 2008.

Nath, P., Kale, S. J., & Bhushan, B. (2019). *A CRC title, part of the Taylor & Francis imprint, a member of the Taylor & Francis Group, the academic division of T&F Informa, plcNon-thermal processing of foods*. Boca Raton: CRC Press, Taylor & Francis.

Ong, M. K., Kazi, F. K., Forney, C. F., & Ali, A. (2013). Effect of gaseous ozone on papaya anthracnose. *Food and Bioprocess Technology, 6*(11), 2996–3005. Available from https://doi.org/10.1007/s11947-012-1013-4.

Osman, K. A. (2015). Production of date palm fruits free of acaricides residues by ozone technology as post-harvest treatment. *Journal of Food Science and Technology, 52*(6), 3322–3335. Available from https://doi.org/10.1007/s13197-014-1398-3.

Oyom, W., Li, Y.-C., Prusky, D., Zhang, Z., Bi, Y., & Tahergorabi, R. (2022). Recent advances in postharvest technology of Asia pears fungi disease control: A review. *Physiological and Molecular Plant Pathology, 117*, 101771. Available from https://doi.org/10.1016/j.pmpp.2021.101771.

Ozkan, R., Smilanick, J. L., & Karabulut, O. A. (2011). Toxicity of ozone gas to conidia of Penicillium digitatum, Penicillium italicum, and Botrytis cinerea and control of gray mold on table grapes. *Postharvest Biology and Technology, 60*(1), 47–51. Available from https://doi.org/10.1016/j.postharvbio.2010.12.004.

Öztekin, S., Zorlugenç, B., & Zorlugenç, F. K. (2006). Effects of ozone treatment on microflora of dried figs. *Journal of Food Engineering, 75*(3), 396–399. Available from https://doi.org/10.1016/j.jfoodeng.2005.04.024.

Palou, L., Crisosto, C. H., Smilanick, J. L., Adaskaveg, J. E., & Zoffoli, J. P. (2002). Effects of continuous 0.3 ppm ozone exposure on decay development and physiological responses of peaches and table grapes in cold storage. *Postharvest Biology and Technology, 24*(1), 39–48. Available from https://doi.org/10.1016/S0925-5214(01)00118-1.

Palou, L., Crisosto, C. H., Smilanick, J. L., Adaskaveg, J. E., & Zoffoloi, J. P. (2001). Evaluation of the effect of ozone exposure on decay development and fruit physiological behavior. *Acta Horticulturae, 553*(553), 429–430. Available from https://doi.org/10.17660/ActaHortic.2001.553.99.

Pandiselvam, R., Subhashini, S., Banuu Priya, E. P., Kothakota, A., Ramesh, S. V., & Shahir, S. (2019). Ozone based food preservation: A promising green technology for enhanced food safety. *Ozone: Science and Engineering, 41*(1), 17–34. Available from https://doi.org/10.1080/01919512.2018.1490636, http://www.tandf.co.uk/journals/titles/01919512.asp.

Pascual, A., Llorca, I., & Canut, A. (2007). Use of ozone in food industries for reducing the environmental impact of cleaning and disinfection activities. *Trends in Food Science and Technology, 18*(1), S29–S35. Available from https://doi.org/10.1016/j.tifs.2006.10.006.

Patil, S., Bourke, P., Frias, J. M., Tiwari, B. K., & Cullen, P. J. (2009). Inactivation of Escherichia coli in orange juice using ozone. *Innovative Food Science and Emerging Technologies, 10*(4), 551–557. Available from https://doi.org/10.1016/j.ifset.2009.05.011.

Patil, S., Valdramidis, V. P., Cullen, P. J., Frias, J., & Bourke, P. (2010). Inactivation of Escherichia coli by ozone treatment of apple juice at different pH levels. *Food Microbiology, 27*(6), 835–840. Available from https://doi.org/10.1016/j.fm.2010.05.002.

Patil, S., Valdramidis, V. P., Tiwari, B. K., Cullen, P. J., & Bourke, P. (2011). Quantitative assessment of the shelf life of ozonated apple juice. *European Food Research and Technology*, *232*(3), 469–477. Available from https://doi.org/10.1007/s00217-010-1416-2, http://link.springer.de/link/service/journals/00217/index.htm.

Paul, A., Radhakrishnan, M., Anandakumar, S., Shanmugasundaram, S., & Anandharamakrishnan, C. (2020). Disinfestation techniques for major cereals: A status report. *Comprehensive Reviews in Food Science and Food Safety*, *19*(3), 1125–1155. Available from https://doi.org/10.1111/1541-4337.12555.

Pinto, L., Palma, A., Cefola, M., Pace, B., D'Aquino, S., Carboni, C., & Baruzzi, F. (2020). Effect of modified atmosphere packaging (MAP) and gaseous ozone pre-packaging treatment on the physico-chemical, microbiological and sensory quality of small berry fruit. *Food Packaging and Shelf Life*, *26*. Available from https://doi.org/10.1016/j.fpsl.2020.100573, http://www.journals.elsevier.com/food-packaging-and-shelf-life/.

Porto, Y. D., Trombete, F. M., Freitas-Silva, O., de Castro, I. M., Direito, G. M., & Ascheri, J. L. R. (2019). Gaseous ozonation to reduce aflatoxins levels and microbial contamination in corn grits. *Microorganisms*, *7*(8). Available from https://doi.org/10.3390/microorganisms7080220, https://www.mdpi.com/2076-2607/7/8/220/pdf.

Prabakaran, M., Selvi, S., Merinal, A., & Panneerselvam. (2012). Effect of ozonation on selected properties of starches. *Pelagia Research Library Advances*, *3*, 299–302.

Prado, J., & Esplugas, S. (1999). Comparison of different advanced oxidation processes involving ozone to eliminate atrazine. *Ozone: Science and Engineering*, *21*(1), 39–52. Available from https://doi.org/10.1080/01919519908547258, http://www.tandf.co.uk/journals/titles/01919512.asp.

Proctor, A. D., Ahmedna, M., Kumar, J. V., & Goktepe, I. (2004). Degradation of aflatoxins in peanut kernels/flour by gaseous ozonation and mild heat treatment. *Food Additives and Contaminants*, *21*(8), 786–793. Available from https://doi.org/10.1080/02652030410001713898.

Prudente, A. D., Jr., & King, J. M. (2002). Efficacy and safety evaluation of ozonation to degrade aflatoxin in corn. *Journal of Food Science*, *67*(8), 2866–2872. Available from https://doi.org/10.1111/j.1365-2621.2002.tb08830.x.

Pérez, A. G., Sanz, C., Ríos, J. J., Olías, R., & Olías, J. M. (1999). Effects of ozone treatment on postharvest strawberry quality. *Journal of Agricultural and Food Chemistry*, *47*(4), 1652–1656. Available from https://doi.org/10.1021/jf980829l.

Qiang, Z., Demirkol, O., Ercal, N., & Adams, C. (2005). Impact of food disinfection on beneficial biothiol contents in vegetables. *Journal of Agricultural and Food Chemistry*, *53*(25), 9830–9840. Available from https://doi.org/10.1021/jf051359f.

Raila, A., Lugauskas, A., Steponavičius, D., Railiene, M., Steponavičiene, A., & Zvicevičius, E. (2006). Application of ozone for reduction of mycological infection in wheat grain. *Annals of Agricultural and Environmental Medicine*, *13*(2), 287–294. Available from http://www.aaem.pl/pdf/13287.pdf.Lithuania.

Rodoni, L., Casadei, N., Concellón, A., Chaves Alicia, A. R., & Vicente, A. R. (2010). Effect of short-term ozone treatments on tomato (Solanum lycopersicum L.) fruit quality and cell wall degradation. *Journal of Agricultural and Food Chemistry*, *58*(1), 594–599. Available from https://doi.org/10.1021/jf9029145.

Rodrigues, A. A. Z., de Queiroz, M. E. L. R., Faroni, L. R. D. 'A., Prates, L. H. F., Neves, A. A., de Oliveira, A. F., ... Zambolim, L. (2021). The efficacy of washing strategies in the elimination of fungicide residues and the alterations on the quality of bell peppers. *Food Research International*, *147*, 110579. Available from https://doi.org/10.1016/j.foodres.2021.110579.

Russel, A. D., Hugo, W. B., & Avliffe, G. A. J. (1999). Principles and practice of disinfection, preservation and sterilization. *Blackwell Science*.

Santis, D. D., Garzoli, S., & Vettraino, A. M. (2021). Effect of gaseous ozone treatment on the aroma and clove rot by fusarium proliferatum during garlic postharvest storage. *Heliyon*, *7*(4).

Sarig, P., Zahavi, T., Zutkhi, Y., Yannai, S., Lisker, N., & Ben-Arie, R. (1996). Ozone for control of post-harvest decay of table grapes caused by Rhizopus stolonifer. *Physiological and Molecular Plant Pathology*, *48*(6), 403–415. Available from https://doi.org/10.1006/pmpp.1996.0032, http://www.elsevier.com/inca/publications/store/6/2/2/9/3/2/index.htt.

Sautter, C. K., Brackmann, A., de, R., Anese, O., Weber, A., Rizzatti, M. R., & Pavanello, E. P. (2011). Controle da podridão-parda e características físico-químicas de pêssegos 'Magnum' submetidos a tratamentos pós-colheita com elicitores abióticos. *Revista Ceres*, *58*(2), 172–177. Available from https://doi.org/10.1590/s0034-737x2011000200007.

Scussel, V. M., Giordano, B. N., Simao, V., Manfio, D., Galvao, S., & Rodrigues, M. N. F. (2011). Effect of oxygen-reducing atmospheres on the safety of packaged shelled Brazil nuts during storage. *International Journal of Analytical Chemistry*, *2011*, 1–9. Available from https://doi.org/10.1155/2011/813591.

Selma, M. V., Ibáñez, A. M., Cantwell, M., & Suslow, T. (2008). Reduction by gaseous ozone of Salmonella and microbial flora associated with fresh-cut cantaloupe. *Food Microbiology*, *25*(4), 558–565. Available from https://doi.org/10.1016/j.fm.2008.02.006.

Sharpe, D., Fan, L., McRae, K., Walker, B., MacKay, R., & Doucette, C. (2009). Effects of ozone treatment on botrytis cinerea and sclerotinia sclerotiorum in relation to horticultural product quality. *Journal of Food Science*, *74*(6), M250–M257. Available from https://doi.org/10.1111/j.1750-3841.2009.01234.x.

Sheng, L., Shen, X., Su, Y., Xue, Y., Gao, H., Mendoza, M., ... Zhu, M.-J. (2022). Effects of 1-methylcyclopropene and gaseous ozone on Listeria innocua survival and fruit quality of Granny Smith apples during long-term commercial cold storage. *Food Microbiology*, *102*, 103922. Available from https://doi.org/10.1016/j.fm.2021.103922.

Silva, S. B., Luvielmo, M. De. M., Geyer, M. C., & Prá, I. (2011). Potencialidades Do Uso Do Ozônio No Processamento de Alimentos. *Semina: Ciencias Agrarias*, *32*(2), 659–682.

Singh, N., Singh, R. K., Bhunia, A. K., & Stroshine, R. L. (2002). Efficacy of chlorine dioxide, ozone, and thyme essential oil or a sequential washing in killing Escherichia coli O157:H7 on lettuce and baby carrots. *LWT – Food Science and Technology*, *35*(8), 720–729. Available from https://doi.org/10.1006/fstl.2002.0933, http://www.elsevier.com/inca/publications/store/6/2/2/9/1/0/index.htt.

Sripong, K., Uthairatanakij, A., & Jitareerat, P. (2022). Impact of gaseous ozone on microbial contamination and quality of fresh-cut durian. *Scientia Horticulturae*, *294*, 110799. Available from https://doi.org/10.1016/j.scienta.2021.110799.

Strano, M. C., Timpanaro, N., Allegra, M., Foti, P., Pangallo, S., & Romeo, F. V. (2021). Effect of ozonated water combined with sodium bicarbonate on microbial load and shelf life of cold stored clementine (Citrus clementina Hort. ex Tan.). *Scientia Horticulturae*, *276*, 109775. Available from https://doi.org/10.1016/j.scienta.2020.109775.

Talbot, P., Martinelli, L., Talvy, S., Chauveheid, E., & Haut, B. (2012). Ozone inactivation of resistant microorganisms: Laboratory analysis and evaluation of the efficiency of plants. *Water Research*, *46*(18), 5893–5903. Available from https://doi.org/10.1016/j.watres.2012.07.052.

Tiwari, B. K., Brennan, C. S., Curran, T., Gallagher, E., Cullen, P. J., & O' Donnell, C. P. (2010). Application of ozone in grain processing. *Journal of Cereal Science, 51*(3), 248–255. Available from https://doi.org/10.1016/j.jcs.2010.01.007.

Tiwari, B. K., Muthukumarappan, K., O' Donnell, C. P., & Cullen, P. J. (2008). Modelling colour degradation of orange juice by ozone treatment using response surface methodology. *Journal of Food Engineering, 88*(4), 553–560. Available from https://doi.org/10.1016/j.jfoodeng.2008.03.021.

Tiwari, B. K., O' Donnell, C. P., Brunton, N. P., & Cullen, P. J. (2009). Degradation kinetics of tomato juice quality parameters by ozonation. *International Journal of Food Science & Technology, 44*(6), 1199–1205. Available from https://doi.org/10.1111/j.1365-2621.2009.01946.x.

Tiwari, B. K., O'Donnell, C. P., Patras, A., Brunton, N., & Cullen, P. J. (2009a). Anthocyanins and color degradation in ozonated grape juice. *Food and Chemical Toxicology, 47*(11), 2824–2829. Available from https://doi.org/10.1016/j.fct.2009.09.001.

Trinetta, V., Vaidya, N., Linton, R., & Morgan, M. (2011). A comparative study on the effectiveness of chlorine dioxide gas, ozone gas and e-beam irradiation treatments for inactivation of pathogens inoculated onto tomato, cantaloupe and lettuce seeds. *International Journal of Food Microbiology, 146*(2), 203–206. Available from https://doi.org/10.1016/j.ijfoodmicro.2011.02.014.

Trombete, F. M., Freitas-Silva, O., Saldanha, T., Venâncio, A. A., & Fraga, M. E. (2016). Ozone against mycotoxins and pesticide residues in food: Current applications and perspectives. *International Food Research Journal, 23*(6), 2545–2556. Available from http://www.ifrj.upm.edu.my/23%20(06)%202016/(34).pdf.

Trombete, F. M., Porto, Y. D., Freitas-Silva, O., Pereira, R. V., Direito, G. M., Saldanha, T., & Fraga, M. E. (2017). Efficacy of ozone treatment on mycotoxins and fungal reduction in artificially contaminated soft wheat grains. *Journal of Food Processing and Preservation, 41*(3). Available from https://doi.org/10.1111/jfpp.12927, http://www.blackwellpublishers.co.uk/journal_new.asp?site = 1.

Tzortzakis, N., Borland, A., Singleton, I., & Barnes, J. (2007). Impact of atmospheric ozone-enrichment on quality-related attributes of tomato fruit. *Postharvest Biology and Technology, 45*(3), 317–325. Available from https://doi.org/10.1016/j.postharvbio.2007.03.004.

Tzortzakis, N., Singleton, I., & Barnes, J. (2008). Impact of low-level atmospheric ozone-enrichment on black spot and anthracnose rot of tomato fruit. *Postharvest Biology and Technology, 47*(1), 1–9. Available from https://doi.org/10.1016/j.postharvbio.2007.06.004.

USEPA. (1999). Alternative disinfectants and oxidants guidance manual.

Vurma, M., Pandit, R. B., Sastry, S. K., & Yousef, A. E. (2009). Inactivation of Escherichia coli O157:H7 and natural microbiota on spinach leaves using gaseous ozone during vacuum cooling and simulated transportation. *Journal of Food Protection, 72*(7), 1538–1546. Available from https://doi.org/10.4315/0362-028x-72.7.1538.

Wang, L., Fan, X., Sokorai, K., & Sites, J. (2019). Quality deterioration of grape tomato fruit during storage after treatments with gaseous ozone at conditions that significantly reduced populations of Salmonella on stem scar and smooth surface. *Food Control, 103*, 9–20. Available from https://doi.org/10.1016/j.foodcont.2019.03.026.

Wen, G., Liang, Z., Xu, X., Cao, R., Wan, Q., Ji, G., … Huang, T. (2020). Inactivation of fungal spores in water using ozone: Kinetics, influencing factors and mechanisms. *Water Research, 185*, 116218. Available from https://doi.org/10.1016/j.watres.2020.116218.

Whangchai, K., Saengnil, K., & Uthaibutra, J. (2006). Effect of ozone in combination with some organic acids on the control of postharvest decay and pericarp browning of longan fruit. *Crop Protection, 25*(8), 821–825. Available from https://doi.org/10.1016/j.cropro.2005.11.003.

Williams, R. C., Sumner, S. S., & Golden, D. A. (2005). Inactivation of Escherichia coli O157:H7 and salmonella in apple cider and orange juice treated with combinations of ozone, dimethyl dicarbonate, and hydrogen peroxide. *Journal of Food Science, 70*(4), M197–M201. Available from https://doi.org/10.1111/j.1365-2621.2005.tb07188.x.

World Health Organization. (2018). Promote health, save lives, serve the vulnerable. https://www.who.int/.

Xu, Y., Ji, J., Wu, H., Pi, F., Blaženović, I., Zhang, Y., & Sun, X. (2019). Untargeted GC-TOFMS-based cellular metabolism analysis to evaluate ozone degradation effect of deoxynivalenol. *Toxicon, 168*, 49–57. Available from https://doi.org/10.1016/j.toxicon.2019.06.022.

Xu, Y., Wang, Y., Ji, J., Wu, H., Pi, F., Zhang, Y., & Sun, X. (2019). Chemical and toxicological alterations of zearalenone under ozone treatment. *Food Additives and Contaminants - Part A Chemistry, Analysis, Control, Exposure and Risk Assessment, 36*(1), 163–174. Available from https://doi.org/10.1080/19440049.2018.1547425, http://www.tandfonline.com/toc/tfac20/current.

Young, J. C., Zhu, H., & Zhou, T. (2006). Degradation of trichothecene mycotoxins by aqueous ozone. *Food and Chemical Toxicology, 44*(3), 417–424. Available from https://doi.org/10.1016/j.fct.2005.08.015.

Yu, H., Neal, J. A., & Sirsat, S. A. (2018). Consumers' food safety risk perceptions and willingness to pay for fresh-cut produce with lower risk of foodborne illness. *Food Control, 86*, 83–89. Available from https://doi.org/10.1016/j.foodcont.2017.11.014.

Zhang, Z.-Y., Liu, X.-J., & Hong, X.-Y. (2007). Effects of home preparation on pesticide residues in cabbage. *Food Control, 18*(12), 1484–1487. Available from https://doi.org/10.1016/j.foodcont.2006.11.002.

Zhu, F. (2018). Effect of ozone treatment on the quality of grain products. *Food Chemistry, 264*, 358–366. Available from https://doi.org/10.1016/j.foodchem.2018.05.047.

Zorlugenç, B., Kiroğlu Zorlugenç, F., Öztekin, S., & Evliya, I. B. (2008). The influence of gaseous ozone and ozonated water on microbial flora and degradation of aflatoxin B1 in dried figs. *Food and Chemical Toxicology, 46*(12), 3593–3597. Available from https://doi.org/10.1016/j.fct.2008.09.003.

C H A P T E R

14

Pulsed electric field processing: food pasteurization, tissue treatment, and seed disinfection

Gulsun Akdemir Evrendilek

College of Engineering, Department of Electrical and Computer Engineering, Boston University, MA, United States

14.1 Introduction

Pulsed electric fields (PEFs) treatment at the different magnitude of electric field strength, frequency, and duration is mainly applied to biological cell membrane to induce electrical breakdown defined as electroporation or electropermeabilization (Pagán & Mañas, 2006; Zimmermann, Pilwat, Beckers, & Riemann, 1976; Zimmermann, 1986) and to provide electrofusion or electroporation on biological membrane (Pagán & Mañas, 2006). Formation of pores on the membrane of cells and organelles is also explained by electroporation cause instabilities in cell structure (Chang, 1992; Pakhomov, Miklavčič, & Markov, 2010; Weaver & Chizmadzhev, 1996). Reversible or irreversible electroporation caused by the magnitude of applied electric field intensity may induce the formation of transient or permanent pores (Barba et al., 2015; Lebovka, Bazhal, & Vorobiev, 2000; Rols, Dahhou, Mishra, & Teissie, 1990; Tsong, 1991). Pores can be resealed after removal of electrical fields and transient pores ensure the survival of the electrically stimulated recipient cells. This use of electric field pulses is useful for injecting exogenous molecules such as drugs, proteins, RNA, or DNA into cells without causing deterioration of cellular functions. Higher magnitude of electric field strength, on the other hand, causes an intense electroporation that electropermeabilization. Irreversible electroporation causes cell membrane disintegration resulting in permanent cell membrane damage or cell lysis. Cell lysis meaning cell destruction form the basis for the food processing method for the food industry as an alternative to thermal processing (Zimmermann, 1986).

There is a critical threshold value called transmembrane potential (U_m) and generally 0.5–1.5 V should be exceeded for the electric field strength to cause electroporation across a membrane (Weaver & Chizmadzhev, 1996). U_m value changes depending on the cell's physical properties such as size and shape. It is directly proportional to the size of spherical cells. U_m for cellular tissues with large cells ($\approx$30–60 μm) ranged from 100 to 500 V/cm (Lebovka et al., 2000), whereas it changes from 3 to 10 kV/cm for small microbial cells ($\approx$1–10 μm) (Lebovka et al., 2000). The applied magnitude of the electric fields causing cell electroporation is the main driving force for microbial inactivation (Weaver & Chizmadzhev, 1996) because significantly increased transmembrane potential resulting in structural changes in the membranes are due to the electroporation resulting membrane polarization. Accumulation of negative and positive charges within the cell at the membrane areas closest to the cathode and anode cause membrane polarization (Chang, 1992). Electroporation is a complex mechanism as pore formation, expansion, and subsequent shrinkage and resealing or cell lysis still remain controversial. Several theories with the help of the studies conducted with model systems such as protoplasts or liposomes, and on large eukaryote cells, such as erythrocytes, chloroplasts, green alga cells, sea urchin eggs, hamsters ovary cells, etc. Pagán & Mañas (2006) explained the reversible electroporation and/or the

Food Packaging and Preservation.
DOI: https://doi.org/10.1016/B978-0-323-90044-7.00014-8

electrical membrane breakdown (Chang, 1992; Kinosita et al., 1992; Newman, Sprafke, Boldt, & Wolf, 1992; Tsong, 1991; Zimmermann, 1986; Zimmermann et al., 1976).

The effect of PEF on microbial inactivation is depended on different factors and inactivation is explained by different theories; but among them, the most accepted theory is the so-called "electromechanical instability theory" (Ho & Mittal, 1996; Weaver & Chizmadzhev, 1996; Zimmermann, 1986; Zimmermann et al., 1976). According to this theory, compare to cell cytoplasm, the cell membrane due to its low dielectric constant can act like capacitor and the potential difference between the cell membrane and its surrounding environment causes accumulation of free charges at both inside and outside of the membrane surfaces. Charge accumulation causes the generation of a transmembrane potential (U_m) of about 10 mV. Transmembrane potential of cell membrane under normal conditions provided by proton motive force is used to regulate a variety of energy-linked processes such as the entrance of certain substrates into the cell against a concentration gradient; maintaining the proper intracellular pH and cells turgor as well as turning flagella, and driving a reverse flow of electrons through the respiratory chain to reduce NAD when the supply of NADH2 is inadequate; generating ATP, etc. (Neidhardt, Ingraham, & Schaechter, 1990; Pagán & Mañas, 2006). However, applying an external electric field to the cell introduces protons to the environment where cells are suspended and this situation causes an increase in U_m due to the accumulation of free charges at both sides of the membrane surface. The addition of free ions at opposite charges attracts each other causing membrane compression and resulting in reduction in membrane thickness. Although free ions cause decrease in membrane thickness, viscoelastic forces oppose the electro-compression of the membrane. The opposite direction of these complex interactions causes tension on the cell membrane, and when U_m reaches approximately 1 V, the electrocompressive forces exceed the viscoelastic properties of the membrane causing membrane breakdown and leading to pore formation. Formation of pores, their size, and the numbers of pores are affected by applied electric field properties such as the magnitude of the electric field and application time. The level of applied electric field intensity that causes membrane breakdown is called threshold or critical electric field strength (E_c). If the applied electric field is lower than that of the critical electric field or the treatment, time is not enough to create compulsive forces on cell membrane; both the size and the number of the formed pores are low. Termination of PEF treatment at this point causes reversible permeabilization of the membrane because the cell membrane is able to restore its structure and functionality. However, if applied PEF treatment is intense (electric field strength is higher than critical electric field and/or treatment time is long enough), both the number of pores and their size increase causing mechanical disruption or irreversible permeabilization of the cell (Pagán & Mañas, 2006; Zimmermann, 1986). PEF inactivation of cell may also be explained by molecular reorientation of lipid bilayer or the protein channels as pore formation is related to these layers. Electropermeabilization may also be caused by the reorientation of phospholipids within the two monolayers of the membrane having conformational changes in membrane structure and losing its semipermeable barrier function resulting in the inactivation of the cell (Tsong, 1991). Structural modifications are also observed in the dynamic lipid fraction of the cell membrane by the applied electric field, and the formation of hydrophobic and hydrophilic pores in the lipid fraction are observed with applied electric fields (Tsong, 1991).

Even though PEF treatment is classified as nonthermal food processing technology, applied electric field resulted in the formation of localized joule heating causes the formation of hydrophilic pores in the lipid bilayer. Moreover, depending on the transmembrane potential; protein channels, pores, and pumps in cell membranes and their functionality may be adversely affected (Tsong, 1991). Protein channel formation to provide membrane transport is usually observed at around 50 mV which is lower than U_m. However, the application of other modifications or joule heating cause denaturation of channel protein, and thus irreversible pore formation. If joule heating is applied above critical U_m, spontaneous pores, because of the structural defects, are observed (Tsong, 1991). These approaches for cell inactivation might not be certain for all biological cells, but are correct for eukaryotic cells is because the cytoplasmic membrane is the only barrier separating the cytoplasm from the environment in most eukaryotic cells (Neidhardt et al., 1990). But these mechanisms may change in bacteria and yeasts due to differences in cytoplasmic membrane. When structural differences between Gram-negative and Gram-positive bacteria are considered, mechanism of action must be different between Gram-negatives and positives. For example, Gram-positive bacteria have very thick cell wall consisting of peptidoglycan layers and teichoic acids outside the cell membrane, and applied electric field must cause pore formation of structural breakdown in the cell wall. Gram-negative bacteria, on the other hand, have a thin layer of cell wall and outer membrane that is also different from eukaryotic cells. Yeasts having external cell wall are also affected differently by external electric fields (Pagán & Mañas, 2006).

14.2 Pulsed electric fields applications for food processing

Molds, yeast, or bacteria, as well as spores, are the target microorganisms for PEF processing on food preservation. Prevention of their growth and their destruction are the major concerns not only for thermal food preservation technologies but also for PEF. Thus, the growing interest in liquid food processing by PEF has evolved over the last decades to produce safer products having equal or better preservation of nutritional, sensory, and quality properties with extended shelf life (Barba et al., 2015). PEF processing studies with bench-scale systems (Fig. 14.1) are found successful mostly in fruit juices due to their low viscosity and high acidity properties. The conductivity of the fruit juices also makes them suitable for PEF processing. Applied electric current is transferred to foods by means of free ions, and fruit juices due to their acidity contain free hydrogen ions that make them suitable for PEF processing. Therefore, most of the studies are performed with fruit juices to inactivate various microorganisms including foodborne pathogens, food spoilage bacteria and plant pathogenic bacteria, inactivation of enzymes, changes in the nutritive value such as vitamins, antioxidant properties, phenolic and anthocyanin compounds, and minerals as well as sensory properties and shelf life extension (Evrendilek & Varzakas, 2015).

Although different microorganisms in different fruit juices and blends are subjected to inactivation studies by PEF, the most studied ones are aerobic microorganisms, yeasts such as *Saccharomyces cerevisiae* (Elez-Martinez, Escola-Hernandez, Soliva-Fortuny, & Martin-Belloso, 2005a; Molinari, Pilosof, & Jagus, 2004), molds (El-Hag, Jayaram, & Griffiths, 2006; Min, Jin, Min, Yeom, & Zhang, 2003a; Yeom, Streaker, Zhang, & Min, 2000) and *Lactobacillus brevis* (Elez-Martinez et al., 2005a) in orange juice, *Lactobacillus plantarum, Escherichia coli, Rhodotorula rubra* (Selma, Salmeron, Valero, & Fernandez, 2004), *L. plantarum* and *Lactobacillus enterocolitica* (Gomez, Garcia, Alvarez, Raso, & Condon, 2005) in orange-carrot juice mix, *E. coli* O157:H7, *Staphylococcus aureus, Listeria monocytogenes, Pseudomonas syringae* subs. *syringae, Erwinia carotovora, Botrytis cinerea*, and *Penicillium expansum* in sour cherry juice (Altuntas, Evrendilek, Sangun, & Zhang, 2010), in apricot (Evrendilek, Altuntas, Sangun, & Zhang, 2013) and in peach nectar (Altuntas, Evrendilek, Sangun, & Zhang, 2011), *E. coli* O157:H7, *Salmonella* Enteritidis

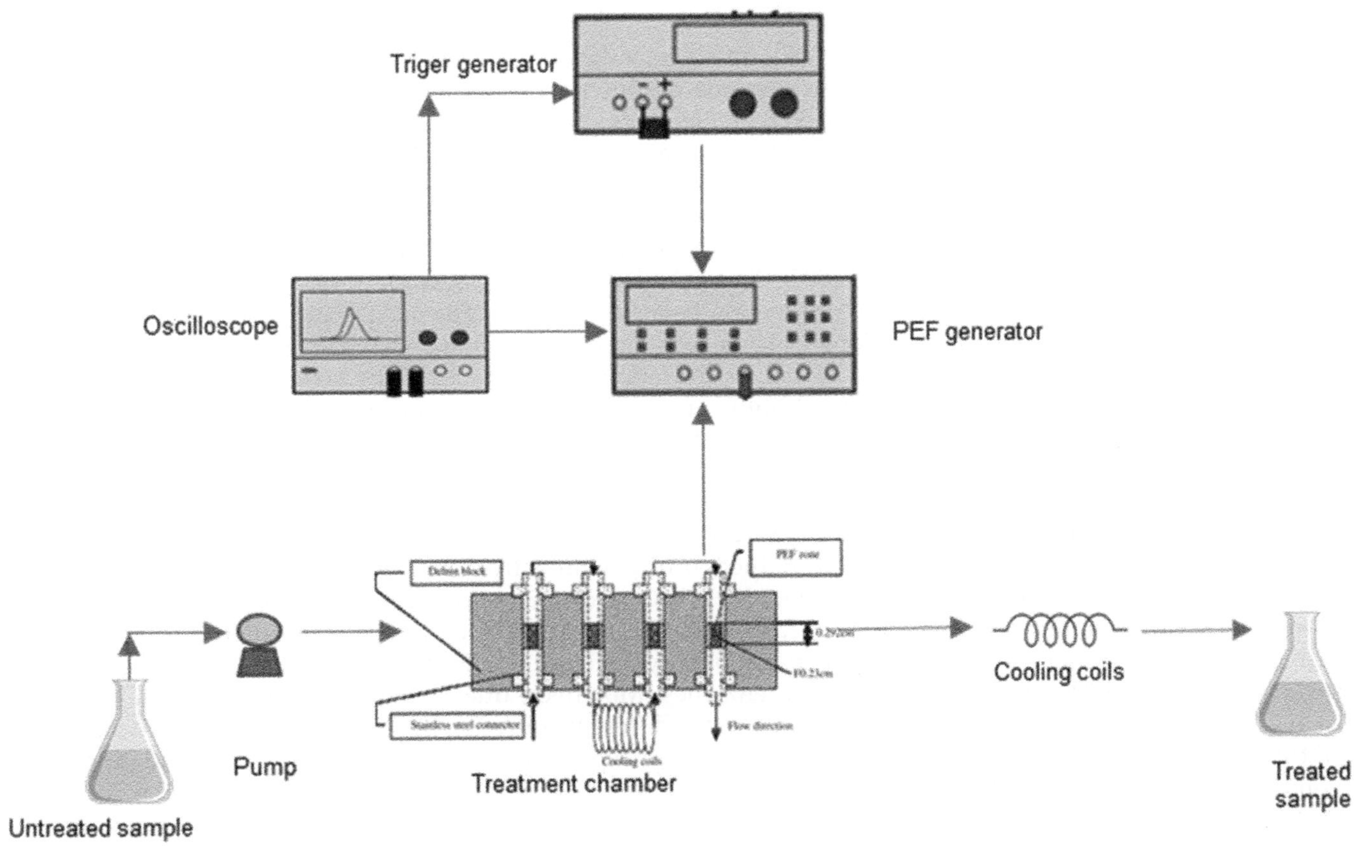

FIGURE 14.1 Bench-scale pulsed electric fields processing system.

and *L. monocytogenes* in melon and watermelon juice (Mosqueda-Melgar, Raybaudi-Massilia, & Martin-Belloso, 2012) and *E. coli* O157:H7 in pomegranate juice (Evrendilek, 2009) are the some recent studies involves inactivation of different microorganisms by PEF in juices or blends. Later studies conducted with microbial inactivation include inactivation of *S. aureus* in skim milk (Evrendilek, Zhang, & Richter, 2004a), *L. innocua* in raw skim milk (Guerrero-Beltran, Sepulveda, Gongora-Nieto, Swanson, & Barbosa-Cánovas, 2010), *L. monocytogenes* in skim milk gel (Fleischman, Ravishankar, & Balasubramaniam, 2004) and *L. innocua* and *Zygosaccharomyces bailii* in cheese whey (Dave, Walkling-Ribeiro, Rodríguez-González, Griffiths, & Corredig, 2012). More challenging studies involve the processing of liquid eggs by PEF due to their high protein content, low denaturation temperature of the protein fraction, and viscosity. Inactivation of *E. coli* O157:H7 (Amiali, Ngadi, Raghavan, & Smith, 2004) in liquid dialyzed egg and *S.* Entritidis in egg white (Hermawan, Evrendilek, Dantzer, Zhang, & Richter, 2004) are the two distinct studies of egg products processed by PEF. Moreover, fermented drinks such as beer with inactivation of *L. plantarum, Pediococcus damnosus, Bacillus subtilis, Saccharomyces uvarum, R. rubra* and *S. cerevisiae* (Evrendilek, Li, Dantzer, & Zhang, 2004b), *Dekkera anomala, Dekkera bruxellensis, Lactobacillus hilgardii and L. plantarum* (Puertolas, Lopez, Condon, Raso, & Alvarez, 2009), red wine with the inactivation of *E. coli* O157:H7, *Candida lipolytica, S. cerevisiae, Hansenula anomala*, and *Lactobacillus bulgaricus* (Abca & Evrendilek, 2015) are the some of the successful examples.

In order for PEF to be commercially available for pasteurization, the efficacy of inactivation should be proven for different pathogens in juices as well as against microbial contamination in different fruit juice matrices (Barba et al., 2015). For the inactivation of foodborne pathogens, a "5-log reduction standard" is required and this criterion is established based on the recommendations by the National Advisory Committee on Microbiological Criteria for Foods as required by the FDA for juice blends. Testing and re-testing of inactivation studies have proven that PEF meet the criteria for pathogen inactivation (Barba et al., 2015).

14.3 Pulsed electric fields applications for enzyme inactivation

Similar to microbial inactivation, different enzymes are subjected to PEF treatment for inactivation. Although some or complete inactivation is obtained by PEF, the mechanism of enzyme inactivation by PEF is still unclear. Enzymes are complex proteins whose activity depends on different factors, and their catalytic activity relies on the native configuration of their active site as well as the conformation of the surrounding proteins. Enzymes are polar molecules and they have charged regions due to charges in amino acids that are affected by applied external electric fields (Laberge, 1998; Martín-Belloso & Elez-Martínez, 2005). Complex network of non-covalent (Van der Waals forces, hydrogen bonding, electrostatic forces, ion pairing, and interaction hydrophobic effect), as well as covalent interactions (disulfide disulfide bonds), provide stability as well as the catalytic activity of the enzymes (Martín-Belloso & Elez-Martínez, 2005; Wong, 1995). Electric field application introducing free charges to the medium where enzymes are present change ion balance and thus cause disintegration of catalytic activity. Charge separation due to changes in the ionic balance leads to conformational changes ending up unfolding and denaturation, breaking of covalent bonds as well as oxidation-reduction reactions between sulfide groups and disulfide bonds (Barsotti & Cheftel, 1999). In addition, applied electric fields can cause changes in conformational state of proteins through charge, dipole, or induced dipole chemical reactions (Tsong & Astunian, 1986). The applied electric field also may lead the formation of structural modifications by affecting charged groups that resulted in loss of activity due to the difficulty of assembling the substrate with the active site (Castro, Swanson, Barbosa-Cánovas, & Zhang, 2001; Ho, Mittal, & Cross, 1997; Tsong, 1990; Vega-Mercado, Powers, Barbosa-Cánovas, & Swanson, 1995). For example, the reduction of papain activity by PEF is found to be related to the loss of α-helix structure (Yeom, Zhang, & Dunne, 1999). Structural differences among the enzymes cause differentiation in the inactivation or magnitude of reduction of enzyme activity. For example, the differences between the inactivation of polyphenol oxidase (PPO) and pectin methyl esterase (PME) might be due to the relevant differences between the molecular structures of these enzymes. Moreover, due to the huge differences in molecular size and structures displayed among enzymes, PPO from different sources may show different sensitivity during PEF treatment (Giner et al., 2002). Inactivation of alkaline phosphatase (ALP) is related to the degradation of the secondary structure in addition to the alteration of the entire globular configuration by PEF (Castro et al., 2001). Structural changes as the factor for inactivation in PEF-treated pepsin, lysozyme, and papain enzymes are also observed (Yang & Zhang, 2003). Compare to continuous PEF systems, batch PEF systems provide conformational changes in the lipase enzyme molecule and these changes lead to inhibition of enzymatic activity. PEF affects the forces involved in maintaining the three-dimensional structure of lipase causing the denaturation of the enzyme (Bendicho, Estela, Giner, Barbosa-Cánovas, & Martín, 2002a).

Inactivation of the enzymes is also greatly dependent on the process parameters. PEF effectiveness for enzyme inactivation is higher with increased electric field strength and treatment time (Bendicho et al., 2002a; Bendicho, Barbosa-Cánovas, & Martín, 2003a; Giner et al., 2000; Giner et al., 2002; Giner, Gimeno, Palomes, Barbosa-Cánovas, & Martín, 2003; Ho et al., 1997; Min et al., 2003a; Vega-Mercado et al., 1995; Yeom et al., 1999; Yeom, Zhang, & Chism, 2002). In addition, pulse width is an important PEF-processing factor affecting enzyme activity as an increase in pulse width will lead to higher levels of enzyme inactivation when the number of pulses and electric field strength are kept constant. Studies conducted with tomato PME (Giner et al., 2000), peach PPO (Giner et al., 2002), peroxidase (POD) (Elez-Martínez, Aguiló-Aguayo, Barbosa-Cánovas, & Martín-Belloso, 2003a), orange juice PME (Elez-Martínez, Suárez-Recio, Espachs-Barroso, Barbosa-Cánovas, & Martín-Belloso, 2003b; Espachs-Barroso, Barbosa-Cánovas, & Martín-Belloso, 2003) verify the effect of pulse width on enzyme inactivation. Frequency is also found important for enzyme inactivation is because more inactivation is obtained at higher frequencies for protease inactivation when similar treatments in terms of accumulated time and field strength are applied (Bendicho et al., 2003a; Bendicho, Barbosa-Cánovas, & Martín, 2003b). The pulse polarity is another imported factor reported to affect PEF effectiveness for inactivating enzymes, and different results have been reported depending on the enzyme. Although no difference in tomato inactivation is observed between mono- and bi-polar pulses (Giner, Gimeno, Barbosa-Cánovas, & Martín, 2001), greater PME inactivation in orange juice is observed by bipolar pulses than that of monopolar pulses (Elez-Martínez et al., 2003a). On the contrary, inactivation of the commercial pectic enzyme complex (CPC) is higher for the monopolar than for the bipolar treatments, and the maximum reduction in enzymatic activity is reported as 80% after PEF treatment with monopolar pulses of 8μs at a frequency of 400 Hz (Espachs-Barroso et al., 2003). Inactivation is also affected by pulse wave shape. While orange PME inactivation is less than 10% after PEF treatment of 25 kV/cm for 184 ms with square wave pulses (Yeom et al., 2002), tomato PME inactivation is 93.8% after PEF treatment of 24 kV/cm for 8 ms with exponential decay pulses (Giner et al., 2000), revealing that inactivation of PME is greater when the enzyme is subjected to exponential decay pulses.

PEF processing temperature is also an important factor for enzyme inactivation because higher temperature provides more enzyme inactivation. For example, increase in PME inactivation from 0% to 83.2% is obtained with an increasing the temperature from 30°C to 61.9°C (Yeom et al., 2002). A 47% difference between residual activity values of lipoxygenase (LOX) is obtained after PEF treatment at 10°C (62.2%) and 50°C (15.4%) with 20 kV/cm for 60 μs (Min, Min, & Zhang, 2003b). An increase in the process temperature provides higher rates of inactivation of ALP and lactoperoxidase (Van Loey, Verachtert, & Hendrickx, 2002). Although most of the studies have reported a positive correlation between the increased treatment temperature and enzyme inactivation, it is reported that an increase in temperature did not affect the inactivation of ALP (Castro et al., 2001).

The sensitivity of the enzymes to PEF is also structure dependent. Enzymes with different structural confirmations reveal that under the same PEF treatment, the most resistant enzyme is ALP followed by peroxidase and the most sensitive one is lipase (Giner et al., 2001). Product parameters such as pH, conductivity, and medium composition may cause differentiation in enzyme inactivation (Giner et al., 2001). Although the effect of conductivity is not clear; higher electric conductivity of the treatment medium provides less inactivation of PPO, and medium conductivity does not show any influence on the stability of both PPO and POD under the same PEF processing parameters (Van Loey et al., 2002). While changes in pH do not influence the stability of PPO to PEF treatment, lowering the pH to 4 causes a decrease in the stability of POD processed by PEF (Giner et al., 2001).

Studies about the effect of the fat content of the medium treated by PEF have controversial results. While higher inactivation of ALP is obtained when the milk fat content is lower (Castro et al., 2001), several researchers report that fat content in milk has no effect inactivation of ALP (Grahl & Märkl, 1996; Ho et al., 1997; Van Loey et al., 2002). Effect of the protein content of the medium on enzyme inactivation reveals different results, in which protein content of the medium show no significant influence on the sensitivity of PPO, POD, and LOX enzymes to PEF (Van Loey et al., 2002), whereas protease inactivation by PEF is more effective in a medium with proteins (Bendicho, Barbosa-Cánovas, & Martín, 2002b).

14.4 Pulsed electric field applications for plant tissue treatment

Compare to processing of liquid food, electroporation efficiency in biological plant tissues or suspension of cells is a rather complicated task (Kanduser & Miklavčič, 2008) because of the requirements of homogeneous distribution of the local electric fields, distribution of cell sizes and shapes, their state and local solute concentrations, local electrical conductivity, etc. (Canatella, Black, Bonnichsen, McKenna, & Prausnitz, 2004; Pucihar, Kotnik, Kanduser, & Miklavčič, 2001; Pucihar, Kotnik, Valic, & Miklavčič, 2006; Pucihar, Kotnik, Teissie, & Miklavčič, 2007). Increase in the electric conductivity, permeability of the whole sample, and solute local diffusivity are observed in cellular tissue after PEF treatment (Fincan & Dejmek, 2002; Lebovka, Praporscic, & Vorobiev, 2004). Tissue damage is mainly affected by electric field strength and the treatment time (Jeyamkondan, Jayas, & Holley, 1999; Knorr, Geulen, Grahl, & Sitzmann, 1994; Knorr, Angersbach, Eshtiaghi, Heinz, & Lee, 2001; Vorobiev, Jemai, Bouzrara, Lebovka, & Bazhal, 2004). Decrease in the fraction of electroporated cells causes an increase in cell density (Kanduser & Miklavčič, 2008); however, the changes in the electrical conductivity due to electroporation process should be taken into account in tissue treatment (Ben Ammar, Lanoiselle, Lebovka, Van Hecke, & Vorobiev, 2011; Corovic et al., 2013).

Biological membranes are characterized by the main components of lipid species, protein channels, and pumps which show phase transitions within the temperature range of 20°C−55°C (Exerova & Nikolova, 1992; Mouritsen & Jergensen, 1997). PEF-induced damage and electroporation caused by PEF on membrane are temperature dependent (Heimburg, 1998); and the temperature at the range of 20°C−50°C provides the breakdown in a single membrane (Zimmermann, 1986). Joule overheating on the membrane surface causes electroporation (primary damage processes) or electroosmotic transfer through the pores (secondary damage processes). Although the PEF effect on membranes is explained well, the physical mechanisms of the thermal effects influencing electroporation are not well understood yet on the cell because damage mechanisms are more complex than those of membranes (Weaver & Chizmadzhev, 1996).

Reversible or irreversible damage on cell membrane of cellular tissue is also dependent on PEF parameters, especially on treatment time. If the treatment time exceeds considerably the characteristic time, the rupture covers the membrane surface completely and it becomes irreversible (Weaver & Chizmadzhev, 1996). If treatment time is shorter than characteristic time, then cell membranes and electroporation may be partially reversible (Gabriel & Teissie, 1995, 1999), and the membrane damages may vanish. The time required for the cell to reseal is within the values of 1 second−100 seconds (DeBruin & Krassowska, 1999; Knorr, 1999; Neumann, Kakorin, & Toensing, 1999). The resealing ability of cell membranes, on the other hand, depends on the electric field intensity and the treatment time. Complete or partially complete resealing can be achieved at moderate treatment conditions under low magnitude of electric field strength or shorter treatment time. Hidden damage can be observed a day and even later after the PEF treatment (Teissie, Eynard, Gabriel, & Rols, 1999). Besides cell membrane damage, additional moisture released from the damaged cells in tissue is another secondary effect related to PEF-induced structural changes such as the diffusion migration, osmotic flow, long-lasting moisture, and air redistribution processes in tissue (Labuza & Hyman, 1998; Lebovka, Bazhal, & Vorobiev, 2001). Conductivity of a sample, on the other hand, is not affected for some time after the PEF treatment (Lebovka, Bazhal, & Vorobiev, 2002; Lebovka, Praporscic, & Vorobiev, 2003).

Electroporation or electroosmotic transfer on plant tissue by PEF provides enhancement of diffusion extraction, osmotic treatment, pressing extraction, drying, and freezing (Barba et al., 2015). Besides, increase in the release of some bioactive compounds is also quantified after PEF application. For example, increase in antioxidant capacity of apple juice due to an enhancement in the release of phenolics from the mash into the juice is obtained after PEF treatment of apple mash at 3 kV/cm with the specific energy input set at 10 kJ/kg (Schilling et al., 2008). Moreover, increase in the concentration of total native polyphenols in the juices by virtue of PEF treatment of apple mash (0.65 kV/cm, 23.2 ms, 32 kJ/kg) is also reported (Schilling et al., 2008). An increase in polyphenol content in subsequent juice by 15% is provided by PEF pretreatment (0.4 kV/cm, 100 ms, 15 kJ/kg) from white grape (Grimi, Lebovka, Vorobiev, & Vaxelaire, 2009). Similar results are revealed for the PEF treatment of grapes. PEF pretreatment on grapes provides enhancement of the release of the major anthocyanin found in Pinot Noir, that is, malvidin-3-O-glucoside (+224%) (Leong, Burritt, & Oey, 2016). A 61% increase in the content of total phenolic and a 19% increase in vitamin C, as well as a 31% of increase in antioxidant, are reported in grape juices treated by PEF as pretreatment of the grapes. The antioxidant potential of juices is improved by PEF pretreatments of grapes and the highest PEF energy treatment is the one yielding the best results for total phenolics, anthocyanins, and stilbenes (Elez-Martínez, Odriozola-Serrano, Oms-Oliu, Soliva-Fortuny, & Martín-Belloso, 2017).

Although the processing of oil-rich tissues by PEF treatment is a great challenge, studies reveal that it can be used to enhance the extraction of oil from tissues, such as maize, olives, and rapeseed with new treatment chamber and PEF system design (Fig. 14.2). A 14.7% increase in phytosterol yield in maize germ oil is obtained by PEF treatment of maize germ at 7.3 kV/cm and 120 pulses (91.4 kJ/kg) (Guderjan, Töpfl, Angersbach, & Knorr, 2005).

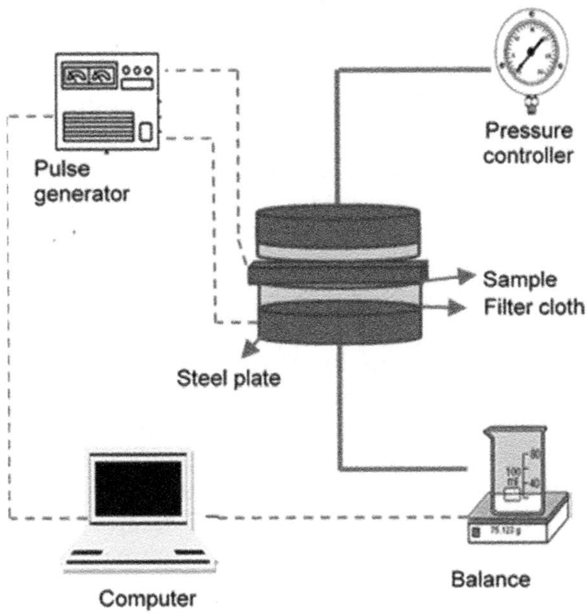

FIGURE 14.2 Bench-scale pulsed electric fields treatment system for plant tissue treatment.

PEF treatment of rapeseeds yields higher concentrations of total antioxidants, tocopherols, polyphenols, and phytosterols in oil. PEF treatments at different intensities (up to 2 kV/cm and 5.22 kJ/kg) on Arbequina olive paste in reference to olive oil extraction at different malaxation times (0 minute, 15 minutes, and 30 minutes) and temperatures (15°C and 26°C) reveals that the concentration of the main pigments in virgin olive oils (chlorophylls and carotenoids) is higher for the control. While the concentration of phenolic compounds that are related to the oxidative stability of olive oils is higher for the control than for the PEF-treated olive, the α-tocopherol content of olive oil is slightly higher when olives are PEF treated (Abenoza et al., 2013).

Mass and heat transfer rates between the cells and their surroundings can be increased by PEF due to the permeabilization of vegetable tissues. Vitamin C contents in PEF-treated (1.4 kV/cm, 0.8 ms) apples are found similar to those of untreated samples after osmotic dehydration (Taiwo, Angersbach, Ade-Omowaye, & Knorr, 2001). Application of PEF at different field strengths (0.5, 1.0, 2.0 kV/cm corresponding to 12, 48, and 192 J/kg, per pulse, respectively) and pulse numbers (2 to 50) to apple slices as a pretreatment to detect their influence on osmotic dehydration shows that using higher field strength resulted in lower amounts of vitamin C, whereas the effect of higher pulse number is not distinct on vitamin C retention in the samples (Taiwo, Angersbach, & Knorr, 2003). The vitamin C content of *Actinidia kolomikta* after PEF treatment of 5 kV/cm electric field strength for 150 seconds treatment time and subsequent hot air drying remains unchanged. Even though drying time is increased to up to 3 hours, the vitamin C content of the fruit remains unchanged (Lamanauskas, Satkauskas, Bobinaite, & Viskelis, 2015).

The possibility of enzyme re-activation by low-intensity PEF technology for stress induction and accumulation of bioactive compounds in plants due to the permeabilization phenomenon has gained great attention recently. It has been indicated that pore formation in the cell membrane is induced by the application of an electric field of about 1 kV/cm creating a potential difference of about 200 mV to 1 V across the cytoplasmic membrane for a long enough period (microseconds to milliseconds) (Weaver, 2000). Some studies have shown that reversible membrane permeabilization induces the generation of radical oxygen species (ROS) within plant cells and changes in cell metabolites accumulation (Cai et al., 2011; Gómez Galindo, Wadso, Vicente, & Dejmek, 2008). It is indicated that ROS is a part of the endogenous signal components required for elicitor-induced synthesis of secondary metabolites, which are widely believed to be part of the defense response of plants to stress (Shohael et al., 2006). Metabolite profiling provides insight into tissue response to stress induced by external stimulation to characterize stress response is used after reversible permeabilization of potato tissue (Galindo et al., 2009). PEF treatment at 1.2 kV/cm electric field strength and 30 pulses of tomato provides higher individual polyphenol and carotenoid contents after refrigeration at 4°C for 24 with the greatest increase in chlorogenic (152%), caffeic acid-O-glucoside (170%), and caffeic acids (140%). PEF treatment at 1.2 kV/cm and 5 pulses led to a maximum increase of a-carotene, 9- and 13-cis-lycopene by 93, 94, and 140%, respectively (Vallverdú-Queralt et al., 2013a).

It is a common application to combine PEF with other processing and/or preservation technologies to enhance the effect of combined technologies. While PEF is often combined with mild heat to increase inactivation efficacy or extraction efficacy, it is often combined with low heat to extend shelf life during storage. Thus, the combination of moderate and high-intensity PEF treatments could be used as a strategy for producing tomato juices with higher content of phenolic compounds and carotenoids. In fact, tomatoes subjected to treatments of moderate intensity provide an enhancement of 63%−65% in 15-cis-lycopene, 25% in chlorogenic acid, and 52% in naringenin-7-O-glucoside in juices (Vallverdú-Queralt et al., 2012; 2013b). It is also revealed that PEF can be an external stimulus to enhance secondary metabolite biosynthesis from plant cell cultures.

Comparing to the control, treatment of soy plant tissue culture with 20−50 pulses of 1.3 kV/cm PEF provides an increase in the daidzein and genistein contents by 20% and 21%, respectively (Guderjan, Elez-Martínez, & Knorr, 2007). Isoflavonoid content of 7-day-old soy plant tissue culture is increased by the application of 1.6 kV/cm electric field strength (Gueven and Knorr, 2011). The potential of PEF to increase phenylalanine ammonia-lyase (PAL) activity in plant cell cultures, the key enzyme for secondary metabolite synthesis from tomato cell culture, is another use of this technology. Due to the ability of the sub-culturing of enhancing the cell structure and improved the cell membrane permeability, PEF treatment combined with the sub-culturing technique provides an increase in PAL activity (Gürsul, Gueven, Grohmann, & Knorr, 2016).

14.5 Pulsed electric field applications for seed treatment

To increase agricultural yield, different agricultural pesticides are widely used, but this achievement has often been accompanied by the degradation and destruction of the natural environment. In addition to passing on genetic materials, seeds also transmit different pathogens such as bacteria, fungi, and viruses, and these seed-borne pathogens, and most commonly fungi, have been largely controlled by chemicals, or by treatments such as heat, electrons, natural antifungal products or biological control agents (Mancini and Romanazzi, 2014). Although the earliest report regarding application of electricity in relation to biological objects goes back to the 1746s and was done by Maimbray (Edinburgh), not much data was collected until recently. He observed the stimulating role of electricity upon the growth and flowering of myrtle plants (Nollet, 1749). In 1747, Abbot Nollet discovered the accelerated rates of germination and overall growth when cultivated under charged electrodes (Nollet, 1749). He also applied electrostatic machines and Leyden jars for electrotherapy. Increase in the yield for electrically pretreated potatoes in a galvanic cell was also observed (Ross, 1844). The growth stimulation (up to 70%) of crops such as potatoes, carrots, and celery by the electrical discharge treatment was observed later in 1885 (Sitzmann, Vorobiev, & Lebovka, 2016). It is reported at the beginning of the 20th century that application of the electrical discharge allowed accelerating the ripening of cucumber fruits with a 17% increase in yield (Newman, 1911; Priestley, 1907). The effects of various electric fields (electrostatic fields, microwave radiation, electrical discharges, electric shocks, and directly applied electric currents) on plants growth, seed germination, living organisms, and weed control were also intensively studied and the term "lethal electrotropism" is suggested to characterize an observed deterioration of plants during continuous exposure to the electric fields of the leaf tips (Murr, 1964).

On top of the studies reported in the literature, a recent effort is made to design and construct a bench-scale PEF system consisting of a loading unit, a seed treatment chamber, and a high-voltage pulse generator for the PEF treatment of seeds with different physical properties (Fig. 14.3). The seeds are feed from the hopper to the treatment chamber having two vertical plate electrodes to which high voltage pulses are supplied. The gap distance between two electrodes is adjusted with a small motor according to the size of the seed and electrodes are designed as 300-mm high. The height of the electrodes determines the duration of the free-fall of the seeds (treatment time). The high-voltage pulse generator includes an oscillator microprocessor with an integrated frequency counter, a low-voltage power pulse generator to the (dis)charge of the capacitor, the high-voltage transformer (transformer ratio 1:45), and a control unit. An unipolar rectangular pulse with 1.2 μs-wide and with a frequency of 10−500 Hz is supplied by the high voltage pulse generator. By using this PEF system, different seed samples are PEF treated with 12 kV/cm electric field strength and two different frequencies of 50 and 200 Hz. Results reveal the different degree of inactivation on indigenous microflora of total mold and yeast and total mesophilic aerobic bacteria on cabbage (A), parsley (B), onion (C), lettuce (D), wheat (E), garden rocket (F) and tomato (G) (Figs. 14.4 and 14.5). In general, increase in frequency, and thus,

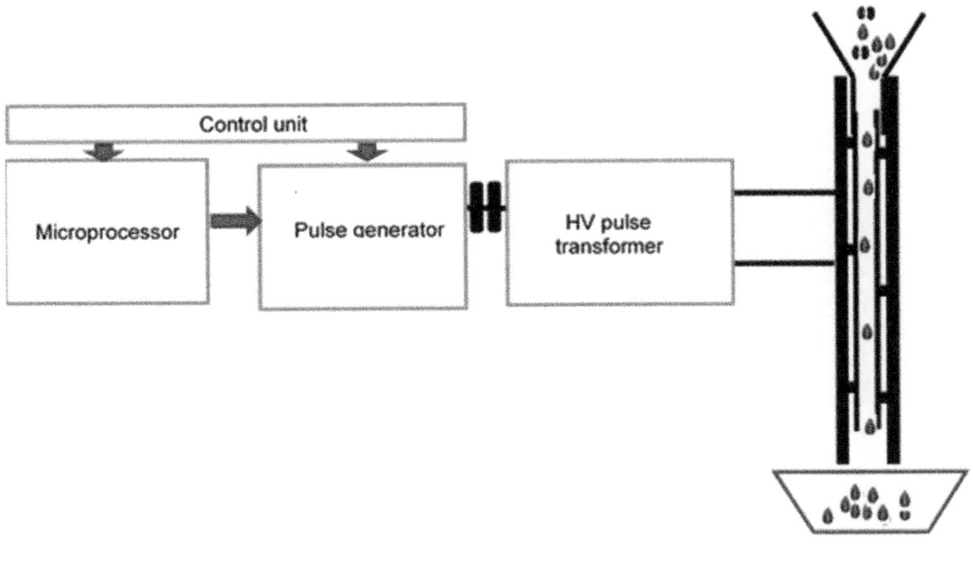

FIGURE 14.3 Bench-scale pulsed electric fields treatment system for seed.

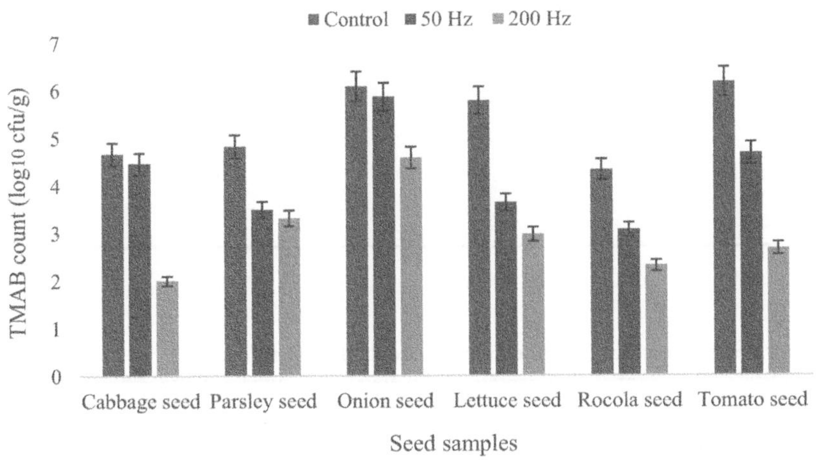

FIGURE 14.4 Inactivation of total aerobic mesophilic bacteria by pulsed electric fields system for seeds of cabbage, parsley, onion, lettuce, garden rocket, and tomato.

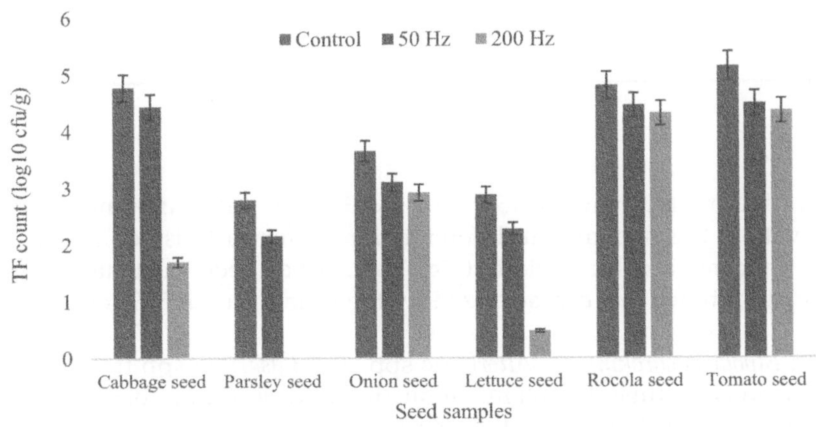

FIGURE 14.5 Inactivation of total fungi by pulsed electric fields system for seeds of cabbage, parsley, onion, lettuce, garden rocket, and tomato.

energy provides more inactivation on both TF and TMAB in all samples. Except for the garden rocket, the initial number of TMAB is found higher than that of the TF. Inactivation of TF is found higher than that of the TMAB in all samples except for garden rocket for both 50 and 200 Hz and tomato for 200 Hz frequency (Figs. 14.4 and 14.5) (Evrendilek and Tanasov, 2017).

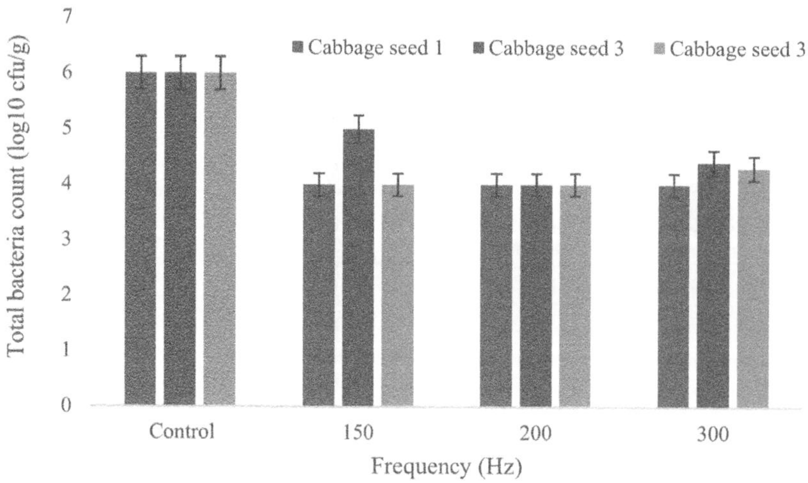

FIGURE 14.6 Inactivation of total aerobic bacteria on cabbage seeds with different maturity fractions.

TABLE 14.1 Fungal decontamination, germination energy and germination of pulsed electric fields -treated barley.

Corresponding Test		Untreated seeds	PEF treatment (250 Hz)	PEF treatment (290 Hz)
Fungal contamination (% Infected seedlings)	*Bipolaris sorokiniana*	20	3.3	8.7
	Alternaria spp.	90	58	5.5
	Fusarium spp	18	3.6	2.6
Germination energy (GE), %		69	70	33
Germination (G), %		83	86	42

TABLE 14.2 Fungal decontamination, germination energy, and germination of pulsed electric fields-treated wheat.

Corresponding Test		Untreated seeds	PEF treated seeds
Fungal contamination (% Infected seedlings)	*Drechslera graminea*	10.7	3.1
	Fusarium graminearum	7.5	2.0
Germination energy (GE), %		80	93
Germination (G), %		93	98

Moreover, PEF treatment of cabbage seeds at different frequencies (150, 200, and 300 Hz) with different maturity fractions reveals different amounts of inactivation. It is shown that differentiation on maturity is not as effective as PEF for the inactivation of soil-borne pathogenic aerobic bacteria (Fig. 14.6). It has been calculated that PEF does not have any adverse effect on germination time (T_{50}) characterized by the time to reach 50% of maximum germination.

Studies with winter barley contaminated with *Bipolaris sorokiniana, Alternaria* spp, and *Fusarium* spp treated by 250 and 290 Hz frequencies provide great reductions in fungal contamination (infected seedlings percentage) with increased germination energy and germination at 250 Hz treatment (Table 14.1). Similarly, PEF treatment of *Drechslera graminea* and *Fusarium graminearum* provides a significant amount of reduction in fungal contamination percentage (infected seeds) and an increase in germination energy from 80% to 93% Germination itself is measured as 93% and 98% for untreated and PEF-treated seeds, respectively (Table 14.2). PEF treatment of barley samples with 120, 200, and 300 Hz frequencies enables to reduce *F. graminearum* seeds' internal infection (%) significantly (Fig. 14.7).

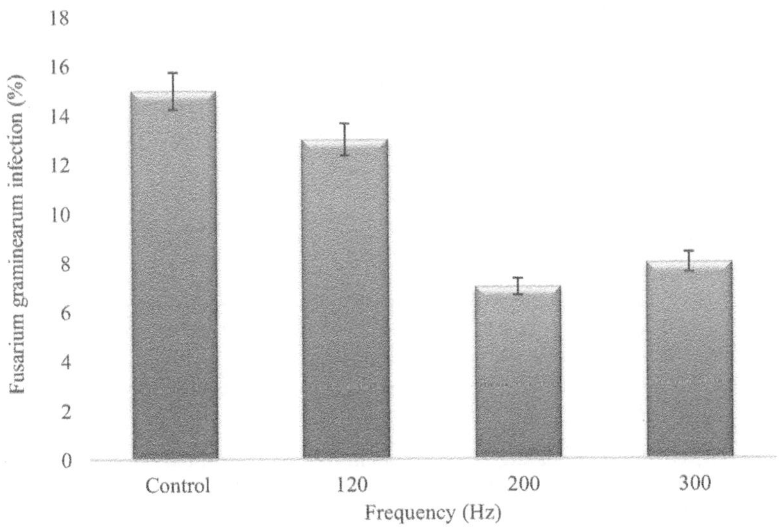

FIGURE 14.7 Reduction of the *Fusarium graminearum* seeds internal infection (%) by pulsed electric fields.

14.6 Conclusions

Green technologies with excluding or minimizing the use of chemicals causing some chronic diseases in humans, environmental pollution, soil, and water are gaining much attention, and this trend caused an increased interest in nonthermal technologies and especially PEF since its wide and environmentally friendly applications. No doubt PEF technology has lots of opportunities and a broad range of application on different processes that makes it one of the most promising technologies. With the improvement of system and equipment design, more and more applications of PEF technology may replace most of the common technologies on food processing, extraction of bioactive compounds as well as oil from different seeds, freeze-drying properties, seed disinfection, etc. One of the biggest hurdles for this technology, on the other hand, is to expand the variations in equipment design and validation of the data collected. Thus, studies need to focus on more equipment design both on bench and pilot scale and comparative collaboration studies with validation of collected data.

References

Abca, E. E., & Evrendilek, G. A. (2015). Processing of red wine by pulsed electric fields with respect to quality parameters. *Journal of Food Processing and Preservation, 39*(6), 758–767. Available from https://doi.org/10.1111/jfpp.12285.

Abenoza, M., Benito, M., Saldaña, G., Álvarez, I., Raso, J., & Sánchez-Gimeno, A. C. (2013). Effects of pulsed electric field on yield extraction and quality of olive oil. *Food and Bioprocess Technology, 6*, 1367–1373. Available from https://doi.org/10.1007/s11947-012-0817-6.

Altuntas, J., Evrendilek, G. A., Sangun, M. K., & Zhang, H. Q. (2010). Effects of pulsed electric field processing on the quality and microbial inactivation of sour cherry juice. *International Journal of Food Science and Technology, 45*, 899–905. Available from https://doi.org/10.1111/j.1365-2621.2010.02213.x.

Altuntas, J., Evrendilek, G. A., Sangun, M. K., & Zhang, H. Q. (2011). Processing of peach nectar by pulsed electric fields with respect to physical and chemical properties and microbial inactivation. *Journal of Food Process Engineering, 34*, 1506–1522. Available from https://doi.org/10.1111/j.1745-4530.2009.00549.x.

Amiali, M., Ngadi, M. O., Raghavan, V. G. S., & Smith, J. P. (2004). Inactivation of *Escherichia coli* O157:H7 in liquid dialyzed egg using pulsed electric fields. *Food and Bioproducts Processing, 82*(2), 151–156. Available from https://doi.org/10.1205/0960308041614936.

Ben Ammar, J., Lanoiselle, J.-L., Lebovka, N. I., Van Hecke, E., & Vorobiev, E. (2011). Impact of a pulsed electric field on damage of plant tissues: Effects of cell size and tissue electrical conductivity. *Journal of Food Science, 76*(1), 90–97. Available from https://doi.org/10.1111/j.1750-3841.2010.01893.x.

Barba, F. J., Parniakov, O., Pereira, S. A., Wiktor, A., Grimi, N., Boussetta, N., ... Vorobiev, E. (2015). Current applications and new opportunities for the use of pulsed electric fields in food science and industry. *Food Research International, 77*, 773–798. Available from https://doi.org/10.1016/j.foodres.2015.09.015.

Barsotti, L., & Cheftel, J. C. (1999). Food processing by pulsed electric fields. II. Biological aspects. *Food Review International, 15*(2), 181–213. Available from https://doi.org/10.1080/87559129909541186.

Bendicho, S., Barbosa-Cánovas, G.V., & Martín, O. (2002b). *Reduction of protease activity in simulated milk ultrafiltrate by high intensity pulsed electric fields.* Institute of Food Technologists Annual Meeting. Anaheim, California, USA. Extended abstracts. Abstract no: 91E-19.

Bendicho, S., Barbosa-Cánovas, G. V., & Martín, O. (2003a). Reduction of protease activity in simulated milk ultrafiltrate by continuous flow high intensity pulsed electric field treatments. *Journal of Food Science, 68*(3), 952–957. Available from https://doi.org/10.1111/j.1365-2621.2003.tb08270.x.

Bendicho, S., Barbosa-Cánovas, G. V., & Martín, O. (2003b). Reduction of protease activity in milk by continuous flow high-intensity pulsed electric field treatments. *Journal of Dairy Science, 86*, 697–703.

Bendicho, S., Estela, C., Giner, J., Barbosa-Cánovas, G. V., & Martín, O. (2002a). Effects of high intensity pulsed electric field and thermal treatments on a lipase from *Pseudomonas fluorescens*. *Journal of Dairy Science, 85*, 19–27. Available from https://doi.org/10.3168/jds.S0022-0302(02)74048-4.

Cai, Z., Riedel, H., Min, N., Kütük, O., Mewis, I., Jäger, H., ... Smetanska, I. (2011). Effects of pulsed electric field on secondary metabolism of *Vitis vinifera* L. cv. Gamay Fréaux suspension culture and exudates. *Applied Biochemistry and Biotechnology, 164*, 443–453. Available from https://doi.org/10.1007/s12010-010-9146-2.

Canatella, P. J., Black, M. M., Bonnichsen, D. M., McKenna, C., & Prausnitz, M. R. (2004). Tissue electroporation: Quantification and analysis of heterogeneous transport in multicellular environments. *Biophysical Journal, 86*, 3260–3268. Available from https://doi.org/10.1016/S0006-3495(04)74374-X.

Castro, A. J., Swanson, B. G., Barbosa-Cánovas, G. V., & Zhang, Q. H. (2001). Pulsed electric field modification of milk alkaline phosphatase activity. In G. V. Barbosa-Cánovas, & Q. H. Zhang (Eds.), *Pulsed electric fields in food processing. Fundamental aspects and applications* (pp. 65–82). Lancaster: Technomic Publishing Company Inc, ISBN: 9781566767835.

Chang, D. C. (1992). Structure and dynamics of electric field-induced membrane pores as revealed by rapid-freezing electron microscopy. In D. C. Chang, B. M. Chassy, J. A. Saunders, & A. E. Sowers (Eds.), *Guide to electroporation and electrofusion* (pp. 9–28). California: Academic Press, ISBN: 0323145019.

Corovic, S., Lackovic, I., Sustaric, P., Sustar, T., Rodic, T., & Miklavčič, D. (2013). Modeling of electric field distribution in tissues during electroporation. *Biomedical Engineering Online, 12*(16), 1–27. Available from https://doi.org/10.1186/1475-925X-12-16.

Dave, A., Walkling-Ribeiro, M., Rodríguez-González, O., Griffiths, M. W., & Corredig, M. (2012). Effect of PEF and UV and their combination on selected microorganisms and physico-chemical properties in whey. *Journal of Animal Science Suppl. 90(3)/Journal of Dairy Science Supplement, 95*(2), 168.

DeBruin, K. A., & Krassowska, W. (1999). Modeling electroporation in a single cell. I. Effects of field strength and rest potential. *Biophysics Journal, 79*, 670–679. Available from https://doi.org/10.1016/S0006-3495(99)76973-0.

Elez-Martinez, P., Escola-Hernandez, J., Soliva-Fortuny, R. C., & Martin-Belloso, O. (2005a). Inactivation of *Lactobacillus brevis* in orange juice by high-intensity pulsed electric fields. *Food Microbiology, 22*(4), 311–319. Available from https://doi.org/10.1016/j.fm.2004.09.005.

Elez-Martínez, P., Aguiló-Aguayo, I., Barbosa-Cánovas, G. V., & Martín-Belloso, O. (2003a). Inactivation of peroxidase in orange juice processed by high intensity pulsed electric fields. *Proceedings Workshop on Nonthermal Food Preservation*, 40–42, Wageningen, Netherlands.

Elez-Martínez, P., Odriozola-Serrano, I., Oms-Oliu, G., Soliva-Fortuny, R., & Martín-Belloso, O. (2017). Effects of pulsed electric fields processing strategies on health-related compounds of plant-based foods. *Food Engineering Reviews, 9*(3), 213–225. Available from https://doi.org/10.1007/s12393-017-9162-x.

Elez-Martínez, P., Suárez-Recio, M., Espachs-Barroso, A., Barbosa-Cánovas, G. V., & Martín-Belloso, O. (2003b). High intensity pulsed electric field inactivation of pectin methyl esterase in orange juice. *XII World Congress of Food Science and Technology*, Chicago, USA. Extended abstracts.

El-Hag, A. H., Jayaram, S. H., & Griffiths, M. W. (2006). Inactivation of naturally grown microorganisms in orange juice using pulsed electric fields. *IEEE Transaction of Plasma Science, 34*(4), 1412–1415. Available from https://doi.org/10.1109/TPS.2006.878382.

Espachs-Barroso, A., Barbosa-Cánovas, G. V., & Martín-Belloso, O. (2003). Microbial and enzymatic changes in fruit juice induced by high-intensity pulsed electric fields. *Food Reviews International, 19*(3), 253–273.

Evrendilek, G.A. (2009). Pulsed electric field processing of pomegranate juice. In: *International conference on bio and food electrotechnologies* (pp. 267–270), 22–23 October, Compiegne, France.

Evrendilek, G. A., Altuntas, J., Sangun, M. K., & Zhang, H. Q. (2013). Apricot nectar processing by pulsed electric fields. *International Journal of Food Properties, 16*(1), 216–227. Available from https://doi.org/10.1080/10942912.2010.551307.

Evrendilek, G. A., Li, S., Dantzer, W. R., & Zhang, Q. H. (2004b). Pulsed electric field processing of beer: Microbial, sensory and quality analyses. *Journal of Food Science, 69*(8), 228–232. Available from https://doi.org/10.1111/j.1365-2621.2004.tb09892.x.

Evrendilek, G. A., & Tanasov, I. (2017). Configuring pulsed electric fields to process seeds: An innovative method of seed disinfection. *Seed Science and Technology, 45*(1), 72–80. Available from https://doi.org/10.15258/sst.2017.45.1.13. (9).

Evrendilek, G. A., & Varzakas, T. (2015). Pulsed electric fields. In T. Varzakas, & C. Tzia (Eds.), *Handbook of food processing: Food preservation* (pp. 469–509). CRC Press, ISBN 9781498721752.

Evrendilek, G. A., Zhang, Q. H., & Richter, E. R. (2004a). Application of pulsed electric fields to skim milk inoculated with *Staphylococcus aureus*. *Biosystems Engineering, 87*, 137–144. Available from https://doi.org/10.1016/j.biosystemseng.2003.11.005.

Exerova, D., & Nikolova, A. (1992). Phase transitions in phospholipid foam bilayers. *Langmuir: The ACS Journal of Surfaces and Colloids, 8*, 3102–3108. Available from https://doi.org/10.1021/la00048a041.

Fincan, M., & Dejmek, P. (2002). In situ visualization of the effect of a pulsed electric field on plant tissue. *Journal of Food Engineering, 55*(3), 223–230.

Fleischman, G. J., Ravishankar, S., & Balasubramaniam, V. M. (2004). The inactivation of *Listeria monocytogenes* by pulsed electric field (PEF) treatment in a static chamber. *Food Microbiology, 21*, 91–95. Available from https://doi.org/10.1016/S0740-0020(03)00015-7.

Gabriel, B., & Teissie, J. (1995). Control by electric parameters of short- and long-term cell death resulting from electropermeabilization of Chinese hamster ovary cells. *Biochimia et Biophysica Acta, 1266*, 171–178. Available from https://doi.org/10.1016/0167-4889(95)00021-J.

Gabriel, B., & Teissie, J. (1999). Time courses of mammalian cell electropermeabilization observed by millisecond imaging of membrane property changes during the pulse. *Biophysics, 1*(76), 2158–2165. Available from https://doi.org/10.1016/S0006-3495(99)77370-4.

Galindo, F., Dejmek, P., Lundgren, K., Rasmusson, A., Vicente, A., & Moritz, T. (2009). Metabolomic evaluation of pulsed electric field induced stress on potato tissue. *Planta, 230*, 469–479. Available from https://doi.org/10.1007/s00425-009-0950-2.

Giner, J., Gimeno, V., Barbosa-Cánovas, G., & Martín, O. (2001). Effects of pulsed electric field processing on apple and pear polyphenoloxidases. *Food Science and Technology International, 7*(4), 339–345.

Giner, J., Gimeno, V., Espachs, A., Elez, P., Barbosa-Cánovas, G. V., & Martín, O. (2000). Inhibition of tomato (*Licopersicon esculentum* Mill.) pectin methylesterase by pulsed electric fields. *Innovative Food Science and Emerging Technologies*, *1*(1), 57–67. Available from https://doi.org/10.1016/S1466-8564(00)00003-5.

Giner, J., Gimeno, V., Palomes, M., Barbosa-Cánovas, G. V., & Martín, O. (2003). Lessening polygalacturonase activity in a commercial enzyme preparation by exposure to pulsed electric fields. *European Food Research and Technology*, *217*, 43–48. Available from https://doi.org/10.1007/s00217-003-0688-1.

Giner, J., Ortega, M., Mesegué, M., Gimeno, V., Barbosa-Cánovas, G. V., & Martín, O. (2002). Inactivation of peach polyphenoloxidase by exposure to pulsed electric fields. *Journal of Food Science*, *67*(4), 1467–1472. Available from https://doi.org/10.1111/j.1365-2621.2002.tb10307.x.

Gomez, N., Garcia, D., Alvarez, I., Raso, J., & Condon, S. (2005). A model describing the kinetics of inactivation of *Lactobacillus plantarum* in a buffer system of different pH and in orange and apple juice. *Journal of Food Engineering*, *70*, 7–14. Available from https://doi.org/10.1016/j.jfoodeng.2004.09.007.

Grahl, T., & Märkl, H. (1996). Killing of microorganisms by pulsed electric fields. *Applied Microbiology and Biotechnology*, *45*, 148–157.

Grimi, N., Lebovka, N., Vorobiev, E., & Vaxelaire, J. (2009). Effect of a pulsed electric field treatment on expression behavior and juice quality of Chardonnay grape. *Food Biophysics*, *4*, 191–198. Available from https://doi.org/10.1007/s11483-009-9117-8.

Guderjan, M., Elez-Martínez, P., & Knorr, D. (2007). Application of pulsed electric fields at oil yield and content of functional food ingredients at the production of rapeseed oil. *Innovative Food Sciences and Emerging Technologies*, *8*, 55–62. Available from https://doi.org/10.1016/j.ifset.2006.07.001.

Guderjan, M., Töpfl, S., Angersbach, A., & Knorr, D. (2005). Impact of pulsed electric field treatment on the recovery and quality of plant oils. *Journal of Food Engineering*, *67*, 281–287. Available from https://doi.org/10.1016/j.jfoodeng.2004.04.029.

Guerrero-Beltran, J. A., Sepulveda, D. R., Gongora-Nieto, M. M., Swanson, B., & Barbosa-Cánovas, G. B. (2010). Milk thermization by pulsed electric fields (PEF) and electrically induced heat. *Journal of Food Engineering*, *100*(1), 56–60. Available from https://doi.org/10.1016/j.jfoodeng.2010.03.027.

Gueven, A., & Knorr, D. (2011). Isoflavonoid production by soy plant callus suspension culture. *Journal of Food Engineering*, *103*(3), 237–243. Available from https://doi.org/10.1016/j.jfoodeng.2010.10.019.

Gómez Galindo, F., Wadso, L., Vicente, A., & Dejmek, P. (2008). Exploring metabolic responses of potato tissue induced by electric pulses. *Food Biophysics*, *3*, 352–360. Available from https://doi.org/10.1007/s11483-008-9086-3.

Gürsul, I., Gueven, A., Grohmann, A., & Knorr, D. (2016). Pulsed electric fields on phenylalanine ammonia lyase activity of tomato cell culture. *s. Journal of Food Engineering*, *188*, 66–76. Available from https://doi.org/10.1016/j.jfoodeng.2016.05.007.

Heimburg, T. (1998). Mechanical aspects of membrane thermodynamics. Estimation of the mechanical properties of lipid membranes close to the chain melting transition from calorimetry. *Biochimica et Biophysica Acta*, *1415*, 147–162. Available from https://doi.org/10.1016/S0005-2736(98)00189-8.

Hermawan, N., Evrendilek, G. A., Dantzer, W. R., Zhang, Q. H., & Richter, E. R. (2004). Pulsed electric field treatment of liquid whole egg inoculated with *Salmonella enteritidis*. *Journal of Food Safety*, *24*(2004), 71–85. Available from https://doi.org/10.1111/j.1745-4565.2004.tb00376.x.

Ho, S. Y., & Mittal, G. S. (1996). Electroporation of cell membranes: A review. *Critical Reviews in Biotechnology*, *6*, 349–362. Available from https://doi.org/10.3109/07388559609147426.

Ho, S. Y., Mittal, G. S., & Cross, J. D. (1997). Effects of high field electric pulses on the activity of selected enzymes. *Journal of Food Engineering*, *31*, 69–84. Available from https://doi.org/10.1016/S0260-8774(96)00052-0.

Jeyamkondan, S., Jayas, D. S., & Holley, R. A. (1999). Pulsed electric field processing of foods: A review. *Journal of Food Protection*, *62*, 1088–1096. Available from https://doi.org/10.4315/0362-028X-62.9.1088.

Kanduser, M., & Miklavčič, D. (2008). Electroporation in biological cell and tissue: An overview. In E. Vorobiev, & N. Lebovka (Eds.), *Electrotechnologies for extraction from food plants and biomaterials*. Springer Sciencer + Business Media, LLC. Available from https://doi.org/10.1007/978-0-387-79374-0, 1 1.

Kinosita, K., Hibino, M., Itoh, H., Shigernori, M., Hirano, K., Kirino, Y., & Hayakawa, T. (1992). Events of membrane electroporation visualized on a time scale from microsecond to seconds. In D. C. Chang, B. M. Chassy, J. A. Saunders, & A. E. Sowers (Eds.), *Guide to electroporation and electrofusion* (pp. 47–62). California: Academic Press, ISBN: 0080917275.

Knorr, D. (1999). Novel approaches in food-processing technology: New technologies for preserving foods and modifying function. *Current Opinions in Biotechnology*, *10*, 485–491. Available from https://doi.org/10.1016/S0958-1669(99)00015-4.

Knorr, D., Angersbach, A., Eshtiaghi, M. N., Heinz, V., & Lee, D.-U. (2001). Processing concepts based on high intensity electric field pulses. *Trends in Food Science and Technology*, *12*, 129–135. Available from https://doi.org/10.1016/S0924-2244(01)00069-3.

Knorr, D., Geulen, M., Grahl, T. A., & Sitzmann, W. (1994). Food application of high electric field pulses. *Trends in Food Science and Technology*, *5*, 71–75. Available from https://doi.org/10.1016/0924-2244(94)90240-2.

Laberge, M. (1998). Intrinsic protein electric fields: Basic non-covalent interactions and relationship to protein-induced stark effects. *Biochimia et Biophysica Acta*, *1386*, 305–350. Available from https://doi.org/10.1016/S0167-4838(98)00100-9.

Labuza, T. P., & Hyman, C. R. (1998). Moisture migration and control in multi-domain foods. *Trends in Food Science and Technology*, *9*, 47–55. Available from https://doi.org/10.1016/S0924-2244(98)00005-3.

Lamanauskas, N., Satkauskas, S., Bobinaite, R., & Viskelis, P. (2015). Pulsed electric field (PEF) impact on *Actinidia kolomikta* drying efficiency. *Journal of Food Process Engineering*, *38*, 243–249. Available from https://doi.org/10.1111/jfpe.12161.

Lebovka, N. I., Bazhal, M. I., & Vorobiev, E. (2000). Simulation and experimental investigation of food material breakage using pulsed electric field treatment. *Journal of Food Engineering*, *44*, 213–223. Available from https://doi.org/10.1016/S0260-8774(00)00029-7.

Lebovka, N. I., Bazhal, M. I., & Vorobiev, E. (2001). Pulsed electric field breakage of cellular tissues: Visualization of percolative properties. *Innovative Food Science & Emerging Technologies*, *2*, 113–125. Available from https://doi.org/10.1016/S1466-8564(01)00024-8.

Lebovka, N. I., Bazhal, M. I., & Vorobiev, E. (2002). Estimation of characteristic damage time of food materials in pulsed electric fields. *Journal of Food Engineering*. Available from https://doi.org/10.1016/S0260-8774(01)00220-5.

Lebovka, N. I., Praporscic, I., & Vorobiev, E. (2003). Enhanced expression of juice from soft vegetable tissues by pulsed electric fields: Consolidation stages analysis. *Food Engineering, 59*, 309–317. Available from https://doi.org/10.1016/S0260-8774(02)00472-7.

Lebovka, N. I., Praporscic, I., & Vorobiev, E. (2004). Effect of moderate thermal and pulsed electric field treatments on textural properties of carrots, potatoes and apples. *Innovative Food Science & Emerging Technologies, 5*(1), 9–16. Available from https://doi.org/10.1016/j.ifset.2003.12.001.

Leong, S. Y., Burritt, D. J., & Oey, I. (2016). Evaluation of the anthocyanin release and health promoting properties of Pinot Noir grape juices after pulsed electric fields. *Food Chemistry, 196*, 833–841. Available from https://doi.org/10.1016/j.foodchem.2015.10.025.

Mancini, V., & Romanazzi, G. (2014). Seed treatments to control seedborne fungal pathogens of vegetable crops. *Pest Management Science, 70*, 860–868.

Martín-Belloso, O., & Elez-Martínez, P. (2005). Enzymatic inactivation by pulsed electric fields. In D.-W. Sun (Ed.), *Emerging technologies for food processing* (pp. 155–181). Academic Press.

Min, S., Jin, Z. T., Min, S. K., Yeom, H., & Zhang, Q. H. (2003a). Commercial-scale pulsed electric field processing of orange juice. *Journal of Food Science, 68*, 1265–1271. Available from https://doi.org/10.1111/j.1365-2621.2003.tb09637.x.

Min, S., Min, S. K., & Zhang, Q. H. (2003b). Inactivation kinetics of tomato juice lipoxygenase by pulsed electric fields. *Journal of Food Science, 68*(6), 1995–2001. Available from https://doi.org/10.1111/j.1365-2621.2003.tb07008.x.

Molinari, P., Pilosof, A. M. R., & Jagus, R. J. (2004). Effect of growth phase and inoculum size on the inactivation of *S. cerevisiae* in fruit juices by pulsed electric fields. *Food Research International, 37*(8), 793–798. Available from https://doi.org/10.1016/j.foodres.2004.03.010.

Mosqueda-Melgar, J., Raybaudi-Massilia, R. M., & Martin-Belloso, O. (2012). Microbiological shelf life and sensory evaluation of fruit juices treated by high-intensity pulsed electric fields and antimicrobials. *Food and Bioproducts Processing, 90*, 205–214. Available from https://doi.org/10.1016/j.fbp.2011.03.004.

Mouritsen, O. G., & Jergensen, K. (1997). Small-scale lipid-membrane structure: Simulation versus experiment. *Current Opinion in Structural Biology, 7*, 518–527. Available from https://doi.org/10.1016/S0959-440X(97)80116-9, org/.

Murr, L. E. (1964). Mechanism of plant cell damage in an electrostatic field. *Nature, 201*, 1305–1306.

Neidhardt, F. C., Ingraham, J. L., & Schaechter, M. (Eds.), (1990). *Physiology of the bacterial bell: A molecular approach*. Sunderland, MA: Sinauer Associates. Available from https://doi.org/10.1016/0168-9525(91)90427-R.

Neumann, E., Kakorin, S., & Toensing, K. (1999). Fundamentals of electroporative delivery of drugs and genes. *Mini-review. Bioelectrochemistry and Bioenerg, 48*, 3–16. Available from https://doi.org/10.1016/S0302-4598(99)00008-2.

Newman, E., Sprafke, A., Boldt, E., & Wolf, H. (1992). Biophysical considerations of membrane electroporation. In D. C. Chang, B. M. Chassy, J. A. Saunders, & A. E. Sowers (Eds.), *Guide to electroporation and electrofusion* (pp. 61–82). California: Academic Press, ISBN: 0080917275.

Newman, J. E. (1911). Electricity as applied to agriculture. *Electrician., 66*, 915.

Nollet, J. A. (1749). *Recherches sur les Causes Particulières des Phénomènes Electriques*. Paris: chez les Frères Guérin.

Pagán, R., & Mañas, P. (2006). Fundamental aspects of microbial membrane electroporation Fundamental aspects of microbial membrane electroporation. In J. Raso, & V. Heinz (Eds.), *Pulsed electric fields technology for the food industry* (pp. 73–94). New York, NY: Springer Science & Business Media, LLC, ISBN: 038731122X.

Pakhomov, A. G., Miklavčič, D., & Markov, M. S. (2010). *Advanced electroporation techniques in biology and medicine* (pp. 71–97). Boca Raton: CRC Press, ISBN: 1439819076.

Priestley, J. H. (1907). The effect of electricity upon plants. *Proceedings of the Bristol Naturalists' Society Series, 1*(3), 192–203. (1).

Pucihar, G., Kotnik, T., Kanduser, M., & Miklavčič, D. (2001). The influence of medium conductivity on electropermeabilization and survival of cells in vitro. *Bioelectrochemistry (Amsterdam, Netherlands), 54*(2), 107–115. Available from https://doi.org/10.1016/S1567-5394(01)00117-7.

Pucihar, G., Kotnik, T., Teissie, J., & Miklavčič, D. (2007). Electropermeabilization of dense cell suspensions. *European Biophysics Journal, 36*(3), 173–185. Available from https://doi.org/10.1007/s00249-006-0115-1.

Pucihar, G., Kotnik, T., Valic, B., & Miklavčič, D. (2006). Numerical determination of transmembrane voltage induced on irregularly shaped cells. *Annals of Biomedical Engineering, 34*(4), 642–652. Available from https://doi.org/10.1007/s10439-005-9076-2.

Puertolas, E., Lopez, N., Condon, S., Raso, J., & Alvarez, I. (2009). Pulsed electric fields inactivation of wine spoilage yeast and bacteria. *International Journal of Food Microbiology, 130*(1), 49–55. Available from https://doi.org/10.1016/j.ijfoodmicro.2008.12.035.

Rols, M. P., Dahhou, F., Mishra, K. P., & Teissie, J. (1990). Control of electric field induced cell membrane permeabilization by membrane order. *Biochemistry, 29*, 2960–2966. Available from https://doi.org/10.1021/bi00464a011.

Ross, W. (1844). Galvanic experiments on vegetation. U.S. Patent office report, 27, 370–373.

Schilling, S., Toepfl, S., Ludwig, M., Dietrich, H., Knorr, D., Neidhart, S., ... Carle, R. (2008). Comparative study of juice production by pulsed electric field treatment and enzymatic maceration of apple mash. *European Food Research and Technology, 226*, 1389–1398. Available from https://doi.org/10.1007/s00217-007-0669-x.

Selma, M. V., Salmeron, M. C., Valero, M., & Fernandez, P. S. (2004). Control of *Lactobacillus plantarum* and *Escherichia coli* by pulsed electric fields in MRS broth, Nutrient broth and orange-carrot juice. *Food Microbiology, 21*, 519–525. Available from https://doi.org/10.1016/j.fm.2003.12.004.

Shohael, A. M., Ali, M. B., Yu, K. W., Hahn, E. J., Islam, R., & Paek, K. Y. (2006). Effect of light on oxidative stress, secondary metabolites and induction of antioxidant enzymes in *Eleutherococcus senticosus* somatic embryos in bioreactor. *Process in Biochemistry, 41*, 1179–1185. Available from https://doi.org/10.1016/j.procbio.2005.12.015.

Sitzmann, W., Vorobiev, E., & Lebovka, N. (2016). Applications of electricity and specifically pulsed electric fields in food processing: Historical backgrounds. *Innovative Food Science and Emerging Technologies, 37*, 302–311. Available from https://doi.org/10.1016/j.ifset.2016.09.021.

Taiwo, K. A., Angersbach, A., Ade-Omowaye, B. I. O., & Knorr, D. (2001). Effects of pretreatments on the diffusion kinetics and some quality parameters of osmotically dehydrated apple slices. *Journal of Agriculture and Food Chemistry, 49*, 2804–2811. Available from https://doi.org/10.1021/jf0009798.

Taiwo, K. A., Angersbach, A., & Knorr, D. (2003). Effects of pulsed electric field on quality factors and mass transfer during osmotic dehydration of apples. *Journal of Food Process Engineering, 26*, 31–48. Available from https://doi.org/10.1111/j.1745-4530.2003.tb00588.x.

Teissie, J., Eynard, N., Gabriel, B., & Rols, M. P. (1999). Electropermeabilization of cell membranes. *Advances in Drug Delivery Reviews, 35*, 3–19. Available from https://doi.org/10.1016/S0169-409X(98)00060-X.

Tsong, T. Y. (1990). Electrical modulation of membrane proteins: Enforced conformational oscillations and biological energy and signal transductions. *Annual Reviews in Biophysics and Chemistry, 19*, 83–106. Available from https://doi.org/10.1146/annurev.bb.19.060190.000503.

Tsong, T. Y. (1991). Electroporation of cell membranes. *Biophysical Journal, 24*, 271–295. Available from https://doi.org/10.1016/S0006-3495(91)82054-9.

Tsong, T. Y., & Astunian, R. D. (1986). Absorption and conversion of electric field energy by membrane bound ATPases. *Bioelectric and Bioenergetics, 15*, 457–476. Available from https://doi.org/10.1016/0302-4598(86)85034-6.

Vallverdú-Queralt, A., Odriozola-Serrano, I., Oms-Oliu, G., Lamuela-Raventós, R. M., Elez-Martínez, P., & Martín-Belloso, O. (2012). Changes in the polyphenol profile of tomato juices processed by pulsed electric fields. *Journal of Agriculture and Food Chemistry, 60*, 9667–9672. Available from https://doi.org/10.1021/jf302791k.

Vallverdú-Queralt, A., Odriozola-Serrano, I., Oms-Oliu, G., Lamuela-Raventós, R. M., Elez-Martínez, P., & Martín-Belloso, O. (2013b). Impact of high-intensity pulsed electric fields on carotenoids profile of tomato juice made of moderate-intensity pulsed electric field treated tomatoes. *Food Chemistry, 141*(3), 3131–3138. Available from https://doi.org/10.1016/j.foodchem.2013.05.150.

Vallverdú-Queralt, A., Oms-Oliu, G., Odriozola-Serrano, I., Lamuela-Raventós, R. M., Martín-Belloso, O., & Elez-Martínez, P. (2013a). Metabolite profiling of phenolic and carotenoid contents in tomatoes after moderate-intensity pulsed electric field treatments. *Food Chemistry, 136*, 199–205. Available from https://doi.org/10.1016/j.foodchem.2012.07.108.

Van Loey, A., Verachtert, B., & Hendrickx, M. (2002). Effects of high electric field pulses on enzymes. *Trends in Food Science and Technology, 12*, 94–102. Available from https://doi.org/10.1016/S0924-2244(01)00066-8.

Vega-Mercado, H., Powers, J. R., Barbosa-Cánovas, G. V., & Swanson, B. G. (1995). Plasmin inactivation with pulsed electric fields. *Journal of Food Science, 60*, 143–1146. Available from https://doi.org/10.1111/j.1365-2621.1995.tb06310.x.

Vorobiev, E., Jemai, A. B., Bouzrara, H., Lebovka, N. I., & Bazhal, M. I. (2004). Pulsed electric field assisted extraction of juice from food plants. In G. V. Barbosa-Canovas, & M. P. Cano (Eds.), *Novel food processing technologies* (pp. 105–130). New York: Marcel Dekker, ISBN: 9780824753337 - CAT# DK1128.

Weaver, J. C. (2000). Electroporation of cells and tissues. *IEEE Transactions of Plasma Science, 28*, 24–33. Available from https://doi.org/10.1109/27.842820.

Weaver, J. C., & Chizmadzhev, Y. A. (1996). Theory of electroporation: A review. *Bioelectrochemistry and Bioenergetics, 41*, 135–160. Available from https://doi.org/10.1016/S0302-4598(96)05062-3.

Wong, D. W. S. (1995). Tailoring enzyme: Structures and functions. In D. W. S. Wong (Ed.), *Food enzymes: Structure and mechanisms* (pp. 17–36). New York: Chapman & Hall, ISBN: 978-1-4757-2349-6.

Yang, R.J., & Zhang, Q.H. (2003). Effects of high voltage pulsed electric fields on activity and structure of enzymes. In: *Institute of food technologists annual meeting*. Chicago, USA. Extended abstracts. Abstract no:103-2.

Yeom, H. W., Streaker, C. B., Zhang, Q. H., & Min, D. B. (2000). Effects of pulsed electric fields on the quality of orange juice and comparison with heat pasteurization. *Journal of Agriculture and Food Chemistry, 48*(10), 4597–4605. Available from https://doi.org/10.1021/jf000306p.

Yeom, H. W., Zhang, Q. H., & Chism, G. W. (2002). Inactivation of pectin methyl esterase in orange juice by pulsed electric fields. *Journal of Food Science, 67*(6), 2154–2159. Available from https://doi.org/10.1111/j.1365-2621.2002.tb09519.x.

Yeom, H. W., Zhang, Q. H., & Dunne, P. (1999). Inactivation of papain by pulsed electric fields in a continuous system. *Food Chemistry, 67*, 53–59. Available from https://doi.org/10.1016/S0308-8146(99)00109-0.

Zimmermann, U. (1986). Electric breakdown, electropermeabilization and electrofusion, Reviews of Physiology. *Biochemistry and Pharmacology, 105*, 196–256. Available from https://doi.org/10.1007/BFb0034499.

Zimmermann, U., Pilwat, G., Beckers, F., & Riemann, F. (1976). Effects of external electrical fields on cell membranes. *Bioelectrochemistry and Bioenergetics, 3*(1), 58–83. Available from https://doi.org/10.1016/0302-4598(76)85007-6.

CHAPTER

15

Pulsed ultraviolet light for microbial inactivation and its applications for food decontamination

Gulsad Uslu-Senel[1] *and Ali Demirci*[2]

[1]Department of Environmental Engineering, Firat University, Elazığ, Turkey [2]Department of Agricultural & Biological Engineering, The Pennsylvania State University, University Park, PA, United States

15.1 Introduction

UV light has been used as a bactericidal agent from as early as 1928 (Miller, Jeffrey, Mitchell, & Elasri, 1999). High-energy ultraviolet (UV)-light (200–320 nm) has a significant germicidal effect on bacteria and viruses (Kowalski, 2009; Wan et al., 2020). UV light is a portion of the electromagnetic spectrum ranging from 100 to 400 nm wavelengths and has the potential to induce damage to the DNA by forming thymine-thymine dimers (Bank, John, Schmehl, & Dratch, 1990; Miller et al., 1999). These dimmers prevent the microorganisms from transcription and replication, which leads to the cell death (Miller et al., 1999). UV photons can excite base pairs in DNA, leading to improper transcription that is necessary for survival (Harm, 1980). UV light can be applied via one of three modes, namely continuous UV light (conventional one), UV-LED, and pulsed UV light.

Recently, the UV light-emitting diode (UV-LED), a p-n junction diode, has emerged as a small, robust, and environmentally friendly (mercury-free) UV source (Cole & Bradley, 2021; Keshavarzfathy, Malayeri, Mohseni, & Taghipour, 2020; Vilhunen, Särkkä, & Sillanpää, 2009). This type of LED emits UV radiation in a narrow spectrum and has no warm-up time (Vilhunen et al., 2009), which makes it suitable for use as a pulse source for disinfection processes (Keshavarzfathy et al., 2020; Song, Mohseni, & Taghipour, 2016; Song, Taghipour, & Mohseni, 2019). The disinfection capacity of UV-LED increases when its peak wavelength approaches the maximum absorption spectra of a microorganism (Keshavarzfathy et al., 2020; Song et al., 2019).

Therefore, UV-LEDs are an alternative source of UV light suitable for food industry applications, drinking water, and wastewater disinfection. The cost of UV-LEDs, which have a compact and robust design, has recently decreased substantially due to technological advances. Output power is increasingly efficient. Additionally, the average life span of UV-LEDs in comparison to UV lamps used in industry today is much longer (D'Souza, Yuk, Khoo, & Zhou, 2015; Hamamoto et al., 2007; Hinds, O'Donnell, Akhter, & Tiwari, 2019; Hinds, Charoux, Akhter, O'Donnell, & Tiwari, 2020). Moreover, heat emissions from UV-LEDs are far lower than UV lamps, thus they are more suitable for food treatment applications. Also since UV-LEDs do not need a warm-up time and thus consume less energy (Akgün & Ünlütürk, 2017).

On the other hand, the pulsed UV light provides higher instantaneous energy intensities than continuous UV light and hence is more effective in inactivating microorganisms. Pulsed UV light is produced by accumulating the energy in a capacitor and releasing it as a short-duration pulse (in the order of micro seconds) to magnify the power greatly by Xenon lamps (Xenon, 2003). These flash lamps radiate a broadband emission light (approximately between 100 and 1100 nm) with a considerable amount of light in the UV spectrum (<400 nm). Pulsed UV (PUV)

Food Packaging and Preservation.
DOI: https://doi.org/10.1016/B978-0-323-90044-7.00015-X

disinfection is a relatively new technology that has been accepted by both the European Union (EU) and Food & Drug Administration (FDA) in the U.S. as a static microbial disinfection technology for the food industry (Elmnasser et al., 2007). Pulsed light is one such technology, which achieves mostly surface microbial decontamination via the emittance of a broad-spectrum electromagnetic radiation delivered in a series of short, but high-intensity pulses (Cassar, Mills, & Demirci, 2022). Pulsed UV light technology is an alternative to thermal treatment for killing pathogenic and spoilage microorganisms in foods, including bacteria, yeasts, molds, and viruses. It has received considerable attention as a novel approach for decontaminating food, packaging, water, and air. The efficacy of pulsed UV light has been documented both for inactivating food-borne microorganisms in suspension as well as on food surfaces. This new technology has also been used for treating waterborne pathogens, reducing disinfection by-products, and improving water quality in drinking water systems.

However, the bactericidal effect of pulsed UV light is still limited by its low degree of penetration. Therefore, it is well suited for the inactivation of microorganisms on solid surfaces, but its applicability in fluid foods is rather limited to transparent liquids. Still, the penetration depth is the main limitation of pulsed UV-light treatment, depending on the optical properties of the medium (Demirci & Panico, 2008). The experimental investigations of Krishnamurthy, Demirci, and Irudayaraj (2004) on the inactivation of *Staphylococus aureus* in milk matrix and of Hillegras and Demirci (2003) on the inactivation of *Clostridium sporogenes* in honey demonstrated that the overall decontamination effect of pulsed-light treatment decreases with the sample depth due to optical opaqueness of the matrix fluid. The effectiveness of pulsed UV light is strongly influenced by the interaction of the substrate with the incident light. Therefore, the treatment is more effective on smooth, nonreflecting surfaces or in liquids that are free of suspended particulates. Despite such limitations, there are unique opportunities for this technology, ranging from water disinfection to the manufacture of safe clear fruit juices, surface treatment of foods and food contact materials, or the terminal antimicrobial treatment of foods packaged in UV transparent materials.

This chapter will provide an overview of the conventional UV light and UV-LED system, fundamental principles of pulsed UV light and its applications in solid and liquid foods and water, available designs and operation of UV systems, process and control parameters, mechanisms and efficacy of pulsed UV light microbial inactivation, examples of applications, and effects on nutritional and quality parameters.

15.2 Conventional UV systems

15.2.1 *UV light system*

UV light disinfection systems have traditionally used low-pressure mercury (LPM) lamps as the source of germicidal radiation. LPM lamps, at pressures of 100–200 Pa, provide several emission lines in the visible and two lines in the UV region at wavelengths of 185 and 254 nm. Most of the output power occurs at 254 nm (40%–50%), and since other lines, including the 185 nm UV line, are outside the region for microbial inactivation, LPM lamps are considered to provide monochromatic radiation as far as disinfection is concerned. Medium-pressure mercury UV lamps are now often used as an alternative to low-pressure lamps. These medium-pressure lamps use mercury pressures of hundreds of kPa, provide increased power, and emit several broad lines over the germicidal wavelength range between 200 and 300 nm. Although the medium-pressure lamp is less electrically efficient than the low-pressure lamp, it emits a substantially higher light intensity and thus provides a given UV germicidal dose in a much shorter irradiation time. The UV radiation from both types of mercury lamps is emitted continuously (Gomez-Lopez, Koutchma, & Linden, 2012; Wang, MacGregor, Anderson, & Woolsey, 2005).

UV light is used in a broad range of antimicrobial applications including disinfection of water and air and disinfection of food preparation surfaces and food containers (Wang et al., 2005). UV light treatment of food is a nonthermal method for microbial inactivation that is free of chemicals and waste effluents, which makes it ecologically friendly. It does not produce by-products (Gomez-Lopez et al., 2012). UV light is a form of electromagnetic radiation that extends from 100 to 400 nm in the electromagnetic spectrum between X-rays and visible light. There are four classes of UV radiation: UV-A (315 to 400 nm), UV-B (280 to 315 nm), UV-C (200 to 280 nm), and vacuum UV (100 to 200 nm) (Meulemans, 1986; Miller et al., 1999). UV disinfection occurs at the germicidal wavelengths of 200 to 300 nm, in the UV-B and UV-C ranges of the electromagnetic spectrum. The main mechanism of UV disinfection is the absorption of UV radiation and the subsequent photochemical breakdown of an organism's DNA or RNA. In order for UV disinfection to be effective, the target organism must absorb energy from the appropriate range of the electromagnetic spectrum. DNA and RNA, and thus all living cells containing either as their genome, absorb radiation in the range of 200 to 300 nm, with a peak of approximately 260 nm (Meulemans, 1986). Krishnamurthy (2006) reported that energy absorption at 254 nm is an indicator of DNA absorption. Indeed, it is important to know which portion of the energy is available at 254 nm, and that contributes to the photochemical changes in the DNA. Only 2% of the total broadband energy is available in the 254 nm wavelength region in the pulsed UV light system.

UV-light treatment is a cost-effective nonthermal emerging technology and is recently revealed in many studies to be promising enough to ensure microbial safety of fruit juices and maintain their nutritional characteristics (Falguera, Pagán, Garza, Garvín, & Ibarz, 2011; Keyser, Muller, Cilliers, Nel, & Gouws, 2008; Tran & Farid, 2004; Uysal-Pala & Kırca-Toklucu, 2011). However, physicochemical and optical characteristics of liquid food products, such as color, turbidity, vitamin C content, viscosity, and absorbance coefficient, are known as essential factors of efficacy of UV-C light on microbial inactivation and are helpful for the UV-C reactor design and performance optimization (Uysal-Pala & Kırca-Toklucu, 2013).

However, conventional UV systems produce continuous UV light with power dissipation in the range of 100 to 1000 W. Generating these high-intensity energy density levels with continuous UV light can be costly to the user, which demands that systems be designed for the maximum conversion and collection efficiency of UV radiation. Therefore, pulsed UV systems have been proposed and designed, which can dissipate many megawatts of electrical power in the light source. Therefore, a modest energy input can yield high peak power dissipation (MacGregor, Turnbull, Tuema, & Farish, 1997; McDonald et al., 2000).

15.2.2 UV-LED system

LED is a semiconductor diode that emits light when an electric current flows through it due to combining of electrons and holes to release energy in the form of photons. The wavelength of released photons depends on the energy difference between electrons and holes. UV-LEDs produce UV radiations in the wavelength ranging from 210 nm to visible light depending on these semiconductors bandgap. Commercially available UV-LEDs can emit photons in the range of 280 to 340 nm. UV-LED unit that operates in the range of 210 to 365 nm is formed by aluminum nitride, gallium nitride, and intermediate alloys where 265 nm is recognized as the peak absorption of DNA (Sunita, Sharma, Kaukab, & Mishra, 2022). In recent years, UV-LED has been developed with many advantages such as low cost, energy-efficient, long life, and easy control of emission without use of mercury (Hamamoto et al., 2007), and UV-LED has a number of features that do not exist in conventional mercury-based UV lamps. One such feature is the ability to turn the radiation on and off with a high frequency, which enables pulsed irradiation with flexible pulse patterns by UV-LED. Another particular feature of UV-LED is its variety of radiation profiles (Keshavarzfathy et al., 2020).

UV-LEDs are made of semiconductor materials, which are similar to visible light LEDs (Khan, Shatalov, Maruska, Wang, & Kuokstis, 2005; Song et al., 2019). Hence, various wavelengths are viable on UV-LEDs by using different materials, such as 235 nm by doped diamond (Koizumi, Watanabe, Hasegawa, & Kanda, 2001), 215 nm by boron nitride (Watanabe, Taniguchi, & Kanda, 2004), and 210 nm by aluminum nitride (Taniyasu, Kasu, & Makimoto, 2006). The wave length of the commercial UV-LED is around 365 nm, UVA (315–400 nm). Unlike conventional UV mercury lamps, the wavelength diversity of UV-LEDs not only provides the chance to select a particular wavelength for a specific purpose but also offers a unique opportunity to selectively combine multiple wavelengths for tailored polychromatic radiation (Song et al., 2019).

15.2.3 Pulsed UV light system

PUV is an emerging decontamination technology, which offers significant advantages over conventional UV technology. Pulsed UV light technology is generally considered a surface decontamination technique for solid and opaque liquid foods, which involves using inert-gas flash lamps to produce short-duration (microseconds) light pulses (Dhar, Basak, & Chakraborty, 2022; Patel, Dhar, & Chakraborty, 2023). The generation of pulsed UV light involves storing electrical energy in a capacitor for a short duration, (milliseconds), and releasing the energy in pulses (John & Ramaswamy, 2018; Mahendran et al., 2019). The pulsed electrical energy is passed through a lamp filled with an inert gas such as xenon, argon, krypton, or other inert gases resulting in the ionization of the gas in the lamp, which leads to the production of a spectrum light in the wavelength ranging from UV to infrared. It targets DNA of the microorganisms and inactivates the DNA repair system thereby irreparable damage and then death occurs (Condon, Alvarez, & Gayan, 2014; Elmnasser et al., 2007; Oner, 2017).

The light pulses are typically at the rate of 1-20 pulses per second with a pulse width of 300 ns to 1 ms. The pulsed UV light typically has the wavelength of 100–1100 nm, but the majority of energy emitted by the lamp is produced in the UV region (Baranda, Barranco, & de Marañón, 2012; Charles, Vidal, Olive, Filgueiras, & Sallanon, 2013; Dunn, Ott, & Clark, 1995; Elmnasser et al., 2007; Gomez-Lopez, Ragaert, Debevere, & Devlieghere, 2007; Krishnamurthy, Demirci, & Irudayaraj, 2007; Murugesan, Orsat, & Lefsrud, 2012; Oms-Oliu, Aguiló-Aguayo, Martín-Belloso, & Soliva-Fortuny, 2010). The pulsed UV light technology could represent an excellent alternative or complement to conventional thermal or chemical destruction, thus ensuring safe foods with satisfactory nutritional and organoleptic qualities (Jean, Morales-Rayas, Anoman, & Lamhoujeb, 2011).

15.2.3.1 Pulsed UV lamp designs

Two types of lamps produce pulsed UV light: the flash lamp and the surface discharge (SD) lamp (Bohrerova, Shemer, Lantis, Impellitteri, & Linden, 2008). Historically, xenon flash lamps have been used commonly as photographic strobe lights. They are also used as light sources for pumping atoms in lasers to excited states to result in the emission of monochromatic light. Pulsed lamps (PL) must be respected from the viewpoint that they contain very high pressure that can rupture explosively. Furthermore, the quartz envelope is dangerously hot (150°C) during operation. Therefore, a cooling system is required to prevent excessive heating of the lamp. PL should be handled only with protective gloves and safety glasses, and only with welder's type goggles when fired up. Very intense UV and visible light as well as IR can cause eye damage. In an SD lamp, plasma is produced by a high-power electrical discharge along the surface of the dielectric substrate inside an envelope containing xenon gas (Schaefer, 2004).

A flash lamp consists of a sealed envelope, usually of quartz filled with a mixture of gases, mainly xenon, and electrodes, which convey electrical current to the gas. They may be formed in virtually any shape, including flat, linear, spiral, serpentine, and helical, to match cure zone shapes (Fig. 15.1). Proper reflectors and lens optics are also important for system optimization.

Pulsed flash lamps differ significantly and approximate quasi-black-body radiation. The temperature rises as a function of current density. The heating is due to ohmic dissipation that is proportional to the resistivity times the square of the current density (Elmer & Griese, 2012; Linden, Shin, & Sobsey, 2001). Higher current density shifts the emitted spectrum to lower wavelengths (i.e., higher frequencies). Thus, application of PL may need some additional engineering considerations. Nonetheless, pulsed lamps offer significant advantages, because their parameters can be adjusted to optimize the radiation spectrum to match the specific application.

Pulsed xenon flash lamp technology has received attention for a wide range of disinfection and remediation applications (Hillegras & Demirci, 2003; Krishnamurthy, Demirci, & Irudayaraj, 2008; Liang, Min, Davis, Green, & Remer, 2003; MacGregor et al., 1998; McDonald et al., 2000; Mofidi et al., 2001; Rowan et al., 1999; Sharma & Demirci, 2003; Weber & Scheible, 2003; Wekhof, 2003). Furthermore, some reports indicate possible increases in inactivation rates of microbes and reduction rates of organic chemicals from the pulsed nature of the light (Grapperhaus, Schaefer, & Linden, 2005; Hillegras & Demirci, 2003; Sharma & Demirci, 2003; Wekhof, 2003).

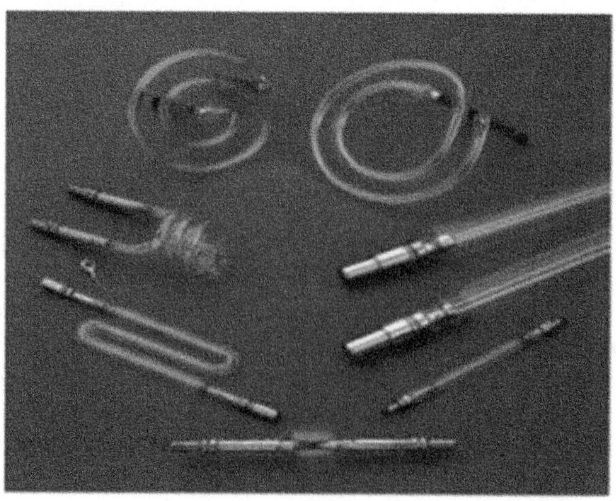

FIGURE 15.1 Commercially available pulsed UV lamp types and shapes (Xenon Corp., Wilmington, MA).

However, quantitative UV spectra, flashlamp efficiency, and performance comparisons with mercury UV lamps have generally not been reported (Linden & Mofidi, 2003). As a result, evaluations of pulsed lamp performance are often qualitative and without the data needed for quantitative comparisons.

According to the demands of the pulse light sterilization, the lamp needs to meet the conditions as follows: strong brightness, quick flash exposure, and be ability to manufacture desired shapes. The pulse xenon lamp has many advantages to meet these conditions. It can start quickly and its light spectrum is close to the sunlight and pretty stable (Elmer & Griese, 2012). In addition, it can produce strong light in a very short time and has owned high brightness, high efficiency, simple structure, and wide applicability. Moreover, considering the convenience of the installation and operation, the tubular pulse xenon lamp is commonly used. For a given pulsed UV lamp, the following parameters can be calculated:

15.2.3.1.1 The maximum allowable input energy of the pulse xenon lamp

The working voltage (V) of the lamp is 800 ~ 3000 V. If the input energy of the pulse xenon lamp exceeds the maximum, it will explode. The value is connected with the structure of the lamp, the pressure of xenon gas, the sealing technology, and the discharging duration. For the pure capacitor discharge circuit, the energy can be calculated by Eq. (15.1) (Zhang, Sun, Ma, & Zheng, 2012):

$$\mathbf{Emax} = 12ld\sqrt{t} \tag{15.1}$$

where E is energy (J), t is pulse width (□s), l is length of the arc (cm), and d is inner diameter of the tube.

With this equation, the maximum allowable input energy could be estimated by the above formula with the given length of the arc l and inner diameter of the tube lamp d. The input energy of the lamp must be less than 30% of the maximum allowable input energy while working (Zhang et al., 2012). Otherwise, the lamp will be damaged easily.

15.2.3.1.2 The calculation of the pulse power

For the rectangular, triangular pulse, the relation of the P and Pp

$$\overline{P}/P_p = r/T \tag{15.2}$$

The average power of the pulse light is the ratio of the output of every single pulse energy E to the pulse repetition period T; the peak power of the pulse light (P_p) is the power of a single pulse in the duration, that is, the ratio of the single pulse output energy E to the pulse width r, and it is the maximum of the light pulse output power (Zhang et al., 2012). When the lamp power is constant, the pulse light intensity is positively correlated with the pulse width. Therefore, the pulse light intensity can be changed by changing the pulse width.

15.2.3.2 Energy levels and wavelength spectra of various pulsed UV lamps

Pulsed UV light also provides a mercury-free treatment that does not produce any hazardous by-products and is environmentally friendly. The most important factor determining the inactivation efficacy of pulsed UV light is the energy dose (or fluence) on the sample. Additionally, the composition of the emitted light spectrum, the distance of the sample from the light source, the thickness, color, opacity, viscosity, and product flow conditions of the liquid samples, as well as the presence of particulate material, the inoculum size, the geometry of the treatment cell and the number of lamps are critical parameters that should be optimized in order to obtain the maximum effectiveness of the treatment along with minimum product alteration (Choi, Cheigh, Jeong, Shin, & Chung, 2010; Gomez-Lopez et al., 2007; Krishnamurthy et al., 2007; Pataro et al., 2011).

Pulsed UV light is a broad-band spectrum in the wavelength range 100−1,100 nm, which includes the UV, visible, and infrared spectra with approximately 54% of energy in the UV range (Fig. 15.2). The spectrum of pulsed UV light resembles the spectrum of sunlight, but is momentarily up to 20,000 times as intense in the visible light spectrum (Krishnamurthy et al., 2007). It is commonly believed that the UV light component of pulsed UV light is responsible for most of the biocidal effects since it is capable of causing thymine-thymine dimers. However, the infrared component of pulsed UV light also contributes significantly to biocidal effect by producing localized heating effects (Krishnamurthy et al., 2007). Krishnamurthy, Tewari, Irudayaraj, and Demirci (2010) identified the damage caused by pulsed UV light treatment to *Staphylococcus aureus* using Fourier-transformed infrared spectroscopy and transmission electron microscopy (TEM). A 5-s treatment with pulsed UV light resulted in severe damage to the *S. aureus* cells. Cell wall damage, cytoplasmic membrane shrinkage, and internal cellular structure collapse were observed. Consequently, cellular contents

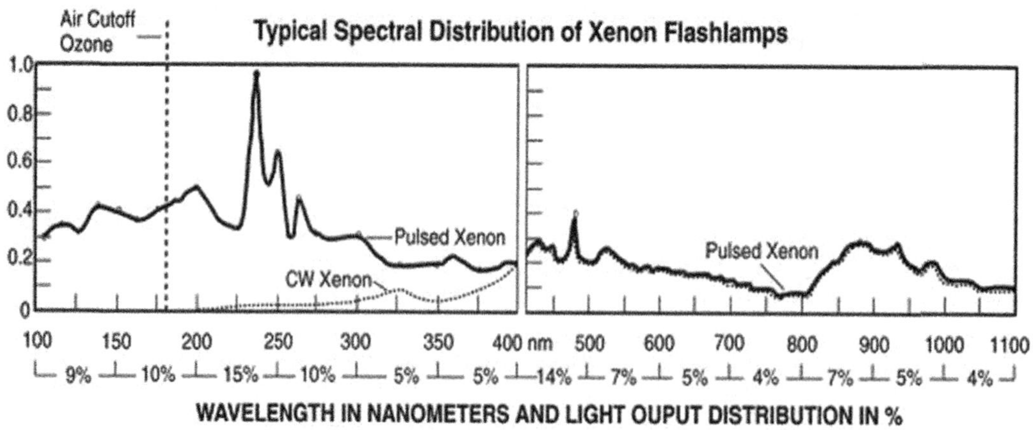

FIGURE 15.2 Spectral distribution of pulsed UV light (Xenon, 2003).

leaked from the cytoplasm leading to cell death. They also confirmed that some of the microbial cells were inactivated without any structural damage, indicating the photochemical effect of pulsed UV light.

It should be pointed out that UV absorbance and transmittance at 254 nm are important design parameters for conventional UV systems containing low-pressure mercury (LPM) lamps and high-pressure mercury (HPM) lamps. UV radiation is a form of electromagnetic radiation that extends from 100 to 400 nm in the electromagnetic spectrum, which is between X-rays and visible light. There are four classes of UV radiation in this range: UV-A (315 to 400 nm), UV-B (280 to 315 nm), UV-C (200 to 280 nm), and vacuum UV (100 to 200 nm). UV disinfection occurs at the germicidal wavelengths of 200 to 300 nm, in the UV-B and UV-C ranges of the electromagnetic spectrum, and peaks at 254 nm (Meulemans, 1986). However, broadband UV lamps, such as medium pressure mercury (MPM) lamps or PL, provide the broadband wavelength including the germicidal UV region up to infrared region.

15.3 Photoreactivation and dark repair post-UV, UV-LED, and PUV treatment

Photoreactivation was first known as a biological recovery phenomenon (Sutherland, 1981). It is a highly efficient process with one dimer split for every blue-light photon absorbed (Sutherland, 1981) and It is manifested as a biological effect by a reduction of UV-induced biological damage by light (300–600 nm) administered after UV (220–320 nm) exposure.

Photoreactivation, the process in which the enzyme photolyase uses the energy from sunlight to split pyrimidine dimers, is easy to demonstrate. When you irradiate cells with UV-C from a germicidal lamp, most of the damage to DNA is in the form of pyrimidine dimers. Adjacent pyrimidine bases (thymine and cytosine) in the DNA become joined by covalent bonds formed between their aromatic rings. Repair of DNA in bacteria following UV disinfection can cause reactivation of inactivated bacteria and negatively impact the efficiency of the UV disinfection process. If the cells are kept in the dark, the excision, error-prone, or recombination repair systems will fix many of the dimers.

UV light is a highly effective antimicrobial treatment and can be applied in a broad range of disinfection and sterilization applications, with pulsed UV light technology becoming increasingly attractive due to its rapid energy delivery and inactivation rates. Microbial inactivation as a result of UV irradiation is almost entirely attributable to photo biochemical reactions that are induced within the exposed microorganisms.

For UV disinfection, DNA or RNA is the key target in a variety of organisms (Hijnen, Beerendonk, & Medema, 2006; Wen, Deng, Wan, Xu, & Huang, 2019). Two major classes of mutagenic DNA lesions induced by UV irradiation are cyclobutane-pyrimidine dimers (CPD) and 6–4 photoproducts (6–4 PPs). Between them, the CPDs are the most abundant and make up around 75% of the UV-induced DNA damage products (Sinha & Hader, 2002). But many microorganisms have developed efficient DNA repair mechanisms in order to counteract the lethal effects of UV-induced damages (Lazarova, Savoye, Janex, Blatchley, & Pommepuy, 1999), known as photoreactivation, in which photolyase enzyme specifically binds to the CPDs and reverses the damage using the energy of light (310–480 nm) (Harm, 1980; Sinha & Hader, 2002). The other is dark repair, which instead replaces the damaged DNA with new, undamaged nucleotides independent of light (Sinha & Hader, 2002; Song et al., 2019). These repair processes are unfavorable because the UV disinfection efficiency could be reduced (Guo, Hu, Bolton, & El-Din, 2009; Sommer, Lhotsky, Haider, & Cabaj, 2000). Photoreactivation is a highly efficient process with one dimer split for every blue-light photon absorbed (Block, 1991; Sinha & Hader, 2002; Zimmer & Slawson, 2002).

Although comparative inactivation of UV-LEDs was obtained, limited studies considered the reactivation processes including photoreactivation and dark repair. And this would not be enough to design the UV-LED systems to meet the required pathogen reduction. UV-LEDs emitting at 265 and 280 nm and combined with different intensities were tested for inactivation of *E. coli* and the potential synergistic effects of multiple wavelengths on the germicidal UV dose and repair.

Fitzhenry, Clifford, Rowan, and Val del Rio (2021) evaluated the efficacy of PUV for water disinfection under flow-through conditions. Bacterial pathogens of interest in the food and water/wastewater applications, namely *Escherichia coli*, *S. aureus*, and *Listeria innocua* (surrogate for *L. monocytogenes*) were used to investigate the potential for photoreactivation and/or dark repair post-PUV flow-through disinfection. A continuous-flow low-pressure UV was also analyzed under similar experimental conditions. Bacterial inactivation via flow-through PUV was dependent on energy output with *E. coli* exhibiting the greatest sensitivity to PUV treatment (5.3 $\log_{10}$ inactivation after treatment at 1539 mJ/cm^2- output in UV range <300 nm); *L. innocua* exhibited the highest PUV resistance (3.0 $\log_{10}$ inactivation after treatment at 1539 mJ/cm^2− output in UV range <300 nm) under similar treatment conditions. Greater photoreactivation occurred at lower PUV outputs for both *S. aureus* and *E. coli* after flow-through PUV treatment. Thus, exposure of treated bacteria to natural light immediately post flow-through PUV treatment, should be avoided to minimize photoreactivation. The LPUV demonstrated the inactivation of all bacteria below the limit of detection (1 CFU/mL) and inhibited the occurrence of photoreactivation.

15.4 Microbial inactivation by UV, UV-LED, and pulsed UV light

The microbial inactivation caused by UV light occurs through several mechanisms depending on the wavelengths applied during the treatment. The inactivation can be achieved indirectly or directly by absorbing UV light incident on the microbial cell (Sunita et al., 2022). UV light microbial inactivation occurs through various mechanisms that are dependent on the wavelengths applied in treatment and can be achieved directly by absorption of the incident light on microbial cell DNA initiating lesion formation (depending on wavelength exposure) or indirectly due to the generation of reactive oxygen species by the interaction of radiation with cellular chromophores acting as photosensitizers (Brem, Guven, & Karran, 2017; Gayán, Condón, & Álvarez, 2014; Sunita et al., 2022). The key mechanism by which UV radiation inactivates microorganisms is largely by the formation of lesions that interfere with DNA replication. The UVC wavelength spectrum, in particular the 254−258 nm range, is the peak absorption range for DNA in cells, and thus it is the most lethal range for microorganisms (Gayán et al., 2014). The absorption of UVC radiation induces the formation of DNA photoproducts, namely CPDs and pyrimidine 6−4 pyrimidone (6−4PP) photoproducts, which inhibit transcription and replication. The UVA range is not as efficiently absorbed by native DNA and, therefore, does not induce severe damage by dimer formation. However, this range may still produce secondary photoreactions of existing DNA photoproducts or damage DNA via indirect photosensitization reactions (Hamamoto et al., 2007; Sinha & Hader, 2002). The indirect germicidal properties of UV radiation are related to membrane damage, growth delay, or DNA damage by the production or increase in reactive oxygen species such as superoxide anion radicals ($O_2^{\cdot}-$), hydroxyl radicals (OH), hydrogen peroxide (H_2O_2), and oxygen (O_2) (Hamamoto et al., 2007). The three main pathways in which microorganisms' DNA is repaired are: reverse damage repair, excision repair (precell replication), and tolerating damage pathways (activated once replication has started or immediately afterward) (Gayán et al., 2014).

Many studies have discussed the mechanism of microorganism inactivation by UV-LEDs, and various mechanisms have been proposed. Much like the conventional mercury-based UV lamps, UVC radiation has proven to be efficient for microorganism inactivation, and most studies on UV-LED disinfection used UVC-LEDs. UVC radiation is believed to have direct germicidal effects by acting directly on the DNA of microorganisms, leading to the formation of pyrimidine dimers and preventing them from reproducing without intermediate steps (Li et al., 2019; Pelvan Akgün & Ünlütürk, 2017). However, direct damage to DNA might be reparable by DNA-repair mechanisms, such as photo-reactivation and dark repair (Nebot Sanz, Salcedo Dávila, Andrade Balao, & Quiroga Alonso, 2007; Song et al., 2016; Wen et al., 2019). Since DNA repair is undesirable for microorganism inactivation, it is necessary to weaken or prevent the repair. DNA repair might be prevented by damaging the repair enzymes, which are proposed to be more vulnerable to high UV intensities (Sommer et al., 1998).

Moreover, the absorption spectrum of protein has a peak around 280 nm, which might help damage repair enzymes and prevent DNA repair (Kalisvaart, 2004).

On the other hand, the mechanism of pulsed UV illumination by xenon lamps with high energy and a broad spectrum has been widely studied for inactivation and food decontamination. Yet, its effect is not well understood, and there is little research on pulsed illumination by UV-LEDs. However, initial studies suggest that the mode of action of the pulsed UV light might be due to the coexistence of different mechanisms. The main mechanism of UV disinfection is the absorption of UV radiation and the subsequent photochemical breakdown of an organism's DNA by forming thymine-thymine dimers. In order for UV disinfection to be effective, the target organism must absorb energy from the appropriate range of the electromagnetic spectrum. DNA and RNA, and thus all living cells containing either as their genome, absorb radiation in the range of 200 to 300 nm, with a peak of approximately 254 nm (Miller et al., 1999; Rowan et al., 1999).

Furthermore, the lethality of pulsed light may be attributed to its rich broad spectrum UV content, its short duration, high peak power, and the ability to regulate the pulse duration and frequency output of flash lamps (Dunn et al., 1995; Takeshita et al., 2003). As a substantial portion of the pulsed light, spectrum covers UV light, it is considered that UV plays a vital role in microbial cell inactivation. The primary target cell of pulsed light in photochemical mechanism is nucleic acid as DNA is the target cell for these UV wavelengths (Chang et al., 1985; Miller et al., 1999). UV light absorbed by the conjugated carbon-carbon double bonds in proteins and nucleic acids induces the antimicrobial effect as it changes the DNA and RNA structures. The bactericidal effect is attributed to the high-energy short-wave UV-C range. In the UV-C range of 250–260 nm, alterations in DNA take place due to pyrimidine dimers mainly thymine dimers.

Even though DNA damage has still been identified as the primary lethal factor for pulsed UV, microbial inactivation by pulsed UV light has additional effects that may explain its higher efficiency on a per fluence basis. Therefore, the recent studies suggest pulsed UV light inactivates microorganisms not only by photochemical reactions but also by other mechanisms associated with the use of high-intensity pulses. Microbial cells become inactive once UV-induced photochemical reactions have taken place, but the cell structures remain intact after UV treatment. However, pulsed UV light also damages the cell structures of some of the pulsed UV-treated cells. These effects can be summarized by pulsed light via photophysical, photochemical, and photothermal mechanisms (Hwang, Lee, Min, & Chung, 2023; Krishnamurthy, 2006). These mechanisms are discussed as follows:

15.4.1 *Photo-chemical mechanism*

The photochemical effects of UV light on some living beings, including microorganisms, can be reversed by illumination with longer wavelengths, especially visible light, a repairing mechanism called photo-reactivation (Cleaver, 2003). Pulsed UV light is produced by accumulating the energy in a capacitor and releasing it as a short-duration pulse to magnify the power greatly. The inactivation is mainly caused by chemical changes in the DNA and RNA. Thymine-thymine dimer formation is the major photochemical change attributed to microbial inactivation. There may also be other minor chemical bond formations and/or breakages in microorganisms (Hwang et al., 2023).

Some researchers explain their results based on the photochemical effect. For example, (Rowan et al., 1999) achieved inactivation with less than 1°C rise in temperature of the bulk solution. Therefore, they concluded that the lethality can be attributed to the photo-chemical action of the shorter UV wavelengths. Wang et al. (2005) supported the hypothesis that the photo-chemical effect produced as a consequence of the UV absorption by DNA is the major cause of microorganism inactivation by pulsed UV light.

15.4.2 *Photo-thermal mechanism*

The lethal action of pulsed UV light can also be due to a photothermal effect. The contribution of the photothermal effect may be considered minimal in liquid foods because microorganisms can easily absorb pulsed UV light energy in clear liquids so that the treatment time will be short, and consequently, the temperature increase should be minimal (Krishnamurthy et al., 2010).

There may be a significant localized temperature increase during longer-duration pulsed UV light treatment. As the heating rate of the bacterial cell and surrounding media are different, localized heating of bacteria occurs, which leads to cell death. The action is primarily photo-thermal in the absence of UV wavelengths,

TABLE 15.1 Comparison of pulse and continuous UV-light.

Parameter	Pulsed UV light	Continuous UV light
Wavelength	100–1100 nm	254 nm
Instantaneous energy	Magnified several thousand-folds	Less
Inactivation mechanism	Damage to cells by photochemical changes	Photochemical DNA damage by thymine dimer formation
Natural cooling of lamp	Enables lamp to cool	No
Mercury	Provides mercury-free alternative	Commonly used as source
Inactivation energy	Up to 4 times increased inactivation efficiency as compared to continuous UV light	Normal
Temperature increase during UV treatment	Significant temperature increase due to infrared	No significant temperature increase

the light pulses being simply a fast method of transferring large amounts of thermal energy to the product (Mertens & Knorr, 1992). Hiramoto (1984) proposed that the rays absorbed by *Aspergillus niger* are expected to heat the mold instantaneously, providing a sort of thermal sterilization.

15.4.3 Photo-physical mechanism

A third inactivation mechanism has recently been proposed by Krishnamurthy et al. (2010), which suggested that the structural damage in microbial cells is attributed to the constant disturbance caused by the high-energy pulses. It has been observed in *S. aureus* cells that damage due to pulsed UV light consisted of cell leakage, lack of cell wall, cytoplasmic membrane shrinkage, and collapse of internal structures (Krishnamurthy et al., 2010). Similarly, structural damage caused by pulsed UV light was identified in *Saccharomyces cerevisiae* cells, which was higher than that caused by a monochromatic UV lamp. This damage consisted of a loss of cell membrane permeability detected by a higher concentration of eluted protein and microscopic observation of raised and expanded vacuoles, cell membrane distortion, and change of cell shape (Takeshita et al., 2003). Because of constant disturbance caused by the high-energy pulses, structural damage to bacteria may occur. Therefore, the effectiveness of pulsed UV light treatment can be improved by optimizing the pulse width and the number of pulses.

Pulsed UV light produces greater instantaneous energy than continuous UV light, but produces the same amount of energy overall, which can result in shortened treatment times (Gomez-Lopez et al., 2007; Krishnamurthy et al., 2007). Table 15.1 illustrates the major differences between pulsed UV light and continuous UV light. The pulsed UV light was found to be more effective than continuous UV in the destruction of microorganisms. This is explained by the instant energy proliferated numerous times in shorter pulse width. Therefore, pulsed UV has a higher penetration depth and may be more effective than continuous UV light (Krishnamurthy et al., 2004). Because of this large instant energy level and better penetration as a result of that, pulsed UV light was found to be more effective than continuous UV in the inactivation of microorganisms. Furthermore, both the pulsed UV light and the continuous wave UV disinfection processes are, in fact, essentially photo-chemical effects of UV light, but pulsed UV can provide more than a photo-chemical effect as explained above.

15.5 Applications of UV, UV-LED, and pulsed UV light for food safety

While UV treatment can achieve desired microbial inactivation rates, it can also, like any other food processing technology, induce chemical changes in foods depending on treatment parameters such as wavelength, exposure time, and food matrix. The primary concern with photochemical reactions in food is the oxidation of key nutrients. Oxidation can occur due to direct reaction of UV radiation with target nutrients including proteins, lipids, and micronutrients. In the majority of cases, the oxidation of proteins and lipids is caused due to the formation of oxygen (O_2) during photochemical reactions, initiating a chain reaction (Davies, 2003; Ghnimi, Budilarto, & Kamal-Eldin, 2017). The application of UV light is well established for disinfection of food preparation surfaces and for air and water treatment (Koutchma, Forney, & Moraru, 2009). However, due to undesirable characteristics of current commercial UV radiation sources (low- and medium-pressure mercury lamps) such as overheating, high-energy costs, the risk of mercury contamination to food or food surfaces, and limited control over wavelength emission, alternative sources including LEDs have been investigated. UV-LEDs have shown promise for food safety applications using various wavelengths, including water disinfection, surface decontamination, and liquid/solid food applications.

Lui, Roser, Corkish, Ashbolt, and Stuetz (2016) reported that UVC LEDs at 270 nm inactivated *E. coli* and *E. faecalis*, achieving a 5-log CFU/mL reduction in 3 min and 18 min, respectively. UVA LEDs (365, 385, 405 nm) were also investigated, however, a far longer duration (3 hour) was required to achieve 5-log reductions at these wavelengths.

Shin, Kim, Kim, and Kang (2016) applied UVC radiation (275 nm) to known foodborne pathogens (*E. coli* O157: H7, *Salmonella enterica* serovar *Typhimurium*, and *L. monocytogenes*) on solid culture media. Inactivation of 6, 6, and 5-log10 CFU/mL reductions for *E. coli*, *S.* Typhimurium, and *L. monocytogenes*, respectively, were observed after 5 minutes exposure to UVC light with 1.67 mJ/cm^2. They found that *L. monocytogenes* was the most UV-resistant bacteria.

Li et al. (2010) investigated the application of pulsed UVA-LED radiation at 365 nm with a reported intensity of 0.28 mW/cm^2 on biofilms that were formed in vitro. It was reported in this study that after 5 minutes of treatment over 90% of the viable microorganisms were inactivated.

Elmnasser et al. (2008) investigated the effect of pulsed light (xenon lamp) on lipid oxidation and possible interaction between lipids and proteins in whole milk, by means of frontal fluorescence (before and after pulsed UV treatment). No difference between treated and untreated samples was observed. The oxidative effects of PUV light in ham slices were investigated by Wambura and Verghese (2011) over a 14-day period. They observed that all samples decreased in oxidative stability over time, but oxidative stability decreased more rapidly in those samples treated with PUV light. Their study highlights the need for more long-term oxidative stability studies on UV-treated foods to further investigate this area.

The effect of UV light on nutrients has also been reported. A study (Bhat, 2016) investigated the impact of UV radiation on tomato juice quality. Significant increases in phenolic content were observed after treatment (60 minutes). In a study by Pan and Zu (2012), the vitamin C content in pineapple juice decreased significantly after UV treatment with monochromatic UV light. While vitamin C content decreased in all samples including the control during storage, it was observed that UV-treated samples declined more rapidly during storage. Akgün and Ünlütürk (2017) investigated the effect of UV light on vitamin C and reported that vitamin C content (2.22 mg/L) in apple juice samples treated with UV-LEDs degraded due to oxidation after long exposure times.

Ma et al. (2012) reported that the degradation of monomeric anthocyanins of *Schisandra chinensis* fruit increased when the samples were exposed to UV light for a long time (9 hours). The decrease in the lightness observed in this study may have been caused by the browning reaction products formed at the end of the monomeric degradation.

Due to the advantages explained so far, pulsed UV light can be used for sanitization, decontamination, and sterilization of liquid and solid foods as well as smooth, dry surfaces such as aluminum, paper, glass, medical devices, and packaging materials. For solid foods, the challenge is the penetration of pulsed UV, which limits the inactivation of microorganisms hidden in crevices. Therefore, Laguna-Solar, Piña, MacDonald, and Bolkan (2006) reported that inactivation of superficial fungal populations on fruits required higher energy levels for inactivation (2–6 fold) than those present on a solid culture medium. For liquid foods, the opaqueness of the fluids limits the pulsed UV reaching the microbial cells. Decontamination of food packaging materials is necessary to ensure microbial safety of food products and extend shelf life, and it is particularly important in aseptic applications (Ringus & Moraru, 2013).

Liu et al. (2021) reported that intense pulsed light treatment could inactivate seven foodborne Gram-positive bacteria, including *S. aureus*, *Bacillus cereus*, *Listeria monocytogenes*, *Enterococcus faecium*, *Bacillus thuringiensis*, *Lactobacillus brevis*, and *Pediococcus acidilactici*.

Zhang et al. (2022) showed that pulsed light treatment is a feasible method to slow down spoilage process in large yellow croaker by inactivating the growth of microorganisms. The study found that 500 J/pulse could lead to lipid oxidation and produce some other odors, but 300 J/pulse pulsed light treatment could achieve the purpose of sterilization without affecting the quality of the yellow croaker. Through the verification of various indicators, the shelf life of the large yellow croaker could be extended by 2–3 days after 300 J/pulse pulsed light treatment at 4°C compared to the control.

Ringus and Moraru (2013) investigated pulsed UV light inactivation of *L. innocua* on food packaging materials of different surface roughness and reflectivity. Their study showed that pulsed UV light is an effective decontaminating method for *L. innocua* cells on five types of food packaging materials. Treatments at a fluence level of about 8 J/cm^2 resulted in a reduction of *L. innocua* ranging from over 3 log CFU for complex, multilayered packaging materials, to over 7 log reduction on high-density or low-density polyethylene (LDPE). Based on Weibull predictions, significantly higher inactivation may be reached by increasing the fluence for the polyethylene-based materials, but not for the complex materials. Other noteworthy observations include the effectiveness of pulsed UV light treatment through transparent polyethylene packaging and the fact that highly reflective and rougher surfaces result in less inactivation than less reflective, smoother materials. This is an important result, consistent with previous works of (Keklik, Demirci, & Puri, 2009, 2010) demonstrating pulsed UV light inactivation of bacteria through LDPE, which suggests that pulsed UV light could potentially be used as a terminal., "through packaging" treatment, when LDPE is used as a packaging material. Therefore, pulsed UV needs to be evaluated for each type of solid and liquid foods. Some of the research for pulsed UV in the literature applied to solid and liquid foods are summarized below:

Jean et al. (2011) investigated inactivation of hepatitis A virus (HAV) and norovirus surrogate in suspension and on food-contact surfaces using pulsed UV light. This study was conducted to evaluate the inactivation of murine norovirus (MNV-1) and HAV by pulsed UV light. Their study showed that short-duration pulses represent an excellent alternative for inactivation of food-borne viruses. This technology could be used to inactivate viruses in drinking water or on food-handling surfaces.

Pulsed UV light has been used for commercial-scale food packaging materials but there is little evidence of large-scale industrial use as a treatment for food. However, emerging applications of pulsed UV light include decontamination of freshly cut, ready-to-eat fruit and vegetables, meat, and fish along with decontamination of associated packaging materials. It is recommended that foods be treated in packaging film based on noncrystalline or low-crystalline polymers such as polystyrene, polyethylene terephthalate, polyvinyl chloride, and polypropylene (Marangoni Junior, Cristianini, & Anjos, 2020).

Woodling and Moraru (2005) demonstrated that surface topography has a complex influence on the efficacy of pulsed UV light treatment for surface microbial reduction. A high degree of surface roughness allows some hiding of microbial cells inside surface details, yet it can also allow for a relatively uniform 2-dimensional distribution of the cells on the treated surface, which favors treatment uniformity. On the other hand, a smooth surface does not necessarily warrant a high level of microbial inactivation by PL treatment if that surface is highly hydrophobic and thus leads to cell clustering. The interaction between light and substrate, such as light reflection and absorption, also plays a role in the inactivation process.

Lagunas-Solar, Piña, MacDonald, and Bolkan (2006) investigated the potential of pulsed UV light technology as an alternative to the use of contact pesticides for the surface disinfection of fresh fruits. More than 5 log reductions in plant (fungal) pathogens on fresh fruits, including apples, kiwi, lemon, nectarines, oranges, peaches, pears, raspberries, and grapes, were obtained in less than 10 seconds. However, in naturally infected or inoculated (sprayed) fruits, a fraction of the inoculum may penetrate into the epidermis or locate in injured tissue in crevices or in surface irregularities. Similarly, the following section will summarize the applications of pulsed UV light on solid foods. Recently, Mohammadi et al. (2023) investigated the described behavior of three primary wines with the presence of epicatechin and acetaldehyde treated with pulse light technology, which would provide a better understanding of the effect of pulse light technology on the wine color. Their results showed that the loss of anthocyanins while applying the pulse light follows the first-order reaction model. It was also demonstrated that the degradation of anthocyanins in pulse light treatment is dependent on the number of hydroxyl or methoxyl group substituents on their B-ring.

15.5.1 *Solid foods*

A limited number of studies have investigated the potential of LED-UV systems for inactivation of food pathogens in solid foods. Foods studied investigated include meat products (Haughton et al., 2012), dairy products (Kim, Kim, & Kang, 2015), and fruits and vegetables (Aihara et al., 2014). While these results are promising, further studies investigating the inactivation capabilities of LEDs on solid foods should be carried out to determine suitable treatments and investigate possible effects of UV-LEDs on food quality. Aihara et al., 2014 reported a 3-log reduction in *E. coli* after 90 minutes treatment with UVA (365 nm), minimal changes in tissue weight and vitamin C or nitrite/nitrate content were observed.

Current literature on solid food applications is relatively low, mainly dealing with the pulsed UV light decontamination of solid and semisolid products such as vegetables (Gomez-Lopez, Devlieghere, Bonduelle, & Debevere, 2005a), fruit (Marquenie, Geeraed, et al., 2003; Marquenie, Michiels, Van Impe, Schrevens, & Nicola, 2003), food powers and seeds (Fine & Gervais, 2004; Jun, Irudayaraj, Demirci, & Geiser, 2003; Sharma & Demirci, 2003), dairy products (Dunn et al., 1991), fish (Ozer & Demirci, 2006), and infant powder milk (Choi et al., 2010). Krishnamurthy et al. (2004), investigated the use of pulsed UV light to inactivate *S. aureus* in a buffer solution and on agar seeded plates. They found a 7 to 8 log reduction of *S. aureus* on seeded agar plates and in buffer solution at treatment times less than 5 seconds without significant temperature increase. Sharma and Demirci (2003) exposed alfalfa seeds inoculated with *E. coli* O157:H7 to pulsed UV light. They found that when a seed layer thickness of 1.02 mm was treated for 30 seconds, a 4.80 log reduction was achieved. For salmon fillets, Ozer and Demirci (2006) reported inactivation rates of 0.86 log CFU/g for *E. coli* after applying 90 pulses at 5 cm from the lamp, and 0.74 log for *L. monocytogenes* with 180 pulses at 8 cm distance, although no information on the energy received by the sample was provided. On the other hand, Ozer and Demirci observed a maximum of 1.09 and 1.02 log reduction of *E. coli* O157:H7 and *L. monocytogenes* on raw salmon fillet skins after a 60-s treatment at a distance of 8 cm from the lamp, with a lamp output of 5.6 J/cm^2 at the surface of the strobe (Ozer & Demirci, 2006).

Hierro, Ganan, Barroso, and Fernández (2012) investigated pulsed UV light treatment for the inactivation of selected pathogens and the shelf-life extension of beef and tuna carpaccio. They found that the application of pulsed UV light to beef and tuna carpaccio at the highest fluences assayed (8.4 and 11.9 J/cm^2) can reasonably improve the microbial safety of the product, achieving inactivation rates of approximately one log. On the other hand, Keklik et al., 2010 reported a 0.9–1.1 log reduction of *Salmonella* typhimurium in vacuum-packaged chicken breast when total doses of 4.0, 8.2, and 16.3 J/cm^2 were applied.

Fine and Gervais (2004) applied a pulsed-UV light for dried food powders such as wheat flour and black pepper and suggested that pulsed UV light can be applied to inactivate *S. cerevisiae* in dried food powders. Stermer, Lasater-Smith, & Brasington (1987) investigated the effect of pulsed UV on inactivation of bacteria in lamb meat. With 4×10^{-3} J/cm^2 energy, 99.9% of the naturally occurring flora of lamb (mostly *Pseudomonas, Micrococcus*, and *Staphylococcus* species) was inactivated.

Ganan, Hierro, Hospital, Barroso, and Fernández (2013) investigated and evaluated the efficacy of pulsed UV light for the inactivation of *L. monocytogenes* and *S. enterica* on two RTE dry-cured meat products (salchichón and loin) in view of a future application of this technology to improve their microbiological safety and shelf-life. This study demonstrates that pulsed UV light is effective to reduce *L. monocytogenes* and *S. typhimurium* on the surface of RTE dry-cured meat products without changing their sensory properties.

Hierro, Manzano, Ordóñez, de la Hoz, and Fernández (2009) performed scanning electron microscopy of an eggshell, revealing that damaged cuticles can create shadow zones and pores where bacteria can hide from pulsed UV light. Thus, they pointed out the importance of an intact cuticle in the efficacious treatment of shell eggs using pulsed UV light. Keklik et al., 2010 reported that the complete inactivation of *Salmonella* Enteritis on eggshells was observed after 20 minutes pulsed UV treatment at 23.6 J/cm^2, which did not significantly chance the albumen height, eggshell strength, or cuticle presence. They showed a maximum inactivation of 2.4 log CFU/g for *S.* Typhimurium applying a total energy of 60–67 J/cm^2 (54% UV light), either vacuum-packaged or not. The authors also reported a 0.9–1.1 log CFU/cm^2 reduction of *S.* Typhimurium in the vacuum-packaged chicken breast when total doses of 4.0, 8.2, and 16.3 J/cm^2 were applied.

Lasagabaster, Arboleya, and De Marañón (2011) also investigated the impact of pulsed UV light technology on the safety and quality of eggs by studying *Salmonella* inactivation on eggshells as well as sensory and rheological properties on egg white. Results obtained in the study show that short treatment time and low fluences are sufficient to inactivate *Salmonella* on eggshells (5 log reduction). Therefore, pulsed UV light processing of eggs could be achieved at low temperatures and under dry conditions.

McLeod et al. (2018) reported that *S. aureus* on chicken fillet surface was reduced by 1.3–3.0 log CFU/cm^2 with PUV light (1.25–18.0 J/cm^2). Rajkovic, Tomasevic, De Meulenaer, and Devlieghere (2017) obtained a log reduction of 1.14 CFU/g for *S. aureus* on one side of a fermented meat sausage slice, which was treated with PUV light at 15 J/cm^2 after 30 minutes following the inoculation.

Keklik, Elik, Salgin, Demirci, and Koçer (2019) demonstrated that *S. aureus* is more resistant to PUV light than *E. coli* O157:H7 on kashar cheese. The most favorable treatment condition, which was selected based on the highest reductions in both microorganisms obtained in the shortest time with no visual alterations on kashar cheese, was identified as the 45 seconds–13 cm treatment. After the PUV treatment at this condition, log reductions of about 1.62 and 3.02 (CFU/cm^2) were achieved for *S. aureus* and *E. coli* O157:H7, respectively, on kashar cheese. In terms of quality changes, there was no significant change in the lipid oxidation, pH, and moisture content of kashar cheese after the favorable treatment. However, PUV treatments including the favorable treatment changed the color parameters of kashar cheese significantly.

Keklik (2019) reported that pastirma (Turkish meat jerky) slices inoculated on top surface with *S. aureus* were subjected to pulsed UV light for 5, 15, 25, 35, and 45 seconds at 5, 8, and 13 cm from the quartz window. Although the 5 cm/45 s treatment (72.3 J/cm^2) yielded a maximum reduction of 2.99 log CFU/cm^2 for *S. aureus*, this treatment changed the color, moisture, and thiobarbituric acid-reactive substances (TBARS) values of pastirma significantly ($P < .05$).

Holck, Liland, Drømtorp, Carlehö, and McLeod (2018) UV-C and pulsed UV light were investigated for their ability to reduce *L. monocytogenes* in salmon. Cold-smoked and raw salmon were spiked with a mix of ten *L. monocytogenes* strains (10^4 CFU/sample) and subsequently exposed to UV-C light (0.0075–0.6 J/cm^2) or high-intensity pulsed UV light (1.3–10.8 J/cm^2). Reductions of *L. monocytogenes* on smoked salmon were 0.7–1.3 log, depending on the fluence. Corresponding reductions for raw salmon muscle side and skin side were 0.2–0.9 log and 0.4–1.1 log, respectively. Generally, reductions using UV-C and pulsed UV light were within the same range, but with some treatments statistically different. *L. monocytogenes* surviving UV treatments on smoked and raw salmon grew at the same rate as controls during storage at 4°C.

Karaoglan, Keklik, and Develi Isıklı (2019) investigated the maximum degradation of anthocyanin after PUV light treatments, which was found to be about 63%. The anthocyanin degradation, brightness, yellow and blue color (%) increased, while red color (%) decreased with longer treatment time and shorter distance.

Holck et al. (2018) investigated the efficiency of continuous UV-C (254 nm) and pulsed UV light for reducing the viability of *Salmonella enteritidis*, *L. monocytogenes*, and enterohemorrhagic *E. coli* on eggs was thoroughly compared. They found that bacterial cells were exposed to UVC light at fluences from 0.05 to 3.0 J/cm^2 (10 mW/cm^2 for 5 to 300 seconds) and pulsed UV light at fluences from 1.25 to 18.0 J/cm, resulting in reductions ranging from 1.6 to 3.8 log, depending on conditions used. Using UV-C light, it was possible to achieve higher reductions at lower fluences compared with pulsed UV light.

Pulsed UV can be useful to inactivate *E. coli* O157:H7 and *Salmonella* spp. on raspberries and strawberries. In Raspberry, 3.9 and 3.4 log reductions occurred in *E. coli* and *Salmonella* spp. at 72 and 59.2 J/cm^2. On strawberries, 2.1 and 2.8 log reductions occurred in *E. coli* and *Salmonella* spp. at 25.7 and 34.2 J/cm^2, respectively (Bialka & Demirci, 2008). Rajkovic et al. (2010)demonstrated that pulsed UV light decontamination treatment can be successfully applied for reduction of *L. monocytogenes* and *E. coli* O157:H7 on a stainless-steel meat contact surfaces, such as a slicing knife. They showed that the decontamination effectiveness of pulsed UV light on the meat contact surface depends on the type of meat product that was in the prior contact with the surface itself and is also a function of time between the contamination and the pulsed UV light treatment.

Hwang, Yee, and Chung (2021) targeted the direct treatment of uninoculated microorganisms on seeds and powdered foods. In addition, a pilot-scale IPL device (rather than a laboratory-scale one) that can handle large sample quantities was devised. Moreover, not only the changes in the numbers of microorganisms but also the changes in the physical properties of the samples that are strongly related to sensory characteristics were measured.

Cardwell et al. (2023) investigated to compare the concentrations of vitamin D_2 in dried button mushrooms exposed to the same dose of pulsed UV radiation either before or after air-drying. In addition, to assess the effect of air-drying separately on the generation of D vitamers, concentrations of D vitamers in irradiated dried mushrooms were compared to those found in irradiated fresh mushrooms exposed to the same dose of pulsed UV radiation. They reported that the potential effect of the sequence of irradiating and drying on the concentration of vitamin D_2 could provide useful information to growers about the optimal sequence of drying and irradiating when producing vitamin D-enhanced dried mushrooms.

15.5.2 Liquid foods

In the last 25 years, pulsed UV light technology research has been conducted by world-wide leading research groups as an alternative to the traditional thermal pasteurization of liquid foods. In addition, there are extremely limited data available on the application of pulsed UV for the decontamination of liquid foods other than clear water, such as fruit juices (Sauer & Moraru, 2009), infant foods (Choi et al., 2010), and milk (Krishnamurthy et al., 2007). Furthermore, to date, only a few studies with the processing of liquid food in a flow-through chamber by using pulsed UV light have been reported (Artíguez, Lasagabaster, & de Marañón, 2011; Krishnamurthy et al., 2007; Pataro et al., 2011).

Krishnamurthy et al. (2008) evaluated pulsed UV light treatment for milk and milk foam decontamination. The inactivation of *S. aureus* in milk and milk foam by pulsed UV light at different distances from the UV source with various dosages of pulsed UV was studied with pulsed UV at the intensity of 1270 mJ/cm^2. The complete inactivation occurred at distances between 8 and 10.5 cm from the UV source for the duration of 180 seconds, while the destruction decreased with the increase of distance and reduction of exposure time for milk.

Hillegras and Demirci (2003) also treated honey, which was inoculated with *Clostridium supergenes, by* pulsed UV. They found that the increase in the number of pulses increased the destruction. They also found that reducing the distance between honey and UV sources increased the destruction. Even though most of the spores were destroyed, the UV treatment was unable to complete the destruction. This suggested that the pulsed UV has limitations in penetrating the honey.

Artíguez et al. (2011) investigated factors affecting microbial inactivation by pulsed UV light in a continuous flow-through unit for liquid products treatment. They demonstrated the high potential of pulsed UV light for microbial inactivation in flow through treated liquid systems. It clearly seems that pulsed UV light technology could be potentially adapted to a commercial setting for the continuous treatment of liquid products. Although further research is needed to corroborate the effectiveness of pulsed UV light for liquid products preservation/decontamination, this technology appears as a promising nonthermal process that could be applied as an alternative to pasteurization in order to improve the microbial safety and quality of this kind of products (e.g., liquid food products and beverages). They also investigated the high potential of pulsed UV light for microbial inactivation in a flow-through liquid systems. It clearly seems that pulsed UV light technology could be potentially adapted to a commercial setting for continuous treatment of liquid products. Uesugi, Woodling, and Moraru (2007) obtained a 6.04 log reduction of *L. innocua* in Butterfield's phosphate buffer (BPB) exposed to 13.3 J/cm^2 of pulsed UV light. Krishnamurthy et al. (2007) obtained up to 7 log reduction *S. aureus* in milk flowing at a rate of 20 mL/min.

Sauer and Moraru (2009) obtained 2.5–2.7 maximum log reductions with a fluence of 13 J/cm^2 for *E. coli* and *E. coli* O157:H7, respectively, in commercial-type clarified apple cider under static conditions. Palgan et al. (2011) reported 5 log reductions in commercial apple juice and one log reduction in commercial orange juice by applying a pulsed UV light fluence of 28 J/cm^2 for *E. coli* K12 DSM 1607 also in static conditions. On the other hand, Ferrario, Alzamora, and Guerrero (2013) reported that 20 seconds of pulsed UV light treatment (24 J/cm^2) demonstrated similar inactivation of *E. coli* in commercial orange juice but somewhat less inactivation in commercial apple juice (3.5 log reduction).

Pataro et al. (2011) investigated the efficacy of pulsed UV light treatments of *L. innocua E. coli* inoculated in apple and orange juices in a range of energy dosages from 1.8 to 5.5 J/cm^2. A laboratory scale continuous flow pulsed UV light system was set up for the experiments. They found that *E. coli* cells displayed greater susceptibility to light pulses than *L. innocua*. Sauer and Moraru (2009) studied the effectiveness of pulsed UV light treatment for the inactivation of *E. coli* in liquid at different levels of clarity such as BPB, tryptic soy broth (TSB), apple juice, and apple cider.

Qi et al. (2023) investigated that the degradation rate of aflatoxins in pulsed light-treated apple juice was determined by high-performance liquid chromatography (HPLC). Subsequently, liquid chromatography-quadruple time-of-flight mass spectrometry (LC-qTOF/Ms) was used to identify degradation products of AFB1, aflatoxin B_2 (AFB_2), aflatoxin G_1 (AFG_1), and aflatoxin G_2 (AFG_2). Finally, the effects of pulsed light on the quality of apple juice were evaluated. They demonstrated the necessity for the application of pulsed light to degrade the mycotoxins in beverages.

Basak, Jha, and Chakraborty (2023) investigated the potential of pulsed light in the pasteurization of tender coconut water (TCW). The initial counts of *E. coli, B. cereus,* and *L. monocytogenes* in the inoculated TCW were 7.00, 9.14, and 7.8 $\log_{10}$ CFU/mL, respectively. For a pulsed light fluence of 465 J/cm^2, *E. coli, B. cereus,* and *L. monocytogenes* exhibited a log reduction of 5.12, 2.97, and 3.40, respectively. *B. cereus* and *L. monocytogenes* exhibited greater resistance than *E. coli* in the TCW. Peroxidase (POD) was more sensitive to pulsed light treatments than polyphenoloxidase in TCW.

15.5.3 *Decontamination of water and wastewater*

Water safety is a significant issue worldwide, particularly in developing counties and rural areas (Song et al., 2016). Water is easily polluted by major pollutant sources such as biological pollutants, which include various pathogenic microorganisms, easily causing the spread and epidemic of water-borne diseases (Wang, Zhu, Chai, & Gu, 2009). UV irradiation has drawn considerable attention in water disinfection and food industries, especially in water treatment applications (Baykuş, Akgün, & Unluturk, 2021; Kim, MylesT, & Maric, 2017; Li et al., 2019). Current UV water disinfection systems consist of low and MPM lamps emitting monochromatic and polychromatic light, respectively (Song et al., 2016). Conventional UV disinfection systems are generally based on low-pressure (LP) mercury UV lamps that emit nearly monochromatic irradiation at ~254 nm, primarily absorbed by nucleic acids in microorganisms (Pousty, Mamane, Cohen-Yaniv, & Bolton, 2022). The conventional way to generate pulsed UV is using a xenon lamp, which has shown application potential in water disinfection and food decontamination. UV disinfection can effectively inactivate most microorganisms in water and possesses a number of merits such as operational simplicity, no chemical addition, and little by-products formation. On the other hand, UV-LEDs are a relatively new UV light source for water treatment. UV-LEDs emit radiation at different wavelengths, and because of their small size, they are flexible in reactor design; optics can control radiation patterns; they have a very short turn-on time; they require a low voltage, and are effective in inactivation of microorganisms at various wavelengths and fluence rate (Pousty et al., 2022). Pulsed UV has yet to be deployed by the water/wastewater sector. This is partly due to a lack of robust, independently validated data for submerged or flow-through treatment applications (Fitzhenry et al., 2021). Newly emerging UV technologies such as pulsed UV light and UV-LEDs have drawn interest in the water/wastewater sector due to advantages such as shorter contact times and compact size (Uslu, Demirci, & Regan, 2015, 2016).

Although UV-LED disinfection has become more widespread in recent years, it is a novel approach for water and wastewater treatment. The effectiveness of UV-LEDs in microbial inactivation depends largely on the wavelength and the species of microorganisms. There is still a debate on whether there is a synergism in microbial inactivation when UV-LEDs at two/multiple wavelengths are used simultaneously for water disinfection. Chevremont, Boudenne, Coulomb, and Farnet (2013) have reported that the combined use of 280 and 365 nm UV-LEDs provided two times higher log-inactivation than the sum of each individually. Nakahashi et al. (2014) have also reported such synergism when 254 and 365 nm UV-LEDs are simultaneously used for *Vibrio parahaemolyticus* inactivation. However, Oguma, Kita, Sakai, Murakami, and Takizawa (2013) did not observe any synergistic effect in *E. coli* inactivation under the simultaneous irradiation from 265, 285, and 310 nm UV-LEDs.

A study carried out by Beck et al. (2017) also investigated the potential of dual-wavelength synergy, using 260 and 280 nm (both in the UVC range) LEDs and water samples inoculated with *E. coli*, Ms2 coliphage, Adenovirus 2, and *Bacillus pumilus* spores. While inactivation was observed at both LED wavelengths, no synergistic effects were observed from combining 260 and 280 nm LED compared to the sum of individual wavelengths acting independently. Nguyen, Suwan, Koottatep, and Beck (2019) investigated a decentralized domestic wastewater treatment system using UV-LEDs. Domestic wastewater was pretreated by a low-cost treatment system with an inclined settler and a sand filter prior to feeding a novel flow-through UV-LED reactor. They found that at an inlet flow rate of 30 L/h, the COD, total suspended solids (TSS), and turbidity of the effluent were 17.7, 3.0 mg/L, and 3.9 NTU, respectively. They show that UV transmittance at 285 nm was enhanced from 29.1% to 70.4%, improving the influent quality for UV-LED disinfection.

Pulsed UV systems have also been used for drinking water as well as wastewater. Lee et al. (2008) reported that the minimum dose of pulsed UV light required for the effective reduction of *Cryptosporidium parvum* (2 log) in water was 15 mJ/cm^2, which was achieved by 5 seconds at 30 cm from the 2 kV light source that operated at 10 pulses/s. Kim, Kim, Hong, Xu, and Lee (2011) investigated the inactivation of *E. coli* in water by pulsed UV. They showed that the sterilization effect of pulsed UV radiation depends on a function of the applied voltage and the number of pulses.

Demirci and Krishnamurthy (2006) investigated the pulsed UV light to inactivate *Bacillus subtilis* spores in deionized water using a flow-through pulsed UV chamber. The results showed the complete inactivation of *B. subtilis* spores ($>5.5\ \log_{10}$ reduction) for all the flow rates evaluated (2, 4, 6, 8, 10, and 14 L/min). They clearly show the potential of pulsed UV light to be used for water disinfection effectively. However, there have been not many studies on pulsed UV disinfection for drinking water and wastewater.

Huffman, Slifko, Salisbury, and Rose (2000) investigated the inactivation of *Klebsiella terrigena*, poliovirus and rotavirus, and *C. parvum* inoculated in water with pulsed UV light at 250 mJ/cm^2. Two pulses achieved a more than 7 log reduction of *K. terrigena* population. A more than 4 log reduction was obtained for polio and rotavirus and *C. parvum* populations.

Pulsed UV light for treating waterborne pathogens, reducing disinfection by-products, and improving water quality in drinking water systems can be used as a novel technology. The enhancement is achieved by rapidly delivering high-energy pulses of broad-spectrum light to the water. An important motivation for using pulsed UV technologies is the requirement to reduce or avoid disinfection by-products (DBPs) as well as the cost, hazards of chlorine, and related chemicals. DPBs are formed when natural organic matter (NOM) is exposed to a disinfectant. Also, chlorine forms trihalomethanes (THM's), ozone forms bromate ions, and chlorine dioxide forms chlorite ions (Amy, Tan, & Davis, 1991; Chow, Fabris, Drikas, & Holmes, 2004; Owen et al., 1995; Singer, 1999). Although it has been known for almost a century that pulsed UV light can kill bacteria, it was universally presumed that pulsed UV had no efficacy against larger pathogens.

In conventional UV disinfection applications, a fluence of 40-mJ/cm^2 is sufficient to deactivate *B. subtilis* spores. In a study carried out by Huffman et al. (2000), water treated with pulsed UV light at 0.25 J/cm^2 (flow rate 15.4 L/min) gave a reduction of >7.4 log for *K. terrigena* after two pulses. These results show that transparency of the media allowed successful microbial inactivation using this technology.

Sharifi-Yazdi and Darghahi (2006) demonstrated that bacterial population in water was reduced by 8 log reduction for *Pseudomonas aeruginosa* after 11 pulses, while the spores of *Bacillus megaterium* was the most resistant and only 4 log reduction achieved after 50 pulses. Therefore, the results of this study demonstrated that pulsed UV light technology could be used as an effective method for the inactivation of pathogenic bacteria in different environments such as drinking water.

Development of pulsed UV has recently also received attention as a potentially novel strategy for the treatment of wastewater and decontaminating wastewater effluent. Baranda et al. (2012) investigated pesticide atrazine photo-degradation in water by pulsed UV light technology. Atrazine was chosen for their study since it is widely used and frequently detected in shallow groundwater (Worthing & Hance, 1991) causing environmental concern because of its low biodegradation. They found that the pulsed UV light degradation process of atrazine is fast, inducing more than 99% of the pesticide degradation in a very short period of exposure time. Moreover, the same degradation rate constant was found for the broad range of initial atrazine concentrations tested (1−1000 mg/L). According to the degradation efficacy, the short treatment time, and the lack of chloro derivatives, pulsed UV light technology appears as a promising process to reduce atrazine contamination in water.

Other studies have also reported that the presence of organic matter such as humic acid may reduce the efficacy of UV for inactivating microorganisms due to the attenuation of light by the former particles, thereby reducing microbial exposures. King, Hoefel, Daminato, Fanok, and Monis (2008) identified increased dissolved organic content in environmental waters with decreased solar inactivation of *C. parvum* oocysts. However, Cantwell et al. (2008) reported that did not significantly affect the inactivation performance of PUV for treating *C. parvum* oocysts suspended in water.

Uslu et al. (2015) investigated decontamination of synthetic municipal wastewater effluent (SMWE) by pulsed UV light treatment. They demonstrated that complete inactivation was obtained after 15-seconds treatment at 8 cm shelf height and 30 mL solution volume for *E. coli* and 15-seconds treatment at 8 cm shelf height and 15 mL solution volume for *B. subtilis*. In addition, a 2.5%−25% (v/v) *E. coli* or *B. subtilis* inoculum was added to the SMWE and treated by pulsed UV light for 5−45 seconds. Results showed that pulsed UV treatment at the optimum conditions increased suspended solids removal by 26.5% and 21.45% for *E. coli* and *B. subtilis*, respectively. Therefore, these results clearly show that pulsed UV light can be used to inactivate microorganisms in a short time. They also investigated the SMWE with 10% *E. coli* or *B. subtilis* inoculum was analyzed for COD and TOC before and after pulsed UV treatment at the determined optimum conditions. They showed that pulsed UV light treatment resulted in COD reductions of 16.4% and 15.8% for *E. coli* and *B. subtilis*, respectively. TOC removal was determined as 17.3% and 16.0% for *E. coli* and *B. subtilis*, respectively. Also, results indicated that TOC and COD decreased by 22.9% and 29.4%, respectively, for cell-free SMWE (uninoculated) after 15-seconds pulsed UV treatments.

The protection by particles is directly related to the number, distribution of sizes, bacterial density, and chemical composition of the particles. These factors contribute to an increased UV demand in wastewater (De la Cruz et al., 2012). The effect of pulsed UV light on inactivating microorganisms depends on a number of factors, including treatment time, TSS, and organic load (TOC or COD), with high concentrations of TSS and organic contaminants in water decreasing the disinfection capability (Uslu at al., 2015). They found the suspended solid concentration increased with increasing initial microbial inoculum percentage from 2.5% to 25% (v/v), 1.7 to 15 mg/L, and 1.3 to 14 mg/L for untreated and treated *E. coli*, respectively and from 2.8 to 16 mg/L and 2 to 16 mg/L for untreated and pulsed UV treated *B. subtilis*, respectively. On the other hand, increasing the initial microbial population generally caused a decrease in the suspended solids removal, and the maximum suspended solids removal was 26.5% for *E. coli* and 21.4% for *B. subtilis* at 2.5% initial inoculum additions.

Uslu et al. (2016) also evaluated the inactivation of *E. coli* and *B. subtilis* via flow-through PUV disinfection in synthetic wastewater with the addition of "multiple pass" analysis whereby the sample was recirculated back under the PUV lamp for a second disinfection step (Uslu et al., 2016). Their results showed complete inactivation with a 10 L/min flow rate for *E. coli* and 6 L/min flow rate for *B. subtilis* using one-pass. For two-pass treatment, complete inactivation for *E. coli* and *B. subtilis* was observed with a 16 L/min flow rate and 10 L/min flow rate, respectively. Furthermore, the PUV treatment resulted in significant COD and TOC reductions (Uslu et al., 2016).

15.6 Limitations to the UV, UV-LED, AND pulsed UV light system

The application of UV light to water, food contact surfaces, and medical equipment for microbial inactivation is widely employed. To date, UV disinfection sources employed are primarily low-pressure and medium-pressure mercury lamps, emitting monochromatic and polychromatic light, respectively. Despite the widespread use of mercury lamps, there are multiple drawbacks associated with their use including high energy consumption, large size, which limits reactor design, high heat emission, and the presence of mercury. UV light tends to penetrate only 50–300 nm into tissue. The reason behind this is its inability to pass through physical barriers. Thus to achieve complete destruction of bacteria, a complete surface should be exposed to help in the stimulation of the defense mechanism against particular organisms. It was seen that waxing shields the bacteria from the rays of UV and hence reduces the performance of UV-C light. If UV-C light is used within the range of 330–480 nm, photoreactivation of the cells will result in the increase of viable microorganisms. To avoid such limitations, the treated products should be refrigerated or should be stored in dark places.

Another limitation of UV light is due to undesirable characteristics of current commercial UV radiation sources (low- and medium-pressure mercury lamps) such as overheating, high-energy costs, the risk of mercury contamination to food or food surfaces, and limited control over wavelength emission. Moreover, while mercury lamps are widely used commercially, their use in the food industry poses some risks to consumers due to the potential exposure of food products to toxic mercury (Hamamoto et al., 2007). In addition to this, the accumulation of mercury waste is undesirable as it can have damaging effects on the environment or human health if not disposed of correctly.

LEDs have potential for use as a highly efficient UV decontamination technology. Recent advances in semiconductor development have resulted in UV-LEDs becoming more widely available. UV-LEDs emit monochromatic light, which enables customized UV-LED disinfection systems at specific wavelengths to be developed. The application of UV-LEDs for disinfection purposes has been studied in recent years, particularly with respect to water disinfection systems. The main challenges in the adoption of UV-LEDs in the food industry are the low output power of LEDs, low external quantum efficiency, and high cost. Moreover, the application of UV light to solid food remains problematic as UV light will only penetrate up to several millimeters depending on the optical properties of the food (Choudhary & Bandla, 2012).

The efficiency of pulsed UV light depends on microbial exposure. Generally, in nontransparent media pulsed UV light, it can only be used as a surface treatment for the first 2 mm (Wallen, May, Rieger, Holloway, & Cover, 2001). Limited pulsed UV light efficiency because of a greater shadow effect has been observed in food products. Indeed, microorganisms may penetrate via crevices or irregularities present on the food surface, or through the epidermis of a vegetable (Gomez-Lopez, Devlieghere, Bonduelle, & Debevere, 2005b; Ozer & Demirci, 2006).

The mercury lamps used in UV light treatments are another source of difficulty. A leakage of mercury from the lamp and its sleeve would pose major health and environmental hazards. However, pulsed UV lamps do not carry this risk because they use inert gases such as xenon. Xenon flash lamps are also more environmentally friendly than UV lamps because they do not use mercury. However, manufacturing pulsed UV lamps can be more expensive.

Sample heating is perhaps the most important limiting factor of pulsed UV light for practical applications. Heat can originate from the absorption of light by food or by lamp heating. When studying the inactivation of *A. niger* spores on corn meal Jun et al. (2003) found that some experimental factor settings resulted in sample temperatures of 120°C, which might have been due to the large amount of heat generated by the lamps, even though their device had a cooling system. Heating also limited the treatment of alfalfa seeds (Sharma & Demirci, 2003), grated carrots (Gomez-Lopez et al., 2005a), and raw salmon fillets (Ozer & Demirci, 2006). It was also demonstrated to occur in honey (Hillegras & Demirci, 2003) and agar (Gomez-Lopez et al., 2005b).

Another limitation of pulsed UV light treatments is the possibility of shadowing occurring when microorganisms readily absorb the rays, as in the case of *A. niger*, and are present one upon another. This makes the organisms in the lower layers very hard to destroy in contrast to those in the upper layer (Hiramoto, 1984), although the use of relatively high peak powers can overcome the shadowing effect.

The superficial absorption of light should be regarded in view of how deep the light can penetrate into the food, that is, the superficial character of the pulsed UV light decontamination process should not be considered limited to an infinitesimal superficial layer of the food because some degree of inactivation can occur below it.

15.7 Concluding remarks and future trends

UV light has long been recognized for its ability to inactivate microorganisms and is widely used in various applications, including food and water treatment. Recently, the development of UV-LEDs and pulsed UV technology has opened up new possibilities for more effective and efficient microbial inactivation in various industries. However, more research is needed to optimize UV-LED systems for microbial inactivation in food and water, explore the effects of wavelength diversity and pulsed illumination, and understand the impact on food quality and nutrition. Pulsed UV technology is a fast and environmentally friendly alternative for food decontamination and water disinfection. However, limitations in opaque and nonuniform surfaces and the impact on nutritional consequences and possible formation of toxic by-products need to be addressed. Future studies should investigate the impact on chemical contaminants and the potential for bio-fortification. Pulsed UV light offers advantages over conventional UV light and could benefit the food sector. The success of pulsed UV technology for liquids depends on the correct alignment of pulsed UV source parameters, and research is needed to establish its suitability for fluid and solid food applications. Continued research and development will be instrumental in the commercial application of these technologies.

References

Aihara, M., Lian, X., Shimohata, T., Uebanso, T., Mawatari, K., Harada, Y., ... Takahashi, A. (2014). Vegetable surface sterilization system using uva light-emitting diodes. *Journal of Medical Investigation, 61*(3–4), 286–290. Available from https://doi.org/10.2152/jmi.61.285, https://www.jstage.jst.go.jp/article/jmi/61/3.4/61_285/_pdf.

Akgün, M. P., & Ünlütürk, S. (2017). Effects of ultraviolet light emitting diodes (LEDs) on microbial and enzyme inactivation of apple juice. *International Journal of Food Microbiology, 260*, 65–74.

Amy, G. L., Tan, L., & Davis, M. K. (1991). The effects of ozonation and activated carbon adsorption on trihalomethane speciation. *Water Research, 25*, 191–202.

Artíguez, M. L., Lasagabaster, A., & de Marañón, I. M. (2011). Factors affecting microbial inactivation by Pulsed Light in a continuous flow-through unit for liquid products treatment. *Procedia Food Science, 1*, 786–791. Available from https://doi.org/10.1016/j.profoo.2011.09.119.

Bank, H. L., John, J., Schmehl, M. K., & Dratch, R. J. (1990). Bactericidal effectiveness of modulated UV light. *Applied and Environmental Microbiology, 56*(12), 3888–3889. Available from https://doi.org/10.1128/aem.56.12.3888-3889.1990.

Baranda, A. B., Barranco, A., & de Marañón, I. M. (2012). Fast atrazine photodegradation in water by pulsed light technology. *Water Research, 46*(3), 669–678. Available from https://doi.org/10.1016/j.watres.2011.11.034, http://www.elsevier.com/locate/watres.

Basak, S., Jha, T., & Chakraborty, S. (2023). Pasteurization of tender coconut water by pulsed light treatment: Microbial safety, enzymatic inactivation, and impact on physicochemical properties. *Innovative Food Science & Emerging Technologies, 84*. Available from https://doi.org/10.1016/j.ifset.2023.103302.

Baykuş, G., Akgün, M. P., & Unluturk, S. (2021). Effects of ultraviolet-light emitting diodes (UV-LEDs) on microbial inactivation and quality attributes of mixed beverage made from blend of carrot, carob, ginger, grape and lemon juice. *Innovative Food Science & Emerging Technologies, 67*. Available from https://doi.org/10.1016/j.ifset.2020.102572.

Beck, S. E., Ryu, H., Boczek, L. A., Cashdollar, J. L., Jeanis, K. M., Rosenblum, J. S., ... Linden, K. G. (2017). Evaluating UV-C LED disinfection performance and investigating potential dual-wavelength synergy. *Water Research, 109*, 207–216. Available from https://doi.org/10.1016/j.watres.2016.11.024, http://www.elsevier.com/locate/watres.

Bhat, R. (2016). Impact of ultraviolet radiation treatments on the quality of freshly prepared tomato (Solanum lycopersicum) juice. *Food Chemistry, 213*, 635–640. Available from https://doi.org/10.1016/j.foodchem.2016.06.096.

Bialka, K. L., & Demirci, A. (2008). Efficacy of pulsed UV-light for the decontamination of *Esherichia coli* O157:H7 and *Salmonella spp.* on raspberries and strawberries. *Journal of Food Science, 73*(5), 201–207.

Block, S. S. (1991). *Disinfection, sterilisation and preservation* (4th ed.). Philadelphia: Lea & Febiger.

Bohrerova, Z., Shemer, H., Lantis, R., Impellitteri, C. A., & Linden, K. G. (2008). Comparative disinfection efficiency of pulsed and continuous-wave UV irradiation technologies. *Water Research, 42*(12), 2975–2982. Available from https://doi.org/10.1016/j.watres.2008.04.001, http://www.elsevier.com/locate/watres.

Brem, R., Guven, M., & Karran, P. (2017). Oxidatively-generated damage to DNA and proteins mediated by photosensitized UVA. *Free Radical Biology and Medicine, 107*, 101–109. Available from https://doi.org/10.1016/j.freeradbiomed.2016.10.488.

Cardwell, G., Bornman, J. F., James, A. P., Daly, A., Strobel, N., Jakobsen, J., & Black, L. J. (2023). Effect of air-drying on the generation of vitamin D2 and 25-hydroxyvitamin D2 by pulsed UV irradiation in button mushroom (Agaricus bisporus). *Journal of Food Composition and Analysis, 115*. Available from https://doi.org/10.1016/j.jfca.2022.105034, http://www.elsevier.com/inca/publications/store/6/2/2/8/7/8/index.htt.

Cassar, J. R., Mills, E. W., & Demirci, A. (2022). Characterization of pulsed light for microbial inactivation. *Journal of Food Engineering, 334*. Available from https://doi.org/10.1016/j.jfoodeng.2022.111152, http://www.sciencedirect.com/science/journal/02608774.

Chang, J. C. H., Ossoff, S. F., Lobe, D. C., Dorfman, M. H., Dumais, C. M., Qualls, R. G., & Johnson, J. D. (1985). UV inactivation of pathogenic and indicator microorganisms. *Applied and Environmental Microbiology, 49*(6), 1361–1365. Available from https://doi.org/10.1128/aem.49.6.1361-1365.1985.

Charles, F., Vidal, V., Olive, F., Filgueiras, H., & Sallanon, H. (2013). Pulsed light treatment as new method to maintain physical and nutritional quality of fresh-cut mangoes. *Innovative Food Science & Emerging Technologies, 18*, 190–195. Available from https://doi.org/10.1016/j.ifset.2013.02.004.

Chevremont, A.-C., Boudenne, J.-L., Coulomb, B., & Farnet, A.-M. (2013). Impact of watering with UV-LED-treated wastewater on microbial and physico-chemical parameters of soil. *Water Research, 47*(6), 1971–1982. Available from https://doi.org/10.1016/j.watres.2013.01.006.

Choi, M. S., Cheigh, C. I., Jeong, E. A., Shin, J. K., & Chung, M. S. (2010). Nonthermal sterilization of Listeria monocytogenes in infant foods by intense pulsed-light treatment. *Journal of Food Engineering, 97*(4), 504–509. Available from https://doi.org/10.1016/j.jfoodeng.2009.11.008.

Choudhary, R., & Bandla, S. (2012). Ultraviolet pasteurization for food industry. *International Journal of Food Science and Nutrition Engineering, 2*(1), 12–15. Available from https://doi.org/10.5923/j.food.20120201.03.

Chow, C.W. K., Fabris, R., Drikas, M., &Holmes, M. (2004). The impact of organic character on water treatment plant performance. In *NOM research: Innovations and applications for drinking watertreatment*. Australia: Victor Harbor.

Cleaver, J. E. (2003). Excision repair – Its bacterial beginnings. *DNA Repair, 2*(11), 1273–1287. Available from https://doi.org/10.1016/S1568-7864(03)00136-8, http://www.elsevier.com/locate/dnarepair.

Cole, H. T., & Bradley, R. A. (2021). Impact of surface reflection on microbial inactivation in a UV LED treatment duct. *Chemical Engineering Science, 230*, 116204.

Condon, S., Alvarez, I., & Gayan, E. (2014). Non-thermal processing: Pulsed UV light. *Encyclopedia of Food Microbiology, 3*, 974–981, Second Edition.

De la Cruz, N., Giménez, J., Esplugas, S., Grandjean, D., de Alencastro, L. F., & Pulgarín, C. (2012). Degradation of 32 emergent contaminants by UV and neutral photo-fenton in domestic wastewater effluent previously treated by activated sludge. *Water Research, 46*(6), 1947–1957. Available from https://doi.org/10.1016/j.watres.2012.01.014.

Davies, M. J. (2003). Singlet oxygen-mediated damage to proteins and its consequences. *Biochemical and Biophysical Research Communications, 305*(3), 761–770. Available from https://doi.org/10.1016/S0006-291X(03)00817-9, http://www.sciencedirect.com/science/journal/0006291X.

Demirci, A., & Krishnamurthy, K. (2006). Disinfection of water by flow-through Pulsed ultraviolet light sterilisation system. *Ultrapure Water Journal, 24*(1), 35–40.

Demirci, A., & Panico, L. (2008). Pulsed ultraviolet light. *Food Science and Technology International, 14*, 443–446.

Dhar, R., Basak, S., & Chakraborty, S. (2022). Pasteurization of fruit juices by pulsed light treatment: A review on the microbial safety, enzymatic stability, and kinetic approach to process design. *Comprehensive Reviews in Food Science and Food Safety, 21*(1), 499–540. Available from https://doi.org/10.1111/1541-4337.12864.

Dunn, J., Clark, R.W., Asmus, J.F., Pearlman, J.S., Boyer, K., Pairchaud, F., & Hofmann, G. (1991). *Methods for preservation of foodstuffs*. Alexandria, VA: US Patent and Trademark Office. US patent 5 034 235.

Dunn, J., Ott, T., & Clark, W. (1995). Pulsed-light treatment of food and packaging. *Food Technology, 49*(9), 95–98.

D'Souza, C., Yuk, H. G., Khoo, G. H., & Zhou, W. (2015). Application of light-emitting diodes in food production, postharvest preservation, and microbiological food safety. *Comprehensive Reviews in Food Science and Food Safety, 14*, 719–740.

Elmer, W., & Griese, Jr. (2012). *Pulsed UV technology is another curing alternative*. Folcroft, PA: Cork Industries, Inc. 19032 610.522.9500 F 610.522.9659.

Elmnasser, N., Dalgalarrondo, M., Orange, N., Bakhrouf, A., Haertle, T., & Federighi, M. (2008). Effect of pulsed-light treatment on milk proteins and lipids. *Journal of Agricultural and Food Chemistry, 56*(6), 1984–1991.

Elmnasser, N., Guillou, S., Leroi, F., Orange, N., Bakhrouf, A., & Federighi, M. (2007). Pulsed-light system as a novel food decontamination technology: A review. *Canadian Journal of Microbiology, 53*(7), 813–821. Available from https://doi.org/10.1139/w07-042.

Falguera, V., Pagán, J., Garza, S., Garvín, A., & Ibarz, A. (2011). Ultraviolet processing of liquid food: A review. Part 2: Effects on microorganisms and on food components and properties. *Food Research International, 44*(6), 1580–1588. Available from https://doi.org/10.1016/j.foodres.2011.03.025.

Ferrario, M., Alzamora, S. M., & Guerrero, S. (2013). Inactivation kinetics of some microorganisms in apple, melon, orange and strawberry juices by high intensity light pulses. *Journal of Food Engineering, 118*(3), 302–311. Available from https://doi.org/10.1016/j.jfoodeng.2013.04.007.

Fine, F., & Gervais, P. (2004). Efficiency of pulsed UV light for microbial decontamination of food powders. *Journal of Food Protection, 67*, 787–792.

Fitzhenry, K., Clifford, E., Rowan, N., & Val del Rio, A. (2021). Bacterial inactivation, photoreactivation and dark repair post flow-through pulsed UV disinfection. *Journal of Water Process Engineering, 41*. Available from https://doi.org/10.1016/j.jwpe.2021.102070.

Ganan, M., Hierro, E., Hospital, X. F., Barroso, E., & Fernández, M. (2013). Use of pulsed light to increase the safety of ready-to-eat cured meat products. *Food Control*, 32(2), 512–517. Available from https://doi.org/10.1016/j.foodcont.2013.01.022.

Gayán, E., Condón, S., & Álvarez, I. (2014). Biological aspects in food preservation by ultraviolet light: A review. *Food and Bioprocess Technology, 7*(1), 1–20. Available from https://doi.org/10.1007/s11947-013-1168-7.

Ghnimi, S., Budilarto, E., & Kamal-Eldin, A. (2017). The new paradigm for lipid oxidation and insights to microencapsulation of omega-3 fatty acids. *Comprehensive Reviews in Food Science and Food Safety, 16*(6), 1206–1218. Available from https://doi.org/10.1111/1541-4337.12300.

Gomez-Lopez, V. M., Devlieghere, F., Bonduelle, V., & Debevere, J. (2005a). Intense light pulses decontamination of minimally processed vegetables and their shelf-life. *International Journal of Food Microbiology, 103*, 79–89.

Gomez-Lopez, V. M., Devlieghere, F., Bonduelle, V., & Debevere, J. (2005b). Factors affecting the inactivation of micro-organisms by intense light pulses. *Journal of Applied Microbiology, 99*(3), 460–470. Available from https://doi.org/10.1111/j.1365-2672.2005.02641.x.

Gomez-Lopez, V. M., Koutchma, T., & Linden, K. (2012). Ultraviolet and pulsed light processing of fluid foods. In P. J. Cullen, B. K. Tiwari, & V. Valdramidis (Eds.), *Novel thermal and non-thermal technologies for fluid foods* (pp. 185–223). Philadelphia PA: Elsevier.

Gomez-Lopez, V. M., Ragaert, P., Debevere, J., & Devlieghere, F. (2007). Pulsed light for food decontamination: A review. *Trends Food Science Technology, 18*, 464–473.

Grapperhaus, M., Schaefer, R.B., & Linden, K. (2005). Modeling of a new UV test cell for evulation of lamp intensity effects on water treatment. In *CDROM Proceedings of the 3rd International Congress on Ultraviolet Technologies*, Whistler, BC, Canada: International Ultraviolet Association.

Guo, M., Hu, H., Bolton, J. R., & El-Din, M. G. (2009). Comparison of low- and medium-pressure ultraviolet lamps: Photoreactivation of Escherichia coli and total coliforms in secondary effluents of municipal wastewater treatment plants. *Water Research, 43*(3), 815–821. Available from https://doi.org/10.1016/j.watres.2008.11.028, http://www.elsevier.com/locate/watres.

Hamamoto, A., Mori, M., Takahashi, A., Nakano, M., Wakikawa, N., Akutagawa, M., … Kinouchi, Y. (2007). New water disinfection system using UVA light-emitting diodes. *Journal of Applied Microbiology, 103*(6), 2291–2298. Available from https://doi.org/10.1111/j.1365-2672.2007.03464.x.

Harm, W. (1980). *Biological effects of ultraviolet radiation* (p. 229) Cambridge: Cambridge University Press.
Haughton, P. N., Grau, E. G., Lyng, J., Cronin, D., Fanning, S., & Whyte, P. (2012). Susceptibility of Campylobacter to high intensity near ultraviolet/visible 395 ± 5nm light and its effectiveness for the decontamination of raw chicken and contact surfaces. *International Journal of Food Microbiology, 159*(3), 267–273. Available from https://doi.org/10.1016/j.ijfoodmicro.2012.09.006.
Hierro, E., Ganan, M., Barroso, E., & Fernández, M. (2012). Pulsed light treatment for the inactivation of selected pathogens and the shelf-life extension of beef and tuna carpaccio. *International Journal of Food Microbiology, 158*(1), 42–48. Available from https://doi.org/10.1016/j.ijfoodmicro.2012.06.018.
Hierro, E., Manzano, S., Ordóñez, J. A., de la Hoz, L., & Fernández, M. (2009). Inactivation of Salmonella enterica serovar Enteritidis on shell eggs by pulsed light technology. *International Journal of Food Microbiology, 135*(2), 125–130. Available from https://doi.org/10.1016/j.ijfoodmicro.2009.07.034.
Hijnen, W. A. M., Beerendonk, E. F., & Medema, G. J. (2006). Inactivation credit of UV radiation for viruses, bacteria and protozoan (oo)cysts in water: A review. *Water Research, 40*(1), 3–22. Available from https://doi.org/10.1016/j.watres.2005.10.030.
Hillegras, S. L., & Demirci, A. (2003). Inactivation of *Clostridium sporogenes* in clover honey by pulsed UV-light treatment. *Agriculture Engineering International: The CIGR Journal of Scientific Research and Development, V*, Manuscript FP 03 009.
Hinds, L. M., Charoux, C. M. G., Akhter, M., O'Donnell, C. P., & Tiwari, B. K. (2020). Effectiveness of a novel UV light emitting diode based technology for the microbial inactivation of Bacillus subtilis in model food systems. *Food Control, 114*. Available from https://doi.org/10.1016/j.foodcont.2019.106910, https://www.journals.elsevier.com/food-control.
Hinds, L. M., O'Donnell, C. P., Akhter, M., & Tiwari, B. K. (2019). Principles and mechanisms of ultraviolet light emitting diode technology for food industry applications. *Innovative Food Science and Emerging Technologies, 56*. Available from https://doi.org/10.1016/j.ifset.2019.04.006, http://www.elsevier.com/locate/ifset.
Hiramoto, T. (1984). *Method of sterilisation*. US Patent number 4464336.
Holck, A. L., Liland, K. H., Drømtorp, S. M., Carlehö, G. M., & McLeod, A. (2018). Comparison of UV-C and pulsed UV light treatments for reduction of salmonella, listeria monocytogenes, and enterohemorrhagic Escherichia coli on eggs. *Journal of Food Protection, 81*(1), 6–16. Available from https://doi.org/10.4315/0362-028X.JFP-17-128, http://jfoodprotection.org/doi/pdf/10.4315/0362-028X.JFP-17-128.
Huffman, D. E., Slifko, T. R., Salisbury, K., & Rose, J. B. (2000). Inactivation of bacteria, virus and Cryptosporidium by a point-of-use device using pulsed broad spectrum white light. *Water Research, 34*(9), 2491–2498. Available from https://doi.org/10.1016/S0043-1354(00)00014-2, http://www.elsevier.com/locate/watres.
Hwang, H. J., Lee, J. Y., Min, S. C., & Chung, M. S. (2023). Application of intense pulsed light – Plasma – Ultraviolet combined system on granular and powdered foods for microbial inactivation. *LWT, 174*, 114447. Available from https://doi.org/10.1016/j.lwt.2023.114447.
Hwang, H. J., Yee, S. Y., & Chung, M. S. (2021). Decontamination of powdery foods using an intense pulsed light (IPL) device for practical application. *Applied Sciences, 11*(4), 1518.
Jean, J., Morales-Rayas, R., Anoman, M.-N., & Lamhoujeb, S. (2011). Inactivation of hepatitis A virus and norovirus surrogate in suspension and on food-contact surfaces using pulsed UV light (pulsed light inactivation of food-borne viruses). *Food Microbiology, 28*(3), 568–572. Available from https://doi.org/10.1016/j.fm.2010.11.012.
John, D., & Ramaswamy, H. S. (2018). Pulsed light technology to enhance food safety and quality: A mini-review. *Current Opinion In Food Science, 23*, 70–79.
Jun, S., Irudayaraj, J., Demirci, A., & Geiser, D. (2003). Pulsed UV-light treatment of corn meal for inactivation of Aspergillus niger spores. *International Journal of Food Science and Technology, 38*(8), 883–888. Available from https://doi.org/10.1046/j.0950-5423.2003.00752.x.
Kalisvaart, B. F. (2004). Re-use of wastewater: Preventing the recovery of pathogens by using medium-pressure UV lamp technology. *Water Science and Technology, 50*(6), 337–344. Available from https://doi.org/10.2166/wst.2004.0393.
Karaoglan, H. A., Keklik, N. M., & Develi Isıklı, N. (2019). Degradation kinetics of anthocyanin and physicochemical changes in fermented turnip juice exposed to pulsed UV light. *Journal of Food Science and Technology, 56*(1), 30–39. Available from https://doi.org/10.1007/s13197-018-3434-1, http://www.springerlink.com/content/121580/.
Keklik, N. M., Demirci, A., & Puri, V. M. (2009). Inactivation of *Listeria monocytogenes* on unpackaged and vacuum-packaged chicken frankfurters using pulsed UV-light. *Journal of Food Science, 74*(8), M431–M439.
Keklik, N. M. (2019). Treatment of pastirma with pulsed UV light: Modeling of Staphylococcus aureus inactivation and assessment of quality changes. *Food Science and Technology International, 26*(2), 185–198.
Keklik, N. M., Demirci, A., & Puri, V. M. (2010). Decontamination of unpackaged and vacuum-packaged boneless chicken breast with pulsed ultraviolet light. *Poultry Science, 89*(3), 570–581. Available from https://doi.org/10.3382/ps.2008-00476.
Keklik, N. M., Elik, A., Salgin, U., Demirci, A., & Koçer, G. (2019). Inactivation of Staphylococcus aureus and Escherichia coli O157:H7 on fresh kashar cheese with pulsed ultraviolet light. *Food Science and Technology International, 25*(8), 680–691. Available from https://doi.org/10.1177/1082013219860925, http://www.sagepub.com.
Keshavarzfathy, M., Malayeri, A. H., Mohseni, M., & Taghipour, F. (2020). UV-LED fluence determination by numerical method for microbial inactivation studies. *Journal of Photochemistry and Photobiology A: Chemistry, 392*. Available from https://doi.org/10.1016/j.jphotochem.2020.112406.
Keyser, M., Muller, I. A., Cilliers, F. P., Nel, W., & Gouws, P. A. (2008). Ultraviolet radiation as a non-thermal treatment for the inactivation of microorganisms in fruit juice. *Innovative Food Science and Emerging Technologies, 9*(3), 348–354. Available from https://doi.org/10.1016/j.ifset.2007.09.002.
Khan, M. A., Shatalov, M., Maruska, H. P., Wang, H. M., & Kuokstis, E. (2005). III-nitride UV devices. *Japanese Journal of Applied Physics, Part 1: Regular Papers and Short Notes and Review Papers, 44*(10), 7191–7206. Available from https://doi.org/10.1143/JJAP.44.7191, http://jjap.ipap.jp/link?JJAP/44/7191/pdf.
Kim, H. J., Kim, D. J., Hong, J. T., Xu, G. C., & Lee, D. G. (2011). Sterilisation of *Escherichia coli* based on Nd: YAG resonator with a pulsed xenon flashlamp. *Journal of Electrical Engineering & Technology, 6*(2), 275–279.
Kim, S. J., Kim, D. K., & Kang, D. H. (2015). Using UVC light-emitting diodes at wavelengths of 266 to 279 nanometers to inactivate foodborne pathogens and pasteurize sliced cheese. *Applied and Environmental Microbiology, 82*(1), 11–17.

Kim, S., MylesT, D., & Maric, R. (2017). Performance of a high temperature polymer electrolyte membrane fuel cell with low catalyst loading produced by reactive spray deposition technology. *ECS Transactions*, *66*, 24.

King, B. J., Hoefel, D., Daminato, D. P., Fanok, S., & Monis, P. T. (2008). Solar UV reduces Cryptosporidium parvum oocyst infectivity in environmental waters. *Journal of Applied Microbiology*, *104*(5), 1311–1323. Available from https://doi.org/10.1111/j.1365-2672.2007.03658.x.

Koizumi, S., Watanabe, K., Hasegawa, M., & Kanda, H. (2001). Ultraviolet emission from a diamond pn junction. *Science (New York, N.Y.)*, *292* (5523), 1899–1901. Available from https://doi.org/10.1126/science.1060258.

Koutchma, T., Forney, L. J., & Moraru, C. I. (2009). *Ultraviolet light in food technology: Principles and applications* (pp. 1–279). Canada: CRC Press. Available from https://www.taylorfrancis.com/books/e/9781420059519.

Kowalski, W. (2009). *Ultraviolet germicidal irradiation handbook: UVGI for air and surface disinfection*. Berlin Heidelberg: Springer. Available from 10.1007/978-3-642-01999-9.

Krishnamurthy, K., Demirci, A., & Irudayaraj, J. (2004). Inactivation of *Staphylococcus aureus* by pulsed UV-light sterilisation. *Journal of Food Protection*, *67*, 1027–1030.

Krishnamurthy, K. (2006). *Decontamination of milk and water by pulsed UV-light and infrared heating* (Ph.D. dissertation). University Park, PA: Pennsylvania State University.

Krishnamurthy, K., Demirci, A., & Irudayaraj, J. (2007). Inactivation of *Staphylococcus aureus* in milk using flow-through pulsed UV-light treatment system. *Journal of Food Science*, *72*(7), M233–M239.

Krishnamurthy, K., Demirci, A., & Irudayaraj, J. (2008). Inactivation of Staphylococcus aureus in milk and milk foam by pulsed UV-light treatment and surface response modeling. *Transactions of the ASABE*, *51*(6), 2083–2090.

Krishnamurthy, K., Tewari, J. C., Irudayaraj, J., & Demirci, A. (2010). Microscopic and spectroscopy evaluation of inactivation of *Staphylococcus aureus* by pulsed UV light and infrared heating. *Food Bioprocess Tech*, *3*(1), 93–104.

Laguna-Solar, M. C., Piña, C., MacDonald, J. D., & Bolkan, L. (2006). Development of pulsed UV light processes for surface fungal disinfection of fresh fruits. *Journal of Food Protection*, *69*, 376–384.

Lagunas-Solar, M. C., Piña, C., MacDonald, J. D., & Bolkan, L. (2006). Development of pulsed UV light processes for surface fungal disinfection of fresh fruits. *Journal of Food Protection*, *69*(2), 376–384.

Lasagabaster, A., Arboleya, J. C., & De Marañón, I. M. (2011). Pulsed light technology for surface decontamination of eggs: Impact on Salmonella inactivation and egg quality. *Innovative Food Science and Emerging Technologies*, *12*(2), 124–128. Available from https://doi.org/10.1016/j.ifset.2011.01.007.

Lazarova, V., Savoye, P., Janex, M. L., Blatchley, E. R., & Pommepuy, M. (1999). Advanced wastewater disinfection technologies: State of the art and perspectives. *Water Science and Technology*, *40*(4-5), 203–213. Available from https://doi.org/10.2166/wst.1999.0593.

Lee, S. U., Joung, M., Yang, D. J., Park, S. H., Huh, S., Park, W. Y., & Yu, J. R. (2008). Pulsed-UV light inactivation of Cryptosporidium parvum. *Parasitology Research*, *102*(6), 1293–1299. Available from https://doi.org/10.1007/s00436-008-0908-5.

Liang, S., Min, J. H., Davis, M. K., Green, J. F., & Remer, D. S. (2003). Use of pulsed-UV processes to destroy NDMA. *Journal - American Water Works Association*, *95*(9), 121–131. Available from https://doi.org/10.1002/j.1551-8833.2003.tb10459.x.

Linden, K.G., & Mofidi, A.A. (2003). *Disinfection efficiency and dose measurement of polychromatic UV light: High quality water-monitoring and treatment*. AWWA RF Report 90931F. London, UK: IWA Publishing.

Linden, K. G., Shin, G., & Sobsey, M. D. (2001). Comparative effectiveness of UV wavelengths for the inactivation of Cryptosporidium parvum oocysts in water. *Water Science and Technology*, *43*(12), 171–174. Available from https://doi.org/10.2166/wst.2001.0731.

Li, X., Cai, M., Wang, L., Niu, F., Yang, D., & Zhang, G. (2019). Evaluation survey of microbial disinfection methods in UV-LED water treatment systems. *Science of the Total Environment*, *659*, 1415–1427.

Li, J., Hirota, K., Yumoto, H., Matsuo, T., Miyake, Y., & Ichikawa, T. (2010). Enhanced germicidal effects of pulsed UV-LED irradiation on biofilms. *Journal of Applied Microbiology*, *109*, 2183–2190.

Liu, Z., Hu, S., Soteyome, T., Bai, C., Liu, J., Wang, Z., ... Xu, Z. (2021). Intense pulsed light for inactivation of foodborne gram-positive bacteria in planktonic cultures and bacterial biofilms. *LWT - Food Science and Technology*, *152*, 112374.

Lui, G. Y., Roser, D., Corkish, R., Ashbolt, N. J., & Stuetz, R. (2016). Point-of-use water disinfection using ultraviolet and visible light-emitting diodes. *Science of the Total Environment*, *553*, 626–635. Available from https://doi.org/10.1016/j.scitotenv.2016.02.039, http://www.elsevier.com/locate/scitotenv.

MacGregor, S. J., Rowan, N. J., McIlvaney, L., Anderson, J. G., Fouracre, R. A., & Farish, O. (1998). Light inactivation of food-related pathogenic bacteria using a pulsed power source. *Letters in Applied Microbiology*, *27*(2), 67–70. Available from https://doi.org/10.1046/j.1472-765x.1998.00399.x.

MacGregor, S. J., Turnbull, S. M., Tuema, F. A., & Farish, O. (1997). Factors affecting and methods of improving the pulse repetition frequency of pulse-charged and DC-charged high-pressure gas switches. *IEEE Transactions on Plasma Science*, *25*(2), 110–117. Available from https://doi.org/10.1109/27.602480.

Mahendran, R., Ramanan, K. R., Barba, F. J., Lorenzo, J. M., López-Fernández, O., Munekata, P. E. S., ... Tiwari, B. K. (2019). Recent advances in the application of pulsed light processing for improving food safety and increasing shelf life. *Trends in Food Science and Technology*, *88*, 67–79. Available from https://doi.org/10.1016/j.tifs.2019.03.010, http://www.elsevier.com/wps/find/journaldescription.cws_home/601278/description#description.

Marangoni Junior, L., Cristianini, M., & Anjos, C. A. R. (2020). Packaging aspects for processing and quality of foods treated by pulsed light. *Journal of Food Processing and Preservation*, *44*(11), 14902.

Marquenie, D., Michiels, C. W., Van Impe, J. F., Schrevens, E., & Nicola, B. N. (2003). Pulsed white light in combinations with UV- C and heat to reduce storage rot of strawberry. *Postharvest Biology and Technology*, *28*, 455–461.

Marquenie, D., Geeraed, A. H., Lammertyn, J., Soontjens, C., Van Impe, J. F., Michiels, C. W., & Nicolaï, B. M. (2003). Combinations of pulsed white light and UV-C or mild heat treatment to inactivate conidia of *Botrytis cinerea* and *Monilia fructigena*. *International Journal of Food Microbiology*, *85*, 185–196.

Ma, C., Yang, L., Yang, F., Wang, W., Zhao, C., & Zu, Y. (2012). Content and color stability of anthocyanins isolated from Schisandra chinensis fruit. *International Journal of Molecular Sciences*, *13*(12), 14294–14310. Available from https://doi.org/10.3390/ijms131114294.

McDonald, K. F., Curry, R. D., Clevenger, T. E., Unklesbay, K., Eisenstark, A., Golden, J., & Morgan, R. D. (2000). Comparison of pulsed and continuous ultraviolet light sources for the decontamination of surfaces. *IEEE Transactions on Plasma Science, 28*(5), 1581–1587. Available from https://doi.org/10.1109/27.901237.

McLeod, A., Hovde Liland, K., Haugen, J. E., Sørheim, O., Myhrer, K. S., & Holck, A. L. (2018). Chicken fillets subjected to UV-C and pulsed UV light: Reduction of pathogenic and spoilage bacteria, and changes in sensory quality. *Journal of Food Safety, 38*(1). Available from https://doi.org/10.1111/jfs.12421, http://www.wiley.com/bw/editors.asp?ref = 0149-6085&site = 1.

Mertens, B., & Knorr, D. (1992). Developments of nonthermal processes for food preservation. *Food Technology, 46*, 124–133.

Meulemans, C. C. E. (1986). *The basic principles of UV-sterilisation of water. Ozone + ultraviolet water treatment, Aquatec Amsterdam* (pp. B.1.1–B.1.13). Paris: International Ozone Association.

Miller, R., Jeffrey, W., Mitchell, D., & Elasri, M. (1999). Bacterial responses to ultraviolet light. *ASM News, 65*(8), 534–541.

Mofidi, A. A., Baribeau, H., Rochelle, P. A., De Leon, R., Coffey, B. M., & Green, J. F. (2001). Disinfection of cryptosporidium parvum with polychromatic UV light. *Journal / American Water Works Association, 93*(6), 95–109. Available from https://doi.org/10.1002/j.1551-8833.2001.tb09229.x, http://onlinelibrary.wiley.com/journal/10.1002/(ISSN)1551-8833.

Mohammadi, X., Matinfar, G., Mandal, R., Singh, A., Fiutak, G., Kitts, D. D., & Pratap Singh, A. (2023). Kinetics of anthocyanin condensation reaction in model wine solution under pulsed light treatment. *Food Chemistry, 405*. Available from https://doi.org/10.1016/j.foodchem.2022.134600, http://www.elsevier.com/locate/foodchem.

Murugesan, R., Orsat, V., & Lefsrud, M. (2012). Effect of pulsed ultraviolet light on the total phenol content of elderberry (Sambucus nigra) fruit. *Food and Nutrition Sciences, 03*(06), 774–783. Available from https://doi.org/10.4236/fns.2012.36104.

Nakahashi, M., Mawatari, K., Hirata, A., Maetani, M., Shimohata, T., Uebanso, T., … Takahashi, A. (2014). Simultaneous irradiation with different wavelengths of ultraviolet light has synergistic bactericidal effect on Vibrio parahaemolyticus. *Photochemistry and Photobiology, 90*(6), 1397–1403. Available from https://doi.org/10.1111/php.12309.

Nebot Sanz, E., Salcedo Dávila, I., Andrade Balao, J. A., & Quiroga Alonso, J. M. (2007). Modelling of reactivation after UV disinfection: Effect of UV-C dose on subsequent photoreactivation and dark repair. *Water Research, 41*(14), 3141–3151. Available from https://doi.org/10.1016/j.watres.2007.04.008.

Nguyen, T. M. H., Suwan, P., Koottatep, T., & Beck, S. E. (2019). Application of a novel, continuous-feeding ultraviolet light emitting diode (UV-LED) system to disinfect domestic wastewater for discharge or agricultural reuse. *Water Research, 153*, 53–62. Available from https://doi.org/10.1016/j.watres.2019.01.006, http://www.elsevier.com/locate/watres.

Oguma, K., Kita, R., Sakai, H., Murakami, M., & Takizawa, S. (2013). Application of UV light emitting diodes to batch and flow-through water disinfection systems. *Desalination, 328*, 24–30. Available from https://doi.org/10.1016/j.desal.2013.08.014.

Oms-Oliu, G., Aguiló-Aguayo, I., Martín-Belloso, O., & Soliva-Fortuny, R. (2010). Effects of pulsed light treatments on quality and antioxidant properties of fresh-cut mushrooms (Agaricus bisporus). *Postharvest Biology and Technology, 56*(3), 216–222. Available from https://doi.org/10.1016/j.postharvbio.2009.12.011.

Oner, M. E. (2017). Inactivation of *Salmonella enteritidis* on almonds by pulsed light treatment. *Academic Food, 15*(3), 242–248.

Owen, D., Amy, G., Chowdhury, Z., Paode, R., McCoy, G., & Viscosil, K. (1995). NOM characterization and tertability. *Journal of American Water Works Association, 87*, 46–63.

Ozer, N. P., & Demirci, A. (2006). Inactivation of *Escherichia coli* O157:H7 and *Listeria monocytogenes* inoculated on raw salmon fillets by pulsed UV-light treatment. *International Journal of Food Science & Technology, 40*, 1–7.

Palgan, I., Caminiti, I. M., Muñoz, A., Noci, F., Whyte, P., Morgan, D. J., … Lyng, J. G. (2011). Effectiveness of high intensity light pulses (HILP) treatments for the control of Escherichia coli and Listeria innocua in apple juice, orange juice and milk. *Food Microbiology, 28*(1), 14–20. Available from https://doi.org/10.1016/j.fm.2010.07.023.

Pan, Y. G., & Zu, H. (2012). Effect of UV-C radiation on the quality of fresh-cut pineapples. *Procedia Engineering, 37*, 113–119.

Pataro, G., Muñoz, A., Palgan, I., Noci, F., Ferrari, G., & Lyng, J. G. (2011). Bacterial inactivation in fruit juices using a continuous flow Pulsed Light (PL) system. *Food Research International, 44*(6), 1642–1648. Available from https://doi.org/10.1016/j.foodres.2011.04.048.

Patel, A. M., Dhar, R., & Chakraborty, S. (2023). Pulsed light, microwave, and infrared treatments of jaggery: Comparing the microbial decontamination and other quality attributes. *Food Control, 149*. Available from https://doi.org/10.1016/j.foodcont.2023.109695.

Pelvan Akgün, M., & Ünlütürk, S. (2017). Effects of ultraviolet light emitting diodes (LEDs) on microbial and enzyme inactivation of apple juice. *International Journal of Food Microbiology, 260*, 65–74.

Pousty, D., Mamane, H., Cohen-Yaniv, V., & Bolton, J. R. (2022). Ultraviolet actinometry Determination of the incident photon flux and quantum yields for photochemical systems using low-pressure and ultraviolet light-emitting diode light sources. *Journal of Environmental Chemical Engineering, 10*, 107947.

Qi, L., Ma, Y., Cai, R., Li, Y., Wang, R., Yue, T., … Wang, Z. (2023). Degradation of aflatoxins in apple juice by pulsed light and the analysis of their degradation products. *Food Control, 148*. Available from https://doi.org/10.1016/j.foodcont.2023.109648.

Rajkovic, A., Tomasevic, I., De Meulenaer, B., & Devlieghere, F. (2017). The effect of pulsed UV light on Escherichia coli O157:H7, Listeria monocytogenes, Salmonella Typhimurium, Staphylococcus aureus and staphylococcal enterotoxin A on sliced fermented salami and its chemical quality. *Food Control, 73*, 829–837. Available from https://doi.org/10.1016/j.foodcont.2016.09.029.

Rajkovic, A., Tomasevic, I., Smigic, N., Uyttendaele, M., Radovanovic, R., & Devlieghere, F. (2010). Pulsed UV light as an intervention strategy against Listeria monocytogenes and Escherichia coli O157:H7 on the surface of a meat slicing knife. *Journal of Food Engineering, 100*(3), 446–451. Available from https://doi.org/10.1016/j.jfoodeng.2010.04.029.

Ringus, D. L., & Moraru, C. I. (2013). Pulsed ligh inactivation of *Listeria innocua* on food packaging materials of different surface roughness and reflectivity. *Journal of Food Engineering, 114*(3), 331–337.

Rowan, N. J., MacGregor, S. J., Anderson, J. G., Fouracre, R. A., McIlvaney, L., & Farish, O. (1999). Pulsed-light inactivation of food-related microorganisms. *Applied and Environmental Microbiology, 65*(3), 1312–1315. Available from https://doi.org/10.1128/aem.65.3.1312-1315.1999.

Sauer, A., & Moraru, C. I. (2009). Inactivation of *Escherichia coli* ATCC 25922 and *Escherichia coli* O157:H7 in apple juice and apple cider, using pulsed light treatment. *Journal of Food Protection, 72*, 937–944.

Schaefer, R.B. (2004). *Patent: Surface discharge lamp and system*. US Patent 6,724,134 B1.
Sharifi-Yazdi, M. K., & Darghahi, H. (2006). Inactivation of pathogenic bacteria using pulsed UV-light and its application in water disinfection and quality control. *Acta Medica Iranica, 44*, 305–308.
Sharma, R. R., & Demirci, A. (2003). Inactivation of *Escherichia coli* O157:H7 on inoculated alfalfa seeds with pulsed ultraviolet light and response surface modeling. *Journal of Food Science, 68*, 1448–1453.
Shin, J. Y., Kim, S. J., Kim, D. K., & Kang, D. H. (2016). Fundamental characteristics of deep-UV light-emitting diodes and their application to control foodborne pathogens. *Applied and Environmental Microbiology, 82*(1), 2–10. Available from https://doi.org/10.1128/AEM.01186-15, http://aem.asm.org/content/82/1/2.full.pdf.
Singer, P. C. (1999). Humic substances as precursors for potentially harmful disinfection by-products. *Water Science and Technology, 40*(9), 25–30. Available from https://doi.org/10.2166/wst.1999.0434.
Sinha, R. P., & Hader, D. P. (2002). UV-induced DNA damage and repair: A review. *Photochemical & Photobiological Sciences: Official Journal of the European Photochemistry Association and the European Society for Photobiology, 1*, 225–236.
Sommer, R., Lhotsky, M., Haider, T., & Cabaj, A. (2000). UV inactivation, liquid-holding recovery, and photoreactivation of Escherichia coli O157 and other pathogenic Escherichia coli strains in water. *Journal of Food Protection, 63*(8), 1015–1020. Available from https://doi.org/10.4315/0362-028x-63.8.1015.
Song, K., Mohseni, M., & Taghipour, F. (2016). Application of ultraviolet light-emitting diodes (UV-LEDs) for water disinfection: A review. *Water Research, 94*, 341–349. Available from https://doi.org/10.1016/j.watres.2016.03.003.
Song, K., Taghipour, F., & Mohseni, M. (2019). Microorganisms inactivation by wavelength combinations of ultraviolet light-emitting diodes (UV-LEDs). *Science of the Total Environment, 665*, 1103–1110. Available from https://doi.org/10.1016/j.scitotenv.2019.02.041.
Stermer, R. A., Lasater-Smith, M., & Brasington, C. F. (1987). Ultraviolet radiation - An effective bactericide for fresh meat. *Journal of Food Protection, 50*(2), 108–111. Available from https://doi.org/10.4315/0362-028X-50.2.108.
Sunita, T., Sharma, A. P. M., Kaukab, S., & Mishra, A. (2022). Light-based processing technologies for food. *Current developments in biotechnology and bioengineering: Advances in food engineering* (pp. 183–218). India: Elsevier. Available from https://www.sciencedirect.com/book/9780323911580, 10.1016/B978-0-323-91158-0.00004-1.
Sutherland, B. M. (1981). Photoreactivation in bacteria and in skin. *Journal of Investigative Dermatology, 77*(1), 91–95. Available from https://doi.org/10.1111/1523-1747.ep12479267.
Takeshita, K., Shibato, J., Sameshima, T., Fukunaga, S., Isobe, S., Arihara, K., & Itoh, M. (2003). Damage of yeast cells induced by pulsed light irradiation. *International Journal of Food Microbiology, 85*(1-2), 151–158. Available from https://doi.org/10.1016/S0168-1605(02)00509-3, http://www.elsevier.com/locate/ijfoodmicro.
Taniyasu, Y., Kasu, M., & Makimoto, T. (2006). An aluminium nitride light-emitting diode with a wavelength of 210 nanometres. *Nature, 441* (7091), 325–328. Available from https://doi.org/10.1038/nature04760.
Tran, M. T. T., & Farid, M. (2004). Ultraviolet treatment of orange juice. *Innovative Food Science & Emerging, 5*, 495–502.
Uesugi, A. R., Woodling, S. E., & Moraru, C. I. (2007). Inactivation kinetics and factors of variability in the pulsed light treatment of Listeria innocua cells. *Journal of Food Protection, 70*(11), 2518–2525. Available from https://doi.org/10.4315/0362-028X-70.11.2518, http://www.ingentaconnect.com/content/iafp/jfp.
Uslu, G., Demirci, A., & Regan, J. M. (2016). Disinfection of synthetic and real municipal wastewater effluent by flow-through pulsed UV-light treatment system. *Journal of Water Process Engineering, 10*, 89–97. Available from https://doi.org/10.1016/j.jwpe.2016.02.004, http://www.journals.elsevier.com/journal-of-water-process-engineering/.
Uslu, G., Demirci, A., & Regan, J. M. (2015). Efficacy of pulsed UV-light treatment on wastewater effluent disinfection and suspended solid reduction. *Journal of Environmental Engineering (United States), 141*(6). Available from https://doi.org/10.1061/(ASCE)EE.1943-7870.0000912, http://ascelibrary.org/toc/joeedu/current.
Uysal-Pala, C., & Kırca-Toklucu, A. (2011). Effect of UV-C light on anthocyanin content and other quality parameters of pomegranate juice. *Journal of Food Composition and Analysis: An Official Publication of the United Nations University, International Network of Food Data Systems, 24* (6), 790–795.
Uysal-Pala, Ç., & Kırca-Toklucu, A. (2013). Microbial, physicochemical and sensory properties of UV-C processed orange juice and its microbial stability during refrigerated storage. *LWT - Food Science and Technology, 50*, 426–431.
Vilhunen, S., Särkkä, H., & Sillanpää, M. (2009). Ultraviolet light-emitting diodes in water disinfection. *Environmental Science and Pollution Research, 16*(4), 439–442. Available from https://doi.org/10.1007/s11356-009-0103-y.
Wallen, R. D., May, R., Rieger, K., Holloway, J. M., & Cover, W. H. (2001). Sterilisation of a new medical device using broad spectrum pulsed light. *Biomedical Instrumentation & Technology, 35*, 323–330.
Wambura, P., & Verghese, M. (2011). Effect of pulsed ultraviolet light on quality of sliced ham. *LWT – Food Science and Technology, 44*, 2173–2179.
Wang, T., MacGregor, S. J., Anderson, J. G., & Woolsey, G. A. (2005). Pulsed ultra-violet inactivation spectrum of *Escherichia coli*. *Water Research, 39*, 2921–2925.
Wang, S., Zhu, H. P., Chai, Z. N., & Gu, C. L. (2009). Safety assessment of the international drinking-water. *Chinese Food Research and Development, 30*(11), 182–185.
Wan, Q., Wen, G., Cao, R., Xu, X., Zhao, H., Li, K., ... Huang, T. (2020). Comparison of UV-LEDs and LPUV on inactivation and subsequent reactivation of waterborne fungal spores. *Water Research, 173*. Available from https://doi.org/10.1016/j.watres.2020.115553.
Watanabe, K., Taniguchi, T., & Kanda, H. (2004). Direct-bandgap properties and evidence for ultraviolet lasing of hexagonal boron nitride single crystal. *Nature Materials, 3*(6), 404–409. Available from https://doi.org/10.1038/nmat1134.
Weber, E., & Scheible, K. (2003). Pulsed-UV unit may inactivate biological agents. *Journal - American Water Works Association, 95*, 34–46.
Wekhof, A. (2003). *Proceedings of the second sterilisation of packaged pharmaceutical solutions, packaging and surgical tools with pulsed UV light*, July 9–11, Vienna, Austria.
Wen, G., Deng, X., Wan, Q., Xu, X., & Huang, T. (2019). Photoreactivation of fungal spores in water following UV disinfection and their control using UV-based advanced oxidation processes. *Water Research, 148*, 1–9. Available from https://doi.org/10.1016/j.watres.2018.10.028.

Woodling, S. E., & Moraru, C. I. (2005). Influence of surface topography on the effectiveness of pulsed light treatment for the inactivation of *Listeria innocua* on stainless-steel surfaces. *Journal of Food Science, 70*(7), m345–m351.

Worthing, C. R., & Hance, R. J. (1991). *The pesticide manual: A world compendium* (9th ed.). Farnham, Surrey, UK: The British Crop Protection Council.

Xenon. (2003). *Sterilisation and decontamination using high energy light*. Woburn, MA: Xenon Corporation.

Zhang, B., Sun, B., Ma, F., & Zheng, H. (2012). The design and theoretical analysis of major components of pulse light sterilisation equipment. *Procedia Engineering, 37*, 260–267.

Zhang, J., Zhou, G., Ji, S., Zou, L., Liang, J., Walayat, N., ... Ding, Y. (2022). Effect of pulse light on the quality of refrigerated (4°C) large yellow croaker (*Pseudosciaena crocea*). *LWT, 167*, 113855.

Zimmer, J. L., & Slawson, R. M. (2002). Potential repair of *Escherichia coli* DNA following exposure to UV radiation from both medium- and low-pressure UV sources used in drinking water treatment. *Applied and Environmental Microbiology, 68*(7), 3293–3299.

C H A P T E R

16

Microwave heating in food processing

Laura Principato and Giorgia Spigno

DiSTAS Department for Sustainable Food Process, Università Cattolica del Sacro Cuore, Piacenza, Italy

16.1 Introduction

Microwave (MW) heating belongs to the so-called nonconventional thermal processing technologies that have been developed or are still under investigation with the main aim of overcoming some typical problems associated with conventional thermal technologies. These problems include cost and environmental aspects, but first, the thermal impact on food quality, particularly in terms of loss of nutritional quality. In fact, irrespective of the main purpose of the heating process (preservation, development of taste and flavor, changes in structure to obtain a specific texture), a certain level of thermal damage will occur. Food technologists and engineers have, then, been working in the direction of minimizing as most as possible such thermal impact following a concept of total food quality, which cannot consider only the compulsory safety prerequisite but also the nutritional and sensorial quality of the final product.

This chapter provides an overview of the basic mechanisms of MW heating, with a focus on its possible applications in the food sector (pasteurization and sterilization, baking and cooking, heating and tempering, freezing, drying, and extraction), underlining, for each application, the main limits of conventional thermal technologies and the potential advantages of MW treatment, together with its drawbacks. Moreover, to overcome typical limits related to MW application, especially nonuniform temperature distribution, the concept of hybrid/combination approach, also called MW-assisted food processing technologies, will be described. These technologies integrate the advantages of MW energy with unconventional processes (e.g., ultrasonication, ohmic heating, electron irradiation, freezing, osmotic dehydration, infrared heating, and vacuum processing) to enhance product quality and process efficiency and reduce time and operational costs of conventional processes (e.g., drying, extraction, baking, pasteurization/sterilization, tempering) (Verma, Mahanti, Thakur, Chakraborty, & Srivastav, 2020, p. 25).

This chapter represents, with permission, an update of the work by Spigno (2016, p. 137).

16.2 Principles of microwave heating

MW heating, as radiofrequency heating, is referred to as dielectric heating and it is based on the use of electromagnetic waves of certain frequencies to generate heat in a material. MW heating is, then, an electro-heat technology, such as radiofrequency and ohmic heating (OH), in which heating is generated volumetrically throughout a product rather than having to rely on the slow conduction of heat through its surface (Regier & Schubert, 2001, p. 178).

The electromagnetic spectrum between frequencies of 300 MHz and 300 GHz is represented by MWs, but for MW heating applications, only some discrete frequency bands have been set aside from telecommunication applications for industrial, scientific, and medical (the so-called ISM) applications. The ISM bands are defined by the International Telecommunication Union-Radiocommunication sector in the radio regulations, but individual countries' use of the bands designated in these sections may differ due to variations in national radio regulations. Typically, MW food processing uses the 2450 and 915 MHz bands, of which the 2450 MHz band is used for home

Food Packaging and Preservation.
DOI: https://doi.org/10.1016/B978-0-323-90044-7.00016-1

ovens, and both are used in industrial heating; however, the 915 MHz band should not be used in countries outside Region 2 (Americas, Greenland and some of the eastern Pacific Islands), while outside the United States, frequencies of 433.92, 896 and 2375 MHz are also used.

MW heating relies on the dielectric properties of the food materials, in particular, the relative dielectric constant ε' and the relative dielectric loss ε'' which, together provide an indication of the electrical insulating ability of the material (Singh & Heldman, 2009, p. 147), where foods are very poor insulators resulting in the absorption of a large fraction of the energy when placed in a MW field with a consequent instantaneous heating.

The relative dielectric constant is the ratio of the permittivity of the material to that of free space or vacuum, and it defines the interaction of a non- or low-conducting material with an electric field and its ability to store electrical energy; while the relative (always to that of vacuum) dielectric loss denotes the ability of the material to dissipate the electrical energy. The two properties are correlated by the loss tangent (tanδ), also called dissipation factor (D) in Eq. (16.1):

$$\varepsilon'' = \varepsilon' \tan\delta \tag{16.1}$$

The loss tangent is frequency dependent and provides an indication of how well the material can be penetrated by an electrical field and how it dissipates electrical energy as heat.

A more in-depth mathematical description of the physical principles behind MW heating can be found in other references (Regier & Schubert, 2001, p. 178; Singh & Heldman, 2009, p. 147), while for the aims of this chapter, only a simplified approach will be presented.

When the food is placed in an electric field of strength E (V/cm) generated from an energy source with frequency f (Hz or s^{-1}), the MW energy will be converted into heat, resulting in a certain rate of heat generation per unit of volume, also called power dissipation (P_D, W/cm^3), according to Eq. (16.2):

$$P_D = 2\pi\varepsilon_o f E^2 \varepsilon' \tan\delta \tag{16.2}$$

where ε_0 is the constant vacuum permittivity (8.85×10^{-14} Fm^{-1}). It follows that the heating rate will be higher at a higher frequency (2450 MHz than 915 MHz), increasing the E, and is, of course, dependent on the food dielectric loss.

The simplified Eq. (16.2) can be integrated with the conventional models for transient heat transfer to predict heat transfer in the food material. However, it is generally quite impossible to estimate the dielectric properties of a food based on its composition and only measured values can act as starting values for accurate calculations. The food ingredients showing the strongest interaction with MW are water and salts. This is because the interaction between an electric field and a dielectric material causes polarization, due to a displacement of the charge centers or an orientation of dipoles. When the applied electric field is time-dependent (e.g., at the MW frequency of 2450 MHz the polarity alternates rapidly at 2.45×10^9 cycles per second), the mass and the corresponding inertia of the dipoles to rotate or the charge centers to displace, causing a difference in phase between the applied field and the resulting polarization, an effect that is reflected in the permittivity value. At the typical frequencies of MW heating, the origin of dielectric loss is due to two main mechanisms:

- The ion conductivity is dependent on the ion concentrations, valences, and mobility. Due to their inherent charge, the ions will move and collide converting the kinetic energy into thermal energy and temperature increase.
- The dipole-orientation is for which polar molecules (such as water) will orient themselves according to the polarity of the electric field, rotating to maintain alignment with the changing polarity. The molecules' rotation generates friction with the surrounding medium and then heat.

Both the effects are frequency dependent, but also temperature dependent in a not easily predictable manner (Regier & Schubert, 2001, p. 178). For example, for water, heating results in a decreasing loss factor value; for salt solutions, in a certain temperature and concentration range, a stable loss factor can be observed; for ethanol and cellulose a thermal runaway effect occurs above 50°C; for melting ice, a sudden increase of 3000 times of the ε'' occurs with consequent water boiling.

To further complicate things, the penetration depth of MW also needs to be considered. In fact, it has been stated above that in the MW process heating is generated volumetrically throughout the product and does not have to rely on the slow conduction of heat through its surface. However, the MWs penetrate the material until a certain penetration depth, where they are absorbed by the material and converted into heat. The generated heat is, then, transferred to the inner material layers through conventional heat transfer mechanisms: conduction for solids and convection for liquids.

The penetration depth can be defined as the distance from the surface of a dielectric material where the incident power is decreased to 1/e of the incident power, but also in terms of half-power depth and then, as the distance from the surface where the incident power is decreased to one-half of the incident power. Lambert's expression for power absorption (Singh & Heldman, 2009, p. 147) is defined by Eq. (16.3):

$$P_d = P_o e^{-2\alpha' d} \tag{16.3}$$

where P_0 is the incident power, P_d is the incident power at the penetration depth d and α' is the attenuation factor. According to the first definition of d, this will be equal to $1/(2\,\alpha')$ or $0.347/\alpha'$ based on the second definition.

The attenuation factor is, on its turn, dependent on the material dielectric properties (ε' and tan δ) and directly related to the frequency of the MW field as described in Eq. (16.4):

$$\alpha' = \frac{f 2\pi}{c}\left[\frac{\varepsilon'}{2}\left(\sqrt{1+\tan^2\delta}-1\right)\right]^{1/2} \tag{16.4}$$

where c is the vacuum speed of light.

It comes that the penetration depth will be higher for the systems operating at 915 MHz than those working at 2450 MHz, while the power dissipation (Eq. 16.2) will be higher at 2450 MHz.

An important consideration related to these aspects is that MW heating is strictly dependent on the sample size. Sample size should be always larger than the penetration depth, otherwise, significant amounts of the electric field will reach the back face of the sample to be reflected, so that the electric field in the sample will be increased many folds leading to dramatically increased heating rate and danger of sample overheating. This is not likely to occur in industrial processing equipment treating large sample sizes (in this case, the electric field and therefore power absorption are greater at the incident face, they decay exponentially into the sample and the total power absorption is nearly size-independent), but MW can also be applied for the cooking, bleaching, and drying of small food pieces; so in these cases, the sample size-heating correlation should be carefully taken into account, such as when scaling-up laboratory scale results. Besides the size, also the sample shape can greatly influence the sample heating and temperature distribution inside it, as discussed in Section 16.3.1.

As a final remark for this section, MW heating modeling (which is indeed required for process control, monitoring, and optimization) results are very complex, as for the other electro-heating technologies, since it must properly include and combine the description of the physical problems of electromagnetism, heat and mass transfer to predict the temperature distribution and the MW power absorption inside the food material. Despite the inevitable simplifications that are introduced, the final models still require powerful computers and specific numerical methods to be solved. The method of finite differences in time domain and the finite element method are among the most used and the reader is referred to Regier and Schubert (2001, p. 178) and Chandrasekaran, Ramanathan, and Basak (2013) for an in-depth analysis.

16.3 Food applications of microwave heating

Various applications of MW heating have been developed for food processing (such as sanitization, cooking, drying, tempering, and defrosting) and food-related sectors (such as analytical methods and extraction of bioactive compounds). A rather complete review has been recently published on MW food processing (Chandrasekaran et al., 2013) together with other review papers and handbooks or books, which witness the ever-increasing interest by researchers and industry in MWs exploitation in the food sector (Datta & Ramaswamy, 2001; Lorence & Pesheck, 2009; Ramaswamy & Tang, 2008; Shaheen, El-Massry, El-Ghorab, & Anjum, 2012, p. 3; Schubert & Regier, 2005; Sumnu & Sahin, 2005, p.419; Vadivambal & Jayas, 2010).

The aim of this chapter is to discuss, from a theoretical point of view, the main limitations of conventional food processes which MW was meant to solve and then, bring some practical examples showing the actual success or failure of MW in achieving this.

16.3.1 *Pasteurization and sterilization*

Pasteurization and sterilization are treatments aimed to produce safe foods destroying pathogenic microorganisms, along with edible foods inactivating enzymes or destroying microorganisms that do not cause any health concern but spoil the food making it unacceptable for the consumer. Pasteurization is a relatively mild heat treatment (typical processing temperatures range from 63°C to 90°C) with the purpose of destroying selected vegetative microbial species or inactivating the enzymes. As a consequence, pasteurized foods need to be contained and stored under refrigeration, with or without chemical additives or modified atmosphere packaging, and show a typical shelf-life from several days to several months. This treatment is generally applied to liquid products and minimally processed vacuum-packaged foods (sous vide products).

Sterilization is a more severe treatment to destroy microorganisms of both spoilage and public health concerns. Since it must guarantee the removal of also sporogenic species, the reference temperature for this treatment is 121°C (250°F), even though it is strictly dependent on pH, therefore acid foods, such as tomato derivatives, can be sterilized also at temperatures below 100°C. Sterilization is combined with vacuum sealing, hermetic packaging, and moderate storage temperature. Conventional pasteurization and sterilization are positive thermal treatments for which food technologists must aim to minimize the impact on food quality. According to the 1977 FDA definition of the "*minimal thermal process*," it is necessary to scientifically determine the minimum required thermal treatment (in terms of specific temperature—time combinations) to ensure the destruction of microorganisms of public health concern. The T-time combination depends on the process target (microorganism or enzyme), its thermal properties (in terms of decimal reduction time, *D*, and thermal resistance, *Z*), its initial content (N_i), and desired final content (N_f). Combining the first and second Bigelow's laws, the required treatment time at a reference temperature T_0 (called thermal death time, *TDT*, or often lethality, *F*) and an equivalent treatment at a different temperature T, can be calculated as reported in Eqs. (16.5 and 16.6):

$$\mathrm{TDT}_{T_0} = \log\left(\frac{N_i}{N_f}\right) D_{T_0} \tag{16.5}$$

$$\mathrm{TDT}_{T} = \mathrm{TDT}_{T_0} \cdot 10^{\frac{T_0 - T}{Z}} \tag{16.6}$$

This minimal treatment needs to be assured for the whole food sample. Product temperature is, first, not constant during the process (which comprises a heating phase, a holding phase, and a cooling phase); second, it is not uniform through the sample. Therefore, the temperature distribution in the sample and in the equipment needs to be assessed to localize the so-called coldest point, and then the thermal history (in terms of temperature trend) of this points needs to be monitored during the process so that the obtained values can be properly processed to calculate the real process time required to achieve a target TDT_{T_0} or to convert the applied process time into the equivalent TDT_{T_0}. Different approaches are possible (the graphical method or the Ball formula method), but their explanation is out of the scope of this chapter and can be found in many other texts (Ramaswamy & Marcotte, 2006; Valentas, Rotstein, & Singh, 1997).

The same concepts can be applied to the food quality attributes, such as vitamins in primis, color compounds, proteins, and other bioactive compounds. Luckily, it has been observed that many quality attributes show both longer decimal reduction times and higher thermal resistances than microbial populations. This means that if, based on Eqs. (16.5) and (16.6), a shorter treatment at a higher temperature is carried out to be equivalent in terms of microbial lethality, the quality attributes will be damaged to a less extent, which is the concept at the basis of the high-temperatures-short-times (HTST) processes.

In conventional thermal processes where conventional heat exchangers are used, heat transfers basically through conduction and conductive mechanisms. It is therefore evident that the food sample starts heating in the external layers close to the heating medium and that solid foods heat up slower than liquid foods (due to the smaller conductive heat-transfer coefficients compared to the convective ones). Furthermore, solid samples need to be treated only after being packaged and in batch or semi-continuous retorts, which worsens the conditions in terms of heating/cooling speed and temperature inhomogeneity. On the other hand, low-viscosity liquid samples, or even more viscous products if specific heat exchangers are used (such as scraped surface heat exchangers), can be treated before packaging (aseptic processing) using continuous heat exchangers (plate or tubular heat exchangers), which allow to make the liquid flow in thin layers minimizing the temperature variation inside the sample and maximizing the heating velocity. Even in the case of after-packaging-treatment, introduction of container movement can greatly improve the convective heat transfer coefficient of the liquid and improve its temperature homogeneity. Finally, the worst conditions are encountered with products consisting of mixtures of liquids and solids (such as vegetable-meat soups and sauces) which must be treated inside containers and where the two phases heat up at greatly different rates.

Following the above premises, it can be summarized that to reduce the thermal impact on food quality, the thermal treatment should:

- Be as fastest as possible (both in the heating and cooling phase)

- Result in a homogenous treatment in terms of temperature distribution inside the food material
- Be associated with other preservation means to reduce the working temperature or duration, according to the hurdle technologies principle (see the above comment about sterilization of acid food, or, for example, combination with chemical preservatives, unfavorable pressure, redox potential, atmosphere composition) (Leistner & Gould, 2002)
- Be substituted by a nonthermal treatment with microbial lethal effects (e.g., high-pressure processing, pulsed electric fields, irradiation, ultrasounds)

Remaining in the field of pure thermal treatments, whereas the first two characteristics can be reached also with conventional technologies if liquid samples are treated in an aseptic process (see the above considerations and the direct contact steam infusion or steam injection heat exchanger), it is almost impossible with solid packaged food and liquid-solid mixtures.

MW heating was thought to be able to solve these problems since it is potentially:

- associated also with nonthermal lethal effects on microorganisms
- homogenous through the entire product volume
- very rapid also for solid phases

The first assumption has been mainly demonstrated not to occur. A very interesting IFT report for the IFT, Report of the Institute of Food Technologists for the Food and Drug Administration of the U.S. Department of Health and Human Services (2000) contains a detailed review of literature works dealing with the assessment of the proposed nonthermal inactivation mechanisms: selective heating, electroporation, cell membrane rupture, and magnetic field coupling. In agreement with the later review by Chandrasekaran et al. (2013), literature results for different microbial species have been inconsistent, therefore at the MW frequencies nonthermal effects are to be excluded. There might be, anyway, an intensification of thermal effects due to MWs, but these have not yet been elucidated. Analogously to what has been reported in MW-assisted solvent extraction of bioactive compounds from plant materials, sudden heating can cause localized hot spots and pressure build-up with related cell disruption.

Anyway, regarding microbial lethality, the only actual recognized effect of MW heating is a thermal effect, therefore successful pasteurization or sterilization needs to be demonstrated as it is done in conventional thermal processes. This can be done, whenever possible (homogenous solid or liquid foods), identifying the coldest point, monitoring the temperature in such location during the process, elaborating the data to calculate the corresponding delivered lethality, and validating the process with microbial tests. If it is not possible to correctly identify the coldest point (e.g., with inhomogeneous products or solid-liquid mixtures), the process needs to be directly evaluated with microbial tests. And, here come the difficulties related to MW processing.

Monitoring temperature distribution inside the food material is not an easy task since traditional thermocouples are not compatible with MWs because their metallic wires change the field distribution. Fiber optic thermometers, more delicate and expensive, are to be used, giving anyway a poor local resolution (Regier & Schubert, 2001, p. 178). The best solution would be the magnetic resonance imaging (MRI) technique which allows good spatial resolution for both surface temperature and temperature distribution throughout the product, it is noninvasive but is extremely expensive to be applied for ordinary measurements at an industrial scale. However, Luan, Tang, Pedrow, Liu, and Tang (2013) showed that in a single mode, 915 MHz MW-assisted thermal sterilization (MATS) system (a 22 m long pilot-scale system that includes four sections: preheating, MW heating, holding, and cooling to simulate continuous industrial processes), it can be possible to use mobile metallic temperature sensors to monitor the temperature of packaged since they did not influence the general heating pattern. A temperature increase close to the sensor occurs causing a local modification of heat pattern, but this can be avoided if the metallic temperature sensor is placed perpendicular to the electric field component.

Prediction of temperature distribution with simulation models (Chandrasekaran et al., 2013) or real temperature measurement inside the MW heated product, can show that it is very difficult to have a homogenous treatment and that, moreover, the position of the coldest point can vary during the process (Vadivambal & Jayas, 2010). The main risk is that some points remain at lower temperatures receiving an inadequate lethality, while others overheat. These are almost the same risks of conventional treatment, with also many variables influencing temperature distribution in MW heating: sample size, shape, and composition; sample position in the MW equipment; type of MW equipment and irradiation mode. Many of these variables influence temperature distribution and heating rate also in conventional thermal processes, but while in these processes the relations variable-effect is well-known and predictable, they are not in MW processes. It follows an attempt to summarize all the factors that can influence temperature distribution and their action mechanisms.

The sample size, shape, and composition are all tightly correlated. As previously commented, MWs penetrate from the outside and their absorption is exponentially reduced advancing towards the center of the product depending on the attenuation factor and, therefore, on the MW frequency and product dielectric properties. It comes that, in principle, the temperature will rise in the external layers and then in the more internal ones, similarly to conventional processes. However, if the sample size is small related to the MW penetration depth, all the thickness of the product can almost uniformly heat up while, if the sample size is too small a reflection of the electric field at the back face will occur causing overheating. Besides the sample size, its shape is fundamental and can have a dramatic effect on the final heating profile. A well-known example is that of exploding in-shell raw eggs inside MW ovens. Due to the cm-dimensions and the convex surface, the MWs are refracted at the surface and focused to the center causing an overheating of this zone (the "in-shell" eggs phenomenon has been studied by different researchers and Kumar and Sanavullah (2011) identified the locations and conditions in a MW oven to cook them without explosion). If the shape is cylindrical, the upper layers behind the surface heat up and with increasing temperature the power absorption capacity increases leading to less penetration of the waves and the migration of hottest locations from interior to the surface (IFT, Report of the Institute of Food Technologists for the Food and Drug Administration of the U.S. Department of Health and Human Services, 2000). This phenomenon is observed if the cylinder is located in vertical position and MWs are applied from the top. If the position is changed to horizontal or if the MW system allows for uniform wave penetration from all directions, then the hot spot will be at the center. The study by Basaran-Akgul, Basaran, and Rasco (2008) showed that tissue orientation in beef meat had a greater effect on dielectric loss values at lower frequencies and that penetration depth tended to be greater when the direction of propagation was perpendicular to the muscle fiber.

In-package MW pasteurization of beef frankfurters (horizontally placed in the cavity) showed that samples experienced shrinkage in the longitudinal direction and expansion in the radial direction, both by approximately 10%, which was probably attributable to the "edge-heating effect" during MW heating (Huang & Sites, 2007).

Moreover, since in MW systems, the air surrounding the sample is transparent to MW and stays cold, the surface will tend to cool (evaporation of water at the surface can further decrease the surface temperature) which is exactly the opposite of what happens in conventional thermal processes. Things change when corners are present, such as in slab/brick shapes. Corners can be either the hottest or the coldest points depending on the type of MW system, single mode, or multimode (details are provided in Section 17.7).

Regarding the food composition, water and salts are the components, showing the strongest interactions with MWs. It turns out that foods with a higher water activity (it is the free water and not the bound water that can interact with MW) and salt content will heat more in MW systems. There might also be the possibility that a solid phase heats up as much as or even more than a liquid one. The salt composition can be exploited to enhance product heating. For example, Wang, Zhang, and Mujumdar (2010) optimized the addition levels of different ingredients (NaCl, sucrose, and sodium glutamate) by response surface methodology to increase the drying rate and shorten drying time (with a MW freeze dryer) of an instant vegetable soup. Fat and oil show low dielectric properties but have low specific heats, therefore they can heat well in an MW field. Proteins have dielectric properties changing after denaturation, and they increase since denaturation brings to an increase in the free water of the system, such as happens when starch gelatinizes. Simple sugars have higher dielectric properties compared to more complex carbohydrates.

The knowledge of dielectric properties of food components and of their dependence on temperature and wave frequency (Datta, Sumnu, & Raghavan, 2005, p. 501) is fundamental in predicting and understanding the MW heating of a product.

The review by Vadivambal and Jayas (2010) reports some practical examples of inhomogeneous temperature distribution for different types of products treated with MW: ready-to-eat meals (such as spaghetti in a tray and a bolognese meat sauce on the top showing different temperature distribution for single-mode and multimode tunnel); different kinds of meats (for example, chicken revealing different temperature in vent, breast, legs, wings and also depending on the breast up- or down-position in the oven, with up to 61°C temperature difference between different parts of the carcass) and grains. Peng, Tanga, Jiaoa, Bohneta, and Barrett (2013) investigated the importance of knowing the dielectric properties of tomatoes to properly develop an MW sanitization treatment, finding a decrease of dielectric constant but an increase of dielectric loss with increasing temperature; lower dielectric losses at 2450 MHz than at 915 MHz; an increase in loss factors and reduction in penetration depth with salt addition; no difference in dielectric constant but difference in loss factor for the different tomato pericarp, locular and placental tissues related to their different ionic content.

Koskiniemi, Truong, Simunovic, and McFeeter (2011) improved the heating uniformity of in-pack pasteurized acidified vegetables with a 915 MHz continuous MW system, by proper addition of NaCl and citric acid solution and rotation of the cups during the treatment. Also, for mashed potatoes it was shown that dielectric loss factors and constants decrease with frequency, that dielectric loss factors increased with temperature and salt content, but dielectric constants were not significantly affected, and, that power penetration depth increased with moisture content and decreased with temperature, frequency, and salt content (Guan, Cheng, Wang, & Tang, 2004). Sakiyan, Sumnu, Sahin, and Meda (2007) studied the dielectric constant and loss factor of cake samples which were shown to be dependent on formulation, baking time, and temperature. In particular, the increase in baking time and temperature decreased the dielectric constant and loss factor of all formulations (due to moisture content reduction), while fat content was shown to increase dielectric constant and loss factor.

Many other papers can be found on the assessment of dielectric properties of different foods under MW for thermal purposes (Al-Muhtaseb, Hararah, Megahey, McMinn, & Magee, 2010; Brinley, Truong, Coronel, Simunovic, & Sandeep, 2008; Kumar, Coronel, Simunovic, & Sandeep, 2008; Kumar, Coronel, Simunovic, Truong, & Sandeep, 2007), together with attempts of using food simulants and estimation procedures that could avoid complex experimental measurements of dielectric properties (Curet, Rouaud, & Boillereaux, 2014).

In the case of homogenous liquid samples, MW penetration and heating of the first layers under the surface will generate natural convective motions and heat will then transfer to the remaining part of the sample. As for conventional treatment, stirring the sample will increase the convective heat transfer coefficients improving temperature homogeneity. If no stirring is provided, it can be useful to cycle the MW irradiation (application of successive cycles consisting in on and off radiation periods) and exploit the off period for heat distribution inside the sample. The maximum achievable temperature in the liquid is, anyway, limited by its specific boiling point, therefore, again as in conventional treatments, pressure needs to be increased if higher temperatures need to be obtained. Continuous fluid pasteurization and sterilization with tubes intersecting waveguides have been investigated since it would be an ideal solution (Abdullah et al., 2013; Campañone & Zaritzky, 2010; Coronel, Simunovic, Sandeep, Cartwright, & Kumar, 2008; Kumar, Coronel, Simunovic, & Sandeep, 2007; Nguyen, Choi, Lee, & Jun, 2013; Zhu, Kuznetsov, & Sandeep, 2007).

The addition of susceptors (MW absorbing materials) can be exploited to enhance the heating of weak MW-absorbing portions of foods since, if opportunely placed, they will heat, and the heat will be transferred by conduction and radiation to the food product. On the opposite, introduction of metallic foils which reflect MW (consumers will know what happens if aluminum containers are used in MW ovens) can reduce the excessive heating of some food portions allowing for a more homogenous treatment. These considerations are particularly useful for solid foods, ready-to-eat meals made of different ingredients, and for packaged foods. MW application for the treatment of packed food is one of the most interesting applications, provided, of course, that packaging material transparent to MW is used.

Addition of susceptors to packaging material makes it a type of active packaging. One of the most well-known active packaging for MW foods is that for popcorn (Risch, 2009). Even though the first paper bag of MW popcorn was sold in 1971, it became a popular product only in the mid-1980s when a metalized PET film, acting as a susceptor, was laminated between two layers of paper in a position so that it lies on the floor of the MW oven. The metalized film is thin enough to absorb MW energy (instead of reflecting it), heat to temperatures of 200°C or higher, and then provide the required energy to get the corn kernels to pop. The same technology is used for other MW products, including pizza, hand-held sandwiches, and French fries for which it helps the surface to dry and enhances browning and crisping.

Contrary to liquid samples, the maximum temperature in solids is not limited by evaporation so that they can overheat in a very short time (thermal runaway effect), therefore in this case, it is even more important to regulate the power irradiation with appropriate cycling or, generally, it is better to work at lower powers for longer times. For sure, MW heating provides an extremely fast heating rate which, if opportunely exploited considering the above considerations, can provide the desired lethal effect with a reduced thermal impact on quality attributes.

Finally, problems associated with an uneven distribution of the electromagnetic field in the cavity where the product is irradiated can be also responsible for an uneven heating of the product. In domestic oven, the turntable has been conceived to partly solve this problem, while the problem is less important in industrial ovens. In any case, after the heating phase, the product must be rapidly cooled down as in conventional pasteurization or sterilization.

To overcome these limitations, new researches on in-package decontamination processes are carried out by coupling MW heating with other technologies such as UV light, electrons irradiation, irradiation, ozone, and cold plasma (Guillard, Mauricio-Iglesias, & Gontard, 2010). Washington State University with the support of the US Department of Defense Dual Use Science and Technology developed a MW-assisted process in combination with a circulating water surface heating system, called MATS (Tang, 2015). The process involves immersing the packaged food in pressurized hot water and simultaneously heating with MWs at 915 MHz. This frequency can penetrate deeply into the water at 80°C–120°C, reducing reflection and refraction of MWs at the interface between water and foods (which possess similar dielectric constants), thus reducing or eliminating edge heating of the foods. The first pilot-scale system was built in batch mode to prove this design concept. This system has been used for various foods. Sun, Tang, and Powers (2007) used this technology on asparagus demonstrating a greater antioxidant activity and better color preservation compared to the products processed by conventional method. Moreover, Guan, Plotka, Clark, and Tang (2002) confirmed the effect of MATS on the microbial sterilization of macaroni and cheese. A second-generation pilot-scale system was built to simulate a continuous process and, more recently, some protocols, such as sterilization of mashed potatoes in trays and salmon fillet in sauce in pouches, were approved by US FDA (Tang, 2015). This process can be used also for pasteurization. Peng et al. (2017) demonstrated how MW-assisted pasteurization (MAP) system reduced the processing time of precooked carrots by half for $F_{90°C} = 3$ minutes, and by 2/3 for $F_{90°C} = 10$ minutes compared with hot water system. Moreover, MAP contributed to a lower thermal degradation effect on color, with observed lower values of ΔE and higher values of a*. No significant differences in texture and pectin methylesterase (PME) activity, as well as in carotenoids content in both the pasteurization methods were detected.

Other combination technologies under study show great synergistic effects in microbial decontamination or sterilization. Maktabi, Watson, and Parton(2011) achieved a significant reduction in bacterial count (*Pseudomonas fragi* and *Escherichia coli*) using a sequence of UV, laser, and MW heat treatments. According to their study, the overall microbial reduction was higher than that achieved with individual treatments. Moreover, the order of the treatment also had a significant effect: the most effective treatment was laser treatment, followed by MW, and finally UV.

MW energy can be also used to induce plasma formation for microbial decontamination. The efficacy of MW-powered cold plasma treatment (MCPT) was tested on both vegetative forms of microorganisms (Won, Lee, & Min, 2017) and spores (Kim, Oh, Won, Lee, & Min, 2017). Won et al. (2017) showed that nitrogen (N2)—MCPT at MW power of 900 W for 10 minutes, resulted in the highest inhibition of *Penicillium italicum* improving storability of mandarins at temperatures of 4°C and 25°C without affecting fruit physiochemical characteristic or appearance. Similar results were found by Kim et al. (2017) on the reduction of target microbial spores, that is, *Bacillus cereus* spores, *Aspergillus brasiliensis*, and *E. coli* of onion powder at storage temperatures of 4°C and 25°C. High MW density-cold plasma treatment (HMCPT) could reduce the initial counts by 2.1 log spores/cm^2, 1.6 log spores/cm^2, and 1.9 CFU/cm^2, respectively, without significantly affecting either the color or the quercetin content of onion powder, when compared with samples not subjected to MCPT.

Another interesting application is the combination of MW and electron-bean irradiation. Electron beam irradiation is known as "electronic pasteurization" and it is effective also in sterilization process. The main drawback related to industrial scale-up is the high application dose required which might cause deleterious effects (Chizoba Ekezie, Sun, Han, & Cheng, 2017). The integration of MW was intensively studied as an effective way to decrease the electron dose without reducing the efficacy of sterilization. Martin et al. (2005) reported comparative results from separate and combined e-beam and MW irradiation treatments to food products. The results showed that the electron dose level and the time for complete sterilization can effectively be reduced using combined treatment. The synergistic sterilization method is based on the plausible assumption that MW energy could change the sensitivity of microorganisms to electron beam irradiation meanwhile acting on temperature distribution, leading to different effects for the same irradiation dose. Nevertheless, the results demonstrated the prospective application of MW and e-beam treatments as a synergistic sterilization method for the decontamination of food products.

16.3.2 *Baking and cooking*

Due to the problems related to the demonstration of the achievement of an equivalent pasteurization or sterilization treatment with MW heating, the latter has been mainly applied in other thermal food processes where safety is not the primary concern, such as cooking, thawing, re-heating (which are also the main domestic exploitations of MW ovens), drying and extraction. MW baking of bread and other starchy foods has been investigated by many authors, with the main objective of accelerating the process, increasing the industrial throughput, and reducing the length of industrial ovens and the related production costs. However, many quality problems are observed in MW-baked products and often the final products do not meet consumer acceptance due to their unacceptable quality (Demirkesen, Sumnu, & Sahin, 2013; Serventi, Sachleben, & Vodovotz, 2011; Sumnu, 2001). The explanation has to be found in the phenomena that occur during baking (Mondal & Datta, 2008), and the intrinsic characteristics of MW heating that have been previously underlined.

Baking of bread and similar products in conventional oven (where heat transfer is mainly by convection through the hot air and radiation from the hot oven surfaces) is a lengthy process during which many physical-chemical and biological phenomena take place leading to a final product with appreciated internal structure porosity, external crispy and brownish crust and definite flavor profile. The two main components of flour, starch and proteins, are involved in the development of the porosity and structure, where starch gelatinization and gluten network (for gluten flours, such as wheat) contribute to the volume increase and gas retention (for leavened products). Reducing sugars and nitrogen compounds (mainly free amino acids) react in the Maillard reaction to give browning and complex aromatic compounds. Sugars' caramelization contributes to the crust's color. Other ingredients such as eggs, fats, shortenings, and emulsifiers interact with the starch and protein component to eventually increase softness, compensate weak or absent gluten, and to retard starch retrogradation in the finished product. The latter is the main cause of the so-called staling phenomenon for which baked products become hard during storage and which involves a crystallization and rearrangement of starch molecules. All the above-cited processes are accompanied by water evaporation and weight loss, and are characterized by specific temperature ranges, activation energy, and kinetics.

In the baking of cookies, the main problem is the difficulty in getting uniform cooking and, in particular in drying the center part which tends to remain more humid for the same reasons that will be explained in Section 8.3.4. The consequence is that the process needs to be carried out at lower temperatures for long time in very long baking tunnels. For cookies, MW can provide selective heating of the inner wet layers reducing the baking time and improving the cooking uniformity (typical biscuit thickness allows for proper MW penetration).

For both bread-type and biscuit products, the problem of cold air in MW oven inevitably prevents the formation of a brownish and crunchy crust (which is the same problem encountered in the MW cooking of many other dishes, such as meats and lasagne), while the excessive speed of the process can prevent the correct flavor production and negatively impair the modifications of the starch and protein fractions. For example, water migration from the inside to the outer surface can be kinetically and energetically favored compared to starch gelatinization. The starch gelatinization is particularly important for the final quality of bakery products. One of the most important drawbacks of MW-only baking is insufficient starch gelatinization due to the short baking period. In an almost analog way, moisture gradients from the interior to the surface induce a weak interaction between water and proteins leading to poor gluten plasticization and tough texture. Serventi et al. (2011) showed that the addition of specific ingredients with water-holding capacity and high polar lipids (soy flour and soymilk powder) can enhance gluten plasticization and improve water retention in microwavable pocket-type flat doughs with evident textural improvements. For the above-commented problems, MW is always used in combination with conventional heating (generally as the finishing step) or other techniques for baking and cooking applications (Bilgen, Coşkuner, & Karababa, 2004; Keskin, Oztürk, Sahin, Koksel, & Sumnu, 2005). Use of active packaging with susceptors, modification of formulation, or the new approach proposed by Albert, Salvador, and Fiszman (2012) (an edible susceptor consisting of a film of alginate gel with high salt concentration) can compensate for the lack of crunchiness.

Currently, coupling infrared (IR) and MW heating systems have found a large interest, especially in bakery products. IR heating is known to be effective for surface heating, but it possesses very poor penetration power. On the other hand, MW's greater penetration depth and volumetric heat transfer are suitable for internal heating (Chizoba Ekezie et al., 2017). Combining IR heating with MW energy can be therefore very profitable. Keskin, Sumnu, and Sahin (2004) studied the effect of halogen lamp-MW combination baking revealing the additional advantages in browning and crisping on surface. Moreover, MW heating was found to be the dominant mechanism in terms of affecting weight loss and texture development during baking. MW-infrared heating was found suitable also for several gluten-driven and gluten-free bakery application. Ozkoc, Sumnu, and Sahin (2009) tested IR-MW combination in breads baking evaluating the effect of guar, κ-carrageenan, and xanthan/guar gum blend on the macro and microstructure of the final product. The pores of breads baked in IR-MW combination oven resulted in coalescence of the gas cells to form channels for each formulation, leading to their being no longer spherical. Moreover, the starch granules in conventionally baked breads were more distorted and seen as a continuous sheet of gelatinized starch. On the other hand, granular residues and continuous starch structures were observed in IR-MW combination heating. However, in another study, it was demonstrated that the addition of fat enhanced starch gelatinization in MW and infrared-MW combination compared to conventional ovens (Ozkoc, Sumnu, Sahin, & Turabi, 2009). MW-infrared combination (MW-IR) applications were found suitable also for gluten-free systems. Ozkoc and Seyhun (2015) obtained gluten-free breads made from frozen dough with different gums and flaxseed concentrations. After frozen storage at 20°C for 10 days, the thawed samples were fermented and subsequently baked in MW-IR oven. The results indicated that the addition of the guar gum and 5% flaxseed to the formulation provided the best results on MW-IR baked gluten-free bread with a softer texture, higher volume, and darker color. Similar results were found for cake application. Sánchez-Pardo et al. (2008) observed that there was insufficient starch gelatinization in MW-baked cakes (degree of gelatinization ranging from 55% to 78% depending on formulation) compared to conventionally baked cakes (degree of gelatinization from 85% to 93%). However, Sakiyan et al. (2011) demonstrated that fat addition can improve cake baking using MW and IR-MW combinations compared to conventional heating. Specifically, the higher gelatinization degree highlighted by differential scanning calorimetry and rapid visco analyzer analysis was related to the enhancement of the dielectric properties of the samples with higher fat content. Other positive results were reported by Ozkahraman, Sumnu, and Sahin (2016) for legume cakes. MW-IR baking guarantees better characteristic texture, as well as larger volumes with desirable surface color, than conventional oven. In a similar manner, Şakıyan (2015) exploited MW-IR ovens for the production of soy-based cakes to obtain hardness, weight loss, color difference, and specific volume similar to cakes baked in a conventional oven. Hybrid systems are intensively studied in baking to gain beneficial synergistic effects and solve the issues arising from the MW application.

MW or MW and conventional combination techniques are especially employed for cooking fruit and vegetables. MW can help in time-saving meanwhile preserving some heat-sensitive nutritional components of the products, more information about the effect of MW on the sensorial quality of these kinds of product are provided in Section 16.3.9.2.

16.3.3 Heating and tempering

Consumers often use their domestic MW ovens to re-heat packed ready-to-eat meals, to re-heat precooked foods and to thaw frozen foods, since MW heating offers an extremely rapid solution for each of these applications compared to conventional convective ovens.

From an industrial point of view, tempering and thawing are widely used MW applications. These two operations are often required as pretreatments of raw materials that are stored in a frozen state before being processed as ingredients in other more complex food products. For example, blocks of frozen meats are to be tempered (that is brought to temperatures just below the melting point of ice) to be cut and milled without excessive thermal damage (such as oxidation of the fat fraction). In the case of large food pieces, thawing/tempering may require very long times (in the order of days) if carried at refrigerated temperature (the low temperature limits dangerous microbial growths) and even longer if carried out in warm rooms. Furthermore, in the latter case, the conditions are favorable for pathogens proliferation, and there is a non-negligible drip loss with detrimental effects on the juiciness and water retention capacity of the product and economic loss for the producer. MW heating can dramatically decrease the time required for defrosting large blocks of food (to minutes or hours depending on the size) with consequent advantages: the possibility for the producer to respond rapidly to any changes in demand; better quality of the final product; minimization of any biological growth and related health hazard (MW can also be used at low air temperatures); minimization of the room space required for the operation.

Also, for this application, the main problem associated with MW is the heating uniformity and the control of the end temperature. As already explained, there can be an uncontrolled thermal runaway in thawing (Akkari, Chevallier, & Boillereaux, 2006; Farag, Lyng, Morgan, & Cronin, 2008; Liu, Wang, & Sakai, 2005) due to the important gap between the dielectric properties of water in solid and liquid phases. Cyclic MW thawing and use of a multi-source multi-mode cavity can help in reducing and controlling the phenomenon Taher and Farid (2001).

It was aforementioned how temperature uniformity is especially critical when solid-liquid mixtures are involved. For heating/tempering applications in these cases, MW has been combined with infrared (IF) or OH to obtain better performance. OH is an electro-heating technique involving the conversion of electrical energy to thermal energy by the application of an alternating current across a food material (Shim, Lee, & Jun, 2010). The combination of MW and OH techniques presents the possibility of selecting the heating of food particles in a liquid medium without considering their electrical conductivity. In this process, the liquid phase is heated via electric current, which can enhance heating uniformity. Choi, Nguyen, Lee, and Jun (2011) investigated the feasibility of eliminating temperature lag for carrot cubes in 0.1% NaCl medium via the synergistic effect of MW and OH combination mode. OH and MWs were applied separately and in combination. During separate OH and

MW heating, thermal lag between the liquid and carrot cubes were 18.9°C and 11.2°C, respectively. On the other hand, when a combined two-step heating technique was employed, temperature difference between the carrot cubes and liquid eventually dropped to less than 2°C at the completion of the process. Further optimization was performed by Nguyen et al. (2013) and Lee, Choi, Kim, and Jun (2015). In the first study, a continuous MW and ohmic combination heater was created and tested on a multi-component food system comprising 0.2%−0.5% salt solution and carrot particles. The assisted process eliminated thermal lags between the solid and liquid phases, as compared to MW and OH techniques separate use. Moreover, the energy conversion of the combined system rose up to 12.8% more than in individual processes (Nguyen et al., 2013). Similarly, Lee et al. (2015) tested a dual cylindrical ohmic and MW combination heater for minimizing thermal dropback during heating of particulate foods (potato particles, chicken and NaCl solution at different ratios). Again, when the OH and MW combination heater was employed, temperature gaps were less than 3.08°C, assuming a possible variation in electrical conductivity for different food components led by the synergistic effect between MW and OH.

MW-IR heating has also been proposed as an efficient and applicable tempering technique. For example, Seyhun, Ramaswamy, Sumnu, Sahin, and Ahmed (2009) used this combination to temper frozen potato puree. The finding was that an increase in temperature and frequency caused significant changes in dielectric constant (ε') and loss factor (ε'') of the samples. In addition, tempering duration was substantially reduced increasing the power levels. Finally, MW-IR heating improved product quality and increased energy efficiency when compared to conventional heating techniques prospecting a quite profitable industrial application.

16.3.4 *Drying*

Drying is, for sure, the most investigated food application of MW heating, since the latter can potentially solve many drawbacks of conventional hot air drying (Feng, Yin, & Tang, 2012; Holtz, Ahrné, Rittenauer, & Rasmuson, 2010; Jangam, 2011; Zhang, Jiang, & Lim, 2010). Drying is a well-known preservation technique used for production of dried fruit and vegetable products (starting from solid raw materials), of instant powders (such as instant tea, coffee, vegetable soups, fruit and vegetable products and milk, starting from a liquid) and of many types of fried and non-fried snacks.

Conventional drying of solids is based on the use of hot air, on convection heat transfer from the air to the product surface, and on conduction heat transfer inside the product, in addition to mass transfer of water from the inside to the outer layer and from the outer layer to the surrounding air. The process can be described in terms of a succession of different steps characterized by a decreasing water evaporation rate. In the first step, which does not always occur and is generally of short duration, the evaporation rate is constant since the product surface is homogenously wet; in the second step, the surface is not uniformly wet anymore and the evaporation rate starts decreasing (this is the so-called falling rate) since water needs to diffuse from the inside to the outer surface; in the third step, the evaporation rate further decreases and water migration is more and more difficult because in the meantime different structural changes have occurred in the product. A crust layer (case hardening) has formed on the surface, while in the interior, the porosity has progressively diminished together with a volume shrinkage. These factors lead to a very slow and long drying process to get the final required residual moisture level, but also to a quality loss in terms of damage of heat-sensitive components, browning, and slow rehydration rates (dried products often require a rehydration step before use, or, simply, rehydration contributes to the mouth feel sensorial quality). Conventional air drying is further dependent on a complex relationship between the hygroscopic properties of air (relative humidity and temperature) and the product to be dried (based on the characteristic isothermal water desorption curves). Technological solutions have been implemented in conventional dryers to improve the process in terms of speed and product quality preservation (forced convection, pneumatic dryers, drum dryers, fluidized bed dryers, foam drying, puff-drying), even though the above-cited problems remain. Freeze-drying can be applied to dry particularly heat-sensitive products, but it is a high-energy-consuming technology. Emerging drying technologies are under investigation, including heat pump-assisted drying, low-pressure superheated steam drying, pulse combustion spray drying, pulsed and ultrasound-assisted osmotic dehydration, as well as novel gas–particle contactors such as impinging streams and pulsed fluidized beds, multistage drying, intermittent drying, and the use of hybrid drying technologies—which combine advantages of different dryers without some of their limitations (Chua, Mujumdar, & Chou, 2005; Jangam, 2011).

MW-assisted drying belongs to these emerging and alternative technologies. The main advantage is to achieve fast drying rates, especially in the falling rate, being, as the RF drying, intrinsically self-leveling with more energy being dissipated in wetter regions than in drier ones. Furthermore, the water vapor generated inside causes an internal pressure gradient which forces the water outside and prevents the shrinkage of the structure, preserving the product porosity or, in some cases, even increasing it. Using MW heating alone for drying is characterized by the same problems underlined for pasteurization and sterilization, such as uneven heating, possible overheating of corners or edges depending on the food shape and size; therefore, it is commonly applied in combination with other techniques (Zhang et al., 2010): MW-assisted air drying (in which MW can be applied at the beginning, in the middle or in the final stage of drying); MW-assisted vacuum drying, and MW-assisted freeze-drying (in which MW can be applied concurrently or after freeze-drying) (Chandrasekaran et al., 2013; Duan, Zhang, Mujumdar, & Wang, 2010). Recently, MW-IR (Si et al., 2016) and MW-assisted ultrasound (Szadzińska, Kowalski, & Stasiak, 2016) have also been accounted for drying applications. When air and vacuum are applied, their action as carriers of evaporated moisture contributes to a more homogeneous and faster drying (Andres et al., 2019). Specifically, in MW-assisted air-drying the combination of the airflow and MW guarantees both the removal of outside moisture of the product and an intensive moisture evaporation and transportation towards the food surface reducing drying time, and thus, energy requirements, while improving product quality. Nistor, Seremet Ceclu, Andronoiu, Rudi, and Botez (2017) investigated the MW-assisted hot air process evaluating the effect of several heat mechanisms like free convection (50°C, 60°C, 70°C) and combination of free convection and forced convection (40°C) on physical and chemical properties of red beetroot. The finding was that an increase in drying rate and a decrease in drying time was obtained by increasing the drying temperature followed by MW heating (350 W). The combined drying methods were more effective than the convective drying (CD) methods, especially when the combination free convection (60°C) followed by forced convection (40°C) was coupled with MW. Also Zielinska, Sadowski, and Błaszczak (2015) evaluated the effect of hot air drying (HAD) + MWD and HAD on the quality of frozen/thawed and raw blueberries. The HAD + MWD drying revealed the reduction in drying time by 80% compared to conventional process. Moreover, the textural properties like hardness, chewiness, cohesiveness, and gumminess are significantly reduced in MW-assisted drying than other drying techniques.

Nowadays, MW-assisted vacuum (MWVD) drying has been intensively investigated due to the advantages derived from combining the volumetric heat absorption provided by MWs with an increase of drying kinetics at reduced pressure. The absence of air prevents oxidation and thus, the color, flavor and texture are minimally affected. As vacuum drying requires less temperature, high vitamin contents are retained compared to convective drying (Chizoba Ekezie et al., 2017). Zhao et al. (2017) studied the effect con MWVD on lotus seeds. They reported that increasing the power density from 10–20 W/g in MWVD led to a decrease in drying time by 56.52%, postulating that high power densities increase the effective moisture diffusivity due to the increase of kinetic energy of water molecules. Moreover, the application of higher MW powers caused the reduction of both non-enzymatic and enzymatic reactions. Zielinska and Michalska (2016) evaluated the quality factors of frozen/thawed blueberries after dehydration using hot air convective drying (HACD), MVD or a combination of HACD and MVD. According to the obtained results, blueberries dried using HACD at 90°C with MVD showed better retention of anthocyanins and higher DPPH (2,2-diphenyl-1-picrylhydrazyl) radical scavenging activity or antioxidant power, when compared to other employed methods. Another outstanding outcomes resulting from MW-assisted vacuum drying application is that the quality of MWVD dried products is similar to than with freeze-drying (FD) (Andres et al., 2019). Monteiro, Carciofi, Marsaioli, and Laurindo (2015) incorporated vacuum condition in a domestic MW oven to dry slices of banana, grape, tomato and carrot. They showed that this technology was feasible to produce dried fruits and vegetables with characteristics like those produced from a FD process. Furthermore, MWVD assured shorter processing time reaching up to 20 minutes compared to typical 14 hours–16 hours for FD process. Some attempts have been also performed to evaluate the impact of MWs during freeze-drying. In this sense, it was reported an additional decrease of energy consuming and freeze-drying time when MWs power is used to supply the heat of sublimation in sea cucumber coated with nanosilver (Duan, Zhang, Li, & Mujumdar, 2008), but significant change in color and browning in mushroom drying (Duan, Liu, Ren, Liu, & Liu, 2016). However, this technology did not find any industrial application due to its high costs and a small market such for freeze-dried food products (Duan et al., 2010).

Some studies have also paid attention to the MW-assisted infrared drying. Recently, Öztürk, Şakıyan, and Özlem Alifakı (2017) examined the correlation between dielectric properties and MW-IR heating on the drying characteristics of banana and kiwi fruit obtaining notable time saving (up to 98%) and lower final moisture contents for both samples when compared with conventional drying. A further development was carried out by Si et al. (2016). They compared the effect of MW-vacuum IR and infrared drying on quality of raspberries. Different MW powers and vacuum pressures were analyzed in terms of the drying kinetics. MW-vacuum IR drying reduced the processing time by 44.44% compared to IR drying. In addition, product quality yielded 2.4 times superior crispness value, 25.63% higher rehydration properties, 17.55% better anthocyanin retention and 21.21% higher DPPH radical-scavenging activity than IR in optimum conditions.

Finally, also MWs-assisted ultrasound drying has been considered efficient for drying applications due to the possibility to conduct the process at lower temperatures than traditional methods, reducing the probability of oxidation or degradation. Szadzińska et al. (2016) showed that the convective process combined with MWs and ultrasound, significantly saved energy and drying time of strawberries by approximately 50% and 94%, respectively. This result was achieved by coupling the "heating effect" of MW with the "vibration effect" of ultrasound. In another study, the effect of the hybrid process was tested on green pepper quality, showing a nearly 80% reduction in energy consumption, shorter drying time of 80 minutes, higher rehydration ratio, and increased retention of up to 70% for color and vitamin C (Szadzińska, Łechtańska, Kowalski, & Stasiak, 2017). Similar findings were reported by Kowalski, Pawłowski, Szadzińska, Łechtańska, and Stasiak (2016) for raspberries dried using convective drying, assisted by ultrasound and MW irradiation. A time saving up to 79% and minimal changes to the sensorial or nutritional profile of raspberries were obtained. Based on these studies, it can be foreseen that the combined application of different energy sources is profitable for efficient drying of food materials.

16.3.5 *Freezing*

Food freezing processes are strongly affected by phase transition and modification of water occurring inside the final product. The size of the ice crystals is especially critical for the quality of the frozen food: large crystal dimensions can lead to irreversible damage to the cellular structure which may impact texture and color of the product after defrosting. (Andres et al., 2019). Among emerging technologies developed to control the crystallization process, the influence of the electrical rather than the magnetic radiation on the water was reported to have a great impact on water molecules due to their polarity and slightly ionic characteristics. Concerning the electromagnetic effects on water, the presence of electromagnetic fields (EMF), the attempts to reorient the water molecules, and some hydrogen bonds of the network may be broken. Furthermore, it was suggested that lower frequency MWs (915 MHz vs 2450 MHz) or radiofrequency radiation (RF) and even extremely low-frequency fields (ELF) (3–300 Hz) have significant and lasting effects on liquid water (Xanthakis et al., 2018). The application of electromagnetic field has been identified as a possible driver to reduce ice crystals dimensions. Xanthakis, Le-Bail, and Ramaswamy (2014) applied MWs during the freezing of pork meat observing a 62% decrease in the average ice crystal size compared to conventional freezing. The same results were found by Jackson, Ungan, Critser, and Gao (1997) who employed a cryo-protectant (ethylene glycol) in conjunction with 2.45 GHz MW radiation to shape the size and formation of ice crystals in tissues, and by Hanyu, Ichikawa, and Matsumoto (1992) who could reduce ice dimensions in squid retina freezing at 83K and in the presence of MW irradiation. In another study, Xanthakis, Le-Bail, Shrestha, Ahrne, and Bernard (2015) better clarified the mechanism stating that the presence of an electromagnetic field induces nucleation and decreases the degree of supercooling during the freezing process. This is an important finding since the size and number of crystal ice have a great effect on the shelf-life of frozen foods. In fact, during frozen storage, the ice crystals enlarge, leading to irreversible tissues damages and quality loss of food products. It is therefore essential to get a high number of small ice crystals during the freezing process to improve product preservation.

16.3.6 *Extraction applications*

MW-assisted extraction (MAE) has been proposed as an innovative extraction technique for recovering active ingredients from plant materials (Chan, Yusoff, Ngoh, & Kung, 2011; Dahmoune et al., 2013; Li, Skouroumounis, Elsey, & Taylor, 2011; Spigno & De Faveri, 2009). One of the main advantages of MAE compared to conventional solid-liquid extraction is the significant reduction of the processing time (from several hours to minutes) and a high yield of active substances. Moreover, MAE is capable of performing a selective/localized heat transfer because of the high absorption of MW energy by water contained inside plant cellular tissues with possible enhancement of cellular disruption. Ever since the first introduction of MAE for plant extracts, process parameters such as the nature of solvent, solvent-to-solid ratio, MW power, and temperature have been subject of intensive study on different matrixes with a particular emphasis on the extraction of anthocyanins, total phenolics, phenolic acids, flavonoids and related compounds associated with antioxidant properties.

Different factors influence the extraction process: irradiation power and time, vessel geometry and agitation, number of irradiation cycles, solvent-to-solid ratio, sample size, type of solvent, and its dielectric properties, and with a primary influence of the temperature reached in the sample on the extraction yield and possible thermal degradation effects of the extracted compounds.

This particular MW application is strongly related to the food sector since the recovered bioactive compounds can be used as natural additives or ingredients in novel food formulations, but also because it can be exploited in fruit juice production to enhance the juice yield, in the grape maceration phase of the wine-making process to increase the color and phenolic compounds extraction, or still in the industrial production of herbal and tea infusions (Carew, Sparrow, Curtin, Close, & Dambergs, 2013; Spigno & De Faveri, 2009).

Recent developments have been focused on the application of the MW-assisted ultrasound process (MAUE) as a novel method for fast and efficient extraction. MW facilitates solvent penetration by molecular and ionic destabilization. On the other hand, ultrasound promotes cavitation, mechanical function, and thermal effect, disrupting cell walls and providing better extraction efficiency due to the enhancement of the release of compounds from plant bodies (Andres et al., 2019). In an extraction of pectin from sugar beet pulp (SBP), Peng et al. (2015) used the ultrasound/MW-assisted acid method as alternative to traditional acid-based technique. The hybrid procedure showed higher yield of pectin from SBP, better emulsifying activity, and increased oxidative stability and viscosity without affecting the protein-polysaccharide structure of pectin. Similarly, Zeng et al. (2015) evaluated the yield of MAE, MAUE, and hot water extraction (HWE) of polysaccharides from citrus fruit. MAUE yield was 9.15% ± 0.13%, increasing up to 76.64%, 128.18%, and 405.52% over MAE, ultrasonic-assisted extraction, and HWE, respectively. Similar results were found for glucosinates from cabbage (Pongmalai, Devahastin, Chiewchan, & Soponronnarit, 2015), trans-resveratrol from tree peony seed (Chen et al., 2016), juglone from walnut green husk (Xu, Yang, & Fu, 2016), and pectin from pomelo peels (Liu, Qiao, Yang, Gu, & Yang, 2017). Due to the high efficiency and dramatically short extraction time, it is anticipated that MAUE has a great potential for commercial applications, with a cost-effective process, fast sample preparation, and new strategy for process intensification.

16.3.7 *Microwaves impact on food quality*

Food quality is the result of a complex interaction between different quality aspects: safety (including biological, chemical, and physical hazards), sensorial (including structure, texture, color, flavor, and taste), and nutritional (including both healthy and anti-nutritional compounds). Convenience and cost factors can also greatly contribute to consumer-perceived quality. MWs can both positively and negatively influence each of these quality aspects. In general, some reports suggest that in MWs foods retention of nutrients and bioactive compounds, such as vitamins and antioxidant molecules, is improved because treatment time is shortened. However, other studies suggest that nutrient retention in MW processing is not much greater than that in other conventional processes. Since we have been speaking of MW *heating*, in most cases, the impact is simply related to the reached temperature and the time taken, therefore knowledge of the product heating profile, calculation of the cumulative undergone thermal effect and comparison with the kinetics of the involved reactions will explain or predict the final effect. In other cases, the change is due to phenomena related to the sudden heating, which can typically occur in MW heating, therefore, for example, flavors can be lost with escaping water vapor, or the release of intracellular components can be enhanced by localized pressure build-up and consequent tissue damage.

The later section describes some literature works that cannot, of course, exhaustively represent the state of the art but that aim to bring significant examples of MW-related effects on safety, sensorial and nutritional properties for different food categories (starchy foods, fruit and vegetables, and animal products).

16.3.8 *Safety properties*

16.3.8.1 *Biological hazards*

In Section 17.3.1, it was already discussed about the fact that the achieved lethal effect of MW heating is of thermal origin. However, the question of whether nonthermal effects of MW radiation existed has always been the focus of debate and research efforts are still focused on the attempt of demonstrating and defining the nonthermal effects of MW treatments (Benlloch-Tinoco, Martínez-Navarrete, & Rodrigo, 2014; Xue, Xu, Jiang, Zhu, & Dong, 2013). However, from the current research, the existence or not of the MW nonthermal effects is still not ready to make a conclusion. However, verification of such nonthermal effects is a difficult task due to the complexity of biological systems; the difficulty to control the temperature during the experiment; the lack of rationality and logic during experiment design. That's why FDA has advised to neglect any nonthermal effect in the validation of pasteurization/sterilization treatment and to rely only on the cumulative thermal lethality. Reliable thermal processing calculations require an understanding of the thermal inactivation kinetics of microorganisms throughout the heating process and in each part of the treated sample. The found microbial inactivation kinetics must be coupled with the heat transfer phenomena to calculate the survival of bacteria, especially in the MW nonuniform heating process, which is the most important limiting factor associated with MW pasteurization/sterilization (Hamoud-Agha, Curet, Simonin, & Boillereaux, 2013).

Benlloch-Tinoco et al. (2014) assessed the suitability of the pasteurization unit (PU) concept to compare the thermal load of MW and conventional kiwifruit puree pasteurization treatments (to inactivate 90% of peroxidase enzyme and reduce more than 5 $\log_{10}$ cycles of *Listeria monocytogenes*). Conventional treatment was carried out in a thermostatic bath at 97°C for 30 minutes, while two MW treatments with a domestic oven were tested: 1000 W-200 seconds and 900 W-225 seconds on 500 g of product preheated to 25°C and then, kept in a glass beaker without stirring and registering the time-temperature history of the sample. Even though the number of PU obtained in MW-treated puree at the hottest spot (1.79 minutes and 7.9 minutes for the 1000 W and 900 W treatment, respectively) was considerably higher than the PU obtained at the coldest spot (0.046 minute and 0.0027 minute for the 1000 and 900 W treatment), it was lower than the PU in conventional heating (19.27 minutes in the coldest point). Since both the conventional and MW-treated samples showed the same level of POD inactivation and *L. monocytogenes* reduction, the authors postulate some contributory nonthermal effects associated with MW.

Also, the results obtained by Lin and Ramaswamy (2011) seemed to confirm the existence of enhanced thermal effects from MW. The authors compared the phosphatase inactivation kinetics in milk under continuous flow MW and conventional heating conditions, observing that it occurred much faster under MW heating. In particular, based on thermal profiles, they calculated the decimal death time (D) and the thermal resistance (Z). The first parameter was one order of magnitude lower under MW heating, while the second was only slightly lower (4.9°C compared to 5.2°C).

MWs have also been successfully used to control insects in stored cereals and cereal products (Vadivambal, Jayas, & White, 2007) with the advantage that no chemical residues are left in the foods. For such applications, the challenge is to determine the mortality of the target insects at their different development stages, at different moisture contents, the effect of MW on germination and the evaluation of any detrimental effects on the flour quality. This is particularly critic for wheat since, depending on the treatment conditions, researchers have reported contrasting results: loss of the structural and functional characteristics of protein-gluten, improvement in the baking quality but also no significant difference in the quality of grain and flour protein, flour yield, flour ash, and loaf volume of wheat treated with MW energy.

Pande, Mishra, and Singh (2012) applied MW treatment for the disinfestation of stored green gram seed, evaluating also some physical-chemical characteristics (grain size, grain hardness, and nutritional quality). It was found that under optimized treatment conditions (808 W, 80 seconds), it was possible to increase insect mortality also reducing moisture content and nutritional factor (phytic acid) with a consequent improvement of protein digestibility and preserving the green color. However, the MW treatment caused a certain cell disruption, swelling of the grain and reduction of hardness.

16.3.8.2 Packaging-related chemical hazards

One of the greatest advantages of MW heating is that it can be applied to packed-foods, especially to ready-to-heat meals. Food packaging materials contain chemical additives and compounds (plasticizers, antioxidants, unreacted monomers, inks, etc.) that could migrate into the foodstuffs inside and thus be ingested by the consumer. Unfortunately, MW heating aggravates molecules movement and decreases the activation energy, and consequently, the chemical migration from the packaging material would be different from that during conventional heating (Bhunia, Sablani, Tang, & Rasco, 2013; Duan, Chen, Zhu, Zhang, & Zhang, 2013; Guillard et al., 2010; Huang, Zhu, Chen, Li, & Li, 2014; Lorence & Pesheck, 2009). Microwavable materials do not constitute a barrier for MW that can thus reach the food. Although they are not expected to be heated by MW, heat transfer from the food can rise their temperature. Temperature increase is especially critical in case susceptors are used. As already reported for the MW popcorn container, susceptors are made of a paperboard sheet covered by aluminum and protected by a PET layer. Local temperatures on the susceptor surface could reach 350°C to 450°C, higher values than possible with conventional heating. At these temperatures, migration is enhanced but also modification of material structure or degradation reactions of materials components with the production of breakdown products can occur.

Huang et al. (2014) found an accelerated migration of plasticizer (triacetin) from a water-resistant hydrophobic starch ester film into milk during MW treatment (film samples immersed in milk and heated from 30°C at

2°C/min for different heating times from 15 minutes to 35 minutes) compared to a control (sample film immersed in milk at 30°C). The phenomenon was related to both structural changes of the film, as well to a higher permeation of water and/or fat molecules from the milk into the amorphous regions of the polymer (worst results were obtained with whole milk due to its higher affinity with triacetin and starch ester film).

Bradley et al. (2010) compared the migration of melamine from melaware articles (melamine-formaldehyde plastics) into various food simulants and foods themselves, under different time and temperature test conditions and in different MW heating tests. The effect of MW heating was the most dramatic, where the peak temperature achieved (>70°C) counterbalances the shorter contact times (6–14 times lower) than in the other contact tests.

Moreover, the study by Alin and Hakkarainen (2013) revealed a faster migration of cyclic oligomers from PET food packaging into the food simulant ethanol and isooctane (10% ethanol had to be added to isooctane which is nonpolar and could not be heated by MW), compared to conventional heating at the same temperature (1 hour at 80°C and 4 hours at 100°C), even though the overall migration was below the limit set by the European Commission in all studied food simulants (water, 3% acetic acid, 10% ethanol, and isooctane/ethanol). The effect of MW was more significant at 80°C and PET chains underwent transesterification with ethanol during MW heating to a minor degree, resulting in the formation and migration of diethyl terephthalate.

The same authors (Alin & Hakkarainen, 2013) also investigated the effect of MW heating on the migration from two MWable polypropylene copolymer packaging, one white and one clear and transparent. Migration of packaging antioxidants (Irgafos 168 and Irganox 1010) and of their degradation products, was assessed. Tests were carried out at 80°C, with MW and conventional heating. When fatty food simulants were used (isooctane and ethanol), antioxidants degraded giving higher specific migration during MW heating compared to conventional heating.

Luckily, Ehlert, Beumer, and Groot (2008) showed that for eighteen different brands of polycarbonate baby bottles sold in Europe, during three MW-heating cycles (each cycle was carried out in a MW oven operated at maximum power 1000 W, boiling was achieved after 1 minutes 54 seconds and maintained for 1 minutes), MW radiation had no effect on migration of bisphenol A into water and that the specific migration level set by EC was never overpassed.

In some cases, the shorter processing times allowed by MW systems can better preserve the properties of packaging materials, compared to conventional processes. This is what was observed by Mokwena, Tang, Dunne, Yang, and Chow (2009) regarding the deterioration of oxygen barrier properties of multilayer EVOH films, which was lower after MW sterilization than after retort sterilization.

16.3.8.3 Other chemical hazards

Many chemical compounds present in the raw materials or generated during processing can represent a potential hazard to human health. For example, fish and derived products are objects of concern and investigation due to the potential high content of heavy metals. Published studies have shown that the heavy metal concentration of fish can be affected by processing or cooking methods. Ersoy (2011) compared the MW cooking (2450 MHz, 4 minutes without specification of power) with oven baking (200°C, 15 minutes), grill oven (200°C, 10 minutes), and frying (200°C, 4 minutes) of African catfish fillets in terms of residual content of Pb, Ni, As, Cd, Cr, and Co. The author concluded that MW cooking, baking, and grilling were suitable methods, but important data related to the moisture content of raw and cooked fish prevent a correct evaluation of metals loss. Almost the same study had been previously carried out by Ersoy, Yanar, Küçükgülmez, and Çelik (2006) on sea bass fillet, with slightly different cooking conditions. Again, the metal concentrations before and after cooking were compared on a wet weight basis, so it could be hypothesized that since moisture should decrease during cooking, a lower concentration after cooking reveals a metal loss. This occurred only for Pb in MW-cooked fish, but the explanation was not provided. In principle, heavy metals cannot be degraded by cooking and their reduction could be only achieved through leaching with fish juices during the cooking process.

Fish can also be a source of pollutants and veterinary drugs. Mitrowska, Posyniak, and Zmudzki (2007) studied the persistence of a fungicide/ectoparasicide (malachite green, MG) and its major metabolite (leucomalachite green, LMG) in carp muscles after different cooking methods: MW (2450 MHz/800 W for 0.25 minute, 0.5 minute, 1 minute); baking in a gas oven at 180°C for 5 minutes, 10 minutes and 15 minutes; boiling in water at 100°C for 5 minutes, 10 minutes and 15 minutes. In the boiled product also the cooking water was analyzed, in the other two cases loss of juices was not observed. The measured concentrations were combined with the product weight to properly calculate the mass reduction. Degradation of MG was observed for all the cooking methods, while the metabolite LMG could be degraded only with microwaving. Since the internal temperature of the meat did not rise above 100°C and did not differ from that achieved in other cooking methods, it was postulated an effect linked to electromagnetic radiation.

Polycyclic aromatic hydrocarbons (PAHs) can be generated during the heating at high temperature (such as grilling) of high-protein foods, such as meat and fish. Farhadian, Jinap, Hanifah, and Zaidul (2011) investigated the influence of some pretreatments on the formation of some PAHs in charcoal-grilled meat (beef filet and chicken breast and leg with skin, all chopped in small pieces and marinated). The pretreatments were MW preheating at 180°C for 60 seconds and 40 seconds for beef and chicken (no details on MW power are provided); steam cooking at 100°C for 4 minutes and 2 minutes for beef and chicken; wrapping with a layer of aluminum foil and a banana leaf. All the applied pretreatments were effective in reducing the analyzed PAHs, with differences between chicken and beef. Regarding the pre-cooking by steam or MW, the effect is probably due to the lower required time for grilling, and then the lower rate of pyrolysis, and the lower levels of deposition and penetration of smoke components into the meat samples from fat drippings that fall on the hot coals. However, combination of pre-cooking and wrapping was required to reduce the formation of one PHAs (fluoranthene).

Furanic compounds (as HMF and furfural) are food toxicants that are generated during the advanced stages of Maillard reactions and caramelization which, on the other hand, are essential for the development of the aroma, taste, and color surface of the baked products. In the study by Petisca, Henriques, Pérez-Palacios, Pinho, and Ferreira (2013), model cakes were cooked in a MW oven (1250 W for 1.5 minutes, 2 minutes, and 2.5 minutes), or in a traditional and in a steam oven (at 160°C for 20 minutes, 40 minutes and 60 minutes). In principle, HMF and furfural formation should be enhanced at low moisture level and at severe thermal treatments. In the MW, HMF is compared after 2 minutes, and furfural after 2.5 minutes when higher moisture loss occurs burning the inside of the cake. The levels were close to that measured in the cakes cooked in the traditional oven, while the steam oven was the best in reducing furanic compounds formation probably due to the reduction of water evaporation on the surface preventing sugar crystallization. Furthermore, MWs cakes possessed a poor volatile profile.

16.3.9 Sensorial and nutritional properties

Tables 16.1–16.4 report some literature examples illustrating the main effect due to MW processing on sensorial and nutritional aspects of different food categories: starchy foods, vegetables, fruit, and animal products. Wherever possible, based on the provided information, the power density applied in the MW treatment, as W/g or W/mL, was calculated and reported.

TABLE 16.1 Influence of microwave (MW) processing on different quality aspects of starchy foods.

Product	MW treatment	Conventional Comparison treatment	Quality aspect	MW effect	References
Puffed snack potato cubes (from blanched 8, 10, 12 mm cubes)	• MSD (2450 MHz, 2 W/g, air at 25°C) • MSD with air at 60°C • Tandem MSD: air at 60°C until 60% moisture content, then MSD with air at 25°C	None	Expansion ratio, breaking force, RR	Tandem MSD: highest expansion ratio and rehydration ratio, and lower breaking force (higher crispness).	Yan et al. (2010)
Re-structured purple-fleshed sweet potato granules	• MVD (4 W/g), 30′ • MFD, 270′ • MSD (air 80°C, 2.5 W/g), 32′	None	Texture (penetration force), color, anthocyanin level, sensory evaluation (appearance, flavor, color, crispness)	MSD: lowest penetration force, color more purple, highest anthocyanins retention. Similar sensorial appreciation for MFD and MSD, lowest for MVD. Best crispness for MFD.	Liu, Zhang, and Mujumdar (2012)
Banana/potato restructured chips	• MFD, 2450 MHz 0–2000 W, 2–3 W/g, 250–300′ • MVD, 0–200 W, 1 W/g, 60′	None	RR, color, texture (hardness and fracturability)	Samples darker with ↑ W/g and overheating in MFD (during melting). Harder texture and higher fracturability with MVD and loss of shape. Better RR with MFD	Jiang, Zhang, Mujumdar, and Lim (2011)

(Continued)

TABLE 16.1 (*Continued*)

Product	MW treatment	Conventional Comparison treatment	Quality aspect	MW effect	References
Potato Puree	MW: Three MW power levels (30%, 40%, and 50%) and MWP 706 W MW-IR: Three upper halogen lamps, power levels (10%, 20%, and 30%) and MWP 706 W	None	Heat capacity and dielectric Properties and tempering time	Decreased tempering time, dielectric constant and loss factor exponentially rose with increase in temperature in MW-IR	Seyhun et al. (2009)
Soy Cake	Two halogen lamps each at infrared power (IRP) = 1500 W, MW power (MWP) = 700 W, 9.5′	CO 24′, 175°C.	Baking time, volume, texture	Shorter baking duration, up to 1.37 m^3/kg increase in specific volume and a lower hardness value of 0.1646 N.	Şakıyan (2015)
Gluten-free breads with chestnut and rice flours	MW-IR oven, 9 min at 200 W (0.5 W/g), with beakers containing water to limit moisture loss during baking.	CO, 200°C, 25′	Macro and microstructure of bread	Puffing effect of MW-IR could not improve the porosity of rice crumb, but it did in chestnut-rice bread due to the higher fiber content	Demirkesen et al. (2013)
Vacuum-packed Potato Omelette	MW(2450 MHz) was performed at four power levels (300 W–150 s; 450 W–90 s, 600 W–60 s and 800 W–45 s)	None	*Salmonella Enteridis* sterilization	4.8 log reduction (safety standard) with 800 W for 45 s.	Valero, Cejudo, and García-Gimeno (2014)
Legume cake	Two halogen lamps each at infrared power (IRP) = 1500 W, MW power (MWP) = 700 W, 4′	CO 24′, 175°C.	Volume, texture, gelatinization degree, microstructure, Hardness, volume, color	Softer texture, more voluminous cakes ($p < 0.05$) with less gelatinization and larger pores.	Ozkahraman et al. (2016)
Pound cakes (sweet bread)	MW 240 W, 5′ (2.2 W/g)	CO, 180°C, 40′	Moisture, total resistant starch, gelatinized starch	MW cakes: lower moisture; higher total resistant starch; lower gelatinization degree	Sánchez-Pardo, Ortiz-Moreno, García-Zaragoza, Necoechea-Mondragón, & Chanona-Pérez (2012)
Cake with different levels of fat or fat mimetics	MW 706 W (50% power), 120–180′ MW-IR oven: MW 706 W (50% power) + IR 1500 W (60% power bottom/top), 210–330″	CO, 175°C, 22–28″	Starch gelatinization level (DSC, RVA)	Increasing fat determines decreasing in gelatinized starch for CO and increasing for both MW and MW-IR. MW-IR possesses values similar to conventional baking (up to 90%)	Sakiyan et al. (2011)
Gluten-free bread with different gums/ flaxseed formulation	Thawing 4°C + Fermentation + baking IR (50% power top and 60% bottom) + MW (20% power), 20′ and 30″	None	weight loss, specific volume, hardness, and color over 72 h	guar gum-5% flaxseed has a lower hardness, higher specific volume, and acceptable color than the other samples over 72 h.	Ozkoc and Seyhun (2015)

CO, Convective oven; *HAD*, hot air drying; *IR*, infrared; *MAD*, MW-assisted drying; *MCWC*, MW-circulation water combination; *MSD*, MW-enhanced spouted bed drying; *MFD*, MW-assisted freeze-drying; *MVD*, MW-assisted vacuum drying; *RH*, relative humidity; *RR*, rehydration rate; ↓ decreased/decreasing; ↑ increased/increasing.

Updated with permission from Spigno, G. (2016). Microwave processing: Impact on food product quality attributes. In A. K. Jaiswal (Ed.), Food processing technologies: Impact on product attributes *(pp. 137–180). CRC Press. https://doi.org/10.1201/9781315372365.*

16.3.9.1 Starchy foods

As described in Section 16.3.2, MW heating of starchy foods can lead to important quality problems related to physical-chemical change of the starch component but also of the protein content, with important consequences on the texture of final products. Also, the absence of crust and adequate surface browning can neutralize the

TABLE 16.2 Influence of microwave (MW) processing on sensorial and nutritional quality of vegetable products.

Product	MW treatment	Conventional Comparison treatment	Quality aspect	MW effect	References
Red Beetroot	HAD + MW in two stage HAD in free convection at different t/T (50°C for 210 min, 60°C for 150 min, and at 70°C for 120 min). HAD in forced convection at 40°C + MW at 315 W (second stage).	HAD in free convection at different t/T (50°C for 510 min, 60°C for 390 min, and at 70°C for 300 min).	betalain variance, and polyphenol, microstructure changes (SEM)	A strong thermal shock, provided by convection at 60 degrees followed by forced convection + MW 315 W, leads to a better preservation of bioactive compounds content (0.631 ± 0.0042 mg/g of betacyanin and 0.795 ± 0.0019 mg/g betaxanthin) when compared to convection at 50, 60 and 70°C.	Nistor et al. (2017)
Dried tomato	• MHAD, 2450 MHz, 40°C, fixed power density (1.5 W/g 100 min; 2 W/g 50′; 2.5 W/g 35′) • 40°C, with power control (3 W/g 170′; 5 W/g 110′; 7 W/g 60′)	None	Color, texture, internal structure, lycopene	Power density control: avoided formation of black spots (due to uneven heating); color close to fresh tomato; better texture (reduced shrinkage and higher porosity); less pronounced isomerization and oxidation of lycopene.	Koné et al. (2013)
Eggplant	MW-IR: two IR lamp 1500 W (10% and 20% power) and MW 630 W (30%, 40%, and 50% power)	HAD 50°C with an air velocity 1.5 m/s	color, rehydration ratio, shrinkage, microstructure, and poresize distribution	Increased effective diffusivity, decreased drying time, that is, 22 to 12 min, 54% decrease in shrinkage values and 37% higher rehydration ratio in MW-IR	Aydogdu, Sumnu, and Sahin (2015)
Asparagus cream	MW pasteurization at different energy densities 1200–3100 J/g (from real power × irradiation time/ sample weight)	None	Lipid fraction oxidation (PV, induction time by Rancimat, K232, K270 nm)	Lipid fraction weakly affected only when low powers are applied, probably due to the protective role of natural occurring antioxidants (e.g., tocopherols, poliphenols) not degraded at low power (low T)	Giuliani, Bevilacqua, Corbo, and Severini (2010)
Carrot in salt solution	MW, 2450 MHz, 900 W Ohmic Heating (OH) (input voltage 0–480 V and output current rated up to 20 A) MW + OH	None	Temperature uniformity in heating of solid/ liquid mixture, energy efficiency	MW + OH shows no significant temperature difference between solid and liquid phases and higher energy efficiency	Nguyen et al. (2013)
Green Pepper	HAD + Ultrasound (100 W) HAD + Ultrasound (200 W) HAD + MW (100 W) HAD + MW(100 W) + Ultrasound (200 W)	HAD, 54°C, air flow velocity = 2 m/s,	drying kinetics, energy consumption, total color change, and vitamin C	HAD + MW shows lowest drying time but hybrid system HAD + MW + U guarantees lower energy consumption, color retention and better rehydration. HAD + UV has higher vitamin C retention	Szadzińska et al. (2017)
Dried garlic cloves	MW, 100 g in MW oven 8 s, 14 s, 20 s	• HAD, 1 h, 2 h, 3 h at 45°C, 50°C, 55°C, 60°C, 70°C, 75°C • FD • FBD, 1, 2 h at 50°C, 55°C, 60°C, 65°C, 70°C	Volatile oils, sensory analysis	Tips of the cloves burnt after 20 s. Skins shrank and cloves are less hard. Loss of aroma and volatiles. Highest volatile oil retention after FD	Puranik, Srivastava, Mishra, and Saxena (2012)
Dried ginger	MW, 2450 MHz, 60 W (0.6 W/g), 320′	• HAD, 50°C, 60°C, 70°C • FD, 29 h • VD, 60°C, 490′	Volatiles	The most favorite drying method based on cluster analysis of aroma compounds (including those of fresh ginger and those novel formed during the process)	Ding et al. (2012)

(*Continued*)

TABLE 16.2 (*Continued*)

Product	MW treatment	Conventional Comparison treatment	Quality aspect	MW effect	References
Dried ginger	MW, 2450 MHz, 385, 525, 660, 800 W (48, 40, 32, 25′)	HAD, 50°C, 12 h	Color, essential oils, TP, and AOC (DPPH, FRAP, reducing power) of ginger extracts	Loss of essential oils and AOC at lower powers; higher color difference at higher powers (less brightness, increase in redness). Retention of maximum volatile fraction at the highest power. ↑ TP yield (probably due to cellular dmage)	Kubra and Rao (2012a, 2012b)
Dlower of *Eucommia ulmoides* (drying for tea making)	2450 MHz, 140, 240, 480, 560 or 700 W (4.7–23 W/g) for 1, 2, 3 and 4′	• HAD, 60°C, 90°C, 120°C, 150°C or 180°C for 1 h • Pan-drying, at 90°C, 120°C, 150°C or 180°C until typical flavor	Functional constituents: TF, cholorogenic acid, aucubin, geniposidic acid, tea flavor	Higher retention of functional constituents. Higher content at higher powers and shorter times (enough to deactivate the enzymes responsible for the destruction of the functional components but not to thermally degrade the latter). Intact shape of the stamens of the flower, color greenish and the liquor of the flower tea was fresh with a strong typical flavor	Dong, Ma, Fu, and Guo (2011)
Green asparagus (water blanching)	MW, 900 W (4.5 W/g), 30 s + cooling to T_{room} + WB (70°C, 80°C, 90°C)	WB, 70°C–360′, 80°C–250′, 90°C–160′	Kinetics of AA degradation and of POD inactivation	For some parts of asparagus (due to different tissue structures): ↑ E_a of AA degradation (the reaction becomes more T-sensitive) ↓ E_a of POD inactivation (the reaction becomes less T sensitive)	Zheng and Lu (2011)
Non-peeled potatoes (white, yellow, red, purple-fleshed) cooking	MW, 750 W, 10′ (cubes 1.5 cm)	• Boiling, 15′ • Oven, 180°C, 45′	AA, cholorogenic acid, TAC, steroidal glycoalkaloids	Intermediate ↓ AA; highest ↓ cholorogenic acid; lowest ↓ steroidal glycoalkaloids; intermediate ↑ TAC	Lachman et al. (2013)
Button Mushrooms	MFD in two mode: – fixed cavity pressure (1 kW and 1.2 kW MW power) – fixed pover level 1.2 kW (50 Pa and 500 Pa cavity pressure)	None	Browing kinetics Color, TP, PPO, moisture content, residual sugar and vitamin C	Moisture content, PPO activity and VC content have significant influences on predominantly enzymatic browning behavior. MW power must be changed during process to minimized both enzymatic and non-enzymatic browning.	Duan et al. (2016)
Onion Powder	MW + Plasma: MW (2.45 GHz, 900 W) at two densities of 170 and 250 mW · m-2. Plasma: Helium (He) flow rate of 165 at 1 L/min, Plasma generation powers 400, 474, 650, 826, 900 W time 10, 14, 25, 36, 40′	None	*B. cereus, A. brasiliensis, and E. coli O157:H7* growth, color, quercitin, antioxidant activity	High-density MW at 400 W for 40 min, determined as the optimum conditions for all strains not affecting the color, antioxidant activity, or quercetin concentration of the powder during storage at both temperatures.	Kim et al. (2017)

AA, Ascorbic acid; *AOC*, antioxidant capacity; *BW*, boiling water; *FBD*, fluidized bed drying; *FD*, freeze-drying; *HAD*, hot air drying; *MVD*, MW-assisted vacuum drying; *MHAD*, MW hot air drying; *POD*, peroxidise; *PV*, peroxide value; *RH*, relative humidity; *RR*, rehydration rate; *TF*, total flavonoids; *TP*, total phenols; *TAC*, total anthocyanins; *VD*, vacuum drying; *WB*, water blanching; ↓ decreased/decreasing; ↑ increased/increasing.

Updated with permission from Spigno, G. (2016). Microwave processing: Impact on food product quality attributes. In A. K. Jaiswal (Ed.), Food processing technologies: Impact on product attributes *(pp. 137–180). CRC Press. https://doi.org/10.1201/9781315372365.*

TABLE 16.3 Influence of microwave (MW) processing on sensorial and nutritional quality of fruit products.

Product	MW treatment	Conventional Comparison treatment	Quality aspects	Main MW effects	References
Apple puree powder	• MFMD, 180 W, air 40°C	• FMD 60°C • FD, 26 h	Sorption isotherms, T_g	No effect of drying method on sorption isotherms and T_g	Jakubczyk, Ostrowska-Ligeza, and Gondek (2010)
Raspberry	• Drying temperature 70°C, IRP = 675 W • MRP different power (400, 600, 800 W) and Vacuum (45, 65, 85 kPa)	None	Color, rehydration property, texture, anthocyanin retention	600 W at 85 kPa shows up to 55.56% time savings with 25.63% higher rehydration ratio, 2.4 times higher crispness value, and 17.55% higher anthocyanin retention.	Si et al. (2016)
Raspberry	• MW vacuum drying (MWVD) • HAD 60°C or 90°C + MWVD	• HAD, 60°C and 90°C	drying kinetics, color, ABTS, TMA, TP, Texture	HAD (90°C)↑ TP, HAD(90°C) + MWVD, ↑ TMA and ABTS with ↓ drying time	Zielinska and Michalska (2016)
Grapefruit Juice	• MP, 900 W (45 W/mL), 30 s (final T: 80°C)	• CP, 95°C, 80 s to reach 80°C and 11 s at 80°C	Particle size distribution, flow behavior, density, turbidity, color, sensory evaluation, AA, total vitamin C, organic acids, TP and single flavonoids, AOA (DPPH test)	MW juice: higher turbidity, darker and purer in color. At sensorial test: the most turbid, viscous, bitter with distinctive grapefruit taste (distinctive grapefruit taste for CP juice). Preservation of AA. Higher retention of AA, Vit.C, TP and flavonoids (probably due to a more intense enzymes inactivation), and AOC during storage.	Igual, Contreras, Camacho, & Martínez-Navarrete (2014), Igual, García-Martínez, Camacho, and Martínez-Navarrete (2010), Igual, García-Martínez, Camacho, and Martínez-Navarrete (2011)
Mandarin (microbial decontamination)	MW + plasma at different condition: – nitrogen (N2)-CPT at 1000 mL/min, $P = .7$ kPa, MWP = 400, 650, and 900 W, $t = 2$, 5, and 10 min, respectively; – helium(He)-CPT and CPT using N2-oxygen (O_2) mixture (N2:O_2 = 4:1), at 1000 mL/min, $P = .7$ kPa, MWP = 900 W, $t = 10$ min.	None	*Penicillium italicum* inhibition, CO_2, weight loss, color, SSC, TA, maturity index, and pH, ascorbic acid, DPPH, TPC	Nitrogen (N2) at 900 W for 10 min ↑ inhibition of P. italicum, TPC and antioxidant activity but did not significantly affect CO_2 generation, weight loss, content of soluble solids, titratable acidity, pH, ascorbic acid concentration (flesh), or surface color during storage at 4°C and 25°C.	Won et al. (2017)
Orange juice (thawing of frozen juice)	800 W, 20 s	• T_{room} • Refrigerated T	Color, provitamin A, carotenoids (content & bioaccessibility), AOC (ABTS test),	↓ in provitamin A, carotenoids, AOC. ↑ bioaccessibility. Highest color difference compared to fresh juice.	Stinco, Fernández-Vázquez, Heredia, Meléndez-Martínez, and Vicario (2013)

(Continued)

TABLE 16.3 (*Continued*)

Product	MW treatment	Conventional Comparison treatment	Quality aspects	Main MW effects	References
Grapefruit jam	Precooking at 900 W (1.8 W/g, 90°C) or 1000 W (2 W/g, 95°C) + cooking at 900 W (100°C) or 1000 W (105°C) for 10′	Conventional: Precooking at 85°C for 10′ + cooking at 95°C–100°C for 20′	Consistency (Bostwick), rheological properties (Hershel-Bulkley model), color	MW jam: lower pH may be due to increased pectin solubilization, confirmed by higher consistency at 900 W and jelly consistency at 1000 W	Igual et al. (2010)
Banana and Kiwifruit	Upper and lower halogen lamps each at IRP = 600 W, MWP = 320 or 420 W	HAD 70°C, 480′	Moisture content, color, and drying time	98% time saving, that is, 0.75 min–2.25 min required to reduce their moisture content to final values compared to HAD	Öztürk et al. (2017)
Dried Apricots	2450 MHz, 2 kW (2.2 W/g), 60°C–18 h, or 70°C–12 h	HAD, 60°C–21 h, 70°C–15 h	Color, β-carotene, AOC (DPPH)	Higher ↓ β-carotene (due to higher T), lower ↓ AOC, reduced darkening	Albanese, Cinquanta, Cuccurullo, and Di Matteo (2013)
Dried Apricots	• MD, 100 W (0.7–0.8 W/g) (2.1 h) • MHAD, 100 W and air at 40°C (2.2 h)	HAD, 40°C–60°C (13–11 h)	Ascorbic acid (AA), vit. C, vit. E, total carotenoids, color, puncture test	Lower loss of AA, lower release of Vit. C, higher Vit. E (for MD), lower loss of carotenoids. More deformable and fragile structure, gummier texture more difficult to cut (higher pectin solubilization and puffing effect). Smallest color difference compared to fresh apricots for MHAD.	García-Martínez, Igual, Martín-Esparza, and Martínez-Navarrete (2013)
Dried Strawberry	MHAD, 100 W (0.2 W/g), and air at 40°C (9.5 h)	HAD, 40°C (37 h)	Color, puncture test, anthocyanins	↑ sample lightness (surface discoloration at higher T); ↑ anthocyanins degradation. Higher mechanical rigidity and puncture resistance.	
Dried Goldenberry	• MW, 2450 MHz, 160 W (0.8 W/g), 380′ • MHAD, 160 W + air 75°C, 150′ • MHAD, 160 W + air 100°C, 110′	• HAD, 75°C–510′, 100°C–240′	Color, TP, AOC (DPPH)	MW treatment allowed for the lowest ↓ AOC and the best color	İzli, Yıldız, Ünal, Işık, and Uylaşer (2014)
Dried Jujube fruit	700 W, 4′, until 2.5% moisture	• Sun drying, 3 weeks until 22% moisture • FD, 48 h, until 0.2% moisture • HAD, 70°C, 8 h, until 2.2%	TP and single phenolics, AOC (ABTS), β-carotene, α-tocopherol,	Higher content of catechin, procatechui acid and epicatechin, ↓ in all other phenolics, same AOC as FD samples (which were the best)	Gao, Wu, Wang, Xu, and Du (2012)

(*Continued*)

TABLE 16.3 (*Continued*)

Product	MW treatment	Conventional Comparison treatment	Quality aspects	Main MW effects	References
Dried Cranberry	MVD, 1.8 kW (1.5 W/g) for 30 to 35′, followed by 1.2 kW (1 W/g) for 20 to 25′	• FD, 5 to 10 days • HAD, 55°C, 60 to 65 h	Total porosity and size distribution, AOC (ORAC, ABTS), anthocyanins	Highest porosity and pore size; higher retention of anthocyanins than HAD, but lower than FD; higher AOC (ABTS) than HAD, comparable with FD; higher AOC (ORAC) than both HAD and FD	Leusink et al. (2010)
Dried Banana	• MVD 400, 700 and 1000 W	• FD 14 h–16 h	Drying kinetics, color, microstructure, texture	dried fruits with characteristics similar (crisp and crunch) to those produced from a freeze-drying process, in much shorter process times, for example, 20 min against the 14 h–16 h, typical of freeze-drying processes.	Monteiro et al. (2015)

AA, Ascorbic acid; *AOA*, antioxidant activity; *CP*, conventional pasteurization; *FD*, freeze-drying; *FMD*, foam mat drying; *HAD*, hot air drying; *IR*, infrared; *MAD*, MW-assisted drying; *MCWC*, MW-circulation water combination; *MFMD*, MW-assisted foam drying; *MSD*, MW-enhanced spouted bed drying; *MFD*, MW-assisted freeze-drying; *MVD*, MW-assisted vacuum drying; *MHAD*, MW hot air drying; *MP*, MW pasteurization; *PME*, pectin methyl-esterase; *RH*, relative humidity; *RR*, rehydration rate; *Tg*, glass transition temperature; *TP*, total phenols; ↓ decreased/decreasing; ↑ increased/increasing.

Updated with permission from Spigno, G. (2016). Microwave processing: Impact on food product quality attributes. In A. K. Jaiswal (Ed.), Food processing technologies: Impact on product attributes *(pp. 137–180). CRC Press. https://doi.org/10.1201/9781315372365.*

time reduction advantage of the MW process, unless it is used in combination with other techniques and special ingredients formulation (Table 16.1).

A few studies have also been found about the possible influence of MW on nutritional aspects related to glycemic index and allergenic potential. García-Zaragoza, Sánchez-Pardo, Ortiz-Moreno, and Bello-Pérez (2011) showed that the glycemic index (calculated from hydrolysis index values by the Granfeldt method) of bread baked in a two-cycle MW toaster was lower than that of the product baked in a conventional oven. Leszczynska, Łącka, Szemraj, Lukamowicz, and Zegota (2003) found a change in the reactivity (due to chemical and conformational changes in the structure) of commercial gliadins and wheat flours exposed to MW irradiation, with a variable effect depending on the dose of applied energy. Commercial gliadins showed an increase in immune reactivity from zero at the highest dose (900 kJ/g) up to 210% at the lowest dose (400 kJ/g). The increase in reactivity was lower for the gliadins contained in the wheat flour, with the maximum immune response at 100 kJ/g and zero at 500 kJ/g. Therefore, MW heating could not be applied for the elimination of the allergic properties of wheat gliadins.

16.3.9.2 Vegetable and fruit products

The review by Vadivambal and Jayas (2007) summarizes the main changes in the quality of MW-treated agricultural products, with reference to drying applications.

In all the cases of MW heating of vegetable and fruit products, together with clear advantages in terms of time-saving, reduction of degradation of nutritional components, reduction of browning, and improvement in texture (higher porosity and faster rehydration rate for dried products), there can be important drawbacks linked, for example, to a low retention of flavors (volatile compounds) and essential oils and a reduced formation of desired aroma compounds, which is particularly critic in the drying of spices. Furthermore, the positive effects are strongly dependent on the MW conditions, such as the applied power density (W/g or W/mL) and the resulting temperature distribution in the sample. Therefore, some works report significant differences between the MW treatment and other comparison treatments with better or worst performance for the MW one, while

TABLE 16.4 Influence of microwave (MW) processing on sensorial and nutritional quality of animal products.

Product	MW treatment	Conventional Comparison treatment	Quality aspects	Main MW effects	References
Freezing of fish and meat	MAF long pulse low power (LPLP: 8 s/6 W–8 s /0 W) and short pulse high power (SPHP: 4 s /12 W–4 s/0 W), as well as freezing under low constant MW power (3 W) + freezing at −25°C and −32°C	Cod freezing at −25°C and −32 degrees	Microstructure analysis and freezable water content.	For pork tenderloin that long pulsed low power MW radiation was able to influence beneficially the size of the formed ice crystals and reduce the freeze damage of the food matrix. The results obtained for fish showed that the application of constant MW power during freezing can lead to the elimination of the intracellular crystals.	Xanthakis et al. (2018)
Chicken meat with potato cubes in salt solution (heating)	MW = 2450 MHz; 900 W OH = maximum frequency of 20 kHz, on/off duty cycle of 0.2, and a max current of 100 Amps MW + OH	None	Heating patterns and thermal Lag, temperature uniformity	The maximum temperature differences between particles and solution obtained by individual MW and ohmic heating were 7.1 ± 1.7 and 11.9°C ± 2.9°C, respectively, while the combination heating exhibited little significant temperature gaps (maximum difference <3.08°C)	Lee et al. (2015)
Cooked grass carp fillet	MW, 4′ at 100% power	• WB, 98°C, 12′ • S, 12′ • G (250°C) 12′ • PF, 180°C, 8′ • DF, 180°C, 5′	Lipid oxidation (peroxide value, AV analysis), fatty acid profile, free amino acids	↑ polar lipids as grilling, pan frying, deep-frying; ↑ oxidation (AV value)	Zhang et al. (2013)
Milk pasteurization	• 500 W (7.1 W/mL) 60 s (final T 75°C) • 500 W 70 s (final T 85°C)	Water bath 75°C–60 s, 85°C–70 s	Salt partitioning, soluble Ca and Pi, coagulation time, and curd characteristics	Lower ionic Ca content (60 s treatment) but recovered after 24 h; longer coagulation time and slower curd formation rate.	De Lafuente, Olano, and Juärez (2002)

DF, Deep-fat frying; *G*, grilling; *OB*, oven baking; *PF*, pan-frying; *S*, steaming; *WB*, water boiling; ↓ decreased/decreasing; ↑ increased/increasing.
Updated with permission from Spigno, G. (2016). Microwave processing: Impact on food product quality attributes. In A. K. Jaiswal (Ed.), Food processing technologies: Impact on product attributes *(pp. 137–180). CRC Press. https://doi.org/10.1201/9781315372365.*

others report a substantial identity (Tables 16.2 and 16.3). For example, Lemmens et al. (2009) could detect no unambiguous effect of the heating technique applied for the treatment of carrot pieces (blanching at low and high temperature; conventional-, MW-, ohmic-heating before conventional cooking) on some quality-related parameters (peroxidise and pectinmethylesterase activity; degree of methoxylation; texture; β-carotene content), because it was the time/temperature conditions to determine the effect. Also, some published papers report the need to optimizing the MW treatment in terms of irradiation time to better preserve the sensorial and nutritional quality, obviously because different combinations of irradiation time and power lead to different temperatures and possible thermal degradation effects (Dadali & Ozbek, 2009; Sharifian, Modarres-Motlagh, Komarizade, & Nikbakht, 2013).

Many other works not included in the tables report that MW processing (alone or in combination with other technologies) resulted in better retention of the quality of untreated original material, including functional constituents and related antioxidant capacity, together with superior sensory characteristics compared to conventional thermal treatments (Leusink, Kitts, Yaghmaee, & Durance, 2010; Li, Cheng, & Liu, 2010; Oszmiański, Wojdyło, & Kolniak, 2007; Padda & Picha, 2008; Piasek et al., 2011; Stea, Johansson, Jägerstad, & Frølich, 2006; Sun et al., 2009).

On the contrary, other works not reported in the tables indicated that MW treatment alone or combined was not the best one in preserving the nutritional component (Gustafson, Yousef, Grusak, & Lila, 2012; Ni et al., 2014). The result is also strongly dependent on the vegetable product (Jiménez-Monreal, García-Diz, Martínez-Tomé, Mariscal, & Murcia, 2009).

MW treatment can also reveal to be efficient in the removal of antinutritional factors, such as trypsin inhibitors in potatoes (Sasi Kiran & Padmaja, 2003), phytic acid in grains (Pande et al., 2012), alpha-amylase inhibitors in potato (Rekha & Padmaja, 2002) (even though the presence of such inhibitor in the diet can be both negative since can lead to conditions like celiac disease, and positive since could be used in the treatment of obesity). The review by Verma, Kumar, Mukul, and Dwived (2012) reports on the possibility of exploiting thermal treatments to reduce or eliminate the allergenic potential of legumes which is linked to their high protein content. During the thermal treatment, conformational changes in protein structure occur which can bring either reduction or enhancement in protein allergenicity. The final effect is dependent on the type of protein, the reached temperature, duration of the treatment, and interaction with the food matrix; therefore, it is impossible to extend a priori the positive effect found for MW heating on soybean allergens, to other legumes.

16.3.9.3 Animal products

It has been already spoken in Section 16.3.3 about the MW advantage on the quality in case of tempering and defrosting of raw materials that need to be further processed. Table 16.4 reports some research works aimed to investigate the effect of MW heating on the quality of the lipid, protein, and flavor fractions. Depending on the product and on the applied conditions, results can indicate MW either as one of the best or worst treatments. Fish represents an important dietary source of high-quality lipids and proteins, therefore many studies have addressed the influence of different cooking methods on the chemical composition of these components. Regarding the comparison between MW cooking and other conventional methods, contrasting results have been found.

The study carried out by Ünal Şengör, Alakavuk, and Yasemin Tosun (2013) showed that grilling and oven baking were the best cooking methods concerning omega-3/omega-6 ratio compared to the MW and steam cooking methods of Atlantic salmon, while Türkkan, Cakli, and Kilinc (2008) found that the cooking method has a considerable effect on fatty acid composition of seabass, with baking and MW cooking the best methods for healthy eating (compared to frying) with reference to the content of n-3 PUFA. However, neither conventional nor MW reheating of sea bream fillets (cooked with different methods) after frozen storage for 4 months had a detrimental effect on the fatty acid profile and its stability (Ozogul, Ozyurt, & Boga, 2009). Again, Naseri, Rezaei, Moieni, Hosseni, and Eskandari (2010) showed that among different precooking methods (steaming, oven-baking, and MW-cooking) of silver carp fillets, steam-cooking was the best one for retaining nutritional constituents. In particular, the free fatty acid content of the fillets did not change by the three different methods, but MW-cooking marginally affected the fatty acid composition and increased the value of thiobarbituric acid (index of lipid oxidation). Regarding the protein quality (total protein content, amino acid content and score, total –SH group content, and protein solubility) of frozen and cooked sardine fillets, García-Arias, Alvarez-Pontes, García-Fernández, and Sánchez-Muniz (2003) concluded that a slow defrosting process is preferable than a quick one obtained with MW, even though with MW there is a lower weight loss and change in proximate composition due to reduced lixiviation.

16.4 Industrial processing implementation of microwave technology

Unfortunately, as with many other innovations in the food sector, MW food application originated from war contexts, since the first magnetron (the system that generates the MW) was built to produce an MW source to power radar sets for the British military during World War II. Later in 1945, the American electrical engineer Spencer, after some fortuitous observations, applied for a patent that was issued in 1950: it was called "method of treating food stuffs" and described for the first time a MW closed oven.

Household MW ovens have enormously diffused all over the world and they are often combined with additional conventional heating equipment (hot air) to achieve browning and crisping effects during cooking.

Industrial applications have been continuously increasing, driven by the potential advantages for product quality that have been discussed in this chapter, but, above all, by the potential economic, practical, and environmental advantages that MW technology could provide.

These benefits originate from the following considerations:

- The reduction of processing times allows for increased throughputs.

- Process lines can be shortened (especially in cooking and drying applications) and generally, the required space for equipment is reduced (industrial ovens operating at 915 MHz are larger than those operating at 2450 MHz but they generate MWs with a larger penetration depth).
- The heating can be controlled very precisely thanks to a switch on/off modality.
- The efficiency of conversion of electrical energy into heat is generally high (even though it is reported that for domestic MW ovens electric energy conversion is evaluated to be approximately 50% and efficiency of MW heating could go up to 65% at the best) (Nguyen et al., 2013).

However, industrial implementation has been greatly limited due to the following main drawbacks:

- The temperature distribution in a MW heated food product is often uneven; more than one cold spot may exist in changing positions during the process; hot spots may also be present at different positions; prediction of heating patterns is difficult (more for multi-mode MW systems than for single-mode systems) due to complex required mathematical models. These factors have particularly limited the MW application for sterilization and pasteurization purposes.
- High investment and operating costs.
- Electrical energy is the most expensive energy form, therefore only a punctual energy cost analysis can properly compare the costs of conventional and MW systems (Pardo & Zufía, 2012).

A MW system consists of a MW source (magnetrons), a system to transfer the generated MW (waveguides and transmission lines can be used), a system to apply the MW to the food (the same waveguides or near-field, single-mode or multi-mode applicators can be used). More details on these components can be found in other references (Hebbar & Rastogi, 2012, p. 369; Regier & Schubert, 2001, p. 178). Many companies can nowadays be found providing industrial MW equipment for food applications, mainly for drying, baking, cooking, tempering and defrosting: RFHIC Corporation; MAX Industrial Microwave; Ferrite Microwave Technologies; AMTek Microwaves; Shangai Nasan Industry Co., Ltd; Industrial Microwave Systems Ltd; Sairem Microwave and radiofrequency; M.I.T SRL, Microwave Industrial Technology; Bi. Elle srl Microwave Technology; Yantai Care Microwave System Co. Ltd. These are only some of the companies that nowadays produce industrial MW equipment for the food sectors.

References

Abdullah, S. A., Lee, S. H., Cho, I. K., Li, Q. X., Jun, S., & Choi, W. (2013). Pasteurization of kava juice using novel continuous flow microwave heating technique. *Food Science and Biotechnology*, *22*(4), 961–966.

Akkari, E., Chevallier, S., & Boillereaux, L. (2006). Observer-based monitoring of thermal runaway in microwaves food defrosting. *Journal of Process Control*, *16*(9), 993–1001.

Albanese, D., Cinquanta, L., Cuccurullo, G., & Di Matteo, M. (2013). Effects of microwave and hot-air drying methods on colour, β-carotene and radical scavenging activity of apricots. International Journal of Food Science & Technology, *48*(6), 1327–1333.

Albert, A., Salvador, A., & Fiszman, S. M. (2012). A film of alginate plus salt as an edible susceptor in microwave able food. *Food Hydrocolloid*, *27*(2), 421–426.

Alin, J., & Hakkarainen, M. (2013). Combined chromatographic and mass spectrometric toolbox for fingerprinting migration from PET tray during microwave heating. *Journal of Agricultural and Food Chemistry*, *61*(6), 1405–1415.

Al-Muhtaseb, A. H., Hararah, M. A., Megahey, E. K., McMinn, W. A. M., & Magee, T. R. A. (2010). Dielectric properties of microwave-baked cake and its constituents over a frequency range of 0.915–2.450 GHz. *Journal of Food Engineering*, *98*(1), 84–92.

Andres, A., Reyes, R.D., Sansano, M., Alcañiz, D., Heredia, A., & Reyes, E.D. (2019). Innovative microwave technologies for food drying processes. In: *IDS'2018 – 21st international drying symposium*, pp. 11–14.

Aydogdu, A., Sumnu, G., & Sahin, S. (2015). Effects of microwave-infrared combination drying on quality of eggplants. *Food and Bioprocess Technology*, *8*(6), 1198–1210.

Basaran-Akgul, N., Basaran, P., & Rasco, B. A. (2008). Effect of temperature (−5 to 130°C) and fiber direction on the dielectric properties of beef semitendinosus at radio frequency and microwave frequencies'. *Journal of Food Science, Vol.73*(6), 243–249.

Benlloch-Tinoco, M., Martínez-Navarrete, N., & Rodrigo, D. (2014). Impact of temperature on lethality of kiwifruit puree pasteurization by thermal and microwave processing. *Food Control*, *35*(1), 22–25.

Bhunia, K., Sablani, S. S., Tang, J., & Rasco, B. (2013). Migration of chemical compounds from packaging polymers during microwave, conventional heat treatment, and storage. *Comprehensive Reviews in Food Science and Food*, *12*(5), 523–545.

Bilgen, S., Coşkuner, Y., & Karababa, E. (2004). Effects of baking parameters on the white layer cake quality by combined use of conventional and microwave ovens. *Journal of Food Processing and Preservation*, *28*(2), 89–102.

Bradley, E. L., Castle, L., Day, J. S., Ebner, I., Ehlert, K., Helling, R., … Pfaff, K. (2010). Comparison of the migration of melamine from melamine-formaldehyde plastics ('melaware') into various food simulants and foods themselves. *Food Additives & Contaminants: Part A: Chemistry, Analysis, Control, Exposure & Risk Assess.*, *27*(12), 1755–1764.

Brinley, T. A., Truong, V. D., Coronel, P., Simunovic, J., & Sandeep, K. P. (2008). Dielectric properties of sweet potato purees at 915 MHz as affected by temperature and chemical composition. *International Journal of Food Properties*, *11*(1), 158–172.

Campañone, L. A., & Zaritzky, N. E. (2010). Mathematical modeling and simulation of microwave thawing of large solid foods under different operating conditions. *Food and Bioprocess Technology*, *3*(6), 813–825.

Carew, A. L., Sparrow, A. M., Curtin, C. D., Close, D. C., & Dambergs, R. G. (2013). Microwave maceration of pinot noir grape must: Sanitation and extraction effects and wine phenolics outcomes. *Food and Bioprocess Technology*, *7*(4), 954–963.

Chan, C. H., Yusoff, R., Ngoh, G. C., & Kung, F. W. (2011). Microwave-assisted extractions of active ingredients from plants. *Journal of Chromatography A*, *1218*, 6213–6225.

Chandrasekaran, S., Ramanathan, S., & Basak, T. (2013). Microwave food processing – A review. *Food Research International*, *52*, 243–261.

Chen, F., Zhang, X., Zhang, Q., Du, X., Yang, L., Zu, Y., & Yang, F. (2016). Simultaneous synergistic microwave–ultrasonic extraction and hydrolysis for preparation of trans-resveratrol in tree peony seed oil-extracted residues using imidazolium-based ionic liquid. *Industrial Crops and Products*, *94*, 266–280.

Chizoba Ekezie, F. G., Sun, D. W., Han, Z., & Cheng, J. H. (2017). Microwave-assisted food processing technologies for enhancing product quality and process efficiency: A review of recent developments. *Trends in Food Science and Technology*, *67*, 58–69.

Choi, W., Nguyen, L. T., Lee, S. H., & Jun, S. (2011). A microwave and ohmic combination heater for uniform heating of liquid-particle food mixtures. *Journal of Food Science*, *76*, 9.

Chua, K. J., Mujumdar, A. S., & Chou, S. K. (2005). Intermittent drying of bioproducts – An overview. *Bioresource Technology*, *90*(3), 285–295.

Coronel, P., Simunovic, J., Sandeep, K. P., Cartwright, G. D., & Kumar, P. (2008). Sterilization solutions for aseptic processing using a continuous flow microwave system. *Journal of Food Engineering*, *85*(4), 528–536.

Curet, S., Rouaud, O., & Boillereaux, L. (2014). Estimation of dielectric properties of food materials during microwave tempering and heating. *Food and Bioprocess Technology*, *7*(2), 371–384.

Dadali, G., & Ozbek, B. (2009). Kinetic thermal degradation of vitamin C during microwave drying of okra and spinach. *International Journal of Food Sciences and Nutrition*, *60*(1), 21–31.

Dahmoune, F., Boulekbache, L., Moussi, K., Aoun, O., Spigno, G., & Madani, K. (2013). Valorization of Citrus limon residues for the recovery of antioxidants: Evaluation and optimization of microwave and ultrasound application to solvent extraction. *Industrial Crops and Products*, *50*, 77–87.

Datta, A. K., & Ramaswamy, C. A. (2001). *Handbook of microwave technology for food processing*. New York, USA: Marcel Dekker, Inc.

Datta, A. K., Sumnu, G., & Raghavan, G. S. V. (2005). Dielectric properties of foods. In M. A. Rao, S. H. Syed Rizvi, & A. K. Datta (Eds.), *Engineering properties of foods* (3rd Ed, pp. 501–565). Boca Raton, FL, US: CRC Press.

Demirkesen, I., Sumnu, G., & Sahin, S. (2013). Image analysis of gluten-free breads prepared with chestnut and rice flour and baked in different ovens. *Food and Bioprocess Technology*, *6*(7), 1749–1758.

Ding, S. H., An, K. J., Zhao, C. P., Li, Y., Guo, Y. H., & Wang, Z. F. (2012). Effect of drying methods on volatiles of Chinese ginger (*Zingiber officinale* Roscoe). *Food and Bioproducts Processing*, *90*(3), 515–524.

Dong, J., Ma, X., Fu, Z., & Guo, Y. (2011). Effects of microwave drying on the contents of functional constituents of *Eucommia ulmoides* flower tea. *Industrial Crops and Products.*, *34*(1), 1102–1110.

Duan, F., Chen, M. Q., Zhu, Y., Zhang, H., & Zhang, J. (2013). Study the migration process of chemical substances through the packaging/food interface during microwave treatment. *Mathematical Problems in Engineering*.

Duan, X., Liu, W. C., Ren, G. Y., Liu, L. L., & Liu, Y. H. (2016). Browning behavior of button mushrooms during microwave freeze-drying. *Drying Technology*, *34*(11), 1373–1379.

Duan, X., Zhang, M., Li, X., & Mujumdar, A. S. (2008). 'microwave freeze drying of sea cucumber coated with nanoscale silver. *Drying Technology*, *26*(4), 413–419.

Duan, X., Zhang, M., Mujumdar, S., & Wang, R. (2010). Trends in microwave-assisted freeze drying of foods. *Drying Technology*, *28*(4), 444–453.

Ehlert, K. A., Beumer, C. W. E., & Groot, M. C. E. (2008). Migration of bisphenol A into water from polycarbonate baby bottles during microwave heating. *Food Additives & Contaminants: Part A: Chemistry, Analysis, Control, Exposure & Risk Assess*, *25*(7), 904–910.

Ersoy, B. (2011). Effects of cooking methods on the heavy metal concentrations of the African catfish (*Clarias gariepinus*). *Journal of Food Biochemistry*, *35*(2), 351–356.

Ersoy, B., Yanar, Y., Küçükgülmez, A., & Çelik, M. (2006). Effects of four cooking methods on the heavy metal concentrations of sea bass fillets (*Dicentrarchus labrax* Linne, 1785). *Food Chemistry*, *99*(4), 748–751.

Farag, K. W., Lyng, J. G., Morgan, D. J., & Cronin, D. A. (2008). Dielectric and thermophysical properties of different beef meat blends over a temperature range of -18 to +10°C. *Meat Science*, *79*(4), 740–747.

Farhadian, A., Jinap, S., Hanifah, H. N., & Zaidul, I. S. (2011). Effects of meat preheating and wrapping on the levels of polycyclic aromatic hydrocarbons in charcoal-grilled meat. *Food Chemistry*, *124*(1), 141–146.

Feng, H., Yin, Y., & Tang, J. (2012). Microwave drying of food and agricultural materials: Basics and heat and mass transfer modeling. *Food Engineering Reviews*, *4*(2), 89–106.

Gao, Q. H., Wu, C. S., Wang, M., Xu, B. N., & Du, L. J. (2012). Effect of drying of jujubes (*Ziziphus jujuba* Mill.) on the contents of sugars, organic acids, α-tocopherol, β-carotene, and phenolic compounds. *Journal of Agricultural and Food Chemistry*, *60*(38), 9642–9648.

García-Arias, M. T., Alvarez-Pontes, E., García-Fernández, M. C., & Sánchez-Muniz, F. J. (2003). Freezing/defrosting/frying of sardine fillets. Influence of slow and quick defrosting on protein quality. *Journal of the Science of Food and Agriculture*, *83*(6), 602–608.

García-Martínez, E., Igual, M., Martín-Esparza, M. E., & Martínez-Navarrete, N. (2013). Assessment of the bioactive compounds, color, and mechanical properties of apricots as affected by drying treatment. *Food and Bioprocess Technology*, *6*(11), 3247–3255.

García-Zaragoza, F. J., Sánchez-Pardo, M. E., Ortiz-Moreno, A., & Bello-Pérez, L. A. (2011). In vitro starch digestibility and expected glycemic index of pound cakes baked in two-cycle microwave-toaster and conventional oven. *International Journal of Food Sciences and Nutrition*, *61* (7), 680–689.

Giuliani, R., Bevilacqua, A., Corbo, M. R., & Severini, C. (2010). Use of microwave processing to reduce the initial contamination by *Alicyclobacillus acidoterrestris* in a cream of asparagus and effect of the treatment on the lipid fraction. *Innovative Food Science and Emerging Technologies*, *11*, 328–334.

Guan, D., Cheng, M., Wang, Y., & Tang, J. (2004). Dielectric properties of mashed potatoes relevant to microwave and radio-frequency pasteurization and sterilization Processes'. *Journal of Food Science, 69*(1), 30–37.

Guan, D., Plotka, V. C. F., Clark, S., & Tang, J. (2002). Sensory evaluation of microwave treated macaroni and cheese. *Journal of Food Processing and Preservation, 26*(5), 307–322.

Guillard, V., Mauricio-Iglesias, M., & Gontard, N. (2010). Effect of novel food processing methods on packaging: Structure, composition, and migration properties. *Critical Reviews in Food Science and Nutrition, 50*, 969–988.

Gustafson, S. J., Yousef, G. G., Grusak, M. A., & Lila, M. A. (2012). Effect of postharvest handling practices on phytochemical concentrations and bioactive potential in wild blueberry fruit. *Journal of Berry Research, 2*(4), 215–227.

Hamoud-Agha, M. M., Curet, S., Simonin, H., & Boillereaux, L. (2013). Microwave inactivation of *Escherichia coli* K12 CIP 54.117 in a gel medium: Experimental and numerical study. *Journal of Food Engineering, 116*(2), 315–323.

Hanyu, Y., Ichikawa, M., & Matsumoto, G. (1992). An improved cryofixation method: Cryoquenching of small tissue blocks during microwave irradiation. *Journal of Microscopy, 165*(2), 255–271.

Hebbar, H. U., & Rastogi, N. K. (2012). Microwave heating of fluid foods. In P. J. Cullen, B. K. Tiwari, & V. Valdramidis (Eds.), *Novel thermal and non-thermal technologies for fluid foods* (pp. 369–409). Amsterdam, Netherlands: Elsevier. Inc.

Holtz, E., Ahrné, L., Rittenauer, M., & Rasmuson, A. (2010). Influence of dielectric and sorption properties on drying behaviour and energy efficiency during microwave convective drying of selected food and non-food inorganic materials. *Journal of Food Engineering, 97*(2), 144–153.

Huang, C., Zhu, J., Chen, L., Li, L., & Li, X. (2014). Structural changes and plasticizer migration of starch-based food packaging material contacting with milk during microwave heating. *Food Control, 36*(1), 55–62.

Huang, L., & Sites, J. (2007). Automatic control of a microwave heating process for in-package pasteurization of beef frankfurters. *Journal of Food Engineering, 80*, 226–233.

IFT, Report of the Institute of Food Technologists for the Food and Drug Administration of the U.S. Department of Health and Human Services (2000). Kinetics of microbial inactivation for alternative food processing technologies. Available from https://www.fda.gov/files/food/published/Kinetics-of-Microbial-Inactivation-for-Alternative-Food-Processing-Technologies.pdf.

Igual, M., Contreras, C., Camacho, M. M., & Martínez-Navarrete, N. (2014). Effect of thermal treatment and storage conditions on the physical and sensory properties of grapefruit juice. *Food and Bioprocess Technology, 7*(1), 191–203.

Igual, M., Contreras, C., & Martínez-Navarrete, N. (2010). Non-conventional techniques to obtain grapefruit jam. *Innovative Food Science and Emerging Technologies, 11*, 335–341.

Igual, M., García- Martínez, M., Camacho, M. M., & Martínez-Navarrete, N. (2010). Effect of thermal treatment and storage on the stability of organic acids and the functional value of grapefruit juice. *Food Chemistry, 118*, 291–299.

Igual, M., García-Martínez, E., Camacho, M. M., & Martínez-Navarrete, N. (2011). Changes in flavonoid content of grapefruit juice caused by thermal treatment and storage. *Innovative Food Science and Emerging Technologies, 12*(2), 153–162.

İzli, N., Yıldız, G., Ünal, H., Işık, E., & Uylaşer, V. (2014). Effect of different drying methods on drying characteristics, color, total phenolic content and antioxidant capacity of goldenberry (*Physalis peruviana* L.). *International Journal of Food Science & Technology, 49*(no.1), 9–17.

Jackson, T. H., Ungan, A., Critser, J. K., & Gao, D. (1997). Novel microwave technology for cryopreservation of biomaterials by suppression of apparent ice formation. *Cryobiology, 34*(4), 363–372.

Jakubczyk, E., Ostrowska-Ligeza, E., & Gondek, E. (2010). Moisture sorption characteristics and glass transition temperature of apple puree powder. *International Journal of Food Science & Technology, 45*(12), 2515–2523.

Jangam, S. V. (2011). 'An overview of recent developments and some R&D challenges related to drying of foods. *Drying Technology, 29*(12), 1343–1357.

Jiang, H., Zhang, M., Mujumdar, A. S., & Lim, R. X. (2011). Comparison of the effect of microwave freeze drying and microwave vacuum drying upon the process and quality characteristics of potato/banana re-structured chips. *International Journal of Food Science & Technology, 46* (3), 570–576.

Jiménez-Monreal, A. M., García-Diz, L., Martínez-Tomé, M., Mariscal, M., & Murcia, M. A. (2009). Influence of cooking methods on antioxidant activity of vegetables. *Journal of Food Science, 74*(3), 97–103.

Keskin, S. O., Oztürk, S., Sahin, S., Koksel, H., & Sumnu, G. (2005). Halogen lamp- microwave combination baking of cookies. *European Food Research and Technology, 220*(5–6), 546–551.

Keskin, S. Ö., Sumnu, G., & Sahin, S. (2004). Usage of enzymes in a novel baking process. *Nahrung – Food, 48*(2), 156–160.

Kim, J. E., Oh, Y. J., Won, M. Y., Lee, K. S., & Min, S. C. (2017). Microbial decontamination of onion powder using microwave-powered cold plasma treatments. *Food Microbiology, 62*, 112–123.

Koné, K. Y., Druon, C., Gnimpieba, E. Z., Delmotte, M., Duquenoy, A., & Laguerre, J. C. (2013). Power density control in microwave assisted air drying to improve quality of food. *Journal of Food Engineering, 119*(4), 750–757.

Koskiniemi, C. B., Truong, V. D., Simunovic, J., & McFeeter, R. F. (2011). Improvement of heating uniformity in packaged acidified vegetables pasteurized with a 915 MHz continuous microwave system'. *Journal of Food Engineering, 105*, 149–160.

Kowalski, S. J., Pawłowski, A., Szadzińska, J., Łechtańska, J., & Stasiak, M. (2016). High power airborne ultrasound assist in combined drying of raspberries. *Innovative Food Science and Emerging Technologies, 34*, 225–233.

Kubra, I. R., & Rao, L. J. M. (2012a). Effect of microwave drying on the phytochemical composition of volatiles of ginger. *International Journal of Food Science & Technology, 47*(1), 53–60.

Kubra, I. R., & Rao, L. J. R. (2012b). Microwave drying of ginger (Zingiber officinale Roscoe) and its effects on polyphenolic content and antioxidant activity. *International Journal of Food Science & Technology, 47*(11), 2311–2317.

Kumar, P., Coronel, P., Simunovic, J., & Sandeep, K. P. (2007). Feasibility of aseptic processing of a low-acid multiphase food product (salsa con queso) using a continuous flow microwave system. *Journal of Food Science, 72*(3), 121–124.

Kumar, P., Coronel, P., Simunovic, J., & Sandeep, K. P. (2008). Thermophysical and dielectric properties of salsa con queso and its vegetable ingredients at sterilization temperatures. *International Journal of Food Properties, 11*(1), 112–126.

Kumar, P., Coronel, P., Simunovic, J., Truong, V. D., & Sandeep, K. P. (2007). Measurement of dielectric properties of pumpable food materials under static and continuous flow conditions. *Journal of Food Science, 72*(4), 177–183.

Kumar, R. S., & Sanavullah, M. Y. (2011). Theoretical and experimental study of cooking regions for shell eggs in a domestic microwave oven. *IEE Transactions*, 306–310.

Lachman, J., Hamouz, K., Musilová, J., Hejtmánková, K., Kotíková, Z., Pazderů, K., ... Cimr, J. (2013). Effect of peeling and three cooking methods on the content of selected phytochemicals in potato tubers with various color of flesh. *Food Chemistry*, *138*(2–3), 1189–1197.

De Lafuente, M. A., Olano, A., & Juärez, M. (2002). Mineral balance in milk heated using microwave energy. *Journal of Agricultural and Food Chemistry*, *50*, 2274–2277.

Lee, S. H., Choi, W., Kim, C. T., & Jun, S. (2015). Development of a dual cylindrical microwave and ohmic combination heater for minimization of thermal lags in the processing of particulate foods. *LWT – Food Science and Technology*, *63*(2), 1220–1228.

Leistner, L., & Gould, G. W. (2002). *Hurdle technologies combination treatments for food stability, safety and quality*. New York, USA: Springer Science + Business Media.

Lemmens, L., Tibäck, E., Svelander, C., Smout, C., Ahrné, L., Langton, M., ... Hendrickx, M. (2009). Thermal pretreatments of carrot pieces using different heating techniques: Effect on quality related aspects. *Innovative Food Science and Emerging Technologies*, *10*, 522–529.

Leszczynska, J., Łącka, A., Szemraj, J., Lukamowicz, J., & Zegota, H. (2003). The effect of microwave treatment on the immunoreactivity of gliadin and wheat flour. *European Food Research and Technology*, *217*(5), 387–391.

Leusink, G. J., Kitts, D. D., Yaghmaee, P., & Durance, T. (2010). Retention of antioxidant capacity of vacuum microwave dried cranberry. *Journal of Food Science*, *75*(3), 311–316.

Li, Y., Skouroumounis, G. K., Elsey, G. M., & Taylor, D. K. (2011). Microwave-assistance provides very rapid and efficient extraction of grape seed polyphenols. *Food Chemistry*, *129*, 570–576.

Li, Z., Cheng, Y., & Liu, S. (2010). Quality kinetics of microwave and pasteurised processed tomato juice. *Transations of the Chinese Society of Agricultural Engineering*, *26*(7), 335–340.

Lin, M., & Ramaswamy, H. S. (2011). Evaluation of phosphatase inactivation kinetics in milk under continuous flow microwave and conventional heating conditions. *International Journal of Food Properties*, *14*(1), 110–123.

Liu, C. M., Wang, Q. Z., & Sakai, N. (2005). Power and temperature distribution during microwave thawing, simulated by using Maxwell's equations and Lambert's law. *International Journal of Food Science & Technology*, *40*(1), 9–21.

Liu, P., Zhang, M., & Mujumdar, A. S. (2012). Comparison of three microwave-assisted drying methods on the physiochemical, nutritional and sensory qualities of re-structured purple-fleshed sweet potato granules. *International Journal of Food Science & Technology*, *47*(1), 141–147.

Liu, Z., Qiao, L., Yang, F., Gu, H., & Yang, L. (2017). Brönsted acidic ionic liquid based ultrasound-microwave synergistic extraction of pectin from pomelo peels. *International Journal of Biological Macromolecules*, *94*, 309–318.

Lorence, M. W., & Pesheck, P. S. (2009). *Development of packaging and products for use in microwave ovens*. Sawston, England: Woodhead Publishing.

Luan, D., Tang, J., Pedrow, P. D., Liu, F., & Tang, Z. (2013). Using mobile metallic temperature sensors in continuous microwave assisted sterilization (MATS) systems. *Journal of Food Engineering*, *119*, 552–560.

Maktabi, S., Watson, I., & Parton, R. (2011). Synergistic effect of UV, laser and microwave radiation or conventional heating on E. coli and on some spoilage and pathogenic bacteria. *Innovative Food Science and Emerging Technologies*, *12*(2), 129–134.

Martin, D. I., Margaritescu, I., Cirstea, E., Togoe, I., Ighigeanu, D., Nemtanu, M. R., ... Iacob, N. (2005). Application of accelerated electron beam and microwave irradiation to biological waste treatment. *Vacuum*, *77*(4), 501–506.

Mitrowska, K., Posyniak, A., & Zmudzki, J. (2007). The effects of cooking on residues of malachite green and leucomalachite green in carp muscles. *Analytica Chimica Acta*, *586*(1–2 Spec Iss.), 420–425.

Mokwena, K. K., Tang, J., Dunne, C. P., Yang, T. C. S., & Chow, E. (2009). Oxygen transmission of multilayer EVOH films after microwave sterilization. *Journal of Food Engineering*, *92*(3), 291–296.

Mondal, A., & Datta, A. K. (2008). Bread baking – A review. *Journal of Food Engineering*, *86*(4), 465–474.

Monteiro, R. L., Carciofi, B. A. M., Marsaioli, A., & Laurindo, J. B. (2015). How to make a microwave vacuum dryer with turntable. *Journal of Food Engineering*, *166*, 276–284.

Naseri, M., Rezaei, M., Moieni, S., Hosseni, H., & Eskandari, S. (2010). Effect of different precooking methods on chemical composition and lipid damage of silver carp (*Hypophthalmichthys molitrix*) muscle. *International Journal of Food Science & Technology*, *45*(10), 1973–1979.

Nguyen, L. T., Choi, W., Lee, S. H., & Jun, S. (2013). Exploring the heating patterns of multiphase foods in a continuous flow, simultaneous microwave and ohmic combination heater. *Journal of Food Engineering*, *116*(1), 65–71.

Ni, Q., Zhang, Y., Xu, G., Gao, Q., Gong, L., & Zhang, Y. (2014). Influence of harvest season and drying method on the antioxidant activity and active compounds of two bamboo grass leaves. *Journal of Food Processing and Preservation*, *38*(4), 1565–1576.

Nistor, O. V., Seremet (Ceclu), L., Andronoiu, D. G., Rudi, L., & Botez, E. (2017). Influence of different drying methods on the physicochemical properties of red beetroot (Beta vulgaris L. var. Cylindra). *Food Chemistry*, *236*, 59–67.

Oszmiański, J., Wojdyło, A., & Kolniak, J. (2007). Effect of L-ascorbic acid, sugar, pectin and freeze-thaw treatment on polyphenol content of frozen strawberries. *LWT – Food Science and Technology*, *42*(2), 581–586.

Ozkahraman, B. C., Sumnu, G., & Sahin, S. (2016). Effect of different flours on quality of legume cakes to be baked in microwave-infrared combination oven and conventional oven. *Journal of Food Science and Technology*, *53*(3), 1567–1575.

Ozkoc, S. O., & Seyhun, N. (2015). Effect of gum type and flaxseed concentration on quality of gluten-free breads made from frozen dough baked in infrared-microwave combination oven. *Food and Bioprocess Technology*, *8*(12), 2500–2506.

Ozkoc, S. O., Sumnu, G., & Sahin, S. (2009). The effects of gums on macro and micro-structure of breads baked in different ovens. *Food Hydrocolloids*, *23*(8), 2182–2189.

Ozkoc, S. O., Sumnu, G., Sahin, S., & Turabi, E. (2009). Investigation of physicochemical properties of breads baked in microwave and infrared-microwave combination ovens during storage. *European Food Research and Technology*, *228*(6), 883–893.

Ozogul, Y., Ozyurt, G., & Boga, E. K. (2009). Effects of cooking and reheating methods on the fatty acid profile of sea bream treated with rosemary extract. *Journal of the Science of Food and Agriculture*, *89*(9), 1481–1489.

Öztürk, S., Şakıyan, Ö., & Özlem Alifakı, Y. (2017). Dielectric properties and microwave and infrared- microwave combination drying characteristics of banana and kiwifruit. *Journal of Food Process Engineering*, *40*, 3.

Padda, M. S., & Picha, D. H. (2008). Phenolic composition and antioxidant capacity of different heat-processed forms of sweetpotato cv. 'Beauregard'. *International Journal of Food Science & Technology, 43*(8), 1404–1409.

Pande, R., Mishra, H. N., & Singh, M. N. (2012). Microwave drying for safe storage and improved nutritional quality of green gram seed (Vigna radiata). *Journal of Agricultural and Food Chemistry, 60*(14), 3809–3816.

Pardo, G., & Zufía, J. (2012). Life cycle assessment of food-preservation technologies. *Journal of Cleaner Production, 28*, 198–207.

Peng, J., Tang, J., Luan, D., Liu, F., Tang, Z., Li, F., & Zhang, W. (2017). Microwave pasteurization of pre-packaged carrots'. *Journal of Food Engineering, 202*, 56–64.

Peng, J., Tanga, J., Jiaoa, Y., Bohneta, S. G., & Barrett, D. M. (2013). Dielectric properties of tomatoes assisting in the development of microwave pasteurization and sterilization processes'. *LWT – Food Science and Technology, 54*(2), 367–376.

Peng, X. Y., Mu, T. H., Zhang, M., Sun, H. N., Chen, J. W., & Yu, M. (2015). Optimisation of production yield by ultrasound-/microwave -assisted acid method and functional property of pectin from sugar beet pulp. *International Journal of Food Science and Technology, 50*(3), 758–765.

Petisca, C., Henriques, A. R., Pérez-Palacios, T., Pinho, O., & Ferreira, I. M. P. L. V. O. (2013). Study of hydroxymethylfurfural and furfural formation in cakes during baking in different ovens, using a validated multiple-stage extraction-based analytical method. *Food Chemistry, 141*(2), 3349–3356.

Piasek, A., Kusznierewicz, B., Grzybowska, I., Malinowska-Pańczyk, E., Piekarska, A., Azqueta, A., ... Bartoszek, A. (2011). The influence of sterilization with EnbioJet® microwave Flow Pasteurizer on composition and bioactivity of aronia and blue-berried honeysuckle juices. *Journal of Food Composition and Analysis, 24*, 80–888.

Pongmalai, P., Devahastin, S., Chiewchan, N., & Soponronnarit, S. (2015). Enhancement of microwave-assisted extraction of bioactive compounds from cabbage outer leaves via the application of ultrasonic pretreatment. *Separation and Purification Technology, 144*, 37–45.

Puranik, V., Srivastava, P., Mishra, V., & Saxena, D. C. (2012). Effect of different drying techniques on the quality of garlic: A comparative study. *American Journal of Food Technology, 7*(5), 311–319.

Ramaswamy, H., & Marcotte, M. (2006). *Food processing principles and applications*. Boca Raton, FL, US: CRC Press.

Ramaswamy, H., & Tang, J. (2008). Microwave and radiofrequency heating. *Food Science and Technology International, Vol.14*(5), 423–427.

Regier, M., & Schubert, H. (2001). Microwave processing. In P. Richardson (Ed.), Thermal technologies in *food* processing (pp. 178–207). Cambridge: Woodhead Publishing Limited.

Rekha, M. R., & Padmaja, G. (2002). Alpha-amylase inhibitor changes during processing of sweet potato and taro tubers. *Plant Foods for Human Nutrition, 57*(3), 285–294.

Risch, S. J. (2009). Food packaging history and innovations. *Journal of Agricultural and Food Chemistry, 57*(18), 8089–8092.

Sakiyan, O., Sumnu, G., Sahin, S., & Meda, V. (2007). 'Investigation of dielectric properties of different cake formulations during microwave and infrared-microwave combination baking. *Journal of Food Science, 72*(4), 205–213.

Sakiyan, O., Sumnu, G., Sahin, S., Meda, V., Koksel, H., & Chang, P. (2011). A study on degree of starch gelatinization in cakes baked in three different ovens. *Food and Bioprocess Technology, 4*(7), 1237–1244.

Şakıyan, Ö. (2015). Optimization of formulation of soy-cakes baked in infrared-microwave combination oven by response surface methodology. *Journal of Food Science and Technology, 52*(5), 2910–2917.

Sánchez-Pardo, M. E., Ortiz-Moreno, A., García-Zaragoza, F. J., Necoechea-Mondragón, H., & Chanona-Pérez, J. J. (2012). Comparison of pound cake baked in a two cycle microwave-toaster oven and in conventional oven. *LWT - Food Science and Technology, 46*(1), 356–362.

Sánchez-Pardo, M. E., Ortiz-Moreno, A., Mora-Escobedo, R., Bello-Pérez, A., Yee-Madeira, H., & Ramos-López, G. (2008). Effect of baking method on some characteristics of starch in pound cake crumbs. *Journal of the Science of Food and Agriculture, 88*(2), 207–213.

Sasi Kiran, K., & Padmaja, G. (2003). Inactivation of trypsin inhibitors in sweet potato and taro tubers during processing. *Plant Foods for Human Nutrition, 58*(2), 153–163.

Schubert, H., & Regier, M. (2005). *The microwave processing of foods*. Cambridge, England: Woodhead Publishing Limited.

Serventi, L., Sachleben, J., & Vodovotz, Y. (2011). Soy addition improves the texture of microwavable par-baked pocket-type flat doughs. *Journal of Thermal Analysis and Calorimetry, 106*(1), 117–121.

Seyhun, N., Ramaswamy, H., Sumnu, G., Sahin, S., & Ahmed, J. (2009). Comparison and modeling of microwave tempering and infrared assisted microwave tempering of frozen potato puree. *Journal of Food Engineering, 92*(3), 339–344.

Shaheen, M. S., El-Massry, K. F., El-Ghorab, A. H., & Anjum, F. M. (2012). Microwave applications in thermal food processing. In W. Cao (Ed.), *The development and application of microwave heating* (pp. 3–16). Intech Openscience.

Sharifian, F., Modarres-Motlagh, A., Komarizade, M. H., & Nikbakht, A. M. (2013). Color change analysis of fig fruit during microwave drying. *International Journal of Food Engineering, 9*(1), 107–114.

Shim, J. Y., Lee, S. H., & Jun, S. (2010). Modeling of ohmic heating patterns of multiphase food products using computational fluid dynamics codes. *Journal of Food Engineering, 99*(2), 136–141.

Si, X., Chen, Q., Bi, J., Yi, J., Zhou, L., & Wu, X. (2016). Infrared radiation and microwave vacuum combined drying kinetics and quality of raspberry'. *Journal of Food Process Engineering, 39*(4), 377–390.

Singh, R. P., & Heldman, D. R. (2009). In R. P. Singh, & D. R. Heldman (Eds.), *Heat transfer in food processing' in* Introduction to food engineering (4th Ed., pp. 247–401). Amsterdam: Academic Press.

Spigno, G. (2016). Microwave processing: Impact on food product quality attributes. In A. K. Jaiswal (Ed.), *Food processing technologies: Impact on product attributes* (pp. 137–180). Boca Raton: CRC Press.

Spigno, G., & De Faveri, D. M. (2009). Microwave-assisted extraction of tea phenols: A phenomenological study. *Journal of Food Engineering, 93*, 210–217.

Stea, T. H., Johansson, M., Jägerstad, M., & Frølich, W. (2006). Retention of folates in cooked, stored and reheated peas, broccoli and potatoes for use in modern large-scale service systems. *Food Chemistry, 101*(3), 1095–1107.

Stinco, C. M., Fernández-Vázquez, R., Heredia, F. J., Meléndez-Martínez, A. J., & Vicario, I. M. (2013). Bioaccessibility, antioxidant activity and color of carotenoids in ultrafrozen orange juices: Influence of thawing conditions. *LWT – Food Science and Technology, 53*(2), 458–463.

Sumnu, G. (2001). A review on microwave baking of foods. *International Journal of Food Science & Technology, 36*(2), 117–127.

Sumnu, G., & Sahin, S. (2005). Recent developments in microwave heating. In D. W. Sun (Ed.), Emerging technologies for *food* processing (pp. 419–444). Amsterdam: Academic Press, Elsevier. Inc.

Sun, J., Elena, G. G., Peng, H. X., Guo, Q., Yang, B., & Zhu, E. J. (2009). Variation in content of monomeric phenolics during the processing of grape seed and skin flours. *Animal Production Science, 49*(2), 170–175.

Sun, T., Tang, J., & Powers, J. R. (2007). Antioxidant activity and quality of asparagus affected by microwave -circulated water combination and conventional sterilization. *Food Chemistry, 100*(2), 813–819.

Szadzińska, J., Kowalski, S. J., & Stasiak, M. (2016). Microwave and ultrasound enhancement of convective drying of strawberries: Experimental and modeling efficiency. *International Journal of Heat and Mass Transfer, 103*, 1065–1074.

Szadzińska, J., Łechtańska, J., Kowalski, S. J., & Stasiak, M. (2017). The effect of high power airborne ultrasound and microwaves on convective drying effectiveness and quality of green pepper. *Ultrasonics Sonochemistry, 34*, 531–539.

Taher, B. J., & Farid, M. M. (2001). Cyclic microwave thawing of frozen meat: Experimental and theoretical investigation. *Chemical Engineering Progress, 40*(4), 379–389.

Tang, J. (2015). Unlocking potentials of microwaves for food safety and quality. *Journal of Food Science, 80*(8), 1776–1793.

Türkkan, A. U., Cakli, S., & Kilinc, B. (2008). Effects of cooking methods on the proximate composition and fatty acid composition of seabass (*Dicentrarchus labrax*, Linnaeus, 1758). *Food and Bioproducts Processing, 86*(3), 163–166.

Ünal Şengör, G. F., Alakavuk, D. U., & Yasemin Tosun, Ş. (2013). Effect of cooking methods on proximate composition, fatty acid composition, and cholesterol content of atlantic salmon (Salmo salar). *Journal of Aquatic Food Product Technology, 22*(2), 160–167.

Vadivambal, R., & Jayas, D. S. (2007). Changes in quality of microwave-treated agricultural products-a review. *Biosystems Engineering, 98*(1), 1–16.

Vadivambal, R., & Jayas, D. S. (2010). Non-uniform temperature distribution during microwave heating of food materials – A review. *Food and Bioprocess Technology, 3*, 161–171.

Vadivambal, R., Jayas, D. S., & White, N. D. G. (2007). Wheat disinfestation using microwave energy. *Journal of Stored Products Research, 43*(4), 508–514.

Valentas, K. J., Rotstein, R., & Singh, P. (1997). *Handbook of food engineering practice*. Boca Raton, FL, US: CRC Press.

Valero, A., Cejudo, M., & García-Gimeno, R. M. (2014). Inactivation kinetics for salmonella enteritidis in potato omelet using microwave heating treatments. *Food Control, 43*, 175–182.

Verma, A. K., Kumar, S., Mukul, D., & Dwived, P. D. (2012). Impact of thermal processing on legume allergens. *Plant Foods for Human Nutrition, 67*, 430–441.

Verma, D. K., Mahanti, N. K., Thakur, M., Chakraborty, S. K., & Srivastav, P. P. (2020). Microwave heating: Alternative thermal process technology for food application. In P. P. Srivastav, D. K. Verma, A. R. Patel, & A. R. Al-Hilphy (Eds.), Emerging thermal and nonthermal technologies in *food* processing (pp. 25–67). Boca Raton: CRC Press.

Wang, R., Zhang, M., & Mujumdar, A. S. (2010). Effect of food ingredient on microwave freeze drying of instant vegetable soup. *LWT – Food Science and Technology, 43*(7), 1144–1150.

Won, M. Y., Lee, S. J., & Min, S. C. (2017). Mandarin preservation by microwave-powered cold plasma treatment. *Innovative Food Science and Emerging Technologies, 39*, 25–32.

Xanthakis, E., Huen, J., Eliasson, L., Jha, P. K., Le-Bail, A., & Shrestha, M. (2018). 'Evaluation of microwave assisted freezing (MAF) impact on meat and fish matrices. *Refrigeration Science and Technology*, 176–181.

Xanthakis, E., Le-Bail, A., & Ramaswamy, H. (2014). Development of an innovative microwave assisted food freezing process. *Innovative Food Science and Emerging Technologies, 26*, 176–181.

Xanthakis, E., Le-Bail, A., Shrestha, M., Ahrne, L., & Bernard, J. P. (2015). FREEZEWAVE – Innovative and low energy microwave assisted freezing process for high quality foods. *Refrigeration Science and Technology*, 2580–2586.

Xu, M. H., Yang, X. Y., & Fu, M. R. (2016). 'Combined ultrasonic and microwave method for juglone extraction from walnut green husk (Juglans nigra). *Waste and Biomass Valorization, 7*(5), 1159–1166.

Xue, D., Xu, B., Jiang, H., Zhu, Y., & Dong, Y. (2013). The non-thermal effects of microwave food processing. *Journal of Chinese Institute of Food Science and Technology, 13*(4), 143–148.

Yan, W., Zhang, M., Huang, L. L., Tang, J., Mujumdar, A. S., & Sun, J. C. (2010). Study of the optimisation of puffing characteristics of potato cubes by spouted bed drying enhanced with microwave. *Journal of the Science of Food and Agriculture, 90*(8), 1300–1307.

Zeng, H., Zhang, Y., Lin, S., Jian, Y., Miao, S., & Zheng, B. (2015). Ultrasonic-microwave synergistic extraction (UMSE) and molecular weight distribution of polysaccharides from Fortunella margarita (Lour.) Swingle. *Separation and Purification Technology, 144*, 97–106.

Zhang, J., Wu, D., Liu, D., Fang, Z., Chen, J., Hu, Y., & Ye, X. (2013). Effect of cooking styles on the lipid oxidation and fatty acid composition of grass carp (*Ctenopharynyodon idellus*) fillet. *Journal of Food Biochemistry, 37*(2), 212–219.

Zhang, M., Jiang, H., & Lim, R. X. (2010). Recent developments in microwave-assisted drying of vegetables, fruits, and aquatic products-drying kinetics and quality considerations. *Drying Technology, 28*(11), 1307–1316.

Zhao, Y., Jiang, Y., Zheng, B., Zhuang, W., Zheng, Y., & Tian, Y. (2017). Influence of microwave vacuum drying on glass transition temperature, gelatinization temperature, physical and chemical qualities of lotus seeds. *Food Chemistry, 228*, 167–176.

Zheng, H., & Lu, H. (2011). Effect of microwave pre-treatment on the kinetics of ascorbic acid degradation and peroxidase inactivation in different parts of green asparagus (*Asparagus officinalis* L.) during water blanching. *Food Chemistry, 128*, 1087–1093.

Zhu, J., Kuznetsov, A. V., & Sandeep, K. P. (2007). Mathematical modelling of continuous flow microwave heating of liquids (effects of dielectric properties and design parameters). *International Journal of Thermal Sciences, 46*(4), 328–341.

Zielinska, M., & Michalska, A. (2016). Microwave-assisted drying of blueberry (Vaccinium corymbosum L.) fruits: Drying kinetics, polyphenols, anthocyanins, antioxidant capacity, color and texture. *Food Chemistry, 212*, 671–680.

Zielinska, M., Sadowski, P., & Błaszczak, W. (2015). Freezing/thawing and microwave-assisted drying of blueberries (*Vaccinium corymbosum* L.). *LWT – Food Science and Technology, 62*(1), 555–563.

Index

Note: Page numbers followed by "*f*" and "*t*" refer to figures and tables, respectively.

N

CPI Antony Rowe
Eastbourne, UK
June 12, 2024